Lecture Notes in Computer Science 3116

Commenced Publication in 1973
Founding and Former Series Editors:
Gerhard Goos, Juris Hartmanis, and Jan van Leeuwen

Editorial Board

Charles Rattray   Savitri Maharaj
Carron Shankland (Eds.)

Algebraic Methodology and Software Technology

10th International Conference, AMAST 2004
Stirling, Scotland, UK, July 12-16, 2004
Proceedings

 Springer

Volume Editors

Charles Rattray
Savitri Maharaj
Carron Shankland
University of Stirling
FK9 4LA Stirling, UK
E-mail: {c.rattray,s.maharaj,c.shankland}@cs.stir.ac.uk

Library of Congress Control Number: 2004108198

CR Subject Classification (1998): F.3-4, D.2, C.3, D.1.6, I.2.3, I.1.3

ISSN 0302-9743
ISBN 3-540-22381-9 Springer-Verlag Berlin Heidelberg New York

Springer-Verlag is a part of Springer Science+Business Media

springeronline.com

© Springer-Verlag Berlin Heidelberg 2004
Printed in Germany

Typesetting: Camera-ready by author, data conversion by Olgun Computergrafik
Printed on acid-free paper   SPIN: 11019428   06/3142   5 4 3 2 1 0

Preface

This volume contains the proceedings of AMAST 2004, the 10th International Conference on Algebraic Methodology and Software Technology, held during July 12–16, 2004, in Stirling, Scotland, UK. The major goal of the AMAST conferences is to promote research that may lead to the setting of software technology on a firm, mathematical basis. This goal is achieved by a large international cooperation with contributions from both academia and industry. The virtues of a software technology developed on a mathematical basis have been envisioned as being capable of providing software that is (a) correct, and the correctness can be proved mathematically, (b) safe, so that it can be used in the implementation of critical systems, (c) portable, i.e., independent of computing platforms and language generations, and (d) evolutionary, i.e., it is self-adaptable and evolves with the problem domain.

Previous AMAST meetings were held in Iowa City (1989, 1991, 2000), Twente (1993), Montreal (1995), Munich (1996), Sydney (1997), Manaus (1999), and Reunion Island (2002), and contributed to the AMAST goals by reporting and disseminating academic and industrial achievements within the AMAST area of interest. During these meetings, AMAST attracted an international following among researchers and practitioners interested in software technology, programming methodology and their algebraic and logical foundations.

For AMAST 2004 there were 63 submissions of overall high quality, authored by researchers from Australia, Canada, China, the Czech Republic, Denmark, France, Germany, India, Iran, Israel, Italy, Korea, Portugal, Spain, Taiwan, The Netherlands, Turkey, the UK, and the USA. All submissions were thoroughly evaluated, and an electronic program committee meeting was held to discuss the reviewers' reports. The program committee selected 35 papers to be presented. This volume includes these papers, and abstracts or papers of invited lectures given by Roland Backhouse, Don Batory, Michel Bidoit, Muffy Calder, Bart Jacobs, and John-Jules Meyer.

We heartily thank the members of the program committee and all the referees for their care and time in reviewing the submitted papers, and all the institutions that supported AMAST 2004: the Edinburgh Mathematical Society, the Engineering and Physical Sciences Research Council, the London Mathematical Society, and the Formal Aspects of Computing Science specialist group of the British Computer Society.

May 2004

Charles Rattray
Savitri Maharaj
Carron Shankland

Program Committee Chairs

AMAST 2004 was organized by the following team at the department of Computing Science and Mathematics, University of Stirling, UK.

C. Rattray
S. Maharaj
C. Shankland

Steering Committee

E. Astesiano (Italy)
R. Berwick (USA)
M. Johnson (Australia)(chair)
Z. Manna (USA)
M. Mislove (USA)
A. Nijholt (The Netherlands)
M. Nivat (France)

C. Rattray (UK)
T. Rus (USA)
G. Scollo (Italy)
M. Sintzoff (Belgium)
J. Wing (USA)
M. Wirsing (Germany)

Program Committee

V.S. Alagar (USA)
G. Barthe (France)
M. Bidoit (France)
R. Bland (UK)
P. Blauth Menezes (Brazil)
G. v. Bochmann (Canada)
C. Brink (South Africa)
M. Broy (Germany)
M.L. Bujorianu (UK)
C. Calude (New Zealand)
C. Choppy (France)
R.G. Clark (UK)
A. Fleck (USA)
M. Frias (Argentina)
J. Goguen (USA)
N. Halbwachs (France)
A. Hamilton (UK)
A. Haxthausen (Denmark)
P. Henderson (UK)
M. Hinchey (USA)
A. Lopes (Italy)

M. Mislove (USA)
P. Mosses (Denmark)
M. Nesi (Italy)
R. De Nicola (Italy)
A. Nijholt (The Netherlands)
F. Orejas (Spain)
D. Pavlovic (USA)
T. Rus (USA)
S. Schneider (UK)
G. Scollo (Italy)
S. Seidman (USA)
D. Smith (USA)
C. Talcott (USA)
A. Tarlecki (Poland)
K.J. Turner (UK)
J. van de Pol (The Netherlands)
P. Veloso (Brazil)
L. Wallen (UK)
K. Williamson (USA)
M. Wirsing (Germany)

Referees

C. Adams	M.J. Healy	M Rappl
V.S. Alagar	P. Henderson	C. Rattray
P. Baillot	M. Hinchey	G. Reggio
G. Barthe	J. Hodgkin	E. Ritter
S. Berghofer	M. Huisman	N. Schirmer
M. Bidoit	M. Huth	S. Schneider
L. Blair	S.B. Jones	G. Scollo
G.v. Bochmann	J. Kleist	R. Segala
C. Brink	H.H. Løvengreen	S.B. Seidman
H. Bruun	S. Maharaj	C. Shankland
M.L. Bujorianu	F. Mancinelli	R. Sharp
M.C. Bujorianu	P.E. Martínez López	D. Smith
M. Calder	I.A. Mason	M. Spichkova
C. Calude	I. Mastroeni	M. Strecker
N. Catano	F. Mehta	A. Suenbuel
M. Cerioli	M. Mislove	C. Talcott
C. Choppy	P.D. Mosses	A. Tarlecki
B. Daou	M. Nesi	H. Treharne
M. Debbabi	H. Nilsson	K.J. Turner
G. Dufay	K. Ogata	M. Valero Espada
N. Evans	F. Orejas	L. Wallen
A. Fleck	J. Ouaknine	K.E. Williamson
M. Fränzle	V. de Paiva	T. Wilson
M. Frias	D. Pavlovic	M. Wirsing
R. Giacobazzi	P. Pelliccione	J. Worrell
J. Goguen	L. Petrucci	H. Yahyaoui
N. Halbwachs	M. Poel	J. Zwiers
R.R. Hansen	J. van de Pol	

Sponsoring Institutions

Edinburgh Mathematical Society
Engineering and Physical Sciences Research Council
Formal Aspects of Computing Science Specialist Group of the British Computer Society
London Mathematical Society

Table of Contents

Invited Speakers

Contributed Talks

Algebraic Approaches to Problem Generalisation

Roland Backhouse

School of Computer Science and Information Technology, University of Nottingham,
Nottingham NG8 1BB, England
rcb@cs.nott.ac.uk

Abstract. A common technique for solving programming problems is
to express the problem in terms of solving a system of so-called "simultaneous" equations (a collection of equations in a number of unknowns
that are often mutually recursive). Having done so, a number of techniques can be used for solving the equations, ranging from simple iterative techniques to more sophisticated but more specialised elimination
techniques.

A stumbling block for the use of simultaneous equations is that there
is often a big leap from a problem's specification to the construction
of the system of simultaneous equations; the justification for the leap
almost invariably involves a *post hoc* verification of the construction.
Thus, whereas methods for solving the equations, once constructed, are
well-known and understood, the process of constructing the equations is
not.

In this talk, we present a general theorem which expresses when the solution to a problem can be expressed as solving a system of simultaneous
equations. The theorem exploits the theory of Galois connections and
fixed-point calculus, which we briefly introduce. We give several examples of the theorem together with several non-examples (that is, examples
where the theorem is not directly applicable). The non-examples serve
two functions. They highlight the gap between specification and simultaneous equations – we show in several cases how a small change in
the specification leads to a breakdown in the solution by simultaneous
equations – and they inform the development of a methodology for the
construction of the equations.

Application of the technique in the case of the more challenging problems depends crucially on finding a suitable generalisation of the original
problem. For example, the problem of finding the *edit distance* between a
word and a context-free language is solved by computing the edit distance
between each segment of the given word and the language generated by
each nonterminal in the given grammar.

A focus of the talk is the use of Conway's factor theory [Con71] in generalising a class of problems we call "bound" problems. Since its publication in 1971, Conway's work has been largely ignored, but its relevance
to program analysis has recently been observed by Oege de Moor and
his colleagues [MDLS02,SdML04]. We show how factor theory underpins
De Moor's work as well as the well-known Knuth-Morris-Pratt pattern
matching algorithm [KMP77]. We also speculate on how further study
of factor theory might have relevance to a broader class of problems.
This talk is based on [Bac04], where further details can be found.

C. Rattray et al. (Eds.): AMAST 2004, LNCS 3116, pp. 1–2, 2004.
© Springer-Verlag Berlin Heidelberg 2004

References

[Bac04]   Roland Backhouse. Regular algebra applied to language problems. *Submitted for publication in Journal of Logic and Algebraic Programming*, 2004.

[Con71]   J.H. Conway. *Regular Algebra and Finite Machines*. Chapman and Hall, London, 1971.

[KMP77]   D.E. Knuth, J.H. Morris, and V.R. Pratt. Fast pattern matching in strings. *SIAM Journal of Computing*, 6:325–350, 1977.

[MDLS02]   O. de Moor, S. Drape, D. Lacey, and G. Sittampalam. Incremental program analysis via language factors. Available from
http://web.comlab.ox.ac.uk/work/oege.de.moor/pubs.htm, 2002.

[SdML04]   Ganesh Sittampalam, Oege de Moor, and Ken Friis Larssen. Incremental execution of transformation specifications. In *POPL'04*, 2004.

A Science of Software Design

Don Batory

Department of Computer Sciences
University of Texas at Austin
Austin, Texas 78746
batory@cs.utexas.edu

Abstract. Underlying large-scale software design and program synthesis are simple and powerful algebraic models. In this paper, I review the elementary ideas upon which these algebras rest and argue that they define the basis for a science of software design.

1 Introduction

I have worked in the areas of program generation, software product-lines, domain specific languages, and component-based architectures for over twenty years. The emphasis of my research has been on large-scale program synthesis and design automation. The importance of these topics is intuitive: higher productivity, improved software quality, lower maintenance costs, and reduced time-to-market can be achieved through automation.

Twenty years has given me a unique perspective on software design and software modularity. My work has revealed that large scale software design and program synthesis is governed by simple and powerful algebraic models. In this paper, I review the elementary ideas on which these algebras rest. To place this contribution in context, a fundamental problem in software engineering is the abject lack of a science for software design. I will argue that these algebraic models can define the basis for such a science.

I firmly believe that future courses in software design will be partly taught using domain-specific algebras, where a program's design is represented by a composition of operators, and design optimization is achieved through algebraic rewrites of these compositions. This belief is consistent with the goal of AMAST. However, I suspect that *how* I use algebras and their relative informality to achieve design automation is unconventional to the AMAST community. As a background for my presentation, I begin with a brief report on the 2003 Science of Design Workshop.

2 NSF's Science of Design Workshop

In October 2003, I attended a *National Science Foundation (NSF)* workshop in Airlie, Virginia on the "Science of Design" [11]. The goal of the workshop was to determine the meaning of the term "Science of Design". NSF planned to start a program with this title and an objective was to determine lines of research to fund. There were 60 attendees from the U.S., Canada, and Europe. Most were from the practical side of

C. Rattray et al. (Eds.): AMAST 2004, LNCS 3116, pp. 3–18, 2004.
© Springer-Verlag Berlin Heidelberg 2004

software engineering; a few attendees represented the area of formal methods. I was interested in the workshop to see if others shared my opinions and experiences in software design, but more generally, I wanted to see what a cross-section of today's Software Engineering community believed would be the "Science of Design". In the following, I review a few key positions that I found particularly interesting.

Richard Gabriel is a Distinguished Engineer at Sun Microsystems and one of the architects of Common Lisp. He described his degree in creative writing – in particular, poetry – and demonstrated that it was far more rigorous in terms of course work than a comparable degree in Software Engineering (of which software design was but a small part). He advocated that students should be awarded degrees in "Fine Arts" for software design. I was astonished: I did not expect to hear such a presentation at a *Science* of Design workshop. Nevertheless, Gabriel reinforced the common perception that software design is indeed an art, and a poorly understood art at that.

Carliss Balwin is a Professor at the Harvard Business School. She argued that software design is an instance of a much larger paradigm of product design. She observed that the processes by which one designs a table, or a chair, or an auditorium, are fundamentally similar to that of designing software. Consequently, software design has firm roots in economic processes and formalisms. Once again, I was not expecting to hear such a presentation at a *Science* of Design workshop. And again, I agreed with her arguments that software design can be viewed as an application of economics.

Did the workshop bring forth the view of is software design as a *science*? I did not see much support for this position. Attendees were certainly using science and scientific methods in their investigations. But I found little consensus, let alone support, for software design as a science. The most memorable summary I heard at the workshop was given by Fred Brookes, the 1999 ACM Turing Award recipient. He summarized the conclusions of his working group as "We don't know what we're doing, and we don't know what we've done!".

The results of the workshop were clear: if there is to be a science of software design, it is a very long way off. In fact, it was questionable to consider software design a "science". Although I do not recall hearing this question posed, it seemed reasonable to ask if design is engineering[1]. For example, when bridges are designed, there is indeed an element of artistry in their creation. But there is also an underlying science called physics that is used to determine if the bridge meets its specifications. So if software design is engineering, then what is the science that underlies software design? Again, we are back to square one.

After many hours of thought, I realized that the positions of Gabriel and Baldwin were consistent with my own. Software design is an art as Gabriel argued, *but not always*. Consider the following: designing the first automobiles was an art – it had never been done before, and required lots of trial and error. Similarly, designing the first computer or designing the first compiler were also works of art. There were no assembly lines for creating these products and no automation. What made them possible was craftsmanship and supreme creativity. Over time, however, people began building variants of these designs. In doing so, they learned answers to the important questions of *how* to design these products, *what* to design, and most importantly, *why* to do it in a particular way. Decision making moved from subjective justifications to

---

[1]   Thanks to Dewayne Perry for this observation.

quantitative reasoning. I am sure you have heard the phrase "we've done this so often, we've gotten it down to a science". Well, that *is* the beginnings of a science.

A distinction that is useful for this paper is the following: given a specification of a program and a set of organized knowledge and techniques, if "magic" (a.k.a. inspiration, creativity) is needed to translate the specification into a program's design, then this process is an *art* or an *inexact science*. However, if it is purely a mechanical process by which a specification is translated into a design of an efficient program, then this process follows an *exact* or *deterministic science*.

Creating one-of-a-kind designs will always be an art and will never be the result of an exact or deterministic science, simply because "magic" is needed. Interestingly, the focus of today's software design methodologies is largely on creating one-of-a-kind products. The objective is to push the envelope on a program or component's capabilities, relying on the creativity and craftsmanship of its creators – and not automation. In contrast, I believe that an exact science for software design lies in the mechanization and codification of well-understood processes, domain-expertise, and design history. We have vast experiences building particular kinds of programs, we know the *how*, the *what*, and the *why* of their construction. We want to automate this process so that there is no magic, no drudgery, and no mistakes. The objective of this approach is *also* to push the envelope on a program or component's capability but *with emphasis on design automation*. That is, we want to achieve the same goals of conventional software development, *but from a design automation viewpoint*.

The mindset to achieve higher levels of automation is unconventional. It begins with a declarative specification of a program. This specification is translated into a design of an efficient program, and then this design is translated to an executable. To do all this requires significant technological advances. First, how can declarative specifications of programs be simplified so that they can be written by programmers with, say, a high-school education? This requires advances in *domain-specific languages*. Second, how can we map a declarative specification to an efficient design? This is the difficult problem of *automatic programming*; all but the most pioneering researchers abandoned this problem in the early 1980's as the techniques that were available at that time did not scale [1]. And finally, how do we translate a program's design to an efficient executable automatically? This is *generative programming* [9]. Simultaneous advances on all three fronts are needed to realize significant benefits in automation.

To do all this seems impossible, yet an example of this futuristic paradigm was realized *over 25 years ago*, around the time that others were giving up on automatic programming. The work was in a significant domain, and the result had a revolutionary impact on industry. The result: *relational query optimization (RQO)* [12].

Here's how RQO works: an SQL query is translated by a parser into an inefficient relational algebra expression. A query optimizer optimizes the expression to produce a semantically equivalent expression with better performance characteristics. A code generator translates the optimized expression into an efficient executable. SQL is a prototypical declarative domain-specific language; the code generators were early examples of generative programming, and the optimizer was the key to a practical solution to automatic programing.

In retrospect, relational database researchers were successful because they automated the development of query evaluation programs. These programs were hard to write, harder to optimize, and even harder to maintain. The insight these researchers had was to create an *exact* or *deterministic science* to specify and optimize query evaluation programs. In particular, they identified the fundamental operators that comprised the domain of query evaluation programs, namely relational algebra. They represented particular programs in this domain by expressions (i.e., compositions of relational operators). And they used algebraic identities to rewrite, and thus optimize, relational algebra expressions.

RQO is clearly an interesting paradigm for automated software development [5]. I cannot recall others ever proposing to generalize the RQO paradigm to other domains. The reason is clear: the generalization is not obvious. It is possible, and in the next sections, I show how.

3 Feature Oriented Programming

Feature Oriented Programming (FOP) originated from work on product-line architectures. The goal is to declaratively specify a program by the features that it is to have, where a feature is some characteristic that can be shared by programs in a domain. So program P1 might have features X, Y, and Z, while program P2 has features X, Q, and R. Features are useful because they align with requirements: customers know their requirements and can see how features satisfy requirements.

Interestingly, feature specifications of products are quite common. (It just isn't common for software). The Dell web site, for example, has numerous web pages where customers can declaratively specify the features they want on their PCs. A Dell web page is a declarative DSL; clicking the check boxes and selecting items from pull-down menus is the way declarative specs are written. By sending Dell a check for the computed amount, that customized PC will be delivered in days. Similarly, ordering a customized meal at a restaurant involves choosing items from a menu; this too is a familiar form of declarative specifications. Neither customizing PCs or ordering customized meals requires an advanced technical degree. We want the same for software.

GenVoca is a simple and powerful algebraic model of FOP. GenVoca is based on the idea of *step-wise refinement*, which is an ancient methodology for building software by progressively adding details [14]. The novelty of GenVoca is that it scales the concept of refinement. That is, instead of composing hundreds or thousands microscopic program rewrites called *refinements*, GenVoca scales refinements so that they each encapsulate an individual feature. A complete program is synthesized by composing a few feature refinements. (Warning: I am using the term "refinement" in its common *object-oriented (OO)* usage, namely to elaborate or extend. In contrast, "refinement" has a different meaning in algebraic specifications – it means to elaborate but *not* extend a program's functionality. "Extension" is a more appropriate term. Henceforth, I use the term "extension", but beware that papers on FOP use the term "refinement" instead).

A GenVoca model of a domain is a set of operators that defines an algebra. Each operator implements a feature. We write:

```
M = { f, h, i, j }
```

to mean model M has operators (or features) f, h, i, and j. One or more of these operators are *constants* that represent base programs:

```
f                // an application with feature f
h                // an application with feature h
```

The remaining operators are *functions* which represent program extensions:

```
i(x)              // adds feature i to application x
j(x)              // adds feature j to application x
```

The design of an application is a named expression called an *equation*:

```
app1 = i(f)       // application with features i and f
app2 = j(h)       // application with features j and h
app3 = i(j(h  ))  // application with features i,j,h
```

The family of programs that can be created from a model is its *product-line*. To simplify notation, we henceforth write `i(j(h))` as $i \bullet j \bullet h$, where $\bullet$ denotes function composition.

A GenVoca expression represents the *design* of a program. Such expressions (and hence program designs) can be automatically optimized. This is possible because a function represents both a feature *and* its implementation. That is, there can be different functions with different implementations of the *same* feature. For example, suppose function k_1 adds feature k with implementation #1 to its input program, while function k_2 adds feature k with implementation #2. When an application requires the use of feature k, it is a problem of *expression optimization* to determine which implementation of k is best (e.g., provides the best performance). Of course, more complicated rewrite rules can be used. Thus, it is possible to design efficient software automatically (i.e., find an expression that optimizes some criteria) given a set of declarative constraints for a target application. An example of this kind of automated reasoning – which is exactly the counterpart to relational query optimization – is [6].

The program synthesis paradigm of GenVoca is straightforward. Figure 1 depicts a program P that is a package of four classes (`class1`–`class4`). These classes are synthesized by composing features X, Y, and Z. X encapsulates a fragment of `class1`–`class3`, which is shown in a solid color. Y extends `class1`–`class3` and introduces `class4`, which is shown in horizontal stripes. Z extends all four classes, and is shown in checker-board. Thus features encapsulate fragments of classes. Composing features yields packages of fully-formed classes.

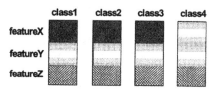

Fig. 1. Package P = Z•Y•X

My colleagues and I have had considerable success using GenVoca for creating product-lines for database systems [7], network protocols [7], data structures [6], avionics [2], extensible Java and Scheme compilers [3, 10], and program verification tools [13]. The next section briefly explains how code synthesis is performed.

3.1 Implementation Details

Extension and composition are very general concepts that can be implemented in many different ways. The core approach relies on inheritance to express method and class extensions. Figure 2a shows method **A()** whose body sequentially executes statements **x, y,** and **z**. Figure 2b declares an extension of this method to be **Super.A()** followed by statement **w**. **Super.A()** says invoke the method's original definition. The composite method is Figure 2c; it was produced by substitution (a.k.a. macro-expansion): that is, **Super.A()** was replaced with the original body of **A()**.

```
      void A() {              void A() {              void A() {
          x; y; z;                Super.A(); w;            x; y; z; w;
(a)   }                  (b)  }                  (c)  }
```

Fig. 2. Method Definition and Extension

Class extensions are similarly familiar. Figure 3a shows a class **P** that has three members: methods **A()**, **B()**, and data member **C**. Figure 3b shows an extension of **P**, which encapsulates extensions to methods **A()** and **B()** and adds a new data member **D**. The composition of the base class and extension is Figure 3c: composite methods **A()** and **B()** are present, plus the remaining members of the base and extension.

```
                                                  class P {
  class P {            refines class P {              void A(){x;y;z;w;}
    void A(){ x;y;z; }     void A(){ Super.A();w; }   void B(){q;r;t;}
    void B(){ r;t; }       void B(){ q;Super.B(); }   int C;
    int C;                 String D;                  String D;
  }            (a)     }                    (b)   }                    (c)
```

Fig. 3. Method Definition and Extension

One way to implement the above is to use subclassing: that is, make Figure 3b a subclass of Figure 3a, where the semantics of the subclass equals that of Figure 3c. Another way is to use substitution (in-place modification) as we have illustrated. There are many other ways to realize these ideas with or without inheritance.

4   AHEAD

Algebraic Hierarchical Equations for Application Design (AHEAD) is the successor to GenVoca [3]. It embodies ideas that have revolutionized my thinking on program synthesis. In particular, AHEAD shows how step-wise development scales to the synthesis of multiple programs and multiple representations, and that software design has an elegant algebraic structure that is expressible as nested sets of expressions. The following sketches the basic ideas of AHEAD.

4.1   Multiple Program Representations

Generating code for individual programs is not sufficient. Today's systems are not individual programs but groups of collaborating programs such as client-servers and tool suites of integrated development environments. Further, systems themselves are not solely defined by source code. Architects routinely use many knowledge representations to express a system's design, such as process models, UML models, makefiles, or formal specifications.

That a program has many representations is reminiscent of Platonic forms. That is, a program is a form. Shining a light on this program casts a shadow that defines a representation of that program in a particular language. Different light positions cast different shadows, exposing different details or *representations* of that program. For example, one shadow might reveal a program's representation in Java, while another an HTML document (which might be the program's design document). There are class file or binary representations of a program, makefile representations, performance models, and so on. A program should encapsulate all of its artifacts or projections.

In general, suppose program P encapsulates artifacts A_p, B_p, and C_p, where the meaning of these artifacts is uninterpreted. I express this relationship algebraically as:

$$P = \{\ A_p,\ B_p,\ C_p\ \}$$

where set notation denotes encapsulation. Members of a set are called *units*.

4.2 Generalize Extensions

Adding a new feature to a program may change any or all of its representations. For example, if a new feature F is added to program P, one would expect changes in P's code (to implement F), documentation (to document F), makefiles (to build F), formal properties (to characterize F), performance models (to profile F), and so on.

In general, suppose feature F changes artifacts A and B (where A_f and B_f denote the specifications of these changes) and adds new artifact D_f. I say F encapsulates A_f, B_f, and D_f, and write this relationship algebraically as:

$$F = \{\ A_f,\ B_f,\ D_f\ \}$$

4.3 Generalize Composition

Given P and F, how is $F \bullet P$ computed? The answer: composition is governed by rules of inheritance. Namely, all units of the parent (inner or right-hand-side) feature are inherited by the child (outer or left-hand-side) feature. Further, units with the same name (ignoring subscripts) are composed pairwise with the parent term as the inner term:

$$
\begin{aligned}
F \bullet P \ &= \ \{\ A_f,\ B_f,\ D_f\ \} \ \bullet \ \{\ A_p,\ B_p,\ C_p\ \} \\
&= \ \{\ A_f \bullet A_p,\ B_f \bullet B_p,\ C_p,\ D_f\ \}
\end{aligned}
\tag{1}
$$

Stated another way, $F \bullet P$ is computed by composing corresponding artifacts and the correspondence is made by name. Thus, the A artifact of $F \bullet P$ is produced by $A_f \bullet A_p$ – the original artifact A_p extended by A_f. Similarly, the B artifact of $F \bullet P$ is $B_f \bullet B_p$ – the original artifact B_p extended by B_f. Artifacts C and D of $F \bullet P$ correspond to their original definitions. (1) defines the *Law of Composition*: it tells us how *composition distributes over encapsulation*.

Readers may recognize Figure 3 to be a particular example of this law. P is the base class of Figure 3a, encapsulating members A, B, and C. F is the class extension of Figure 3b, encapsulating members A, B, and D. The composition F•P – an illustration of (1) – is Figure 3c. More on this in the next section.

You will see shortly that the Law of Composition applies at all levels of abstraction and can be made to apply to all artifacts. Figure 4 is an example of the latter. Figure 4a is a grammar of a language that sums integers. Figure 4b shows a grammar extension that adds the minus operation. In particular, a new token MINUS is added to the grammar and the Operator production is extended with the MINUS rule. (The phrase Super.Operator says substitute the right-hand-side of the original Operator production). Figure 4b shows the composite grammar. Each token and production corresponds to individual terms in the Law of Composition.

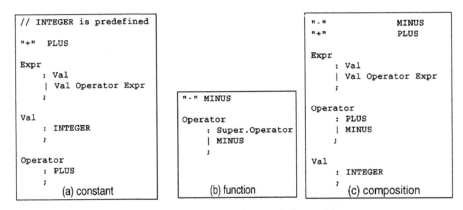

Fig. 4. Grammars, Extensions, and Composition

4.4 Generalize Modularity

A *module* is a containment hierarchy of related artifacts. Figure 5a shows that a class is a 2-level containment hierarchy that encapsulates a set of methods and fields. An interface is also a 2-level containment hierarchy that encapsulates a set of methods and constants. A package is a 3-level containment hierarchy encapsulating a set of classes and interfaces. A J2EE EAR file is a 4-level hierarchy that encapsulates a set of packages, deployment descriptors, and HTML files.

In general, a module hierarchy can be of arbitrary depth and can contain arbitrary artifacts. This enables us to define a module that encapsulates multiple programs. Figure 5b shows a system to encapsulate two programs, a client and a server. Both programs have code, UML, and HTML representations with sub-representations (e.g., code has Java files and binary class files, UML has state machines and class diagrams). Thus, a module allows us to encapsulate all needed representations of a system.

Module hierarchies have simple algebraic representations as nested sets of constants and functions. Figure 6a shows package K to encapsulate class1 and class2, class1 encapsulates method mth1 and field fld1. class2 encapsulates mth2 and mth3. The corresponding set notation is shown in Figure 6b.

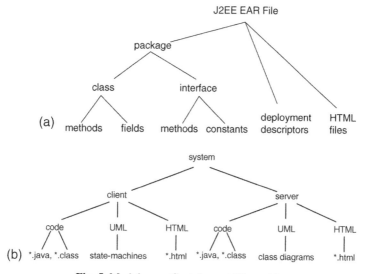

Fig. 5. Modules are Containment Hierarchies

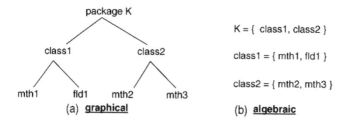

Fig. 6. Modules and Nested Sets

4.5   Generalize GenVoca

A GenVoca model is a set of constants and functions. An AHEAD model is also a set of constants and functions, but now a constant represents a hierarchy that encapsulates the representations of a base program. An AHEAD function is a hierarchy of extensions – that is, it is a containment hierarchy that can add new artifacts (e.g., new Java and HTML files), and can also refine/extend existing artifacts. When features are composed, corresponding program representations are composed. If the representations of each feature are consistent, then their composition is consistent. Thus consistent representations of programs can be synthesized though composition; this is exactly what is needed.

4.6   Implementation Details

We implement module hierarchies as directory hierarchies. Figure 7a shows our algebraic representation of a module, and Figure 7b shows its directory representation.

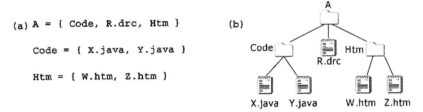

(a) A = { Code, R.drc, Htm }   (b)

    Code = { X.java, Y.java }

    Htm = { W.htm, Z.htm }

Fig. 7. Corresponding Algebraic and Directory Representations

Feature composition is directory composition. That is, when features are composed, their corresponding directories are folded together to produce a directory whose structure is isomorphic to the feature directories that were composed. For example, the X.java file of C = B•A in Figure 8 is produced by composing the corresponding X.java files of B and A.

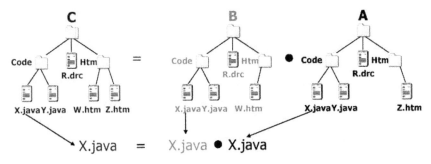

Fig. 8. Composition of Feature Directories

Our implementation is driven by purely algebraic manipulation. We evaluate an expression by alternately expanding nonterminals and applying the Law of Composition:

$$C = B \bullet A$$
$$= \{\ Code_B,\ R.drc_B,\ Htm_B\ \} \bullet \{\ Code_A,\ R.drc_A,\ Htm_A\ \}$$
$$= \{\ Code_B \bullet Code_A,\ R.drc_B \bullet R.drc_A,\ Htm_B \bullet Htm_A\ \}$$
$$= \{\ X.java_B,\ Y.java_B\ \} \bullet \{\ X.java_A,\ Y.java_A\ \},$$
$$\quad R.drc_B \bullet R.drc_A,\ \{\ W.htm_B\ \} \bullet \{\ Z.htm_A\ \}\ \}$$
$$= \{\ \{\ X.java_B \bullet X.java_A,\ Y.java_B \bullet Y.java_A\ \},$$
$$\quad R.drc_B \bullet R.drc_A,\ \{\ W.htm_B,\ Z.htm_A\ \}\}$$

The result is a nested set of expressions. Each expression tells us how to synthesize an artifact of the target system. That is, the X.java artifact of feature C is computed by $X.java_B \bullet X.java_A$; the Y.java artifact of C is computed by $Y.java_B \bullet Y.java_A$, the R.drc artifact of C is computed by $R.drc_B \bullet R.drc_A$, and so on. Thus, there is a simple interpretation for every computed expression, and there is a direct mapping of the nested set of expressions to the directory that is synthesized.

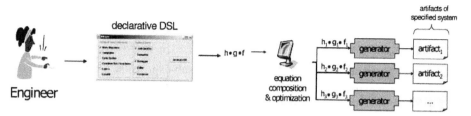

Fig. 9. Program Synthesis Paradigm of AHEAD

Figure 9 illustrates the AHEAD paradigm. An engineer defines a system by declaratively specifying the features it is to have, typically using some GUI-based DSL. The DSL compiler translates the specification into an AHEAD expression. This expression is then expanded and optimized, producing a nested set of expressions. Each expression is typed – expressions that synthesize Java files are distinguishable from expressions that synthesize grammar files – and is submitted to a type-specific generator to synthesize that artifact. The set of artifacts produced are consistent with respect to the original declarative specification. AHEAD is a generalization of the Relational Query Optimization paradigm.

A common question that is asked is: can you realistically design features so that they can be composed? Absolutely. It's easy – this is what software product-lines are all about. Features or components are designed to be composable and compatible. Composability and compatibility are properties that don't happen by magic or by accident; they are premeditated. Some of you may recall the old chip catalogs of the early 1970s, where all the chips in the catalog were designed to be compatible – they worked off of the same voltage, impedance, etc. Chips built by another manufacturer often were not compatible. A more familiar example today are disks for PCs. There are all sorts of disk manufacturers now that have an incredible line of products. These disks are compatible because they meet SCSI or IDE standards. (Recall that prior to plug-and-play standards, adding a disk to a PC required a high-paid technician to do the installation). The same ideas apply to software.

5 Cultural Enrichment

It is beyond the scope of this paper to show a detailed example of these ideas in action. For those interested, please consult [4]. In this section, I explain a simple result that illustrates AHEAD algebras and an elementary optimization. Then I discuss the breadth of the AHEAD framework.

5.1 A Simple Result

Have you ever wondered what an algebraic justification of object-oriented sub-classing (e.g., inheritance) would be and why inheritance is fundamental to software designs? Here's an answer: when a program is synthesized, an expression is generated for every class of that program. Suppose the program has classes **A**, **B**, and **C** with the following expressions:

$$A = Z \bullet Y \bullet X \bullet W$$
$$B = Q \bullet Y \bullet X \bullet W$$
$$C = E \bullet Y \bullet X \bullet W$$

Observe that all three classes share a common sub-expression Y•X•W. Instead of redundantly evaluating this expression, one can eliminate common sub-expressions by *factoring*:

$$F = Y \bullet X \bullet W$$
$$A = Z \bullet F$$
$$B = Q \bullet F$$
$$C = E \bullet F$$

Whenever a new equation F is created, a new class F is generated. The relationship between classes F and A, B, C is indicated in the revised expressions: the expressions for A, B, C reference and thus *extend* F. Code generators materialize F as a class with A, B, and C as subclasses. That is, F contains the data members and methods that are common to its subclasses. *Just as common sub-expression elimination is fundamental to algebra, inheritance hierarchies are fundamental to object-oriented designs, because they are manifestations of the same concept.* Interestingly, the process of promoting common methods and data members of subclasses into a superclass is called *refactoring* in OO parlance, whereas in algebra it is called *factoring*. Not only are the concepts identical, the names are almost identical too.

5.2 Even More Generality

I have concentrated so far on *domain-specific* operators – constants and functions – whose compositions define programs within a target domain. But there are many operators in the AHEAD universe that are not domain-specific. I illustrate some with an example.

In the current implementation of AHEAD, all code is written in Java that has been extended with refinement constructs (e.g., "**refines**" and "**Super**"). This language is called Jak (short for "Jakarta"). To compile these classes, the Jak files are translated to their Java file equivalents using the **jak2java** tool. Next, the Java files are translated to class files using the **javac** tool. These are *derivation steps* that can be expressed algebraically: Let P be a program that is synthesized from some AHEAD equation. P encapsulates a set of Jak files. Let P′ define the set that additionally contains the corresponding Java and class files. The relationship of P and P′ is expressed as:

$$P' = \text{javac}(\ \text{jak2java}(\ P\ )\ ) \tag{2}$$

That is, the **jak2java** tool is an operator that elaborates or extends containment hierarchies by translating every Jak file to its corresponding Java file. Similarly, the **javac** tool is an operator that elaborates containment hierarchies by translating every Java file to its corresponding class file(s).

There are many other operators on containment hierarchies, such as **javadoc** (that derives HTML-documentation files from Java files), **javacc** (that derives Java files that implement a parser from.jj files), and **jar** (a Java archive utility). In short, common tools that programmers use today are operators that derive and thus add new representations to containment hierarchies. Readers will recognize equation (2) to be an algebraic specification of a makefile. It is *much* easier to write and understand (2) than its corresponding `ant` declaration in XML. Thus, it should be possible to generate low-level makefiles from equational specifications like (2). Further, if the makefile expression is particularly complicated, it may be possible also to *optimize* the expression automatically (e.g., perform `javac` operator before `javadoc`) prior to makefile generation. Doing so would further relieve programmers of the burden of manually writing and optimizing these files.

Software design involves many activities that involves many representations of a program (e.g., analysis, implementation, description). Given the ability to compose and derive, you can do just about anything. AHEAD unifies these activities in an algebraic setting. Not only will this simply program specifications, it will also make specifications more amenable to automatic optimization.

6 Why Does AHEAD Work?

I'm sure that some of you, upon reading this paper, will wonder what the big deal is; the ideas are straight from an introductory computer science course (a.k.a. "CS 101"). You are right about the simplicity, but you may have forgotten the state (and abject lack) of science in software design (e.g., Section 2). As computer scientists we understand CS 101 concepts, but I claim that we do *not* know how they are applied to *software design*. Software engineers do *not* think about software designs in terms of algebras or anything (as far as I can tell) dealing with mathematics. They certainly do not think of design in terms of automation. Their reward structure for software development is also screwed up: complexity is appreciated; simplicity and elegance are dismissed. Not only is this bad science, this is bad engineering.

My students and I have synthesized the AHEAD tool suite, which is in excess of 250K Java LOC, from elementary equational specifications. We know of no technical limit on the size of system that can be synthesized. The obvious questions are: why can AHEAD be this simple? What are we giving up? And why does it work as well as it does?

There are many answers; here is one. I have seen the following idea proposed by different people in three consecutive decades, starting in late 1970s. Suppose a pair of independently written packages Q and R are to be composed. Composition is accomplished in the following way: corresponding classes in Q and R are identified. That is, class X in Q corresponds to class Y in R, and so on for every relevant class. Then corresponding methods are paired. That is, method M in class X of Q corresponds to method N in class Y of R, and so on for all relevant methods. And finally, corresponding parameters are matched: parameter 2 of method M in class X of Q corresponds to parameter 3 of method N in class Y of R, etc. Given these relationships, tools can be built to compose Q and R, and examples can show that indeed a package with the functionalities of Q and R can be created.

This approach has many problems. First, it does not scale in general. What happens if Q and R both have 1000 classes, 10,000 methods, and 20,000 parameters? Who can define (or have the patience to define) the required correspondences? How does one validate these correspondences? Note the hidden complexity of this approach: every concept (class, method, parameter) has *two* names: the name used in package Q and that used in R. This means that programmers have to remember *twice* the amount of information than they need to. If *k* additional packages are to be composed, then programmers must know *k* different names for the *same* concept. The concept is the essence of this problem; the different arbitrary names add *accidental complexity* [8]. In general, programmers have a hard time remembering all these accidental details.

Second, it does not work in general. Just because the names in packages can be aligned syntactically does not imply semantic alignment. That is, there is no guarantee that corresponding methods agree on their semantics. If there is disagreement, this approach to composition and software synthesis fails. The only recourse is to define an adaptor that performs the semantic translation – if in fact such an adaptor can be written.

In my experience, software reuse and software composition succeeds because of a premeditated design. Stated another way, components are composable *only* if they were *designed* with the other in mind. *This axiom is crucial to scalability*. It is very easy (and wrong) to assume that just because a small example works where this axiom does not hold, more complicated examples will work just as well.

AHEAD's contribution is basic: it shows how simple ideas and common software development tools can be unified and elegantly captured as an algebra that expresses the essential tasks of software design and scalable software generation as mathematics. I reject axioms that accept accidental complexity as fundamental. As a consequence, AHEAD provides a clean solution to a core design problem where class, method, and parameter names are standardized, and so too are their semantics. This enables us to build simple tools and attack problems of automated design on scale. Our work relies on a common engineering technique: arbitrary and random complexity are eliminated *by design*. AHEAD does indeed have correspondence mappings; they are implicit, and hence easy to write and easy to understand[2].

The following quote from Fred Brooke's 1987 "No Silver Bullet" paper [8] helps to put the above arguments into context:

> The complexity of software is an essential property. Descriptions of software that abstract away its complexity often abstract away its essence...[3] Software people are not alone in facing complexity. Physicists deal with terribly complex objects even at the "fundamental" particle level. The physicist labors on in a firm faith that there are unifying principles to be found ... Einstein argued that there must be simplified explanations of nature, because God is not capricious or arbitrary.

> No such faith comforts the software engineer. Much of the complexity that he must master is arbitrary complexity, forced without rhyme or reason, because they were designed by different people, rather than by God.

Not quite. Just like physicists, I too believe there is an underlying simplicity in software. Programmers are geniuses in making simple things look complicated. If software were truly complicated, people could not program and architects could not design. There *must* be an underlying simplicity. The challenge is not making things

---

[2]   Our belief is that once the core problems are understood, one can begin to relax assumptions within the AHEAD framework itself to address progressively broader design problems.

[3]   The exception, of course, is modularity; hiding the details is what modularity is all about.

complicated, but revealing their essential simplicity. That is what AHEAD does and this is why it scales.

Further, it does not take God to eliminate complexity. All it requires is common sense and the use of common engineering techniques. This has worked for many non-software domains; I and others are showing that it can work for software as well.

7   Conclusions

Just as the structure of matter is fundamental to chemistry and physics, so too must the structure of software be fundamental to Computer Science. By structure, I mean: What are modules? And how are modules composed to build larger modules? Unfortunately, the structure of software is not yet well-understood. Software design, which is the process to define the structure of an application, is an art. As long as it remains an art, our abilities to automate software design and make software development a true engineering discipline are limited.

In short, software design is in desperate need for a science of design. Such a science, I have argued, must be intimately related to automated design. That is, a science of software design will not arise from having virtuoso engineers use their creativity and craftsmanship to create one-of-a-kind products. A science of design must arise from domains where the software development process is mature or reasonably well-understood and where developers can mechanize the process of creating successive variants of programs. Because today's models of software design are not aimed at automation, progress towards a science is understandably lacking.

I believe a science of software design is indeed possible, and have argued that the seminal work on Relational Query Optimization (RQO) is a powerful paradigm that can be emulated in other domains. RQO showed how declarative specifications can be translated into efficient query evaluation programs automatically. The core reason why RQO was successful is because it expressed the design of query evaluation programs algebraically. I presented FOP as a generalization of the RQO paradigm. I believe FOP could be a basis for a science of software design for many reasons:

- it raises the level of modularity from "code" to "design-related-artifacts",
- program design and optimization are expressed algebraically (and thus are ideal for automatic manipulation),
- it allows us to reason about applications in terms of their features (as architects do), and
- it is based on a few simple ideas whose applicability shows no apparent bounds.

Historically, science has progressed by leaps of intuition followed by many years of community debate before the real contributions of a work are understood. The science of software design will be no different; such a science is indeed years away. AHEAD has the right "look and feel" for particular aspects of software design, but that is a far cry from having the software development community accept and appreciate its possibilities. FOP does seem to be a step in the right directions of simplicity, mathematical elegance, and support for automation. These criteria, more than anything else, are the metrics by which advances in software design should be measured.

Acknowledgements

I gratefully acknowledge help from Vicki Almstrum, Dewayne Perry, Chetan Kapoor, Chris Lengauer, and Carron Shankland on improving earlier drafts of this paper.

References

1. R. Balzer, "A Fifteen-Year Perspective on Automatic Programming", *IEEE Transactions on Software Engineering*, November 1985.
2. D. Batory, L. Coglianese, et al., "Creating Reference Architectures: An Example from Avionics", *Symposium on Software Reusability*, Seattle Washington, April 1995.
3. D. Batory, J.N. Sarvela, and A. Rauschmayer, "Scaling Step-Wise Refinement", *International Conference on Software Engineering*, 2003.
4. D. Batory, J. Liu, J.N. Sarvela, "Refinements and Multi-Dimensional Separation of Concerns", *ACM SIGSOFT 2003 (ESEC/FSE2003)*.
5. D. Batory, "The Road to Utopia: A Future for Generative Programming", Dagstuhl on Domain-Specific Program Generation, Springer *LNCS* #3016, 1-17.
6. D. Batory, et al., "Design Wizards and Visual Programming Environments for GenVoca Generators", *IEEE Trans. Software Engineering*, May 2000, 441-452.
7. D. Batory and S. O'Malley. "The Design and Implementation of Hierarchical Software Systems with Reusable Components". *ACM TOSEM*, 1(4):355-398, October 1992.
8. F.P. Brookes, "No Silver bullet: Essence and Accidents of Software Engineering", *IEEE Computer*, April 1987, pp. 10-19.
9. K. Czarnecki and U. Eisenecker. *Generative Programming: Methods, Techniques, and Applications*. Addison-Wesley, 2000.
10. S. Krishnamurthi, "Linguistic Reuse", Ph.D. Rice University, 2001.
11. National Science Foundation, "Science of Design: Software-Intensive Systems", Workshop Proceedings, October 2003.
12. P. Selinger, M.M. Astrahan, D.D. Chamberlin, R.A. Lorie, and T.G. Price, "Access Path Selection in a Relational Database System", *ACM SIGMOD 1979*, 23-34.
13. R.E.K. Stirewalt and L.K. Dillon, "A Component-Based Approach to Building Formal Analysis Tools", *International Conference on Software Engineering*, 2001, 57-70.
14. N. Wirth, "Program Development by Stepwise Refinement", *CACM* Vol. 14, No. 4, April 1971, 221-227.

Glass Box and Black Box Views of State-Based System Specifications

Michel Bidoit[1] and Rolf Hennicker[2]

[1] Laboratoire Spécification et Vérification (LSV), CNRS & ENS de Cachan, France
[2] Institut für Informatik, Ludwig-Maximilians-Universität München, Germany

Abstract. In this talk we will discuss some logical foundations for the semantics of state-based system specifications. In particular we will discuss the need for both:

- a "glass box view" corresponding to the implementor's contract. In this case the underlying paradigm is that the semantics of a specification should be as loose as possible in order to capture all its correct realizations.
- a "black box view" corresponding to the user point of view, and in particular to the properties the user of a realization of a specification can observe or rely upon.

Of course both views should be formally related, and we will explain how this can be done by means of appropriate institution morphisms.

If time permits we will also consider how these ideas could be extended to specifications of object-oriented systems.

C. Rattray et al. (Eds.): AMAST 2004, LNCS 3116, p. 19, 2004.

Abstraction for Safety, Induction for Liveness

Muffy Calder

Department of Computing Science
University of Glasgow
Glasgow, Scotland

1 Introduction

It is very natural to reason about concurrent, communicating systems using model-checking techniques. But, it is not possible to show, by model-checking, that results scale to systems of arbitrary size. Namely, you cannot use model-checking techniques to prove results about about infinite families of about networks of arbitrary size: this is the *parameterised model checking problem* (PMCP) which is, in general undecidable [1].

But in some subclasses of systems PMCP is decidable. We consider two interesting subclasses: proving indexed safety properties by abstraction, and unindexed liveness properties by induction. The latter is suitable when the individual processes degenerate. In both cases we develop some general techniques for state based specification with asynchronous communication, and then illustrate the techniques with a number of examples using the specification language Promela and the model-checker Spin. We discuss under what circumstances these approaches are tractable, and speculate how the techniques can be extended to encompass larger (sub)classes of systems.

References

1. Krzysztof R. Apt and Dexter C. Kozen. Limits for automic verification of finite-state concurrent systems. *Information Processing Letters,* 22:307-309, 1986

C. Rattray et al. (Eds.): AMAST 2004, LNCS 3116, p. 20, 2004.

Counting Votes with Formal Methods

Bart Jacobs

in collaboration with
Engelbert Hubbers, Joseph Kiniry, and Martijn Oostdijk

Security of Systems Group
Department of Computer Science, University of Nijmegen
P.O. Box 9010, 6500 GL Nijmegen, The Netherlands
www.cs.kun.nl/sos

This abstract provides some background information about the electronic voting experiment that is planned in the Netherlands for the European Elections of 2004, and about our own involvement in the infrastructure for this experiment. The talk will elaborate further about the computer security issues involved, especially with respect to the use of formal methods for vote counting software.

Remote Voting

Since the late 1990s voting in the Netherlands proceeds largely via voting machines. These are dedicated computers that record and store votes. These machines are under control of local governments, who put them up in voting stations on election days. These voting machines (and all new versions of them) have undergone independent evaluation before being admitted. However the internal mechanics is secret. Also, the evaluation reports are not public. Nevertheless, at the time of introduction, these machines were uncontroversial. They have been used in several elections, without causing problems. Currently, such machines are the subject of much discussion, see for instance [2].

In 2002 the parliament of the Netherlands adopted a temporary law that went a step further than voting via computers at voting stations. The law allows experimentation with what is called location-independent voting. It has resulted in plans to allow voting via internet and phone in the 2004 elections for the European Parliament. This involves a one-time, limited experiment, largely intended to explore the possibilities and to gather experience with the required techniques and procedures.

Low-Tech Approach

These electronic elections are set up for expatriats. They already have the possibility to participate in elections via voting by (ordinary) mail. To keep things simple, the approach in the electronic elections is modeled after this voting by mail. Hence, participants in the electronic elections are required to register explicitly in advance. Upon registration, they have to submit a copy of their pasport and provide a self-chosen pin-code, for authentication.

The whole organisation is fairly low-tech, and involves various codes for voter identification & authentication, and for candidate selection. In total, thousand different ballots (with different candidate codes) will be distributed randomly, in order to ensure confidentiality.

C. Rattray et al. (Eds.): AMAST 2004, LNCS 3116, pp. 21–22, 2004.
© Springer-Verlag Berlin Heidelberg 2004

The complicated registration procedure thus prevents national adoption of this approach. And because there is no national electronic identity card (yet), more high tech, crypto-based authentication procedures are not an option.

Organisation

The plans for these electronic elections have been elaborated by the Ministry of Internal Affairs, mostly in 2003. There has been an open bidding to build the software for these elections, and to run it as a service. This bidding has been won by LogicaCMG. Also, a panel of independent experts has been set-up, for feedback. The main advice from this panel[1] was to run the project as open as possible, and to compartimentalise it maximally, so that fraud is difficult without cooperation of several parties. The Ministry owns the copyright on the software, and organises its own evaluations – again by several parties. For instance, our group has participated in an evaluation of the robustness of the webservers, during an experiment in nov. 2003. Also, the intention is to make the source code available for inspection[2], but it will probably not appear on the internet, like earlier in Australia[3].

Vote Counting

Late into the project the Ministry decided to open another separate, much smaller bid for the counting of the votes. It has been won by our group, on the basis of a proposal that involves annotating the Java source code with correctness assertions from the Java Modeling Language JML [1]. These annotations are checked both with the run-time assertion checker [4] and with the newest version of the Extended Static Checker, ESC/Java 2 [3], developed in part by our group. Details will be in the talk.

References

1. L. Burdy, Y. Cheon, D. Cok, M. Ernst, J. Kiniry, G.T. Leavens, K.R.M. Leino, and E. Poll. An overview of JML tools and applications. In Th. Arts and W. Fokkink, editors, *Formal Methods for Industrial Critical Systems (FMICS'03)*, number 80 in Elect. Notes in Theor. Comp. Sci. Elsevier, Amsterdam, 2003.
2. D.L. Dill, B. Schneier, and B. Simons. Voting and technology: Who gets to count your vote. *Commun. ACM*, 46(8):29–31, 2003.
3. ESC/Java2. Open source extended static checking for Java version 2 (ESC/Java 2) project. Security of Systems Group, Univ. of Nijmegen
   www.cs.kun.nl/sos/research/escjava/.
4. G.T. Leavens, Y. Cheon, , C. Clifton, C. Ruby, and D.R. Cok. How the design of JML accommodates both runtime assertion checking and formal verification. In F. de Boer, M. Bonsangue, S. Graf, and W.-P. de Roever, editors, *Formal Methods for Components and Objects (FMCO 2002)*, number 2852 in Lect. Notes Comp. Sci., pages 262–284. Springer, Berlin, 2003.

---

[1] Which included the current author.

[2] At the time of writing the details are still unclear, but inspection will probably require an explicit request/registration.

[3] See www.elections.act.gov.au/EVACS.html

Agent-Oriented Programming: Where Do We Stand?

(Invited Talk)

John-Jules Charles Meyer

Institute of Information and Computing Sciences,
Utrecht University,
The Netherlands

In the last decade or so the subject of agent technology has been getting an ever increasing interest in both the fields of software engineering and artificial intelligence [17]. (Intelligent) agents are software (or hardware) entities that display a certain degree of *autonomy* while operating in an environment (possibly inhabited by other agents) that is not completely known by the agent and typically is changing constantly. Agents possess properties like *reactiveness*, *proactiveness* and *social behavior*, often thought of as being brought about by mental or cognitive attitudes involving knowledge, beliefs, desires, goals, intentions,..., in the literature often referred to as '*BDI attitudes*' (for beliefs, desires, intentions).

The concept of an autonomous agent is especially important if the agent is situated in an environment that is inhabited by other agents as well. This brings us to the study of *multi-agent systems*, which incorporates such issues as communication, coordination, negotiation and distributed reasoning/problem solving and task execution (e.g. distributed / cooperative planning and resource allocation). As such the area is part of the field of distributed AI. An important subfield is that of the investigation of agent *communication languages* that enable agents to communicate with other agents in a standardized and high-level manner. Applications of agent technology are numerous (at least potentially), and range from intelligent personal assistants in various contexts to cognitive robots and e-commerce, for example [9].

Historically, after philosophical investigations into the essence of agency (in particular [2]) and several logical proposals of describing the behavior of intelligent agents (e.g. [4, 12, 8], researchers started investigating how these agents could be realized by means of special agent architectures and dedicated agent programming languages such as AGENT_0 ([16]).

We think that especially the idea of developing special agent-oriented programming languages is very interesting and important since this enables us in principle to carry agent concepts through all the way from analysis via design to implementation. In this way one really can use the 'cognitive' concepts employed in the analysis also in the implementation since agent programming languages are equipped with these notions as well (without having to code these notions in some more or less ad hoc way!). That is to say, in theory, since in practice the match is not yet perfect (for instance, since some cognitive notions are lacking in AO programming languages). We believe that we should pursue what Shoham

C. Rattray et al. (Eds.): AMAST 2004, LNCS 3116, pp. 23–26, 2004.

has begun and aim for AO languages that possess the same notions as in the formal/logical descriptions of these agents. (We ourselves have recently extended our own language 3APL with the notion of *declarative goal*, which is generally missing in AO languages [6]). Although agent technology, and agent-oriented programming (AOP) in particular, thus is a very promising field, there are still a number of problems to be overcome, before it could really settle as an established programming paradigm like object orientation is nowadays! We mention a few other issues that should be dealt with.

First of all, proposals for languages should come with a *formal semantics* in order to define language constructs in a precise and unambiguous manner. Our claim is that for this we should reuse concepts and techniques from 'traditional' programming as much as possible: e.g. we advocate the use of structural operational semantics and notions and techniques from process algebra. Fortunately, there are already a number of papers going in this direction, e,g. [10, 11], apart from our own [7, 15].

Also the *correctness* issue should be taken more seriously. Multiagent systems are complicated, as all distributed systems, and to be sure the systems behave as they should, we must be able to verify their correctness in a formal analytical way. (In our view this, too, requires a formal semantics of the language(s) involved!) I believe the correctness issue for agents is still in its infancy. Interest in it is growing since one is getting to realize that it is needed to get reliable systems (cf. the initiative of NASA to enhance the study of formal methods for agent-based systems [13, 14]), but the subfield is certainly not as developed as for e.g. object-oriented programming (OOP). The correctness / verification / specification issue is of course also related to the gap observed in the literature between high-level specifications in agent logics on the one hand and the behavior of implemented practical systems on the other.

Of course, also with respect to the issue of verification it would be unwise to start all over again: we should reuse a lot of theory and tools for non-agent programming styles (in particular OOP; we may think here of Hoare logic, dynamic logic and forms of temporal logic), but it is also clear that crucial adaptations are necessary if one wants to take typical agent-oriented notions such as, for example, BDI and autonomy for individual agents, and roles and normative behavior in agent societies into account as well, so that ideally ideas from traditional programming should be merged with the formalisms and logics that were especially devised for agents (e.g. BDI logic [12]). AOP is not simply an instance or slight variant of OOP...! AOP has its special features that have to be treated. Particular attention should be given to relating the behavior of complex MASs and that of the constituent individual agents. More or less autonomous agents should be restrained within a MAS or agent society by norms enforced by (protocols within) institutions, for example. Implemented institution protocols should be proven correct with respect to the norms they are supposed to enforce.

Finally, and perhaps most importantly, to guide users of the AOP paradigm we need a proper methodology for programming agents in which the typical aspects of agents are captured. Having an appropriate methodology for building agents falls within the area of agent-oriented software engineering (AOSE). Actually, AOSE has two quite distinct meanings:

- the engineering of (general) software using agent concepts in analysis & design phases
- the engineering of software written in an agent-oriented programming language

We think the former is too ambitious and not feasible in general, while the latter is much more restricted and feasible with the advantage that one may hope to use similar concepts in the analysis, the design as well as the implementation of agents, as mentioned before (cf. [5]). We believe that "methodologies for MAS development should assist the developer in making decisions about those aspects of the analysis, design *and implementation*, that are critical for MASs, namely, *social and cognitive concepts (e.g. norms and goals)*" ([5]). Although a number of methodologies for AOP have been proposed in the literature (e.g. Gaia[18] and Tropos[3]), these still suffer from drawbacks and omissions, such as not supporting the implementation phase or not being usable for the important class of *open* systems in which agents may enter and leave the agent society. Therefore, in our view a truly agent-oriented approach is still called for!

Acknowledgments

Many of the ideas expressed in this talk originated from discussions and research done with (former) members of my group, in particular Frank de Boer, Mehdi Dastani, Frank Dignum, Virginia Dignum, Rogier van Eijk, Koen Hindriks, Joris Hulstijn, Wiebe van der Hoek, Meindert Kroese, Cees Pierik, and last but certainly not least, Birna van Riemsdijk.

References

1. F.S. de Boer, C. Pierik, R.M. van Eijk & J.-J. Ch. Meyer, Coordinating Agents in OO, in: *Objects, Agents, and Features* (M.D. Ryan, J.-J. Ch. Meyer & H.-D. Ehrich, eds.), LNCS 2975, Springer, 2004.
2. M.E. Bratman, *Intentions, Plans, and Practical Reason*, Harvard University Press, Massachusetts, 1987.
3. J. Castro, W. Kolp & J. Mylopoulos, Towards Requirements-driven Information Systems Engineering: the TROPOS project, *Information Systems* 27, 2002, pp. 365–389.
4. P.R. Cohen & H.J. Levesque, Intention is Choice with Commitment, *Artificial Intelligence* 42, 1990, pp. 213–261.
5. M. Dastani, J. Hulstijn F. Dignum & J.-J. Ch. Meyer, Issues in Multiagent System Development, to appear in Proc. AAMAS'04, New York, 2004.

6. M. Dastani, B. van Riemsdijk, F. Dignum & J.-J. Ch. Meyer, A Programming Language for Cognitive Agents: Goal-Directed 3APL, in: em Proceedings of the First Workshop on Programming Multiagent Systems: Languages, frameworks, techniques, and tools (ProMAS03), (M. Dastani, J. Dix, A. El Fallah-Seghrouchni & D. Kinny, eds.), Melbourne, 2003, pp. 9-15.

7. R.M. van Eijk, F.S. de Boer, W. van der Hoek & J.-J. Ch. Meyer, Process Algebra for Agent Communication: A General Semantic Approach, in: Communication in Multiagent Systems - Agent Communication Languages and Conversation Policies (M.-Ph. Huget, ed.), LNCS 2650, Springer-Verlag, Berlin, 2003, pp. 113 - 128.

8. W. van der Hoek, B. van Linder & J.-J. Ch. Meyer, An Integrated Modal Approach to Rational Agents, in: *Foundations of Rational Agency* (M. Wooldridge & A. Rao, eds.), Applied Logic Series 14, Kluwer, Dordrecht, 1998, pp. 133–168.

9. N.R. Jennings & M.J. Wooldridge, *Agent technology: Foundations, Applications, and Markets*, Springer, Berlin, 1997.

10. A.F Moreira, R. Vieira & R.H. Bordini, Operational Semantics of Speech-Act Based Communication in AgentSpeak, in: Proc. EUMAS 2003 (M. d'Inverno, C. Sierra & F. Zambonelli, eds.), Oxford, 2003.

11. A. Omicini, A. Ricci & M. Viroli, Formal Specification and Enactment of Security Policies through Agent Coordination Contexts, in: Proc. EUMAS 2003 (M. d'Inverno, C. Sierra & F. Zambonelli, eds.), Oxford, 2003.

12. A.S. Rao & M.P. Georgeff, Modeling Rational Agents within a BDI-Architecture, in *Proceedings of the Second International Conference on Principles of Knowledge Representation and Reasoning (KR'91)* (J. Allen, R. Fikes & E. Sandewall, eds.), Morgan Kaufmann, 1991, pp. 473–484.

13. J.L. Rash, C.A. Rouff, W. Truszkowski, D. Gordon, M.G. Hinchey (eds.),, *Proc. First Goddard Workshop on) Formal Approaches to Agent-Based Systems (FAABS 2000)*, LNAI 1871, Springer, Berlin/Heidelberg, 2001,

14. M.G. Hinchey, J.L. Rash, W.F. Truszkowski, C. Rouff & D. Gordon-Spears (eds.), *Formal Approaches to Agent-Based Systems (Proc. FAABS 2002)*, LNAI 2699, Springer, Berlin, 2003.

15. M.B. van Riemsdijk, F.S. de Boer & J.-J. Ch. Meyer, Semantics of Plan Revision in Intelligent Agents, in this volume, 2004.

16. Y. Shoham, Agent-Oriented Programming, *Artificial Intelligence* 60(1), 1993, pp. 51–92.

17. M.J. Wooldridge & N.R. Jennings (eds.), *Intelligent Agents*, Springer, Berlin, 1995.

18. M.J. Wooldridge, N.R. Jennings & D. Kinny, The Gaia Methodology for Agent-Oriented Analysis and Design, *Autonomous Agents and Multi-Agent Systems* 3(3), 2000, pp. 285–312.

On Guard: Producing Run-Time Checks from Integrity Constraints

Michael Benedikt and Glenn Bruns

Bell Labs, Lucent Technologies

Abstract. Software applications are inevitably concerned with data integrity, whether the data is stored in a database, files, or program memory. An *integrity guard* is code executed before a data update is performed. The guard returns "true" just if the update will preserve data integrity. The problem considered here is how integrity guards can be produced automatically from data integrity constraints. We seek a solution that can be applied in general programming contexts, and that leads to efficient integrity guards. In this paper we present a new integrity constraint language and guard generation algorithms that are based on a rich object data model.

1 Introduction

Every programmer understands the issue of data integrity. In database applications, updates must be checked before they are applied to prevent data corruption. In object-oriented programs, method parameters must be checked to ensure that object attributes are not corrupted. In networking, updates to configuration data and routing tables must be checked.

Ideally, data integrity would be ensured in advance through program verification using static analysis, model checking, or theorem proving. However, large programs and unbounded data values are problematic for model checkers, while theorem provers generally require human guidance. More importantly, many programs accept data from the environment, and this data cannot be checked before run time, even in principle. An alternative to preventing bad updates is to monitor at run-time for the presence of corrupted data. An issue with this approach is how data that has been found to be corrupted can be "repaired" (see e.g., [1]).

In this paper we focus on another run-time approach: the *integrity guard*. An integrity guard is a program that is executed just before a data update is performed. If the guard returns "true" then the update is guaranteed to cause no data corruption. It is desirable that a guard be *exact* – it should return true if and only if the update will not corrupt the data. A guard should also be efficient – it should not interfere with application performance requirements. Guards are often used with constraints that are invariants. When this is the case it is desirable that a guard be *incremental*: it can assume that prior to the update the data is valid. The *guard generation problem* is to automatically

C. Rattray et al. (Eds.): AMAST 2004, LNCS 3116, pp. 27–41, 2004.

generate an exact, incremental, and efficient guard from a data schema, a data update language, and set of data integrity constraints.

Much work on the guard generation problem comes from the database community, where it is known as the "integrity constraint maintenance problem". A related and more general problem is "view maintenance" – recomputing the result of a query as the input data is updated.

Our goal is to make guard generation useful for general-purpose programming. The database community has concentrated on the guard generation problem primarily for the relational data model and for constraint languages whose expressiveness subsumes first-order logic (see Section 5 for details). However, common programming data structures can be hard to model relationally. And in a general programming context, we wish to generate imperative guards, not relational queries. Furthermore, guards generated from arbitrary first-order constraints may have prohibitive execution cost.

Our approach is to use an object data model in which program data structures can be naturally expressed. We use an XPath-based constraint language that can express classical functional and inclusion dependencies and data restriction constraints, but with expressiveness limited enough to ensure that generated guards are inexpensive to evaluate (in particular, much less expensive than for traditional relational query languages). By using XPath we also gain in readability and user-familiarity. Finally, our architecture for guard generation allows guards to be produced for a wide range of data infrastructure.

Thus the main contributions of our work are a) a constraint language that is expressive while having advantages for generation and evaluation of guards b) algorithms for generating low-cost guards and c) a modular implementation framework that allows guard generation in multiple data storage paradigms. The language and algorithms we describe have been implemented as part of the Delta-X system at Lucent. This system is used to generate production code in several complex network management applications.

The paper is organized as follows. Section 2 presents our data model, update model, and constraint language, and gives properties of the constraints that will be useful in guard generation. Section 3 defines formally the incremental precondition problem in general, and presents the high-level structure of our procedures. Section 4 gives a detailed look at guard-generation. Section 5 overviews related work and discusses the implementation and applications of the framework, including experience and open issues.

2   Specifying Data Integrity

In this section we present a simple object data model, a set of data update operations, and our language for expressing constraints.

2.1   Data Model

A *data signature* is a set $A = \{a_1 \ldots a_n\}$ of unary function symbols, where each symbol a_i in A is either of *integer type* or *node type*, and a set *Classes* of class names. An *tree t* of this signature has the form

$$t = (nodes, class, I, child)$$

where $nodes$ is a finite set, $class$ is a mapping from $nodes$ to $Classes$, I is a function interpreting members of A of integer type as functions from $nodes$ to the integers, and members of A of node type as functions from $nodes$ to $nodes$. Function $child$ maps each node n to a sequence $m_1 \ldots m_k$ of distinct nodes, such that the binary relation defined by $m \in child(n)$ forms a tree. For a_i a symbol of A we define $type(a_i)$ to be the integers if a_i has integer type, and $nodes$ if a_i has node type.

This data model is a simplified object model – nodes represent objects, function symbols in A represent attributes, and members of $Classes$ represent class names. The model is simplified in that our model captures only attributes of integer and pointer type. Also, we capture only a weak notion of class in our model, as all objects share the same set of attributes. The absence of an attribute in a class can be modeled through the use of distinguished attribute values. Finally, we do not model class inheritance.

Running Example. Figure 1 shows an example tree that describes a network configuration for routing software in a telecommunication system. The tree represents regions in the network, with attributes storing the minimum and maximum call capacity of each region and the children of a region representing the regions' neighbors. In the figure each box represents an object, with the object's class at the top and its attributes listed below. The system's call-processing component reads this data in order to route calls, while the provisioning component updates the data in response to commands from a technician.

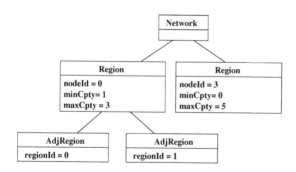

Fig. 1. Data for Running Example

We provide two basic update operations on trees, each of which is parameterized. A create operation $Create(n, C, U)$ for tree t has as parameters a node n of t, a class name C of $Classes$, and elements $U =< u_1, \ldots, u_n >$, with each u_i in $type(a_i)$. The effect of applying $Create(n, C, U)$ to t is that a fresh node n' is added to $nodes(t)$, function $class$ is updated so that $class(n') = C$, each function a_i is updated so that $a_i(n') = u_i$, and function $child$ is extended so that $child(n)$ has additional, final element n'. A delete operation $Delete(n, C)$ for t

has as parameters a leaf node n of t and a class name C of *Classes*. The effect of applying $Delete(n, C)$ to t is that node n of class C is removed from $nodes(t)$.

2.2   Constraint Language

We present a two-tiered language, consisting of an *expression language* and an *assertion language*. The expression language allows one to define sets of nodes in a data tree, while the assertion language consists of statements about sets of nodes.

Our expression language is based on XPath [5], an XML standard that allows one to express queries on an XML document. Evaluating an XPath expression on a tree node yields a set of nodes, or a set of values, or a boolean. In Delta-X we use a simplified form of XPath having the following abstract syntax, where $\propto$ ranges over $\{=, <, \leq, >, \geq\}$, $axis$ ranges over $\{\leftarrow, \leftarrow^*, \rightarrow, \rightarrow^*, \downarrow, \downarrow^*, \uparrow, \uparrow^*\}$, a ranges over attribute names, C ranges over class names, and k ranges over integers:

$$E ::= \epsilon \mid / \mid @a \mid axis \mid C \mid E_1 \cup E_2 \mid [q] \mid E_1/E_2$$
$$q ::= E \mid E_1 \propto E_2 \mid E \propto k \mid class = C \mid q_1 \wedge q_2 \mid q_1 \vee q_2$$

We now briefly describe, in the style of [20], the meaning of expressions by defining the set $E(n)$ of nodes "returned" by expression E relative to a node n of tree t. Expression ϵ returns $\{n\}$. Expression / returns the root node of t. Expression $@a$ returns the singleton set containing the value of attribute a of n. Expression $axis$ returns the nodes related to n by the axis (for example, $\downarrow$ returns the children of a node, $\downarrow^*$ the descendants, and $\rightarrow$ the right siblings). Expression C returns the children in class C of a node. Expression $E_1 \cup E_2$ returns the union of $E_1(n)$ and $E_2(n)$. Expression $[q]$ returns $\{n\}$ if *qualifier* q returns false, and $\emptyset$ otherwise. Expression E_1/E_2 returns the functional composition of E_1 and E_2 (in other words, the union of what E_2 returns relative to each of the nodes in the set E_1 denotes).

A qualifier denotes a node predicate. Qualifier E holds of a node n in tree t iff $E(n)$ is non-empty. Qualifier $E_1 \propto E_2$ holds of n iff $E_1(n)$ contains an element n_1 and $E_2(n)$ contains an element n_2 such that $n_1 \propto n_2$. Similarly, $E_1 \propto k$ holds of n iff $E_1(n)$ contains an element n_1 such that $n_1 \propto k$. Qualifier $class = C$ holds of n iff $class(n) = C$. Qualifiers $q_1 \wedge q_2$ and $q_1 \vee q_2$ provide logical conjunction and disjunction.

Assertions are built up from expressions. Our assertion language supports three types of assertions (or "constraints"). A *restriction constraint* $Never(E)$ asserts that for every node in a data tree, E returns empty on that node.

A *referential constraint* $Reference(E) : Source(E_1 \ldots E_n), Target(F_1 \ldots F_n)$ asserts that for every node n, such that E evaluated at the root satisfies n, there is another node n' in the tree such that for each i, $E_i(n)$ has nonempty intersection with $F_i(n')$.

A *key constraint* $Key(E) : Fields(F_1 \ldots F_n)$ asserts that for any two distinct nodes n_1, n_2 returned by E at the root of a data tree, there is some i such that

$F_i(n_1)$ is distinct from $F_i(n_2)$: that is, a node can be uniquely identified by the values of $F_1, \ldots, F_n$.

We write L_{XP} for the language obtained from this assertion language by taking expressions from XPath. L_{XP} allows one to express a wide range of program invariants.

Example. Returning to the network configuration example, one can express that the minimum capacity of a region must be above the maximum capacity of adjacent regions by

$$Never([class = Region \land$$
$$@minCpty \leq AdjRegion/@regionId/@maxCpty])$$

One can express that a region ID uniquely identifies a region node by

$$Key(\downarrow^* / Region) : Fields(@regionId)$$

One can express that the region ID of every adjacent region points to some node ID of a region by

$$Reference(\downarrow^* / AdjRegion) : Source(@regionId), Target(@nodeId)$$

L_{XP} is strictly more expressive then the standard key and referential constraints of relational and XML data. On the other hand, evaluation of the language is tractable:

Theorem 1. *The problem of evaluating a constraint $\phi \in L_{XP}$ on a tree t can be solved in polynomial time in $|\phi|, |t|$, on a machine that can iterate through t with unit cost for each child, sibling, or parent navigation step.*

The proof follows from the fact that evaluating a XPath expression E can be done in time $|E||t|^2$. This quadratic bound requires a refinement of an argument in [8]. For now we sketch a simpler argument that gives a bound of $|E||t|^3$, and which will be relevant to our later algorithms. An XPath expression can be translated in linear time to a first-order formula in which every subformula has at most three free variables. The evaluation of such formulae on data trees is well known to be in cubic time ([11]) using a bottom-up algorithm. The polynomial bounds now follow easily. A more detailed look at this translation and its target logic is presented in Section 4.

A L_{XP} constraint is called *local* if the initial expression is of the form $\downarrow^* / E$, and all other expressions do not involve navigation of axes or following node-valued (i.e. pointer) attributes. The second and third example constraints above are local. For local constraints, we can get extremely efficient bounds on verification: linear for *Never*, quadratic for all others.

3   Computing Guards from Integrity Constraints

Imagine a system in which condition ϕ currently holds and must continue to hold. To ensure that a state update will not take the system to a state not satisfying ϕ, we want to perform a check such that 1) the check will succeed iff applying the update would leave the system in a state satisfying ϕ, and 2) the check will be inexpensive to perform. We now formalize the problem of finding such a check.

For simplicity we will for a moment treat preconditions in a purely semantic way. Assume a set S of states. Let $op : S \rightarrow S$ be a state update operation and let $\phi : S \rightarrow Bool$ be a state predicate. The *weakest precondition* [7] of ϕ and op, written $wp(op, \phi)$, is the predicate that holds of s exactly when $\phi(op(s))$ holds, for all s in S. A predicate ψ is an *incremental precondition* of ϕ and op if $\phi(s) \Rightarrow (\psi(s) \Leftrightarrow wp(op, \phi))$ holds for all s in S. In other words, ψ is a predicate that acts like $wp(op, \phi)$ for all states in which ϕ holds. Trivially, $wp(op, \phi)$ is an incremental precondition for op and ϕ, and so is the predicate $\neg\phi \lor wp(op, \phi)$. We are interested in incremental preconditions that can be evaluated cheaply, and generally we can expect no relationship between the logical strength of a predicate and its cost. For example, $wp(op, \phi)$ is not necessarily more costly than $\neg\phi \lor wp(op, \phi)$.

In putting these ideas into practice we need languages to describe updates and properties of states. In our case we actually need two property languages: a specification language in which to describe constraints, and an executable target language in which to describe incremental preconditions. Our computational problem is then: given a constraint in the specification language and an operator in the update language, compute a minimal cost incremental precondition in the target language. An issue is whether an incremental precondition exists in the target language for each constraint in the specification language (see [2]).

In Delta-X the specification language is L_{XP} and the update language is the one defined in Section 2, consisting of create and delete operations. These updates have parameters that are instantiated at runtime; we want to generate incremental preconditions with the same parameters.

In producing guards (i.e. incremental preconditions in an executable target language) from L_{XP} constraints, we work in multiple steps using intermediate languages. Delta-X supports several target language and environments (e.g. Java code storing objects within main memory, or C++ code storing objects within a relational database), so we translate constraints to guards via an intermediate language named DIL. This language has the basic control structures of an imperative language plus data types based on our data model.

Another intermediate language is called for because DIL is unsuited to the simplification of guards. We first produce the guards in a first-order logic on trees. This logic, called FOT, is powerful enough to express both the constraints of L_{XP} and the generated guards. The basic precondition generation algorithms are easy to express as transformations on FOT, and as a declarative language it provides a good framework for simplification.

Hence we compute guards in four steps, as shown in Figure 2. We first translate a L_{XP} constraint ϕ into a formula ψ_0 of FOT (see Section 4.1). Next we generate an incremental precondition of ψ_0 within FOT, and then simplify to obtain formula ψ_1. In the next step we translate ψ_1 into a DIL program, which is finally translated into a program in the target programming language. Before going into these steps in detail, we state some results about the algorithm as a whole.

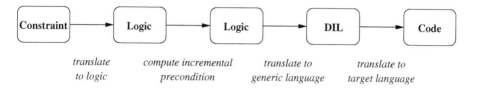

Fig. 2. Code-generation Scheme

Theorem 2. *For any L_{XP} constraint and any update operation, let ψ be the result of the algorithm shown in Figure 2.*

- *ψ runs in time polynomial in the data (where atomic operations of the target language are assumed to have unit cost). In fact, we can give the same bounds for ψ as we can for evaluation of constraints in Theorem 1 – e.g. quadratic in $|t|$ for Never constraints.*
- *If ϕ is a local constraint, then ψ can be evaluated in: constant time for a Never constraint, linear time for a Key constraint, and linear time for a Reference constraint for Create operations.*

On the other hand, there are limits to what one can hope to accomplish for any incremental precondition algorithm:

Theorem 3. *The problem of determining, given a ϕ in L_{XP} and an update operation, whether or not there is an incremental precondition of ϕ with a given quantifier rank, is undecidable. If ϕ consists of only Never constraints, then the problem is decidable but NP-Hard.*

The proof uses the undecidability of the implication problem for XML keys and foreign keys, the decidability of existential first-order logic over graphs, and the intractability of conjunctive query satisfaction.

4   Algorithms

4.1   Translating Constraints to Logic

The first step of our translation is from L_{XP} constraints to *FOT* formulas. The syntax of *FOT* formulas and terms is as follows, where C ranges over *Classes*, k over integers, *axis* over $\{\downarrow, \downarrow^*, \uparrow^*, \rightarrow, \rightarrow^*\}$, and $\propto$ over $\{=, <, \leq, >, \geq\}$:

$$\phi ::= \forall x \in C : \phi \mid \exists x \in C : \phi \mid \phi_1 \wedge \phi_2 \mid \phi_1 \vee \phi_2 \mid \phi_1 \Rightarrow \phi_2 \mid \neg\phi \mid$$
$$class(t) = C \mid t_1 \, axis \, t_2 \mid t_1 \propto t_2$$
$$t ::= x \mid t.a \mid k \mid parent(t)$$

A formula is interpreted relative to an instance t as follows. The logical symbols have their expected meaning. Formula $class(t) = C$ holds if t represents a node of class C. Formula $t_1 \downarrow t_2$ holds if the node represented by t_1 is a child of the node represented by t_2. Other formulae of the form $t_1\, axis\, t_2$ are interpreted similarly. For example, for axis $\downarrow^*$, t_1 must be a descendent of t_2, and for axis $\rightarrow$, t_1 must be the right sibling of t_2. Formula $t_1 = t_2$ holds if the terms represent the same integer values or the same node values. Other formulae of the form $t_1 \propto t_2$ are interpreted similarly.

The term x is a node variable, the term $t.a$ represents the a attribute of the node represented by t, the term k represents integer value k, and the term $parent(t)$ represents the parent of the node represented by t.

We translate a constraint in L_{XP} to a closed formula of FOT in two stages. First, the XPath expressions within the constraint are each translated into open formulas. Second, the constraint itself is translated, with the XPath-related formulas used as parameters.

Example. The *Never* constraint of our running example translates to:

$$\forall x \in Region : \ \forall y \in AdjRegion : \ y \downarrow x \Rightarrow$$
$$\forall z \in Region\ y : regionId = z.nodeId \Rightarrow z.maxCpty < x.minCpty$$

In translating an XPath expression to FOT, we obtain a formula containing free variables x and y. The formula represents a function from an input node x to an output node y. Because of space limitations we cannot present details of the translation. One issue is that an XPath expression can return nodes of different classes. Since our logic provides only for quantification over nodes of a single class, translation requires us to bound the set of classes in which a variable can occur. We do a static analysis to conservatively estimate which classes a particular variable may range over. The analysis can be made more precise in the presence of schema information, such as a DTD.

This translation satisfies two properties. First, ϕ has no subformula containing more than three free variables. As explained in the proof of Theorem 1, this guarantees low evaluation complexity. Second, ϕ has *limited-alternation*. This means that no universal quantifier is found within the scope of a existential quantifier, no negation symbols are present, and in an implication $\psi_1 \rightarrow \psi_1$ the formula ψ_1 contains no universal quantifications and no implications. Precondition algorithms are much simpler for this class.

The translation of assertions is a straightforward implementation of the semantics. This translation produces formulas that are limited-alternation, while our translation of expressions produces formulas that are both limited-alternation and within the three-variable fragment.

4.2   Computing Incremental Preconditions

We now present our incremental precondition algorithm, which accepts as input an update operation and a specification in FOT, and produces an incremental precondition as a formula of FOT.

$$f_{wp}(op, \phi_1 * \phi_2) \overset{\text{def}}{=} f_{wp}(op, \phi_1) * f_{wp}(op, \phi_2) \quad (* \text{ a prop. connective})$$
$$f_{wp}(op, \exists x \in C : \phi) \overset{\text{def}}{=} \exists x \in C : f_{wp}(op, \phi) \vee f_{wp}(op, \phi[x/p])$$
$$f_{wp}(op, \exists x \in D : \phi) \overset{\text{def}}{=} \exists x \in D : f_{wp}(op, \phi[x/p])$$
$$f_{wp}(op, \forall x \in C : \phi) \overset{\text{def}}{=} \forall x \in C : f_{wp}(op, \phi) \wedge f_{wp}(op, \phi[x/p])$$
$$f_{wp}(op, \forall x \in D : \phi) \overset{\text{def}}{=} \forall x \in D : f_{wp}(op, \phi[x/p])$$
$$f_{wp}(op, p \downarrow t) \overset{\text{def}}{=} t = n$$
$$f_{wp}(op, t \downarrow p) \overset{\text{def}}{=} false$$
$$f_{wp}(op, p \downarrow^* t) \overset{\text{def}}{=} n \downarrow^* t$$
$$f_{wp}(op, t \downarrow^* p) \overset{\text{def}}{=} p = t$$
$$f_{wp}(op, t \rightarrow p) \overset{\text{def}}{=} t \downarrow n \wedge \bigwedge_d \forall x \in D : (x \neq t \wedge x \downarrow n) \Rightarrow x \rightarrow t$$
$$f_{wp}(op, p \rightarrow t) \overset{\text{def}}{=} false$$
$$f_{wp}(op, t \rightarrow^* p) \overset{\text{def}}{=} t \downarrow n$$
$$f_{wp}(op, p \rightarrow^* t) \overset{\text{def}}{=} p = t$$
$$f_{wp}(op, axis(t_1, t_2)) \overset{\text{def}}{=} axis(t_1, t_2) \quad (t_i \neq p)$$
$$f_{wp}(op, t_1 \; op \; t_2) \overset{\text{def}}{=} t_1 \; op \; t_2$$
$$f_{wp}(op, t) \overset{\text{def}}{=} t \quad (t \text{ a term or boolean constant})$$

Fig. 3. Weakest Precondition Calculation for Operation $op = Create(n, C, U)$

$$f_{wp}(op, \exists x \in C : \phi) \overset{\text{def}}{=} \exists x \in C : x \neq n \wedge f_{wp}(op, \phi)$$
$$f_{wp}(op, \exists x \in D : \phi) \overset{\text{def}}{=} \exists x \in D : f_{wp}(op, \phi)$$
$$f_{wp}(op, \forall x \in C : \phi) \overset{\text{def}}{=} \forall x \in C : x \neq n \Rightarrow f_{wp}(op, \phi)$$
$$f_{wp}(op, \forall x \in D : \phi) \overset{\text{def}}{=} \forall x \in D : f_{wp}(op, \phi)$$

Fig. 4. Weakest Precondition Calculation for Operation $op = Delete(n, C)$

As a first step towards incremental precondition generation we have a simple inductive algorithm $f_{wp}(op, \phi)$ that calculates a weakest precondition for ϕ under op. Weakest preconditions can be thought of as a default case for incremental integrity checking. In fact, one can obtain an incremental precondition by computing $f_{wp}(op, \phi)$ and then simplifying, using ϕ as an axiom. Instead, we begin with a set of rules that generate incremental checks directly, thus ensuring that the most important simplifications are performed. The algorithms for computing $f_{wp}(op, \phi)$ for Create and Delete operations are shown in Figures 3 and 4. Only the cases that differ from those of Figure 3 are shown in Figure 4. We write $\phi[x/p]$ for the formula obtained by substituting term p for free occurrences of variable x in ϕ. In the figures, p is a parameter for the created object, C the class of p and D an arbitrary class other than C.

Figures 5 and 6 show the incremental precondition algorithms for Create and Delete operations. Again, the Delete case includes only rules differing from the Create case. These rules are valid for the limited-alternation fragment of FOT only. For example, in both the Create and Delete case we use the fact that since limited-alternation formulas are Π_2, there will be no universal quantifiers nested

$$\Delta(op, \phi) \stackrel{\text{def}}{=} true \quad (\phi \text{ does not contain } c)$$

$$\Delta(op, \exists x \in D : \phi) \stackrel{\text{def}}{=} true \quad (d \text{ any class, including } c)$$

$$\Delta(op, \forall x \in C : \phi) \stackrel{\text{def}}{=} \forall x \in C : \Delta(op, \phi) \wedge f_{wp}(op, \phi[x/p])$$

$$\Delta(op, \forall x \in D : \phi) \stackrel{\text{def}}{=} \forall x \in D : \Delta(op, \phi)$$

$$\Delta(op, \phi_1 \wedge \phi_2) \stackrel{\text{def}}{=} \Delta(op, \phi_1) \wedge \Delta(op, \phi_2)$$

$$\Delta(op, \phi_1 \vee \phi_2) \stackrel{\text{def}}{=} (\Delta(op, \phi_1) \wedge \Delta(op, \phi_2)) \vee f_{wp}(op, \phi_1 \vee \phi_2)$$

$$\Delta(op, \phi_1 \Rightarrow \phi_2) \stackrel{\text{def}}{=} f_{wp}(op, \phi_1) \Rightarrow \Delta(op, \phi_2)$$

$$\Delta(op, \phi) \stackrel{\text{def}}{=} \phi \quad (\phi \text{ atomic})$$

Fig. 5. Incremental Precondition Calculation for Operation $op = Create(n, C, U)$

$$\Delta(op, \exists x \in C : \phi) \stackrel{\text{def}}{=} CBF(op, \exists x \in C : \phi) \Rightarrow f_{wp}(op, \phi)$$

$$\Delta(op, \exists x \in D : \phi) \stackrel{\text{def}}{=} true \quad (d \neq c)$$

$$\Delta(op, \forall x \in C : \phi) \stackrel{\text{def}}{=} \forall x \in C : x \neq n \Rightarrow \Delta(op, \phi)$$

$$\Delta(op, \forall x \in D : \phi) \stackrel{\text{def}}{=} \forall x \in D : \Delta(op, \phi)$$

$$\Delta(op, \phi_1 \Rightarrow \phi_2) \stackrel{\text{def}}{=} f_{wp}(op, \phi_1) \Rightarrow \Delta(op, \phi_2)$$

$$CBF(op, \exists x \in C : \phi) \stackrel{\text{def}}{=} \phi[x/n] \quad (c \text{ does not appear free in } \phi)$$

$$CBF(op, \exists x \in C : \phi) \stackrel{\text{def}}{=} (\exists x \in C : x \neq n \wedge CBF(op, \phi)) \vee CBF(op, \phi[x/n])$$
$$(c \text{ appears free in } \phi)$$

$$CBF(op, \exists x \in D : \phi) \stackrel{\text{def}}{=} \exists x \in D : CBF(op, \phi)$$

$$CBF(op, \phi_1 \vee \phi_2) \stackrel{\text{def}}{=} CBF(op, \phi_1) \vee CBF(op, \phi_2)$$

$$CBF(op, \phi_1 \wedge \phi_2) \stackrel{\text{def}}{=} CBF(op, \phi_1) \vee CBF(op, \phi_2)$$

$$CBF(op, \phi) \stackrel{\text{def}}{=} \phi \quad (\phi \text{ atomic})$$

Fig. 6. Incremental Precondition Calculation for Operation $op = Delete(n, C)$

inside existential quantifiers. In the case for implication in Figure 5, we use that if $\phi_1 \Rightarrow \phi_2$, then ϕ_1 must contain only universal quantifiers.

The algorithm for deletion uses an auxiliary function $CBF(op, \phi)$ ("can become false"), defined for every formula without universal quantifiers, which returns true whenever operation op can cause ϕ to change from true to false. We present a basic algorithm for computing *CanBecomeFalse*, but algorithms based on more precise static analyses are possible. Indeed, much more precise, albeit ad-hoc, algorithms are used in our implementation.

Example. In our example constraint, for the operation op of creating a new AdjRegion, the algorithm will produce the incremental precondition:

$$\forall x \in Region. \ p \downarrow x \Rightarrow$$
$$\forall z \in Region. \ p.regionId = z.nodeId \Rightarrow z.maxCpty < x.minCpty$$

Theorem 4. *Algorithm $f_{wp}(op, \phi)$ computes the weakest precondition of ϕ and op, while algorithm $\Delta(op, \phi)$ computes an incremental precondition of ϕ and op.*

One can also verify that for local constraints the output of these algorithms meets the complexity bound of Theorem 2. For general constraints the claims we can make are weaker. First of all, there are cases where the logical complexity of an incremental precondition is higher than that of the original constraint. For example, in a constraint of the form $\exists x \in C : \phi$, a delete operation $Delete(n, C)$ may yield the precondition $\phi(n) \Rightarrow \exists x \in C : x \neq n \wedge \phi(x)$, which is logically more complex but should yield better performance in the "average case". However, one can show that the worst-case running time for evaluation of the precondition can never be more than a constant factor above that for evaluation of the original constraint. This follows because f_{wp} preserves the structure of formulas, hence preserves the running-time bounds, while the running time of Δ is at worst linear in that of f_{wp}.

4.3   Logical Simplification

The Delta-X simplifier is a rewrite system that takes a formula of FOT as input and produces a logically equivalent formula of FOT as output. The quantifier depth and maximum number of free variables of subformulas does not increase through simplification. Indeed, because maximum quantifier depth is a factor that relates strongly to performance of the generated code, a main goal of simplification is to reduce quantifier depth.

The rewrite rules of the simplifier are based on laws of first-order logic and the domain of trees. The following are a few sample rules:

$$\exists x \in A : \phi_1 \vee \phi_2 \rightsquigarrow \phi_1 \vee \exists x \in A : \phi_2 \ (x \text{ not free in } \phi_1) \tag{1}$$

$$x \neq t \vee \phi \rightsquigarrow x \neq t \vee \phi[x/t] \ (x \text{ not free in } t) \tag{2}$$

$$t \downarrow x \rightsquigarrow x = parent(t) \tag{3}$$

Rule 1 captures a validity of first-order logic. Rule 2 is a demodulation rule, allowing one term to be substituted for an equal term. Rules 3 is domain-specific; it says that t is a child of x just when x is the parent of t. Additional rules would be present if schema information were available. For example, rules would capture the relationships given by a class hierarchy.

Example. Figure 7 shows how our system simplifies the example precondition of Section 4.2. The subformula $class(parent(p)) \neq Region$ could be simplified to *false* if schema information showed that the parent of p must be of class Region.

To get a feeling for the cost savings achieved by the simplifier, we generated preconditions from the 83 constraints in the Delta-X regression test suite, which are modelled after constraints in Delta-X applications. The cost of a precondition was computed as n^d, where n is the number of nodes per class, and d is the maximum loop nesting depth in the code generated for the precondition. Assuming 1000 nodes/class, the cost before simplification was 9.6×10^6, and the cost afterward was 3.5×10^6 – a savings of about 63%. Although the cost savings are good for this constraint set, the simplifier lacks robustness. We have had to make extensions to the rule set as new constraint examples appear.

$$\forall x \in Region: \ p \downarrow x \Rightarrow$$
$$\forall z \in Region: \ p.regionId = z.nodeId \Rightarrow z.maxCpty < x.minCpty$$
$$\rightsquigarrow \forall x \in Region: \ \neg p \downarrow x \ \vee$$
$$\forall z \in Region: \ p.regionId \neq z.nodeId \vee z.maxCpty < x.minCpty$$
$$\rightsquigarrow \forall x \in Region: \ parent(p) \neq x \ \vee$$
$$\forall z \in Region: \ p.regionId \neq z.nodeId \vee z.maxCpty < x.minCpty$$
$$\rightsquigarrow \forall x \in Region: \ parent(p) \neq x \ \vee$$
$$\forall z \in Region: \ p.regionId \neq z.nodeId \vee z.maxCpty < parent(p).minCpty$$
$$\rightsquigarrow \forall z \in Region: \ p.regionId \neq z.nodeId \vee z.maxCpty < parent(p).minCpty \vee$$
$$\forall x \in Region: \ parent(p) \neq x$$
$$\rightsquigarrow \forall z \in Region: \ p.regionId \neq z.nodeId \vee z.maxCpty < parent(p).minCpty \vee$$
$$class(parent(p)) \neq Region$$

Fig. 7. Simplification of the Example Precondition

Our rewriting system currently uses a bottom-up rewriting strategy. To increase the power of our system, we are experimenting with alternative rewriting strategies. We are also considering the use of third-party systems (e.g., Simplify [14], PVS [16]), but have yet to find a system that meets our needs. Most systems are geared towards theorem proving rather than simplification (e.g., they use skolemization, which does not preserve logical equivalence). There are licensing issues with most academic systems, and many commercial systems are targeted to special domains, such as hardware design. Finally, we require that simplification be directed by our cost function.

4.4   Translating Logic to Code

We translate formulas of our logic into code via the Delta-X Imperative Language (*DIL*). The translation of *DIL* to popular imperative languages is straightforward, so here we describe only the translation from logic to *DIL*.

DIL looks roughly like C++ and Java. It has classes, methods, statements, and expressions. The types, expressions, and basic statements of the language relate directly to the data model we use. For example, there are expressions to get the attribute of a node and to get the parent of a node. Sequencing, looping, and conditional statements are provided as control structures. However, only special-purpose looping statements are provided: for iterating over all objects of a class, or over all children of a node.

The translation to *DIL* takes a formula of our logic, plus a boolean variable, and produces a statement such that the value assigned by the statement to the variable is exactly the value our semantics dictates for the formula. Terms are translated similarly, except that the computed value need not be boolean.

Example. The following shows the DIL code produced from a formula similar to the simplified precondition of Figure 7.

```
// forall z in Region:
//     p.region <> z.nodeId or z.maxCpty < parent(p).minCpty
```

```
okay := true
for all z in Region while okay {
    // p.region <> z.nodeId or a.maxCpty < parent(p).minCpty
    okay := p.region <> z.nodeId
    if (!(okay)) {
        okay := a.maxCpty < parent(p).minCpty
    }
}
```

The translation of logic to DIL works in a bottom-up fashion on the structure of the formula. The details are mostly straightforward, because most terms of the logic correspond closely to expressions of DIL. However, in the translation of quantifiers one can tradeoff space for time, so we define two translation methods. In the *iterative method*, a quantifier is translated directly to a loop construct of DIL, as in the example above. In the *tabular method*, a quantifier is translated to a mapping represented as a table. The mapping obtained from the universally quantified formula in the example above takes as input a value for free variable p, and produces a boolean value as output.

The iterative translation method produces code that in the worst case requires dk space and n^k time, where d is the maximum size of data values and $O(n^k)$ time, n is the number of nodes in the tree, and k is the maximum quantifier depth of the formula. The tabular method produces code with worst case time and space $O(n^v)$, where v is the maximum number of free variables found in any subformula of the formula.

5   Related Work and Discussion

The relational data model is the basis of much work on integrity constraint maintenance. [15, 3] deal with a rich class of constraints on relational databases, with an emphasis on static verification of transformational programs. Runtime approaches using relational calculus (equivalent in expressiveness to first-order logic) as the specification language include [10, 18, 9]. [12] surveys the problem, dealing with questions such as which additional data structures are to be maintained. [10] gives algorithms that can be used to provide a solution to the integrity constraint maintenance problem for relational calculus. The maintenance of arbitrary first-order constraints is problematic because the evaluation of first-order formulas has PSPACE-complete combined complexity. For this reason the incremental evaluation of such powerful languages has not become a standard part of commercial database management systems.

Richer than the relational model is the object data model, in which program data structures can be captured in a natural way. [13] studies constraint maintenance within an object-oriented database, again using a language that can capture all first-order constraints. Due to the richness of the language [13] cannot provide efficient exact guards, and hence looks for ways to provide weaker guarantees.

There is much recent work on the hierarchical XML data model, which lies in expressiveness between the relational and object models. XML constraint languages have evolved from DTDs, which express purely structural properties, to XML Schema [19], which extends DTDs with further structural features as well as the key and foreign key constraints of [4]. Yet more expressive languages [6] include both structural and data-oriented features. [17] presents integrity constraint algorithms for a subset of XML Schema dealing only with the tree structure, not with data values. To our knowledge no work on incremental constraint checking for more expressive XML constraint languages exists.

The specification language of the Lucent version of Delta-X differs in several ways from the one presented here. The most significant difference is that node expressions allow just the child axis, and that only key constraints must be of the form $Key(\downarrow^* /C)\ldots$ for C a classname. On the other hand, Delta-X allows a generalization of *Never* constraints that restricts the cardinality of a node expression — *Never* asserts that this cardinality is 0; the general version allows any integer upper or lower bound. The production version can make use of schemas specifying for each class C the possible child classes. The Delta-X simplifier allows these additional constraints to be exploited in guard generation – the schema is read in along with the constraints, and a set of simplification rules are dynamically generated. The schema for these applications requires the class containment hierarchy to be acyclic, and hence implies a fixed bound on the depth of the object hierarchy. With this restriction the use of the descendant relation in constraints becomes unnecessary. The absence of sibling axes is due to the fact that in these data-oriented applications the sibling ordering is arbitrary. In addition to *Create* and *Delete*, Delta-X supports a *Modify* update operation on trees, which allows the modification of selected attribute values.

Acknowledgements

Lucent engineers Robin Kuss, Amy Ng, and Arun Sankisa were involved in the inception of the Delta-X project and influenced many aspects of Delta-X, especially the design of the constraint language. Lucent engineers Julie Gibson and James Stuhlmacher contributed to the design of the constraint language and the *DIL* language. We thank Kedar Namjoshi and Nils Klarlund for helpful comments and discussion.

References

1. Marcelo Arenas, Leopoldo Bertossi, and Jan Chomicki. Consistent query answers in inconsistent databases. In *PODS*, pages 68–79, 1999.
2. Michael Benedikt, Timothy Griffin, and Leonid Libkin. Verifiable properties of database transactions. *Infomation and Computation*, 147:57–88, 1998.
3. Véronique Benzaken and Xavier Schaefer. Static integrity constraint management in object-oriented database programming languages via predicate transformers. In *ECOOP '97*, 1997.
4. Peter Buneman, Susan Davidson, Wenfei Fan, Carmem Hara, and WangChiew Tan. Keys for XML. In *WWW 10*, 2001.

5. James Clark and Steve DeRose. *XML Path Language (XPath)*. W3C Recommendation, November 1999. `http://www.w3.org/TR/xpath`.
6. Alin Deutsch and Val Tannen. Containment for classes of XPath expressions under integrity constraints. In *KRDB*, 2001.
7. E.W. Dijkstra. *A Discipline of Programming*. Prentice-Hall, 1976.
8. Görg Gottlob, Chistroph Koch, and Reinhard Pichler. Efficient algorithms for processing XPath queries. In *VLDB*, 2002.
9. Tim Griffin and Howard Trickey. Integrity maintenance in a telecommunications switch. *IEEE Data Engineering Bulletin, Special Issue on Database Constraint Management*, 1994.
10. Timothy Griffin, Leonid Libkin, and Howard Trickey. An improved algorithm for the incremental recomputation of active relational expressions. *TKDE*, 9:508–511, 1997.
11. Martin Grohe. Finite variable logics in descriptive complexity theory. *Bulletin of Symbolic Logic*, 4, 1998.
12. Ashish Gupta and Inderpal Mumick. *Materialized Views: Techniques, Implementations, and Applications*. MIT Press, 1999.
13. H.V. Jagadish and Xiaolei Qian. Integrity maintenance in an object-oriented database. In *VLDB*, 1992.
14. K. Rustan M. Leino and Greg Nelson. An extended static checker for Modula-3. In *Compiler Construction: 7th International Conference*, 1998.
15. William McCune and Lawrence Henschen. Maintaining state constraints in relational databases: A proof theoretic basis. *Journal of the ACM*, 36(1):46–68, 1989.
16. S. Owre, S. Rajan, J.M. Rushby, N. Shankar, and M.K. Srivas. PVS: Combining specification, proof checking, and model checking. In *CAV*, pages 411–414, 1996.
17. Yannis Papakonstantinou and Victor Vianu. Incremental validation of XML documents. In *ICDT*, 2003.
18. Xiaolei Qian. An effective method for integrity constraint simplification. In *ICDE*, pages 338–345, 1988.
19. Henry S. Thompson, David Beech, Murray Maloney, and Noah Mendelsohn. *XML Schema Part 1: Structures*. W3C Working Draft, April 2000. `http://www.w3.org/TR/xmlschema-1/`.
20. Philip Wadler. Two Semantics for XPath. Technical report, Computing Sciences Research Center, Bell Labs, Lucent Technologies, 2000.

Behavioural Types and Component Adaptation*

Antonio Brogi[1], Carlos Canal[2], and Ernesto Pimentel[2]

[1] Department of Computer Science, University of Pisa, Italy
brogi@di.unipi.it
[2] Department of Computer Science, University of Málaga, Spain
{canal,ernesto}@lcc.uma.es

Abstract. Component adaptation is widely recognised to be one of the crucial problems in Component-Based Software Engineering. The objective of this paper is to set a formal foundation for the adaptation of heterogeneous components that present mismatching interaction behaviour. The proposed adaptation methodology relies on: (1) the inclusion of behavioural types in component interfaces, to describe the interaction behaviour of components, (2) a simple high-level notation for adaptor specification, to express the intended connection between component interfaces, and (3) a formal definition of adaptor, a component-in-the-middle capable of making two components interoperate successfully according to a given specification.

1 Introduction

Component adaptation is widely recognised to be one of the crucial problems in Component-Based Software Engineering [4, 5, 10]. The capability of adapting off-the-shelf software components to work properly within larger applications is a must for the development of a true component marketplace, and for component deployment in general [3]. The need for component adaptation is also motivated by the ever-increasing attention devoted to developing extensively interacting distributed systems, consisting of large numbers of heterogeneous components.

Available component-oriented platforms address software interoperability at the signature level by means of Interface Description Languages (IDLs), a sort of *lingua franca* for specifying the functionality offered by heterogeneous components. While IDLs allow to overcome signature mismatches, there is no guarantee that the components will interoperate correctly, as mismatches may occur because of differences in the interaction behaviour of the components involved [11].

The objective of this paper is to set a formal foundation for the *adaptation* of heterogeneous components that may present mismatching interaction behaviour. The notion of *adaptor* was originally introduced in [13] to name a component-in-the-middle aimed at enabling the successful interoperation of two components presenting mismatching behaviour.

---

* This work has been partly supported by the project NAPOLI funded by the Italian Ministry of Instruction, University and Research (MIUR), and the projects TIC2002-4309-C02-02 and TIC2001-2705-C03-02 funded by the Spanish Ministry of Science and Technology (MCYT).

C. Rattray et al. (Eds.): AMAST 2004, LNCS 3116, pp. 42–56, 2004.

In our previous work [2], we have developed a formal methodology for component adaptation that supports the successful interoperation of heterogeneous components presenting mismatching interaction behaviour. As pointed out in [13], the first step needed to overcome behavioural mismatch is to let behaviour information be explicitly represented in **component interfaces**. Process algebras feature a very expressive description of interaction protocols, and enable sophisticated analyses of concurrent systems. For these reasons, their use for the specification of component interfaces and for the analysis of component compatibility has been widely advocated (e.g. [1, 7, 9]). On the other hand, a serious drawback of employing process algebras is the inherent complexity of verification procedures, which inhibits their usability in practice.

Therefore, a suitable trade-off between expressiveness and effectiveness of verification is needed to reason about component protocols. To this end, we propose the notion of *session* for the description of component interfaces. Intuitively speaking, sessions feature a modular projection of component behaviour both in space and in time. On the one hand, each session describes a partial view of the behaviour of a component (w.r.t. another component that will be attached to it), thus partitioning the component behaviour into several facets or roles. On the other hand, each session describes a (possibly finite) connection, thus partitioning the full life of a component into a sequence of sessions.

From a technical viewpoint, we will use *session types* (firstly defined in [6]) to describe component sessions as true types. The ultimate objective of employing session types is to provide a basic means to describe complex interaction behaviour with clarity and discipline at a high-level of abstraction, together with a formal basis for analysis and verification. Session types are supported by a rigorous type discipline, thus featuring a powerful type checking mechanisms of component behaviour. Moreover, the use of types – instead of processes – to describe behaviour features the possibility of describing recursive behaviour while maintaining the analysis tractable.

The second ingredient necessary to address formally the issue of component adaptation is a suitable formalism to express **adaptor specifications**. Indeed, as shown in [2], separating adaptor specification and adaptor derivation permits the automation of the error-prone, time-consuming task of manually constructing a detailed implementation of a correct adaptor. The desired adaptation will be expressed by simply defining a set of (possibly non-deterministic) correspondences between the actions provided by the two components to be adapted. As we will see, the distinguishing aspect of the notation used is that it produces a high-level, partial specification of the adaptor.

In a third step, we define formally the notion of **adaptor** in terms of an adaptor specification and of the session types of the components involved. Finally, we prove that adaptors guarantee the safe interaction of the components adapted, in the sense that they will never deadlock during an interaction session.

The rest of the paper is organized as follows. Sect. 2 is devoted to introduce session types for typing component behaviour. After defining a process calculus to denote component protocols, a type system is introduced and the notion of type compatibility is presented. In Sect. 3 adaptor specifications and the formal definition of adaptor are introduced, and the session-safety result is established. Finally some concluding remarks are drawn in Sect. 4.

2   Typing Component Behaviour

Process algebras have been widely used to specify software systems. In particular, they have been often employed to describe the interactive behaviour of software components [1, 7, 9]. The advantage of having these formal specifications of components is two-fold. First, component behaviour is unambiguously fixed and it can be animated with a convenient interpreter tool. Second, it is possible to analyze a number of liveness and security properties such as safe composition or replaceability in complex systems. In spite of the usefulness of process algebras for component description, they present an important drawback due to the complexity of the decision procedures to verify the mentioned properties. In order to cut off this complexity, we have applied to our context the notion of *session type* introduced in [6].

Session types present some important features that distinguish them from processes written in a general process algebra like the π-calculus:

- session types abstract from data values, referring to the corresponding data types instead;
- sessions are limited to diadic communications between two components;
- mobility is expressed by means of explicit *throw/catch* actions, and since sessions are diadic, once a process throws a session, it cannot use it anymore;
- no mixed alternatives are allowed: input and output actions cannot be combined in a non-deterministic choice.

It is worth noting that these restrictions are not relevant limitations in our context, as we will show. On the contrary, they make session types a calculus much more tractable than other studied alternatives, like the π-calculus or CSP. A thorough discussion of the advantages of employing session types instead of other concurrency formalisms is reported in [12]. Under this approach, a process is viewed as a collection of sessions, each one being a chain of diadic interactions. Each session is designated by a private session name, through which interactions belonging to that session are performed. The use of diadic sessions for the specification of software interaction allows a modular description of complex systems. The objective is to provide a basic means to specify complex interaction behaviour with clarity and discipline at a high-level of abstraction, together with a formal basis for analysis and verification.

Throughout the paper, we will use both a session type description language and a process calculus $\mathcal{L}$. The former will be used to type the behaviour of components (and will be exposed in component interfaces), while the latter will be used to refer to (and exemplify) the actual implementation of components.

2.1   A Process Calculus for Component Description

In this section we present the process calculus for describing component implementations. It is a variant of that used in [6]. Apart from some simplifications in the notation, the main difference is that we allow the alternative composition of output actions (somehow equivalent to an *if-then-else* construct), and not only of input actions as in [6]. We give also a transition system for the calculus, not present in [6]. The syntax of the process calculus $\mathcal{L}$ is defined as follows:

$$P ::= 0 \mid act.P \mid \sum_i k!m_i.P_i \mid \sum_i k?m_i.P_i \mid P_1 \parallel P_2 \mid A(\tilde{x}\tilde{k})$$

$$act ::= x!request(k) \mid x?accept(k) \mid k!throw(k') \mid k?catch(k')$$

where 0 represents the empty process, P, P_i denote a process, A is a process identifier, x denotes a link name, k and k' denote session names, $\tilde{\cdot}$ denotes a sequence of names, and m_i denotes a message, syntactically composed by a message selector and a sequence of data arguments.

For any process identifier $A(\tilde{x}\tilde{k})$ there must be a unique defining equation $A(\tilde{x}\tilde{k}) = P$. Then, $A(\tilde{y}\tilde{h})$ behaves like $P\{\tilde{y}/\tilde{x}, \tilde{h}/\tilde{k}\}$. Defining equations provide recursion, since P may contain any process identifier, even A itself.

We consider two kinds of actions in the process calculus $\mathcal{L}$: output actions $(k!m_i)$, where a message m_i is sent through a session k, and input actions $(k?m_i)$, where a message m_i is received through a session k. There are four special message selectors: *request*, *accept*, *throw*, and *catch*. All of them require a single argument representing a session name. A *request* output action issued on a link name x waits for the acceptance (*accept*) of a session on this link. When these two actions synchronize, a new session is created linking the processes where the interaction was performed. Similarly, *throw* and *catch* are complementary actions, too. In this case, an existing session (name) can be moved from a process (where the *throw* action is made) to another one (where a *catch* action is performed to capture the session). These last two actions permit to change dynamically the topology of the system. Notice that link names are only used to create sessions (via *request/accept* actions), while all other interactions take place on session names.

The transition relation described in Fig. 1 defines the operational semantics of $\mathcal{L}$. There are both labelled and unlabelled transitions, the latter corresponding to silent actions. The four initial rules describe the behaviour of actions concerning session manipulations. Rules (REQ) and (ACC) model session creation, while (THR) and (CTH) model session transmission. Both *accept* and *catch* are binding actions, receiving a new session name not occurring in $fn(P)$ (free names of P). Rule (SUM) defines the behaviour of a sum of (either input or output) actions $\lambda_i = k?m_i$ or $\lambda_i = k!m_i$, respectively, which is modelled by the usual non-deterministic choice, assuming that the choice is made locally for output actions, and globally for input actions. To model the parallel composition of processes we have four different transition rules. Rules (PAR$_{open}$) and (PAR$_{throw}$) model session opening and session throwing, respectively, whereas (SYNC) models the synchronous exchange of input and output messages. Rule (PAR) describes the standard interleaving of parallel compositions. Finally, rule (DEF) models process definition. Note that the label λ in these last two rules may be empty.

$$\text{REQ}: \frac{}{x!request(k).P \xrightarrow{x!rqt(k)} P} \qquad \text{THR}: \frac{}{k!throw(k').P \xrightarrow{k!thw(k')} P}$$

$$\text{ACC}: \frac{}{x?accept(k).P \xrightarrow{x?acp(h)} P\{h/k\}} \quad (h \notin fn(P) \setminus \{k\})$$

$$\text{CTH}: \frac{}{k?catch(k').P \xrightarrow{k?cth(h)} P\{h/k'\}} \quad (h \notin fn(P) \setminus \{k'\})$$

$$\text{SUM}: \frac{}{\sum_{i=1}^{n} \lambda_i.P_i \xrightarrow{\lambda_j} P_j} \quad (j = 1 \cdots n)$$

$$\text{PAR}_{open}: \frac{P \xrightarrow{x!rqt(k)} P' \quad Q \xrightarrow{x?acp(k)} Q'}{P \parallel Q \longrightarrow P' \parallel Q'}$$

$$\text{PAR}_{throw}: \frac{P \xrightarrow{k!thw(k')} P' \quad Q \xrightarrow{k?cth(k')} Q'}{P \parallel Q \longrightarrow P' \parallel Q'}$$

$$\text{SYNC}: \frac{P \xrightarrow{k!m} P' \quad Q \xrightarrow{k?m} Q'}{P \parallel Q \longrightarrow P' \parallel Q'} \qquad \text{PAR}: \frac{P \xrightarrow{\lambda} P'}{P \parallel Q \xrightarrow{\lambda} P' \parallel Q}$$

$$\text{DEF}: \frac{P\{\tilde{y}/\tilde{x}, \tilde{h}/\tilde{k}\} \xrightarrow{\lambda} P'}{A(\tilde{y}\tilde{h}) \xrightarrow{\lambda} P'} \quad (A(\tilde{x}\tilde{k}) = P)$$

Fig. 1. Transition system for $\mathcal{L}$.

Additionally to the transition system in Fig. 1, we assume also standard commutativity and associativity axioms for choice and parallel composition operators.

Throughout this paper we will use a simplified FTP system to illustrate the different aspects of our proposal. Suppose that the system is composed of a server, with whom clients interact for opening FTP sessions, and a set of n daemons, each one responsible for handling one client session:

$$System(client, daemon) = Server(client, daemon) \parallel \prod_{i=1}^{n} Daemon(daemon)$$

Suppose also that the specification in $\mathcal{L}$ of the *Server* component is:

```
Server(client,daemon) =
 client?accept(a). a?user(usr). a?pass(pwd).
 ( a!rejected("User unauthorized"). Server(client,daemon)
  + a!rejected("Service unavailable"). Server(client,daemon)
  + a!connected!(). daemon!request(b). a!throw(b). Server(client,daemon) )
```

The FTP `Server` component declares two link names (`client` and `daemon`). The first one is used for its interaction with its clients, while the second is used for interacting with the daemons. When a request is received on link `client`, a new session `a` is created, for handling the interaction between the system and this specific client. Then, the client identifies itself, providing its name and password. As a result, the server may either reject (message `rejected`), or accept (message `connected`) the connection. In the latter case, the server requests on link `daemon` a session b from one of its daemons, and throws the newly created session b to the client. Session a ends now, and the server returns to its original state, ready for attending a new client request.

On the other hand, the daemon component can be represented as follows:

```
Daemon(server) = server?accept(b). Daemon2(server,b)
Daemon2(server,b) = ( b?get(filename).
                      ( b!result(data). Daemon2(server,b)
                      + b!nosuchfile(). Daemon2(server,b) )
                    + b?quit(). Daemon(server) )
```

The specification above shows how once a new session b is established with the server, the daemon waits for different user commands (here, only get is considered, the rest being alike), finishing with a quit command, after which the session ends, and the daemon is ready for accepting a new session request. After each get command, either the corresponding data (result) or an error (nosuchfile) is returned. Notice how the daemon remains unaware of the fact that the session b initially requested by the server is afterwards sent to the client, which is the one who actually performs get and quit commands.

However, our interest is not focused on using a process calculus like $\mathcal{L}$ for describing the behaviour of software components, but rather in *typing* this behaviour for establishing the correct interaction among the corresponding components. This is the objective of the next section.

2.2 Typing System

Whereas the type system defined in [6] deals both with data and session types, without loss of generality we shall omit data arguments in process definitions and message arguments. This simplification is not relevant, and our typing system could be easily extended to deal also with data. We will denote by *TExp* the set of type expressions constructed by following grammar:

$$\alpha ::= 0 \mid \perp \mid !\alpha \mid ?\alpha \mid !(\alpha).\beta \mid ?(\alpha).\beta \mid ! \sum_i t_i.\alpha_i \mid ? \sum_i t_i.\alpha_i \mid \Lambda$$

where α, α_i, β are type expressions, and Λ is a type identifier (we assume that for each identifier Λ exists a unique defining equation $\Lambda = \alpha$). The constant type 0 represents inaction's type, and $\perp$ denotes a specific type indicating that no further connection is possible at a given session. In other words, if a session k has a type $\perp$ in a process, then k is not offered by this process as an open session. Type expressions $!\alpha$ and $?\alpha$ correspond to *request* and *accept* primitives, respectively, whereas $!(\alpha)$ and $?(\alpha)$ correspond to *throw* and *catch*. The expression t_i denotes the type associated to a message (which will coincide with the message selector, since we abstract from message arguments). Then, the type $? \sum_i t_i.\alpha_i$ denotes the branching behaviour given by a process which is waiting with several options, and which behaves as type α_i if the i-th action is selected. Similarly, $! \sum_i t_i.\alpha_i$ denotes the complementary selecting behaviour, w.r.t. output actions.

Given a type α where $\perp$ does not occur, we define its dual type $\bar{\alpha}$, as follows:

$$\begin{array}{llll}
\overline{?\alpha} = !\bar{\alpha} & \overline{?(\alpha).\beta} = !(\alpha).\bar{\beta} & \overline{!\sum_i t_i.\alpha_i} = ?\sum_i t_i.\bar{\alpha_i} & \overline{0} = 0 \\
\overline{!\alpha} = ?\bar{\alpha} & \overline{!(\alpha).\beta} = ?(\alpha).\bar{\beta} & \overline{?\sum_i t_i.\alpha_i} = !\sum_i t_i.\bar{\alpha_i} &
\end{array}$$

The dual of a given type denotes the complementary behaviour of the original type, and $\bar{\bar{\alpha}} = \alpha$. It is obvious that the composition of a type and its dual is successful, in the sense that the corresponding processes will execute without deadlocks [6], eventually terminating in empty processes. However, imposing such a condition seems too restrictive in the context of real software components.

In [12] a notion of type compatibility is defined in terms of a subtyping relation and type duality. Intuitively speaking, a session type α is a subtype of β if α can be used in any context where β is used, and no error occurs in the session. Basically, this means that α should have more – or equal – branchings (input alternatives), and less – or equal – selections (output alternatives). Based on this subtyping relation, α is said compatible with β, denoted by $\alpha \bowtie \beta$, if α is a subtype of the dual of β. Defined this way, compatibility is a sufficient condition to ensure successful composition of the corresponding processes. (More details about these subtyping and compatibility relationships can be found in [12].)

The typing system for the calculus $\mathcal{L}$ is shown in Fig. 2, and deals with sequents of the form: $\Theta; \Gamma \vdash P \triangleright \Delta$, which means: "under the current environment, given by Θ and Γ, the process P has a typing Δ". As in [6], Θ denotes a *basis*, which is a mapping from process names to the types in *TExp* of the corresponding arguments (links and sessions); while the *sorting* $(t \in)\Gamma$ stores types for links, which are expressed by means of sorts $\langle \alpha, \bar{\alpha} \rangle$. A sort of this form represents a pair of complementary interactions which are associated with a link name: one starting with *accept*, and the other one starting with *request*. Each of them correspond to the type of certain session in the *typing* Δ, which is a partial mapping from session names to types. Given a typing (or sorting or basis) Ξ, we write $\Xi \cdot k : \alpha$ to denote the mapping $\Xi \bigcup \{k : \alpha\}$ provided that $k \notin dom(\Xi)$.

Using session types for describing component behaviour makes it possible to determine when two components can interact safely. This analysis will be done in terms of the compatibility of the typings of the components. Indeed, the notion of compatibility in [12], previously mentioned, can be naturally extended to typings. When two typings, Δ_1 and Δ_2, are compatible ($\Delta_1 \bowtie \Delta_2$), their composition ($\Delta_1 \odot \Delta_2$) is defined as a new typing given by:

$$(\Delta_1 \odot \Delta_2)(k) = \begin{cases} \perp & \text{if } k \in dom(\Delta_1) \cap dom(\Delta_2) \\ \Delta_1(k) & \text{if } k \in dom(\Delta_1) \setminus dom(\Delta_2) \\ \Delta_2(k) & \text{if } k \in dom(\Delta_2) \setminus dom(\Delta_1) \end{cases}$$

As we have already said, the typing system in Fig. 2 is similar to that provided in [6], but adapted to the process calculus $\mathcal{L}$. The first two rules define the sort associated to a link x, on which an *accept* or *request* is made. Notice that the sort for x is a pair composed by the session type α and its dual, α being the derived type for the session opened on x. Rule $\mathcal{T}_{\text{CTH}}$ types a *catch* action, assigning to

$$\mathcal{T}_{\text{ACC}} : \quad \frac{\Theta; \Gamma \vdash P \triangleright \Delta \cdot k : \alpha}{\Theta; \Gamma, x : \langle \alpha, \bar{\alpha} \rangle \vdash x?accept(k).P \triangleright \Delta \cdot k :?\alpha}$$

$$\mathcal{T}_{\text{REQ}} : \quad \frac{\Theta; \Gamma \vdash P \triangleright \Delta \cdot k : \bar{\alpha}}{\Theta; \Gamma, x : \langle \alpha, \bar{\alpha} \rangle \vdash x!request(k).P \triangleright \Delta \cdot k :!\alpha}$$

$$\mathcal{T}_{\text{THR}} : \quad \frac{\Theta; \Gamma \vdash P \triangleright \Delta \cdot k : \beta}{\Theta; \Gamma \vdash k!throw(k').P \triangleright \Delta \cdot k :!(\alpha).\beta \cdot k' : \alpha}$$

$$\mathcal{T}_{\text{CTH}} : \quad \frac{\Theta; \Gamma \vdash P \triangleright \Delta \cdot k : \beta \cdot k' : \alpha}{\Theta; \Gamma \vdash k?catch(k').P \triangleright \Delta \cdot k :?(\alpha).\beta}$$

$$\mathcal{T}_{\text{IN}} : \quad \frac{\Theta; \Gamma \vdash P_1 \triangleright \Delta \cdot k : \alpha_1 \quad \cdots \quad \Theta; \Gamma \vdash P_n \triangleright \Delta \cdot k : \alpha_n}{\Theta; \Gamma \vdash \sum_{i=1}^{n} k?m_i.P_i \triangleright \Delta \cdot k : ? \sum_{i=1}^{n} m_i.\alpha_i}$$

$$\mathcal{T}_{\text{OUT}} : \quad \frac{\Theta; \Gamma \vdash P_1 \triangleright \Delta \cdot k : \alpha_1 \quad \cdots \quad \Theta; \Gamma \vdash P_n \triangleright \Delta \cdot k : \alpha_n}{\Theta; \Gamma \vdash \sum_{i=1}^{n} k!m_i.P_i \triangleright \Delta \cdot k : ! \sum_{i=1}^{n} m_i.\alpha_i}$$

$$\mathcal{T}_{\text{PAR}} : \quad \frac{\Theta; \Gamma \vdash P \triangleright \Delta \quad \Theta; \Gamma \vdash Q \triangleright \Delta'}{\Theta; \Gamma \vdash P \parallel Q \triangleright \Delta \odot \Delta'} \quad (\Delta \bowtie \Delta')$$

$$\mathcal{T}_{\text{INACT}} : \quad \frac{}{\Theta; \Gamma \vdash 0 \triangleright \Delta}$$

$$\mathcal{T}_{\text{DEF}} : \quad \frac{\Theta \cdot A : \tilde{t}\tilde{\alpha}; \Gamma \cdot \tilde{x} : \tilde{t} \vdash P \triangleright \Delta \cdot \tilde{k} : \tilde{\alpha}}{\Theta \setminus A; \Gamma, \tilde{y} : \tilde{t} \vdash A(\tilde{y}\tilde{h}) \triangleright \Delta \cdot \tilde{h} : \tilde{\alpha}} \quad (A(\tilde{x}\tilde{k}) = P)$$

$$\mathcal{T}_{\text{VAR}} : \quad \frac{}{\Theta \cdot A : \tilde{t}\tilde{\alpha}; \Gamma, \tilde{y} : \tilde{t} \vdash A(\tilde{y}\tilde{h}) \triangleright \Delta \cdot \tilde{h} : \tilde{\alpha}}$$

Fig. 2. Typing system for the calculus $\mathcal{L}$.

the captured session k' a type α, which is provided by the rule $\mathcal{T}_{\text{THR}}$ in the corresponding *throw* action. Rules $\mathcal{T}_{\text{IN}}$ and $\mathcal{T}_{\text{OUT}}$ define the expected type for a sum of input and output actions, respectively, where with abuse of notation, we still use m_i to denote the type of the message m_i. Rule $\mathcal{T}_{\text{PAR}}$ defines the synchronization among processes having compatible types; the resulting type is given by their composition. Finally, the rules $\mathcal{T}_{\text{DEF}}$ and $\mathcal{T}_{\text{VAR}}$ define the types for process definitions, where the information accumulated on the basis Θ about process variables may be used to type recursive definitions. We assume that the range of Δ in $\mathcal{T}_{\text{INACT}}$ and $\mathcal{T}_{\text{VAR}}$ contains only 0 and $\perp$.

If a type sequent $\Theta; \Gamma \vdash A(\tilde{x}\tilde{k}) \triangleright \Delta$ is derivable from the typing system, we say that the process A is *typable*, and its type, denoted by $[A]$, is determined by the types associated to each link x of A in Γ, and to each session k of A in Δ. We write $[A]_x$ to denote the session type for x in the process A. Given a link x of a typable component A, we will denote by k_x the session that A opens on link x. Then, we have that $[A]_x$ is the session type α (respectively, $\bar{\alpha}$) such that $x : \langle \alpha, \bar{\alpha} \rangle$ is in Γ, and the session k_x has type α (respectively, $\bar{\alpha}$) in Δ. Thus, from the point of view of A, the type of a link x is the type of the session k_x opened on that link. Hence, we usually write $[A]_x$ and $[A]_{k_x}$ interchangeably.

Let us show now which are the types corresponding to the FTP system of our example. Starting from the daemon component, its type $[Daemon]$ is given by the type DAEMON_b, assigned accordingly to the typing system in Fig. 2 to the session b that a daemon establishes with the server:

```
DAEMON_b =  ? DAEMON2_b
DAEMON2_b = ?(get.  !(result.DAEMON_b + nosuchfile.DAEMON_b) + quit.0)
```

On the other hand, the type of the server component [Server] defines two session types: SERVER_a for the session a it establishes with the client, and SERVER_b for the session b it establishes with the daemons:

```
SERVER_a = ? ?user. ?pass.
                !( rejected. 0 + connected. !(SERVER_b). 0 )
```

and, from rule $\mathcal{T}_{THR}$, we have that SERVER_b is precisely the type dual of DAEMON_b (and notice that by definition of $\odot$ their composition is $\bot$). Consequently, the type for the whole FTP system, obtained composing the typings of the server and the daemons, is:

$$[System] = \{ \ a : [Server]_a, \ b : [Server]_b \odot [Daemon]_b \ \} = \{ \ a : \texttt{SERVER\_a}, \ b : \bot \ \}.$$

We can see several interesting differences between the specification in $\mathcal{L}$ of the components previously shown and their corresponding session types above. First of all, session types describe the behaviour of the component during a single session (that is, from the point where a new session name is created in an *accept/request*, to the moment where this session is no more used, and its name is lost). Hence, session types describe usually a finite pattern of actions (as it can be seen in particular in the session types of the server component). Second, session types separate the interleaving of actions performed on different session names, allowing a modular description of interactions. On the contrary, the process algebra specification describes the full behaviour of the components, interleaving actions from different sessions (as in the case of the server component), and is usually recursive to an *accept* or *request* action.

Now, we could write the session type of an FTP client component perfectly compliant (that is, dual) with the server and the daemon described so far, ensuring successful interaction among them. However, our objective is to deal with component adaptation, and for this reason, a more interesting (although very simple) client is represented below (on the left the component, on the right the corresponding type):

```
Client(server) = server!request(c).      CLIENT_c = !
                     c!login(usr,pwd).              !login.
                     c!download(filename).          !download.
                     c?data(filedata). 0            ?data. 0
```

It is easy to see that the above client is not compatible with our FTP system (and obviously the corresponding components would deadlock when composed in parallel). Indeed, the more relevant differences between them are:

- The name of the messages used in both types are different, though we could guess some correspondences between them (e.g., get and download).

- Also the protocols of the components are different. For instance, the `DAEMON_b` type recurs until a `quit` message is received, while the client just performs a single `download` command.
- The client ignores the throwing of a daemon session from the server. Instead, its behaviour is represented by a single session.
- Finally, the client unwisely ignores any error message coming from the server.

3   Adaptor Specification

In this section, we introduce a simple, high-level notation for describing the intended correspondence among the functionality of two components being adapted. This correspondence is specified by means of a mapping that establishes a number of rules relating messages of both components. This specification will be then used for the automatic construction of an adaptor that mediates the interaction of the two components.

A detailed description of the notation for adaptor specification can be found in [2]. Here, we will show only how we can use it for accommodating the differences between the types `SERVER_a` and `CLIENT_c` above. The intended adaptation between them may be represented by the following specification:

```
S = { !login(usr,pwd)        <> ?user(usr), ?pass(pwd);   // r1
                             <> !rejected(msg);           // r2
                             <> !connected();             // r3
      !download(filename)    <> ?get(filename);           // r4
      !download(filename)    <> ;                         // r5
      ?data(filedata)        <> !result(filedata);        // r6
      ?data("No such file")  <> !nosuchfile();            // r7
      ?data("Not connected") <> ;                         // r8
                             <> ?quit(); }                // r9
```

where the actions of the client are represented in the left terms while those of the server (and its daemons) are on the right.

The specification `S` establishes a correspondence between messages in both components. For instance, the `login` output message in the client is mapped to a pair of `user` and `pass` actions in the server, as indicated in the first rule. Instead, server's messages `rejected`, `connected` or `quit` have no correspondence in the other part (rules **r2**, **r3**, and **r9**). Finally, some other messages, like client's `data` may correspond to different actions in the server, like `result` (rule **r6**), `nosuchfile` (rule **r7**), or even to no action at all (rule **r8**).

An adaptor specification defines the properties that the behaviour of an adaptor component must satisfy [2]. Each rule in a specification can be translated into a term in the process calculus $\mathcal{L}$. For instance, for the rule **r1**, we have:S
`R1(l,r) = l?login(usr,pwd).( r!user(usr).0 || r?pass(pwd).0 || R1(l,r) )`
meaning that if the adaptor accepts a message `login` from the component at its left (represented by the session `l`), then it will eventually perform one action `user` and one action `pass` in its interaction with the component at its right.

Given an adaptor specification, we say that a process A satisfies it, if A fulfills the rules of the specification. Formally, this means that the process A is simulated by the parallel composition of the rules of the specification [2].

The specification S above provides a minimal description of an adaptor that plays the role of "component-in-the-middle" between the FTP system and its client, mediating their interaction. The ultimate goal of such a specification is to obtain an adaptor, a component both satisfying the specification, and providing the required adaptation between the components being adapted.

Formally, the notion of adaptor is introduced as follows.

Definition 1 (Adaptor). *Let α_P and β_Q be sessions types for two components P and Q, respectively, and let S be an adaptor specification. A process $A(l, r)$ is an adaptor for α_P and α_Q under S iff:*

1. *$A(l, r)$ satisfies S, and*
2. *$[A]_l \bowtie \alpha_P$ and $[A]_r \bowtie \alpha_Q$.*

The adaptor A is a process with two session types, ($[A]_l$ and $[A]_r$) – one for each component to be adapted –, compatible with the corresponding session types α_P and α_Q. The two conditions a process has to satisfy to be an adaptor ensure that: (i) the process follows the adaptation pattern given by the adaptor specification, and (ii) the parallel composition of the adaptor with the components P and Q is "safe", as we will illustrate later.

Given an adaptor specification, and the session types of the components to be adapted, an automatic procedure [2] deploys a concrete adaptor (not a type, but an actual component, here represented by a process in $\mathcal{L}$), that both satisfies the specification, and adapts the mismatching behaviour of the actual components represented by those session types. For instance, for the previous specification S, and the session types of the FTP client and server (CLIENT_c and SERVER_a, respectively), a possible result of the generation procedure is the adaptor:

```
Adaptor(l,r) =
  l?accept(x). r!request(y). x?login(usr,pwd). y!user(usr). y!pass(pwd).
  ( y?connected(). y?(z). x?download(filename). z!get(filename).
                ( z?result(filedata). x!data(filedata). z!quit(). 0
                + z?nosuchfile(). x!data("No such file"). z!quit(). 0 )
  + y?rejected(msg). x?download(filename). x!data("Not connected"). 0 )
```

The adaptor above has two link names l and r for interacting with its left and right counterparts (the client and the server, respectively). In the first two actions, a session is opened on each of these links, and then the adaptor begins to interact with the client and the server following the behaviour represented in their session types, and according to the correspondence between messages indicated in the specification S. Thus, the initial client's message login is transmitted to the server by means of one user and one pass message (rule r1 in S). If the server replies with a connected message (without correspondence in the client, as per rule r3), then a daemon session z will be accepted by the adaptor and the interaction goes on through this session (note how the client remains completely unaware of this fact). The following client's action `download` will be transmitted by a `get` (rule **r4**), and the reply of the daemon (either `result` or `nosuchfile`) will be returned to the client with a `data` message (rules **r6** and **r7**). Anyway, in both branches the client closes its session at this point, and the adaptor also ends by sending the `quit` message (rule **r9**) the server is waiting for. On the contrary, if the server rejects the connection and closes the session, client's message `download` will not be transmitted (rule **r5**). Instead, the client will be replied with a `"Not connected"` indication in a `data` message.

3.1   Safe Composition

As we already mentioned, the conditions to be fulfilled by an adaptor guarantee the safe interaction of the components to be adapted. In order to clarify what we mean by that, we introduce the notion of *session-safety*.

Definition 2 (Session safety). *A process P in $\mathcal{L}$ is* session safe *for a set of links L if for every trace $P \longrightarrow^* E \not\longrightarrow$, we have that either:*

1. *$E \equiv 0$, or*
2. *if $E \xrightarrow{\xi}$ then $\xi = x?acp(k)$ or $\xi = x!rqt(k)$ for some link $x \in L$ and some session k.*

Session safety states that a process does not deadlock in the middle of the computation of a session. In other words, once a session is open, then it will finish without deadlocks. We now prove that the definition of adaptor ensures the conditions for guaranteeing that the interactions are safe.

Proposition 1. *Let P, Q, A be three processes sharing only two sessions (k_l, k_r), such that $[P]_{k_l} \bowtie [A]_{k_l}$, $[Q]_{k_r} \bowtie [A]_{k_r}$, and for every session $k \notin \{k_l, k_r\}$ we have $[P]_k = [A]_k = [Q]_k = \bot$. Assume that $fn(P) \cap fn(Q) = \emptyset$ (i.e. P and Q do not interact each other). Then, for $E = P \parallel A \parallel Q$ we have that either:*

1. *$E \equiv 0$, or*
2. *if $E \not\longrightarrow$, and $P \xrightarrow{\xi}$ or $Q \xrightarrow{\xi}$, then for some link x, $\xi \in \{x!rqt(k_l), x?acp(k_l), x!rqt(k_r), x?acp(k_r)\}$, or*
3. *$E \longrightarrow P' \parallel A' \parallel Q'$ where:*
   (a) $P' = P$, $[A']_{k_l} = [A]_{k_l}$, $[A']_{k_r} \bowtie [Q']_{k_r}$, or
   (b) $Q' = Q$, $[A']_{k_r} = [A]_{k_r}$, $[A']_{k_l} \bowtie [P']_{k_l}$

Proof. Let us suppose $E \not\equiv 0$, $E \not\longrightarrow$, and $P \xrightarrow{\xi}$ (analogously if $Q \xrightarrow{\xi}$). If, for any link x, $\xi \notin \{x!rqt(k_l), x?acp(k_l)\}$, then ξ is one of the following transition labels: $k_l!thw(k)$, $k_l?cth(k)$, $k_l!m$, or $k_l?m$ (for some session k and message m). In any of these cases, as $[P]_{k_l} \bowtie [A]_{k_l}$, the process A would present the corresponding complementary action, and therefore $P \parallel A \longrightarrow$, which contradicts the original hypothesis ($E \not\longrightarrow$). If $E \longrightarrow P' \parallel A' \parallel Q'$, then (as P and Q does not share names) the interaction comes from a synchronization made by A and either P or Q. Let us assume the interaction is produced by the parallel composition of A and P. Then, clearly $Q' = Q$. Moreover, since the only session shared by A and P is k_l, we will get $[A']_{k_r} = [A]_{k_r}$. Finally, it is easy to prove (by structural induction on the transition system of $\mathcal{L}$) that $[A']_{k_l} \bowtie [P']_{k_l}$. Analogously, if the interaction is performed between A and Q.   □

Proposition 1 establishes a correspondence between a property on session types (compatibility) and the way the corresponding processes proceed. The theorem below generalizes this result, and proves that a deadlock-freedom result for the parallel composition of two processes can be deduced from the compatibility of the corresponding session types. Of course, this result is only derivable when processes do not interact (among them or with others) through other sessions.

Theorem 1. *Let P, Q, A be three processes sharing only two links $\{l, r\}$ such that $fn(P) \cap fn(Q) = \emptyset$, and for every session $k \notin \{k_l, k_r\}$ we have $[P]_k = [A]_k = [Q]_k = \bot$. If A is an adaptor for $[P]_l$ and $[Q]_r$ under a certain specification S, then $P \parallel A \parallel Q$ is session safe for $\{l, r\}$.*

Proof. We will prove the following equivalent result. One of the following situations hold, for every trace $P \parallel A \parallel Q \longrightarrow^* E$:

i. $E \equiv 0$, or

ii. if $E \not\longrightarrow$ and $E \xrightarrow{\xi}$, then $\xi \in \{l!rqt(k_l), l?acp(k_l), r!rqt(k_r), r?acp(k_r)\}$, or

iii. $E = P' \parallel A' \parallel Q'$ where $[A']_{k_l} \bowtie [P']_{k_l}$ and $[A']_{k_r} \bowtie [Q']_{k_r}$.

Now, we reason by induction on the trace length. The base case is trivial, because $E = P \parallel A \parallel Q$, and then we can apply previous Proposition. Let us assume the statement if true for every trace with length n (inductive hypothesis). Then, if we consider a $(n+1)$-trace $P \parallel A \parallel Q \longrightarrow^n E' \longrightarrow E$, by applying the inductive hypothesis to the first n transitions, we obtain $E' = P' \parallel A' \parallel Q'$ where $[A']_l \bowtie [P']_l$ and $[A']_r \bowtie [Q']_r$ (note that two first statements of the inductive hypothesis are not applicable because $E' \longrightarrow E$). Therefore, by Proposition 1, $E = P'' \parallel A'' \parallel Q''$ satisfying either $(3.a)$ or $(3.b)$. In both cases, P'', A'' and Q'' satisfy the desired requirements.   □

Notice that the conditions of the theorem ensure that the components being adapted will not deadlock in their interaction with other possible components of the system, and this is the sense of enforcing that the type of every session on links different from l or r is $\bot$. If this were not the case, obviously we could not ensure session-safety since a deadlock occurred in another session would deadlock the whole component, including sessions k_l and k_r.

Returning to our example, the types corresponding to the `Adaptor(l,r)` component above are:

```
ADAPTOR_l = ? ?login. ?download. !data. 0
ADAPTOR_r = ! !user. !pass. (?connected. ?(ADAPTOR_z). 0 + ?rejected. 0)
ADAPTOR_z = !get. ( ?result. !quit. 0  + ?nosuchfile. !quit. 0 )
```

and it can be easily proved both that the adaptor satisfies the specification S and also that its types are compatible with those of the client and the server: `CLIENT_c` $\bowtie$ `ADAPTOR_l` and `ADAPTOR_r` $\bowtie$ `SERVER_a`. Hence, we are under the conditions of Theorem 1, and we can conclude that the system:

$$Client(l) \parallel Adaptor(l, r) \parallel Server(r, s) \parallel \prod_{i=1}^{n} Daemon(s)$$

is session-safe for the links $\{l, r\}$, and now these components are able to interact successfully. Indeed, after the client performs its FTP session, both the client and the adaptor end, and the FTP server and its daemons are again in their original states, all of them expecting to perform an accept action, requested by a new client (that may need a completely different adaptation).

4 Concluding Remarks

The objective of this paper was to set a formal foundation for the adaptation of heterogeneous components that may present mismatching interaction behaviour. The three main ingredients of the methodology of software adaptation that we began to develop in [2] can be synthetised as follows:

- IDL component interfaces are extended with an explicit description of the interaction behaviour of a component;
- the desired adaptation is specified via a simple set of (possibly nondeterministic and partial) correspondences between actions of two components;
- a concrete adaptor is automatically generated (if possible), given its partial specification and the interfaces of two components.

In this paper we improved the methodology with the adoption of *session types* to describe the interaction behaviour of components. As shown in Sect. 2, session types feature a disciplined, modular representation of behaviour at a suitable level of abstraction. Most importantly, their type discipline permits a rigorous type checking of component protocols that paves the way for rigorous analysis.

The adoption of session types also permits to describe recursive component protocols, which could not be fully achieved in [2] where component interfaces could only declare finite, non-recursive interaction patterns followed by a component. It is important to stress that while session types permit to describe recursive protocols, their analysis and verification remains tractable.

One of the main contributions of this paper is the formal definition of the notion of *adaptor*, and the proof that any adaptor for (the session types of) two components guarantees their correct session-wise interaction (Sect. 3).

Session types were originally introduced in [6]. Our treatment of session types however differs from [6] in that we employ a more expressive process algebra (and formally define its operational semantics), while simplifying the sometimes cumbersome notation used for session types in [6]. Moreover, our notion of type compatibility relies on the notion of subtyping introduced in [12], rather than coinciding with the more restrictive notion of type duality used in [6].

One of the advantages of using session types is that they can cope with heterogeneous descriptions of component interfaces. Namely, if the protocols of different components have been expressed using different formalisms, such protocols can be typed into the corresponding session types, providing so an homogeneous way of dealing with the composition of third-part components.

Indeed, we argue that the introduction of behavioural types in component interfaces is necessary to achieve a systematic component-based software development, capable of guaranteeing properties of the resulting systems. The results presented in this paper are a step in this direction. Our plans for future work include the integration of the proposed adaptation methodology with available CBSE environments so as to promote its experimentation and assessment.

Space limitations do not allow a proper comparison of our adaptation methodology with other proposals, including those centering on the introduction of connectors in software architectures (e.g., [8]). A thorough comparison of our adaptation methodology with other proposals can be found in [2], while a detailed comparison between session types and other formalisms is reported in [6].

References

1. R. Allen and D. Garlan. A formal basis for architectural connection. *ACM Transactions on Software Engineering and Methodology*, 6(3):213–49, 1997.
2. A. Bracciali, A. Brogi, and C. Canal. A formal approach to component adaptation. *Journal of Systems and Software, Special Issue on Automated Component-Based Software Engineering*, 2003. (in press). A preliminary version of this paper was published in *Component deployment*, LNCS 2370, pages 185–199. Springer, 2002.
3. A.W. Brown and K.C. Wallnau. The current state of CBSE. *IEEE Software*, 15(5):37–47, 1998.
4. G.H. Campbell. Adaptable components. In *ICSE 1999*, pages 685–686. IEEE Press, 1999.
5. G.T. Heineman. An evaluation of component adaptation techniques. In *ICSE'99 Workshop on CBSE*, 1999.
6. K. Honda, V.T. Vasconcelos, and M. Kubo. Language primitives and type disciplines for structured communication-based programming. In *European Symposium on Programming (ESOP'98)*, volume 1381 of *LNCS*, pages 122–138. Springer, 1998.
7. P. Inverardi and M. Tivoli. Automatic synthesis of deadlock free connectors for COM/DCOM applications. In *ESEC/FSE'2001*. ACM Press, 2001.
8. A. Lopes, M. Wermelinger, and J.L. Fiadeiro. Higher-order connectors. *ACM Transactions on Software Engineering and Methodology*, 12(1):64–104, 2003.
9. J. Magee, S. Eisenbach, and J. Kramer. Modeling Darwin in the π-calculus. In *Theory and Practice in Distributed Systems*, LNCS 938, pages 133–152. 1995.
10. B. Morel and P. Alexander. Automating component adaptation for reuse. In *Proc. IEEE International Conference on Automated Software Engineering (ASE 2003)*, pages 142–151. IEEE Computer Society Press, 2003.
11. A. Vallecillo, J. Hernández, and J.M. Troya. New issues in object interoperability. In *Object-Oriented Technology*, LNCS 1964, pages 256–269. Springer, 2000.
12. A. Vallecillo, V.T. Vasconcelos, and A. Ravara. Typing the behavior of objects and components using session types. *ENTCS*, 68(3), 2003.
13. D.M. Yellin and R.E. Strom. Protocol specifications and components adaptors. *ACM Transactions on Programming Languages and Systems*, 19(2):292–333, 1997.

Towards Correspondence Carrying Specifications

Marius C. Bujorianu and Eerke A. Boiten

Computing Laboratory
University of Kent
Canterbury CT2 7NF, UK
Fax +44 1227 762811
{M.C.Bujorianu,E.A.Boiten}@kent.ac.uk

Abstract. In this work we study the unification of heterogeneous partial specifications using category theory. We propose an alternative to institution morphisms, which we call (abstract) correspondences carrying specifications. Our methodology is illustrated using a categorical specification style inspired by the state-and-operations style of Z as well as a categorical unification procedure.

Keywords: partial specification, viewpoints, UML, Z, LOTOS, category theory.

1 Introduction

The development of large software projects raises specific problems and rich pallets of techniques and methodologies have been proposed in response. Here, we consider the method of *partial specification* or *viewpoints*. A partial specification focuses on a particular aspect or subsystem, with the intention that it is still small enough to be embraced, while still containing complete information about its particular concern or perspective. A system is then described as a collection of partial specifications, each from a different perspective or "viewpoint". Important issues arising from such approaches are *consistency* of the different descriptions, and how specifications or systems can be constructed that simultaneously satisfy all of them (*"unification"*).

Prominent examples of partial specifications system development models include the Reference Model for Open Distributed Processing (RM-ODP) [3], Viewpoints Oriented Software Engineering (VOSE) [16] and object-oriented analysis and design models such as UML [25]. Partial specifications are also the cornerstone of the newly approved standard IEEE 1471 [19]. This standard prescribes the description of software architectures by multiple partial specifications (called views), each one being an instance of a previously defined viewpoint.

A similar style of specification occurs naturally in telecommunication systems, which are often described as collections of *features*. Feature interactions are a real problem for telecom companies – they may be viewed as a consistency issue between partial specifications.

This paper builds upon the partial specification approach for formal methods developed by Bowman, Derrick, Boiten, et al [3, 4, 6, 13]. We investigate

C. Rattray et al. (Eds.): AMAST 2004, LNCS 3116, pp. 57–71, 2004.

the categorical foundation of this approach and develop a topos theoretic formalization, based on the principles set out in [2, 6, 7]. Here, we concentrate on specific ingredients of the partial specification method, in particular on the role of *correspondences*. We define the notion of a *specification style*, and provide a topos theoretic formalization of Z to support the state-and-operations style, emphasizing a constructive categorical technique for unifying partial specifications.

The paper is structured as follows. In the next section we give an overview of partial specification, its main challenges and the most important approaches. We identify the need for a unified and uniform formalization, and suggest category theory as a potential mathematical framework. In Section three we present a small but intricate example through three partial specifications in a variety of formalisms, all describing simple labelled transition systems. We show that the standard categorical semantics of each partial specification's language helps very little in relating these partial specifications. In Section four we propose a solution, inspired by this example. We propose that every partial specification should be accompanied by an alternative, possibly simpler, semantics called *abstract correspondence*. A specification style is then a formally defined abstract correspondence. The next section presents a categorical account of partial specification unification in the state-and-operations style of Z. The Conclusions section ends the paper.

The categorical framework we propose for correspondence carrying specification is largely inspired by Z, but generalises its classical set theoretic basis to a topos theoretic basis. This allows non-classical logic variants of Z to be derived, as well as constructive versions and alternative notions of subtyping (the standard Z type system is a particular instance, called maximal types).

We assume the reader is familiar with specification notations such as Z [12], algebraic specification [17], LOTOS and its structured operational semantics [13]. We make use of formal category theory [20], FCT for short, introduced by S. Mac Lane in 1969 as the branch of mathematics that studies the foundations of category theory using category theory itself [20]. We use *double categories* for modelling structured operational semantics, and the formalization of the state-and-operations style from [7] within the *bicategory* of relations [9]. In Section 5 we make use of internal category theory [5]. The unification construction is *internal* to the topos introduced in Section 4.3.

2   A Quick Tour of Partial Specification

Viewpoint methods have been studied extensively [16], particularly in the context of requirements engineering. Here, we describe a particular variant, using formal methods, which one might call the 'Development Approach', by the Kent group (Derrick, Bowman, Boiten, Steen, Linington [3, 4, 13, 12]). The consistency of partial specifications is initially defined as the existence of a common *implementation*. The term '*development relation*' is used to collect concepts like refinement, implementation, and translation. A sufficient condition for showing the existence of a common implementation is to find a common development. In order to reduce the global consistency check of multiple partial specifications of the same system to an iteration of *binary* consistency checks, additional conditions are required.

The definition of consistency in this approach is [4]: "*A set of partial system specifications are considered to be consistent if there exists a specification that is a development of each of the partial specifications with respect to the identified development relations and the correspondences between the partial specifications*". This common development is called an *integration*. A special case of this is the *least developed integration*. This is called *unification* and all other integrations are developments of it.

The Development approach naturally generalizes to a categorical one. We aim to construct a practical categorical framework for studying the consistency and the unification of partial specification.

Partial specifications can overlap in some common parts of the envisaged system that they describe. Redundancy between a set of partial specifications implies that one partial specification contains information about another one. Thus it is necessary to describe the relationship (*"correspondence"*) between the partial specifications. In the simplest case, two viewpoints refer to the same entity using the same name (and give it the same type), in which case the correspondence is implicitly characterised by equality of names. In complex applications, explicit correspondences between the partial specifications are needed (e.g. correspondences between complex behaviors or types). We need to represent correspondences in the categorical framework. The "interconnections" in algebraic specifications [15] serve to identify commonalities in the partial specifications, and to form the basis of the co-limit diagram that defines the composition. Correspondences appear to play a similar role, which would suggest that they could be modelled as objects (with particular outgoing arrows) as well. In our algebraic approach, correspondences are expressed using relations, defined as a span diagram.

3   Heterogenous Specification

Previous work has concentrated on partial specifications written in the same language. If partial specifications represent different aspects, it is likely that each would use a different language appropriate to its particular concern. Indeed, major application areas like UML require a *heterogeneous* approach.

Here, we present the main example for this paper. It consists of a heterogenous partial specification of a simple system, a 'door'. On the basis of this, we discuss the algebraic semantics of the underlying specification languages.

3.1   An Algebraic Specification View

We use a general purpose algebraic specification notation, with a boolean (meta) type *bool*, based on partial conditional equational logic ($\mathcal{PCEL}$).

Spec AS
sort door
opns init: door
      isopen : door→ bool
      move : door → door
      hello : door ↛ door
var D : door
eqns
        isopen(init) .
        isopen(move(D)) = not(isopen(D)).
        isopen(hello(D)) if isopen(D)
end

The standard (initial) algebraic semantics of AS is an algebra A, whose carrier $| A |$ has two elements, *True* and *False*, obtained by partitioning the set of terms by the predicate *isopen*. A has a total operation *move*, equivalent to negation, and a partial one, *hello*, equivalent to the identity on the value *True*.

The modern categorical semantics of algebraic specification is based on the concept of institutions [18]. An *institution* consists of

- a category of signatures **SIGN**;
- two functors *Form* and *Mod* from **SIGN**, associating to each signature Σ a set of formulae *Form*$[\Sigma]$ and a category of algebras *Mod*$[\Sigma]$, and
- a satisfaction relation between formulae and algebras, for each signature Σ, subject to an invariance axiom.

An institution for $\mathcal{PCEL}$ is presented in detail in [17].

3.2   A Process Algebraic Partial Specification

We describe a simple process in the process algebra LOTOS.

    P := *open*; Q [] **exit**
    Q := *hello*; Q [] *close*;P

Categorical Operational Semantics. The 'natural' semantics of LOTOS is given in terms of structured operational semantics (SOS). The SOS rules have been formalized categorically as natural transformation [26] and structured coalgebras [9]. Both axiomatizations are instances of formal category theory [20]. We propose here a formulation of SOS rules as double cells, derived by applying the tile calculus [9] to LOTOS. To a transition $s \xrightarrow{a} t$ of the lts $(A, T, \to \subseteq T \times A \times T)$ corresponds a double cell tran and to a SOS rule $\dfrac{s_i \xrightarrow{a_i} t_i \, , \, i = 1..n}{s \xrightarrow{b} t}$ (R) corresponds a double cell R

$$
\begin{array}{ccc}
1 & \xrightarrow{a} & 1 \\
s \uparrow & \text{tran} \uparrow t & \\
0 & \xrightarrow[id]{} & 0
\end{array}
\qquad\qquad
\begin{array}{ccc}
1 & \xrightarrow{b} & 1 \\
s \uparrow & \text{R} \uparrow t & \\
n & \xrightarrow[a]{} & n
\end{array}
$$

where $a = a_1 \otimes ... \otimes a_n$ is the product object in the Lawvere theory [9] $\mathbf{TH}_A$ associated with the process signature A. We use objects of $\mathbf{TH}_A$ to model the number of premises in a SOS rule.

The SOS semantics for LOTOS is given by the following set of double cells

$$
\begin{array}{ccc}
1 \xrightarrow{id} 1 & \quad & 1 \xrightarrow{u} 1 \\
a \uparrow \quad act_a \quad \uparrow id & & \backslash a \uparrow \quad res_a \quad \uparrow \backslash a \\
1 \xrightarrow{id} 1 & & 1 \xrightarrow{u} 1
\end{array}
$$

$$
\begin{array}{ccccc}
1 \xrightarrow{u} 1 & \quad 1 \xrightarrow{u} 1 & \quad 1 \xrightarrow{id_1} 1 \\
[] \uparrow \quad []_1 \quad \uparrow \pi_0 & [] \uparrow \quad []_2 \quad \uparrow \pi_1 & exit \uparrow \quad exit \quad \uparrow exit \\
2 \xrightarrow{u \otimes id} 2 & 2 \xrightarrow{id \otimes u} 2 & 0 \xrightarrow{id_0} 0
\end{array}
$$

$$
\begin{array}{cc}
1 \xrightarrow{u} 1 & \quad 1 \xrightarrow{u} 1 \\
|| \uparrow \quad ||_1 \quad \uparrow|| & || \uparrow \quad ||_2 \quad \uparrow|| \\
2 \xrightarrow{u \otimes id} 2 & 2 \xrightarrow{id \otimes u} 2
\end{array}
$$

The SOS rules form an object in a double category $\mathbf{SOS}$, by defining a morphism between the double cell sets $U \in| \mathbf{DC} |$ and $V \in| \mathbf{DD} |$ as a triple (F_h, F_v, f) where $F_h : \mathbf{DC}_h \to \mathbf{DD}_h$ and $F_v : \mathbf{DC}_v \to \mathbf{DD}_v$ are functors and $f : U \to V$ is a function preserving all kinds of identities and compositions. The function f is defined as $f(u) : F_h[x] \to F_h[y] \in \mathbf{DD}_h$ and $F_v[v] : F_v[x] \to F_v[y] \in \mathbf{DD}_v$ for any $u : x \to y \in \mathbf{DC}_h$ and $v : x \to y \in \mathbf{DC}_v$. The composition of SOS rules is modelled by the left adjoint of the forgetful functor from $\mathbf{DCAT}$ to $\mathbf{SOS}$.

For example, the process term $open;\ close;\ exit$ is derived as follows

$$
\begin{array}{ccc}
1 \xrightarrow{open} 1 & \quad 1 \xrightarrow{close} 1 \\
open \uparrow & id^1 \uparrow & \uparrow id^1 \\
1 \xrightarrow{id_1} 1 & 1 \xrightarrow{close} 1 \\
close \uparrow & close \uparrow & \uparrow id^1 \\
1 \xrightarrow{id_1} 1 & 1 \xrightarrow{id_1} 1 \\
exit \uparrow & exit \uparrow & \uparrow exit \\
0 \xrightarrow{id_0} 0 & 0 \xrightarrow{id_0} 0
\end{array}
$$

Composing all double cells one obtains $open;\ close;\ exit \uparrow$
$$
\begin{array}{c}
1 \xrightarrow{open;\ close} 1 \\
\uparrow exit \\
0 \xrightarrow{id_0} 0
\end{array}
$$

that corresponds to the transition $open;\ close;\ exit \xrightarrow{open;\ close} exit$. Observe that composing cells along the columns results in the set T of all possible terms, corresponding to the state of the lts semantics of the LOTOS process, whilst composing double cells along the rows gives the set of all allowable sequences of actions, that corresponds to the trace semantics of the LOTOS process.

Bisimulation can be defined as an equivalence on T following [9], and the resulting lts is drawn below

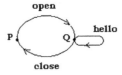

3.3   An Object Oriented View

The next partial specification uses a OCL style notation. We use "@pre" to refer to initial values in postconditions as proposed in [23].

> **let** isopen: **bool**
> **let** oc,cc: **integer**
> **method** reset
>> post: oc = 0 AND cc = 0
>
> **method** open
>> pre: not(isopen)
>> post: oc = oc@pre + 1
>
> **method** close
>> pre: isopen
>> post: cc = cc@pre + 1

This partial specification uses a richer state, a counter for each transition being added. Moreover, the transition 'hello' is missing.

A semantics for OCL, with some institutional flavor, has been proposed in [21]. In this work we make use of a translation of OCL into Z, as in [25].

4   Relating Partial Specifications

In this section we relate the partial specifications given previously. The first way one might try to do this is to relate the specification languages, with some help from category theory. Despite initial optimism inspired by research in the field, many researchers in formal methods found the approach rather unpractical, especially because translations are defined only for a limited set of logics. In this section we argue for a different category theory based methodology to relate the partial specifications. This defines translations locally in a manner specific to the partial specifications involved. We call this methodology *correspondence carrying specification* and it is the main topic of this work.

4.1   Language Translations

The theory of categories appears a natural formalism for defining translations between different formalizations. However, in formal specification we are interested in practical solutions. In this section we investigate to what extent category theory provides useful translations for our example.

Let us consider the first two partial specifications. The 'natural' way to translate one partial specification into another is to use a translation LT between their languages. LT should define a translations between semantics, i.e., use institution morphisms [17]. Such a translation can be constructed in three steps:

1. define an institution for LOTOS;
2. chose a suitable definition of morphism;
3. ensure that the translation LT is related to the two dimensions of the process algebra semantics.

The first aspect seems to be the most difficult. There is no institutional semantics for LOTOS reported in the literature, and the invariance axiom seems difficult to satisfy in LOTOS semantics. A solution would be to first translate LOTOS into a suitable formal notation with an institutional semantics, such as temporal logics [15]. This solution, acceptable from a categorical perspective, is against the spirit of partial specification: partial specification stakeholders must have their freedom respected in choosing their own expertise language. A translation of a partial specification could not be transparent to its own stakeholder and this is the purpose of any support tool (software or theoretical). But this is not possible for our example: the institution $\mathcal{PCEL}$ offers too poor semantics, and this is because there is no distinction between operations constructing states and those denoting transitions. Therefore, the existence of any translation is conditioned by further information from the stakeholder. This (necessary) additional information is formalised in our approach as abstract correspondences in the next (sub)section.

The second aspect, although it looks the easiest, is actually more difficult. Research on institutions has so far failed to agree on a definition of morphism. Software specifiers cannot realistically be expected to make the choice between the various alternatives proposed. The existing approaches to heterogeneous specification based on Grothendieck institutions [22] are not suitable for partial specification: it is not possible to model the concept of correspondence and they are applicable only to those formal notations that admit institutional semantics.

The answer to the third question is related to the particular property of institution morphisms, namely that they define both syntactic and semantic translations such that satisfiability is preserved. For process algebras, syntax is related to SOS rules and not directly to lts or trace semantics; the latter are derived.

An advantage one might get by avoiding complete language translations can be found in the SOS semantics of the process algebraic partial specification. Indeed, SOS semantics is, in fact, a *metalevel* semantics: it does not describe a lts for a process, but a set of rules to construct that lts. This is why institution morphism theory is difficult to apply in our example. The metalevel of semantics is reflected categorically in the extra dimension of semantic category: we have to move to double categories.

The OCL partial specification provides another example of a domain where metalevel semantics is present: UML [2]. Metamodelling is frequently used in UML, where UML itself is used to give semantics of UML [14]. In the absence of a ground, solid formalization of metalevel semantics, metamodelling rather seems to express the lack of semantics. This is why FCT promises to offer a potential for transforming the metamodelling methodology of UML into a formal specification tool.

The most important use of FCT can be obtained when considering the state and operations style of Z. Its categorical formalization is done using FCT, viz. the bicategory of relations $\mathbf{REL_R}$ over a regular category [24] $\mathbf{R}$. The metalevel interpretation is as follows: the states are viewed as objects of a regular category $\mathbf{R}$. State transformation is done by operations, and their formalization is done by considering an extra dimension (i.e. a meta reasoning over states) to $\mathbf{R}$, namely constructing the bicategory $\mathbf{REL_R}$. Considering refinement, modelled as relational inclusion, one gets a three-category.

4.2   Abstract Correspondences

An *abstract correspondence* (AC) is a semantic description of the partial specification, which helps to derive concrete correspondences with respect to another partial specification. AC is a simplification of the standard semantics of the partial specification or may be written in a different language.

The algebraic specification in our example is relatively difficult to relate with the other partial specifications because its standard semantics provides an algebra. In our example, an useful AC associated with the AS partial specification is the lts drawn below

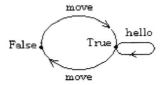

Our alternative to institutions is to use a categorical semantics of the state-and-operations style derived from Z. The above lts semantics can be rigorously specified by considering states with no structure - a simple set of states

$ST \ ::= \ False \mid True$

and operations defining in a compact manner possible transitions, like

_States_____
$s : ST$

_Hello_____
$\Delta States$
$s = True \wedge s' = True$

$$\begin{array}{|l|}\hline \textit{Move} \\\hline \Delta \textit{States} \\\hline (s = \textit{True} \wedge s' = \textit{False}) \vee (s = \textit{False} \wedge s' = \textit{True}) \\\hline \end{array}$$

AC is specific to a partial specification, and not to a language. Two different partial specifications, written in the same specification language, can have very different ACs. However, we can have the situation where an AC can be extended to the whole language. We call such an AC a *specification style*. This is an alternative, simpler (for a specific use) semantics of the language. In our view, the main contribution of the theory of institution to formal logics consists in the introduction of localisation. The satisfaction relation is not defined for the entire language as in most formal logics, but locally, for every signature separately. The global aspects of logics are then recovered by introducing the invariance condition, and most mathematical results of the theory consist in re-establishing globality. We develop a constructive interpretation of this localisation principle as ACs.

For us, the most relevant specification style is the state-and-operations style in Z. The standard semantics of Z offers no operational interpretation of operations: they are just sets. The usual interpretation is that an operation is a relation between the semantics of pre and post states.

4.3   A Categorical State-and-Operations Style for Z

This section establishes the home category of our categorical investigation following a categorical type theory analysis of Z.

The Z Type System. The primitive types in Z are numbers and the *given sets* G (i.e. sorts in algebraic specification terms). Types are constructed from the basic types using type constructors. The type constructors include power set and product. The formalization of types proposed here has Standard Z as a model, but also variants such as *Constructive Z* or multi-valued logic variants of Z. The logical constructors with possibly multiple interpretations are marked using the $-$ decoration.

Standard given sets comprise the singleton type $\imath$, the integers $\mathbb{Z}$, the Booleans $\mathbb{B}$, and any other given sets defined without further information on their structure.
Type constructors are formally defined as: if α and β are types, so are $\mathbb{P}\,\alpha$ (the power set), $\alpha \times \beta$ (product) and $[a_i : \alpha_i]_{i \in I}$ (schema type; all labels a_i must be different).

Terms are defined recursively as follows:

- for every type α we consider countably many variables of type α. We write $x : \alpha$ for a variable x of type α
- $*$ is a term of type $\imath$ (i.e. $\imath = \{*\}$)
- $true : \mathbb{B}$. Moreover, if $a, b, \phi(x) : \mathbb{B}$ (with ϕ -possibly- containing the free variable $x : \alpha$) so are $a \,\bar{\wedge}\, b$, $a \Rightarrow b$, $\bar{\forall}\, x : \alpha \bullet \phi(x)$.

- 0 is term of type $\mathbb{Z}$, and, if $n : \mathbb{Z}$ then so are Sn and $-n$
- if $a : \alpha$ and $b : \beta$ then $< a, b > : \alpha \times \beta$
- if $x_i : \alpha_i$ for all $i \in I$, then $\langle\!\langle a_i == x_i \rangle\!\rangle_{i \in I} : [a_i : \alpha_i]_{i \in I}$
- if $a : \alpha$ and $u : \mathbb{P}\,\alpha$ then $(a \,\bar{\in}\, u) : \mathbb{B}$
- if $\phi(x) : \mathbb{B}$ (possibly) contains the free variable $x : \alpha$ then $\{x : \alpha \mid \phi(x)\} : \mathbb{P}\alpha$

Important *particular terms* can be defined as follows

- *false* $: \mathbb{B}$ is defined as $\bar{\forall}\, a : \mathbb{B}\bullet a$ and $(\bar{\neg}\, a) : \mathbb{B}$ is defined as $a \Rightarrow$ *false*
- $\bar{\exists}\, x : \alpha \bullet \phi(x) : \mathbb{B}$ is defined as $\bar{\forall}\, c : \mathbb{B}\bullet(\bar{\forall}\, x : \alpha \bullet (\phi(x) \Rightarrow c) \Rightarrow c)$
- $(a = b) : \mathbb{B}$ is defined as $\bar{\forall}\, u : \mathbb{P}\alpha\bullet(a \,\bar{\in}\, u \Leftrightarrow b \,\bar{\in}\, u)$
- $\{a\} : \mathbb{P}\alpha$ is defined as $\{x : \alpha \mid x = a\}$ where $a : \alpha$

States
The Z Standard document defines the semantics of a state as a set of bindings (members of a schema type).
The next result allows us to define semantics of a Z type as an object in a topos.

Theorem 1. *Every model of a Z's type system is equivalent with a relational category over a topos, that we denote by* **MOD**.

This result imposes a change in the categorical formalization of the Z standard semantics by adding subtyping. We make the convention that every object of **MOD** denotes a type. Every type in the sense of Z Standard becomes a maximal type in our topos theoretic semantics.
Bindings over a set of identifiers V are modelled as the slice topos of **MOD** over V, denoted, by abuse of notation, as **MOD**.

Refinement. In order to support partial specification, a formal approach must also offer strong support for refinement. This is the key of the success of applying model oriented methods to partial specification. The lack of space stopped us presenting the rich theory of relational refinement, see [12]. The data refinement concept is powerful enough to prove, for example, that the Z schema *Move* is refined by the schemas *Open* and *Close* (defined in the following section). One might ask how such a relational semantics might relate to the behavioural semantics required by process algebra; this is discussed in detail in [11].

5   Unification from Z

In this section, the host category is a slice topos **MOD** over a set comprising all partial specification identifiers V. All category concepts we use in this section are internal [5, Section 8] in this topos.

This section follows the set theoretic approach from [4]. We refer to [12, 4] for the definitions and properties of Z abstract datatypes (ADTs) and relational semantics. Consider the Z partial specifications (ADTs) $A = (StA, InitA, \{OpA\})$ and $B = (StB, InitB, \{OpB\})$, where StA denotes the state space, $InitA$ denotes the initialization of described state machine and the set of operations $\{OpA\}$ specify all possible transitions. Z unification comprises three stages.

5.1   State Unification

The hidden states of the two ADTs can be unified by pushout, using the amalgamation operation developed in algebraic specification.

The correspondence between two types C and D is given as a relation $\rho : A \leftrightarrow B$. The internalization of relations plays a crucial role in unifying the states. The unified state can be viewed as a relation between extended states, namely

$$tot\rho = \rho \cup (StA \setminus dom\_\rho) \times \{\perp_A\} \cup (StB \setminus ran\_\rho) \times \{\perp_B\}$$

This formula shows that state unification is basically relational, and the concept of correspondence plays a dominant role. Basically, every item in a partial specification must be considered as being related with an item from another partial specification. When the last item is not known, then a special element is introduced, the 'unknown', which is considered to be in correspondence with any other item from a different partial specification that is not already linked by the correspondence relation.

Let $\mathbf{C}$ be a category and $A, B, C, D \in |\mathbf{C}|$.

Given morphisms $f_1, ..., f_n : A \to B (n \geq 1)$, and $f : A \to C$ in $\mathbf{C}$ the following arrows $g_1, ..., g_n : C \to D$ and $g : B \to D$, chasing the diagram

$$
\begin{array}{ccc}
A & \overset{f_1}{\underset{f_n}{\rightrightarrows}} & B \\
f \downarrow & & \downarrow g \\
C & \overset{g_1}{\underset{g_n}{\rightrightarrows}} & D
\end{array}
\qquad (RP)
$$

in $\mathbf{C}$ are called *relational pushout* of $(f_1, ..., f_n)$ and f if we have:
(CC) $g\_f_i = g_i\_f$ for every $i = 1, ..., n$ and
(U) For each object D' and morphisms $g', g_1, ..., g_n$ with $g'\_f_i = g'_i\_f$, $(i = 1, ..., n)$ there is a unique morphism $h : D \to D'$ s.t. $h\_g = g'$ and $h\_g_i = g'_i$.

Remark 1. For $n = 1$ the relational pushout is the standard pushout.

Proposition 1. *(Construction of relational pushouts) If* $\mathbf{C}$ *has finite coproducts and (standard) pushouts then* $\mathbf{C}$ *has relational pushouts.*

Corollary 1. MOD *has relational pushouts.*

Important here is the concept of *states consistent with respect to the correspondence* relation. Not all states can be integrated in a consistent way; supplementary constraints are necessary to ensure consistency.

The two state schemas $StA \widehat{=} [a : A \mid Pred_{StA}]$ and $St_2 \widehat{=} [b : B \mid Pred_{StB}]$ are *state consistent with respect to the correspondence* relation $\rho \subseteq A \times B$ iff

$$(a, b) \in \rho \Leftrightarrow (Pred_{StA} \Leftrightarrow Pred_{StB}).$$

Proposition 2. $tot\rho$ *is the relational pushout of StA and StB via their projections from* ρ.

Theorem 2. $tot\rho$ *is the least developed refinement of the states consistent with respect to the correspondence relation ρ.*

For the first two viewpoints in our example, the correspondence ρ consists in the identification of True with Q and False with P. The state space of the OCL partial specification is essentially isomorphic as its add extra components (the counters) which are left unconstrained by the correspondence relation.

5.2   Operation Adaptation

As the new (unified) state was created, operations must be adapted to this. This is usually achieved using injective vocabulary morphisms. In this way all related techniques from the institution literature [17] get a fruitful application field here.

For every state schema S, we consider an injective renaming $verS$ and denote by $revS$ the renaming from $verS$ back to S.

Every operation Op over the state space *State* is renamed (adapted) to an operation $OpAd$ on the unified state space $StU = tot\rho$.

```
┌─ Op ─────────────────────
│ ΔState
│ Decl
├──────────
│ Pred
└──────────────────────────
```

```
┌─ OpAd ─────────────────────────────────
│ ΔStU
│ Decl
├────────────────────────────────────────
│ x ∈ ran verS
│ x′ ∈ ran verS
│ letx == revS x; x′ == revS x′ • Pred
└────────────────────────────────────────
```

5.3   Operations Unification

As in Z, the precondition is only specified implicitly as the satisfiability of the operation, the operation itself serves as its postcondition. Conversely, we can specify the *weakest* postcondition of an operation Op as

wpc $Op ==$ pre $Op \Rightarrow Op$

For every operation Op, we have that $Op \equiv$ pre $Op \wedge$ wpc Op.
In general, unification is given in a category having object operations, acting on a common state space, and arrows $OpA \rightarrow OpB$ refinements given as pairs of implications pre $OpA \Leftarrow$ pre OpB and wpc $OpA \Rightarrow$ wpc OpB.

The categorical product acts contravariant on pre-conditions (i.e. it generates the disjunction of pre-conditions) and covariant on weakest post-conditions (i.e. it generates the conjunction of post-conditions) giving the following unification expression

```
┌─ OpU ───────────────────────────────────────
│ ΔStU
├──────────────────
│ (pre OpA ∨ pre OpB) ∧ (wpc OpA ∧ wpc OpB)
└──────────────────────────────────────────────
```

Remark 2. The unification *InitU* of adapted initializations is their conjunction.

Proposition 3. *Operations unification is a limit.*

The following result, from [4], proves that our procedure provides the unification. The categorical version of data refinement involved will be presented in a subsequent work [8]. The proof in [4] is a set-theoretic one.

Theorem 3. $U = (StU, Init, \{OpU\})$ *is the least developed (data) refinement of A and B.*

Applying this procedure to the partial specifications of example from Section 3, the resulting unification is described below. The reader might observe that no partial specification alone can be taken as the unification.

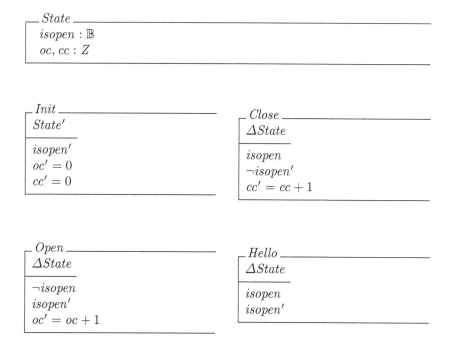

6   Conclusions

The categorical approach to partial specification is a useful initiative. It provides a very rigorous foundation and helps to distinguish many facets and aspects of specification languages. Category theory helps us not only to discover the complexity of translations between formal specification languages, but also to explain it (in our example, by discovering the double dimension of SOS semantics) and to offer alternative solutions.

It is necessary to distinguish between correspondences carrying specifications (CoCS) and semantic frameworks. CoCS adresses the partial specification only.

Semantic frameworks propose universal solutions, by embedding an important ammount of specification languages. These encodings might get very complex and aspects of the original specification might be lost during translation, affecting the unification and the possibility to provide constructive feedback. By contrast, CoCS does not prescribe any universal solution. It provides, for given partial specifications, a simplified common semantics called abstract correspondence. An AC acts like a common semantic framework, but relative to the partial specifications involved. Different partial specifications can have different associated AC. The advantage of using category theory consists in the possibility of providing generic ACs: an abstract AC defined in categorical terms can be instantiated with several models, providing concrete ACs. The AC described in this paper models not only Z, but any specification language that can be axiomatised as specification frames (examples are given in [7]).

Another relevant approach is the similarly sounding paradigm SCS: *specification carrying code* [1]. The main differences between these approaches come from their different purposes. SCS main purposes are centred around code, as cryptography. CoCS is specialised on formal partial specification. However, the two paradigms make often use of the same formal tools, as category theory and refinement.

We propose an extension of this paper to a full categorical framework for partial specification as further work. In [7] first steps were given, by formalizing categorically the (set theoretic) relational framework presented [6].

The detailed presentation of the proofs of results presented in this paper will appear in [8].

The most important conclusion that emerged from our formal experiments, points out the crucial role of correspondences in unification of specifications.

Acknowledgements

The authors want to gratefully acknowledge the contribution of John Derrick in the early stages of this work and thank to Dusko Pavlovic for discussions on specification carrying code and to the anonymous referees for useful comments.

References

1. M. Anlauf and D. Pavlovic: On Specification Carrying Software, its Refinement and Composition, in: H. Ehrig, B.J. Kramer and A.Ertas, eds., *Proceedings of IDPT 2002*, Society for Design and Process Science, 2002.
2. E.A. Boiten and M.C. Bujorianu: Exploring UML Refinement through Unification, in: *Workshop on Critical Systems Development with UML, UML 2003 Conference*, San Francisco, California, USA, 2003.
3. E.A. Boiten, H. Bowman, J. Derrick, P.F. Linington, and M.W.A. Steen: Viewpoint Consistency in ODP. *Computer Networks* 34(3):503–537, 2000.
4. E.A. Boiten, J. Derrick, H. Bowman, M.W.A. Steen: Constructive consistency checking for partial specification in Z. *Science of Computer Programming*, 35(1):29–75, 1999.

5. F. Borceux: *Handbook of Categorical Algebra 1. Basic Category Theory* Cambridge University Press, 1994.
6. H. Bowman, M.W.A. Steen, E.A. Boiten, and J. Derrick: A formal framework for partial specification consistency. *Formal Methods in System Design*, 21:111–166, 2002.
7. M.C. Bujorianu: Integration of Specification Languages Using Viewpoints , in: *IFM 2004: Integrated Formal Methods*, Canterbury, Kent, LNCS 2999, Springer Verlag, 2004.
8. M.C. Bujorianu: *A Categorical Framework for Partial Specification*, forthcoming PhD Thesis, Computing Laboratory, University of Kent, 2004.
9. R. Bruni, J. Meseguer and U. Montanari: Symmetric and Cartesian Double Categories as a Semantic Framework for Tile Logic. *Math Struct in Comp Science* 11:1–33, 1995.
10. A.Corradini, R. Heckel, and U. Montanari: From SOS Specifications to Structured Coalgebras: How to Make Bisimulation a Congruence, in: *Coalgebraic Methods in Computer Science CMCS'99*, ENTCS 19, 1999.
11. J. Derrick and E.A. Boiten: Relational concurrent refinement. *Formal Aspects of Computing*, 15(2):182–214, 2003.
12. J. Derrick and E. Boiten: *Refinement in Z and Object-Z: Foundations and Advanced Applications*. Springer FACIT Series, 2001.
13. J. Derrick, E.A. Boiten, H. Bowman, M. Steen: Viewpoints and Consistency: Translating LOTOS to Object–Z. *Computer Standards and Interfaces*, 21:251–272, 1999.
14. A. Evans, R. France, K. Lano, B. Rumpe: Meta-modelling semantics of UML, in: H. Kilov, ed., *Behavioural Specifications for Businesses and Systems*. Kluwer, 1999.
15. J.L.Fiadeiro and J.F.Costa: Mirror, Mirror in my Hand: a duality between specifications and models of process behaviour. *Mathematical Structures in Computer Science* 6:353–373, 1996.
16. A. Finkelstein and G.Spanoudakis (eds.), *International Workshop on Multiple Perspectives in Software Development (Viewpoints '96)* ACM, 1996.
17. J. Goguen and G. Rosu: Institution Morphisms. *Formal Aspects of Computing* 12(3-5):274–307, 2002.
18. J. Goguen and R. Burstall: Institutions: Abstract Model Theory for Specification and Programming. *Journal of the ACM*, 39(1):95-146, 1992.
19. IEEE Architecture Working Group *IEEE P1471/D5.0 Information Technology - Draft Recommended Practice for Architectural Description*, 1999.
20. J.W. Gray: *Formal Category Theory: Adjointness for 2-Categories*, Lectures Notes in Mathematics 391, Springer Verlag, 1974.
21. P. Kosiuczenko: Formal Redesign of UML Class Diagrams, in: *Practical UML-Based Rigorous Development Methods*, Lecture Notes in Informatics P-7, 2001.
22. T. Mossakowski: Heterogeneous development graphs and heterogeneous borrowing, in: *Fossacs'02*, LNCS 2303, pp. 326–341, 2002.
23. OMG: *Object Constraint Language (OCL) 2.0.* RFI Response Draft (University of Kent, Microsoft e.a.) 2002.
24. D. Pavlovic: Maps II: Chasing diagrams in categorical proof theory, *Journal of the IGPL* 4(2): 1–36, 1996,
25. J.Rumbaugh and I.Jacobson: *The Unified Modeling Language Reference Manual*, Addison Wesley Longman Inc., 1999.
26. D. Turi and G. Plotkin: Towards a Mathematical Operational Semantics, in: *Procs. of IEEE Conference on Logic in Computer Science: LICS'97*, IEEE Press, 1997.

Formalizing and Proving Semantic Relations between Specifications by Reflection*

Manuel Clavel, Narciso Martí-Oliet, and Miguel Palomino

Departamento de Sistemas Informáticos y Programación
Universidad Complutense de Madrid, Spain
{clavel,narciso,miguelpt}@sip.ucm.es

Abstract. This work contains both a theoretical development and a novel application of ideas introduced in [1] for using reflection in formal metareasoning. From the theoretical side, we extend the metareasoning principles proposed in [1] to cover the case of metatheorems about equational theories which are unrelated by the inclusion relation. From the practical side, we apply the newly introduced metareasoning principles to formalize and prove semantic relations between equational theories used in formal specification.

1 Introduction

Intuitively, a *reflective* logic is a logic in which important aspects of its metalogic can be represented at the object level in a consistent way, so that the object-level representations correctly simulate the relevant metalogical aspects. More concretely, a logic is reflective when there exists a theory – that we call *universal* – in which we can represent and reason about all finitely presentable theories in the logic, including the universal theory itself [8, 3]. As a consequence, in a reflective logic, metatheorems involving families of theories can be represented and logically proved as theorems about its universal theory. Of course, one of the advantages of formal metareasoning based on reflection is that it can be carried out using already existing logical reasoning tools.

The use of reflection for formal metareasoning was first proposed in [1], both abstractly and concretely. Abstractly, it proposes a set of requirements for a logic to be used as a reflective metalogical framework. Concretely, it presents membership equational logic [12] as a particular logic that satisfies those requirements. In addition, it provides metareasoning principles to logically prove metatheorems about families of membership equational theories as theorems about its universal theory [9].

This work is both a development and an application of the reflective methodology proposed in [1] for formal metareasoning using membership equational logic. First, we extend the metareasoning principles introduced in [1]. In particular, while [1] only considered metatheorems about membership equational

---

* Research supported by Spanish MCYT Projects MELODIAS TIC2002-01167 and MIDAS TIC2003-01000, and CICYT Project AMEVA TIC2000-0701-C02-01.

C. Rattray et al. (Eds.): AMAST 2004, LNCS 3116, pp. 72–86, 2004.
© Springer-Verlag Berlin Heidelberg 2004

theories which are related by the inclusion relation, our reflective methodology can also deal with metatheorems about theories which are unrelated with respect to inclusion. Thus, our extension increases significantly the applicability of the reflective methodology proposed in [1], as we show with the following case study. As is well-known, equational specifications can be related in different ways, and these relations can be informally formulated as metalogical statements of equational logic. The semantic relations between different equational specifications are in fact key conceptual tools in the stepwise specification methodology, and different techniques and criteria have been proposed to metalogically prove them [10]. We show that some of these semantic relations can be formalized as theorems about the universal theory of membership equational logic and that they can be logically proved in a way that mirrors their corresponding proofs at the metalogical level.

Organization. The paper is organized as follows. First, in Sect. 2 we introduce membership equational logic; the content of this section is standard material borrowed from other papers. Then, in Sect. 3 we formulate some semantic relations between membership equational specifications as metalogical statements, and in Sect. 4 we introduce metareasoning principles for metalogically proving them. Finally, in Sects. 5, 6, and 7 we present our reflective framework, and we put it to work. In particular, we show how a whole class of metalogical statements, that contains those in Sects. 3 and 4, can be represented and logically proved, using reflection, in membership equational logic; we include in an appendix a detailed example of this.

2   Membership Equational Logic

Membership equational logic is an expressive version of equational logic. A full account of the syntax and semantics of membership equational logic can be found in [2, 12]. Here we define the basic notions needed in this paper.

A *signature* in membership equational logic is a triple $\Omega = (K, \Sigma, S)$ with K a set of *kinds*, Σ a K-kinded signature $\Sigma = \{\Sigma_{k_1 \ldots k_n, k}\}_{(k_1 \ldots k_n, k) \in K^* \times K}$, and $S = \{S_k\}_{k \in K}$ a pairwise disjoint K-kinded family of sets. We call S_k the set of *sorts* of kind k. The pair (K, Σ) is what is usually called a many-sorted signature of function symbols; however we call the elements of K *kinds* because each kind k now has a set S_k of associated *sorts*, which in the models will be interpreted as subsets of the carrier for the kind.

The atomic formulae of membership equational logic are either *equations* $t = t'$, where t and t' are Σ-terms of the same kind, or *membership assertions* of the form $t : s$, where the term t has kind k and $s \in S_k$. Sentences are Horn clauses on these atomic formulae, i.e., sentences of the form

$$\forall (x_1, \ldots, x_m).\, A_1 \wedge \ldots \wedge A_n \implies A_0 \,,$$

where each A_i is either an equation or a membership assertion, and each x_j is a K-kinded variable. A theory in membership equational logic is a pair (Ω, E), where E is a finite set of sentences in membership equational logic over the signature Ω. We write $(\Omega, E) \vdash \phi$ to denote that (Ω, E) entails the sentence ϕ.

We employ standard semantics concepts from many-sorted logic. Given a signature $\Omega = (K, \Sigma, S)$, an Ω-*algebra* is a many-kinded Σ-algebra (that is, a K-indexed-set $A = \{A_k\}_{k \in K}$ together with a collection of appropriately kinded functions interpreting the function symbols in Σ) and an assignment that associates to each sort $s \in S_k$ a subset $A_s \subseteq A_k$. An algebra A and a valuation σ, assigning to variables of kind k values in A_k, satisfy an equation $t = t'$ iff $\sigma(t) = \sigma(t')$. We write $A, \sigma \models t = t'$ to denote such a satisfaction. Similarly, $A, \sigma \models t : s$ holds iff $\sigma(t) \in A_s$.

Note that an Ω-algebra is a K-kinded first-order model with function symbols Σ and a kinded alphabet of unary predicates $\{S_k\}_{k \in K}$. We can then extend the satisfaction relation to Horn and first-order formulae ϕ over the atomic formulae in the standard way. We write $A \models \phi$ when the formula ϕ is satisfied for all valuations σ, and then say that A is a model of ϕ. As usual, we write $(\Omega, E) \models \phi$ when all the models of the set E of sentences are also models of ϕ. As expected, the rules of inference for membership equational logic are sound and complete.

Theories in membership equational logic have initial models. This provides the basis for reasoning by induction. In the initial model of a membership equational theory, sorts are interpreted as the smallest sets satisfying the axioms in the theory, and equality is interpreted as the smallest congruence satisfying those axioms. Given a theory (Ω, E), we denote its initial model by $T_{\Omega/E}$. In particular, when $E = \emptyset$ we obtain the term algebra T_Ω, and for X a K-kinded set of variables the free algebra $T_\Omega(X)$. We write $(\Omega, E) \models_i \phi$ to denote that the initial model of the membership equational theory (Ω, E) is also a model of ϕ, that is, that the satisfaction relation $T_{\Omega/E} \models \phi$ holds.

3   Semantic Relations between Specifications

The formalization and proof of certain semantic relations between equational specifications are important aspects of the algebraic specification methodology. In this regard, a classical notion is that of *enrichment*, which is a key conceptual tool in the stepwise specification methodology [10, 11]. We consider the following definition of the enrichment relation between membership equational specifications.

Definition 1. *Let* $R = (\Omega, E)$ *and* $R' = (\Omega', E')$ *be specifications, with* $\Omega = (K, \Sigma, S)$ *and* $\Omega' = (K', \Sigma', S')$, *such that* $R \subseteq R'$ *componentwise. Let* k *be a kind in* K, *and let* s *be a sort in* S_k. *Then* R' *is an* s-*enrichment of* R *if and only if:*

1-a. $\forall t \in T_{\Omega'}. R' \vdash t : s \Longrightarrow \exists t' \in T_\Omega. R \vdash t' : s \land R' \vdash t = t'$
1-b. $\forall t, t' \in T_\Omega. R \vdash t : s \land R \vdash t' : s \land R' \vdash t = t' \Longrightarrow R \vdash t = t'$.

Note that our definition is slightly different from that in [10] – (1-a) and (1-b) correspond, respectively, to their notions of *complete* and *consistent extensions* – since we define the enrichment relation relative to a particular sort. The idea captured by our definition is that each ground term in the specification R' having the sort s can be proved equal to a ground term of the specification R having the sort s, and also that R' does not impose new equalities on ground terms of sort s of the specification R. These properties correspond, respectively, to the *no junk* and *no confusion* properties in Burstall and Goguen's terminology.

The enrichment relation assumes an inclusion between the given specifications. There are other semantic relations, however, that do not require such an inclusion. Consider, for example, the specifications INT_1 and INT_2 below. They are presented using Maude syntax [5, 6], where operators, variables, membership axioms, and equations are introduced, respectively, with the keywords op, var, mb (or cmb for the conditional case), and eq. The Maude system implements membership equational logic (and rewriting logic) and can infer kind information automatically; however, for increased clarity, we have explicitly named kinds, with their associated sort lists inside square brackets following the kind's name. The specifications INT_1 and INT_2 are clearly related since both specify the integer numbers – and, in that sense, they are *interchangeable* –, but neither INT_1 is included in INT_2, nor INT_2 in INT_1.

```
fmod INT1 is                          fmod INT2 is
kind Num[Neg, Nat, Int] .             kind Num[Int] .
op 0 : -> Num .                       op 0 : -> Num .
op s : Num -> Num .                   op s : Num -> Num .
op p : Num -> Num .                   op p : Num -> Num .
var N : Num .                         var N : Num .
--  Nonpositive numbers               --  Integers
mb 0 : Neg .                          mb 0 : Int .
cmb p(N) : Neg if N : Neg .           cmb s(N) : Int if N : Int .
--  Natural numbers                   cmb p(N) : Int if N : Int .
mb 0 : Nat .                          eq p(s(N)) = N .
cmb s(N) : Nat if N : Nat .           eq s(p(N)) = N .
                                      endfm
--  Integers
cmb N : Int if N : Neg .
cmb N : Int if N : Nat .
endfm
```

We propose the following definition for this particular relation between membership equational specifications. For the sake of simplicity, we restrict our definition to specifications with a common set of kinds.

Definition 2. *Let $R = (\Omega, E)$ and $R' = (\Omega', E')$ be specifications, with $\Omega = (K, \Sigma, S)$ and $\Omega' = (K, \Sigma', S')$. Let k be a kind in K, and let s be a sort in $S_k \cap S'_k$. Then R and R' are s-interchangeable if and only if:*

2-a. $\forall t \in T_\Omega. \, R \vdash t : s \implies \exists t' \in T_{\Omega'}. \, R' \vdash t' : s \land R \vdash t = t'$
2-b. $\forall t, t' \in T_{\Omega'}. \, R' \vdash t : s \land R' \vdash t' : s \land R \vdash t = t' \implies R' \vdash t = t'$
2-c. $\forall t \in T_{\Omega'}. \, R' \vdash t : s \implies \exists t' \in T_\Omega. \, R \vdash t' : s \land R' \vdash t = t'$
2-d. $\forall t, t' \in T_\Omega. \, R \vdash t : s \land R \vdash t' : s \land R' \vdash t = t' \implies R \vdash t = t'$.

The idea captured by the above definition is that each ground term in the specification R' (resp. R) having the sort s can be proved equal to a ground term of the specification R (resp. R') having the sort s, and also that R' (resp. R) does not impose new equalities on ground terms of sort s of the specification R (resp. R').

Now, note that to prove properties (1-b), (2-b), and (2-d) we need, in general, to examine the form of the axioms in R and R'. But to prove properties (1-a), (2-a), and (2-c) we can use inductive techniques. In fact, an inductive reasoning principle is proposed in [2] to logically prove property (1-a). However, the lack of an inclusion between R and R' invalidates the use of this principle for proving properties (2-a) and (2-c). We will propose in Sect. 4 an inductive reasoning principle (ind^+) to metalogically prove these properties, and we will show in Sect. 7 how this principle can be transformed, using reflection, into an inductive reasoning principle $\overline{(ind^+)}$ to logically prove them.

4 An Inductive Principle for Metalogically Proving Semantic Relations

In order to provide a simpler and more compact formulation, and soundness proof, of an inductive principle (ind^+) for *metalogically* proving semantic relations between equational specifications, we begin by extending the definitions of terms, atomic formulae, and entailment relation for membership equational logic. Basically, these extensions allow us to metareason on equivalence classes of terms, instead than on concrete terms, which is essential for metareasoning about families of theories which are unrelated with respect to inclusion. However, this change of logical framework is only transitory. We will show in Sect. 7 how the inductive principle (ind^+) can be transformed, using reflection, into an inductive principle $\overline{(ind^+)}$ for *logically* proving, in standard membership equational logic, semantic relations between equational specifications. Of course, since our final goal is to provide principles for carrying out formal metareasoning, we are interested in $\overline{(ind^+)}$ rather than in (ind^+), but we introduce the latter as a technical device to simplify the presentation and proof of the former.

In what follows, let $\mathcal{CR}$ be the class of finite multisets of finitely presentable theories with a common, nonempty set of kinds $\mathcal{R} = \{R_i\}_i = \{(\Omega_i, E_i)\}_{i \in [1..p]}$, with each $\Omega_i = (K, \Sigma_i, S_i)$. We consider multisets instead of lists or sets for technical reasons: it simplifies our definition of the inductive principles (ind^+).

Definition 3. *Given* $\mathcal{R} = \{R_i\}_i = \{(\Omega_i, E_i)\}_{i \in [1..p]} \in \mathcal{CR}$, *with each* $\Omega_i = (K, \Sigma_i, S_i)$, *we define the set* $T_{\Omega_i}^{\mathcal{R}}(X)$ *of* $(\Omega_i, \mathcal{R})$-*terms with K-kinded variables in X as follows:*

- $c \in (T_{\Omega_i}^{\mathcal{R}}(X))_k$ *iff* $c \in (\Sigma_i)_{\lambda,k}$, $k \in K$, *where λ denotes the empty sequence of kinds;*
- $x \in (T_{\Omega_i}^{\mathcal{R}}(X))_k$ *iff* $x \in X_k$, $k \in K$;
- $f(t_1, \ldots, t_n) \in (T_{\Omega_i}^{\mathcal{R}}(X))_k$ *iff* $f \in (\Sigma_i)_{k_1 \ldots k_n, k}$, *and* $t_j \in (T_{\Omega_i}^{\mathcal{R}}(X))_{k_j}$, *for* $j = 1, \ldots, n$,
- $[t]_{R_j} \in (T_{\Omega_i}^{\mathcal{R}}(X))_k$ *iff* $R_j \in \mathcal{R}$ *and* $t \in (T_{\Omega_j}^{\mathcal{R}}(X))_k$.

As we will formalize in Def. 5 below the intended meaning of the term $[t]_{R_j}$, in the particular case that $t \in T_{\Omega_j}$, is the equivalence class of the terms provably equal to t in the theory R_j.

Definition 4. *Given $\mathcal{R} = \{R_i\}_i = \{(\Omega_i, E_i)\}_{i \in [1..p]} \in \mathcal{CR}$, with each $\Omega_i = (K, \Sigma_i, S_i)$, an atomic $(\Omega_i, \mathcal{R})$-formula is either an equation $t = t'$, where t and t' are $(\Omega_i, \mathcal{R})$-terms of the same kind, or a membership assertion of the form $t : s$, where the $(\Omega_i, \mathcal{R})$-term t has kind k and $s \in (S_i)_k$.*

Definition 5. *Given $\mathcal{R} = \{R_i\}_i = \{(\Omega_i, E_i)\}_{i \in [1..p]} \in \mathcal{CR}$, with each $\Omega_i = (K, \Sigma_i, S_i)$, for all theories $R_i \in \mathcal{R}$ and atomic $(\Omega_i, \mathcal{R})$-formulae ϕ, we recursively define the entailment relation $\vdash^{\mathcal{R}}$ as follows:*

- *if there is a position p in ϕ (for an appropriate definition of positions in atomic formulae), and a term $t \in (T_{\Omega_j})_k$, with $R_j \in \mathcal{R}$, such that $[t]_{R_j}$ occupies position p in ϕ, then*

$$R_i \vdash^{\mathcal{R}} \phi \text{ iff } \exists t' \in (T_{\Omega_i})_k \cap (T_{\Omega_j})_k . (R_i \vdash^{\mathcal{R}} \phi[t']_p \wedge R_j \vdash t = t'),$$

  where $\phi[t']_p$ is the replacement operation of the term t' inside the atomic formula ϕ at position p;
- *otherwise, $R_i \vdash^{\mathcal{R}} \phi$ iff $R_i \vdash \phi$.*

According to this definition, a theory R_i entails an atomic $(\Omega_i, \mathcal{R})$-formula ϕ if ϕ can be proved in R_i after recursively replacing all occurrences of terms $[t]_{R_j}$, such that $t \in T_{\Omega_j}$, with appropriate ground terms in the corresponding equivalence classes.

Remark 1. Given $\mathcal{R} = \{R_i\}_i = \{(\Omega_i, E_i)\}_{i \in [1..p]} \in \mathcal{CR}$, with each $\Omega_i = (K, \Sigma_i, S_i)$, for all theories $R_i, R_j \in \mathcal{R}$, atomic $(\Omega_i, \mathcal{R})$-formulae $\phi(x)$, with free variable x of kind k, and terms $t, t' \in (T_{\Omega_i})_k \cap (T_{\Omega_j})_k$, the following statements hold:

- if $R_j \vdash t = t'$, then $R_i \vdash^{\mathcal{R}} \phi([t]_{R_j})$ iff $R_i \vdash^{\mathcal{R}} \phi([t']_{R_j})$;
- $R_i \vdash^{\mathcal{R}} \phi(t)$ iff $R_i \vdash^{\mathcal{R}} \phi([t]_{R_i})$;
- if $R_i \vdash^{\mathcal{R}} \phi(t)$ then $R_i \vdash^{\mathcal{R}} \phi([t]_{R_j})$;
- if $E_j \subseteq E_i$, then $R_i \vdash^{\mathcal{R}} \phi(t)$ iff $R_i \vdash^{\mathcal{R}} \phi([t]_{R_j})$; and
- if E_j does not include any equations, then $R_i \vdash^{\mathcal{R}} \phi(t)$ iff $R_i \vdash^{\mathcal{R}} \phi([t]_{R_j})$.

For example, using these extended definitions of terms, atomic formulae, and entailment relation for membership equational logic, we can express, in a simple and compact way, property (2-c) with respect to INT_2 and INT_1 by the following metalogical statement:

$$\forall t \in (T_{INT_2})_{Num} . (INT_2 \vdash t : Int \implies INT_1 \vdash^{\mathcal{I}} [t]_{INT_2} : Int), \tag{1}$$

where $\mathcal{I}$ is the multiset $\{INT_1, INT_2\}$.

We are now ready to prove in Prop. 1 below an inductive principle (ind^+) for metalogically proving metatheorems about finite multisets of theories $\mathcal{R} = \{R_i\}_i = \{(\Omega_i, E_i)\}_{i \in [1..p]}$ in $\mathcal{CR}$. Since this proposition is rather technical, we informally advance its content.

- The inductive principle (ind^+) can be applied to metalogical statements of the form: "for all terms t of a sort s in a membership equational theory R_i in $\mathcal{R}$, some property P holds." Here, P is a Boolean expression, $bexp(B_1, \ldots, B_p)$, whose propositional variables are instantiated with metalogical statements of the form: "an atomic $(\Omega_j, \mathcal{R})$-formula $\phi([t]_{R_i})$ holds in R_j," with respect to our extended definition of the entailment relation. For example, the metalogical statement (1) belongs to the class of metatheorems to which our inductive principle can be applied.
- The inductive cases generated by the inductive principle (ind^+) are directly derived from the inductive definition of the sort s in the membership equational theory R_i. Therefore, our inductive cases mirror the inductive cases generated by the usual structural induction principle for membership equational theories. For example, the three inductive cases generated by (ind^+) when applied to the metalogical statement (1) correspond to the three cases in the inductive definition of the sort Int in INT_2: namely, 0 is an Int; $s(n)$ is an Int, if n is an Int; and $p(n)$ is an Int, if n is an Int.

In what follows, given a term $u \in T_\Omega^\mathcal{R}(X)$, we denote by $u(x_1, \ldots, x_n)$, or just $u(\vec{x})$, the fact that the variables in u are in the set $\vec{x} = \{x_1, \ldots, x_n\} \subseteq X$. Thus, given a set $\{t_1, \ldots, t_n\}$ of *metavariables*, we denote by $u(\vec{t})$ the simultaneous replacement of x_i by t_i in u, for $i = 1, \ldots, n$. Similarly, given an atomic formula $\phi(\vec{x})$ with free variables in $\vec{x}$, we denote by $\phi(\vec{t})$ the simultaneous replacement of x_i by t_i in ϕ, for $i = 1, \ldots, n$.

Proposition 1. *Let $\mathcal{R} = \{R_i\}_i = \{(\Omega_i, E_i)\}_{i \in [1..p]}$ be a finite multiset of theories in $\mathcal{CR}$, with each $\Omega_i = (K, \Sigma_i, S_i)$. Let s be a sort in some $(S_e)_k$, $e \in [1..p]$ and $k \in K$, and let $C_{[R_e, s]} = \{C_1, \ldots, C_n\}$ be those sentences in E_e that specify s, i.e., those C_i of the form*

$$\forall (x_1, \ldots, x_{r_i}).\, A_1 \wedge \ldots \wedge A_{q_i} \Longrightarrow A_0 , \tag{2}$$

where, for $1 \leq j \leq r_i$, x_j is of kind k_{i_j}, and for some term w of kind k, A_0 is $w : s$.

Then, for all finite multisets of atomic formulae, $\{\phi_l(x)\}_{l \in [1..p]}$, with each $\phi_l(x)$ an atomic $(\Omega_l, \mathcal{R})$-formula with free variable x of kind k, and Boolean expressions $bexp$, the following metalogical statement holds:

$$\psi_1 \wedge \ldots \wedge \psi_n \tag{3}$$
$$\Longrightarrow \forall t \in (T_{\Omega_e})_k.\, (R_e \vdash t : s \Longrightarrow bexp(R_1 \vdash^\mathcal{R} \phi_1([t]_{R_e}), \ldots, R_p \vdash^\mathcal{R} \phi_p([t]_{R_e}))) ,$$

where, for $1 \leq i \leq n$ and C_i in E_e of the form (2), ψ_i is

$$\forall t_1 \in (T_{\Omega_e})_{k_{i_1}} \ldots \forall t_{r_i} \in (T_{\Omega_e})_{k_{i_{r_i}}}.\, [A_1]^\sharp \wedge \ldots \wedge [A_{q_i}]^\sharp \Longrightarrow [A_0]^\sharp$$

and, for $0 \leq j \leq q_i$,

$$[A_j]^\sharp \triangleq \begin{cases} bexp\left(R_1 \vdash^\mathcal{R} \phi_1([u(\vec{t})]_{R_e}), \ldots, R_p \vdash^\mathcal{R} \phi_p([u(\vec{t})]_{R_e})\right) & \text{if } A_j = u : s \\ R_e \vdash A_j(\vec{t}) & \text{otherwise.} \end{cases}$$

The metalogical statement (3) *introduces an inductive metareasoning principle* (*ind*$^+$), *where each* ψ_i *corresponds to an inductive case and the top line in the definition of* $[A_j]^\sharp$ *provides the corresponding induction hypotheses.*

Proof (soundness). Assume that $\psi_1 \wedge \cdots \wedge \psi_n$ holds. We must prove that

$$\forall t \in (T_{\Omega_e})_k.\,(R_e \vdash t:s \Longrightarrow bexp(R_1 \vdash^{\mathcal{R}} \phi_1([t]_{R_e}), \ldots, R_p \vdash^{\mathcal{R}} \phi_p([t]_{R_e})))$$

also holds. Let $t \in (T_{\Omega_e})_k$ be a term such that $R_e \vdash t : s$; we proceed by structural induction on this derivation. If $R_e \vdash t : s$, then there exists a sentence C_i in E_e of the form $\forall(x_1, \ldots, x_{r_i}).\,A_1 \wedge \ldots \wedge A_{q_i} \Longrightarrow A_0$, where, for $1 \leq j \leq r_i$, x_j is of kind k_{i_j}, and for some term w of kind k, A_0 is $w:s$, and a substitution $\sigma : \{x_1, \ldots, x_{r_i}\} \longrightarrow T_{\Omega_e}$, such that

- $R_e \vdash t = \sigma(w)$, and
- $R_e \vdash \sigma(A_j)$, for $1 \leq j \leq q_i$.

In this case, we must prove that $bexp(R_1 \vdash^{\mathcal{R}} \phi_1([t]_{R_e}), \ldots, R_p \vdash^{\mathcal{R}} \phi_p([t]_{R_e}))$ holds, under the inductive hypothesis that, for $1 \leq j \leq q_i$, if $A_j = u_j : s$, then $bexp\big(R_1 \vdash^{\mathcal{R}} \phi_1([\sigma(u_j)]_{R_e}), \ldots, R_p \vdash^{\mathcal{R}} \phi_p([\sigma(u_j)]_{R_e})\big)$ holds. Since, by assumption, ψ_i holds, then it also holds $[A_1]^\sharp_\sigma \wedge \ldots \wedge [A_{q_i}]^\sharp_\sigma \Longrightarrow [A_0]^\sharp_\sigma$, where, for $0 \leq j \leq q_i$,

$$[A_j]^\sharp_\sigma \triangleq \begin{cases} bexp\big(R_1 \vdash^{\mathcal{R}} \phi_1([\sigma(u_j)]_{R_e}), \ldots, R_p \vdash^{\mathcal{R}} \phi_p([\sigma(u_j)]_{R_e})\big) & \text{if } A_j = u_j : s \\ R_e \vdash \sigma(A_j) & \text{otherwise.} \end{cases}$$

Note then that, for $1 \leq j \leq q_i$,

- If $A_j = (u_j : s)$, then $[A_j]^\sharp_\sigma$ holds by induction hypothesis.
- If $A_j \neq (u_j : s)$, then $[A_j]^\sharp_\sigma$ holds by assumption.

Hence, $[A_0]^\sharp_\sigma$, that is, $bexp(R_1 \vdash^{\mathcal{R}} \phi_1([\sigma(w)]_{R_e}), \ldots, R_p \vdash^{\mathcal{R}} \phi_p([\sigma(w)]_{R_e}))$, also holds. Finally, since $R_e \vdash t = \sigma(w)$, by Rem. 1, we have that $bexp(R_1 \vdash^{\mathcal{R}} \phi_1([t]_{R_e}), \ldots, R_p \vdash^{\mathcal{R}} \phi_p([t]_{R_e}))$ as required.   □

As a final remark, note that a case analysis metareasoning principle (*case*$^+$) can be introduced in a way entirely similar to (*ind*$^+$), except of course for the definition of $[A_j]^\sharp$, that will be as follows (see [7] for more details):

$$[A_j]^\sharp \triangleq \begin{cases} bexp\big(R_1 \vdash \phi_1([u(\vec{t})]_{R_e}), \ldots, R_p \vdash \phi_p([u(\vec{t})]_{R_e})\big) & \text{if } j = 0 \\ R_e \vdash A_j(\vec{t}) & \text{otherwise.} \end{cases}$$

5   Reflection in Membership Equational Logic

A logic is reflective when there exists a *universal* theory in which we can represent and reason about all finitely presentable theories in the logic, including the universal theory itself [8, 3]. A universal theory MB-META for membership equational logic was introduced in [9], along with a representation function $(\_\vdash\_)$

that encodes pairs, consisting of a finitely presentable membership equational theory with nonempty kinds and a sentence in it, as sentences in MB-META. The signature of MB-META contains constructors to represent operators, variables, terms, kinds, sorts, signatures, axioms, and theories. In particular, the signature of MB-META includes the sorts Op, Var, Term, TermList, Kind, Sort, and Theory for terms representing, respectively, operators, variables, terms, lists of terms, kinds, sorts, and theories. In addition, it contains three Boolean operators

```
op _::_in_ : [Term] [Kind] [Theory] -> [Bool] .
op _:_in_  : [Term] [Sort] [Theory] -> [Bool] .
op _=_in_  : [Term] [Term] [Theory] -> [Bool] .
```

to represent, respectively, that a term is a ground term of a given kind in a membership equational theory, and that a membership assertion or an equation holds in a membership equational theory. Note that here, and in what follows, we use Maude's convention for naming kinds: kinds are not named but denoted using the name of their sorts enclosed in square brackets.

The representation function $(\_ \vdash \_)$ is defined in [9] as follows: for all finitely presentable membership equational theories with nonempty kinds R and atomic formulae ϕ over the signature of R,

$$\overline{R \vdash \phi} \triangleq \begin{cases} (\overline{t} : \overline{s} \text{ in } \overline{R}) = \texttt{true} \text{ if } \phi = (t:s) \\ (\overline{t} = \overline{t'} \text{ in } \overline{R}) = \texttt{true} \text{ if } \phi = (t = t'), \end{cases}$$

where $\overline{(\_)}$ is a representation function defined recursively over theories, signatures, axioms, and so on. Under this representation function, a term t is represented in MB-META by a ground term $\overline{t}$ of sort Term, a kind k is represented by a ground term $\overline{k}$ of sort Kind, a sort s is represented by a ground term $\overline{s}$ of sort Sort, and a theory R is represented by a ground term $\overline{R}$ of sort Theory. In particular, to represent terms the signature of MB-META contains the constructors

```
op _[_] : [Op] [TermList] -> [Term] .
op nil : -> [TermList] .
op _,_ : [TermList] [TermList] -> [TermList] .
```

and the representation function $\overline{(\_)}$ is defined as follows:

$$\overline{t} \triangleq \begin{cases} \overline{c} & \text{if } t = c \text{ is a constant} \\ \overline{x} & \text{if } t = x \text{ is a variable} \\ \overline{f}[\overline{t_1}, \dots, \overline{t_n}] & \text{if } t = f(t_1, \dots, t_n). \end{cases}$$

For example, the term $s(0)$ of kind Num is represented in MB-META as the term $\overline{s}[\overline{0}]$ of sort Term.

The following propositions state the main properties of MB-META as a universal theory [9]:

Proposition 2. *For all finitely presentable membership equational theories with nonempty kinds* $R = (\Omega, E)$*, with* $\Omega = (K, \Sigma, S)$*, terms* t *in* T_Ω*, and kinds* $k \in K$,

$$t \in (T_\Omega)_k \iff \texttt{MB-META} \vdash (\overline{t} :: \overline{k} \text{ in } \overline{R}) = \texttt{true}.$$

Proposition 3. *For all finitely presentable membership equational theories with nonempty kinds $R = (\Omega, E)$, with $\Omega = (K, \Sigma, S)$, kinds $k \in K$, and ground terms u of the kind* [Term], *if*

$$\text{MB-META} \vdash (u :: \overline{k} \text{ in } \overline{R}) = \text{true},$$

then there is a term $t \in (T_\Omega)_k$ such that $\overline{t} = u$.

Proposition 4. *For all finitely presentable membership equational theories with nonempty kinds $R = (\Omega, E)$, with $\Omega = (K, \Sigma, S)$, terms t in $(T_\Omega)_k$ and sorts s in S_k,*

$$R \vdash t : s \iff \text{MB-META} \vdash (\overline{t} : \overline{s} \text{ in } \overline{R}) = \text{true}.$$

Similarly, for all terms t, t' in $(T_\Omega)_k$,

$$R \vdash t = t' \iff \text{MB-META} \vdash (\overline{t} = \overline{t'} \text{ in } \overline{R}) = \text{true}.$$

For example,

$$\text{MB-META} \vdash \overline{(\text{p(s(p(0)))} : \text{Int in INT2})} = \text{true},$$

but

$$\text{MB-META} \not\vdash \overline{(\text{p(s(p(0)))} : \text{Int in INT1})} = \text{true}.$$

6   Reflection in Extended Membership Equational Logic

To represent and reason about our extended definition of entailment relation, we define a new theory MB-META$^=$ that extends the universal theory MB-META with a binary operator

```
op _in_ : [Term] [Theory] -> [Term] .
ceq (t̄ in R̄) = (t̄' in R̄) if (t̄ = t̄' in R̄) = true .
```

to represent the equivalence class of a term in a membership equational theory.

Proposition 5. *For all finitely presentable membership equational theories with nonempty kinds $R = (\Omega, E)$, with $\Omega = (K, \Sigma, S)$, and terms t, t' in $(T_\Omega)_k$, $k \in K$,*

$$R \vdash t = t' \iff \text{MB-META}^= \vdash (\overline{t} \text{ in } \overline{R}) = (\overline{t'} \text{ in } \overline{R}).$$

Using this operator, we can now define a representation function $\overline{(\_)}$ for terms in the extended class, which satisfies the expected property, as shown in Prop. 6 below. Let $\mathcal{R} = \{R_i\}_i = \{(\Omega_i, E_i)\}_{i \in [1..p]}$ be a finite multiset of theories in $\mathcal{CR}$. Then, for all terms $t \in T_\Omega^{\mathcal{R}}(X)$,

$$\overline{t} \triangleq \begin{cases} \overline{c} & \text{if } t = c \text{ is a constant} \\ \overline{x} & \text{if } t = x \text{ is a variable} \\ \overline{f}[\overline{t_1}, \dots, \overline{t_n}] & \text{if } t = f(t_1, \dots, t_n) \\ (\overline{t'} \text{ in } \overline{R}) & \text{if } t = [t']_R. \end{cases} \tag{4}$$

Proposition 6. *Let* $\mathcal{R} = \{R_i\}_i = \{(\Omega_i, E_i)\}_{i \in [1..p]}$ *be a finite multiset of theories in* $\mathcal{CR}$, *with each* $\Omega_i = (K, \Sigma_i, S_i)$. *Then, for all theories* $R_i \in \mathcal{R}$, *terms* $t \in (T^{\mathcal{R}}_{\Omega_i})_k$ *and sorts* s *in* $(S_i)_k$, $k \in K$,

$$R_i \vdash^{\mathcal{R}} t : s \iff \texttt{MB-META}^= \vdash (\overline{t} : \overline{s} \text{ in } \overline{R_i}) = \texttt{true}.$$

Similarly, for all terms $t, t' \in (T^{\mathcal{R}}_{\Omega_i})_k$, $k \in K$,

$$R_i \vdash^{\mathcal{R}} t = t' \iff \texttt{MB-META}^= \vdash (\overline{t} = \overline{t'} \text{ in } \overline{R_i}) = \texttt{true}.$$

Proof. This proposition is a corollary of Props. 4 and 5.   □

7   An Inductive Principle
for Logically Proving Semantic Relations

We are now ready to prove in Prop. 7 below the main technical result in this paper, namely, that there is a class of metatheorems about membership equational logic theories which can be represented and logically proved as theorems about the initial model of the membership equational theory $\texttt{MB-META}^=$. As a corollary of this proposition we will obtain an inductive reasoning principle $\overline{(ind^+)}$ for *logically* proving metatheorems about families of membership equational theories.

In order to simplify the presentation of the upcoming material, we introduce here some additional notation. Let $\mathcal{R} = \{R_i\}_i = \{(\Omega_i, E_i)\}_{i \in [1..p]}$ be a finite multiset of theories in $\mathcal{CR}$, with each $\Omega_i = (K, \Sigma_i, S_i)$. For all theories $R_i \in \mathcal{R}$ and terms $t \in T^{\mathcal{R}}_{\Omega_i}(X)$, we denote by $\overline{t}^{[X]}$ the reflective representation of t defined in (4), except that now variables $x \in X$ are replaced by variables $\overline{x}^{[X]} = x$ of the kind [Term] [1], and we denote by $\overline{X}^{[X]}$ the set $\overline{X}^{[X]} \triangleq \{\overline{x}^{[X]} \mid x \in X\}$. In addition, for all theories $R_i \in \mathcal{R}$, and membership assertions $t : s$, with t in $T^{\mathcal{R}}_{\Omega_i}(X)$ and s in some $(S_i)_k$,

$$\overline{t : s}^{[R_i, X]} \triangleq (\overline{t}^{[X]} : \overline{s} \text{ in } \overline{R_i}) = \texttt{true},$$

and, similarly, for all equations $t = t'$, with t, t' in $T^{\mathcal{R}}_{\Omega_i}(X)$,

$$\overline{t = t'}^{[R_i, X]} \triangleq (\overline{t}^{[X]} = \overline{t'}^{[X]} \text{ in } \overline{R_i}) = \texttt{true}.$$

We can now define a representation function for metalogical statements, which satisfies the expected property, as shown in Prop. 7 below. Let $\mathcal{R} = \{R_i\}_i = \{(\Omega_i, E_i)\}_{i \in [1..p]}$ be a finite multiset of theories in $\mathcal{CR}$, with each $\Omega_i = (K, \Sigma_i, S_i)$. Let $\{k_1, \dots, k_n\}$ be a finite multiset of kinds, with each k_i in K, let $\vec{x} = \{x_1, \dots, x_n\}$ be a finite set of variables, with each x_i of kind k_i, and let τ be a metalogical statement of the form

$$\forall t_1 \in (T_{\Omega_1})_{k_1}. \ \dots \ \forall t_n \in (T_{\Omega_n})_{k_n}. \ bexp(R_1 \vdash^{\mathcal{R}} \phi_1(\vec{t}), \dots, R_p \vdash^{\mathcal{R}} \phi_p(\vec{t})), \quad (5)$$

---

[1] The key difference between $\overline{t}$ and $\overline{t}^{[X]}$ is that $\overline{t}$ is a *ground term* of sort Term, whereas $\overline{t}^{[X]}$ is a term of kind [Term] with variables of the kind [Term].

where each $\phi_l(\vec{x})$ is an atomic $(\Omega_l, \mathcal{R})$-formula with free variables in $\vec{x}$. Then,

$$\bar{\tau} \triangleq \forall x_1. \ldots \forall x_n. (((x_1 :: \overline{k_1} \text{ in } \overline{R_1}) = \mathbf{true} \wedge \cdots \wedge (x_n :: \overline{k_n} \text{ in } \overline{R_n}) = \mathbf{true}$$
$$\implies bexp(\overline{\phi_1(\vec{x})}^{[R_1,\vec{x}]}, \ldots, \overline{\phi_p(\vec{x})}^{[R_p,\vec{x}]})),$$

where $\{x_1, \ldots, x_n\}$ are now variables of the kind [Term].

Note that the class of metalogical statements of the form (5) includes, for example, all instances of the properties (1-a) in Def. 1, and (2-a) and (2-c) in Def. 2. In particular, the metalogical statement (1) is represented in MB-META$^=$ as the formula

$$\forall \mathtt{N}.((\mathtt{N} :: \overline{\mathtt{Num}} \text{ in } \overline{\mathtt{INT2}} = \mathbf{true})$$
$$\implies (\mathtt{N} : \overline{\mathtt{Int}} \text{ in } \overline{\mathtt{INT2}} = \mathbf{true}) \implies ((\mathtt{N} \text{ in } \overline{\mathtt{INT2}}) : \overline{\mathtt{Int}} \text{ in } \overline{\mathtt{INT1}} = \mathbf{true})),$$

where $\mathtt{N}$ is a variable of the kind [Term].

Proposition 7. *Let* $\mathcal{R} = \{R_i\}_i = \{(\Omega_i, E_i)\}_{i \in [1..p]}$ *be a finite multiset of theories in* $\mathcal{CR}$, *with each* $\Omega_i = (K, \Sigma_i, S_i)$. *For all metalogical statements* τ *of the form (5),* τ *holds iff* MB-META$^=$ $\models \bar{\tau}$.

Proof. We first prove the ($\Rightarrow$)-direction of this proposition. Suppose that τ holds. Let $\sigma : \{x_1, \ldots, x_n\} \longrightarrow T_{\text{MB-META}^=}$ be a substitution such that, for $1 \leq i \leq n$,

$$\text{MB-META}^= \models (\overline{\sigma(x_i) :: \overline{k_i} \text{ in } \overline{R_i}}) = \mathbf{true}. \tag{6}$$

We must prove that

$$\text{MB-META}^= \models \sigma \left(bexp(\overline{\phi_1(\vec{x})}^{[R_1,\vec{x}]}, \ldots, \overline{\phi_p(\vec{x})}^{[R_p,\vec{x}]}) \right).$$

Note that, since $(\overline{\sigma(x_i) :: \overline{k_i} \text{ in } \overline{R_i}})$ is a ground term, (6) implies

$$\text{MB-META}^= \models (\overline{\sigma(x_i) :: \overline{k_i} \text{ in } \overline{R_i}}) = \mathbf{true},$$

which, by completeness of membership equational logic, implies

$$\text{MB-META}^= \vdash (\overline{\sigma(x_i) :: \overline{k_i} \text{ in } \overline{R_i}}) = \mathbf{true}.$$

Thus, by Prop. 3, we know that, for $1 \leq i \leq n$, $\sigma(x_i) = \overline{w_i}$ for some $w_i \in (T_{\Omega_i})_{k_i}$. Note then that

$$\sigma \left(bexp(\overline{\phi_1(\vec{x})}^{[R_1,\vec{x}]}, \ldots, \overline{\phi_p(\vec{x})}^{[R_p,\vec{x}]}) \right) = bexp(\overline{\phi_1(\vec{w})}^{[R_1,\emptyset]}, \ldots, \overline{\phi_p(\vec{w})}^{[R_p,\emptyset]}),$$

and that, by Prop. 6, for $1 \leq l \leq p$, $R_l \vdash^{\mathcal{R}} \phi_l(\vec{w})$ iff MB-META$^=$ $\vdash \overline{\phi_l(\vec{w})}^{[R_l,\emptyset]}$. Since we are assuming that $bexp(R_1 \vdash^{\mathcal{R}} \phi_1(\vec{w}), \ldots, R_p \vdash^{\mathcal{R}} \phi_p(\vec{w}))$ holds, then

$$\text{MB-META}^= \vdash bexp(\overline{\phi_1(\vec{w})}^{[R_1,\emptyset]}, \ldots, \overline{\phi_p(\vec{w})}^{[R_p,\emptyset]}),$$

and, by soundness of membership equational logic,

$$\text{MB-META}^= \models bexp(\overline{\phi_1(\vec{w})}^{[R_1,\emptyset]}, \ldots, \overline{\phi_p(\vec{w})}^{[R_p,\emptyset]}),$$

as required.

The proof of the $(\Leftarrow)$-direction is similar. In particular, consider for any terms $\{w_1, \ldots, w_n\}$ the substitution $\sigma : \{x_1, \ldots, x_n\} \longrightarrow T_{\texttt{MB-META}^=}$ given by $\sigma(x_i) = \overline{w_i}$, for $1 \le i \le n$. $\qquad\qquad\square$

As corollaries of Prop. 7 we can prove the reflective versions of Rem. 1 and Prop. 1, which we will denote, respectively, as Rem. $\overline{1}$, and Prop. $\overline{1}$. Both are obtained by replacing each metalogical statement ϕ in Rem. 1 and Prop. 1 by its logical representation $\overline{\phi}$ in $\texttt{MB-META}^=$. Of course, this is key for our purposes, since it automatically gives us an inductive reasoning principle $(\overline{ind^+})$, and a case analysis reasoning principle $(\overline{case^+})$, for proving metalogical statements about membership equational theories represented as logical statements in $\texttt{MB-META}^=$. Moreover, since Rem. $\overline{1}$ and Prop. $\overline{1}$ mirror their metalogical counterparts, the metalogical proofs based on the latter will also be mirrored by the logical proofs based on the former. As an example of this, we *logically* prove in the appendix, using the induction principle $(\overline{ind^+})$, that INT_2 satisfies property (2-c) with respect to INT_1.

8   Conclusion

The work presented is based on the ideas proposed in [1] for formal metareasoning using reflection. Here we extend the metareasoning principles introduced in [1], increasing their applicability as we show in a case study. The reader can find in [1] a detailed discussion on tradeoffs and limitations of reflective metalogical frameworks, and a survey of related work.

One of the advantages of formal metareasoning based on reflection is that it can be carried out using already existing logical reasoning tools. Our experience shows also that the logical proofs of metatheorems using reflection mirror their standard metalogical proofs. In this regard, we plan to use our results to extend the ITP tool [3, 4], which is an interactive inductive theorem prover for membership equational theories, with metareasoning capabilities, so that it can be used, for example, as a methodological tool for software development.

Acknowledgments

We thank David Basin and José Meseguer for many discussions on using reflection for formal metareasoning. We also thank two anonymous referees for their helpful comments on ways of improving the presentation of our results.

References

1. D. Basin, M. Clavel, and J. Meseguer. Reflective metalogical frameworks. *ACM Transactions on Computational Logic*, 2004. To appear. http://www.acm.org/pubs/tocl/accepted.html.
2. A. Bouhoula, J.-P. Jouannaud, and J. Meseguer. Specification and proof in membership equational logic. *Theoretical Computer Science*, 236:35–132, 2000.

3. M. Clavel. *Reflection in Rewriting Logic: Metalogical Foundations and Metaprogramming Applications*. CSLI Publications, 2000.

4. M. Clavel. The ITP tool's home page. http://geminis.sip.ucm.es/~clavel/itp, 2004.

5. M. Clavel, F. Durán, S. Eker, P. Lincoln, N. Martí-Oliet, J. Meseguer, and J. F. Quesada. Maude: Specification and programming in rewriting logic. *Theoretical Computer Science*, 285:187–243, 2002.

6. M. Clavel, F. Durán, S. Eker, P. Lincoln, N. Martí-Oliet, J. Meseguer, and C. Talcott. Maude Manual (Version 2.1). Manual distributed as documentation of the Maude system. http://maude.cs.uiuc.edu, 2004.

7. M. Clavel, N. Martí-Oliet, and M. Palomino. Formalizing and proving semantic relations between specifications by reflection (extended version). http://geminis.sip.ucm.es./~clavel/pubs/pubs.html, 2004.

8. M. Clavel and J. Meseguer. Axiomatizing reflective logics and languages. In G. Kiczales, editor, *Proc. Reflection'96*, pages 263–288. Xerox PARC, 1996.

9. M. Clavel, J. Meseguer, and M. Palomino. Reflection in membership equational logic, many-sorted equational logic, Horn logic with equality, and rewriting logic. In F. Gadducci and U. Montanari, editors, *Proc. Fourth International Workshop on Rewriting Logic and its Applications*, volume 71 of *Electronic Notes in Theoretical Computer Science*, pages 63–78. Elsevier, 2002. http://geminis.sip.ucm.es/~clavel/pubs/pubs.html.

10. H. Ehrig and B. Mahr. *Fundamentals of Algebraic Specification 1*, volume 6 of *EATCS Monographs on Theoretical Computer Science*. Springer-Verlag, 1985.

11. J. Loeckx, H.-D. Ehrich, and M. Wolf. *Specification of Abstract Data Types*. J. Wiley & Sons and B.G. Teubner, 1996.

12. J. Meseguer. Membership algebra as a logical framework for equational specification. In F. Parisi-Presicce, editor, *Proc. WADT'97*, volume 1376 of *LNCS*, pages 18–61. Springer-Verlag, 1998.

Appendix

We show how Fact 1 below, that is, the representation of the metalogical statement (1) as a logical statement about the initial model of MB-META$^=$, can be logically proved using the inductive principle $(\overline{ind^+})$, along with the reflective properties of membership equational logic. Our proof mirrors at the logical level the metalogical proof of (1), that we omit here for the sake of space limitations; this proof can be found in [7].

Fact 1.

$$\text{MB-META}^= \models \forall \text{N.}(\text{N} :: \overline{\text{Num}} \text{ in } \overline{\text{INT2}} = \text{true}$$
$$\Longrightarrow (\text{N} : \overline{\text{Int}} \text{ in } \overline{\text{INT2}} = \text{true}) \Longrightarrow ((\text{N in } \overline{\text{INT2}}) : \overline{\text{Int}} \text{ in } \overline{\text{INT1}} = \text{true})),$$

where N is a variable of the kind [Term].

Proof. By $(\overline{ind^+})$, we can prove the theorem by showing:

$$\text{MB-META}^= \mathrel{\ventrianglerighteq} ((\overline{0} \text{ in } \overline{\text{INT2}}) : \overline{\text{Int}} \text{ in } \overline{\text{INT1}} = \text{true}) \tag{7}$$

$$\wedge\ \forall \text{N.}(\text{N} :: \overline{\text{Num}} \text{ in } \overline{\text{INT2}} = \text{true}$$

$$\implies ((\text{N in } \overline{\text{INT2}}) : \overline{\text{Int}} \text{ in } \overline{\text{INT1}} = \text{true}) \tag{8}$$

$$\implies ((\overline{\text{s}}[\text{N}] \text{ in } \overline{\text{INT2}}) : \overline{\text{Int}} \text{ in } \overline{\text{INT1}} = \text{true}))$$

$$\wedge\ \forall \text{N.}(\text{N} :: \overline{\text{Num}} \text{ in } \overline{\text{INT2}} = \text{true}$$

$$\implies ((\text{N in } \overline{\text{INT2}}) : \overline{\text{Int}} \text{ in } \overline{\text{INT1}} = \text{true}) \tag{9}$$

$$\implies ((\overline{\text{p}}[\text{N}] \text{ in } \overline{\text{INT2}}) : \overline{\text{Int}} \text{ in } \overline{\text{INT1}} = \text{true})),$$

where N is a variable of the kind [Term]. Note that (7) holds by Prop. 6 (using soundness of membership equational logic). Regarding (8) and (9), their proofs are similar; we show here only the proof of (8). It is a fact[2] that, (8) holds if

$$\text{MB-META}^= \mathrel{\ventrianglerighteq} \forall \text{N.}(\text{N} :: \overline{\text{Num}} \text{ in } \overline{\text{INT1}} = \text{true} \implies (\text{N} : \overline{\text{Int}} \text{ in } \overline{\text{INT1}} = \text{true}) \tag{10}$$

$$\implies ((\overline{\text{s}}[\text{N}] \text{ in } \overline{\text{INT2}}) : \overline{\text{Int}} \text{ in } \overline{\text{INT1}} = \text{true})),$$

which, by Rem. $\overline{1}$, is equivalent to

$$\text{MB-META}^= \mathrel{\ventrianglerighteq} \forall \text{N.}(\text{N} :: \overline{\text{Num}} \text{ in } \overline{\text{INT1}} = \text{true} \implies (\text{N} : \overline{\text{Int}} \text{ in } \overline{\text{INT1}} = \text{true}) \tag{11}$$

$$\implies ((\overline{\text{s}}[\text{N in } \overline{\text{INT1}}] \text{ in } \overline{\text{INT2}}) : \overline{\text{Int}} \text{ in } \overline{\text{INT1}} = \text{true})).$$

To prove (11) we can use again $(\overline{ind^+})$, and reduce its proof to showing:

$$\text{MB-META}^= \mathrel{\ventrianglerighteq} \forall \text{N.}(\text{N} :: \overline{\text{Num}} \text{ in } \overline{\text{INT1}} = \text{true} \implies (\text{N} : \overline{\text{Nat}} \text{ in } \overline{\text{INT1}} = \text{true}) \tag{12}$$

$$\implies ((\overline{\text{s}}[\text{N in } \overline{\text{INT1}}] \text{ in } \overline{\text{INT2}}) : \overline{\text{Int}} \text{ in } \overline{\text{INT1}} = \text{true}))$$

$$\wedge\ \forall \text{N.}(\text{N} :: \overline{\text{Num}} \text{ in } \overline{\text{INT1}} = \text{true} \implies (\text{N} : \overline{\text{Neg}} \text{ in } \overline{\text{INT1}} = \text{true}) \tag{13}$$

$$\implies ((\overline{\text{s}}[\text{N in } \overline{\text{INT1}}] \text{ in } \overline{\text{INT2}}) : \overline{\text{Int}} \text{ in } \overline{\text{INT1}} = \text{true}))$$

The proofs of (12) and (13) are similar; we show here only the proof of (13). By $(\overline{case^+})$ and Rem. $\overline{1}$, we can reduce proving (13) to showing:

$$\text{MB-META}^= \mathrel{\ventrianglerighteq} (\overline{\text{s}}[\overline{0}] \text{ in } \overline{\text{INT2}}) : \overline{\text{Int}} \text{ in } \overline{\text{INT1}} = \text{true} \tag{14}$$

$$\wedge\ \forall \text{N.}(\text{N} :: \overline{\text{Num}} \text{ in } \overline{\text{INT1}} = \text{true} \implies (\text{N} : \overline{\text{Neg}} \text{ in } \overline{\text{INT1}} = \text{true}) \tag{15}$$

$$\implies ((\overline{\text{s}}[\overline{\text{p}}[\text{N}]] \text{ in } \overline{\text{INT2}}) : \overline{\text{Int}} \text{ in } \overline{\text{INT1}} = \text{true}))$$

Note that (14) holds by Prop. 6 (using soundness of membership equational logic). Regarding (15), let $\sigma : \{\text{N}\} \longrightarrow T_{\text{MB-META}=}$ be a substitution such that $(\sigma(\text{N}) :: \overline{\text{Num}} \text{ in } \overline{\text{INT1}} = \text{true})$ and $(\sigma(\text{N}) : \overline{\text{Neg}} \text{ in } \overline{\text{INT1}} = \text{true})$ hold in the initial model of MB-META$^=$. Thus, by Prop. 3, $\sigma(\text{N}) = \overline{N}$ for some term N of kind Num, and, by Prop. 4, $INT_1 \vdash N : Neg$. Finally, note that, since $INT_2 \vdash s(p(N)) = N$, then, by Prop. 6 (using again soundness of membership equational logic),

$$\text{MB-META}^= \mathrel{\ventrianglerighteq} (\overline{\text{s}}[\overline{\text{p}}[\overline{N}]] \text{ in } \overline{\text{INT2}}) : \overline{\text{Int}} \text{ in } \overline{\text{INT1}} = \text{true}.$$

$\square$

---

[2] This fact is an instance of the reflective counterpart of a general metareasoning principle for reducing metalogical statements to a form such that (ind^+) can be applied to them. For the sake of space limitations, we omit here the proposition that states this principle, and its proof, that can be found in [7].

Model-Checking Systems
with Unbounded Variables without Abstraction

Magali Contensin[1] and Laurence Pierre[2]

[1] CMI/Université de Provence
39 Rue Joliot-Curie
13453 Marseille cedex 13, France
Magali.Contensin@cmi.univ-mrs.fr
[2] I3S, Université de Nice
2000 Route des Lucioles, BP 121
06903 Sophia Antipolis cedex, France
Laurence.Pierre@i3s.unice.fr

Abstract. The problem addressed in this paper is the formal verification of temporal properties in the presence of unbounded data types. In that framework, state of the art model-checkers use reduction techniques, e.g. abstraction, to compute finite counterparts of the systems under consideration. The method we present integrates a model-checker for the modal ν-calculus with a theorem prover, it processes unbounded systems without having to reduce them.

1 Introduction

In software engineering, as in hardware design, there is a crucial need for formal verification methods. Testing is one of the main software verification techniques used in practice. However, an exhaustive testing of all execution paths is practically infeasible, testing can thus never be complete and cannot guarantee the absence of errors. Formal methods allow an early integration of verification in the development process, and provide more reliable techniques. They are meant either to check the functionality of the system or to verify properties.

This paper addresses the problem of formally verifying *temporal properties* of high-level specifications that take into account unbounded domains, for instance with integer variables. Keeping these variables under their numeric form engenders an infinite state space, classical model-checking techniques do not apply. Even assuming an encoding that induces a fixed size, model-checking techniques suffer from the state space explosion problem (as the size of the data increases, the state space of the system increases exponentially). Many existing model-checkers perform either explicit or symbolic enumerations of the reachable states of the system, using BDD-based representations and their various enhancements. For systems with arithmetic data or for real-time systems with several clocks, *state space reduction techniques* have to be applied to get tractable problems. Various solutions have been proposed, such as the consideration of *uninterpreted functions* [1, 2], the use of *abstraction techniques* [3–5], the construction of *region*

C. Rattray et al. (Eds.): AMAST 2004, LNCS 3116, pp. 87–101, 2004.
© Springer-Verlag Berlin Heidelberg 2004

graphs [6], the computation of *partial order reductions* [7,8], or the exploitation of *symmetries* [9]. There are various disadvantages with techniques like abstraction: difficulties to build the abstraction, necessity for validating it (do the properties that hold at the abstract level still hold at the concrete level?), difficulties to reason on liveness properties, to express counterexamples, etc.

Deductive approaches implemented in theorem provers do not impose a limit on the size or the complexity of the system and give the opportunity to express and to verify properties over infinite domains. For instance, the simple *induction* technique is efficient to reason on data types like natural numbers. *Combination of model-checking and theorem proving* allows to exploit their complementary strengths to verify large or infinite systems [10]. Among these works, we can mention [11] that proposes a combination of different proof tools, among them the model-checker SMV [12] and the theorem prover HOL [13], through a variety of strategies. In [14], an induction mechanism for natural numbers is integrated in the proof assistant of SMV. The inductive proof uses an abstraction that partitions the natural numbers into a finite number of intervals. A finite model is obtained, that can be processed fully mechanically with SMV.

Our method proposes a combination of a model-checker for the modal ν-calculus with a theorem prover, first theoretical results were reported in [15]. It does not assume any specific encoding of the variables, does not require any state space reduction, and is not limited to combining model-checking and theorem proving techniques via strategies. The algorithm realizes a classical fixed-point computation, and calls on the theorem prover at specific and well-identified steps to process by deduction arithmetic conjectures. This paper presents the methodology and the tool under development.

After presenting in Section 2 the characteristics of the systems that we consider and the foundations of our model-checker, we describe the tool and detail the algorithm in Section 3. Section 4 proposes some related works, and then we conclude with a description of our ongoing and future work.

2   Model-Checking without Abstraction

2.1   Modelling the Systems and Their Properties

Our method is inspired from the one proposed by G.Winskel in [16], which is an algorithm for checking whether a state of a labelled transition system (LTS) satisfies a property in the modal ν-calculus [17]: the models under consideration are of the form $(Proc, \{\overset{\alpha}{\rightarrow} \,|\, \alpha \in \text{labels}\}, V)$, where $Proc$ is a nonempty finite set of *processes* (states) and V is a function from basic assertions a to the subsets $V(a) \subseteq Proc$ of processes satisfying them.

The *models* we consider allow to describe more elaborate systems: transition systems where transitions are not labelled by atoms of a finite set, but by pairs of the form *condition/assignments*, and *condition* and *assignments* can handle unbounded data types. More precisely, the models are tuples of the form $(S, V, I, O, \mathcal{L}, \mathcal{A}, \rightarrow, S_0, P_0)$, where

S is a finite set of symbolic states,

V is a set of variables,

I and O are sets of inputs and outputs that carry boolean or arithmetic data,

$\mathcal{L}$ is a set of labels (transition conditions), that are formulas of the predicate calculus without quantifiers,

$\mathcal{A}$ is a set of assignments of the form $v \leftarrow exp(v_1, ..., v_n)$; they correspond to the actions that are performed on the transitions, and will be interpreted hereinafter as equalities $v = exp(v_1, ..., v_n)$

$\rightarrow \subseteq S \times (\mathcal{L} \times 2^{\mathcal{A}}) \times S$ denotes the transition relation, a notation of the form α/a is used to label the transitions,

$S_0 \in S$ is the initial state, and P_0 are the properties that hold in S_0.

Example. Let us illustrate this definition with a simple version of a *cash withdrawal system*, see Fig. 1 (transition labels of the form $\alpha/-$ mean that there is no assignment). S contains 6 symbolic states, $V = \{code, ok, n\}$, $I = \{inc, cc, codin, take\}$, and $O = \{outc, keep\}$. The inputs inc and $take$ are true resp. when the card is entered and withdrawn, cc is the card code, $codin$ is the submitted code. The outputs $outc$ and $keep$ indicate resp. that the card can be withdrawn or that the machine keeps it. The initial state is S_0, the associated property is $outc = false$. If the customer gives the right code before the third attempt (included), he is allowed to retrieve his card; otherwise the card is kept.

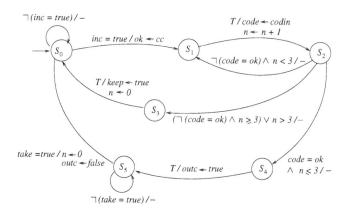

Fig. 1. Behaviour of a cash machine

Our *language of assertions* contains two modal operators $\Diamond$ and $\Box$ (their intuitive meaning is close to the one of the operators introduced in [17]) and a greatest fixed-point operator inspired from the one of [16]:

$$A ::= f \mid T \mid F \mid \neg A \mid A_0 \wedge A_1 \mid A_0 \vee A_1 \mid \langle \psi \rangle A \mid [\psi] A \mid X \mid \nu X \{\vec{r}\} A$$

where f represents "atomic" formulas (formulas without propositional connective, but that can involve predicates symbols), $\psi \in \mathcal{L}$, and $\vec{r}$ is a list of the form

$(r_1, \varphi_1), ..., (r_n, \varphi_n)$, $r_i \in S$, φ_i is a formula made of subformulas of $\mathcal{A}' \cup \mathcal{L}$ where $\mathcal{A}'$ is $\mathcal{A}$ where the assignments are interpreted as equalities.

As in [16] the semantics $[\![A]\!]$ of these assertions is given w.r.t. an environment function ρ, but it is also related to a formula φ that represents the conjunction of the properties that are known to be true in the current state: $[\![A]\!]_{\rho,\varphi}$ is the subset of S in which $\varphi \Rightarrow A$. The definition of this semantics is as follows:

$$[\![f]\!]_{\rho,\varphi} = \begin{cases} S & \text{if } \varphi \Rightarrow f, \\ \emptyset & \text{otherwise} \end{cases}$$

$\varphi \Rightarrow f$ means that the conjunction of the formulas that are known to be true is sufficient to deduce that f is true.

$$[\![T]\!]_{\rho,\varphi} = S$$

$$[\![F]\!]_{\rho,\varphi} = \begin{cases} S & \text{if } \varphi = F, \\ \emptyset & \text{otherwise} \end{cases}$$

$$[\![\neg A]\!]_{\rho,\varphi} = \begin{cases} S & \text{if } \varphi = F, \\ S \setminus [\![A]\!]_{\rho,\varphi} & \text{otherwise} \end{cases}$$

$$[\![A_0 \wedge A_1]\!]_{\rho,\varphi} = [\![A_0]\!]_{\rho,\varphi} \cap [\![A_1]\!]_{\rho,\varphi}$$

$$[\![A_0 \vee A_1]\!]_{\rho,\varphi} = [\![A_0]\!]_{\rho,\varphi} \cup [\![A_1]\!]_{\rho,\varphi}$$

$$[\![\langle \alpha \rangle A]\!]_{\rho,\varphi} = \{p \in S \mid \exists q \in [\![A]\!]_{\rho,\varphi'}, p \xrightarrow{\alpha'/a} q \wedge \alpha' \Rightarrow \alpha \wedge \neg(\varphi \Rightarrow \neg\alpha')\}$$

$[\![\langle \alpha \rangle A]\!]_{\rho,\varphi}$ denotes the set of states p such that there exists a state of $[\![A]\!]_{\rho,\varphi'}$ reachable from p by the transition α'/a with $\alpha' \Rightarrow \alpha$. The condition $\varphi \Rightarrow \neg\alpha'$ allows to detect a contradiction between φ and the transition condition α'; these conditions will be used to prune branches in the exploration tree.

$$[\![[\alpha]A]\!]_{\rho,\varphi} = \{p \in S \mid \forall q, ((p \xrightarrow{\alpha'/a} q \wedge \alpha' \Rightarrow \alpha) \Rightarrow q \in [\![A]\!]_{\rho,\varphi'}) \wedge \neg(\varphi \Rightarrow \neg\alpha')\}$$

$[\![[\alpha]A]\!]_{\rho,\varphi}$ denotes the set of states p such that for all state q, if q is reachable from p by the transition α'/a, with $\alpha' \Rightarrow \alpha$, then q belongs to the set $[\![A]\!]_{\rho,\varphi'}$.

$$[\![X]\!]_{\rho,\varphi} = \rho(X)$$

$$[\![\nu X\{\overrightarrow{r}\}A]\!]_{\rho,\varphi} = \cup\{E \subseteq S \mid E \subseteq \{r_i \mid (r_i, \varphi_i) \in \{\overrightarrow{r}\} \wedge Equiv(\varphi, \varphi_i)\}$$
$$\cup [\![A]\!]_{\rho[E/X],\varphi}\}$$

where the equivalence between φ and φ_i, $Equiv(\varphi, \varphi_i)$, can be roughly expressed as follows: the values of every variable v in φ and φ_i must correspond to comparable expressions (a matching function is used), and the condition parts of φ and φ_i (conjunctions of the α') must be equivalent in the state defined by the equalities induced by the assignments a of φ and φ_i (after eliminating irrelevant terms i.e., terms that involve identifiers that are no longer in use).

$Equiv(\varphi, \varphi_i) = Match\_variables(\varphi, \varphi_i) \wedge$
$\quad (Equalities(\varphi) \wedge Equalities(\varphi_i) \Rightarrow (Conditions(\varphi) \Leftrightarrow Conditions(\varphi_i)))$

Let us see how φ' (the condition associated with q in the semantics of $\Diamond$ and $\Box$) is defined. $\varphi' = Next\varphi(\varphi, \alpha', a, p)$ can be seen as a renaming of $\varphi \wedge \alpha' \wedge a$: the property associated with q is the conjunction of the previous property φ with the condition α' of the transition that is taken (excluding the conditions on the inputs, that become obsolete), and with the assignments a (interpreted as equalities) performed on this transition. We give the definition of $Next\varphi(\varphi, \alpha', a, p)$ below, it will be illustrated on an example in Section 3.2. For each variable v, v' refers to its next value. In the very first step (i.e., $\varphi = P_0 \wedge p = S_0$), φ' corresponds to the conjunction $P_0 \wedge \alpha' \wedge a$ (note that, for a uniform processing in the subsequent steps, each variable must be represented. Hence equalities of the form $v' = v$ are introduced for every variable v that is not assigned in a). Afterwards, φ' is $\varphi \wedge \alpha'$, in which the assignments of a (where the inputs are indexed according to the step number) are incorporated.

$$Next\varphi(\varphi, \alpha', a, p) = \begin{cases} P_0\backslash_I \wedge \alpha'\backslash_I \wedge a' \wedge \bigwedge_{v \in V \text{not assigned in } a} v' = v & \text{if first step} \\ \sigma_1(\varphi, \sigma_2(a, \varphi)) \wedge \sigma_3(\alpha'\backslash_I, \varphi) & \text{otherwise} \end{cases}$$

where

$E\backslash_I$ is the expression E without the subexpressions that involve the inputs,
a' is a where the assignments are interpreted as equalities and the left-hand side operands are primed,
$\sigma_2(a, \varphi) = $ replace in a every input identifier i by i_k (where k is the step number) and every assignment $v = exp$ by the assignment $v' = \sigma(exp, \varphi)$ where $\sigma(exp, \varphi) = $ replace in exp every $v \in V$ by exp_1 if the equality $v' = exp_1$ is in φ,
$\sigma_1(\varphi, A) = $ replace in φ every equality $v' = exp_1$ by the equality $v' = exp_2$ of A, and update the output values,
$\sigma_3(\alpha, \varphi) = \sigma(\alpha, \varphi)$.

2.2   The Model-Checking Procedure

Like [16], we define the semantics of correctness assertions, $[\![p, \varphi \vdash A]\!]$:

$$[\![p, \varphi \vdash A]\!] = \begin{cases} true & \text{if } p \in [\![A]\!]_{\rho, \varphi} \\ false & \text{otherwise} \end{cases}$$

Example. Let us consider the system of Fig. 1. We can express for example the following invariant: when the output $outc$ is true (the card can be retrieved), the number of attempts n is not greater than 3.

Its formulation is
$$S_0, outc = false \vdash \nu Y\{\}(outc = true \Rightarrow n \le 3) \wedge [T]Y$$

Thanks to the definition above and the ones of Section 2.1, we get the following set of reduction rules. The model-checker algorithm (fixed-point computation) corresponds to the application of these rules to rewrite the expression that corresponds to the property to be proven, until $true$ or $false$ is obtained.

$$(f) \quad (p, \varphi \vdash f) \rightarrow \begin{cases} \text{true} & \text{if } \varphi \Rightarrow f \\ \text{false} & \text{otherwise} \end{cases}$$

$$(T) \quad (p, \varphi \vdash T) \rightarrow \text{true}$$

$$(F) \quad (p, \varphi \vdash F) \rightarrow \begin{cases} \text{true} & \text{if } \varphi = F \\ \text{false} & \text{otherwise} \end{cases}$$

$$(\neg) \quad (p, \varphi \vdash \neg A) \rightarrow \begin{cases} \text{true} & \text{if } \varphi = F \\ \neg(p, \varphi \vdash A) & \text{otherwise} \end{cases}$$

$$(\wedge) \quad (p, \varphi \vdash A_0 \wedge A_1) \rightarrow (p, \varphi \vdash A_0) \wedge (p, \varphi \vdash A_1)$$

$$(\vee) \quad (p, \varphi \vdash A_0 \vee A_1) \rightarrow (p, \varphi \vdash A_0) \vee (p, \varphi \vdash A_1)$$

$$(\Diamond) \quad (p, \varphi \vdash \langle \alpha \rangle A) \rightarrow \bigvee_{(q, a, \alpha') \in Reach_{p, \alpha}} (q, \varphi' \vdash A)$$

$$\text{where } Reach_{p, \alpha} = \{(q, a, \alpha') \mid p \xrightarrow{\alpha'/a} q \wedge \alpha' \Rightarrow \alpha \wedge \neg(\varphi \Rightarrow \neg \alpha')\}$$

$$(\Box) \quad (p, \varphi \vdash [\alpha] A) \rightarrow \bigwedge_{(q, a, \alpha') \in Reach_{p, \alpha}} (q, \varphi' \vdash A)$$

$$(\nu) \quad (p, \varphi \vdash \nu X\{\overrightarrow{r}\} A) \rightarrow \begin{cases} \text{true} & \text{if } \exists (p, \varphi_i) \text{ in } \{\overrightarrow{r}\} \mid Equiv(\varphi, \varphi_i) \\ (p, \varphi \vdash A[\nu X\{\overrightarrow{r}, (p, \varphi)\} A \, / \, X]) & \text{otherwise} \end{cases}$$

The theorem prover takes the relay when arithmetic properties have to be checked. These properties are: $\varphi \Rightarrow f$ in the (f) rule, $\varphi = F$ ($=$ is the equivalence) in the (F) and $(\neg)$ rules, $\alpha' \Rightarrow \alpha$ and $\neg(\varphi \Rightarrow \neg \alpha')$ in the $(\Diamond)$ and $(\Box)$ rules, and $Equiv(\varphi, \varphi_i)$ in the (ν) rule. We have chosen the ACL2 prover [18] that supports first order logic without quantifiers, integrates an induction mechanism to reason on inductively defined abstract data types, and imposes no restriction on arithmetic operators. Since the underlying logic is undecidable, the only case in which the algorithm cannot go on and requires user guidance is when the prover cannot decide whether a property is true. It is worth noticing that our methodology can be linked with other theorem provers, the choice of the prover (and of the associated logic) depends on the characteristics of the systems under consideration. In cases where a decidable framework (for instance Presburger arithmetic) suffices, processing the properties above can be fully automated. Let us also remark that the set of data types that can be handled is also a consequence of the choice of the prover. For instance, ACL2 is very powerful with integers, and also allows to reason with floating-point operators.

3   Implementation of This Methodology

3.1   The Prototype Tool

We have implemented a prototype tool that can be used on a network of worstations under any web browser, see a screen dump on Fig. 2. It takes as input

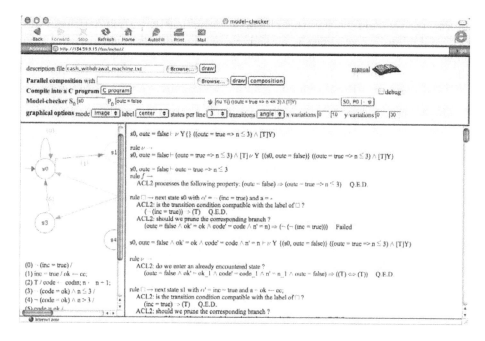

Fig. 2. Screen dump of an execution

simple textual descriptions: the list of inputs, outputs and variables (with their types), followed by the transcription of the transition system (for each transition, the source and destination states, the transition condition and actions). Since this first prototype has only been realized to show the feasibility of the approach, without any optimization, CPU times devoted to graph explorations are not worth being reported (only the ACL2 CPU times will be mentioned).

Using the "draw" button, a *graphical view* of the automaton can be displayed in the left frame; graphical options can be chosen for drawing the picture.

The designer also has the possibility to compute the *parallel composition* of several automata, which is useful when a complex specification is expressed as the composition of subsystems (e.g. different actors involved in a mutual exclusion algorithm, the sender and the receiver in a communication protocol, etc.).

Before running formal verification tools, it can be useful to perform debugging steps by executing the specification. To achieve that goal, it is also possible to *translate into a C program* the original specification file.

The main component of this tool is the *model-checker* described in the previous section. The user has to supply the property p he wants to check, the initial state S_0, and the conjunction of properties P_0 that hold in the initial state and that are relevant for processing p. The tool displays a human readable and concise view of its execution, a debug mode can be chosen to get more detailed information. It writes the name of the rule that is being applied, and the property currently under consideration. Each time ACL2 is used, the tool informs the user and displays the returning status (Q.E.D. in case of success and Failed if failure). A module should be included to allow user guidance when the prover cannot decide. This module, that has not yet been implemented, would allow a more elaborate dialogue between the user and the prover, and prevent the fixed-point algorithm from going on until a definitive answer is obtained.

Example. Let us consider the 2-process version of the famous *Bakery algorithm for mutual exclusion* (see for instance [4]).

shared variables $y1, y2$ initially 0;

```
loop {                              loop {
    NC section;                         NC section;
    y1 ← y2 + 1;                        y2 ← y1 + 1;
    await y2 = 0 ∨ y1 ≤ y2              await y1 = 0 ∨ y2 < y1
    CR section;                         CR section;
    y1 ← 0;                             y2 ← 0;
}                                   }
    (a) Process 1                       (b) Process 2
```

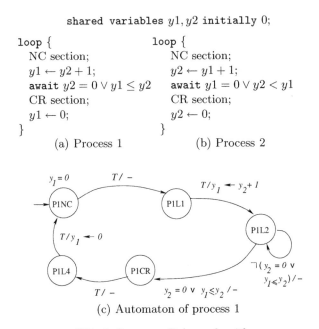

(c) Automaton of process 1

Fig. 3. 2-process Bakery algorithm

Variables y_1 and y_2 are used to determine which process can enter the critical section. Fig. 3(a) and 3(c) give the algorithm and the corresponding automaton for process 1: its starts its execution in state P1NC (non critical section), and can enter the critical section (state P1CR) only if the second process does not make a request ($y_2 = 0$) or if $y_1 \leq y_2$. The second process has a symmetrical behaviour: it can enter the critical section only if the first process does not make a request ($y_1 = 0$) or if $y_2 < y_1$.

Our tool can build the composition of the processes, as represented on Fig. 4, and then model-check this specification. For instance, verifying that mutual exclusion is guaranteed requires the exploration of 11 states (the exploration path is highlighted with heavy lines on Fig. 4), there is only one case in which the (ν) rule prevents from entering a loop (state $P1L4P2L2$) but the ($\square$) rule allows to prune 8 branches (contradiction between the transition condition and the current property). ACL2 processes 25 non-trivial properties (a trivial theorem is for instance $T \Leftrightarrow T$) in a total CPU time of 0.06 s on a Pentium 4.

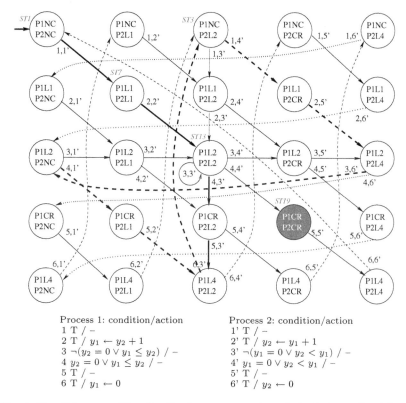

Process 1: condition/action
1 T / –
2 T / $y_1 \leftarrow y_2 + 1$
3 $\neg(y_2 = 0 \lor y_1 \leq y_2)$ / –
4 $y_2 = 0 \lor y_1 \leq y_2$ / –
5 T / –
6 T / $y_1 \leftarrow 0$

Process 2: condition/action
1' T / –
2' T / $y_2 \leftarrow y_1 + 1$
3' $\neg(y_1 = 0 \lor y_2 < y_1)$ / –
4' $y_1 = 0 \lor y_2 < y_1$ / –
5' T / –
6' T / $y_2 \leftarrow 0$

Fig. 4. Parallel composition for the 2-process version of the Bakery algorithm

3.2   Detailed Execution on an Example

Considering the cash withdrawal system of Fig. 1, we now illustrate the algorithm with the following property (this is the one used in the screen dump of Fig. 2):

$$S_0, P_0 \vdash \nu Y\{\}\phi(Y)$$

where $\phi(Y) = ((outc = true \Rightarrow n \leq 3) \land [T]Y)$ and P_0 is $outc = false$.

The algorithm executes as follows. The execution tree is given on Fig. 5, the leaves correspond to states in which the (ν) rule detects that it is not necessary to go on (state already traversed with an equivalent condition).

$$S_0, P_0 \vdash \nu Y\{\}((outc = true \Rightarrow n \leq 3) \land [T]Y)$$

$$\xrightarrow{\nu} S_0, P_0 \vdash (outc = true \Rightarrow n \leq 3) \land [T]\nu Y\{(S_0, P_0)\}\phi(Y)$$

$$\xrightarrow{\land} S_0, P_0 \vdash (outc = true \Rightarrow n \leq 3) \land S_0, P_0 \vdash [T]\nu Y\{(S_0, P_0)\}\phi(Y)$$

$$\xrightarrow{f} true \land S_0, P_0 \vdash [T]\nu Y\{(S_0, P_0)\}\phi(Y)$$

The first part of the conjunction reduces to $true$ as P_0 includes $outc = false$. This corresponds to the elementary ACL2 theorem below (ACL2 uses a Lisp-like syntax, each variable is implicitly universally quantified, *nil* is the counterpart of *false*). For the sake of simplicity, the other ACL2 theorems will not be given (none of them is difficult to prove, but some of them have long statements).

```
(defthm cashmachine1
   (implies (and (booleanp outc) ; outc is a boolean
                 (integerp n) (>= n 0) ; n is a natural number
                 (equal outc nil))
            (implies (equal outc t) (<= n 3)))))
```

The second part is rewritten with the ($\square$) rule, two cases are considered: the loop to state S_0 and the transition to S_1.

$$\xrightarrow{\square} S_0, P_0 \wedge ok' = ok \wedge code' = code \wedge n' = n \vdash \nu Y\{(S_0, P_0)\}\phi(Y) \quad \wedge$$
$$S_1, P_1 \vdash \nu Y(S_0, P_0)\phi(Y)$$

where $P_1 = (P_0 \wedge ok' = cc_0 \wedge code' = code \wedge n' = n)$.
The property φ associated with S_0 is obtained by the function $Next\varphi$. The transition condition on the input *inc* refers to its value in the previous state and is obsolete, hence not included. There is no assignment performed on this transition, $ok' = ok \wedge code' = code \wedge n' = n$ expresses that the variable values are not modified. Similarly, the property P_1 associated with S_1 includes $ok' = cc_0 \wedge code' = code \wedge n' = n$ to express that the only assignment realized on the corresponding transition is $ok \leftarrow cc$ (this is the first time the input cc is consulted, it is numbered 0).

$$\xrightarrow{\nu} true \wedge S_1, P_1 \vdash \nu Y\{(S_0, P_0)\}\phi(Y)$$

the (ν) rule detects that S_0 has already been encountered with a caracteristic property equivalent to the current one, and it stops iterating on this branch.

$$\xrightarrow{\nu} S_1, P_1 \vdash (outc = true \Rightarrow n \leq 3) \wedge [T]\nu Y\{(S_0, P_0), (S_1, P_1)\}\phi(Y)$$
$$\xrightarrow{\wedge, f} true \wedge S_1, P_1 \vdash [T]\nu Y\{(S_0, P_0), (S_1, P_1)\}\phi(Y)$$
$$\xrightarrow{\square} S_2, P_2 \vdash \nu Y\{(S_0, P_0), (S_1, P_1)\}\phi(Y)$$

where $P_2 = (P_0 \wedge ok' = cc_0 \wedge code' = coding_0 \wedge n' = n + 1)$.

$$\xrightarrow{\nu} S_2, P_2 \vdash (outc = true \Rightarrow n \leq 3)$$
$$\wedge [T]\nu Y\{(S_0, P_0), (S_1, P_1), (S_2, P_2)\}\phi(Y)$$
$$\xrightarrow{\wedge, f} true \wedge S_2, P_2 \vdash [T]\nu Y\{(S_0, P_0), (S_1, P_1), (S_2, P_2)\}\phi(Y)$$
$$\xrightarrow{\square} S_1, P_3 \vdash \nu Y\{(S_0, P_0), (S_1, P_1), (S_2, P_2)\}\phi(Y) \wedge$$
$$S_3, P_4 \vdash \nu Y\{(S_0, P_0), (S_1, P_1), (S_2, P_2)\}\phi(Y) \wedge$$
$$S_4, P_5 \vdash \nu Y\{(S_0, P_0), (S_1, P_1), (S_2, P_2)\}\phi(Y)$$

The ($\square$) rule considers the three successor states S_1, S_3 and S_4 of S_2. In the first case, after going through S_1 with the property $P_3 = (P_2 \wedge \neg(coding_0 =$

$cc_0) \wedge (n+1) < 3$), we come back to S_2 with $P_6 = (P_0 \wedge ok' = cc_0 \wedge code' = codin_1 \wedge n' = n+2 \wedge \neg(codin_0 = cc_0) \wedge (n+1) < 3)$. Note that the function $Next\varphi$ updates the variable values, and replaces their names by their symbolic values in the transition conditions when they are integrated.

$$S_2, P_6 \vdash [T]\nu Y\{(S_0, P_0), (S_1, P_1), (S_2, P_2), (S_1, P_3)\}\phi(Y)$$

$$\xrightarrow{\square} S_1, P_7 \vdash \nu Y\{(S_0, P_0), (S_1, P_1), (S_2, P_2), (S_1, P_3)\}\phi(Y) \wedge$$
$$S_3, P_8 \vdash \nu Y\{(S_0, P_0), (S_1, P_1), (S_2, P_2), (S_1, P_3)\}\phi(Y) \wedge$$
$$S_4, P_9 \vdash \nu Y\{(S_0, P_0), (S_1, P_1), (S_2, P_2), (S_1, P_3)\}\phi(Y)$$

– First, let us consider the transition to S_1 with $P_7 = (P_6 \wedge \neg(codin_1 = cc_0) \wedge (n+2) < 3)$. The ($\nu$) rule detects that P_7 and P_3 are equivalent: both of them express that $\neg(code' = ok') \wedge n' < 3$, and the variables $code'$ and n' are of the same form in both of them (a non-constant atom and an additive expression respectively). The exploration stops on this branch.

– In the third case, the transition to S_4 is associated with the property $P_9 = (P_6 \wedge codin_1 = cc_0 \wedge (n+2) \leq 3)$. Then we reach the state S_5 with the property $P_{10} = (outc = true \wedge ok' = cc_0 \wedge code' = codin_1 \wedge n' = n+2 \wedge \neg(codin_0 = cc_0) \wedge (n+1) < 3 \wedge codin_1 = cc_0 \wedge (n+2) \leq 3)$.

$$S_5, P_{10} \vdash outc = true \Rightarrow n \leq 3 \wedge$$
$$S_5, P_{10} \vdash [T]\nu Y\{(S_0, P_0), (S_1, P_1), (S_2, P_2), (S_1, P_3), (S_2, P_6), (S_4, P_9)\}\phi(Y)$$

In the first part of this conjunction, the (f) rule applies. The hypothesis P_{10} indicates that $outc = true$ and $n' \leq 3$, hence the (f) rule returns $true$ because $\varphi \Rightarrow f$ (the variable identifiers are primed in f to coincide with the identifiers in φ). The second conjunct is rewritten with the ($\square$) rule:

$$S_5, P_{10} \vdash [T]\nu Y\{(S_0, P_0), (S_1, P_1), (S_2, P_2), (S_1, P_3), (S_2, P_6), (S_4, P_9)\}\phi(Y)$$
$$\xrightarrow{\square} S_5, P_{10} \vdash \nu Y\{(S_0, P_0), (S_1, P_1), (S_2, P_2), (S_1, P_3), (S_2, P_6), (S_4, P_9)\}\phi(Y) \wedge$$
$$S_0, P_{11} \vdash \nu Y\{(S_0, P_0), (S_1, P_1), (S_2, P_2), (S_1, P_3), (S_2, P_6), (S_4, P_9)\}\phi(Y)$$

The (ν) rule straightforwardly reduces the first conjunct to true because the state S_5 with the property P_{10} as already been explored.

In the case where we reach S_0, P_{11} is ($outc = false \wedge ok' = cc_0 \wedge code' = codin_1 \wedge n' = 0 \wedge \neg(codin_0 = cc_0) \wedge (n+1) < 3 \wedge codin_1 = cc_0 \wedge (n+2) \leq 3)$. The two possible next states are S_0 and S_1. The property associated with S_0 is still P_{11}, thus the (ν) rule decides to stop. Next state S_1 is associated with the property ($outc = false \wedge ok' = cc_1 \wedge code' = codin_1 \wedge n' = 0 \wedge \neg(codin_0 = cc_0) \wedge (n+1) < 3 \wedge codin_1 = cc_0 \wedge (n+2) \leq 3$), the variable ok has received a new input value cc_1. Eliminating the irrelevant hypotheses (hypotheses that concern identifiers that are no longer involved in the expressions assigned to the variables), this property is equivalent to P_1. Therefore, the (ν) rule stops in this case too.

– On the branch that corresponds to the successor (S_3, P_8) of (S_2, P_6), the algorithm stops for similar reasons.

As for the successors (S_3, P_4) and (S_4, P_5) of (S_2, P_2), the algorithm has a behavior that is close to what has been described above. Finally, *true* is returned, which means that the temporal property under consideration holds.

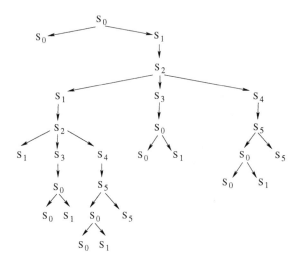

Fig. 5. Execution tree

Fifteen states are actually explored, the (ν) rule prevents from entering loops in 12 states (the leaves); 54 non-trivial properties are submitted to ACL2, for a total CPU time of 0.26 s.

The (ν) rule makes use of the function $Equiv$ to recognize equivalent states and to stop the fixed-point iteration. If φ and φ_i are characterized by equivalent conditions, but present different *constant* values for the corresponding variables, they cannot be considered equivalent. Our strategy is well-suited to *general* problems that do *not* require variables to be given specific initial values. This is the case with this formulation of the cash machine, that specifies a correct behaviour, whatever the initial value of n is.

4   Related Work

4.1   Model-Checking/Theorem-Proving Combined Approaches with Reduction

Most model-checking/theorem proving combined approaches try to reduce the size of the problem before solving it [19]. Some of them exploit the modular structure of the system to decompose the initial problem into smaller sub-problems. Other ones build on abstraction techniques.

Using *modularity*, the proofs are broken down by exploiting the hierarchy of the system. Each finite module is independently checked by a model-checker. The role of the theorem provers is usually to verify that the composition of the verified modules entails the specification. Mocha [20] exploits the hierarchy of the design; it is an interactive environment for the specification, simulation and verification of concurrent systems. Most of the verification work is done by the built-in or external model-checker. In [21], McMillan introduces compositional verification and symmetry reduction in SMV in order to reduce the problem into a set of subgoals discharged by model-checking.

Abstraction techniques reduce the verification of a property of a concrete infinite-state or large system to checking a related property over a finite abstract system. The abstract problem is verified by model-checking, deductive methods are used either to produce it or to verify its correctness. InVeSt [22] belongs to this category of tools, it uses the reduction of the invariance problem to a set of first-order formulas, and fixed-point calculation; it integrates the theorem prover PVS [23] with the model-checker SMV. The approach proposed in [24] uses ACL2 to reduce an infinite-state system through the computation of a quotient structure induced by a well-founded equivalence bisimulation (WEB). The prover is mainly used to verify that the selected equivalence relation is a WEB. The system STeP [4] integrates both abstraction techniques and modularity. The abstractions are deductively justified and algorithmically model-checked.

4.2   Model-Checking as Decision Procedure in a Theorem-Prover

Another current trend is the integration of model-checking modules as *decision procedures within a theorem prover*. The authors of [25] integrate a μ-calculus model-checker as a decision procedure for a fragment of the PVS higher-order logic corresponding to a finite μ-calculus. Moreover, [26] gives a method to build predicate abstraction for the μ-calculus; PVS is used to determine when predicate abstraction is needed.

In [27] the linear temporal logic LTL is embedded in the theorem prover HOL. The translation of LTL formulas into equivalent ω-automata permits the integration of ad hoc symbolic model-checkers as decision procedures in HOL.

5   Conclusion

We have presented a methodology in which a combination of algorithmic and deductive techniques allows to model-check unbounded systems without using state space reduction. A prototype tool has been developed, the role of this first prototype is *not* to measure the efficiency of the algorithm in terms of CPU times, but to demonstrate the feasibility and the interest of the approach.

The technique can be applied to software specifications as well as to some kinds of high-level hardware descriptions. In section 3, we have mentioned its application to two software problems, a mutual exclusion algorithm and the controller of a cash withdrawal system. We are currently considering another class of problems, the communication protocols, e.g. FDDI. Hardware devices made of a control part and a data part are also among the applications of the method presented here. For instance, this tool has also been used to reason on a high-level (behavioural) specification of a circuit that realizes a GCD function, we have proved a loop invariant that ensures the validity of the implementation. The arithmetic theorems involved in this benchmark are more complex than in the examples above. Twenty-one states are traversed by the algorithm, the (ν) rule prevents from entering loops in 21 states, ACL2 processes 111 non-trivial properties in a total CPU time of 2.01 s.

In the current prototype tool, the mechanism that applies the rewrite rules for the state space exploration has not been optimized. Moreover the tool does not give the user the possibility to interact with the prover. This tool is being reimplemented, data structures and the associated functionalities will be improved, and a module for guiding the prover when necessary will be included.

References

1. Hojati, R., Isles, A., Kirkpatrick, D., Brayton, R.: Verification using Uninterpreted Functions and Finite Instantiations. In: Proc. FMCAD'96 (LNCS 1166). (1996)
2. Berezin, S., Biere, A., Clarke, E., Zhu, Y.: Combining Symbolic Model Checking with Uninterpreted Functions for Out-of-Order Processor Verification. In: Proc. FMCAD'98 (LNCS 1522). (1998)
3. Bensalem, S., Lakhnech, Y., Owre, S.: Computing Abstractions of Infinite State Systems Compositionally and Automatically. In: Proc. CAV'98. (1998)
4. Bjørner, N., Browne, A., Colón, M., Finkbeiner, B., Manna, Z., Sipma, H., Uribe, T.: Verifying Temporal Properties of Reactive Systems: A STeP Tutorial. Formal Methods in System Design **16** (2000)
5. Henzinger, T., Jhala, R., Majumdar, R., Sutre, G.: Lazy abstraction. In: Symposium on Principles of Programming Languages. (2002)
6. Du, X., Ramakrishnan, C., Smolka, S.: Real-Time Verification Techniques for Untimed Systems. Electronic Notes in Theoretical Computer Science **39** (2000)
7. Peled, D.: Combining Partial Order Reductions with On-the-Fly Model-Checking. Formal Methods in System Design **8** (1996)
8. Naumovich, G., Clarke, L., Cobleigh, J.: Using partial order techniques to improve performance of data flow analysis based verification. In: Workshop on Program Analysis For Software Tools and Engineering. (1999)
9. Godefroid, P., Sistla, P.: Symmetry and reduced symmetry in model checking. In: Proc. CAV'2001 (LNCS 2102). (2001)
10. Uribe, T.E.: Combinations of model checking and theorem proving. In: Frontiers of Combining Systems. (2000) 151–170
11. Schneider, K., Kropf, T.: A unified approach for combining different formalisms for hardware verification. In: Proc. FMCAD'96. (1996)
12. McMillan, K.: Symbolic Model Checking. Kluwer Academic Pub. (1993)
13. Gordon, M., Melham, T., eds.: Introduction to HOL: A theorem proving environment for higher order logic. Cambridge University Press (1993)
14. McMillan, K.L., Qadeer, S., Saxe, J.B.: Induction in compositional model checking. In: Proc. Computer Aided Verification. (2000) 312–327

15. Contensin, M., Pierre, L.: Combining ACL2 and a ν-calculus Model-checker to Verify System-level Designs. In: Proc. ACM & IEEE International Conference MEMOCODE'03. (2003)
16. Winskel, G.: A note on model-checking the modal ν-calculus. Theoretical Computer Science **83** (1991)
17. Kozen, D.: Results on the propositional μ-calculus. Theoretical Computer Science **27** (1983)
18. Kaufmann, M., Manolios, P., Moore, J.S.: Computer-Aided Reasoning: An Approach. Kluwer Academic Press (2000)
19. Shankar, N.: Combining Theorem Proving and Model Checking through Symbolic Analysis. In: Proc. CONCUR'2000. (2000)
20. Alur, R., Henzinger, T.A., Mang, F.Y.C., Qadeer, S., Rajamani, S.K., Tasiran, S.: MOCHA: Modularity in model checking. In: Proc. CAV'98. (1998)
21. McMillan, K.L.: Verification of infinite state systems by compositional model checking. In: Proc. Charme'99. (1999)
22. Bensalem, S., Lakhnech, Y., Owre, S.: InVeSt: A tool for the verification of invariants. In Hu, A.J., Vardi, M.Y., eds.: Computer-Aided Verification, CAV '98. Volume 1427., Vancouver, Canada, Springer-Verlag (1998) 505–510
23. Crow, J., Owre, S., Rushby, J., Shankar, N., Srivas, M.: A tutorial introduction to PVS. In: Proc. Workshop on Industrial-Strength Formal Specification Techniques. (1995)
24. Manolios, P., Namjoshi, K., Sumners, R.: Linking Theorem Proving and Model-Checking with Well-Founded Bisimulation. In: Proc. Computer Aided Verification, LNCS 1633. (1999)
25. Shankar, N.: PVS: Combining specification, proof checking and model checking. In: Proc. FMCAD'96 (LNCS 1166). (1996)
26. Saidi, H., Shankar, N.: Abstract and model check while you prove. In: Proc. CAV'99 (LNCS 1633). (1999)
27. Schneider, K., Hoffmann, D.W.: A HOL conversion for translating linear time temporal logic to ω-automata. In: Proc. TPHOLs'99. (1999)

A Generic Software Safety Document Generator

Ewen Denney[1] and Ram Prasad Venkatesan[2],[*]

[1] QSS Group, NASA Ames Research Center, Moffett Field, CA, USA
edenney@email.arc.nasa.gov
[2] Dept. of Computer Science, University of Illinois at Urbana-Champaign, IL, USA
rpvenkat@uiuc.edu

Abstract. *Formal certification* is based on the idea that a mathematical proof of some property of a piece of software can be regarded as a certificate of correctness which, in principle, can be subjected to external scrutiny. In practice, however, proofs themselves are unlikely to be of much interest to engineers. Nevertheless, it is possible to use the information obtained from a mathematical analysis of software to produce a detailed textual justification of correctness. In this paper, we describe an approach to generating textual explanations from automatically generated proofs of program *safety*, where the proofs are of compliance with an explicit *safety policy* that can be varied. Key to this is tracing proof obligations back to the program, and we describe a tool which implements this to certify code auto-generated by AutoBayes and AutoFilter, program synthesis systems under development at the NASA Ames Research Center. Our approach is a step towards combining formal certification with traditional certification methods.

1 Introduction

Formal methods are becoming potentially more applicable due, in large part, to improvements in automation: in particular, in automated theorem proving. However, this increasing use of theorem provers in both software and hardware verification also presents a problem for the applicability of formal methods: how can such specialized tools be combined with traditional process-oriented development methods?

The aim of formal certification is to prove that a piece of software is free of certain defects. Yet certification traditionally requires documentary evidence that the software development complies with some process (e.g., DO-178B). Although theorem provers typically generate a large amount of material in the form of formal mathematical proofs, this cannot be easily understood by people inexperienced with the specialized formalism of the tool being used. Consequently, the massive amounts of material that experts can create with these theorem provers remains fairly inaccessible. If you trust a theorem prover, then a proof

---

[*] Ram Prasad Venkatesan carried out this work during a QSS summer internship at the NASA Ames Research Center.

C. Rattray et al. (Eds.): AMAST 2004, LNCS 3116, pp. 102–116, 2004.

of correctness tells that a program is safe, but this is not much help if you want to understand *why*.

One approach is to verbalize high-level proofs produced by a theorem prover. Most of the previous work in this direction has focused on translating low-level formal languages based on natural deduction style formal proofs. A few theorem provers, like Nuprl [CAB+86] and Coq [BBC+97], can display formal proofs in a natural language format, although even these readable texts can be difficult to understand. However, the basic problem is that such proofs of correctness are essentially stand-alone artifacts with no clear relation to the program being verified.

In this paper, we describe a framework for generating comprehensive explanations for *why* a program is safe. Safety is defined in terms of compliance with an explicitly given safety policy. Our framework is generic in the sense that we can instantiate the system with a range of different safety policies, and can easily add new policies to the system.

The safety explanations are generated from the proof obligations produced by a verification condition generator (VCG). The verification condition generator takes as input a synthesized program with logical annotations and produces a series of verification conditions. These conditions are preprocessed by a rewrite-based simplifier and are then proved by an automated theorem prover. Unfortunately, any attempt to directly verbalize the proof steps of the theorem prover would be ineffective as

- the process of simplifying the proof objects makes it difficult to provide a faithful reproduction of the entire proof;
- it is difficult to relate the simplified proof obligations to the corresponding parts of the program.

We claim that it is unnecessary to display actual proof steps – the proof obligations alone provide sufficient insight into the safety of a program. Hence we adopt an approach that generates explanations directly from the verification conditions. Our goals in this paper are:

- using natural language as a basis for safety reports;
- describing a framework in which proofs of safety explicitly refer back to program components;
- providing an approach to merge automated certification with traditional certification procedures.

Related Work. Most of the previous work on proof documentation has focused on translating low-level formal proofs, in particular those given in natural deduction style. In [CKT95], the authors present an approach that uses a proof assistant to construct proof objects and then generate explanations in pseudonatural language from these proof objects. However, this approach is based on a low-level proof even when a corresponding high-level proof was available. The Proverb system [Hua94] renders machine-found natural deduction proofs in natural language using a reconstructive approach. It first defines an intermediate representation called *assertion level* inference rules, then abstracts the machine-found natural deduction proofs using these rules; these abstracted proofs are then verbalized into natural language. Such an approach allows atomic justifications at a higher level of abstraction. In [HMBC99], the authors propose a new approach to text generation from formal proofs exploiting the high-level interactive features of a tactic-style theorem prover. It is argued that tactic steps correspond approximately to human inference steps. None of these techniques, though, is directly concerned with program verification. Recently, there has also been research on providing formal traceability between specifications and generated code. [BRLP98] presents a tool that indicates how statements in synthesized code relate to the initial problem specification and domain theory. In [WBS+01], the authors build on this to present a documentation generator and XML-based browser interface that generates an explanation for every executable statement in the synthesized program. It takes augmented proof structures and abstracts them to provide explanations of how the program has been synthesized from a specification.

One tool which does combine verification and documentation is the PolySpace static analysis tool [Pol]. PolySpace analyzes programs for compliance with fixed notions of safety, and produces a marked-up browsable program together with a safety report as an Excel spreadsheet.

2   Certification Architecture

The certification tool is built on top of two program synthesis systems. Auto-Bayes [FS03] and AutoFilter [WS03] are able to auto-generate executable code in the domains of data analysis and state estimation, respectively. Both systems are able to generate substantial complex programs which would be difficult and time-consuming to develop manually. Since these programs can be used in safety-critical environments, we need to have some guarantee of correctness. However, due to the complex and dynamic nature of the synthesis tools, we have departed from the traditional idea of program synthesis as being "correct by construction" or *process-oriented certification*, and instead adopt a *product-oriented* approach. In other words, we certify the individual programs which are generated by the system, rather than the system itself.

The synthesis tools, themselves, are about 75KLOC, use many "non-trivial" features of Prolog, and are continually being extended by a number of developers. Requiring ongoing verification of the system itself would cripple development. The programs generated, on the other hand, are in a "clean" imperative subset and so it is substantially easier to extend the systems to enable certification of the programs they generate.

Figure 1 gives an overview of the components of the system. The synthesis system takes as input a high-level specification together with a *safety policy*. Low-level code is then synthesized to implement the specification. The synthesizer first generates "intermediate" code which can then be translated to different platforms. A number of target language backends are currently supported. The safety policy is used to *annotate* the intermediate code with mark-up information

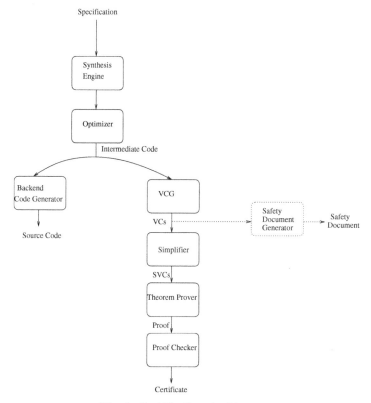

Fig. 1. Certification Architecture

relevant to the policy. These annotations give "local" information, which must then be propagated throughout the code. Next, the annotated code is processed by a Verification Condition Generator (VCG), which applies the rules of the safety policy to the annotated code in order to generate safety conditions (which express whether the code is safe or not). The VCG has been designed to be "correct-by-inspection", that is, sufficiently simple that it is relatively easy to be assured that it correctly implements the rules of the safety logic. In particular, the VCG does not carry out any optimizations, not even reducing substitution terms. Consequently, the verification conditions (VCs) tend to be large and must be preprocessed before being sent to a theorem prover. The preprocessing is done by a traceable rewrite system. The more manageable SVCs are then sent to a first-order theorem prover, and the resulting proof is sent to a proof checker. In the above diagram, the safety documentation extension is indicated using dotted lines.

3   Safety Policies

Formal reasoning techniques can be used to show that programs satisfy certain *safety policies*, for example, memory safety (i.e. they do not access out of bound memory locations), and initialization safety (i.e. undefined or uninitialized variables are not used). Formally, a safety policy is a set of proof rules and auxiliary definitions which are designed to show that programs satisfy a safety property of interest. The intention is that a safety policy enforces a particular safety property, which is an operational characterization that *a program does not go wrong*. The distinction between safety properties and policies is explored in detail in [DF03]. We summarize the important points here.

Axiomatic semantics for (simple) programming languages are traditionally given using Hoare logic [Mit96], where $P \{C\} Q$ means that if precondition, P, holds before the execution of command, C, then postcondition, Q, holds afterwards. This can be read backwards to compute the weakest precondition which must hold to satisfy a given postcondition.

We have extended the standard Hoare framework with the notion of safety properties. [DF03] outlines criteria when a (semantic) safety property can be encoded as an executable safety policy.

We have initially restricted ourselves to safety of array accesses (ensuring that the access is within the array bounds) and safety of variables with respect to initialization (ensuring that all variables which are read have been assigned to). However, the tool has been designed to be generic and we intend to extend it to support other forms of safety.

Hoare logic treats commands as transformations of the execution environment. The key step in formalizing safety policies is to extend this with a "shadow", or *safety environment*. Each variable (both scalar and vector) has a corresponding shadow variable which records the appropriate safety information for that variable. For example, for initialization safety, the shadow variable x_{init} is set to *init* or *uninit* depending on whether x has been initialized or not. In general, there is no connection between the values of a variable and its shadow variables. The semantic definition of a safety property can then be factored into two families of formulas, Safe^- and $\mathsf{Sub}^-(\_)$. A feature of our framework is that the safety of a command can only be expressed in terms of its *immediate* subexpressions. The subscripts give the class of command (assignment, for-loop, etc.), and the superscript lists the immediate subexpressions.

For a given safety policy, for each command, C, of class cl with immediate subexpressions, $e_1 \dots e_n$, $\mathsf{Safe}_{\mathsf{cl}}^{e_1 \cdots e_n}$ expresses the safety conditions on C, in terms of program variables and shadow variables; $\mathsf{Sub}_{\mathsf{cl}}^{e_1 \cdots e_n}(P)$ is a substitution applied to formula P expressing the change C makes to the safety environment.

For example, for the initialization safety policy, the assignment $x := y$ has safety condition, $\mathsf{Safe}_{\mathsf{assign}}^{x,y}$, which is the formula $y_{\mathrm{init}} = init$ (i.e., "y must be initialized") and, for formula P, $\mathsf{Sub}_{\mathsf{assign}}^{x,y}(P)$ is the safety substitution $P[init/x]$ (i.e., "x becomes initialized").

Hence, in our framework, *verifying* the safety of a program amounts to working backwards through the code, applying safety substitutions to compute the safety environment, and accumulating safety obligations while proving that the safety environment at each point implies the corresponding safety obligations. *Explaining* the safety of a program amounts to giving a textual account of why (decl) $$\frac{}{\mathtt{lab}(l, \mathsf{Sub}^x_{\mathtt{decl}}(Q) \wedge \mathsf{Safe}^x_{\mathtt{decl}})\ \{(\mathtt{var}\ x)^l\}\ Q}$$

(adecl) $$\frac{}{\mathtt{lab}(l, \mathsf{Sub}^{x,n}_{\mathtt{adecl}}(Q) \wedge \mathsf{Safe}^{x,n}_{\mathtt{decl}})\ \{(\mathtt{var}\ x\,[n])^l\}\ Q}$$

(assign) $$\frac{}{\mathtt{lab}(l, \mathsf{Sub}^{x,e}_{\mathtt{assign}}(Q) \wedge \mathsf{Safe}^{x,e}_{\mathtt{assign}})\ \{(x\ \mathtt{:=}\ e)^l\}\ Q}$$

(update) $$\frac{}{\mathtt{lab}(l, \mathsf{Sub}^{x,e_1,e_2}_{\mathtt{update}}(Q) \wedge \mathsf{Safe}^{x,e_1,e_2}_{\mathtt{update}})\ \{(x\,[e_1]\ \mathtt{:=}\ e_2)^l\}\ Q}$$

(if) $$\frac{P_1\ \{c_1\}\ Q \quad P_2\ \{c_2\}\ Q}{\mathtt{lab}(\mathtt{if}(l), \mathsf{Sub}^b_{\mathtt{if}}(b \Rightarrow P_1 \wedge \neg b \Rightarrow P_2) \wedge \mathsf{Safe}^b_{\mathtt{if}})\ \{(\mathtt{if}\ b\ \mathtt{then}\ c_1\ \mathtt{else}\ c_2)^l\}\ Q}$$

(while) $$\frac{P\ \{c\}\ I \quad I\&b \Rightarrow \bar{P} \quad I\&\neg b \Rightarrow \bar{Q}}{\mathtt{lab}(\mathtt{wh}(l), \mathtt{inv}(I), \mathsf{Sub}^b_{\mathtt{while}}(I) \wedge \mathsf{Safe}^b_{\mathtt{while}})\ \{(\mathtt{while}\ b\ \mathtt{inv}\ I\ \mathtt{do}\ c)^l\}\ Q}$$

(comp) $$\frac{P\ \{c_1\}\ R \quad R\ \{c_2\}\ Q}{P\ \{c_1; c_2\}\ Q}$$

Fig. 2. Extended Hoare Rules

these implications hold, in terms relating to the safety conditions and safety substitutions.

Our goal, then, is to augment the certification system such that the proof obligations have sufficient information that we can give them a comprehensible textual rendering. We do this by extending the intermediate code to accommodate labels and the VCG to generate verification conditions with labels. We add labels for each declaration, assignment, loop construct and conditional statement by giving them a number in increasing order starting from zero. For loops and conditions, we also add the command type to the label. For example, for loops are given a label `for`*(label)*. Similarly, we also have labels `if`*(label)* and `wh`*(label)*. Figure 2 gives the Hoare rules extended by labels which are implemented by the VCG. The composition rule is the standard one. The conclusions of the other rules, which have the general form $\mathtt{lab}(L, P)\ \{c^l\}\ Q$, when read backwards, say that in order to satisfy the postcondition, Q, the command, c, labeled by l, requires precondition, P, which is given label, L (i.e. the command label, l, and possibly the command type). In the rule for while loops, $\bar{P}$ denotes P with labels removed.

4   Documentation Architecture

In this section, we introduce the general architecture of the safety document generator and discuss the notions of def-use analysis and template composition.

4.1   Document Generator

Figure 3 shows the structure of the document generation subsystem. The synthesized intermediate code is labeled by adding line numbers to the code before

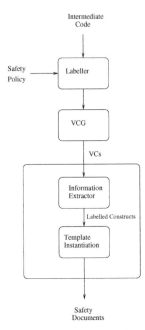

Fig. 3. Document Generation Architecture

it is sent to the VCG. The VCG then produces verification conditions for the corresponding safety policy. These verification conditions preserve the labels by encapsulating them along with the weakest safety preconditions.

Because of the way our safety logic is formulated in terms of immediate subexpressions of commands, we define a *fragment* to be a command "sliced" to its immediate subexpressions. For atomic commands, this is equivalent to the command itself. For compound commands, we will represent this as if b and while b. These are the parts of a program that require an independent safety explanation.

The document generator takes as input the verification conditions generated in this manner and first identifies each fragment that requires explanation (for array safety, any fragments which contain array accesses and for initialization safety, any fragments containing variable reads). It then selects appropriate *explanation templates* from a repository of safety-dependent templates. Text is then generated by instantiating the templates with program fragments and composing the results.

4.2   Def-use Analysis

Since commands can affect the safety of other commands in their effect on the program environment we cannot consider the safety of commands in isolation. In particular, the safety of a command involving a variable x depends on the safety of all previous commands in the program that contain an occurrence of x. Consider the following code:

$\vdots$

(L1) x = 2 ;

$\vdots$

(L2) y = x ;

$\vdots$

(L3) a[y] = 0 ;

Now consider the safety of the expression a[y] = 0 with respect to array bounds. To determine whether this access is safe, we need to ensure that the value held by y is within the array bounds. Now supposing that a is an array of size 10, we need to reason that y is defined from x which in turn is initialized to 2, which is less than 10. Hence we can state that the access is safe. Similarly, if we were analyzing the safety of the same expression with respect to initialization, we would need to convince ourselves simply that y is initialized. Reasoning from y = x alone would be insufficient and incorrect because x could be uninitialized. So we need to convince ourselves that x is also initialized by considering the expression x = 2. In other words, the safety of the expression a[y] = 0 depends on the safety of the fragments y=x and x=2.

To summarize, we trace each variable in a program fragment ϕ to the point where it was first defined or initialized and reason about the safety of all the fragments encountered in the path to obtain a thorough explanation of the safety of ϕ. For a given program fragment ϕ having variables ω, we use $\Omega(\phi)$ to represent the set of all fragments, with their labels, that were encountered while tracking each variable in ω to its origin. We also include ϕ in $\Omega(\phi)$. Strictly speaking, the argument to Ω should be a distinguished occurrence of a fragment within a program, but we will gloss over this.

4.3 Contexts

In addition to tracking variables to their original definition, we also need to find which fragments the fragment under consideration depends on. For example, the safety of an assignment statement appearing within a conditional block also depends on the safety of the conditional expression. Similarly, the safety of statements inside **while** loops depends on the safety of the loop condition. In the case of nested loops and conditional statements, a fragment's safety depends on multiple fragments. To provide complete safety explanations for a program fragment ϕ, we construct a set[1] $\Psi_{sp}(\phi)$ as follows. As above, ϕ is assumed to be distinguished within a given program. We first identify all the fragments ϕ' on which ϕ depends. That is, if ϕ lies within conditional blocks and/or loop blocks, then we include the fragments representing those conditional expressions and/or loop expressions in ϕ'. We will refer to this as the *context* of the fragment, ϕ, and denote it by cxt(ϕ). Since we add special labels to loops and conditional statements, we can easily identify blocks and can easily determine the set ϕ'.

---

[1] In the implementation, this is a tree.

Then, we trace each component and variable in the fragment ϕ and the set of fragments ϕ' to their origin (as explained in the previous section); that is,

$$\Psi_{sp}(\phi) = \cup\{\Omega(\phi') \mid \phi' \in \mathtt{cxt}(\phi)\}.$$

Intuitively, we can view $\Psi_{sp}(\phi)$ as the set of all expressions and program fragments that we need to consider while reasoning about the safety of ϕ with respect to the safety policy sp. Each element in this set is represented as a (*label*, *fragment*) pair.

We now state (without proof) that ϕ is safe if each of the fragments in $\Psi_{sp}(\phi)$ is safe. That is,

$$safe_{sp}(\Psi_{sp}(\phi)) \Rightarrow safe_{sp}(\phi).$$

Here, we use the predicate *safe* to indicate that a set of program fragments are safe with respect to a policy *sp*.

For example, consider the following piece of code in C:

```
(1) x = 5 ;
(2) z = 10 ;
(3) if(x > z)
(4)    y = x ;
    else
(5)    y = z ;
```

The safety of the assignment y = x at line 4 with respect to initialization of variables depends not only on the assignment statement y = x but also on the the conditional fragment if(x > z) so, in this case, for the program fragment y = x, the context would be simply $\{\mathtt{if}(x > z)\}$. We can further deduce that the safety of the conditional statement in turn depends on the two assignment statements x = 5 and z = 10. So, to explain the safety of the expression y = x at line 4, we need to reason about the safety of the fragments if(x > z) , z = 10 and x = 5 at lines 3, 2 and 1 respectively. Hence, $\Psi_{sp}(y = x)$ is the set $\{(4, y = x), (3, \mathtt{if}(x > z)), (2, z = 10), (1, (x = 5))\}$.

4.4   Templates

We have defined a library of templates which are explanation fragments for the different safety policies. These templates are simply strings with holes which can be instantiated by program components to form safety explanations. A program component can be a simple program variable, a program fragment, an expression or a label.

Template Composition and Instantiation: The composition of an explanation for a given program fragment is obtained from the templates defined for a given policy, *sp*. For each fragment, ϕ, we first construct the set $\Psi_{sp}(\phi)$. Then, for each element ψ in $\Psi_{sp}(\phi)$, we find the required template(s), $Temp_{sp}(\psi)$. Next we insert the appropriate program components in the gaps present in the template to form the textual safety explanation. This process is repeated recursively for each fragment in $\Psi_{sp}(\phi)$ and then all the explanations obtained in this way are concatenated to form the final safety explanation. It should be noted that the safety explanations generated for most of these fragments are phrases rather than complete sentences. These phrases are then combined in such a manner that the final safety explanations reflects the data flow of the program.

As an example, consider the following code fragment:

```
(1) var a[10] ;
(2) x = 0 ;
(3) a[x] = 0 ;
```

Here, a is declared to be an array of size 10 at line 1. x is initialized to 0 at line 2 and a[x] is initialized to 0 at line 3.

Considering the safety of the expression a[x] = 0 (ϕ), the set $\Psi_{sp}(\phi)$ is {(3, (a[x] = 0)), (2, (x = 0))}. Now, for each of these program fragments, we apply the appropriate templates for array bounds safety and generate explanations by combining them with the program variables a and x along with their labels. In this case, the safety explanation is:

The access a[x] at line 3 is safe as the term x is evaluated from x = 0 at line 2; x is within 0 and 9; and hence the access is within the bounds of the array declared at line 1.

Now if we were interested in initialization of variables, the set $\Psi_{sp}(\phi)$ is still {(3, (a[x] = 0)), (2, (x = 0))}. However, the template definitions for the same fragments differ and the explanation is:

The assignment a[x] = 0 at line 3 is safe; the term x is initialized from x=0 at line 2.

5   Implementation and Illustration

We now describe an implementation of the safety document generator based on the principles discussed in the previous sections and give an example of how it works for different safety policies.

5.1   Implementation

The process of generating the explanations from the intermediate language can be broadly classified into two phases.

- Labeling the intermediate code and generating verification conditions.
- Scanning the verification conditions and generating explanations.

We scan the verification conditions to identify the different parts of the program that require safety explanations collecting as much information as possible about the different data and variables along the way, and computing the $\Psi_{sp}(\phi)$. Fragments that require safety explanations differ for different safety policies. Since we analyze the verification conditions and not the program, the safety policy has already determined this. For example, for safety with respect to array bounds, the fragments that require explanations would be the array accesses in the program. On the other hand, we need to consider all variable assignments, array accesses and assignments, declarations, and conditional statements for safety with respect to initialization of variables. That is, we consider all fragments where a variable is used and determine whether the variable has been initialized. In addition, we also accumulate information about the program variables, constants and the blocks.

Finally, using the information that we have accumulated during the scanning phase, we generate explanations for why the program is safe. As we have already mentioned, our tool is designed to be generic. Irrespective of the safety policy that we are currently concerned with, the tool analyzes each fragment that requires an explanation, and generates explanations using templates as discussed in the previous section. This approach makes extensions very easy as the introduction of a new safety policy only involves providing definitions of each template for the policy.

5.2   A Simple Example

We now give a (slightly contrived) example of some intermediate code and the corresponding explanations provided by the document generator. The program needs an invariant (not given here) in order to prove its safety for each policy. Our prototype will simply indicate where invariants have been used to prove a verification obligation.

```
0          proc(eg)
    {
1          a[10] : int ;
2          b : int ;
3          c : int ;
4          d : int ;
5          x : int ;

6          b = 1 ;
7          c = 2 ;
8          d = b*b + c*c ;

9          for(i=0;i<10;i++)
      {
10          if(i < 5)
11            a[d+i] = i ;
            else
12            a[2*d-1-i] = i ;
      }
13          x = a[a[5]] ;
    }
```

The explanations generated for safety with respect to array bounds and initialization are given in Figures 4 and 5, respectively. The explanations are only generated if the theorem prover successfully proves all the corresponding verification conditions, and this is noted at the end. Note that we currently perform no symbolic evaluation during the analysis. The safety of the final assignment is proven using the invariant.

Safety Explanations for Array Bounds

The access a[d+i] *at line* 11 *(if the condition* i<5 *at line* 10 *is true) is safe as the term* d *is evaluated from* d=b*b+c*c *at line* 8; *the term* b *is evaluated from* b=1 *at line* 6; *the term* c *is evaluated from* c=2 *at line* 7; *for each value of the loop index* i *from* 0 *to* 9 *at line* 9, *d+i is within* 0 *and* 9; *and hence the access is within the bounds of the array declared at line* 1.

The access a[2*d-1-i] *at line* 12 *(if the condition* i<5 *at line* 10 *is false) is safe as the term* d *is evaluated from* d=b*b+c*c *at line* 8; *the term* b *is evaluated from* b=1 *at line* 6; *the term* c *is evaluated from* c=2 *at line* 7; *for each value of the loop index* i *from* 0 *to* 9 *at line* 9, *2*d-1-i is within* 0 *and* 9; *and hence the access is within the bounds of the array declared at line* 1.

The access a[a[5]] *at line* 13 *is safe; using the invariant for the loop at line* 9 *and the postcondition* i=9+1 *after the loop;* a[5] *is within* 0 *and* 9; *and hence the access is within the bounds of the array declared at line* 1.

[Certified by e-setheo on Mon Mar 15 17:58:28 PST 2004 for array policy.]

Fig. 4. Auto-generated Explanation: *array* safety policy

Safety Explanations for Initialization of Variables

The assignment b=1 *at line* 6 *is safe.*

The assignment c=2 *at line* 7 *is safe.*

The assignment d=b*b+c*c *at line* 8 *is safe; the term* b *is initialized from* b=1 *at line* 6; *the term* c *is initialized from* c=2 *at line* 7.

The loop index i *ranges from* 0 *to* 9 *and is initialized at line* 9.

The conditional expression i<5 *appears at line* 10; *the loop index* i *ranges from* 0 *to* 9 *and is initialized at line* 9.

The assignment a[d+i]=i *at line* 11 *is safe (if the condition* i<5 *at line* 10 *is true); the term* d *is initialized from* d=b*b+c*c *at line* 8; *the term* b *is initialized from* b=1 *at line* 6; *the term* c *is initialized from* c=2 *at line* 7; *the loop index* i *ranges from* 0 *to* 9 *and is initialized at line* 9.

The assignment a[2*d-1-i]=i *at line* 12 *is safe (if the condition* i<5 *at line* 10 *is false); the term* d *is initialized from* d=b*b+c*c *at line* 8; *the term* b *is initialized from* b=1 *at line* 6; *the term* c *is initialized from* c=2 *at line* 7; *the loop index* i *ranges from* 0 *to* 9 *and is initialized at line* 9.

The assignment x = a[a[5]] *at line* 13 *is safe; using the invariant for the loop at line* 9 *and the postcondition* i=9+1 *after the loop.*

[Certified by e-setheo on Mon Mar 15 18:02:24 PST 2004 for init policy.]

Fig. 5. Auto-generated Explanation: *init* safety policy

6   Design Issues

In this section, we present some issues that were analyzed during the design and implementation of the safety document generator and then describe features that we have implemented for flexibility.

6.1   Invariants

To enable the document generator to recognize those parts of the verification conditions which come from loop invariants, we need to specifically label them with labels of the form $\texttt{inv}(I)$. Then, while generating explanations for fragments within loops, we first find if the loop has an explicit invariant. If it does, we check if the fragment shares any variables with the invariant. The idea behind this is that it is always possible that the loop invariant might be completely unrelated to the safety of a fragment within that loop. In such a case, our explanations should not consider the loop invariant. However, if the invariant does (presumably) affect the safety of a fragment, we incorporate it into the explanation using the label giving the line at which the invariant was defined.

6.2   Simplifying the Explanations

We have designed the document generator to be comprehensive. For some of the more complex programs synthesized by AutoBayes, the safety document can run to over a hundred pages.

One technique which we have implemented is to allow users to *slice* the explanation with respect to specific variables or line numbers. The document generator will then provide explanations only for those fragments that fall within the parts of interest (as well as any fragments they depend on).

However, although slicing can be used to focus attention to individual areas, it is still nice to get an overall justification of why a program is safe. One important technique for making the tool scalable (discussed in the conclusion) is to make use of high-level structuring information from a model. Another is to observe that some facts are clearly more important than others. We have implemented a simple heuristic which *ranks* the fragments and displays them based on user request. For instance, initialization of variables to constants can be viewed as a trivial command so the corresponding explanation can be eliminated. We have categorized fragments (in order of increasing priority) in terms of assignments to numeric constants, loop variable initializations, variable initializations, array accesses and – the highest priority – any command involving invariants.

The rationale behind giving explanations involving invariants the highest priority is that invariants are generally used to fill in the trickiest parts of a proof, so are most likely to be of interest.

7   Conclusions and Future Work

The documentation generation system which we have described here builds on our state-of-the-art program synthesis system, and offers a novel combination of synthesis, verification and documentation. We believe that documentation capabilities such as this are essential for formal techniques to gain acceptance.

Although we do not (yet) use any symbolical evaluation, we believe that the simple technique of analyzing the verification conditions provides useful insights. Our plan is to combine the safety documentation with ongoing work on *design documentation*. We currently have a system which is able to document the synthesized code (explaining the specification, how it relates to the generated code, design choices made during synthesis, and so on), either as in-line comments in the code or as a browsable document, but it remains to integrate this with the safety document generator. We intend to let the user choose between various standard formats for the documentation (such as those mandated by DO-178B or internal NASA requirements).

A big problem for NASA is the recertification of modified code. In fact, this can be a limiting factor in deciding whether a code change is feasible or not. For synthesis, the problem is that there is currently no easy way to combine manual modifications to synthesized code with later runs of the synthesis system. We would like to be able to generate documentation which is specific to the changes which have been made.

Finally, we intend to extend our certification system with new policies (including resource usage, unit safety, and constraints on the implementation environment). The two safety policies which we have illustrated this with here are *language-specific* in the sense that the notion of safety is at the level of individual commands in the language. We have also looked at *domain-specific* policies (such as for various matrix properties) where the reasoning takes place at the level of code blocks. This will entail an interesting extension to the document generator, making use of domain-specific concepts, and we are currently redesigning our entire synthesis framework to make use of explicit *domain models*. Clearly, reflecting the high-level structure of the code in the documentation will be important in making this approach scalable.

References

[BBC+97] Bruno Barras, Samuel Boutin, Cristina Cornes, Judicael Courant, Jean-Christophe Filliatre, Eduardo Gimenez, Hugo Herbelin, Gerard Huet, Cesar Munoz, Chetan Murthy, Catherine Parent, Christine Paulin-Mohring, Amokrane Saibi, and Benjamin Werner. The Coq proof assistant reference manual: Version 6.1. Technical Report RT-0203, 1997.

[BRLP98] Jeffrey Van Baalen, Peter Robinson, Michael Lowry, and Thomas Pressburger. Explaining synthesized software. In D. F. Redmiles and B. Nuseibeh, editors, *Proc. 13th IEEE Conference on Automated Software Engineering*, pages 240–248, 1998.

[CAB+86] Robert L. Constable, Stuart F. Allen, H. M. Bromley, W. R. Cleaveland, J. F. Cremer, R. W. Harper, Douglas J. Howe, T. B. Knoblock, N. P. Mendler, P. Panangaden, James T. Sasaki, and Scott F. Smith. *Implementing Mathematics with the Nuprl Development System*. Prentice-Hall, NJ, 1986.

[CKT95]   Y. Coscoy, G. Kahn, and L. Théry. Extracting text from proofs. In
          M. Dezani-Ciancaglini and G. Plotkin, editors, *Proc. Second International
          Conference on Typed Lambda Calculi and Applications, Edinburgh, Scot-
          land*, volume 902, pages 109–123, 1995.
[DF03]   Ewen Denney and Bernd Fischer. Correctness of source-level safety policies.
          In Keijiro Araki, Stefania Gnesi, and Dino Mandrioli, editors, *Proc. FM
          2003: Formal Methods*, volume 2805 of *Lect. Notes Comp. Sci.*, pages 894–
          913, Pisa, Italy, September 2003. Springer.
[DKT93]   Arie Van Deursen, Paul Klint, and Frank Tip. Origin tracking. *Journal of
          Symbolic Computation*, 15(5/6):523–545, 1993.
[FS03]   Bernd Fischer and Johann Schumann. AutoBayes: A system for generating
          data analysis programs from statistical models. *J. Functional Programming*,
          13(3):483–508, May 2003.
[HMBC99] Amanda M. Holland-Minkley, Regina Barzilay, and Robert L. Constable.
          Verbalization of high-level formal proofs. In *AAAI/IAAI*, pages 277–284,
          1999.
[Hua94]   Xiaorong Huang. Proverb: A system explaining machine-found proofs. In
          Ashwin Ram and Kurt Eiselt, editors, *Proc. 16th Annual Conference of the
          Cognitive Science Society, Atlanta, USA*, pages 427–432. Lawrence Erlbaum
          Associates, 1994.
[Mit96]   John C. Mitchell. *Foundations for Programming Languages*. The MIT Press,
          1996.
[Pol]   PolySpace Technologies. `http://www.polyspace.com`.
[WBS+01] Jon Whittle, Jeffrey Van Baalen, Johann Schumann, Peter Robinson,
          Thomas Pressburger, John Penix, Phil Oh, Mike Lowry, and Guillaume
          Brat. Amphion/NAV: Deductive synthesis of state estimation software. In
          Proc. IEEE Conference on Automated Software Engineering, 2001.
[WS03]   Jon Whittle and Johann Schumann. Automating the implementation of
          Kalman filter algorithms, 2003. In review.

Linear Temporal Logic and Z Refinement

John Derrick[1] and Graeme Smith[2]

[1] Computing Laboratory, University of Kent, Canterbury, CT2 7NF, UK
[2] School of Information Technology and Electrical Engineering, The University of
Queensland 4072, Australia

Abstract. Since Z, being a state-based language, describes a system in
terms of its state and potential state changes, it is natural to want to
describe properties of a specified system also in terms of its state. One
means of doing this is to use Linear Temporal Logic (LTL) in which
properties about the state of a system over time can be captured. This,
however, raises the question of whether these properties are preserved
under refinement. Refinement is observation preserving and the state of
a specified system is regarded as internal and, hence, non-observable.
In this paper, we investigate this issue by addressing the following questions. Given that a Z specification A is refined by a Z specification C,
and that P is a temporal logic property which holds for A, what temporal logic property Q can we deduce holds for C? Furthermore, under
what circumstances does the property Q preserve the intended meaning
of the property P? The paper answers these questions for LTL, but the
approach could also be applied to other temporal logics over states such
as CTL and the μ-calculus.

Keywords: Z, refinement, temporal logic, LTL.

1 Introduction

Z [14], like other state-based languages such as B [1] and VDM [9], describes a
system in terms of its state and the changes on this state. A specification typically
comprises a state schema declaring and restricting a set of state variables, an
initial state schema restricting the initial values of the state variables, and a
set of operation schemas detailing possible changes to the state variables with
respect to additional variables representing inputs and outputs.

While invariant properties of the system can be captured directly by the
specification, more complex behavioural properties need to be proved to hold.
Given the emphasis on the state while specifying using Z, it seems natural to
want to also describe desired behavioural properties in terms of the system's
state. Ideal for this purpose are temporal logics which define predicates over
infinite sequences of states. They can be used to specify how the state of the
system evolves over time, in a way that isn't possible directly in the model.

Along with their ability to capture behavioural properties in terms of state,
temporal logics are also commonly used for describing properties in model checking [4]. Hence, investigating their use with Z is an important first step toward

C. Rattray et al. (Eds.): AMAST 2004, LNCS 3116, pp. 117–131, 2004.
© Springer-Verlag Berlin Heidelberg 2004

developing model checking support for the language [13]. The most common temporal logics used in model checking are Linear Temporal Logic (LTL) [8], Computation Tree Logic (CTL) [8] and the μ-calculus [10]. In this paper, we focus on the use of LTL, although the approach we develop could also be used to similarly investigate CTL and the μ-calculus.

The purpose of the definition of refinement in Z is to formalise the development from an abstract to a more concrete specification [6]. It is therefore prudent to ask whether temporal logic properties are preserved under refinement. That is, if a property is proved to hold for a Z specification, will it also hold for a refinement of that specification? This question is complicated by the fact that refinement is defined to preserve *observable* properties. In Z, only the operations and their inputs and outputs are regarded as observable; the state of a specification, to which temporal logic properties refer, is regarded as internal and, hence, non-observable.

In this paper, we develop a general approach for investigating preservation of temporal logic properties under refinement, and apply it to LTL. In Section 2, we provide an overview of refinement in Z and motivate our work with an indicative example of where a temporal property is not preserved. Our general approach to determining temporal logic property preservation is defined in Section 3 and applied to LTL in Section 4. We conclude with a discussion of related work and future directions in Section 5.

2   Refinement

Refinement in Z is defined so that the observable behaviour of a Z specification is preserved. This behaviour is in terms of the operations that are performed and their input and output values. Values of the state variables are regarded as being internal. Hence, they are not observable and properties on them are, in general, not preserved. The reason for regarding them as internal is so that refinement can be used to change the representation of the state of a system. This is referred to as *data refinement*.

The definition of refinement in Z is derived from a relational model into which Z is embedded. The details of this are contained in [6], and if a specification C is a refinement of another specification A we write $A \sqsubseteq C$.

To prove a data refinement one needs to link the states in the abstract and concrete specifications via a relation known as the *retrieve relation*. The retrieve relation shows how a state in one specification is represented in the other. For refinement to be complete, a relation, rather than simply a function, is required [6].

Given such a retrieve relation, simulation rules relating the schemas of the specifications are used to verify the refinement. The standard simulation rules for Z assume that operations have preconditions outside of which they can occur changing the state arbitrarily. Such arbitrary state changes make it difficult to prove any interesting temporal properties however. Hence in our approach, we adopt an alternative view of operations in which they are *blocked*, i.e., cannot occur outside their preconditions.

As well as facilitating the use of temporal logics, the blocking model also makes our approach applicable to variants of Z which adopt this model, such as Object-Z [12]. It is also equivalent to the standard non-blocking model of Z when operation preconditions are *totalised*, i.e., post-states are defined for all pre-states. Such totalisation of preconditions would be necessary in standard Z for the specification to exhibit interesting temporal properties.

As for standard Z refinement, there are two simulation rules for refinement under the blocking model which are together complete, i.e., all possible refinements can be proved with a combination of the rules (in fact, every refinement can be verified with one downward together with one upward simulation).

The first rule, referred to as *downward simulation*, requires that

1. the initial states of the concrete specification are related to abstract initial states,
2. the concrete operations are only enabled in states related to abstract states where the corresponding abstract operations are enabled, and vice versa (i.e., they have equivalent preconditions), and
3. whenever a concrete operation can result in the state change (t, t'), for any abstract state s related to t, the corresponding abstract operation can result in (s, s') such that s' is related to t'. That is, the effect of the concrete operation is consistent with the requirements of the corresponding abstract operation.

It is defined as follows [6]. (Note that the state variables of a schema S after each operation are denoted by S', and, in general, the Z notation is also used for expressing its own metatheory, details of this notation can be found in [6].)

Definition 1. *A Z specification with state schema CState, initial state schema CInit and operations $COp_1 \ldots COp_n$ is a downward simulation of a Z specification with state schema AState, initial state schema AInit and operations $AOp_1 \ldots AOp_n$, if there is a retrieve relation R such that the following hold for all $i : 1..n$.*

    1. $\forall\, CState \bullet CInit \Rightarrow \exists\, AState \bullet AInit \wedge R$
    2. $\forall\, AState;\ CState \bullet R \Rightarrow (pre\,AOp_i \Leftrightarrow pre\,COp_i)$
    3. $\forall\, AState;\ CState;\ CState' \bullet R \wedge COp_i \Rightarrow (\exists\, AState' \bullet R' \wedge AOp_i)$

The second rule, referred to as *upward simulation*, requires that

1. there is an abstract state related to every concrete state,
2. the initial states of the concrete specification are only related to abstract initial states,
3. for every concrete state, there exists an abstract state, such that all operations enabled on the abstract state are also enabled on the concrete state, and
4. whenever a concrete operation can result in the state change (t, t'), for any abstract state s' related to t', the corresponding abstract operation can result in (s, s') where s is related to t.

It is defined as follows [6].

Definition 2. *A Z specification with state schema CState, initial state schema CInit and operations $COp_1 \ldots COp_n$ is an upward simulation of a Z specification with state schema AState, initial state schema AInit and operations $AOp_1 \ldots AOp_n$, if there is a retrieve relation R such that the following hold.*

1. $\forall\, CState \bullet \exists\, AState \bullet R$
2. $\forall\, AState;\ CState \bullet CInit \land R \Rightarrow AInit$
3. $\forall\, CState \bullet \exists\, AState \bullet \forall\, i : 1..n \bullet R \land (pre\, AOp_i \Rightarrow pre\, COp_i)$
4. $\forall\, i : 1..n \bullet \forall\, AState';\ CState;\ CState'$
   $\bullet\ (COp_i \land R' \Rightarrow (\exists\, AState \bullet R \land AOp_i)$

Note that we use a slightly stronger form of upward simulation, in particular, one where the quantification over the operations (that is, i) in condition 3 is after the existential quantification of the abstract state. The reasons for using this form of upward simulation are explored in [2, 7]. In particular, this form ensures that refinement corresponds to failures-divergences refinement in CSP. We need this form here since Lemma 2 does not hold for the weaker form of upward simulation.

Given these definitions, one might assume that properties involving states of the abstract system would hold for the related states in the concrete system. That is, a property referring to abstract state s would hold for concrete state t when t was related to s. This is not the case, as the following example shows.

Consider the following Z specification which, from an initial state $s = 0$, moves nondeterministically to either state $s = 1$ or $s = 2$ via operation $AOp1$ and then to state $s = 3$ or state $s = 4$ depending on its current state via operation $AOp2$.

```
┌─ AState ─────────────────
│ s : 0...4
└───────────────────────
```

```
┌─ AInit ─────────────────
│ AState
├───────────────────────
│ s = 0
└───────────────────────
```

```
┌─ AOp1 ─────────────────
│ ΔAState
├───────────────────────
│ s = 0 ∧ s' ∈ {1, 2}
└───────────────────────
```

```
┌─ AOp2 ─────────────────
│ ΔAState
├───────────────────────
│ s ∈ {1, 2}
│ s = 1 ⇒ s' = 3
│ s = 2 ⇒ s' = 4
└───────────────────────
```

This is refined by the following specification which from an initial state $t = 0$ moves to a state $t = 1$ via operation $COp1$ and then nondeterministically to state $t = 2$ or $t = 3$ via operation $COp2$.

```
┌─ CState ─────────────────
│ t : 0...3
└───────────────────────
```

```
┌─ CInit ─────────────────
│ CState
├───────────────────────
│ t = 0
└───────────────────────
```

$$\begin{array}{|l}
\hline
\;COp1 \rule{5cm}{0.4pt} \\
\;\Delta CState \\
\hline
\;t = 0 \wedge t' = 1 \\
\hline
\end{array}$$

$$\begin{array}{|l}
\hline
\;COp2 \rule{5cm}{0.4pt} \\
\;\Delta CState \\
\hline
\;t = 1 \wedge t' \in \{2,3\} \\
\hline
\end{array}$$

This can be proved to be a refinement using an upward simulation with the following retrieve relation.

$$\begin{array}{|l}
\hline
\;R \rule{8cm}{0.4pt} \\
\;\;AState \\
\;\;CState \\
\hline
\;\;s = 0 \Leftrightarrow t = 0 \\
\;\;s \in \{1,2\} \Leftrightarrow t = 1 \\
\;\;s = 3 \Leftrightarrow t = 2 \\
\;\;s = 4 \Leftrightarrow t = 3 \\
\hline
\end{array}$$

This can be seen more clearly in Figure 1 which depicts the behaviours of the specifications and the retrieve relation; with only the operations visible, both behaviours are identical.

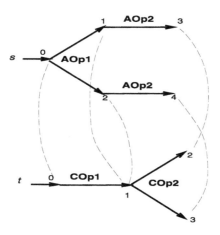

Fig. 1. An example refinement

Some temporal properties that hold for the abstract specification also hold for the related states in the concrete specification. For example, the property "it is always true that when $s = 0$, $s \in \{1,2\}$ in the next state" holds for the abstract specification, and the corresponding property "it is always true that when $t = 0$, $t = 1$ in the next state" holds in the concrete specification.

Note, however, that while the similar property "it is always true that when $s = 1$, $s = 3$ in the next state" holds for the abstract specification, the corresponding property "it is always true that when $t = 1$, $t = 2$ in the next state" does not hold for the concrete specification. Hence, this property is not preserved under this refinement.

This motivates the question as to when a temporal property is preserved by refinement. Does it depend on the property, the nature of the refinement or both? In the next section, we describe a general approach for answering this question for any temporal logic over states.

3   Temporal Structures

The temporal logics LTL, CTL and the μ-calculus are usually interpreted on temporal, or Kripke, structures [8]. Given the set of all atomic predicates AP, a temporal structure $(S, S_0, \textit{Trans}, L)$ comprises

- a set of states S,
- a set of initial states $S_0 \subseteq S$,
- a transition relation $\textit{Trans} \subseteq S \times S$ where $\forall\, s \in S.\, \exists\, s' \in S.(s, s') \in \textit{Trans}$, i.e., $\textit{Trans}$ is total, and
- a labelling function $L : S \to \mathbb{P}\, AP$ mapping each state in S to the atomic propositions which hold in it.

The condition that $\textit{Trans}$ is total is necessary and, together with an assumption that some transition always occurs, ensures that temporal structures continue to progress.

To interpret temporal logic predicates on Z specifications, it is necessary to also adopt this assumption that the specified system progresses, i.e., that its environment always allows some enabled operation to eventually occur. It is also necessary to represent Z specifications as temporal structures.

In order to state temporal properties which refer to inputs and outputs, we need to embed these in the states of the specification without changing its behaviour. This can be done as detailed in [13]. In brief, the type of each input and output when embedded in the state is extended with a value $\perp$ and this value is used for the pre-state value of embedded inputs and post-state value of embedded outputs when they are not declared in a particular schema of the original specification.

The set S of the temporal structure representing a Z specification is the set of bindings of the state schema of the specification. For example, for the abstract specification of Section 2

$$S = \{s \rightsquigarrow 0, s \rightsquigarrow 1, s \rightsquigarrow 2, s \rightsquigarrow 3, s \rightsquigarrow 4\}$$

Similarly, the set S_0 is the set of bindings of the initial state schema. For the same example,

$$S_0 = \{s \rightsquigarrow 0\}$$

The transition relation $\textit{Trans}$ includes mappings $(s, s') : S \times S$ when there exists an operation with binding $s \cup s'$, and mappings $(s, s) : S \times S$ when there does not exist an operation with binding $s \cup s'$ for any $s' : S$. The latter mappings represent stuttering transitions when no other transitions are available. They are necessary for *Trans* to be total. For the example,

$$Trans = \{(s \rightsquigarrow 0, s \rightsquigarrow 1), (s \rightsquigarrow 0, s \rightsquigarrow 2), (s \rightsquigarrow 1, s \rightsquigarrow 3), (s \rightsquigarrow 2, s \rightsquigarrow 4),$$
$$(s \rightsquigarrow 3, s \rightsquigarrow 3), (s \rightsquigarrow 4, s \rightsquigarrow 4)\}$$

where the final two pairs are stuttering transitions.

The labelling function L maps each binding s in S to the set comprising those atomic propositions P over the state variables where s is also a binding of the schema $[AState \mid P]$. For example, since $s \rightsquigarrow 0$ is a binding of $[AState \mid s < 10]$, the proposition $s < 10$ is a member of $L(s \rightsquigarrow 0)$.

3.1 Temporal Structures and Refinement

Given the above representation of Z specifications as temporal structures, we can express that a Z specification A meets a temporal logic property P using the standard notation $A \models P$. This means that the property P holds from all initial states of the specification.

To relate the temporal properties of Z specifications under a refinement, we first prove a lemma on Z refinement.

Lemma 1. *Given Z specifications, A and C, if $A \sqsubseteq C$ verified via a retrieve relation R then $\forall CInit \bullet (\exists AInit \bullet R)$.*

Proof. Since downward and upward simulation are complete, C is either a downward or upward simulation of A, or a combination of the two. It suffices to consider each case separately.

In the former case, we have from Definition 1

$$\forall CState \bullet CInit \Rightarrow (\exists AState \bullet AInit \wedge R)$$

This simplifies to the required condition.

In the latter case, we have from Definition 2

$$\forall CState \bullet \exists AState \bullet R$$

We also have

$$\forall AState; \ CState \bullet CInit \wedge R \Rightarrow AInit$$

Together these properties of upward simulation give us the required condition. □

For all temporal logics over states, the following general theorem holds.

Theorem 1. *Given Z specifications, A and C, and temporal logic property P, if $A \models P$ and $A \sqsubseteq C$ under retrieve relation R then $C \models \exists AState \bullet P \wedge R$.*

Proof. From Lemma 1, we know that for each initial state of C there exists an initial state of A related to it by R. Hence, $C \vDash \exists\, AInit \bullet R$. Also since $A \vDash P$, we know that P is true from all states satisfying $AInit$. Hence, we have $C \vDash \exists\, AInit \bullet P \wedge R$ which implies the required condition.   □

This general theorem provides the basis for our investigations. If a temporal logic property holds for a specification A, we want to know whether a *corresponding property* holds for a refinement C. We define a corresponding property to be one constructed from the same logical operators and with each atomic proposition P replaced by the following representation of it in the concrete state:

$$\exists\, AState \bullet P \wedge R$$

where R is the refinement retrieve relation.

Determining whether this is true reduces to determining whether conjunction and existential quantification distribute through the operators in the temporal logic property in such a way that the resulting property is no stronger than the original. For example, suppose $A \vDash P \wedge Q$. From Theorem 1, we know that

$$C \vDash \exists\, AState \bullet (P \wedge Q) \wedge R$$

Since conjunction distributes through conjunction resulting in an equivalent property (i.e., $(P \wedge R) \wedge (Q \wedge R)$ is equivalent to $(P \wedge Q) \wedge R$), we have

$$C \vDash \exists\, AState \bullet (P \wedge R) \wedge (Q \wedge R)$$

Since existential quantification distributes through conjunction resulting in a weaker property (i.e., $(\exists\, x \bullet P) \wedge (\exists\, x \bullet Q)$ is implied by $\exists\, x \bullet P \wedge Q$), we have

$$C \vDash (\exists\, AState \bullet P \wedge R) \wedge (\exists\, AState \bullet Q \wedge R)$$

If P and Q are atomic propositions then the above property is the concrete property corresponding to the abstract property $P \wedge Q$. Hence, the abstract property is preserved by refinement. If P or Q involve additional operators then we need to repeat the above process on the relevant conjunct or conjuncts above. Note that these conjuncts are of the same form as the concrete property derived from Theorem 1 and so the distribution of conjunction and existential quantification is again required.

In the following section, we investigate whether conjunction and existential quantification distribute through the operators of Linear Temporal Logic (LTL). This shows us when LTL properties are preserved by refinement and throws light on the cases when they are not.

4   Linear Temporal Logic

Linear Temporal Logic (LTL) [8] is defined on *paths*, i.e., infinite sequences of states of a temporal structure where each pair of consecutive states is related by the transition relation of the temporal structure. In this context, $A \vDash P$ means that P holds on all paths originating from initial states of A. Given a path π of A, we also adopt the more specific notations $A, \pi \vDash P$ and $A, \pi_i \vDash P$ for some $i \geqslant 0$. The former means that property P is true on path π and the latter that the property is true on the suffix of path π starting with its ith state, i.e., if $\pi = s_0 s_1 s_2 \ldots s_i s_{i+1} s_{i+2} \ldots$ then $\pi_i = s_i s_{i+1} s_{i+2} \ldots$.

We introduce a further lemma on refinement.

Lemma 2. *Given Z specifications A and C, if $A \sqsubseteq C$ under retrieve relation R then for all paths $\pi^C = t_0 t_1 t_2 \ldots$ of C there exists a path $\pi^A = s_0 s_1 s_2 \ldots$ of A such that each state t_i of π^C is related to the corresponding state s_i of π^A by R.*

Proof. The proof follows by induction.

(i) For state t_0, we know that there exists an abstract state related to it by Lemma 1. Hence, there is a path $\pi^A = s_0 s_1 s_2 \ldots$ of A such that s_0 and t_0 are related by R.

(ii) Assume there exists a path $\pi^A = s_0 s_1 s_2 \ldots$ of A and a $j : \mathbb{N}$ such that for all $i : \mathbb{N}$ where $i \leqslant j$, s_i is related to t_i by R. Since downward and upward simulation are jointly complete, C is either a downward or upward simulation of A, or a combination of the two. It suffices to consider each case separately.

In the former case, we have from Definition 1

$$\forall \, AState; \; CState; \; CState' \bullet R \wedge COp_i \Rightarrow (\exists \, AState' \bullet R' \wedge AOp_i)$$

Hence, for any concrete transition from t_j, there will be a transition from s_j to a state s'_{j+1} such that t_{j+1} and s'_{j+1} are related by R. The definition of a path means there will always be some transition from t_j to t_{j+1}. Hence, there is a path $\pi^{A'} = s_0 s_1 s_2 \ldots s'_{j+1} s'_{j+2} \ldots$ of A such that for all $i : \mathbb{N}$ where $i \leqslant j$, s_i is related to t_i by R and s'_{j+1} is related to t_{j+1} by R.

For upward simulation, we do not require the induction assumption. We simply prove that a path $\pi^{A'} = s'_0 s'_1 s'_2 \ldots$ of A exists such that for all $i : \mathbb{N}$ where $i \leqslant j + 1$, s'_i is related to t_i by R.

We have from Definition 2

$$\forall \, CState \bullet \exists \, AState \bullet R$$

Hence, there will be an abstract state s'_{j+1} related to t_{j+1} by R. Consider a concrete transition from t_j to t_{j+1}. Either this is skip (when no concrete operations are enabled) or the concrete transition is due to a concrete operation COp_i. In the former case the upward simulation applicability condition means no abstract operations are enabled, thus there exists a skip in the abstract state and $t_j = t_{j+1}$, $s_j = s_{j+1}$. In the latter case, we use the following condition from Definition 2

$$\forall \, AState'; \; CState; \; CState' \bullet COp_i \wedge R' \Rightarrow (\exists \, AState \bullet R \wedge AOp_i)$$

Hence, for the concrete transition from t_j to t_{j+1} due to COp_i, there will be a transition from a state s'_j of A to s'_{j+1} such that s'_j is related to t_j by R.

Similarly, for the concrete transition from t_{j-1} to t_j, there will be a transition from a state s'_{j-1} of A to s'_j such that s'_{j-1} is related to t_{j-1} by R. Following this line of reasoning, we can deduce that there exists a path $\pi^{A'} = s'_0 s'_1 s'_2 \ldots$ such that for all $i : \mathbb{N}$ such that $i \leqslant j + 1$, s'_i is related to t_i by R.   □

We also specialise Theorem 1 of Section 3.1 for LTL as follows.

Theorem 2. *Given Z specifications, A and C, and temporal logic property P, if there exists an $i : \mathbb{N}$ such that for all abstract paths π^A, $A, \pi_i^A \vDash P$ and $A \sqsubseteq C$ under retrieve relation R then for all concrete paths π^C, $C, \pi_i^C \vDash \exists\, AState \bullet P \wedge R$.*

Proof. From Lemma 2, we know that for each concrete path $\pi^C = t_0 t_1 t_2 \ldots$ of C there exists an abstract path $\pi^A = s_0 s_1 s_2 \ldots$ of A such that for all $i : \mathbb{N}$, s_i is related to t_i by R. Hence, $C, \pi_i^C \vDash \exists\, AState \bullet R$ where one instance of $AState$ satisfying the existentially quantified predicate is s_i. Also since $A, \pi_i^A \vDash P$, we know that P is true from state s_i. Hence, we have $C, \pi^C \vDash \exists\, AState \bullet P \wedge R$ as required.   □

4.1   Syntax and Semantics

Formulae in LTL are generated from the following rules of syntax.

> atomic propositions are formulae[1]
> if P and Q are formulae, then $\neg P$ and $P \wedge Q$ are formulae
> if P and Q are formulae, then $\mathbf{X} P$ and $P \mathbf{U} Q$ are formulae

The operator $\mathbf{X}$ is read as "next" and $\mathbf{U}$ as "until". The following abbreviations are also commonly used:

$$
\begin{aligned}
P \vee Q &\equiv \neg (\neg P \wedge \neg Q) \\
P \Rightarrow Q &\equiv \neg P \vee Q \\
true &\equiv P \vee \neg P \quad \text{for some } P \\
false &\equiv P \wedge \neg P \quad \text{for some } P \\
\mathbf{F} P &\equiv true\ \mathbf{U}\ P \quad (\text{read "eventually P"}) \\
\mathbf{G} P &\equiv \neg \mathbf{F} \neg P \quad (\text{read "always P"})
\end{aligned}
$$

The semantics of LTL is given in terms of a temporal structure A, path $\pi = s_0 s_1 s_2 \ldots$ of A, and LTL formulae P and Q as follows.

$$
\begin{aligned}
A, \pi &\vDash P && \text{if and only if} && P \text{ is true in } s_0 \\
A, \pi &\vDash \neg P && \text{if and only if} && A, \pi \nvDash P \\
A, \pi &\vDash P \wedge Q && \text{if and only if} && A, \pi \vDash P \text{ and } A, \pi \vDash Q \\
A, \pi &\vDash \mathbf{X} P && \text{if and only if} && A, \pi_1 \vDash P \\
A, \pi &\vDash P \mathbf{U} Q && \text{if and only if} && \exists j.(A, \pi_j \vDash Q) \text{ and } \forall k < j.(A, \pi_k \vDash P)
\end{aligned}
$$

---

[1] LTL, as used in model checking, is usually restricted to atomic propositions of the form $n = v$. We do not require this restriction for our results; any proposition is allowed.

4.2 Property Preservation

Let A and C be Z specifications such that $A \sqsubseteq C$ under retrieve relation R. To determine which LTL properties of A are preserved under Z refinement, we examine the distribution of conjunction and existential quantification through each LTL operator in turn. That distribution through conjunction works was shown in Section 3.1. Of the other operators (surprisingly) negation, rather than one of the temporal operators, turns out to be the most interesting case and so we leave it until last.

Next (X). If $A \vDash \mathbf{X}\, P$ then from the semantics of the next operator, for all abstract paths π^A, we have

$$A, \pi_1^A \vDash P$$

Hence, from Theorem 2 we have, for all concrete paths π^C,

$$C, \pi_1^C \vDash \exists\, AState \bullet P \wedge R$$

Hence, from the semantics of the next operator, we have

$$C \vDash \mathbf{X}\,(\exists\, AState \bullet P \wedge R)$$

Hence, conjunction and existential quantification distribute through the next operator.

Until (U). If $A \vDash P \,\mathbf{U}\, Q$ then from the semantics of the until operator, we know there exists a $j : \mathbb{N}$ such that for all $k : \mathbb{N}$ such that $k < j$, for all abstract paths π^A, we have

$$A, \pi_k^A \vDash P$$

and

$$A, \pi_j^A \vDash Q$$

Hence, from Theorem 2 we have, for all concrete paths π^C,

$$C, \pi_k^C \vDash \exists\, AState \bullet P \wedge R$$

and

$$C, \pi_j^C \vDash \exists\, AState \bullet Q \wedge R$$

Hence, from the definition of the until operator, we have

$$C \vDash (\exists\, AState \bullet P \wedge R) \,\mathbf{U}\, (\exists\, AState \bullet Q \wedge R)$$

That is, conjunction and existential quantification distribute through the until operator.

Since the eventually operator (**F**) is defined in terms of the until operator, conjunction and existential quantification also distribute through it.

Negation ($\neg$)

Negation of a property distributes through conjunction with a retrieve relation. That is, $\neg P \wedge R$ implies $\neg (P \wedge R)$. However, it does not distribute through existential quantification. That is, $\exists x \bullet \neg P$ does not imply $\neg (\exists x \bullet P)$ (since there may be some values of x satisfying $\neg P$ and others satisfying P).

So, in general, LTL properties involving negation are not preserved by refinement. Consider, for example, the following Z specification which performs $Op1$ once and then $Op2$ an infinite number of times.

```
┌─ AState ──────────────          ┌─ AInit ──────────────
│ s : ℕ                           │ AState
└──────────────────────          ├──────────────────────
                                  │ s = 0
                                  └──────────────────────
```

```
┌─ AOp1 ────────────────          ┌─ AOp2 ────────────────
│ ΔAState                         │ ΔAState
├──────────────────────          ├──────────────────────
│ s = 0 ∧ s′ = 1                  │ s ≠ 0 ∧ s′ = s + 1
└──────────────────────          └──────────────────────
```

It is refined by the following specification with identical behaviour.

```
┌─ CState ──────────────          ┌─ CInit ──────────────
│ t : {0, 1}                      │ t = 0
└──────────────────────          └──────────────────────
```

```
┌─ COp1 ────────────────          ┌─ COp2 ────────────────
│ ΔCState                         │ ΔCState
├──────────────────────          ├──────────────────────
│ t = 0 ∧ t′ = 1                  │ t = 1 ∧ t′ = 1
└──────────────────────          └──────────────────────
```

The refinement is a downward simulation under the retrieve relation

```
┌─ R ──────────────────────────────────────────────
│ AState
│ CState
├──────────────────────────────────────────────────
│ s = 0 ⇔ t = 0
│ s ≠ 0 ⇔ t = 1
└──────────────────────────────────────────────────
```

However, while the LTL property $\mathbf{G}\ (s = 1 \Rightarrow \mathbf{X}\ (\neg\ s = 1))$ holds for the abstract specification, the corresponding property $\mathbf{G}\ (t = 1 \Rightarrow \mathbf{X}\ (\neg\ t = 1))$ does not hold for the concrete specification.

This result does not mean that LTL properties involving negation are never preserved by refinement. For example, the property $\mathbf{G}\ (s = 0 \Rightarrow \mathbf{X}\ (\neg\ s = 0))$ holds for the abstract specification above, and the corresponding property $\mathbf{G}\ (t = 0 \Rightarrow \mathbf{X}\ (\neg\ t = 0))$ holds for the concrete specification.

The reason, in this case, is that the retrieve relation R is functional for the state $t = 0$, i.e., there is only one related abstract state. When this is so, $\exists\, AState \bullet \neg\, P \wedge R$ implies that $\neg\, P$ is true from the single instance of $AState$ related to the concrete state. Hence, it is equivalent to $\neg\, (\exists\, AState \bullet P \wedge R)$. That is, negation distributes through existential quantification.

Also, even without the retrieve relation being functional on negated states, some LTL properties involving negation are preserved. For example, disjunction which is defined in terms of negation does distribute through existential quantification as follows.

$$\begin{aligned}
&\exists\, x \bullet P \vee Q \\
\equiv\;& \exists\, x \bullet \neg\, (\neg\, P \wedge \neg\, Q) \\
\equiv\;& \neg\, (\forall\, x \bullet \neg\, P \wedge \neg\, Q) \\
\Rightarrow\;& \neg\, ((\forall\, x \bullet \neg\, P) \wedge (\forall\, x \bullet \neg\, Q)) \\
\equiv\;& \neg\, ((\neg\, (\exists\, x \bullet P)) \wedge (\neg\, (\exists\, x \bullet Q))) \\
\equiv\;& (\exists\, x \bullet P) \vee (\exists\, x \bullet Q)
\end{aligned}$$

In fact, whenever we have an even number of successive negations to distribute, as above, then they will distribute through existential quantification. The proof follows the reasoning of that above. Therefore, as well as the disjunction operator, we have that conjunction and existential quantification distribute through the always operator which is defined by $\mathbf{G}\, P = \neg\, \mathbf{F} \neg\, P$.

Of the operators defined as abbreviations, the only one we haven't considered is implication. Since this is defined by $P \Rightarrow Q = \neg\, P \vee Q$, we have an odd number of successive negations preceding predicate P. For this reason, the operator does not distribute through existential quantification, and hence temporal logic properties involving implication are not, in general, preserved by refinement.

This explains the example at the end of Section 2. The first concrete property can be formulated in LTL as

$$\mathbf{G}\, (s = 0 \Rightarrow \mathbf{X}\, (s \in \{1, 2\}))$$

which despite involving implication is preserved by the refinement since the retrieve relation is functional on $t = 0$ (the concrete state related to s_0).

The second property can be formulated in LTL as

$$\mathbf{G}\, ((s = 1 \Rightarrow (\mathbf{X}\, s = 3))$$

In this case, the retrieve relation is not functional on state $t = 1$ (the concrete state related to $s = 1$) and so the property is not preserved.

5   Conclusion

In this paper, we have shown that all temporal properties P over the state $AState$ of an abstract Z specification A are transformed to properties $\exists\, AState \bullet P \wedge R$ over a concrete Z specification C which is a refinement of A under the retrieve relation R.

This result allowed us to reduce the problem of determining when properties of a given temporal logic are preserved by refinement to an investigation of the distribution of the various operators of the temporal logic through conjunction and existential quantification.

We carried out such an investigation for Linear Temporal Logic (LTL) and discovered that while properties containing only conjunction and the temporal operators "next" and "until" (and, hence, also the derived operator "eventually") are preserved, those involving negation are, in general, not preserved.

We also pointed out two important cases when properties involving negation are preserved. The first is when any negated concrete state is related to only one abstract state. The second is when the number of successive negations in any part of the property is even. The first case is important as it shows that all LTL properties are preserved when the refinement retrieve relation is functional. The second is important as it shows that properties involving certain operators derived using negation, namely, disjunction and the temporal operator "always", are preserved.

The preservation of all LTL properties under a functional retrieve relation agrees with the recent result by Darlot et al. [5]. Our work extends this result by considering non-functional retrieve relations and also a complete definition of refinement (Darlot et al. consider only a variant of downward simulation in their approach).

We used a form of upward simulation that is compatible with CSP refinement, and is slightly stronger than the form often used in Z. The extension of our results to the weaker form of upward simulation is left for future work.

Our work is also closely related to that on abstraction, relevant in model checking as a means to reduce the size of the state space of a specification to be checked. Such techniques are in essence the inverse of refinement. Clarke et al. [3] proved that all properties in the universal fragment of CTL* (i.e., the fragment of CTL*, an extension of CTL, which excludes existential quantification over paths) are true of a specification when they are true of an abstraction of that specification. Loiseaux et al. [11] proved a similar result for the universal fragment of the mu-calculus. Both of these fragments subsume LTL, but there is no contradiction with our result as only functional abstraction relations were considered.

The abstraction results, as well as being limited to functional relations, also only consider variants of downward simulation. It would be interesting, therefore, to extend our results to CTL and the μ-calculus. Not only would this make our approach of using temporal properties with Z more general, and compatible with more model checking techniques, it would also open the possibility of developing more general abstraction techniques for model checking. This extension of our results could readily be achieved using the approach we have presented in this paper.

Acknowledgement

This work was carried out while Graeme Smith was on a visit to the University of Kent funded by a University of Queensland External Support Enabling Grant.

References

1. J.R. Abrial. *The B Book: Assigning Programs to Meaning.* Cambridge University Press, 1996.
2. C. Bolton and J. Davies. A Singleton Failures Semantics for Communicating Sequential Processes. *Formal Aspects of Computing*, 2002. Under consideration.
3. E. Clarke, O. Grumberg, and D. Long. Model checking and abstraction. *ACM Transactions on Programming Languages and Systems*, 16(5):1512–1542, 1994.
4. E. Clarke, O. Grumberg, and D. Peled. *Model Checking.* MIT Press, 2000.
5. C. Darlot, J. Julliand, and O. Kouchnarenko. Refinement preserves PLTL properties. In D. Bert, J.P. Bowen, S. King, and M. Waldén, editors, *International Conference of Z and B Users (ZB2003)*, volume 2651 of *Lecture Notes in Computer Science*, pages 408–420. Springer Verlag, 2003.
6. J. Derrick and E. Boiten. *Refinement in Z and Object-Z, Foundations and Advanced Applications.* Springer-Verlag, 2001.
7. J. Derrick and E.A. Boiten. Relational concurrent refinement. *Formal Aspects of Computing*, 15(1):182–214, November 2003.
8. E. A. Emerson. Temporal and modal logic. In J. van Leeuwen, editor, *Handbook of Theoretical Computer Science*, volume B, pages 996–1072. Elsevier Science Publishers, 1990.
9. C.B. Jones. *Systematic Software Development using VDM.* Prentice Hall, 1986.
10. D. Kozen. Results on the propositional μ-calculus. *Theoretical Computer Science*, 27:333–354, 1983.
11. C. Loiseaux, S. Graf, J. Sifakis, A. Bouajjani, and S. Bensalem. Property preserving abstractions for the verification of concurrent systems. *Formal Methods in System Design*, 6(1), 1995.
12. G. Smith. *The Object-Z Specification Language.* Advances in Formal Methods. Kluwer Academic Publishers, 2000.
13. G. Smith and K. Winter. Proving temporal properties of z specifications using abstraction. In D. Bert, J.P. Bowen, S. King, and M. Waldén, editors, *International Conference of Z and B Users (ZB2003)*, volume 2651 of *Lecture Notes in Computer Science*, pages 408–420. Springer Verlag, 2003.
14. J.M. Spivey. *The Z Notation: A Reference Manual.* Prentice Hall, 2nd edition, 1992.

Formal JVM Code Analysis in JavaFAN

Azadeh Farzan, José Meseguer, and Grigore Roşu

Department of Computer Science,
University of Illinois at Urbana-Champaign
{afarzan,meseguer,grosu}@cs.uiuc.edu

Abstract. JavaFAN uses a Maude rewriting logic specification of the
JVM semantics as the basis of a software analysis tool with competitive
performance. It supports formal analysis of concurrent JVM programs
by means of symbolic simulation, breadth-first search, and LTL model
checking. We discuss JavaFAN's executable formal specification of the
JVM, illustrate its formal analysis capabilities using several case studies,
and compare its performance with similar Java analysis tools.

1 Introduction

There is a general belief in the algebraic specification community that all traditional programming language features can be described with equational specifications [2, 9, 29]. What is less known, or tends to be ignored, is that *concurrency*, which is a feature of almost any current programming language, cannot
be naturally handled by equational specifications, unless one makes deterministic restrictions on how the different processes or threads are interleaved. While
some of these restrictions may be acceptable, as most programming languages
also provide thread or process scheduling algorithms, most of them are unacceptable in practice because concurrent execution typically depends upon the
external environment, which is unpredictable. Rewriting logic [17] extends equational logic with rewriting rules and has been mainly introduced as a *unified
model of concurrency*; indeed, many formal theories of concurrency have been
naturally mapped into rewriting logic during the last decade.

A next natural challenge is to define mainstream concurrent programming
languages in rewriting logic and then use those definitions to build formal analysis tools for such languages. There is already a substantial body of case studies,
of which we only mention [25, 24, 28], backing up one of the key claims of this paper, namely that *rewriting logic can be fruitfully used as a unifying framework for
defining programming languages*. Further evidence on this claim includes modeling of a wide range of programming language features that has been developed
and tested as part of a recent course taught at the University of Illinois [22]. In
this paper we give detailed evidence for a second key claim, namely that rewriting logic specifications can be used *in practice* to build simulators and formal
analysis tools for mainstream programming languages such as Java with competitive performance. Here, we focus on Java's bytecode, but our methodology
is general and can be applied also to the Java source code level and to many
other languages.

C. Rattray et al. (Eds.): AMAST 2004, LNCS 3116, pp. 132–147, 2004.
© Springer-Verlag Berlin Heidelberg 2004

The JavaFAN (Java Formal Analyzer) tool specifies the semantics of the most commonly used JVM bytecode instructions (150 out of the 250 total) as a Maude module specifying a rewrite theory $T_{\mathrm{JVM}} = (\Sigma_{\mathrm{JVM}}, E_{\mathrm{JVM}}, R_{\mathrm{JVM}})$, where $(\Sigma_{\mathrm{JVM}}, E_{\mathrm{JVM}})$ is an equational theory giving an algebraic semantics with semantic equations E_{JVM} to the *deterministic* JVM instructions, whereas R_{JVM} is a set of *rewrite rules*, with concurrent transition semantics, specifying the behavior of all *concurrent* JVM instructions. The three kinds of formal analysis currently supported in JavaFAN are: (1) *symbolic simulation*, where the theory T_{JVM} is executed in Maude as a JVM intepreter supporting fair execution and allowing some input values to be symbolic; (2) *breadth-first search*, where the entire, possibly infinite, state space of a program is explored starting from its initial state using Maude's `search` command to find safety property violations; and (3) *model checking*, where if a program's set of reachable states is finite, linear time temporal logic (LTL) properties are verified using Maude's LTL model checker.

A remarkable fact is that, as we explain in Section 4, even though T_{JVM} gives indeed a *mathematical semantics* to the JVM, it becomes the basis of a formal analysis tool whose performance is *competitive* and in some cases surpasses that of other Java analysis tools. The reasons for this are twofold. On the one hand, Maude [3] is a high-performance logical engine, achieving millions of rewrites per second on real applications, efficiently supporting search, and performing model checking with performance similar to that of SPIN [13]. On the other, the algebraic specification of system states, as well as the equations E_{JVM} and rules R_{JVM}, have been *optimized* for performance through several techniques explained in Section 3.5, including keeping only the dynamic parts of the state explicitly in the state representation, and making most equations and rules *unconditional*. In this regard, rewriting logic's distinction between the equations E_{JVM} and the rules R_{JVM} has a crucial performance impact in drastically reducing the sate space size. The point is that rewriting with the rules R_{JVM} takes place *modulo* the equations E_{JVM}, and therefore only the rules R_{JVM} affect state space size. Our experience in specifying the JVM in rewriting logic is that we gain the best benefits from algebraic (equations) and SOS [20] (Rules) paradigms in a combined way, while being able to distinguish between deterministic and concurrent features in a way not possible in either SOS or algebraic semantics.

Related Work. The different approaches to formal analysis for Java can be classified as focusing on either *sequential* or *concurrent* programs. Our work falls in the second category. More specifically, it belongs to a family of approaches that use a *formal executable specification of the concurrent semantics* of the JVM as a basis for formal reasoning. Two other approaches in precisely this category are one based on the ACL2 logic and theorem prover [15], and another based on a formal JVM semantics and reasoning based on Abstract State Machines (ASM) [23]. Our approach seems complementary to both of these, in the sense that it provides new formal analysis capabilities, namely search and LTL model checking. The ACL2 work is in a sense more powerful, since it uses an inductive theorem prover, but this greater power requires greater expertise and effort.

Outside the range of approaches based on executable formal specification, but somewhat close in the form of analysis, is NASA's Java Path Finder (JPF) [1, 12], which is an explicit state model-checker for Java bytecode based on a modified version of a C implementation of a JVM. Preliminary rough comparisons of JavaFAN and JPF[1] are encouraging, in the sense that we can analyze the same types of JVM programs of the same or even larger size. Other related work includes [21], which proposes an algorithm that takes the bytecode for a method and generates a temporal logic formula that holds iff the bytecode is safe; an off-the-shelf model checker can then be used to determine the validity of the formula. Among the formal techniques for *sequential* Java programs, some related approaches include the work on defensive JVM [5], and the collective effort around the JML specification language and verification tools for sequential Java, e.g. [16, 26].

Another approach to define analysis tools for Java is based on *language translators*, generating simpler language code from Java programs and then analyzing them later. Bandera [6] extracts abstract models from Java programs, specified in different formalisms, such as PROMELA, which can be further analyzed with specialized tools such as SPIN. JCAT [7] also translates Java into PROMELA. [19] presents an analysis tool which translates Java bytecode into C++ code representing an executable version of a model checker. While the translation-based approaches can benefit from abstraction techniques being integrated into the generated code, they inevitably lead to natural worries regarding the correctness of the translations. Unnecessary overhead seems to be also generated, at least in the case of [19]; for example, exactly the same Remote Agent Java code that can be analyzed in 0.3 second in JavaFAN [8] takes more than 2 seconds even on the most optimized version of the tool in [19].

In section 2 we present a brief background on Maude's methodology. A detailed description of our model is given in 3. In Section 4, we present the various kinds of formal analysis done for the Java programs together with the performance results for several case studies. Finally, Section 5 presents the conclusion and future work.

2   Rewriting Logic, Maude and Its Object Methodology

Here we briefly explain our methodology to specify the state of a concurrent system, in this case focusing on the JVM, as a "pool" or "soup" of objects whose interaction is modeled by rewrite rules. As a whole, the specification of the JVM is a *rewrite theory*, that is, a triple $(\Sigma_{\text{JVM}}, E_{\text{JVM}}, R_{\text{JVM}})$, with Σ_{JVM} a signature of operators, E_{JVM} a set of equational axioms, and R_{JVM} a collection of labeled Σ_{JVM}-rewrite rules. The equations describe the *static* structure of the JVM's state space as an algebraic data type, as well as the operational semantics of its deterministic features. The *concurrent transitions* that can occur in different threads are described by the rules R_{JVM}. Arbitrary interleavings of rewrite rules are possible, leading to different concurrent computations of a

---

[1] Authors thank Willem Visser for examples and valuable information about JPF.

multithreaded JVM program. The rewriting rules R_{JVM} are applied *modulo* the equations E_{JVM}. Important equations are those of *associativity, commutativity* and *identity* (ACU) of binary operations – such as the multiset union operation that builds up the "soup" of objects – allowing us to effectively define the state infrastructure of the JVM. Even though we focus on the algebraic definition of the JVM in this paper, the same methodology has been used to define the Java language as well as several other programming languages [8, 22].

Maude [3] supports, executes, and formally analyzes rewriting logic theories, via a series of efficient algorithms for term rewriting, state-space breadth-first search, and linear temporal logic (LTL) model checking. Once the JVM is formally specified as a rewrite theory in Maude, the above provide us with JVM program analysis tools at no additional cost, capable of performing fair interpretation and simulation, potentially infinite state-space exploration for detecting safety violations, as well as LTL model-checking of JVM multithreaded programs. E_{JVM} contains associativity, commutativity and identity equational axioms to represent the concurrent state of the JVM computation as a *multiset* of entities such as the threads, Java classes, Java objects, etc. Following a well-established methodology in rewriting logic, for which Maude provides generic support [3], we call these entities *objects*. To avoid terminology confusion with Java objects, we may sometimes call them *Maude objects*. Unless differently specified, from now on by "object" we mean a "Maude object". Maude supports a fully generic object-oriented specification environment, where one can define classes and then objects as instances of classes. Aiming at a maximum of efficiency for our JVM analysis tools, we decided *not* to use Maude's generic OO meta-level framework and, instead, to define a minimal object infrastructure at the core level. As a consequence, we have dropped the generic definition of classes, "hardwiring" our object types according to the JVM language.

Most of our equations and rules are applied modulo ACU, which in Maude is a highly optimized and efficient process. For example, Figures 3 and 4 present typical object-oriented rewrite rules. An object in a given state is formally represented as a term $\langle O : C \mid a_1 : v_1, \ldots, a_n : v_n \rangle$, where O is the object's name or identifier, C is its class, the a_i's are the names of the object's *attribute identifiers*, and the v_i's are the corresponding *values*.

3 Rewriting Logic Semantics of the JVM

We use Maude to specify the operational semantics of a sufficiently large subset of JVM bytecode. This includes 150 out of 250 bytecode instructions, defined in about 2000 lines of Maude code, including around 300 equations and 40 rewrite rules. We support multithreading, dynamic thread and object creation, virtual functions, recursive functions, inheritance, and polymorphism. Exception handling, garbage collection, native methods and many of the Java built-in libraries are not supported in the current version. The formal semantics of each instruction is defined based on the informal description of JVM in [27]. Section 3.2 explains the operational semantics of the deterministic part of the JVM given by the 300 equations in E_{JVM}, and Section 3.3 discusses the semantics of the concurrent part of JVM specified by the 40 rewrite rules in R_{JVM}.

3.1 Algebraic Representation of the JVM State

Here, we describe the representation of *states* in our model. Our major design goal has been to reduce the size and the number of system states to improve the performance of the formal analysis. The reduction in size has been achieved through separating the *static* and *dynamic* aspects of the program, maintaining only the dynamic part in the system's state. To reduce the number of states, we keep the number of rewrite rules in the specification minimal. A detailed discussion on these optimizations is given in Section 3.5.

The JVM has four basic components: (1) the class space, (2) the thread space, (3) the heap, and (4) the state transition machine, updating the internal state at each step.

In our model, no specific entity plays the role of the state transition system, and the strict separation of the classes, threads, and objects no longer exists. Instead, the state of the JVM is represented as a multiset of objects and messages[2] in Maude [3]. Rewrites (with rewrite rules and equations) model the changes in the state of the JVM.

Elements of the Multiset. Objects in the multiset fall into four categories:

1. Maude objects which represent Java objects,
2. Maude objects which represent Java threads,
3. Maude objects which represent Java classes, and
4. auxiliary Maude objects used mostly for definitional purposes.

Below, we discuss each in detail.

Java Objects are modeled by objects containing the following attributes.

```
< O:JavaObject | Addr:HeapAddress, FieldValues:FieldValues, CName:ClassName, Lock:Lock >
```

The `Addr` attribute refers to the heap address at which the object is stored. Physical heap addresses are employed only because they are used in the bytecode to refer to objects. The `FieldValues` attribute contains all instance fields and their values. Note that a single field may have more than one value, depending on its appearance in more than one class in the hierarchy of superclasses of the Java class from which the object is instantiated. The sort `FieldValues` is a list of pairs, with each pair consisting of a class name and a list. The latter list by itself consists of pairs of field names and field values. Therefore, based on the current class of the object, we can extract the right value for a desired field. The `CName` attribute holds the name of the object's class. The `Lock` attribute holds the lock associated with the object.

---

[2] Messages are used as a method to define the semantics in our model. One can use a somewhat different approach which does not include any messages.

Java Threads are modeled by objects with the following attributes.

```
< T : JavaThread | callStack: CallStack, Status: CallStackStat, ORef: HeapAddress >
```

The `callStack` attribute models the runtime stack of threads in Java. It is a stack of **frames**, where each **frame** models the activation record of a method call. Therefore, at any time, the top frame corresponds to the activation record of the method currently being executed. A **frame** is a tuple defined as follows.

```
op [_,_,_,_,_,_,_] : Int Inst LabeledPgm LocalVars OperandStack SyncFlag ClassName -> Frame .
```

The first component is an integer representing the program counter. The second component is the next instruction of the thread to be executed. The third component is a complete list of the instructions of the method, along with their corresponding offsets. The fourth component is the list of the current values of the local variables of the method. The fifth component contains the current operand stack, which carries instruction arguments and results. The sixth component is a flag indicating whether the call of the current method has locked the corresponding class (**SLOCKED**) or the corresponding object (**LOCKED**) or nothing at all (**UNLOCKED**). The last component represents the class from which the method has been invoked.

The **Status** attribute is a flag indicating the scheduling status of the thread: **scheduled** when the thread is ready to execute the next instruction, or **waiting** otherwise. Examples of threads with **waiting** status include a thread waiting for the completion of a communication it has started in order to get the code of the method being invoked. The **Oref** attribute contains the address of the object to which the thread is associated.

Java Classes. Each class is divided into *static* and *dynamic* parts (see Section 3.5), represented by `JavaClassS` and `JavaClassD` objects respectively. These objects contain the following attributes.

```
< C:JavaClassS| SupClass:ClassName,StaticFields:FlatFNL,Fields:FlatFNL,Methods:MethodList >
< C' : JavaClassD | ConstPool:ConstantPool, Lock:Lock, StaticFieldValues:FieldPairList >
```

The **SupClass** attribute contains the name of the immediate superclass of the class represented. The attribute **StaticFields** is a list of pairs, each pair consisting of a class name along with the list of static field names of that class. The classes in the first components of the pairs in this list are exactly the class represented by this object along with all its ancestors. These lists are compiled in a preprocessing phase. The **Fields** attribute has exactly the same structure as **StaticFields**, but for instance fields. The **ConstPool** attribute models the constant pool in the Java class file. In our model the constant pool is an indexed list containing information about methods, classes, and fields. Bytecode instructions refer to these indices, that the threads use to extract (from the constant pool) the required information to execute the instructions. By looking at the ith entry of the constant pool, we get a **FieldInfo**, which contains a field name and the name of the class the field belongs to, or a **MethodInfo**, which contains the method name, the name of the class the method belongs to, and the number of arguments of the method, or a `ClassInfo`, which only contains a class name. Examples of instructions which refer to the constant pool include,

- `new #5`, which creates a new object of the class whose name can be found in the 5th element of the constant pool, or
- `invokevirtual #3`, which invokes a method whose information (name, class, and number of arguments) can be found at the 3rd entry of the constant pool.

The `Methods` attribute contains a list of tuples, each representing a method. The structure of the tuple is as follows:

```
op {_,_,_,_,_} : MethodName MethodFormals MethodSync LabeledPgm Int -> Method .
```

The tuple components respectively represent the method name, a list of types of formal arguments of the method, a flag indicating whether or not the method is synchronized, the code of the method, and the number of local variables of the method. The `StaticFieldValues` attribute is exactly the same as `FieldValues` already discussed for `JavaObject`, except that this list refers to the values of static fields (which are stored inside the class) as opposed to the values of instance fields (which are stored inside the object). The `Lock` attribute holds the lock associated with the class.

Auxiliary Objects: Several objects in the multiset do not belong to any of the above categories. They have been added for definitional/implementation purposes. Examples include:

1. An *object collecting the outputs* of the threads. This object contains a list of values. When a thread prints a value, it adds this value to the end of this list. Input is assumed to be hardwired in the Java program at the moment.
2. A *heap manager*, that maintains the last address being used on the heap. We do not model garbage collection at the moment. but a modification of the heap manager can add garbage collection to our current JVM definition.
3. A *thread name manager*, that is used to generate new thread names.
4. There are several *Java built-in classes* that had to be apriori defined. The support for input/output, creating new threads, and `wait`/`notify` facilities are among the most important ones. All of these built-in classes have been created separately and are added as part of the initial multiset.

3.2   Equational Semantics of Deterministic JVM Instructions

If a bytecode instruction can be executed locally in the thread, meaning that no interaction with the outside environment is needed, that instruction's semantics can be specified using only equations. The equations specifying the semantics of all these deterministic bytecode instructions form the E_{JVM} part of the JVM's rewrite theory. In this section we present some examples of how deterministic bytecode instructions are modeled in our system. The complete Maude representation and a collection of examples can be found in [8].

```
eq < T : JavaThread | callStack: [PC, iadd, Pgm, LocalVars, (I # J # OperandStack),
                      SyncFlag, ClassName] CallStack, Status: scheduled, ORef : OR >
= < T : JavaThread | callStack: [PC + size(Pgm[PC]), Pgm[PC + size(Pgm[PC])]], Pgm, LocalVars,
    ((I + J) # OperandStack), SyncFlag, ClassName] CallStack, Status: scheduled, ORef: OR > .
```

Fig. 1. The `iadd` instruction.

iadd instruction is executed locally in the thread, and therefore, is modeled by
the equation shown in Figure 1. Values `I` and `J` on top of the operand stack are
popped, and the sum `I + J` is pushed. The program counter is moved forward
by the size of the `iadd` instruction to reach the beginning offset of the next
instruction. The current instruction (which was `iadd` before) is also changed to
be the next instruction in the current method code. Nothing else is changed in
the thread. Many of the bytecode instructions are typed. In this example, by
defining `I` and `J` to be integer variables, we support dynamic type checking as
well. Several dynamic checks of this kind are supported.

Invokevirtual is used to invoke a method from an object. It is among the most
complicated bytecode instructions and its specification includes several equations
and rewrite rules. The equation in Figure 2 is the first part of `invokevirtual`
semantics. One thread, one Java object, and one Java class are involved. When

```
ceq < T : JavaThread | callStack: [PC, invokevirtual(I), Pgm, LocalVars, OperandStack,
                       SyncFlag, ClassName] CallStack, Status: scheduled, ORef: OR >
   < ClassName : JavaClassV | StaticFieldValues: SFV, , Lock: L
                       ConstPool:[I, {J, MethodName, CName]] ConstantPool >
   < O : JavaObject | Addr: K , FieldValues: FV, CName: ClName, Lock: L >
=  < T : JavaThread | callStack: [PC, invokevirtual(I), Pgm, LocalVars, OperandStack,
                       SyncFlag, ClassName] CallStack, Status: waiting, ORef: OR >
   < ClassName : JavaClassV | StaticFieldValues: SFV, Lock: L,
                       ConstPool: [I, {J, MethodName, ClName]] ConstantPool >
   < O : JavaObject | Addr: K, FieldValues: FV, CName: ClName, Lock: L >
   (GetMethod MethodName ofClass ClName ArgSize J forThread T)
if K==int((popLocalVars(J+1,OperandStack))[O]) .
```

Fig. 2. The `invokevirtual` instruction.

the thread reaches the `invokevirtual` instruction, by looking at the reference
on top of the operand stack (`REF(K)`), it figures out from what object it has
to call the method. The method information (see Section 3.1) will be extracted
from the constant pool. The class `ClassName` needs to be involved, since the
constant pool is stored inside this class. The class (`ClName`) is the current[3] class
of the object `O`, therefore the code of the desired method should be extracted
from the constant part of it. The thread will send a message asking for the
code of the method, sending all the information to uniquely specify it. The last
entity before the condition is a message. This message is consumed later and the

---

[3] Note that this can change dynamically.

desired method is sent back to the thread through another message. The thread receives the message, and that is when the invocation is complete. If the method being invoked is a *synchronized* method, the thread has to acquire a lock before the invocation is complete. This then has to be done using a rewrite rule (see Section 3.5).

3.3   Rewriting Semantics of Concurrent JVM Instructions

The semantics of those bytecode instructions that involve interaction with the outside environment is defined using rewrite rules, thus allowing us to explore all the possible concurrent executions of a program. In this section we present the semantics of several concurrent bytecode instructions.

monitorenter (Figure 3) is used to acquire a lock. This makes a change in the shared space between threads, and so has to be specified by a rewrite rule. One Java object and one Java thread are involved. The thread execut-

```
rl [MONITORENTER1] :
     < T : JavaThread | callStack: [PC, monitorenter, Pgm, LocalVars, (REF(K) # OperandStack),
       SyncFlag, ClassName] CallStack, Status: scheduled, ORef: OR >
     < O : JavaObject | Addr: K, Lock: Lock(OIL, NoThread, 0), REST >
=> < T : JavaThread | callStack: [PC + size(Pgm[PC]), Pgm[PC + size(Pgm[PC])], Pgm,
       LocalVars, OperandStack, SyncFlag, ClassName] CallStack, Status: scheduled, ORef : OR >
     < O:JavaObject | Addr: K, Lock: Lock(OIL, T, 1), REST > .
```

Fig. 3. The `monitorenter` instruction.

ing `monitorenter` acquires the lock of the object whose reference is on top of the operand stack (`REF(K)`). The heap address of the object (`K`) is matched with this reference, and the lock of the object is changed to indicate that the object is now locked once by the thread `T` (note that a thread can lock or unlock an object several times). See section 3.4 for a more detailed discussion on locking and unlocking procedures.

getfield is a more complex instruction modeled by the rewrite rule in Figure 4. One thread and two Java classes are involved in this rule. The `I` operand is an index to the constant pool referring to the field information (`[I, {ClName, fieldname}]`), namely, field's name and its corresponding class name. The Java class `ClassName` is needed to extract the constant pool. The Java object `O` is identified by matching its heap address `K` with the reference `REF(K)` on top of the operand stack. On the right hand side of the rule, the thread proceeds to the next instruction, and the value of the indicated field of object `O` is placed on top of the operand stack of the thread (`FV[ClName, FieldName] # OperandStack`).

3.4   Synchronization

We support three means of thread synchronization: (1) `synchronized` sections, (2) `synchronized` methods, and (3) `wait/notifyAll` methods. In this section we explain how these means of synchronization are modeled.

```
rl [GETFIELD] :
    < T : JavaThread | callStack : ([PC, getfield(I), Pgm, LocalVars, REF(K) # OperandStack,
      SyncFlag, ClassName] CallStack),  Status: scheduled, ORef: OR >
    < ClassName : JavaClassV | ConstPool: ([I, {ClName, FieldName}] ConstantPool), REST >
    < O : JavaObject | Addr: K, FieldValues: FV, REST' >
=> < T : JavaThread | callStack: ([PC + size(Pgm[PC]), Pgm[PC + size(Pgm[PC])], Pgm,
      LocalVars, (FV[ClName, FieldName])#OperandStack, SyncFlag, ClassName] CallStack),
      Status: scheduled, ORef: OR >
    < ClassName : JavaClassV | ConstPool : ([I, {ClName, FieldName}] ConstantPool), REST >
    < O : JavaObject | Addr: K, FieldValues: FV, REST' >  .
```

Fig. 4. `getfield` Instruction.

The `synchronized` sections in Java are translated into sections surrounded by `monitorenter` (see Figure 3) and `monitorexit` bytecode instructions. During the execution of both, an object reference is expected to be on top of the operand stack whose corresponding lock is acquired and released respectively. Each Java object is modeled by a Maude object that includes a `Lock` attribute. This attribute has a tuple structure of the following form:

```
op Lock : OidList Oid Int -> Lock .
```

The first component is a list of identifiers of all threads that have waited on this object. This corresponds to `wait` and `notifyAll` methods (see below). The second component shows what thread currently owns the lock of this object (`NoThread` if none). The third component is a counter that shows how many times the owner of the lock has acquired the lock, since each lock can be acquired several times by the same owner.

When a thread encounters the `monitorenter` instruction, it checks whether the lock of the corresponding object is free. If so, the lock is changed to belong to this thread, and the thread can proceed to the critical section. It is also possible that the lock is not free, but has been acquired by the same thread before. In this case, only the counter is increased by one. When the thread finishes the execution of the critical section and reaches the `monitorexit` instruction, it simply decreases the counter by one. If the counter becomes zero, the lock is marked as free.

The `synchronized` methods are modeled in a very similar way. The difference is that, when the method is synchronized, `monitorenter` and `monitorexit` are replaced by method invocation and return, respectively. These methods are modeled through different rewrite rules, since different bytecode instructions are used for them.

Adding support (with little effort) for the `wait` and `notifyAll` methods of the Java built-in class `Object` is an interesting problem that we have solved. Similar to synchronization primitives, `wait` and `notifyAll` are called expecting an object reference on top of the operand stack. The thread (calling these methods) should already own the lock of the object on top of the operand stack. When `wait` is called, the thread releases the lock of the corresponding object, which it must own, and goes to sleep. It will not continue unless notified by another thread. The lock of the object is marked as free, the identifier of the current thread is added to the list of threads waiting on this object (the first component of the lock), and the integer indicating the number of times the thread had locked the corresponding object is stored locally in the sleeping thread, so that it can be recalled when the thread wakes up.

When `notifyAll` is called, a (broadcast) message is created containing the list of all threads which have waited on the corresponding object up to that point. This message will then be consumed by all the threads in this list. Each thread that consumes the message will *try to* wake up. In order to continue their execution, all these threads have to compete to acquire the lock on the specific object, to follow the rest of their executions inside the synchronized section. After the lock becomes available, one thread nondeterministically[4] acquires it.

3.5   Optimizations

Below, we discuss two major optimizations we have applied to decrease the size and number of system states, as well as the size of the state space.

Size of the State. In order to keep the state of the system small, we only maintain the dynamic part of the Java classes inside the system state. Every attribute of Java threads and Java objects can potentially change during the execution, but Java classes contain attributes that remain constant all along, namely, the methods, inheritance information, and field names. This, potentially huge amount of information, does not have to be carried along in the state of the JVM. The attributes of each class are grouped into *dynamic* and *static* attributes. The former group appears in the multiset, and the latter group is kept outside the multiset, in a Maude constant accessed through auxiliary operations.

Rules vs. Equations. Using equations for all deterministic computations, and rules only for concurrent ones leads to great savings in state space size. The key idea is that the only two cases in which a thread interacts with (possibly changes) the outside environment are shared memory access and acquiring locks. Examples of the former include the semantics of the instruction `getfield` (see Section 3.3) where a rule has been used. As an example for the latter case, we refer the reader to semantics of the `monitorenter` instruction (see Section 3.3). Since only the 40 rules in R_{JVM} contribute to the size of the state space, which is basically a graph with states as nodes and rewrite transitions as edges, we obtain a much smaller state space than if all the deterministic bytecode instructions had been specified as rules, in which case 340 rules would be used.

4   Formal Analysis

Using the underlying fair rewriting, search and model checking features of Maude, JavaFAN can be used to formally analyze Java programs in bytecode format. The

---

[4] In our model, but in general various implementations of the JVM use a variety of algorithms to choose the thread. By not committing to any specific deterministic choice approach, our formal analysis can discover subtle violations that may appear in some JVM implementations, but may not show up in others.

Maude's specification of the JVM can be used as an interpreter to simulate fair JVM computations by rewriting. *Breadth-first search analysis* is a semi-decision procedure that can be used to explore all the concurrent computations of a program looking for safety violations characterized by a pattern and a condition. Infinite state programs can be analyzed this way. For finite state programs it is also possible to perform explicit-state model checking of properties specified in linear temporal logic (LTL).

4.1 Simulation

Our Maude specification provides executable semantics for the JVM, which can be used to execute Java programs in bytecode format. This simulator can also be used to execute programs with symbolic inputs. Maude's `frewrite` command provides fair rewriting with respect to objects, and since all Java threads are defined as objects in the specification, no thread ever starves, although no specific scheduling algorithm is imposed. This assumption of fairness (with respect to threads) coincides with real models of JVM with a built-in scheduler, since scheduling algorithms also take the fairness into account. This fairness assumption does not mean that a deadlock is avoided; a *deadlock* in our model is a state in which no more rewrites are possible. The fair rewriting helps us avoid the situations in which a thread stuck in a loop is being executed forever, while other threads that can also be executed are starving.

To facilitate user interaction, the JVM semantics specification is integrated within the JavaFAN tool, that accepts standard bytecode as its input. The user can use `javac` (or any Java compiler) to generate the bytecode. She can then execute the bytecode in JavaFAN, being totally unaware of Maude. We use `javap` as the disassembler on the class files along with another disassembler `jreversepro` [14] to extract the constant pool information that `javap` does not provide.

4.2 Breadth-First Search

Using the simulator (Section 4.1), one can explore only one possible trace (modeled as sequence of rewrites) of the Java program being executed. Maude's `search` command allows exhaustively exploring all possible traces of a Java program. The breadth-first nature of the `search` command gives us a semi-decision procedure to find errors even in infinite state spaces, being limited only by the available memory. Below, we discuss a number of case studies.

Remote Agent. The Remote Agent (RA) is an AI-based spacecraft controller that has been developed at NASA Ames Research Center and has been part of the software component of NASA's Deep Space 1 shuttle. On Tuesday, May 18th, 1999, Deep Space 1's software deadlocked 96 million kilometers away from the Earth and consequently had to be manually interrupted and restarted from ground. The blocking was due to a missing critical section in the RA that had led to a data-race between two concurrent threads, which further caused a deadlock [10, 11]. This real life example shows that even quite experienced programmers can miss data-race errors in their programs. Moreover, these errors are so subtle that they often cannot be exposed by intensive testing procedures, such as NASA's, where more than 80% of a project's resources go into testing. This justifies formal analysis techniques like the ones presented in this paper which could have caught that error.

The RA consists of three components: a Planner that generates plans from mission goals; an Executive that executes the plans; and a Recovery system that monitors RA's status. The Executive contains features of a multithreaded operating system, and the Planner and Executive exchange messages in an interactive manner. Hence, this system is highly vulnerable to multithreading errors. Events and tasks are two major components (see [8] for the code). In order to catch the events that occur while tasks are executing, each event has an associated event counter that is increased whenever the event is signaled. A task then only calls `wait_for_event` in case this counter has not changed, hence, there have been no events since it was last restarted from a call of `wait_for_event`.

The error in this code results from the unprotected access to the variable `count` of the class `Event`. When the value of `event1.count` is read to check the condition, it can change before the related action is taken, and this can lead to a possible deadlock. This example has been extensively studied in [10, 11]. Using the search capability of our system, we also found the deadlock in the same faulty copy in 0.3 seconds. This is while the tool in [19] finds it in more than 2 seconds in its most optimized version[5].

The Thread Game. The Thread Game [18] is a simple multithreaded program which shows the possible data races between two threads accessing a common variable (see [8] for the code). Each thread reads the value of the static variable c twice and writes the sum of the two values back to c. Note that these two readings may or may not coincide. An interesting question is what values can c possibly hold during the infinite execution of the program. Theoretically, it can be proved that all natural numbers can be reached [18].

We can use Maude's search command to address this question for each specific value of N. The search command can find one or all existing solutions (sequences) that lead to get the value N. We have tried numbers up to 1000 where for all of them a solution is found in a reasonable amount time (Table 1).

Table 1. Thread Game Times.

N	50	100	200	400	500	1000
Time(s)	7.2	17.1	41.3	104	4.5m	10.1m

4.3   Model Checking

Maude's model checker is explicit state and supports Linear Temporal Logic. This general purpose rewriting logic model checker can be directly used on the

---
[5] All the performance results given in this section are in seconds on a 2.4GHz PC.

Maude specification of JVM's concurrent semantics. This way, we obtain a model checking procedure for Java programs for free. The user has to specify in Maude the atomic propositions to be used in order to specify relevant LTL properties. We illustrate this kind of model checking analysis by the following examples.

Table 2. Dining Philosophers Times.

Tests	Times(s)
DP(4)	0.64
DP(5)	4.5
DP(6)	33.3
DP(7)	4.4m
DP(8)	13.7m
DP(9)	803.2m
DF(4)	21.5
DF(5)	3.2m
DF(6)	23.9m
DF(7)	686.4m

Dining Philosophers. See [8] for the version of the dining philosophers problem that we have used in our experiments (DP). The property that we have model checked is whether all the philosopher can eventually dine. Each philosopher prints her ID when she dines. Therefore, to check whether the first philosopher has dined, we only have to check if 1 is written in the output list (see Section 3.1 for the output process). The LTL formula can be built based on propositions defined as follows. op `Check : Int -> Prop`, where `Check(N)` will be true at some state if the output list contains all the numbers from 1 to N. In this case, we check the following LTL formula using the `modelCheck`, where `InitialState` is the initial state of the program defined automatically. The formula that we model checked is $\Diamond$Check(n) for n philosophers. The model checker generates counterexamples, in this case a sequence of states that lead to a possible deadlock. The sequence shows a situation in which each philosopher has acquired one fork and is waiting for the other fork. Currently, we can detect the deadlock for up to 9 philosophers (Table 2). We also model checked a slightly modified version of the same program which avoids deadlock (DF). In this case, we can prove the program deadlock-free when there are up to 7 philosophers. This compares favorably with JPF [1,12] which for the same program cannot deal with 4 philosophers.

2-Stage Pipeline implements a pipeline computation (see [8] for the code), where each pipeline stage executes as a separate thread. Stages interact through *connector* objects that provide methods for **adding** and **taking** data. The property we have model checked for this program is related to the proper shutdown of pipelined computation, namely, "the eventual shutdown of a pipeline stage in response to a call to **stop** on the pipeline's input connector". The LTL formula for the property is $\Box$(c1stop $\rightarrow$ $\Diamond$($\neg$stage1return)). JavaFAN model checks the property and returns `true` in 17 minutes (no partial order reduction was used). This compares favorably with the model checker in [19] which without using the partial order reduction performs the task in more than 100 minutes.

5   Lessons Learned and Future Work

We have presented JavaFAN, explained its design, its rewriting logic semantic basis, and its Maude implementation. We have also illustrated JavaFAN's formal analysis capabilities and its performance on several case studies. The main lessons learned are that, using a rewriting logic semantics and a high-performance logical engine as a basis to build software analysis tools for conventional concurrent programs has the following advantages: (1) it is *cost-effective* in terms of amount of work needed to develop such tools; (2) it provides a *generic* technology that can bee applied to many different languages and furthermore the analysis tools for each language come essentially *for free* from the underlying logical engine; and (3) it has *competitive performance* compared to similar software analysis tools tailored to a specific language.

As always, there is much work ahead. On the one hand, in collaboration with Feng Chen support for the Java source code level has also been added to to JavaFAN and we plan to gain more experience and further optimize the tool at both levels. On the other, we plan to extend the range of formal analyses supported by JavaFAN, including, among the others, support for *program abstraction*, to model check finite-state abstractions of infinite-state programs, and for *theorem proving*, using Maude's inductive theorem prover (ITP) [4] as a basis. Furthermore, since the general techniques used in JavaFAN are in fact language-independent, we hope that other researchers find these techniques useful and apply them to develop similar tools for other concurrent languages.

Acknowledgment

Research supported by NASA/NSF grant CCR-0234524 and by ONR grant N00014-02-0715.

References

1. G. Brat, K. Havelund, S. Park, and W. Visser. Model checking programs. In *Automated Software Engineering 2000*, pages 3 – 12, 2000.
2. M. Broy, M. Wirsing, and P. Pepper. On the algebraic definition of programming languages. *ACM Trans. on Prog. Lang. and Systems*, 9(1):54–99, January 1987.
3. M. Clavel, F. Durán, S. Eker, P. Lincoln, N. Martí-Oliet, J. Meseguer, and C. Talcott. *Maude 2.0 Manual*, 2003. http://maude.cs.uiuc.edu/manual.
4. M. Clavel, F. Durán, S. Eker, and J. Meseguer. Building equational proving tools by reflection in rewriting logic. In *Proc. of the CafeOBJ Symposium*, April 1998.
5. R. M. Cohen. The defensive Java Virtual Machine specification. Technical report, Electronic Data Systems Corp, 1997.

6. J. C. Corbett, M. B. Dwyer, J. Hatcliff, S. Laubach, C. S. Păsăreanu, R. Zheng, and H. Zheng. Bandera: extracting finite-state models from Java source code. In *International Conference on Software Engineering*, pages 439 – 448, 2000.

7. C. Demartini, R. Iosif, and R. Sisto. A deadlock detection tool for concurrent Java programs. *Software - Practice and Experience*, 29(7):577 – 603, 1999.

8. A. Farzan, F. Chen, J. Meseguer, and G. Roşu. JavaFAN. `fsl.cs.uiuc.edu/javafan`.

9. J. Goguen and G. Malcolm. *Algebraic Semantics of Imperative Programs*. MIT, 1996.

10. K. Havelund, M. Lowry, S. Park, C. Pecheur, J. Penix, W. Visser, and J. White. Formal analysis of the remote agent before and after flight. In *the 5th NASA Langley Formal Methods Workshop*, 2000.

11. K. Havelund, M. Lowry, and J. Penix. Formal Analysis of a Space Craft Controller using SPIN. *IEEE Transactions on Software Engineering*, 27(8):749 – 765, August 2001. Previous version appeared in Proceedings of the 4th SPIN workshop, 1998.

12. K. Havelund and T. Pressburger. Model checking Java programs using Java PathFinder. *Software Tools for Technology Transfer*, 2(4):366 – 381, April 2000.

13. G. J. Holzmann. The model checker SPIN. *Software Eng.*, 23(5):279 – 295, 1997.

14. Jreversepro 1.4.1. `http://jrevpro.sourceforge.net/`.

15. M. Kaufmann, P. Manolios, and J. S. Moore. *Computer-Aided Reasoning: ACL2 Case Studies*. Kluwer Academic Press, 2000.

16. G. T. Leavens, K. R. M. Leino, E. Poll, C. Ruby, and B. Jacobs. JML: notations and tools supporting detailed design in Java. In *Object Oriented Programming, Systems, and Applications*, pages 105–106, 2000.

17. J. Meseguer. Conditional Rewriting Logic as a Unified Model of Concurrency. *Theoretical Computer Science*, pages 73–155, 1992.

18. J. S. Moore. `http://www.cs.utexas.edu/users/xli/prob/p4/p4.html`.

19. D. Y. W. Park, U. Stern, J. U. Sakkebaek, and D. L. Dill. Java model checking. In *Automated Software Engineering*, pages 253 – 256, 2000.

20. G. D. Plotkin. A structural approach to operational semantics. Technical report, Computer Science Department, Aarhus University, 1981.

21. J. Posegga and H. Vogt. Java bytecode verification using model checking. In *Workshop "Formal Underpinnings of Java" OOPSLA*, October 1998.

22. G. Roşu. Programming Language Design - CS322 Course Notes.

23. R. Stärk, J. Schmid, and E. Börger. *Java and the Java Virtual Machine - Definition, Verification, Validation*. Springer-Verlag, 2001.

24. M. Stehr and C. Talcott. Plan in Maude: Specifying an active network programming language. In *Rewriting Logic and its Applications*, volume 71 of *Electronic Notes in Theoretical Computer Science*, 2002.

25. P. Thati, K. Sen, and N. Martí-Oliet. An executable specification of asynchronous Pi-Calculus semantics and may testing in Maude 2.0. In *Rewriting Logic and its Applications*, volume 71 of *Electronic Notes in Theoretical Computer Science*, 2002.

26. J. van den Berg and B. Jacobs. The LOOP compiler for Java and JML. In *Tools and Algorithms for the Construction and Analysis of Systems*, volume 2031 of *LNCS*, pages 299 – 312, 2001.

27. B. Venners. *Inside The Java 2 Virtual Machine*. McGraw-Hill, 1999.

28. A. Verdejo and N. Martí-Oliet. Executable structural operational semantics in Maude. Manuscript, Dto. Sistemas Informáticos y Programación, Universidad Complutense, Madrid, August 2003.

29. M. Wand. First-order identities as a defining language. *Acta Informatica*, 14:337–357, 1980.

Verifying a Sliding Window Protocol in μCRL

Wan Fokkink[1,2], Jan Friso Groote[1,3], Jun Pang[1],
Bahareh Badban[1], and Jaco van de Pol[1]

[1] CWI, Embedded Systems Group
{wan,pangjun,badban,vdpol}@cwi.nl
[2] Vrije Universiteit Amsterdam, Theoretical Computer Science Group
[3] Eindhoven University of Technology, Systems Engineering Group
J.F.Groote@tue.nl

Abstract. We prove the correctness of a sliding window protocol with
an arbitrary finite window size n and sequence numbers modulo $2n$. We
show that the sliding window protocol is branching bisimilar to a queue
of capacity $2n$. The proof is given entirely on the basis of an axiomatic
theory, and was checked with the help of PVS.

1 Introduction

Sliding window protocols [6] (SWPs) ensure successful transmission of messages
from a sender to a receiver through a medium, in which messages may get lost.
Their main characteristic is that the sender does not wait for an incoming acknowledgement before sending next messages, for optimal use of bandwidth.
This is the reason why many data communication systems include the SWP, in
one of its many variations.

In SWPs, both the sender and the receiver maintain a buffer. In practice the
buffer at the receiver is often much smaller than at the sender, but here we make
the simplifying assumption that both buffers can contain up to n messages. By
providing the messages with sequence numbers, reliable in-order delivery without
duplications is guaranteed. The sequence numbers can be taken modulo $2n$ (and
not less, see [42] for a nice argument). The messages at the sender are numbered
from i to $i + n$ (modulo $2n$); this is called a *window*. When an acknowledgement
reaches the sender, indicating that k messages have arrived correctly, the window
slides forward, so that the sending buffer can contain messages with sequence
numbers $i+k$ to $i+k+n$ (modulo $2n$). The window of the receiver slides forward
when the first element in this window is passed on to the environment.

Within the process algebraic community, SWPs have attracted much attention. We provide a comparison with verifications of SWPs in Section 8, and
restrict here to the context in which the current paper was written. After the
advent of process algebra in the early 80's of last century, it was observed that
simple protocols, such as the alternating bit protocol, could readily be verified. In
an attempt to show that more difficult protocols could also be dealt with, SWPs
were considered. Middeldorp [31] and Brunekreef [4] gave specifications in ACP
[1] and PSF [30], respectively. Vaandrager [43], Groenveld [12], van Wamel [44]
and Bezem and Groote [3] manually verified one-bit SWPs, in which the windows

C. Rattray et al. (Eds.): AMAST 2004, LNCS 3116, pp. 148–163, 2004.

have size one. Starting in 1990, we attempted to prove the most complex SWP from [42] (not taking into account additional features such as duplex message passing and piggybacking) correct using the process algebraic language μCRL [16]. This turned out to be unexpectedly hard, which is shown by the 13 year it took to finish the current paper, and led to significant developments in the realm of process algebraic proof techniques for protocol verification. We therefore consider the current paper as a true milestone in process algebraic verification.

Our first observation was that the external behaviour of the protocol, as given in [42], was unclear. We adapted the SWP such that it nicely behaves as a queue of capacity $2n$. The second observation was that the SWP of [42] contained a deadlock [13, Stelling 7], which could only occur after at least n messages were transmitted. This error was communicated to Tanenbaum, and has been repaired in more recent editions of [42]. Another bug in the μCRL specification of the SWP was detected by means of a model checking analysis. A first attempt to prove the resulting SWP correct led to the verification of a bakery protocol [14], and to the development of the *cones and foci* proof method [19, 9]. This method rephrases the question whether two system specifications are branching bisimilar in terms of proof obligations on relations between data objects. It plays an essential role in the proof in the current paper, and has been used to prove many other protocols and distributed algorithms correct. But the correctness proof required an additional idea, already put forward by Schoone [37], to first perform the proof with unbounded sequence numbers, and to separately eliminate modulo arithmetic.

We present a specification in μCRL of a SWP with buffer size $2n$ and window size n, for arbitrary n. The medium between the sender and the receiver is modelled as a lossy queue of capacity one. We manually prove that the external behaviour of this protocol is branching bisimilar [10] to a FIFO queue of capacity $2n$. This proof is entirely based on the axiomatic theory underlying μCRL and the axioms characterising the data types. It implies both safety and liveness of the protocol (the latter under the assumption of fairness). First, we linearise the specification, meaning that we get rid of parallel operators. Moreover, communication actions are stripped from their data parameters. Then we eliminate modulo arithmetic, using the proof principle CL-RSP [2], which states that each linear specification has a unique solution (modulo branching bisimulation). Finally, we apply the cones and foci technique, to prove that the linear specification without modulo arithmetic is branching bisimilar to a FIFO queue of capacity $2n$. All lemmas for the data types, all invariants and all correctness proofs have been checked using PVS. The PVS files are available via http://www.cwi.nl/~badban/swp.html. Ongoing research is to extend the current verification to a setting where the medium is modelled as a lossy queue of unbounded capacity, and to include duplex message passing and piggybacking.

In this extended abstract we omitted most equational definitions of the data types, most lemmas regarding these data types, part of the invariants and part of the correctness proofs. The reader is referred to the full version of the paper [8], for these definitions and proofs.

2 μCRL

μCRL [16] (see also [18]) is a language for specifying distributed systems and protocols in an algebraic style. It is based on the process algebra ACP [1] extended with equational abstract data types [28]. In a μCRL specification, one part specifies the data types, while a second part specifies the process behaviour.

The data types needed for our μCRL specification of a SWP are presented in Section 3. In this section we focus on the process part of μCRL. Processes are represented by process terms, which describe the order in which the actions from a set $\mathcal{A}$ may happen. A process term consists of action names and recursion variables combined by process algebraic operators. Actions and recursion variables may carry data parameters. There are two predefined actions outside $\mathcal{A}$: δ represents deadlock, and τ a hidden action. These two actions never carry data parameters. $p \cdot q$ denotes sequential composition and $p + q$ non-deterministic choice. Summation $\sum_{d:D} p(d)$ provides the possibly infinite choice over a data type D, and the conditional construct $p \triangleleft b \triangleright q$ with b a data term of sort *Bool* behaves as p if b and as q if $\neg b$. Parallel composition $p \parallel q$ interleaves the actions of p and q; moreover, actions from p and q may also synchronise to a communication action, when this is explicitly allowed by a predefined communication function. Two actions can only synchronise if their data parameters are equal. Encapsulation $\partial_{\mathcal{H}}(p)$, which renames all occurrences in p of actions from the set $\mathcal{H}$ into δ, can be used to force actions into communication. Hiding $\tau_{\mathcal{I}}(p)$ renames all occurrences in p of actions from the set $\mathcal{I}$ into τ. Finally, processes can be specified by means of recursive equations $X(d_1:D_1,\ldots,d_n:D_n) \approx p$, where X is a recursion variable, d_i a data parameter of type D_i for $i = 1,\ldots,n$, and p a process term (possibly containing recursion variables and the parameters d_i). A recursive specification is linear if it is of the form

$$X(d_1:D_1,\ldots,d_n:D_n) \approx \sum_{i=1}^{\ell} \sum_{z_i:Z_i} a_i(e_1^i,\ldots,e_{m_i}^i) \cdot X(d_1^i,\ldots,d_n^i) \triangleleft b_i \triangleright \delta.$$

To each μCRL specification belongs a directed graph, called a labelled transition system, which is defined by the structural operational semantics of μCRL (see [16]). In this labelled transition system, the states are process terms, and the edges are labelled with parameterised actions. Branching bisimulation $\underline{\leftrightarrow}_b$ [10] and strong bisimulation $\underline{\leftrightarrow}$ [33] are two well-established equivalence relations on the states in labelled transition systems. Conveniently, strong bisimulation equivalence implies branching bisimulation equivalence. The proof theory of μCRL from [15] is sound modulo branching bisimulation equivalence, meaning that if $p \approx q$ can be derived from it then $p \underline{\leftrightarrow}_b q$.

The goal of this paper is to prove that the initial state of the forthcoming μCRL specification of a SWP is branching bisimilar to a FIFO queue. We use three proof principles from the literature:

- *Sum elimination* [14] states that a summation over a data type from which only one element can be selected can be removed.
- *CL-RSP* [2] states that the solutions of a linear μCRL specification that does not contain any infinite τ sequence are all strongly bisimilar.

– The *cones and foci* method from [9, 19] rephrases the question whether two linear μCRL specifications $\tau_{\mathcal{I}}(S_1)$ and S_2 are branching bisimilar, where S_2 does not contain actions from some set $\mathcal{I}$ of internal actions, in terms of data equalities. A *state mapping* ϕ relates each state in S_1 to a state in S_2. Furthermore, some states in S_1 are declared to be *focus points*, by means of a predicate *FC*. The *cone* of a focus point consists of the states in S_1 that can reach this focus point by a string of actions from $\mathcal{I}$. It is required that each reachable state in S_1 is in the cone of a focus point. If a number of *matching criteria* are satisfied, then $\tau_{\mathcal{I}}(S_1)$ and S_2 are branching bisimilar.

3   Data Types

In this section, the data types used in the μCRL specification of the SWP are presented: booleans, natural numbers supplied with modulo arithmetic, and buffers.

Booleans and Natural Numbers. *Bool* is the data type of booleans. t and f denote true and false, $\wedge$ and $\vee$ conjunction and disjunction, $\rightarrow$ and $\leftrightarrow$ implication and logic equivalence, and $\neg$ negation. For a boolean b, we abbreviate $b = \mathsf{t}$ to b and $b = \mathsf{f}$ to $\neg b$. Unless otherwise stated, data parameters in boolean formulas are universally quantified.

For each data type D in this paper there is an operation $if : Bool \times D \times D \rightarrow D$ with as defining equations $if(\mathsf{t}, d, e) = d$ and $if(\mathsf{f}, d, e) = e$. Furthermore, for each data type D in this paper one can easily define a mapping $eq : D \times D \rightarrow Bool$ such that $eq(d, e)$ holds if and only if $d = e$ can be derived. For notational convenience we take the liberty to write $d = e$ instead of $eq(d, e)$.

Nat is the data type of natural numbers. 0 denotes zero, $S(n)$ the successor of n, $+$, $\dot{-}$ and $\cdot$ addition, monus (also called proper subtraction) and multiplication, and $\leq$, $<$, $\geq$ and $>$ less-than(-or-equal) and greater-than(-or-equal). Usually, the sign for multiplication is omitted, and $\neg(i = j)$ is abbreviated to $i \neq j$. As binding convention, $\{=, \neq\} > \{\cdot\} > \{+, \dot{-}\} > \{\leq, <, \geq, >\} > \{\neg\} > \{\wedge, \vee\} > \{\rightarrow, \leftrightarrow\}$.

Since the buffers at the sender and the receiver in the sliding window are of size $2n$, calculations modulo $2n$ play an important role. $i|_n$ denotes i modulo n, while $i \, div \, n$ denotes i integer divided by n.

Buffers. The sender and the receiver in the SWP both maintain a buffer containing the sending and the receiving window, respectively (outside these windows both buffers are empty). Let Δ be the set of data elements that can be communicated between sender and receiver. The buffers are modelled as a list of pairs (d, i) with $d \in \Delta$ and $i \in Nat$, representing that position (or sequence number) i of the buffer is occupied by datum d. The data type *Buf* is specified as follows, where $[]$ denotes the empty buffer: $[] :\rightarrow Buf$ and $in : \Delta \times Nat \times Buf \rightarrow Buf$. $q|_n$ denotes buffer q with all sequence numbers taken modulo n. $[]|_n = []$ and $in(d, i, q)|_n = in(d, i|_n, q|_n)$. $test(i, q)$ produces t if and only if position i in q is occupied, $retrieve(i, q)$ produces the datum that resides at position i in buffer q (if this position is occupied), and $remove(i, q)$ is obtained by emptying position

i in buffer q. $release(i, j, q)$ is obtained by emptying positions i up to j excluded in q. $release|_n(i, j, q)$ does the same modulo n:

$$release(i, j, q) \ = \ if(i \geq j, q, release(S(i), j, remove(i, q)))$$
$$release|_n(i, j, q) = if(i|_n{=}j|_n, q, release|_n(S(i), j, remove(i, q)))$$

$next\text{-}empty(i, q)$ produces the first empty position in q, counting upwards from sequence number i onward. $next\text{-}empty|_n(i, q)$ does the same modulo n.

$$next\text{-}empty(i, q) \ = \ if(test(i, q), next\text{-}empty(S(i), q), i)$$
$$next\text{-}empty|_n(i, q) = \begin{cases} next\text{-}empty(i|_n, q|_n) & \text{if } next\text{-}empty(i|_n, q|_n) < n \\ next\text{-}empty(0, q|_n) & \text{otherwise} \end{cases}$$

Intuitively, $in\text{-}window(i, j, k)$ produces t if and only if j lies in the range from i to $k \dot{-} 1$, modulo n, where n is greater than i, j and k.

$$in\text{-}window(i, j, k) = i \leq j < k \lor k < i \leq j \lor j < k < i$$

Lists. The data type *List* of lists is used in the specification of the desired external behaviour of the SWP: a FIFO queue of capacity $2n$. It is specified by the empty list $\langle \rangle :\rightarrow List$ and $in : \Delta \times List \rightarrow List$. $length(\lambda)$ denotes the length of λ, $top(\lambda)$ produces the datum at the top of λ, $tail(\lambda)$ is obtained by removing the top position in λ, $append(d, \lambda)$ adds datum d at the end of λ, and $\lambda{+}{+}\lambda'$ represents list concatenation. Furthermore, $q[i..j\rangle$ is the list containing the elements in buffer q at positions i up to but not including j. An empty position in q, in between i and j, gives rise to an occurrence of the default datum d_0 in $q[i..j\rangle$.

$$q[i..j\rangle = \begin{cases} \langle \rangle & \text{if } i \geq j \\ in(retrieve(i, q), q[S(i)..j\rangle) & \text{if } i < j \land test(i, q) \\ in(d_0, q[S(i)..j\rangle) & \text{if } i < j \land \neg test(i, q) \end{cases}$$

4   Sliding Window Protocol

In this section, a μCRL specification of a SWP is presented, together with its desired external behaviour.

Specification of a Sliding Window Protocol. A sender **S** stores data elements that it receives via channel A in a buffer of size $2n$, in the order in which they are received. **S** can send a datum, together with its sequence number in the buffer, to a receiver **R** via a medium that behaves as lossy queue of capacity one, represented by the medium **K** and the channels B and C. Upon reception, **R** may store the datum in its buffer, where its position in the buffer is dictated by the attached sequence number. In order to avoid a possible overlap between the sequence numbers of different data elements in the buffers of **S** and **R**, no more than one half of the buffers of **S** and **R** may be occupied at any time; these halves are called the sending and the receiving window, respectively. **R** can pass on a datum that resides at the first position in its window via channel D; in that case the receiving window slides forward by one position. Furthermore, **R** can send the sequence number of the first empty position in (or just outside) its window as an acknowledgement to **S** via a medium that behaves as lossy queue of capacity one, represented by the medium **L** and the channels E and F. If **S** receives this acknowledgement, its window slides accordingly.

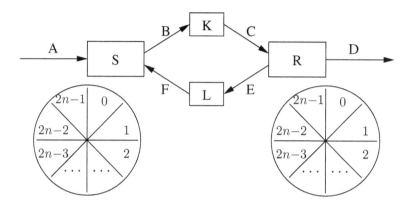

The sender **S** is modelled by the process $\mathbf{S}(\ell, m, q)$, where q is a buffer of size $2n$, ℓ the first position in the sending window, and m the first empty position in (or just outside) the sending window. Data elements can be selected at random for transmission from (the filled part of) the sending window.

$\mathbf{S}(\ell{:}Nat, m{:}Nat, q{:}Buf)$
$\approx \sum_{d:\Delta} r_{\mathrm{A}}(d){\cdot}\mathbf{S}(\ell, S(m)|_{2n}, in(d, m, q)) \triangleleft in\text{-}window(\ell, m, (\ell + n)|_{2n}) \triangleright \delta$
$+ \sum_{k:Nat} s_{\mathrm{B}}(retrieve(k, q), k){\cdot}\mathbf{S}(\ell, m, q) \triangleleft test(k, q) \triangleright \delta$
$+ \sum_{k:Nat} r_{\mathrm{F}}(k){\cdot}\mathbf{S}(k, m, release|_{2n}(\ell, k, q))$

The receiver **R** is modelled by the process $\mathbf{R}(\ell', q')$, where q' is a buffer of size $2n$ and ℓ' the first position in the receiving window.

$\mathbf{R}(\ell'{:}Nat, q'{:}Buf)$
$\approx \sum_{d:\Delta} \sum_{k:Nat} r_{\mathrm{C}}(d, k){\cdot}(\mathbf{R}(\ell', in(d, k, q')) \triangleleft in\text{-}window(\ell', k, (\ell' + n)|_{2n}) \triangleright \mathbf{R}(\ell', q'))$
$+ s_{\mathrm{D}}(retrieve(\ell', q')){\cdot}\mathbf{R}(S(\ell')|_{2n}, remove(\ell', q')) \triangleleft test(\ell', q') \triangleright \delta$
$+ s_{\mathrm{E}}(next\text{-}empty|_{2n}(\ell', q')){\cdot}\mathbf{R}(\ell', q')$

For $i \in \{\mathrm{B}, \mathrm{C}, \mathrm{E}, \mathrm{F}\}$, s_i and r_i can communicate, resulting in c_i.

Finally, the mediums **K** and **L**, which have capacity one, may lose frames between **S** and **R**. The action j indicates an internal choice.

$$\mathbf{K} \approx \sum_{d:\Delta} \sum_{k:Nat} r_{\mathrm{B}}(d, k){\cdot}(j{\cdot}s_{\mathrm{C}}(d, k) + j){\cdot}\mathbf{K}$$
$$\mathbf{L} \approx \sum_{k:Nat} r_{\mathrm{E}}(k){\cdot}(j{\cdot}s_{\mathrm{F}}(k) + j){\cdot}\mathbf{L}$$

The initial state of the SWP is expressed by $\tau_{\mathcal{I}}(\partial_{\mathcal{H}}(\mathbf{S}(0, 0, [\,]) \parallel \mathbf{R}(0, [\,]) \parallel \mathbf{K} \parallel \mathbf{L}))$, where the set $\mathcal{H}$ consists of the read and send actions over the internal channels B, C, E, and F, while the set $\mathcal{I}$ consists of the communication actions over these internal channels together with j.

External Behaviour. Data elements that are read from channel A by **S** should be sent into channel D by **R** in the same order, and no data elements should be lost. In other words, the SWP is intended to be a solution for the linear specification

$$\mathbf{Z}(\lambda:List) \approx \sum_{d:\Delta} r_A(d)\cdot\mathbf{Z}(append(d,\lambda)) \lhd length(\lambda) < 2n \rhd \delta$$
$$+ s_D(top(\lambda))\cdot\mathbf{Z}(tail(\lambda)) \lhd length(\lambda) > 0 \rhd \delta$$

Note that $r_A(d)$ can be performed until the list λ contains $2n$ elements, because in that situation the sending and receiving windows will be filled. Furthermore, $s_D(top(\lambda))$ can only be performed if λ is not empty.

The remainder of this paper is devoted to proving the following theorem.

Theorem 1. $\tau_{\mathcal{I}}(\partial_{\mathcal{H}}(\mathbf{S}(0,0,[]) \parallel \mathbf{R}(0,[]) \parallel \mathbf{K} \parallel \mathbf{L})) \underline{\leftrightarrow}_b \mathbf{Z}(\langle\rangle).$

5   Transformations of the Specification

The starting point of our correctness proof is a linear specification $\mathbf{N}_{mod}$, in which no parallel operators occur. $\mathbf{N}_{mod}$ can be obtained from the μCRL specification of the SWP without the hiding operator, by means of a linearisation algorithm presented in [17]. $\mathbf{N}_{mod}$ contains five extra parameters: $e{:}D$ and $g, g', h, h'{:}Nat$. Intuitively, g (resp. g') equals zero when medium **K** (resp. **L**) is inactive, equals one when **K** (resp. **L**) just received a datum, and equals two if **K** (resp. **L**) decides to pass on this datum. Furthermore, e (resp. h) equals the datum that is being sent from **S** to **R** (resp. the position of this datum in the sending window) while $g \neq 0$, and equals the dummy value d_0 (resp. 0) while $g = 0$. Finally h' equals the first empty position in the receiving window while $g' \neq 0$ and equals 0 while $g' = 0$. Furthermore, data arguments are stripped from communication actions, and these actions are renamed to a fresh action c. For the sake of presentation, we only present parameters whose values are changed.

$$\mathbf{N}_{mod}(\ell{:}Nat, m{:}Nat, q{:}Buf, \ell'{:}Nat, q'{:}Buf, g{:}Nat, e{:}D, h{:}Nat, g'{:}Nat, h'{:}Nat)$$
$$\approx \sum_{d:\Delta} r_A(d)\cdot\mathbf{N}_{mod}(m{:=}S(m)|_{2n}, q{:=}in(d,m,q)) \lhd in\text{-}window(\ell, m, (\ell+n)|_{2n}) \rhd \delta$$
$$+ \sum_{k:Nat} c\cdot\mathbf{N}_{mod}(g{:=}1, e{:=}retrieve(k,q), h{:=}k) \lhd test(k,q) \wedge g = 0 \rhd \delta$$
$$+ j\cdot\mathbf{N}_{mod}(g{:=}0, e{:=}d_0, h{:=}0) \lhd g = 1 \rhd \delta$$
$$+ j\cdot\mathbf{N}_{mod}(g{:=}2) \lhd g = 1 \rhd \delta$$
$$+ c\cdot\mathbf{N}_{mod}(q'{:=}in(e,h,q'), g{:=}0, e{:=}d_0, h{:=}0) \lhd in\text{-}window(\ell', h, (\ell'+n)|_{2n})$$
$$\wedge g = 2 \rhd \delta$$
$$+ c\cdot\mathbf{N}_{mod}(g{:=}0, e{:=}d_0, h{:=}0) \lhd \neg in\text{-}window(\ell', h, (\ell'+n)|_{2n}) \wedge g = 2 \rhd \delta$$
$$+ s_D(retrieve(\ell', q'))\cdot\mathbf{N}_{mod}(\ell'{:=}S(\ell')|_{2n}, q'{:=}remove(\ell', q')) \lhd test(\ell', q') \rhd \delta$$
$$+ c\cdot\mathbf{N}_{mod}(g'{:=}1, h'{:=}next\text{-}empty|_{2n}(\ell', q')) \lhd g' = 0 \rhd \delta$$
$$+ j\cdot\mathbf{N}_{mod}(g'{:=}0, h'{:=}0) \lhd g' = 1 \rhd \delta$$
$$+ j\cdot\mathbf{N}_{mod}(g'{:=}2) \lhd g' = 1 \rhd \delta$$
$$+ c\cdot\mathbf{N}_{mod}(\ell{:=}h', q{:=}release|_{2n}(\ell, h', q), g'{:=}0, h'{:=}0) \lhd g' = 2 \rhd \delta$$

Theorem 2.

$$\tau_{\mathcal{I}}(\partial_{\mathcal{H}}(\mathbf{S}(0,0,[]) \parallel \mathbf{R}(0,[]) \parallel \mathbf{K} \parallel \mathbf{L})) \; \underline{\leftrightarrow} \; \tau_{\{c,j\}}(\mathbf{N}_{mod}(0,0,[],0,[],0,d_0,0,0,0)).$$

The specification of $\mathbf{N}_{nonmod}$ is obtained by eliminating all occurrences of $|_{2n}$ from $\mathbf{N}_{mod}$, and replacing $in\text{-}window(\ell, m, (\ell+n)|_{2n}$ by $m < \ell + n$ and $in\text{-}window(\ell', h, (\ell'+n)|_{2n}$ by $\ell' \leq h < \ell' + n$.

Theorem 3. $\mathbf{N}_{mod}(0,0,[],0,[],0,d_0,0,0,0) \; \underline{\leftrightarrow} \; \mathbf{N}_{nonmod}(0,0,[],0,[],0,d_0,0,0,0).$

The proof of Theorem 2, using a linearisation algorithm [17] and a simple renaming, is omitted. The proof of Theorem 3 is shown in Section 7.1.

6   Properties of Data and Invariants of $\mathbf{N}_{nonmod}$

Lemma 1 collects results on modulo arithmetic related to buffers. Simpler lemmas on modulo arithmetic, buffers, the *next-empty* operation, and lists can be found in the full version of this paper [8]. We use those lemmas without mention.

Lemma 1. *The lemmas below hold for modulo arithmetic related to buffers.*

1. $\forall j{:}Nat(test(j,q) \rightarrow i \leq j < i+n) \wedge i \leq k \leq i+n \rightarrow test(k,q) = test(k|_{2n}, q|_{2n})$
2. $\forall j{:}Nat(test(j,q) \rightarrow i \leq j < i+n) \wedge test(k,q) \rightarrow retrieve(k,q) = retrieve(k|_{2n}, q|_{2n})$
3. $i \leq k < i+n \rightarrow in\text{-}window(i|_{2n}, k|_{2n}, (i+n)|_{2n})$
4. $in\text{-}window(i|_{2n}, k|_{2n}, (i+n)|_{2n}) \rightarrow k+n < i \vee i \leq k < i+n \vee k \geq i+2n$
5. $\forall j{:}Nat(test(j,q) \rightarrow i \leq j < i+n) \wedge test(k,q|_{2n}) \rightarrow in\text{-}window(i|_{2n}, k, (i+n)|_{2n})$

Invariants of a system are properties of data that are satisfied throughout the reachable state space of the system. Lemma 2 collects 9 invariants of $\mathbf{N}_{nonmod}$ that are needed in the correctness proofs in the current paper.

Lemma 2. *The invariants below hold for* $\mathbf{N}_{nonmod}(\ell, m, q, \ell', q', g, e, h, g', h')$.

1. $max\{h', \ell\} \leq next\text{-}empty(\ell', q')$
2. $g \neq 0 \rightarrow h < m$
3. $next\text{-}empty(\ell', q') \leq min\{m, \ell'+n\}$
4. $test(i,q) \leftrightarrow \ell \leq i < m$
5. $\ell \leq m \leq \ell+n \leq \ell'+2n$
6. $g \neq 0 \rightarrow next\text{-}empty(\ell', q') \leq h+n$
7. $g \neq 0 \wedge test(h,q) \rightarrow retrieve(h,q) = e$
8. $g \neq 0 \wedge test(h,q') \rightarrow retrieve(h,q') = e$
9. $\ell \leq i \wedge j \leq next\text{-}empty(i,q') \rightarrow q[i..j\rangle = q'[i..j\rangle$

7   Correctness of $\mathbf{N}_{mod}$

7.1   Equality of $\mathbf{N}_{mod}$ and $\mathbf{N}_{nonmod}$

In this section we present a proof of Theorem 3. It suffices to prove that for all $\ell, m, \ell', h, h' : Nat, \; q, q' : Buf, \; e : \Delta$ and $g, g' \leq 2$,

$$\mathbf{N}_{mod}(\ell|_{2n}, m|_{2n}, q|_{2n}, \ell'|_{2n}, q'|_{2n}, g, e, h|_{2n}, g', h'|_{2n})$$
$$\underline{\leftrightarrow} \; \mathbf{N}_{nonmod}(\ell|_{2n}, m|_{2n}, q|_{2n}, \ell'|_{2n}, q'|_{2n}, g, e, h|_{2n}, g', h'|_{2n})$$

Proof. We show that $\mathbf{N}_{mod}(\ell|_{2n}, m|_{2n}, q|_{2n}, \ell'|_{2n}, q'|_{2n}, g, e, h|_{2n}, g', h'|_{2n})$ is a solution for the defining equation of $\mathbf{N}_{nonmod}(\ell, m, q, \ell', q', g, e, h, g', h')$. Hence, we must derive the following equation.

$$
\begin{aligned}
&\mathbf{N}_{mod}(\ell|_{2n}, m|_{2n}, q|_{2n}, \ell'|_{2n}, q'|_{2n}, g, e, h|_{2n}, g', h'|_{2n}) \\
&\approx \sum_{d:\Delta} r_A(d) \cdot \mathbf{N}_{mod}(m:=S(m)|_{2n}, q:=in(d, m, q)|_{2n}) \triangleleft m < \ell + n \triangleright \delta & (A) \\
&+ \sum_{k:Nat} c \cdot \mathbf{N}_{mod}(g:=1, e:=retrieve(k, q), h:=k|_{2n}) \triangleleft test(k, q) \wedge g = 0 \triangleright \delta & (B) \\
&+ j \cdot \mathbf{N}_{mod}(g:=0, e:=d_0, h:=0) \triangleleft g = 1 \triangleright \delta & (C) \\
&+ j \cdot \mathbf{N}_{mod}(g:=2) \triangleleft g = 1 \triangleright \delta & (D) \\
&+ c \cdot \mathbf{N}_{mod}(q':=in(e, h, q')|_{2n}, g:=0, e:=d_0, h:=0) \triangleleft \ell' \leq h < \ell' + n \wedge g = 2 \triangleright \delta & (E) \\
&+ c \cdot \mathbf{N}_{mod}(g:=0, e:=d_0, h:=0) \triangleleft \neg(\ell' \leq h < \ell' + n) \wedge g = 2 \triangleright \delta & (F) \\
&+ s_D(retrieve(\ell', q')) \cdot \mathbf{N}_{mod}(\ell':=S(\ell')|_{2n}, q':=remove(\ell', q')|_{2n}) \triangleleft test(\ell', q') \triangleright \delta & (G) \\
&+ c \cdot \mathbf{N}_{mod}(g':=1, h':=next\text{-}empty(\ell', q')|_{2n}) \triangleleft g' = 0 \triangleright \delta & (H) \\
&+ j \cdot \mathbf{N}_{mod}(g':=0, h':=0) \triangleleft g' = 1 \triangleright \delta & (I) \\
&+ j \cdot \mathbf{N}_{mod}(g':=2) \triangleleft g' = 1 \triangleright \delta & (J) \\
&+ c \cdot \mathbf{N}_{mod}(\ell:=h'|_{2n}, q:=release(\ell, h', q)|_{2n}, g':=0, h':=0) \triangleleft g' = 2 \triangleright \delta & (K)
\end{aligned}
$$

In order to prove this, we instantiate the parameters in the defining equation of $\mathbf{N}_{mod}$ with $\ell|_{2n}, m|_{2n}, q|_{2n}, \ell'|_{2n}, q'|_{2n}, g, e, h|_{2n}, g', h'|_{2n}$.

$$
\begin{aligned}
&\mathbf{N}_{mod}(\ell|_{2n}, m|_{2n}, q|_{2n}, \ell'|_{2n}, q'|_{2n}, g, e, h|_{2n}, g', h'|_{2n}) \\
&\approx \sum_{d:\Delta} r_A(d) \cdot \mathbf{N}_{mod}(m:=S(m|_{2n})|_{2n}, q:=in(d, m|_{2n}, q|_{2n})) \\
&\qquad\qquad \triangleleft in\text{-}window(\ell|_{2n}, m|_{2n}, (\ell|_{2n} + n)|_{2n}) \triangleright \delta \\
&+ \sum_{k:Nat} c \cdot \mathbf{N}_{mod}(g:=1, e:=retrieve(k, q|_{2n}), h:=k) \triangleleft test(k, q|_{2n}) \wedge g = 0 \triangleright \delta \\
&+ j \cdot \mathbf{N}_{mod}(g:=0, e:=d_0, h:=0) \triangleleft g = 1 \triangleright \delta \\
&+ j \cdot \mathbf{N}_{mod}(g:=2) \triangleleft g = 1 \triangleright \delta \\
&+ c \cdot \mathbf{N}_{mod}(q':=in(e, h|_{2n}, q'|_{2n}), g:=0, e:=d_0, h:=0) \\
&\qquad\qquad \triangleleft in\text{-}window(\ell'|_{2n}, h|_{2n}, (\ell'|_{2n} + n)|_{2n}) \wedge g = 2 \triangleright \delta \\
&+ c \cdot \mathbf{N}_{mod}(g:=0, e:=d_0, h:=0) \\
&\qquad\qquad \triangleleft \neg in\text{-}window(\ell'|_{2n}, h|_{2n}, (\ell'|_{2n} + n)|_{2n}) \wedge g = 2 \triangleright \delta \\
&+ s_D(retrieve(\ell'|_{2n}, q'|_{2n})) \cdot \mathbf{N}_{mod}(\ell':=S(\ell'|_{2n})|_{2n}, q':=remove(\ell'|_{2n}, q'|_{2n})) \\
&\qquad\qquad \triangleleft test(\ell'|_{2n}, q'|_{2n}) \triangleright \delta \\
&+ c \cdot \mathbf{N}_{mod}(g':=1, h':=next\text{-}empty|_{2n}(\ell'|_{2n}, q'|_{2n})) \triangleleft g' = 0 \triangleright \delta \\
&+ j \cdot \mathbf{N}_{mod}(g':=0, h':=0) \triangleleft g' = 1 \triangleright \delta \\
&+ j \cdot \mathbf{N}_{mod}(g':=2) \triangleleft g' = 1 \triangleright \delta \\
&+ c \cdot \mathbf{N}_{mod}(\ell:=h'|_{2n}, q:=release|_{2n}(\ell|_{2n}, h'|_{2n}, q|_{2n}), g':=0, h':=0) \triangleleft g' = 2 \triangleright \delta
\end{aligned}
$$

To equate the eleven summands in both specifications, we obtain a number of proof obligations. Here, we focus on summands A, B, and E.

A $m < \ell + n = in\text{-}window(\ell|_{2n}, m|_{2n}, (\ell|_{2n} + n)|_{2n})$.
    $m < \ell + n \leftrightarrow \ell \leq m < \ell + n$ (Inv. 2.5) $\rightarrow in\text{-}window(\ell|_{2n}, m|_{2n}, (\ell + n)|_{2n})$
    (Lem. 1.3). Reversely, $in\text{-}window(\ell|_{2n}, m|_{2n}, (\ell + n)|_{2n}) \rightarrow m+n < \ell \vee \ell \leq m < \ell + n \vee m \geq \ell + 2n$ (Lem. 1.4) $\leftrightarrow m < \ell + n$ (Inv. 2.5). Since $(\ell + n)|_{2n} = (\ell|_{2n} + n)|_{2n}$, we have $m < \ell + n = in\text{-}window(\ell|_{2n}, m|_{2n}, (\ell|_{2n} + n)|_{2n})$.

B Below we equate the entire summand B of the two specifications. The conjunction $g = 0$ and the argument $g:=1$ of summand B are omitted, because they are irrelevant for this derivation.

By Inv. 2.4 and 2.5, $test(j, q) \to \ell \leq j < \ell + n$. So by Lem. 1.5, $test(k', q|_{2n})$ implies $in\text{-}window(\ell|_{2n}, k', (\ell + n)|_{2n})$. $test(k', q|_{2n})$ implies $k' = k'|_{2n}$, and by Lem. 1.4, $k' + n < \ell|_{2n} \vee \ell|_{2n} \leq k' < \ell|_{2n} + n \vee k' \geq \ell + 2n$. $k' = k'|_{2n} < 2n$ implies $k' + n < \ell|_{2n} \vee \ell|_{2n} \leq k' < \ell|_{2n} + n$.

$$\sum_{k:Nat} c \cdot \mathbf{N}_{mod}(e:=retrieve(k, q), h:=k|_{2n})$$
$$\vartriangleleft test(k, q) \vartriangleright \delta$$
$$\approx \sum_{k:Nat} c \cdot \mathbf{N}_{mod}(e:=retrieve(k, q), h:=k|_{2n})$$
$$\vartriangleleft test(k, q) \wedge \ell \leq k < \ell + n \vartriangleright \delta \qquad \text{(Inv. 2.4, 2.5)}$$
$$\approx \sum_{k:Nat} c \cdot \mathbf{N}_{mod}(e:=retrieve(k|_{2n}, q|_{2n}), h:=k|_{2n})$$
$$\vartriangleleft test(k|_{2n}, q|_{2n}) \wedge \ell \leq k < \ell + n \vartriangleright \delta \qquad \text{(Lem. 1.1, 1.2)}$$
$$\approx \sum_{k':Nat} \sum_{k:Nat} c \cdot \mathbf{N}_{mod}(e:=retrieve(k', q|_{2n}), h:=k')$$
$$\vartriangleleft test(k', q|_{2n}) \wedge \ell \leq k < \ell + n \wedge k' = k|_{2n} \vartriangleright \delta \qquad \text{(sum elimination)}$$
$$\approx \sum_{k':Nat} \sum_{k:Nat} c \cdot \mathbf{N}_{mod}(e:=retrieve(k', q|_{2n}), h:=k')$$
$$\vartriangleleft test(k', q|_{2n}) \wedge k = (\ell \; div \; 2n)2n + k' \wedge \ell|_{2n} \leq k' < \ell|_{2n} + n \wedge k' = k|_{2n} \vartriangleright \delta$$
$$+ \sum_{k':Nat} \sum_{k:Nat} c \cdot \mathbf{N}_{mod}(e:=retrieve(k', q|_{2n}), h:=k')$$
$$\vartriangleleft test(k', q|_{2n}) \wedge k = S(\ell \; div \; 2n)2n + k' \wedge k' + n < \ell|_{2n} \wedge k' = k|_{2n} \vartriangleright \delta$$
$$\approx \sum_{k':Nat} c \cdot \mathbf{N}_{mod}(e:=retrieve(k', q|_{2n}), h:=k')$$
$$\vartriangleleft test(k', q|_{2n}) \wedge \ell|_{2n} \leq k' < \ell|_{2n} + n \wedge k' = k' \vartriangleright \delta$$
$$+ \sum_{k':Nat} c \cdot \mathbf{N}_{mod}(e:=retrieve(k', q|_{2n}), h:=k')$$
$$\vartriangleleft test(k', q|_{2n}) \wedge k' + n < \ell|_{2n} \wedge k' = k' \vartriangleright \delta \qquad \text{(sum elimination)}$$
$$\approx \sum_{k':Nat} c \cdot \mathbf{N}_{mod}(e:=retrieve(k', q|_{2n}), h:=k')$$
$$\vartriangleleft test(k', q|_{2n}) \vartriangleright \delta \qquad \text{(see above)}$$

E $g = 2 \to \ell' \leq h < \ell' + n = in\text{-}window(\ell'|_{2n}, h|_{2n}, (\ell' + n)|_{2n})$.
Let $g = 2$. We have $\ell' \leq next\text{-}empty(\ell', q')$, and by Inv. 2.6 together with $g = 2$, $next\text{-}empty(\ell', q') \leq h + n$, so $\ell' \leq h + n$. Furthermore, by Inv. 2.2 together with $g = 2$, $h < m$, by Inv. 2.5, $m \leq \ell' + 2n$. Hence, $h < \ell' + 2n$. So using Lem. 1.3 and 1.4, it follows that $\ell' \leq h < \ell' + n = in\text{-}window(\ell'|_{2n}, h|_{2n}, (\ell' + n)|_{2n})$.

Equality of other summands can be derived without much difficulty. Hence, we prove that $\mathbf{N}_{mod}(\ell|_{2n}, m|_{2n}, q|_{2n}, \ell'|_{2n}, q'|_{2n}, g, e, h|_{2n}, g', h'|_{2n})$ is a solution for the specification of $\mathbf{N}_{nonmod}(\ell, m, q, \ell', q', g, e, h, g', h')$. By CL-RSP, they are strongly bisimilar.

7.2   Correctness of $\mathbf{N}_{nonmod}$

We prove that $\mathbf{N}_{nonmod}$ is branching bisimilar to the FIFO queue $\mathbf{Z}$ of capacity $2n$ (see Section 4), using the cones and foci method [9].

Let Ξ abbreviate $Nat \times Nat \times Buf \times Nat \times Buf \times Nat \times \Delta \times Nat \times Nat \times Nat$. Furthermore, let $\xi:\Xi$ denote $(\ell, m, q, \ell', q', g, e, h, g', h')$. The state mapping $\phi : \Xi \to List$, which maps states of $\mathbf{N}_{nonmod}$ to states of $\mathbf{Z}$, is defined by:

$$\phi(\xi) = q'[\ell'..next\text{-}empty(\ell', q')\rangle \mathbin{+\!\!+} q[next\text{-}empty(\ell', q')..m\rangle$$

Intuitively, ϕ collects the data elements in the sending and receiving windows, starting at the first position of the receiving window (i.e., ℓ') until the first empty position in this window, and then continuing in the sending window until the first empty position in that window (i.e., m). Note that ϕ is independent of e, g, ℓ, h, g', h'; we therefore write $\phi(m, q, \ell', q')$.

The focus points are those states where either the sending window is empty (meaning that $\ell = m$), or the receiving window is full and all data elements in the receiving window have been acknowledged, meaning that $\ell = \ell' + n$. That is, the focus condition for $\mathbf{N}_{nonmod}(\ell, m, q, \ell', q', g, e, h, g', h')$ is

$$FC(\ell, m, q, \ell', q', g, e, h, g', h') := \ell = m \vee \ell = \ell' + n$$

Lemma 3. *For each $\xi:\Xi$ where the invariants in Lemma 2 hold, there is a $\hat{\xi}:\Xi$ with $FC(\hat{\xi})$ such that $\mathbf{N}_{nonmod}(\xi) \xrightarrow{c_1} \cdots \xrightarrow{c_n} \mathbf{N}_{nonmod}(\hat{\xi})$, where $c_1, \ldots, c_n \in \mathcal{I}$.*

Proof. In case $g \neq 0$ in ξ, by summands C, E and F, we can perform one or two communication actions to a state where $g = 0$. By Inv. 2.3, $next\text{-}empty(\ell', q') \leq \min\{m, \ell'+n\}$. We prove by induction on $\min\{m, \ell'+n\} - next\text{-}empty(\ell', q')$ that for each state ξ' where $g = 0$ and the invariants in Lemma 2 hold, a focus point can be reached.
BASE CASE: $next\text{-}empty(\ell', q') = \min\{m, \ell' + n\}$.
In case $g' \neq 0$ in ξ', by summands I and K, we can perform communication actions to a state where $g' = 0$ and $next\text{-}empty(\ell', q') = \min\{m, \ell' + n\}$. By summands H, J and K we can perform three communication actions to a state $\hat{\xi}$ where $\ell = h' = next\text{-}empty(\ell', q') = \min\{m, \ell' + n\}$. Then $\ell = m$ or $\ell = \ell' + n$, so $FC(\hat{\xi})$.
INDUCTION CASE: $next\text{-}empty(\ell', q') < \min\{m, \ell' + n\}$.
By Inv. 2.1, $\ell \leq next\text{-}empty(\ell', q') < m$. By Inv. 2.4, $test(next\text{-}empty(\ell', q'), q)$. Furthermore, $\ell' \leq next\text{-}empty(\ell, q') < \ell' + n$. Hence, by summands B, D and E from ξ' we can perform three communication actions to a state ξ''. In ξ'', $g := 0$, and in comparison to ξ', m and ℓ' remain the same, while $q' := in(d, next\text{-}empty(\ell', q'), q')$ where d denotes $retrieve(next\text{-}empty(\ell', q'), q)$. Since $next\text{-}empty(\ell', in(d, next\text{-}empty(\ell', q'), q'))$ $= next\text{-}empty(S(next\text{-}empty(\ell', q')), q') > next\text{-}empty(\ell', q')$, we can apply the induction hypothesis to conclude that from ξ'' a focus point can be reached.

Theorem 4. *For all $e:\Delta$, $\tau_{\{c,j\}}(\mathbf{N}_{nonmod}(0, 0, [], 0, [], 0, e, 0, 0, 0)) \underline{\leftrightarrow}_b \mathbf{Z}(\langle\rangle)$.*

Proof. By the cones and foci method we obtain the following matching criteria (cf. [9]). Trivial matching criteria are left out.

$$\begin{cases} \text{I.1}: & \ell' \leq h < \ell' + n \wedge g = 2 \;\rightarrow\; \phi(m, q, \ell', q') = \phi(m, q, \ell', in(e, h, q')) \\ \text{I.2}: & g' = 2 \;\rightarrow\; \phi(m, q, \ell', q') = \phi(m, release(\ell, h', q), \ell', q') \\ \text{II.1}: & m < \ell + n \;\rightarrow\; length(\phi(m, q, \ell', q')) < 2n \\ \text{II.2}: & test(\ell', q') \;\rightarrow\; length(\phi(m, q, \ell', q')) > 0 \\ \text{III.1}: & (\ell = m \vee \ell = \ell' + n) \wedge length(\phi(m, q, \ell', q')) < 2n \;\rightarrow\; m < \ell + n \\ \text{III.2}: & (\ell = m \vee \ell = \ell' + n) \wedge length(\phi(m, q, \ell', q')) > 0 \;\rightarrow\; test(\ell', q') \\ \text{IV}: & test(\ell', q') \;\rightarrow\; retrieve(\ell', q') = top(\phi(m, q, \ell', q')) \\ \text{V.1}: & m < \ell + n \;\rightarrow\; \phi(S(m), in(d, m, q), \ell', q') = append(d, \phi(m, q, \ell', q')) \\ \text{V.2}: & test(\ell', q') \;\rightarrow\; \phi(m, q, S(\ell'), remove(\ell', q')) = tail(\phi(m, q, \ell', q')) \end{cases}$$

I.1 $\ell' \leq h < \ell' + n \wedge g = 2 \rightarrow \phi(m, q, \ell', q') = \phi(m, q, \ell', in(e, h, q'))$.
CASE 1: $h \neq next\text{-}empty(\ell', q'))$.
Let $g = 2$. Since $next\text{-}empty(\ell', in(e, h, q')) = next\text{-}empty(\ell', q')$, it follows that $\phi(m, q, \ell', in(e, h, q')) = in(e, h, q')[\ell'..next\text{-}empty(\ell', q')\rangle ++ q[next\text{-}empty(\ell', q')..m\rangle$.
CASE 1.1: $\ell' \leq h < next\text{-}empty(\ell', q'))$.
$test(h, q')$, so by Inv. 2.8 together with $g = 2$, $retrieve(h, q') = e$. Hence, $in(e, h, q')[\ell'..next\text{-}empty(\ell', q')\rangle = q'[\ell'..next\text{-}empty(\ell', q')\rangle$.

CASE 1.2: $\neg(\ell' \leq h \leq next\text{-}empty(\ell', q'))$.
$in(e, h, q')[\ell'..next\text{-}empty(\ell', q')\rangle = q'[\ell'..next\text{-}empty(\ell', q')\rangle$.
CASE 2: $h = next\text{-}empty(\ell', q')$.
Let $g = 2$. The derivation splits into two parts.
(1) $in(e, h, q')[\ell'..h\rangle = q'[\ell'..h\rangle$.
(2) By Inv. 2.1, $\ell \leq h$, and by Inv. 2.2 together with $g = 2$, $h < m$. Thus, by Inv. 2.4, $test(h, q)$. So by Inv. 2.7 together with $g = 2$, $retrieve(h, q) = e$. Hence,

$$
\begin{aligned}
&in(e, h, q')[h..next\text{-}empty(S(h), q')\rangle \\
&= in(e, in(e, h, q')[S(h)..next\text{-}empty(S(h), q')\rangle) \\
&= in(e, q'[S(h)..next\text{-}empty(S(h), q')\rangle) \\
&= in(e, q[S(h)..next\text{-}empty(S(h), q')\rangle) \qquad \text{(Inv. 2.9)} \\
&= q[h..next\text{-}empty(S(h), q')\rangle
\end{aligned}
$$

Finally, we combine (1) and (2). We recall that $h = next\text{-}empty(\ell', q')$.

$$
\begin{aligned}
&in(e, h, q')[\ell'..next\text{-}empty(\ell', in(e, h, q'))\rangle \\
&{+\!\!+}q[next\text{-}empty(\ell', in(e, h, q'))..m\rangle \\
&= in(e, h, q')[\ell'..next\text{-}empty(S(h), q')\rangle \\
&\quad {+\!\!+}q[next\text{-}empty(S(h), q')..m\rangle \\
&= (in(e, h, q')[\ell'..h\rangle {+\!\!+} in(e, h, q')[h..next\text{-}empty(S(h), q')\rangle) \\
&\quad {+\!\!+}q[next\text{-}empty(S(h), q')..m\rangle \\
&= q'[\ell'..h\rangle {+\!\!+} q[h..next\text{-}empty(S(h), q')\rangle \\
&\quad {+\!\!+}q[next\text{-}empty(S(h), q')..m\rangle \qquad \text{(1), (2)} \\
&= q'[\ell'..h\rangle {+\!\!+} q[h..m\rangle
\end{aligned}
$$

I.2 $g' = 2 \rightarrow \phi(m, q, \ell', q') = \phi(m, release(\ell, h', q), \ell', q')$.
By Inv. 2.1, $h' \leq next\text{-}empty(\ell', q')$.
So $release(\ell, h', q)[next\text{-}empty(\ell', q')..m\rangle = q'[next\text{-}empty(\ell', q')..m\rangle$.
II.1 $m < \ell + n \rightarrow length(\phi(m, q, \ell', q')) < 2n$.
Let $m < \ell + n$. By Inv. 2.3, $next\text{-}empty(\ell', q') \leq \ell' + n$. Hence,

$$
\begin{aligned}
&length(q'[\ell'..next\text{-}empty(\ell', q')\rangle {+\!\!+} q[next\text{-}empty(\ell', q')..m\rangle) \\
&= length(q'[\ell'..next\text{-}empty(\ell', q')\rangle) + length(q[next\text{-}empty(\ell', q')..m\rangle)) \\
&= (next\text{-}empty(\ell', q') \dot{-} \ell') + (m \dot{-} next\text{-}empty(\ell', q')) \\
&\leq n + (m \dot{-} \ell) \qquad \text{(Inv. 2.1)} \\
&< 2n
\end{aligned}
$$

II.2 $test(\ell', q') \rightarrow length(\phi(m, q, \ell', q')) > 0$.
$test(\ell', q')$ yields $next\text{-}empty(\ell', q') = next\text{-}empty(S(\ell'), q') \geq S(\ell')$. Hence,
$length(\phi(m, q, \ell', q')) = (next\text{-}empty(\ell', q') \dot{-} \ell') + (m \dot{-} next\text{-}empty(\ell', q')) > 0$.
III.1 $(\ell = m \vee \ell = \ell' + n) \wedge length(\phi(m, q, \ell', q')) < 2n \rightarrow m < \ell + n$.
CASE 1: $\ell = m$.
Then $m < \ell + n$ holds trivially.
CASE 2: $\ell = \ell' + n$.
By Inv. 2.3, $next\text{-}empty(\ell', q') \leq \ell' + n$. Hence,

$$
\begin{aligned}
&length(\phi(m, q, \ell', q')) \\
&= (next\text{-}empty(\ell', q') \dot{-} \ell') + (m \dot{-} next\text{-}empty(\ell', q')) \\
&\leq ((\ell' + n) \dot{-} \ell') + (m \dot{-} \ell) \qquad \text{(Inv. 2.1)} \\
&= n + (m \dot{-} \ell)
\end{aligned}
$$

So $length(\phi(m, q, \ell', q')) < 2n$ implies $m < \ell + n$.

III.2 $(\ell = m \vee \ell = \ell' + n) \wedge length(\phi(m, q, \ell', q')) > 0 \rightarrow test(\ell', q')$.

CASE 1: $\ell = m$.

Since $m \dot{-} next\text{-}empty(\ell', q') \leq (m \dot{-} \ell)$ (Inv. 2.1) $= 0$, we have
$length(\phi(m, q, \ell', q')) = next\text{-}empty(\ell', q') \dot{-} \ell'$.
Hence, $length(\phi(m, q, \ell', q')) > 0$ yields $next\text{-}empty(\ell', q') > \ell'$, which implies
$test(\ell', q')$.

CASE 2: $\ell = \ell' + n$.

Then by Inv. 2.1, $next\text{-}empty(\ell', q') \geq \ell' + n$, which implies $test(\ell', q')$.

IV $test(\ell', q') \rightarrow retrieve(\ell', q') = top(\phi(m, q, \ell', q'))$.

$test(\ell', q')$ implies $next\text{-}empty(\ell', q') = next\text{-}empty(S(\ell'), q') \geq S(\ell')$.
So $q'[\ell'..next\text{-}empty(\ell', q')\rangle = in(retrieve(\ell', q'), q'[S(\ell')..next\text{-}empty(\ell', q')\rangle)$.
Hence, $top(\phi(m, q, \ell', q')) = retrieve(\ell', q')$.

V.1 $m < \ell + n \rightarrow \phi(S(m), in(d, m, q), \ell', q') = append(d, \phi(m, q, \ell', q'))$.

$$q'[\ell'..next\text{-}empty(\ell', q')\rangle + \!\!+ in(d, m, q)[next\text{-}empty(\ell', q')..S(m)\rangle$$
$$= q'[\ell'..next\text{-}empty(\ell', q')\rangle + \!\!+ append(d, q[next\text{-}empty(\ell', q')..m\rangle)$$
$$= append(d, q'[\ell'..next\text{-}empty(\ell', q')\rangle + \!\!+ q[next\text{-}empty(\ell', q')..m\rangle)$$

V.2 $test(\ell', q') \rightarrow \phi(m, q, S(\ell'), remove(\ell', q')) = tail(\phi(m, q, \ell', q'))$.

$test(\ell', q')$ implies $next\text{-}empty(\ell', q') = next\text{-}empty(S(\ell'), q')$. Hence,

$$remove(\ell', q')[S(\ell')..next\text{-}empty(S(\ell'), remove(\ell', q'))\rangle$$
$$+ \!\!+ q[next\text{-}empty(S(\ell'), remove(\ell', q'))..m\rangle$$
$$= remove(\ell', q')[S(\ell')..next\text{-}empty(S(\ell'), q')\rangle + \!\!+ q[next\text{-}empty(S(\ell'), q')..m\rangle$$
$$= remove(\ell', q')[S(\ell')..next\text{-}empty(\ell', q')\rangle + \!\!+ q[next\text{-}empty(\ell', q')..m\rangle$$
$$= q'[S(\ell')..next\text{-}empty(\ell', q')\rangle + \!\!+ q[next\text{-}empty(\ell', q')..m\rangle$$
$$= tail(q'[\ell'..next\text{-}empty(\ell', q')\rangle + \!\!+ q[next\text{-}empty(\ell', q')..m\rangle)$$

7.3   Correctness of the Sliding Window Protocol

Finally, we can prove Theorem 1.

Proof.

$$\tau_I(\partial_{\mathcal{H}}(\mathbf{S}(0, 0, []) \parallel \mathbf{R}(0, []) \parallel \mathbf{K} \parallel \mathbf{L}))$$
$$\underline{\leftrightarrow} \ \tau_{\{c,j\}}(\mathbf{N}_{mod}(0, 0, [], 0, [], 0, d_0, 0, 0, 0)) \quad \text{(Thm. 2)}$$
$$\underline{\leftrightarrow} \ \tau_{\{c,j\}}(\mathbf{N}_{nonmod}(0, 0, [], 0, [], 0, d_0, 0, 0, 0)) \quad \text{(Thm. 3)}$$
$$\underline{\leftrightarrow}_b \ \mathbf{Z}(\langle\rangle) \quad \text{(Thm. 4)}$$

8   Related Work

Sliding window protocols have attracted considerable interest from the formal verification community. In this section we present an overview. Many of these verifications deal with unbounded sequence numbers, in which case modulo arithmetic is avoided, or with a fixed finite window size. The papers that do treat arbitrary finite window sizes mostly restrict to safety properties.

Infinite window size. Stenning [41] studied a SWP with unbounded sequence numbers and an infinite window size, in which messages can be lost, duplicated or reordered. A timeout mechanism is used to trigger retransmission. Stenning gave informal manual proofs of some safety properties. Knuth [26] examined more general principles behind Stenning's protocol, and manually verified some safety properties. Hailpern [20] used temporal logic to formulate safety and liveness properties for Stenning's protocol, and established their validity by informal reasoning. Jonsson [23] also verified both safety and liveness properties of the protocol, using temporal logic and a manual compositional verification technique.

Fixed finite window size. Richier *et al.* [34] specified a SWP in a process algebra based language Estelle/R, and verified safety properties for window size up to eight using the model checker Xesar. Madelaine and Vergamini [29] specified a SWP in Lotos, with the help of the simulation environment Lite, and proved some safety properties for window size six. Holzmann [21, 22] used the Spin model checker to verify both safety and liveness properties of a SWP with sequence numbers up to five. Kaivola [25] verified safety and liveness properties using model checking for a SWP with window size up to seven. Godefroid and Long [11] specified a full duplex SWP in a guarded command language, and verified the protocol for window size two using a model checker based on Queue BDDs. Stahl *et al.* [40] used a combination of abstraction, data independence, compositional reasoning and model checking to verify safety and liveness properties for a SWP with window size up to sixteen. The protocol was specified in Promela, the input language for the Spin model checker. Smith and Klarlund [38] specified a SWP in the high-level language IOA, and used the theorem prover MONA to verify a safety property for unbounded sequence numbers with window size up to 256. Latvala [27] modeled a SWP using Colored Petri nets. A liveness property was model checked with fairness constraints for window size up to eleven.

Arbitrary finite window size. Cardell-Oliver [5] specified a SWP using higher order logic, and manually proved and mechanically checked safety properties using HOL. (Van de Snepscheut [39] noted that what Cardell-Oliver claims to be a liveness property is in fact a safety property.) Schoone [37] manually proved safety properties for several SWPs using assertional verification. Van de Snepscheut [39] gave a correctness proof of a SWP as a sequence of correctness preserving transformations of a sequential program. Paliwoda and Sanders [32] specified a reduced version of what they call a SWP (but which is in fact very similar to the bakery protocol from [14]) in the process algebra CSP, and verified a safety property modulo trace semantics. Röckl and Esparza [35] verified the correctness of this bakery protocol modulo weak bisimulation using Isabelle/HOL, by explicitly checking a bisimulation relation. Jonsson and Nilsson [24] used an automated reachability analysis to verify safety properties for a SWP with arbitrary sending window size and receiving window size one. Rusu [36] used the theorem prover PVS to verify both safety and liveness properties for a SWP with unbounded sequence numbers. Chkliaev *et al.* [7] used a timed state machine in PVS to specify a SWP in which messages can be lost, duplicated or reordered, and proved some safety properties with the mechanical support of PVS.

References

1. J.A. Bergstra and J.W. Klop. Process algebra for synchronous communication. *Information and Control*, 60(1/3):109–137, 1984.
2. M.A. Bezem and J.F. Groote. Invariants in process algebra with data. In *Proc. CONCUR'94*, LNCS 836, pp. 401–416. Springer, 1994.
3. M.A. Bezem and J.F. Groote. A correctness proof of a one bit sliding window protocol in µCRL. *The Computer Journal*, 37(4):289–307, 1994.
4. J.J. Brunekreef. Sliding window protocols. In S. Mauw and G. Veltink, eds, *Algebraic Specification of Protocols*. Cambridge Tracts in Theoretical Computer Science 36, pp. 71–112. Cambridge University Press, 1993.
5. R. Cardell-Oliver. Using higher order logic for modelling real-time protocols. In *Proc. TAPSOFT'91*, LNCS 494, pp. 259–282. Springer, 1991.
6. V.G. Cerf and R.E. Kahn. A protocol for packet network intercommunication. *IEEE Transactions on Communications*, COM-22:637–648, 1974.
7. D. Chkliaev, J. Hooman, and E. de Vink. Verification and improvement of the sliding window protocol. In *TACAS'03*, LNCS 2619, pp. 113–127. Springer, 2003.
8. W.J. Fokkink, J.F. Groote, J. Pang, B. Badban, and J.C. van de Pol. Verifying a sliding window protocol in µCRL. Technical Report SEN-R0308, CWI, 2003.
9. W.J. Fokkink and J. Pang. Cones and foci for protocol verification revisited. In *Proc. FOSSACS'03*, LNCS 2620, pp. 267–281. Springer, 2003.
10. R.J. van Glabbeek and W.P. Weijland. Branching time and abstraction in bisimulation semantics. *Journal of the ACM*, 43(3):555–600, 1996.
11. P. Godefroid and D.E. Long. Symbolic protocol verification with Queue BDDs. *Formal Methods and System Design*, 14(3):257–271, 1999.
12. R.A. Groenveld. Verification of a sliding window protocol by means of process algebra. Report P8701, University of Amsterdam, 1987.
13. J.F. Groote. *Process Algebra and Structured Operational Semantics*. PhD thesis, University of Amsterdam, 1991.
14. J.F. Groote and H.P. Korver. Correctness proof of the bakery protocol in µCRL. In *Proc. ACP'94*, Workshops in Computing, pp. 63–86. Springer, 1995.
15. J.F. Groote and A. Ponse. Proof theory for µCRL: A language for processes with data. In *Proc. SoSL'93*, Workshops in Computing, pp. 232–251. Springer, 1994.
16. J.F. Groote and A. Ponse. Syntax and semantics of µCRL. In *Proc. ACP'94*, Workshops in Computing, pp. 26–62. Springer, 1995.
17. J.F. Groote, A. Ponse, and Y.S. Usenko. Linearization of parallel pCRL. *Journal of Logic and Algebraic Programming*, 48(1/2):39–72, 2001.
18. J.F. Groote and M. Reniers. Algebraic process verification. In *Handbook of Process Algebra*, pp. 1151–1208. Elsevier, 2001.
19. J.F. Groote and J. Springintveld. Focus points and convergent process operators: A proof strategy for protocol verification. *Journal of Logic and Algebraic Programming*, 49(1/2):31–60, 2001.
20. B.T. Hailpern. *Verifying Concurrent Processes Using Temporal Logic*. LNCS 129, Springer, 1982.
21. G.J. Holzmann. *Design and Validation of Computer Protocols*. Prentice Hall, 1991.
22. G.J. Holzmann. The model checker Spin. *IEEE Transactions on Software Engineering*, 23(5):279-295, 1997.
23. B. Jonsson. *Compositional Verification of Distributed Systems*. PhD thesis, Uppsala University, 1987.

24. B. Jonsson and M. Nilsson. Transitive closures of regular relations for verifying infinite-state systems. In *TACAS'00*, LNCS 1785, pp. 220–234. Springer, 2000
25. R. Kaivola. Using compositional preorders in the verification of sliding window protocol. In *Proc. CAV'97*, LNCS 1254, pp. 48–59. Springer, 1997.
26. D.E. Knuth. Verification of link-level protocols. *BIT*, 21:21–36, 1981.
27. T. Latvala. Model checking LTL properties of high-level Petri nets with fairness constraints. In *Proc. APN'01*, LNCS 2075, pp. 242–262. Springer, 2001.
28. J. Loeckx, H.-D. Ehrich, and M. Wolf. *Specification of Abstract Data Types*. Wiley/Teubner, 1996.
29. E. Madelaine and D. Vergamini. Specification and verification of a sliding window protocol in Lotos. In *Proc. FORTE'91*, IFIP Transactions, pp. 495-510. North-Holland, 1991.
30. S. Mauw and G.J. Veltink. A process specification formalism. *Fundamenta Informaticae*, 13(2):85–139, 1990.
31. A. Middeldorp. Specification of a sliding window protocol within the framework of process algebra. Report FVI 86-19, University of Amsterdam, 1986.
32. K. Paliwoda and J.W. Sanders. An incremental specification of the sliding-window protocol. *Distributed Computing*, 5:83–94, 1991.
33. D.M.R. Park. Concurrency and automata on infinite sequences. In *Proc. 5th GI Conference*, LNCS 104, pp. 167–183. Springer, 1981.
34. J.L. Richier, C. Rodriguez, J. Sifakis, and J. Voiron. Verification in Xesar of the sliding window protocol. In *Proc. PSTV'87*, pp. 235–248. North-Holland, 1987.
35. C. Röckl and J. Esparza. Proof-checking protocols using bisimulations. In *Proc. CONCUR'99*, LNCS 1664, pp. 525–540. Springer, 1999.
36. V. Rusu. Verifying a sliding-window protocol using PVS. In *Proc. FORTE'01*, Conference Proceedings 197, pp. 251-268. Kluwer, 2001.
37. A.A. Schoone. *Assertional Verification in Distributed Computing*. PhD thesis, Utrecht University, 1991.
38. M.A. Smith and N. Klarlund. Verification of a sliding window protocol using IOA and MONA. In *Proc. FORTE/PSTV'00*, pp. 19–34. Kluwer, 2000.
39. J.L.A. van de Snepscheut. The sliding window protocol revisited. *Formal Aspects of Computing*, 7(1):3–17, 1995.
40. K. Stahl, K. Baukus, Y. Lakhnech, and M. Steffen. Divide, abstract, and model-check. In *Proc. SPIN'99*, LNCS 1680, pp. 57–76. Springer, 1999.
41. N.V. Stenning. A data transfer protocol. *Computer Networks*, 1(2):99–110, 1976.
42. A.S. Tanenbaum. *Computer Networks*. Prentice Hall, 1981.
43. F.W. Vaandrager. Verification of two communication protocols by means of process algebra. Report CS-R8608, CWI, Amsterdam, 1986.
44. J.J. van Wamel. A study of a one bit sliding window protocol in ACP. Report P9212, University of Amsterdam, 1992.

State Space Reduction
for Process Algebra Specifications

Hubert Garavel and Wendelin Serwe

INRIA Rhône-Alpes / VASY
655, avenue de l'Europe
F-38330 Montbonnot St Martin, France
{Hubert.Garavel,Wendelin.Serwe}@inria.fr

Abstract. Data-flow analysis to identify "dead" variables and reset them to an "undefined" value is an effective technique for fighting state explosion in the enumerative verification of concurrent systems. Although this technique is well-adapted to imperative languages, it is not directly applicable to value-passing process algebras, in which variables cannot be reset explicitly due to the single-assignment constraints of the functional programming style. This paper addresses this problem by performing data-flow analysis on an intermediate model (Petri nets extended with state variables) into which process algebra specifications can be translated automatically. It also addresses important issues, such as avoiding the introduction of useless reset operations and handling shared read-only variables that children processes inherit from their parents.

1 Introduction

We consider the verification of concurrent systems using *enumerative* (or *explicit state*) techniques, which consist in enumerating all the system states reachable from the initial state.

Among the various approaches to avoid state explosion, it has been known for long (e.g. [11]) that a significant reduction of the state space can be achieved by *resetting* state variables as soon as their values are no longer needed. This avoids to distinguish between states that only differ by the values of so-called *dead variables*, i.e., variables that will no longer be used in the future before they are assigned again. Resetting these variables, as soon as they become useless, to some "undefined" value (usually, a pattern of 0-bits) allows states that would otherwise differ to be considered as identical.

When concurrent systems are described using an imperative language with explicit assignments, it is possible to reset variables by inserting zero-assignments manually in the source program (e.g. [11]). Some languages even provide a dedicated instruction for resetting variables (e.g. [14, §6]). Despite its apparent simplicity, this approach proves to be tedious and error-prone, and it obscures the source program with verification artefacts. Both its correctness and efficiency critically depend on the specifier's skills (resets have to be inserted at all the right places and only these).

C. Rattray et al. (Eds.): AMAST 2004, LNCS 3116, pp. 164–180, 2004.

Moreover, this approach does not apply to value-passing process algebras (i.e., process algebras with data values such as CCS, CSP, LOTOS [13], μCRL, etc.), which use a functional programming style in which variables are initialised only once and cannot be reassigned (thus, reset) later.

This paper addresses these two problems by presenting a general method, which is applicable to process algebras and which allows variables to be reset automatically, in a fully transparent way for the specifier. This method proceeds in two steps.

In a first step, process algebra specifications are translated automatically into an intermediate model with an imperative semantics. This approach was first proposed in [8, 10], which proposed a so-called *network* model consisting of a Petri net extended with state variables, the values of which are consulted and modified when the transitions are executed. This network model is used in the CÆSAR compiler for LOTOS (CÆSAR is distributed as part of the widespread CADP verification toolbox [9]). This paper presents the most recent version of the network model, which adds to the model of [8, 10] the enhancements introduced since 1990 in order to allow state space reductions based on transition compaction and to support the EXEC/CÆSAR framework for rapid prototyping of LOTOS specifications. We believe that this network model is sufficiently general to be used for other process algebras than LOTOS.

In a second step, resets are introduced, not at the source level (process algebraic specifications), but in the intermediate model, by attaching the resets to the transitions of the network.

Various techniques can be used to determine automatically which variables can be reset by which transitions. A simple approach consists in resetting all the variables of a process as soon as this process terminates. This approach was implemented in CÆSAR 4.3 (January 1992) and gives significant reductions[1] for terminating processes (especially at the points corresponding to the sequential composition ("">>") and disabling ("[>") operators of LOTOS, which are detected by analysing the structure of the network model), but not for cyclic (i.e., non-terminating) processes. The XMC model checker uses a similar approach [6], with two minor differences: dead variables are determined by analysing the sequential composition of processes at the source level and are removed from the representation of the state instead of being reset[2].

A more sophisticated approach was studied in 1992–1993 by the first author and one of his MSc students [7] in order to introduce variable resets everywhere it would be possible, including in cyclic processes. A key idea in [7] was the computation of variable resets by means of classical data-flow analysis techniques (precisely, dead variable analysis), such as those used in optimising compilers for sequential languages. An experimental version of CÆSAR implementing this idea

---

[1] For the "rel/REL" reliable atomic multicast protocol, CÆSAR 4.3 generated a state space of 126,223 states and 428,766 transitions in 30 minutes on a DEC Station 5000 with 24 MB RAM, while CÆSAR 4.2 would generate a state space of 679,450 states and 1,952,843 transitions in 9 hours on the same machine.

[2] See the concerns expressed in [12] about the poor efficiency of such a variable-length state representation scheme.

was developed in 1993. Although it gave significant state space reductions, it also happened to produce incorrect results on certain examples, which prevented it from being integrated in the official releases of CÆSAR. The reason for these errors was unknown at that time, but is now understood and addressed in this paper.

The use of data-flow analysis for resetting dead variables was later mentioned in [12] and formalised in [3, 4], the main point of which is the proof that reduction based on dead variable analysis preserves strong bisimulation. Compared to [7], [3, 4] target at the SDL language rather than the LOTOS process algebra, and, instead of the network model, consider a set of communicating automata with state variables that are consulted and assigned by the automata transitions. The main differences between the model of [3, 4] and the network model are twofold.

As regards system architecture, the network model allows concurrent processes to be nested one in another at an arbitrary depth; this is needed for a compositional translation of process algebra specifications in which parallel and sequential composition operators are intertwined arbitrarily – such as the LOTOS behaviour "$B_1 >> (B_2 | | | B_3) >> B_4$" expressing that the execution of process B_1 is followed by the concurrent execution of two processes B_2 and B_3, which, upon termination of both, will be followed by the execution of process B_4. On the contrary, the model of [3, 4] lacks any form of process hierarchy by allowing only a "flat" collection of communicating automata, all activated in the initial state.

As regards interprocess communications, the network model implements the Hoare-style rendezvous mechanism used in process algebras by synchronised Petri net transitions, which allow data exchanges between processes; additionally, concurrent processes may share variables inherited from their parent process(es) – as in the LOTOS behaviour "$G?X:S;(B_1 | | | B_2)$", in which both processes B_1 and B_2 can use variable X of sort S, whose value has been set in their parent process; these shared variables are read-only, in the sense that children processes cannot modify them. On the contrary, the model of [3, 4] relies on FIFO message queues and shared variables that can be arbitrarily read/written by all the processes. As regards shared variables, [3, 4] propose an approach in which variable resets are computed partly at compile-time (when analysing each communicating automaton separately) and partly at run-time (when generating all reachable states of the product automaton). Although it is difficult to figure out how this approach can be implemented in practice – since the authors stand far from algorithmic concerns and since the most recent versions[3] of their IF tool set [5] do not reset shared variables actually – we believe that the communicating automata model used by [3, 4] is not sufficient in itself to express resets of shared variables, so that some extra information (yet to be specified) must be passed from compile-time to run-time. In comparison, the approach presented in this paper can be performed entirely at compile-time and requires no addition to the network model.

This paper is organised as follows. Section 2 presents the network model and its operational semantics. Sections 3 and 4 respectively present the local and

---

[3] Namely, IF 1.0 (dated November 2003) and IF 2.0 (dated March 2003).

global data-flow analyses of [7] for determination of variable resets. Section 5 deals with the particular case of inherited variables, which need careful attention to avoid semantic problems caused by a "naive" insertion of resets. Section 6 reports about experimental results and Sect. 7 gives concluding remarks.

2   Presentation of the Network Model

The network model presented here is based on the definitions of [8, 10], the essential characteristics of which are retained (namely, the Petri net structure with state variables); but it also contains some more recent extensions that proved to be useful.

Formally, a network is a tuple $\langle \mathcal{Q}, \mathcal{Q}_0, \mathcal{U}, \mathcal{T}, \mathcal{G}, \mathcal{X}, \mathcal{S}, \mathcal{F} \rangle$, the components of which will be presented progressively, so as to avoid forward references. We will use the following convention consistently: elements of set $\mathcal{Q}$ (*resp.* $\mathcal{U}$, $\mathcal{T}$, $\mathcal{G}$, $\mathcal{X}$, $\mathcal{S}$, $\mathcal{F}$) are noted by the corresponding capital letter, e.g. Q, Q_0, Q_1, Q', Q'', etc.

Sorts, Functions, and Variables. In the above definition of a network, $\mathcal{S}$ denotes a finite set of *sorts* (i.e., data types), $\mathcal{F}$ denotes a finite set of *functions*, and $\mathcal{X}$ denotes a finite set of *(state) variables*. We note $domain(S)$ the (possibly infinite) set of *ground values* of sort S. Functions take (zero, one, or many) typed arguments and return a typed result. Variables also are typed.

Contexts. To represent the memory containing state variables, we define a *context* C as a (partial) function mapping each variable of $\mathcal{X}$ either to its ground value or to the undefined value, noted "$\perp$". We need 5 operations to handle contexts. For contexts C_1 and C_2, and variables $X_0, \ldots, X_n$, we define the contexts:
- $\{\}$: $X \mapsto \perp$ (i.e., the empty context)
- $\{X_0 \mapsto v\}$: $X \mapsto$ if $X = X_0$ then v else $\perp$
- $C_1 \ominus \{X_0, \ldots, X_n\}$: $X \mapsto$ if $X \in \{X_0, \ldots, X_n\}$ then $\perp$ else $C_1(X)$
- $C_1 \oslash C_2$: $X \mapsto$ if $C_2(X) \neq \perp$ then $C_2(X)$ else $C_1(X)$
- $C_1 \oplus C_2$: $X \mapsto$ if $C_2(X) \neq \perp$ then $C_2(X)$ else $C_1(X)$
  We only use $\oplus$ on "disjoint" contexts, i.e., when $(C_1(X) = \perp) \vee (C_2(X) = \perp)$.

Value Expressions. A *value expression* is a term built using variables and functions: $V ::= X \mid F(V_1, \ldots, V_{n \geq 0})$. We note $eval(C, V)$ the (unique) ground value obtained by evaluating value expression V in context C (after substituting variables with their ground values given by C and applying functions). Because the network is generated from a LOTOS specification that is correctly typed and well-defined (i.e., each variable is initialised before used), evaluating a value expression never fails due to type errors or undefined variables.

Offers. An *offer* is a term of the form: $O ::= !V \mid ?X : S \mid O_1 \ldots O_{n \geq 0}$, meaning that an offer is a (possibly empty) sequence of *emissions* (noted "!") and/or *receptions* (noted "?"). We define a relation noted "$[C, O] \xrightarrow{o} [C', v_1 \ldots v_n]$" expressing that offer O evaluated in context C yields a (possibly empty) list of ground values $v_1 \ldots v_n$ and a new context C' (C' reflects that $?X : S$ binds X to the received value(s)). For given pair $[C, O]$ there might be one or several pairs $[C', v_1 \ldots v_n]$ such that $[C, O] \xrightarrow{o} [C', v_1 \ldots v_n]$, since a reception $?X : S$ generates as many pairs as there are ground values in $domain(S)$.

$$\frac{v = eval(C,V)}{[C, !V] \overset{o}{\to} [\{\}, v]} \quad \frac{v \in domain(S)}{[C, ?X : S] \overset{o}{\to} [\{X \mapsto v\}, v]} \quad \frac{\forall i \in \{1, \dots, n\} \; [C, O_i] \overset{o}{\to} [C_i, v_i]}{[C, O_1 \dots O_n] \overset{o}{\to} [\bigoplus_{i=1}^{n} C_i, v_1 \dots v_n]}$$

Actions. Actions are terms of the form:

$$
\begin{array}{lll}
A ::= & \textbf{none} & \textit{(empty action)} \\
  | & \textbf{when } V & \textit{(condition)} \\
  | & \textbf{for } X \textbf{ among } S & \textit{(iteration)} \\
  | & X_0, \dots, X_{n \geq 0} \texttt{:=} V_0, \dots, V_n & \textit{(vectorial assignment)} \\
  | & \textbf{reset } X_0, \dots, X_{n \geq 0} & \textit{(variable reset)} \\
  | & A_1 \,; A_2 & \textit{(sequential composition)} \\
  | & A_1 \& A_2 & \textit{(collateral composition)}
\end{array}
$$

We define a relation noted "$[C, A] \overset{c}{\to} C'$" expressing that successful execution of action A in context C yields a new context C'. For given pair $[C, A]$ there might be zero, one, or several C' such that $[C, A] \overset{c}{\to} C'$, since a "**when** V" condition may block the execution if V evaluates to false, whereas a "**for** X **among** S" iteration triggers as many executions as there are ground values in $domain(S)$.

$$\frac{}{[C, \textbf{none}] \overset{c}{\to} C} \quad \frac{eval(C,V) = \textsf{true}}{[C, \textbf{when } V] \overset{c}{\to} C} \quad \frac{v \in domain(S) \quad [C, X \texttt{:=} v] \overset{c}{\to} C'}{[C, \textbf{for } X \textbf{ among } S] \overset{c}{\to} C'}$$

$$\frac{C' = C \oslash \bigoplus_{i=0}^{n} \{X_i \mapsto eval(C, V_i)\}}{[C, X_0, \dots, X_n \texttt{:=} V_0, \dots, V_n] \overset{c}{\to} C'} \quad \frac{C' = C \ominus \{X_0, \dots, X_n\}}{[C, \textbf{reset } X_0, \dots, X_n] \overset{c}{\to} C'}$$

$$\frac{[C, A_1] \overset{c}{\to} C' \quad [C', A_2] \overset{c}{\to} C''}{[C, A_1 \,; A_2] \overset{c}{\to} C''} \quad \frac{[C, A_1 \,; A_2] \overset{c}{\to} C'' \quad [C, A_2 \,; A_1] \overset{c}{\to} C''}{[C, A_1 \& A_2] \overset{c}{\to} C''}$$

Gates. In the above definition of a network, $\mathcal{G}$ denotes a finite set of *gates* (i.e., names for communication points). There are two special gates: "τ", the usual notation for the internal steps of a process, and "ε", a powerful artefact (see [8, 10]) allowing the compositional construction of networks for a large class of LOTOS behaviours such as "$B_1 [] (B_2 ||| B_3)$". Although ε deserves a special semantic treatment, this has no influence on the approach proposed in this paper; thus, we do not distinguish ε from "ordinary" gates.

Places and Transitions. In the above definition of a network, $\mathcal{Q}$ denotes a finite set of *places*, $Q_0 \in \mathcal{Q}$ is the *initial place* of the network, and $\mathcal{T}$ denotes a finite set of *transitions*. Each transition T is a tuple $\langle Q_i, Q_o, A, G, O, W, R \rangle$, where $Q_i \subseteq \mathcal{Q}$ is a set of *input places* (we note $in(T) \overset{\Delta}{=} Q_i$), $Q_o \subseteq \mathcal{Q}$ is a set of *output places* (we note $out(T) \overset{\Delta}{=} Q_o$), A is an action, G is a gate, O is a (possibly empty) offer, W is a *when-guard* (i.e., a restricted form of action constructed only with "**none**", "**when**", "$;$", and "$\&$"), and R is a *reaction* (i.e., a restricted form of action constructed only with "**none**", "$:=$", "**reset**", "$;$", and "$\&$").

Markings. As regards the firing of transitions, the network model obeys the standard rules of Petri nets with the particularity that it is one-safe, i.e., each place may contain at most one token – this is due to the so-called *static control contraints* [2, 8, 10], which only allow a statically bounded dynamic creation of processes (for instance, the following behaviour "$B_1 >> (B_2 ||| B_3) >> B_4$" is permitted, whereas recursion through parallel composition is prohibited).

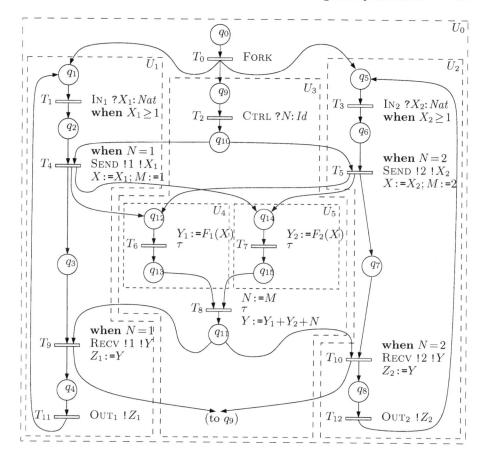

Fig. 1. Example of a network

Therefore, we can define a *marking* M as a subset of the places of the network (i.e., $M \subseteq \mathcal{Q}$). We define the *initial marking* $M_0 \triangleq \{Q_0\}$, which expresses that, initially, only the initial place of the network has one token. We define a relation noted "$[M, T] \overset{m}{\rightarrow} M'$" meaning that transition T can be fired from marking M, leading to a new marking M'. Classically, $[M, T] \overset{m}{\rightarrow} M'$ holds iff $in(T) \subseteq M$ (i.e., all input places of T have a token) and $M' = (M \setminus in(T)) \cup out(T)$ (i.e., tokens move from input to output places).

Units. Contrary to standard Petri nets, which consists of "flat" sets of places and transitions, the places of a network are properly structured using a tree-shaped hierarchy of *units*. The set of units, which is finite, is noted $\mathcal{U}$ in the above definition of a network. To each unit U is associated a non-empty, finite set of places, called the *proper places of* U and noted $places(U)$, such that all sets of proper places $\{places(U) \mid U \in \mathcal{U}\}$ form a partition of $\mathcal{Q}$. Although units play no part in the transition relation "$[M, T] \overset{m}{\rightarrow} M'$" between markings, they satisfy an important invariant: for each marking M reachable from the initial marking M_0 and for each unit U, one has $card(M \cap places(U)) \leq 1$, i.e., there is at most one token among the proper places of U, meaning that each unit models a (possibly inactive) sequential behaviour. This invariant serves both for correctness proofs and compact memory representation of markings.

Units can be nested recursively: each unit U may contain zero, one, or several units, called the *sub-units* of U; this is used to encapsulate sequential or concurrent sub-behaviours. There exists a *root unit* containing all other units. We note "$U' \sqsubseteq U$" the fact that U' is equal to U or transitively contained in U: this relation is a complete partial order, the maximum of which is the root unit. We note $places^*(U) = \bigcup_{U' \sqsubseteq U} places(U')$ the set of places transitively contained in U. For some marking M reachable from M_0, one may have $card\big(M \cap places^*(U)\big) > 1$ in case of concurrency between the sub-units of U; yet, for all units U and $U' \sqsubseteq U$, one has $\big(M \cap places(U) = \emptyset\big) \vee \big(M \cap places(U') = \emptyset\big)$, meaning that the proper places of a unit are mutually exclusive with those of its sub-units.

Variables may be *global*, or *local* to a given unit. We note $unit(X)$ the unit to which variable X is attached (global variables are attached to the root unit). A variable X is said to be *inherited* in all sub-units of $unit(X)$. In a first approximation, we will say that variable X is *shared* between two units U_1 and U_2 iff $\big(U_1 \sqsubseteq unit(X)\big) \wedge \big(U_2 \sqsubseteq unit(X)\big) \wedge (U_1 \not\sqsubseteq U_2) \wedge (U_2 \not\sqsubseteq U_1)$; this definition will be sufficient until Sect. 5, where a refined definition will be given.

Labelled Transition Systems. Finally, the operational semantics of the network model is defined as a *Labelled Transition System* (LTS), i.e., a tuple $\langle \Sigma, \sigma_0, \mathcal{L}, \to \rangle$ where Σ is a set of *states*, $\sigma_0 \in \Sigma$ is the *initial state*, $\mathcal{L}$ is the set of *labels* and $\to \subseteq \Sigma \times \mathcal{L} \times \Sigma$ is the *transition relation*.

The LTS is constructed as follows. Each state of Σ consists of a pair $\langle M, C \rangle$, with M a marking and C a context. The initial state σ_0 is the pair $\langle M_0, \{\} \rangle$, i.e., one token is in the initial place and all variables are undefined initially. Each label of $\mathcal{L}$ consists of a list $G\ v_1 \ldots v_n$, with G a gate and $v_1 \ldots v_n$ a (possibly empty) list of ground values resulting from the evaluation of an offer. A transition (σ_1, L, σ_2) belongs to the "$\to$" relation, a fact which we note "$\sigma_1 \xrightarrow{L} \sigma_2$", iff

$$\frac{[M, T] \xrightarrow{m} M' \quad [C, A] \xrightarrow{c} C' \quad [C', O] \xrightarrow{o} [C'', v_1 \ldots v_n] \quad [C'', (W\,;R)] \xrightarrow{c} C'''}{\langle M, C \rangle \xrightarrow{G\ v_1 \ldots v_n} \langle M', C''' \rangle}$$

The above definition expresses that firing a transition involves several steps, the execution of each must succeed: the action is executed first, then the offer is evaluated, then the when-guard is checked, and the reaction is executed finally. In fact, the actual definition of the transition relation is more complex because there are rules to eliminate ε-transitions from the LTS; as mentioned before, we do not detail these rules here.

Example. Figure 1 gives an example of a network. According to Petri Net graphical conventions, places and transitions are represented by circles and rectangles. Dashed boxes are used to represent units. For each transition, the corresponding action, gate and offer, when-guard, and reaction are displayed (in that order) from top to bottom on the right; we omit every action, when-guard, or reaction that is equal to **none**. The variables attached to U_1 are X_1 and Z_1; those attached to U_2 are X_2 and Z_2; those attached to U_3 are M, N, X, Y, Y_1, and Y_2. Variable X inherited from U_3 is shared between U_4 and U_5.

3   Local Data-Flow Analysis

In the network model, transitions constitute the equivalent of the "basic blocks" used for data-flow analysis of sequential programs. We first analyse the flow of data within each transition taken individually to characterise which variables are accessed by this transition. Our definitions are based on [7] with adaptations to take into account the latest extensions of the network model and to handle networks that already contain "**reset**" actions. We define the following sets by structural induction over the syntax of value expressions, offers, and actions:

- $use_v(V)$ (*resp.* $use_o(O)$, $use_a(A)$) denotes the set of variables consulted in value expression V (*resp.* offer O, action A).
- $def_o(O)$ (*resp.* $def_a(A)$) denotes the set of variables assigned a defined value by offer O (*resp.* action A).
- $und_a(A)$ denotes the set of variables assigned an undefined value (i.e., reset) by action A.
- $use\_before\_def_a(A)$ denotes the set of variables consulted by action A and possibly modified by A later (modifications, if present, should only occur after the variables have been consulted at least once).

$use_v(X) \triangleq \{X\}$ $use_v(F(V_1,\dots,V_n)) \triangleq \bigcup\limits_{i=1}^{n} use_v(V_i)$	$use_o(!V) \triangleq use_v(V)$ $use_o(?X\!:\!S) \triangleq \emptyset$ $use_o(O_1\dots O_n) \triangleq \bigcup\limits_{i=1}^{n} use_o(O_i)$
$und_a(\textbf{reset}\ X_0,\dots,X_n) \triangleq \{X_0,\dots,X_n\}$ $und_a(A_1;A_2) \triangleq (und_a(A_1) \setminus def_a(A_2)) \cup und_a(A_2)$ $und_a(A_1\&A_2) \triangleq und_a(A_1) \cup und_a(A_2)$ otherwise : $und_a(A) \triangleq \emptyset$	$def_o(!V) \triangleq \emptyset$ $def_o(?X\!:\!S) \triangleq \{X\}$ $def_o(O_1\dots O_n) \triangleq \bigcup\limits_{i=1}^{n} def_o(O_i)$
$def_a(X_0,\dots,X_n:=V_0,\dots,V_n) \triangleq \{X_0,\dots,X_n\}$ $def_a(\textbf{for}\ X\ \textbf{among}\ S) \triangleq \{X\}$ $def_a(A_1;A_2) \triangleq (def_a(A_1) \setminus und_a(A_2)) \cup def_a(A_2)$ $def_a(A_1\&A_2) \triangleq def_a(A_1) \cup def_a(A_2)$ otherwise : $def_a(A) \triangleq \emptyset$	$use_a(\textbf{when}\ V) \triangleq use_v(V)$ $use_a(X_0\dots:=V_0\dots) \triangleq \bigcup\limits_{i=0}^{n} use_v(V_i)$ $use_a(A_1;A_2) \triangleq use_a(A_1) \cup use_a(A_2)$ $use_a(A_1\&A_2) \triangleq use_a(A_1) \cup use_a(A_2)$ otherwise : $use_a(A) \triangleq \emptyset$
$use\_before\_def_a(A_1;A_2) \triangleq use\_before\_def_a(A_1) \cup (use\_before\_def_a(A_2) \setminus def_a(A_1))$ $use\_before\_def_a(A_1\&A_2) \triangleq use\_before\_def_a(A_1) \cup use\_before\_def_a(A_2)$ otherwise : $use\_before\_def_a(A) \triangleq use_a(A)$	

Finally, for a transition $T = \langle Q_i, Q_o, A, G, O, W, R \rangle$ and a variable X, we define three predicates, which will be the only local data-flow results used in subsequent analysis steps:

- $use(T, X)$ holds iff X is consulted during the execution of T.
- $def(T, X)$ holds iff X is assigned a defined value by the execution of T, i.e., if X is defined by A, O or R, and not subsequently reset.

– *use_before_def*(T, X) holds iff if X is consulted during the execution of T and possibly modified later (modification, if present, should only occur after X has been consulted at least once).

Formally:

$$use(T, X) \triangleq X \in use_a(A) \cup use_o(O) \cup use_a(W) \cup use_a(R)$$

$$def(T, X) \triangleq X \in \big((def_a(A) \cup def_o(O)) \setminus und_a(R) \big) \cup def_a(R)$$

$$use\_before\_def(T, X) \triangleq X \in \begin{pmatrix} use\_before\_def_a(A) \; \cup \; \big(use_o(O) \setminus def_a(A) \big) \; \cup \\ \big(use\_before\_def_a(W ; R) \big) \setminus \big(def_a(A) \cup def_o(O) \big) \end{pmatrix}$$

Example. For the variable N in the network of Fig. 1, we have: $use(T, N)$ for $T \in \{T_4, T_5, T_8, T_9, T_{10}\}$, $def(T, N)$ for $T \in \{T_2, T_8\}$, and $use\_before\_def(T, N)$ for $T \in \{T_4, T_5, T_9, T_{10}\}$.

4   Global Data-Flow Analysis

Based on local (intra-transition) data-flow predicates, we now perform global (inter-transition) data-flow analysis, the goal being to compute, for each transition $T = \langle Q_i, Q_o, A, G, O, W, R \rangle$ and for each variable X, a predicate $reset(T, X)$ expressing that it is possible to *reset variable X at the end of transition T* (i.e., to append "**reset** X" at the end of A if X is neither defined in O nor used in O, W, and R, or else to append "**reset** X" at the end of R). To be exact, if X is an inherited shared variable, it is not always possible to insert "**reset** X" at the end of every transition T such that $reset(T, X)$; this issue will be dealt with in Sect. 5; for now, we focus on computing $reset(T, X)$.

For sequential programs, the classical approach to global data-flow analysis (e.g. [1]) consists in constructing a *control-flow graph* on which boolean predicates will then be evaluated using fixed point computations. The vertices of the control-flow graph are usually the basic blocks connected by arcs expressing that two basic blocks can be executed in sequence. Since the control-flow graph is a data-independent abstraction, it represents a superset of the possible execution paths, i.e., some paths of the control-flow graph might not exist in actual executions of the sequential program.

A significant difference between sequential programs and our setting is that networks feature concurrency. One could devise a "true concurrency" extension of data-flow analysis by evaluating the boolean predicates, not on control-flow graphs, but directly on Petri nets. Instead, following [7], we adopt an "interleaving semantics" approach that maps concurrency onto a standard control-flow graph, on which the boolean predicates can be evaluated as usual.

To abstract away concurrency from the network model, various possibilities exist, leading to different control-flow graphs. One possibility would be to base the analysis on the graph of reachable markings of the underlying Petri net; this would be accurate but costly to compute, as state explosion might occur. Hence, we choose a stronger abstraction by defining the control-flow graph as the directed graph CFG $= \langle \mathcal{T}, \rightarrow \rangle$, the vertices of which correspond to the transitions of the network and such that there is an arc $T_1 \rightarrow T_2$ iff $out(T_1) \cap in(T_2) \neq \emptyset$.

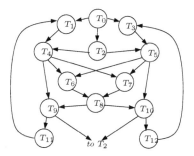

Fig. 2. CFG for Fig. 1

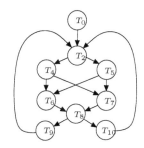

Fig. 3. CFG_N for Fig. 1

Example. The CFG corresponding to the network of Fig. 1 is shown in Fig. 2.

Instead of constructing a unique CFG valid for all variables, [7] suggests to build, for each variable X, a dedicated control-flow graph CFG_X, which is a subset of CFG containing only the execution paths relevant to X (nowadays, this would be called "slicing"). According to [7, § 4.3.3], such a restricted control-flow graph increases the algorithmic efficiency; by our accounts, it also gives more precise data-flow results.

To define CFG_X formally, we need two auxiliary definitions. Let $trans(U) \triangleq \{T \mid (in(T) \cup out(T)) \cap places^*(U) \neq \emptyset\}$ be the set of transitions with an input or an output place in unit U. Let $scope(X)$ be (an upper-approximation of) the set of places through which the data-flow for variable X passes. Following the simple, "syntactic" approximation of $scope(X)$ given in [7], we define $scope(X) \triangleq places^*(unit(X))$ as the set of all places in the unit to which X is attached.

We now define CFG_X as the directed graph $\langle \mathcal{T}_X, \rightarrow_X \rangle$ with the set of vertices $\mathcal{T}_X \triangleq trans(unit(X))$ and such that there is an arc $T_1 \rightarrow_X T_2$ iff $out(T_1) \cap in(T_2) \cap scope(X) \neq \emptyset$. For $T \in \mathcal{T}_X$, we note $succ_X(T) \triangleq \{T' \in \mathcal{T}_X \mid T \rightarrow_X T'\}$ and $pred_X(T) \triangleq \{T' \in \mathcal{T}_X \mid T' \rightarrow_X T\}$.

Example. Figure 3 shows CFG_N for the network of Fig. 1 and variable N; notice that $T_4 \rightarrow T_9$, but not $T_4 \rightarrow_N T_9$.

Following the classical definition of "live" variables (e.g. [1, pages 631–632]), we define, for $T \in \mathcal{T}_X$, the following predicate:

$$live(T, X) \triangleq \bigvee_{T' \in succ_X(T)} use\_before\_def(T', X) \vee (live(T', X) \wedge \neg def(T', X))$$

that holds iff after T it is possible, by following the arcs of CFG_X, to reach a transition T' that uses X before any modification of X. For a given X, the set $\{T \in \mathcal{T}_X \mid live(T, X)\}$ is computed as a backward least fixed point.

We could now, as in [3, 4], define $reset(T, X) \triangleq \neg live(T, X)$. Unfortunately, this simple approach inserts superfluous resets, e.g. before a variable is initialised or at places where a variable has already been reset. For this reason, one needs an additional predicate:

$$available(T, X) \triangleq def(T, X) \vee \left(\bigvee_{T' \in pred_X(T)} (live(T', X) \wedge available(T', X)) \right)$$

that holds iff T can be reached from some transition that assigns X a defined value, by following the arcs of CFG_X and ensuring that X remains alive all along the path. For a given X, the set $\{T \in \mathcal{T}_X \mid available(T, X)\}$ is computed as a forward least fixed point. [7] uses a similar definition without the $live(T', X)$ condition and, thus, only avoids resetting uninitialised variables.

Finally, we define $reset(T, X) \triangleq available(T, X) \wedge \neg live(T, X)$, expressing that a variable can be reset where it is both available and dead.

Example. For the network of Fig. 1 and variable N, we have $\{T \mid live(T, N)\} = \{T_2, T_8\}$ and $\{T \mid available(T, N)\} = \{T_2, T_4, T_5, T_8, T_9, T_{10}\}$. Thus, we can insert "**reset** N" at the end of T_4, T_5, T_9, and T_{10}. Using the definition of [3, 4], one would insert superfluous "**reset** N" at the end of T_0, T_6, and T_7. Using the definition of [7], one would insert superfluous "**reset** N" at the end of T_6 and T_7. Using CFG instead of CFG$_N$ would give $\{T \mid live(T, N)\} = \{T_0 \ldots T_5, T_8 \ldots T_{12}\}$ and $\{T \mid available(T, N)\} = \{T_1 \ldots T_5, T_8 \ldots T_{12}\}$, so that no "**reset** N" at all would be inserted.

5   Treatment of Inherited Shared Variables

Issues when Resetting Shared Variables. Experimenting with the approach of [7], we noticed that systematic insertion of a "**reset** X" at the end of every transition T such that $reset(T, X)$ could produce either incorrect results (i.e., an LTS which is not strongly bisimilar to the original specification) or run-time errors while generating the LTS (i.e., accessing a variable that has been reset).

Example. In the network of Fig. 1, there exists a fireable sequence of transitions $T_0, T_1, T_2, T_4, T_6, T_7$. Although $reset(T_6, X)$ is true, one should not reset X at the end of T_6, because X is used just after in T_7. Clearly, the problem is that T_6 and T_7 are two "concurrent" transitions sharing the same variable X. This was no problem as long as X was only read by both transitions, but as soon as one transition (here, T_6) tries to reset X, it affects the other transition (here, T_7).

So, insertion of resets turns a read-only shared variable into a read/write shared variable, possibly creating read/write conflicts as in a standard reader-writer problem. The sole difference is that resets do not provoke write/write conflicts (concurrent resets assign a variable the same undefined value).

To avoid the problem, a simple solution consists in never resetting inherited shared variables (as in the IF tool set [5]). Unfortunately, opportunities for valuable state space reduction are missed by doing so.

Example. As shown in Fig. 4 (a) and (b), the LTS generated for the LOTOS behaviour "$G?X : \mathbf{bit}; (G_1!X; \mathbf{stop} \mid\mid\mid G_2!X; \mathbf{stop})$" has 9 states if the inherited shared variable X is not reset, and only 8 states if X is reset after firing transitions $G_1!X$ and $G_2!X$ (state space reduction would be more substantial if both occurrences of "**stop**" were replaced by two complex behaviours B_1 and B_2 in which the value of X is not used). Figure 4 (c) shows the incorrect LTS obtained by resetting X to 0 after each transition $G_1!X$ and $G_2!X$.

Duplication of Variables. The deep reason behind the issues when resetting inherited shared variables is that the control-flow graphs CFG and CFG$_X$ defined in Sect. 4 are nothing but approximations. Their definitions follow the place-transition paths in the network, which has the effect of handling similarly

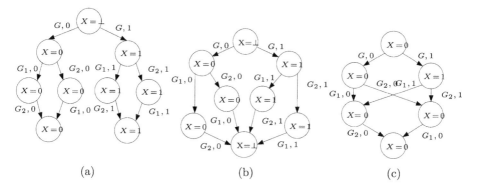

Fig. 4. LTS (a) without reset, (b) with correct resets, and (c) with incorrect resets

nondeterministic choice (i.e., a place with several outgoing transitions) and asynchronous concurrency (i.e., a transition with several output places). Indeed, both LOTOS behaviours "$G;(B_1|||B_2)$" and "$G;(B_1[]B_2)$" have the same CFG. These approximations produce compact control-flow graphs, but are only correct in absence of data dependencies (caused by inherited shared variables) between "concurrent" transitions.

To address the problem, we introduce the notion of *variable duplication*. For an inherited variable X shared between two concurrent behaviours B_1 and B_2, duplication consists in replacing, in one behaviour (say, B_2), all occurrences of X with a local copy X' initialised to X at the beginning of B_2. This new variable X' can be safely reset in B_2 without creating read/write conflicts with B_1. A proper application of duplication can remove all data dependencies between "concurrent" transitions, hence ensuring correctness of our global data-flow analysis approximations. It also enables the desired state space reductions.

Example. In the previous example, duplicating X in B_2 yields the LOTOS behaviour   "$G?X:\mathbf{bit};\mathbf{let}\ X':\mathbf{bit}{=}X\ \mathbf{in}\ (G_1!X;\mathbf{stop}|||G_2!X';\mathbf{stop})$",   in which it is possible to reset X after the $G_1!X$ transition and X' after the $G_2!X'$ transition; this precisely gives the optimal LTS shown on Fig. 4 (b). Note that it is not necessary to duplicate X in B_1.

Instead of duplicating variables at the LOTOS source level, as in the above example, we prefer duplicating them in the network model, the complexity of which has already been reduced by detecting constants, removing unused variables, identifying variables local to a transition, etc. Taking into account that concurrent processes are represented by units, we define the *duplication of a variable* X in a unit U, with $U \sqsubseteq unit(X)$ and $U \neq unit(X)$, as the operation of creating a new variable X' of the same sort as X attached to U (whereas X is attached to $unit(X)$), replacing all occurrences of X in the transitions of $trans(U)$ by X' and adding an assignment "$X':=X$" at the end of all transitions $T \in entry(U)$ such that $live(T, X)$, where $entry(U) \triangleq \{T \in trans(U) \mid in(T) \cap places^*(U) = \emptyset\}$ is the set of transitions "entering" U.

In general, several duplications may be needed to remove all read/write conflicts on a shared variable X. On the one hand, if X is shared between n concurrent behaviours, $(n-1)$ duplications of X may be necessary. On the other hand, each new variable X' duplicating X might itself be shared between concurrent sub-units, so that duplications of X' may also be required.

Concurrency Relation between Units. We now formalise the notion of "concurrent units". Ideally, two units U_i and U_j are concurrent if it exists a reachable state $\langle M, C \rangle$ in the corresponding LTS such that the two sets of places $\big(M \cap places^*(U_i)\big)$ and $\big(M \cap places^*(U_j)\big)$ are both non empty and disjoint (meaning that U_i and U_j are "separate" and simultaneously "active" in marking M). *Example.* In the LOTOS behaviour "$(B_1 | | | B_2) >> (B_3 | | | B_4)$", units U_1 and U_2 corresponding to B_1 and B_2 are concurrent, units U_3 and U_4 corresponding B_3 and B_4 also, but neither U_1 nor U_2 is concurrent with either U_3 or U_4.

Practically, to avoid enumerating all states of the LTS, we need a relation noted "$U_i \| U_j$" that is an upper-approximation of the ideal definition above, i.e., U_i and U_j concurrent implies $U_i \| U_j$. We base our definition on an abstraction function $\alpha : \mathcal{Q} \to \{1, \ldots, N\}$ (N being the number of units in the network) that maps all the proper places of each unit to the same number: $\forall Q \in places(U_i)$: $\alpha(Q) \triangleq i$. We extend α to sets of places by defining $\widehat{\alpha} : \wp(\mathcal{Q}) \to \wp(\{1, \ldots, N\})$ such that $\widehat{\alpha}(\{Q_1, \ldots, Q_n\}) \triangleq \{\alpha(Q_1), \ldots, \alpha(Q_n)\}$. We then use α and $\widehat{\alpha}$ to "quotient" the network, yielding a Petri net with N places numbered from 1 to N, with initial place $\alpha(Q_0)$ (Q_0 being the initial place of the network), and which possesses, for each transition T in the network, a corresponding transition t such that $in(t) \triangleq \widehat{\alpha}\big(in(T)\big)$ and $out(t) \triangleq \widehat{\alpha}\big(out(T)\big)$ – "self-looping" transitions such that $in(t) = out(t)$, as well as transitions identical to another one, can be removed. As the number of units is usually small compared to the number of places, one can easily generate the set $\mathcal{M}$ of all reachable markings for the quotient Petri net. Finally, we define $U_i \| U_j$ iff there exists $M \in \mathcal{M}$ such that both sets $\big(M \cap \widehat{\alpha}(places^*(U_i))\big)$ and $\big(M \cap \widehat{\alpha}(places^*(U_j))\big)$ are not empty and disjoint. Notice that $U_i \| U_j$ implies $U_i \neq U_j$, $U_i \not\sqsubseteq U_j$, and $U_j \not\sqsubseteq U_i$.

Conflicts between Units. For two units U_i and U_j such that $U_i \| U_j$, let $ancestor(U_i, U_j)$ denote the largest unit U such that $U_i \sqsubseteq U$ and $U_j \not\sqsubseteq U$ and let $link(U_i, U_j)$ denote the set of transitions connecting the ancestors of U_i and those of U_j; formally: $link(U_i, U_j) \triangleq trans\big(ancestor(U_i, U_j)\big) \cap trans\big(ancestor(U_j, U_i)\big)$.

To characterise whether two units U_i and U_j are in conflict for variable X according to given values of predicates *use* and *reset*, we define the predicate:

$$conflict(U_i, U_j, X, use, reset) \triangleq U_i \sqsubseteq unit(X) \wedge U_j \sqsubseteq unit(X) \wedge U_i \| U_j \wedge$$
$$\big(\exists T_i \in trans(U_i) \setminus link(U_i, U_j)\big) \ \big(\exists T_j \in trans(U_j) \setminus link(U_i, U_j)\big)$$
$$\big(reset(T_i, X) \wedge use(T_j, X)\big) \vee \big(reset(T_j, X) \wedge use(T_i, X)\big)$$

Intuitively, units U_i and U_j are in conflict for X if there exist two "independent" transitions T_i and T_j likely to create a read/write conflict on X. To avoid irrelevant conflicts (and thus, unnecessary duplications), one can dismiss the transitions of $link(U_i, U_j)$, i.e., the transitions linking the ancestor of U_i with that of U_j, since the potential impact of these transitions on the data-flow for

---

1. compute the relation $U_i \| U_j$ and $link(U_i, U_j)$ (cf. Sect. 5)
2. $VARS := \mathcal{X}$
3. **while** $VARS \neq \emptyset$ **do**
4. **begin**
5. $X := one\_of(VARS)$
6. $VARS := VARS \setminus \{X\}$
7.   **repeat**
8.   **forall** $T \in trans(unit(X))$ **do**
9.   compute $use(T, X)$, $def(T, X)$, and $use\_before\_def(T, X)$ (cf Sect. 3)
10.   **forall** $T \in trans(unit(X))$ **do**
11.   compute $reset(T, X)$ (cf Sect. 4)
12.   compute $conflict(U_i, U_j, X, use, reset)$ and UCG_X (cf Sect. 5)
13.   compute $U := best\_of(UCG_X)$ (cf. Sect. 5)
14.   **if** $U \neq \bot$ **then**
15.   **begin**
16.   duplicate X in U by creating a new variable X'
17. $VARS := VARS \cup \{X'\}$
18.   **end**
19.   **until** $U = \bot$
20. $--$ at this point, there is no more conflict on X
21.   **forall** $T \in trans(unit(X))$ **such_that** $reset(T, X)$ **do**
22.   insert "**reset** X" at the end of T (cf. Sect. 4)
23. **end**

Fig. 5. Complete algorithm

X has already been considered when constructing CFG_X and computing $reset$ – based on the observation that $link(U_i, U_j) \subseteq trans(unit(X))$.

We then define, for given values of predicates use and $reset$, the *unit conflict graph for variable* X, noted UCG_X, as the undirected graph whose vertices are the units of $unit(X)$ and such that there is an edge between U_i and U_j iff $conflict(U_i, U_j, X, use, reset)$.

Complete Algorithm. The algorithm shown in Fig. 5 operates as follows. *VARS* denotes the set of all variables in the network, which might be extended progressively with new, duplicated variables. All the variables X in *VARS* are processed individually, one at a time, in an unspecified order. For a given X, the algorithm performs local and global data-flow analysis, then builds UCG_X. If UCG_X has no edge, X needs not be duplicated and "**reset** X" can be inserted at the end of every transition $T \in trans(unit(X))$ such that $reset(T, X)$. Otherwise, X must be duplicated in one or several units to solve read/write conflicts. This adds to *VARS* one or several new variables X', which will be later analysed as if they were genuine variables of the network (i.e., to insert resets for X' and/or to solve read/write conflicts that may still exist for X'). Everytime a new variable X' is created to duplicate X, data-flow analysis for X and UCG_X are recomputed, as duplication modifies the network by removing occurrences (definitions, uses, and resets) of X and by adding new assignments of the form $X' := X$.

Since each creation of a new variable X' increases the size of the state representation (thus raising the memory cost of model checking), it is desirable to minimise the number of duplications by choosing carefully in which unit(s) X will be duplicated. Based on the observation that duplicating X in some unit U removes from UCG_X all conflict edges connected to U, the problem is similar to the classical NP-complete "vertex cover problem", except that each edge removal provokes the recalculation of UCG_X. To select the unit (noted $best\_of(UCG_X)$) in which X should be duplicated first, we adopt a combination of top-down and greedy strategies by choosing, among the units of UCG_X having at least one edge, outermost ones; if there are several such units, we then choose one having a maximal number of edges; if UCG_X has no edges, $best\_of(UCG_X)$ returns $\bot$.

For a given variable X, the "**repeat**" loop (line 7) terminates because of fixed point convergence of global data-flow analysis and because the cardinal of $U_X = \{U \sqsubseteq unit(X) \mid \exists T \in trans(U) \setminus entry(U) \text{ such that } use(T, X)\}$ strictly decreases at each duplication (line 16). Indeed, U_X is the set of units U in which variable X is used within a transition T that does not "enter" U (i.e., $in(T) \cap places^*(U) \neq \emptyset$). After duplicating X in U, there are no remaining occurrences of X in the transitions of $trans(U) \setminus entry(U)$ and, thus, U is no longer in U_X. While duplication might add assignments $X' := X$ (and consequently, new uses of X) to the transitions T of $entry(U)$ (in fact, only those T such that $live(T, X)$) and, therefore, might add to U_X some new unit(s) U' such that $U \sqsubseteq U' \sqsubseteq unit(X)$, this is not the case actually, as all such units U' already belong to U_X (since $U \in U_X$ and $U \sqsubseteq U' \sqsubseteq unit(X)$ implies $U' \in U_X$).

The outermost "**while**" loop (line 3), which removes one variable X from $VARS$ but possibly inserts new variables X' in this set, also terminates. Let $\delta(U)$ be the nesting depth of unit U in the unit hierarchy, i.e., the number of parent units containing U (the root unit having depth 0). Let $L = \max\{\delta(U) \mid U \in \mathcal{U}\}$ be the maximal nesting depth, and let $\Delta(VARS)$ be the vector $(n_0, \ldots, n_L)$ such that $\forall i, n_i = card\{X \in VARS \mid \delta(unit(X)) = i\}$. At each iteration of the outermost loop, $\Delta(VARS)$ strictly decreases according to the lexicographic ordering on integer vectors of length L, as all variables X' created to duplicate X are attached to units strictly included in $unit(X)$, i.e., $\delta(unit(X')) < \delta(unit(X))$.

6   Experimental Results

We implemented our approach in a prototype version of CÆSAR and compared this prototype with the "standard" version of CÆSAR, which, as mentioned in Sect. 1, already resets variables but in a more limited way (upon process termination only). We performed all our measurements on a Sun Sparc Station 100 with 1.6 GB RAM.

We considered 279 LOTOS specifications (many of which are derived from "real world" applications) for which the entire state space could be generated with the "standard" version of CÆSAR. For 112 examples out of 279 (40%), our approach reduced the state space (still preserving strong bisimulation) by a mean factor of 9.7 (with a maximum of 220) as regards the number of states and a mean factor of 13 (with a maximum of 360) as regards the number of transitions. For none of the other 167 examples did our prototype increase the state space.

Then, we considered 3 new "realistic" LOTOS specifications, for which our prototype could generate the corresponding state space entirely, whereas the "standard" version of CÆSAR would fail due to lack of memory. For one of these examples, the "standard" version stopped after producing an incomplete state space with more than 9 million states, while our prototype generated an LTS with 820 states and 1,500 transitions (i.e., a reduction factor greater than 10^4).

We then extended our set of $279+3$ examples with 17 new, large examples for which our prototype is still unable to generate the entire state space. On these 299 examples, variable duplication occurred in only 28 cases (9.4%) for which it increased the memory size needed to represent a state by 60% on average (most of the increase being due to 2 particular examples; for the 26 remaining ones, the increase is only of 10% on average). However, on all examples for which the "standard" version of CÆSAR could generate the LTS entirely, we measured that, on average, the increased memory cost of state representation was outweighed by the global reduction in the number of states.

As regards execution time, we observed that our approach divides by a factor of 4 the total execution time needed to generate all LTSs corresponding to the 279 examples mentioned above. The cost in time of our algorithm is small (about 4%) and more than outweighed by the resulting state space reductions.

7   Conclusion

This paper has shown how state space reduction based on a general (i.e., not merely "syntactic") analysis of dead variables can be applied to process algebra specifications. Our approach requires two steps.

First, the process algebra specifications are compiled into an intermediate network model based on Petri nets extended with state variables, which can be consulted and modified by actions attached to the transitions.

Then, data-flow analysis is performed on this network to determine automatically where variable resets can be inserted. This analysis generalizes the "syntactic" technique (resetting variables of a process upon its termination) implemented in CÆSAR since 1992. It handles shared read-only variables inherited from parent processes, an issue which so far prevented the approach of [7] from being included into the official releases of CÆSAR.

Compared to related work, our network model features a hierarchy of nested processes, where other approaches are usually restricted to a flat collection of communicating automata. Also, our data-flow analysis uses two passes (backward, then forward fixed points computations) in order to introduce no more variable resets than necessary.

Experiments conducted on several hundreds of realistic LOTOS specifications indicate that state space reduction is frequent (20–40% of examples) and can reach several orders of magnitude (e.g. 10^4). Additionally, state space reduction makes CÆSAR four times faster when processing the complete set of examples.

As regards future work, we can mention some open issues (not addressed in this paper, since they are beyond the scope of the CÆSAR compiler for LOTOS), especially data-flow analysis in presence of dynamic creation/destruction of processes (arising from recursion through parallel composition) and data-flow analysis for shared read/write variables (in which case duplication is no longer possible).

References

1. A. V. Aho, R. Sethi, and J. D. Ullman. *Compilers: Principles, Techniques and Tools.* Addison-Wesley, 1986.
2. G. Ailloud. Verification in ECRINS of LOTOS Programs. In *Towards Practical Verification of LOTOS specifications*, Universiteit Twente, 1986. Technical Report ESPRIT/SEDOS/C2/N48.1.
3. M. Bozga, J.-C. Fernandez, and L. Ghirvu. State Space Reduction based on Live Variables Analysis. In *SAS'99*, LNCS 1694, pages 164–178, Sept. 1999. Springer.
4. M. Bozga, J.-C. Fernandez, and L. Ghirvu. State Space Reduction based on Live Variables Analysis. *Science of Computer Programming*, 47(2–3):203–220, 2003.
5. M. Bozga, J.-C. Fernandez, L. Ghirvu, S. Graf, J.-P. Krimm, and L. Mounier. IF: An Intermediate Representation and Validation Environment for Timed Asynchronous Systems. In *FM'99*. Springer, Sept. 1999.
6. Y. Dong and C. R. Ramakrishnan. An Optimizing Compiler for Efficient Model Checking. In *FORTE'99*, pages 241–256, Beijing, Oct. 1999. Kluwer.
7. J. Galvez Londono. Analyse de flot de données dans un système parallèle. Mémoire de DEA, Institut National Polytechnique de Grenoble and Université Joseph Fourier, Grenoble. Supervised by Hubert Garavel and defended on June 22, 1993 before the jury composed of Hubert Garavel, Farid Ouabdesselam, Claude Puech, and Jacques Voiron.
8. H. Garavel. Compilation et vérification de programmes LOTOS, Thèse de doctorat, Université Joseph Fourier, Grenoble, Nov. 1989.
9. H. Garavel, F. Lang, and R. Mateescu. An Overview of CADP 2001. *EASST Newsletter*, 4:13–24, Aug. 2002. Also INRIA Technical Report RT-0254.
10. H. Garavel and J. Sifakis. Compilation and Verification of LOTOS Specifications. In 10th *International Symposium on Protocol Specification, Testing and Verification*, pages 379–394. IFIP, June 1990.
11. S. Graf, J.-L. Richier, C. Rodríguez, and J. Voiron. What are the Limits of Model Checking Methods for the Verification of Real Life Protocols? In 1st *Workshop on Automatic Verification Methods for Finite State Systems*, LNCS 407, pages 275–285, June 1989.
12. G. J. Holzmann. The Engineering of a Model Checker: The Gnu i-Protocol Case Study Revisited. In 6th *SPIN Workshop*, LNCS 1680, pages 232–244, 1999.
13. ISO/IEC. LOTOS – A Formal Description Technique Based on the Temporal Ordering of Observational Behaviour. International Standard 8807, ISO, Genève, Sept. 1989.
14. R. Melton and D. L. Dill. *Murphi Annotated Reference Manual*, 1996. Release 3.1. Updated by C. Norris Ip and Ulrich Stern. Available at `http://verify.stanford.edu/dill/Murphi/Murphi3.1/doc/User.Manual`.

A Hybrid Logic of Knowledge Supporting Topological Reasoning

Bernhard Heinemann

Fachbereich Informatik, FernUniversität in Hagen
58084 Hagen, Germany
Phone: + 49 2331 987 2714, Fax: + 49 2331 987 319
Bernhard.Heinemann@fernuni-hagen.de

Abstract. We consider a certain, in a way hybrid extension of a system called topologic, which has been designed for reasoning about the spatial content of the idea of knowledge. What we add to the language of topologic are names of both points and neighbourhoods of points. Due to the special semantics of topologic these names do not quite behave like nominals in hybrid logic. Nevertheless, corresponding satisfaction operators can be simulated with the aid of the global modality, which becomes, therefore, another means of expression of our system. In this paper we put forward an axiomatization of the set of formulas valid in all such hybrid scenarios, and we prove the decidability of this logic. Moreover, we argue that the present approach to hybridizing modal concepts for knowledge and topology is not only much more powerful but also much more natural than a previous one.

Keywords: logical frameworks for reasoning, reasoning about knowledge and topology, hybridization, completeness, decidability

1 Introduction

Logics of knowledge and time are today widely accepted formalisms for analysing distributed scenarios[1]. While a temporal component is obviously inherent to such systems, it is less recognized that they have a *spatial* element as well. In fact, knowledge can be viewed as *closeness* and knowledge acquisition as *approximating points* in suitable spaces of sets. In this way *topology* enters the context of evolving knowledge. This was discovered and made precise in the paper [3] first, and drawn up in detail in [4]. The basic issue of these papers is a rather general bi-modal system called *topologic*, comprising two one-place operators K representing *knowledge* and $\square$ representing *effort* (in which *time* is encoded implicitly). The formulas of the underlying language are interpreted in *set spaces* $(X, \mathcal{O}, V)$ consisting of a non-empty set X of points or *states,* a set $\mathcal{O}$ of distinguished subsets of X, which can be taken as *knowledge states* (changing in the course of time) of an agent, and a valuation V determining the states where the atomic propositions are true. The operator K quantifies then across

---

[1] The textbooks [1] and [2] provide comprehensive introductions to this field.

C. Rattray et al. (Eds.): AMAST 2004, LNCS 3116, pp. 181–195, 2004.

any knowledge state, whereas $\Box$ quantifies 'downward' across $\mathcal{O}$. That is, more knowledge, i.e., closer proximity, can be achieved by descending with respect to the set inclusion relation inside $\mathcal{O}$, and just this is modelled by $\Box$.

Focussing on the topological part of knowledge leads us to a new area of application: why not use *topologic* as a basic geometric specification tool? Pursuing this idea causes, however, new expressiveness requirements very soon. The way we follow here in order to meet these requirements is marked in the title of the paper. Two concepts typical of *hybrid logic,* viz *naming* and *evaluating formulas at* distinguished states, are added to the source language[2]. When doing this we have to be careful, because of the semantics of *topologic*. As a result, we get to a 'soft' hybridization of the original system. Nevertheless, our skills with regard to expressiveness can be substantially expanded while the nice properties of *topologic* are preserved.

An 'orthodox' hybrid logic of knowledge and time has been developed already, cf [8]. The system presented in this paper is, however, much more powerful and flexible than that. In particular, the earlier logic can be embedded; see Section 5 below.

The paper is organized as follows. The new language for spaces of knowledge states is introduced in the next section, where a couple of examples too are contained in. These examples show on the one hand how the language can be applied to modelling evolving knowledge and on the other hand which topological properties can be expressed. Section 3 contains an axiom system for the resulting hybridized logic of set spaces. Moreover, we touch on the question of completeness there. In Section 4 we state the decidability of our logic and outline the proof of this main result of the present paper. A comparison with other approaches to 'hybrid topologic' is given in the final section.

It should be mentioned that the system proposed in this paper is related to those we presented at previous AMAST conferences; cf [9] and [10]. While certain variants of the basic logic of set spaces were studied there, a substantial development is yielded in our view by the hybridization carried out here.

2   The Language

First in this section we define precisely the syntax and semantics of the logical language described colloquially above. Second, we give a couple of examples showing how this language can be used for modelling basic frameworks of knowledge acquisition. Finally, we turn to topology and give sample specifications of some fundamental topological notions. In this way the new language proves to be an appropriate core language supporting topological reasoning.

We extend the bi-modal language of *topologic* by adding both two sets of names and a third unary modal operator. The denotation of each of the new

---

[2] The paper [5] is a very readable introduction to hybrid logic; see also [6], Sec. 7.3.
– The origin of hybrid logic from Prior's work on temporal representation and the part hybrid logic plays for today's temporal logic are commented on, eg, in [7].

names is either a unique state or a distinguished set of states, whereas the further modality represents the global one (cf [6], Sec. 7.1).

Let $\text{PROP} = \{p, q, \ldots\}$ be a denumerable set of symbols called *proposition variables*. Moreover, let $\text{N}_{stat} = \{i, j, \ldots\}$ and $\text{N}_{sets} = \{A, B, \ldots\}$ be two further sets of symbols called *names of states* and *names of sets*, respectively. We assume that the sets PROP, N_{stat} and N_{sets} are mutually disjoint. Then, we define the set WFF of *well-formed formulas* over $\text{PROP} \cup \text{N}_{stat} \cup \text{N}_{sets}$ by the rule

$$\alpha ::= p \mid i \mid A \mid \neg\alpha \mid \alpha \wedge \beta \mid K\alpha \mid \Box\alpha \mid A\alpha.$$

The missing boolean connectives $\top, \bot, \vee, \rightarrow, \leftrightarrow$ are treated as abbreviations, as needed. The duals of the modal operators $K, \Box$ and A are denoted $L, \Diamond$ and E, respectively.

Now we give meaning to formulas. To begin with we have to define the domains where formulas are to be interpreted in. In the following we let $\mathcal{P}(X)$ designate the powerset of a given set X.

Definition 1 (Set spaces with names).

1. Let $X \neq \emptyset$ be a set and $\mathcal{O} \subseteq \mathcal{P}(X)$ a set of subsets of X such that $X \in \mathcal{O}$. Then the pair $\mathcal{S} := (X, \mathcal{O})$ is called a *set frame*.
2. Let $\mathcal{S} = (X, \mathcal{O})$ be a set frame. The set of *neighbourhood situations* of $\mathcal{S}$ is $\mathcal{N}_{\mathcal{S}} := \{x, U \mid x \in U \text{ and } U \in \mathcal{O}\}$.
3. Let $\mathcal{S}$ be as above. An $\mathcal{S}$–valuation is a mapping

$$V : \text{PROP} \cup \text{N}_{stat} \cup \text{N}_{sets} \longrightarrow \mathcal{P}(X)$$

   such that $V(i)$ is either $\emptyset$ or a singleton subset of X for every $i \in \text{N}_{stat}$, and $V(A) \in \mathcal{O}$ for every $A \in \text{N}_{sets}$.
4. A triple $\mathcal{M} := (X, \mathcal{O}, V)$, where $\mathcal{S} = (X, \mathcal{O})$ is a set frame and V an $\mathcal{S}$–valuation, is called a *set space with names* (or, in short, an SSN). We say that $\mathcal{M}$ is based on $\mathcal{S}$.

For a given SSN $\mathcal{M}$ we define now the relation of satisfaction, $\models_{\mathcal{M}}$, between neighbourhood situations of the underlying frame and formulas in WFF. (The obvious clauses for negation and conjunction are omitted.)

Definition 2 (Satisfaction and validity).

1. Let $\mathcal{M} = (X, \mathcal{O}, V)$ be an SSN based on $\mathcal{S} = (X, \mathcal{O})$, and x, U a neighbourhood situation of $\mathcal{S}$. Then

$$\begin{aligned}
x, U &\models_{\mathcal{M}} p & &:\Longleftrightarrow x \in V(p) \\
x, U &\models_{\mathcal{M}} i & &:\Longleftrightarrow x \in V(i) \\
x, U &\models_{\mathcal{M}} A & &:\Longleftrightarrow V(A) = U \\
x, U &\models_{\mathcal{M}} K\alpha & &:\Longleftrightarrow y, U \models_{\mathcal{M}} \alpha \text{ for all } y \in U \\
x, U &\models_{\mathcal{M}} \Box\alpha & &:\Longleftrightarrow \forall U' \in \mathcal{O} : (x \in U' \subseteq U \text{ implies } x, U' \models_{\mathcal{M}} \alpha) \\
x, U &\models_{\mathcal{M}} A\alpha & &:\Longleftrightarrow y, U' \models_{\mathcal{M}} \alpha \text{ for all } y, U' \in \mathcal{N}_{\mathcal{S}},
\end{aligned}$$

   where $p \in \text{PROP}$, $i \in \text{N}_{stat}$, $A \in \text{N}_{sets}$, and $\alpha \in \text{WFF}$. In case $x, U \models_{\mathcal{M}} \alpha$ is true we say that α holds in $\mathcal{M}$ at the neighbourhood situation x, U.

2. *Let $\mathcal{M}$ be an SSN. A formula α is called* valid in $\mathcal{M}$ *iff it holds in $\mathcal{M}$ at every neighbourhood situation of the set frame $\mathcal{M}$ is based on. (Manner of writing: $\mathcal{M} \models \alpha$.)*

Note that the meaning of both proposition variables and names of states is regardless of neighbourhoods, thus 'stable' with respect to $\square$.

Because of the second clause in Definition 2.1 a name $i \in \mathrm{N}_{stat}$ reminds one of a nominal as known from hybrid logic; cf [5]. However, due to the two-component semantics just defined we can*not* specify accordingly an accompanying satisfaction operator $@_i$. Solely for *names of neighbourhood situations* such an operator can be mimicked. In fact, formulas of the form $i \wedge A$, where $i \in \mathrm{N}_{stat}$ and $A \in \mathrm{N}_{sets}$, can be taken as appropriate names for elements of $\mathcal{N}_S$, and a satisfaction operator associated with such a name reads then $\mathsf{E}(i \wedge A \wedge \ldots)$. I.e., pairs (i, A) can act like proper nominals in set spaces with names.

The following two examples show that the language just defined is closely related to knowledge acquisition, actually.

Example 1. Suppose that an agent (who is an expert in heraldry) can get to know which countries of the Continent do not belong to EURO-land in the following way: every member of this alliance is represented by an accompanying EURO, there is one toss of these 12 coins, and the agent may turn round one coin after the other. Let

$$\text{A - B - D - 0 - F - 0 - GR - I - IRL - L - NL - 0}$$

be the outcome of the toss, where '0' indicates the (matching) reverse. Then, assuming that the agent knows in advance that no country from Eastern Europe is involved, the following sequence of successive *knowledge states* (i.e., sets of states the agent considers possible) results:

$$\{Y \mid \mathrm{card}(Y) = 5 \text{ and } Y \subseteq \{\mathrm{CH, DK, E, FIN, GB, N, P, S}\}\}$$
$$\supseteq \{Y \mid \mathrm{card}(Y) = 5 \text{ and } Y \subseteq \{\mathrm{CH, DK, FIN, GB, N, P, S}\}\}$$
$$\supseteq \{Y \mid \mathrm{card}(Y) = 5 \text{ and } Y \subseteq \{\mathrm{CH, DK, GB, N, P, S}\}\}$$
$$\supseteq \{\{\mathrm{CH, DK, GB, N, S}\}\}$$

More generally, the agent will always get to know the countries not belonging to EURO-land by such a scenario . Therefore, the formula $\Diamond K \mathsf{non-members}$ holds at every initial situation of an SSN constructed from the above set of countries in an obvious way, where $\mathsf{non-members}$ serves as a name for the knowledge state $\{\{\mathrm{CH, DK, GB, N, S}\}\}$ desired by the agent.

Example 2. Objects that are potentially infinite can often be represented by elements of $\{0, 1\}^\omega$ in a natural way. This is true, in particular, for the real numbers (for which in fact many natural representations exist). An element $\rho \in \mathbb{R}$ is called *computable*, iff there is a program computing a binary stream that encodes a fast-converging Cauchy sequence having limit ρ; see [11]. Now assume that an agent wants to acquire some knowledge about an actually computed $\rho \in \mathbb{R}$. A suitable model is based on the *Cantor space* of all binary streams in such a way that every knowledge state of the agent coincides with a basic open set consisting of all possible prolongations of the output the program has delivered up to any time. Then, if it *is* true that ρ is different from, eg, π, the agent will know this eventually and forever afterwards. Thus the formula $\mathsf{EK}\square\neg\pi$ is valid in every SSN based on the Cantor space (where π names any stream representing this real number).

The connection between knowledge and spaces of shrinking sets, i.e., topological scenarios, is revealed by these examples, in particular.

Concluding this section we take a step further and give reasons for our thesis that the above language is sufficiently expressive to serve as a topological specification language. The examples contained in the following definition and proposition, respectively, concern the basic topological ideas of *separation* and *connectedness*. The use of the global modality is exemplified by expressing the fact that set frames are *generated* in a sense (by any neighbourhood situation x, X).

Definition 3 (Special classes of set frames). *Let $\mathcal{S} = (X, \mathcal{O})$ be a set frame. Then $\mathcal{S}$ is called*

1. separated *iff for all $x, y \in X$ such that $x \neq y$ there exists $U_1, U_2 \in \mathcal{O}$ satisfying $x \in U_1 \setminus U_2$ and $y \in U_2 \setminus U_1$,*
2. connected *iff for all $U, U_1, U_2 \in \mathcal{O}$ such that $U_1 \neq \emptyset \neq U_2$ we have that $U = U_1 \cup U_2$ implies $U_1 \cap U_2 \neq \emptyset$.*

Proposition 1. *Let $\mathcal{S} = (X, \mathcal{O})$ be a set frame. Then*

1. *$\mathcal{S}$ is separated iff $\mathcal{M} \models i \wedge L(j \wedge \neg i) \rightarrow \Diamond K \neg j \wedge K(j \rightarrow \Diamond K \neg i)$ for all SSNs $\mathcal{M}$ based on $\mathcal{S}$ and $i, j \in \mathsf{N}_{stat}$.*
2. *$\mathcal{S}$ is connected iff $\mathcal{M} \models K \Diamond (A \vee B) \rightarrow L(\Diamond A \wedge \Diamond B)$ for all SSNs $\mathcal{M}$ based on $\mathcal{S}$ and $A, B \in \mathsf{N}_{sets}$.*
3. *$\mathcal{M} \models \mathsf{E}(\mathsf{E}\alpha \rightarrow L\Diamond\alpha)$ for all SSNs $\mathcal{M}$ based on $\mathcal{S}$ and $\alpha \in \mathsf{WFF}$.*

Proof. We prove only item 2^3. First, let $\mathcal{S}$ be connected. Let $\mathcal{M}$ be any set space with names based on $\mathcal{S}$, and take any neighbourhood situation x, U of $\mathcal{S}$ such that $x, U \models_{\mathcal{M}} K\Diamond(A \vee B)$, where $A, B \in \mathsf{N}_{sets}$. Then for all $y \in U$ there exists $U_y \in \mathcal{O}$ such that $y, U_y \models A \vee B$. That is, $U_y = V(A)$ or $U_y = V(B)$ due to the definition of the semantics. It follows from this that $U = V(A) \cup V(B)$. Since $U \neq \emptyset$ and $\mathcal{S}$ is assumed to be connected there exists a point $z \in V(A) \cap V(B)$. Thus $z, U \models_{\mathcal{M}} \Diamond A \wedge \Diamond B$, and consequently $x, U \models_{\mathcal{M}} L(\Diamond A \wedge \Diamond B)$, as desired.

Now suppose that $\mathcal{S}$ is *not* connected. Then there $U, U_1, U_2 \in \mathcal{O}$ such that $U_1 \neq \emptyset \neq U_2$ and $U = U_1 \cup U_2$, but $U_1 \cap U_2 = \emptyset$. Take any $\mathcal{S}$–valuation V such that $V(A) = U_1$ and $V(B) = U_2$. Let $\mathcal{M}'$ be the SSN which is based on $\mathcal{S}$ and has $\mathcal{S}$–valuation V. Then, clearly, $x, U \models_{\mathcal{M}'} K\Diamond(A \vee B)$. But since there is no point in the intersection $U_1 \cap U_2$, we have $y, U \models_{\mathcal{M}'} \square\neg A$ or $y, U \models_{\mathcal{M}'} \square\neg B$ for

---

^{3} The proof of item 1 can be done in a similar way; item 3 is due to the fact that $X \in \mathcal{O}$.

all $y \in U$. Therefore, $x, U \models_{\mathcal{M}'} K(\Box \neg A \vee \Box \neg B)$ can be concluded. It follows that the formula schema given above is not valid in every model. This proves the other direction.

It should be remarked that the separation condition from Definition 3.1 equals the $(T\,1)$–Axiom for topological spaces. Moreover, the notion of connectedness from Definition 3.2 means that all non-empty sets $U \in \mathcal{O}$ are connected in the topological sense; cf [12], Ch. I, §11, Sec. 1, Definition 2. – The global modality is notably important for the proof of Theorem 4 below.

3   Completeness

We propose below a sound system of axioms for the logic of set spaces with names. Then, a couple of Gabbay–style proof rules[4] is added. This enables us to prove completeness of the system. It turns out that we obtain even *extended* completeness (in a sense explained below). Note that almost everything that is relevant to the A–free fragment of our language can be found in the paper [13].

The axiom schemata are arranged in three groups. First, the usual axioms for arbitrary set spaces are listed; cf [4].

1. All instances of tautologies.   6. $(p \to \Box p) \wedge (\Diamond p \to p)$
2. $K(\alpha \to \beta) \to (K\alpha \to K\beta)$ 7. $\Box\,(\alpha \to \beta) \to (\Box\alpha \to \Box\beta)$
3. $K\alpha \to \alpha$ 8. $\Box\alpha \to \alpha$
4. $K\alpha \to KK\alpha$ 9. $\Box\alpha \to \Box\Box\alpha$
5. $L\alpha \to KL\alpha$ 10. $K\Box\alpha \to \Box K\alpha,$

where $p \in \mathrm{PROP}$ and $\alpha, \beta \in \mathrm{WFF}$. – The second group of axioms concerns names; cf [13]. (Note the difference between Axiom 14 here and there.)

11. $(i \to \Box i) \wedge (\Diamond i \to i)$ 14. $\mathsf{A}\,(A \wedge L\alpha \to L\beta) \vee \mathsf{A}\,(A \wedge L\beta \to L\alpha)$
12. $i \wedge \alpha \to K(i \to \alpha)$ 15. $K(\Diamond B \to \Diamond A) \wedge L\Diamond B \to \Box\,(A \to L\Diamond B)$
13. $A \to KA$ 16. $K\Diamond A \to A,$

where $i \in \mathrm{N}_{stat}$, $A, B \in \mathrm{N}_{sets}$ and $\alpha, \beta \in \mathrm{WFF}$. – In the final group, each axiom deals with the global modality. In particular, the interplay between A and $K\Box$ is described therein.

17. $\mathsf{A}(\alpha \to \beta) \to (\mathsf{A}\alpha \to \mathsf{A}\beta)$ 19. $\mathsf{A}\alpha \to \mathsf{A}\mathsf{A}\alpha$ 21. $\mathsf{A}\alpha \to K\Box\alpha$
18. $\mathsf{A}\alpha \to \alpha$ 20. $\alpha \to \mathsf{A}\mathsf{E}\alpha$ 22. $\mathsf{E}\,(\mathsf{E}\alpha \to L\Diamond\alpha),$

where $\alpha, \beta \in \mathrm{WFF}$. – Note that the last axiom, which is a 'weak' converse of the penultimate one, equals the formula schema from Proposition 1.3.

Now we define the desired logical system, which we designate **SNG** (indicating a logic for spaces with names and the global modality).

---

[4] This naming is meant to refer to *Gabbay's irreflexivity rule* (cf, eg, [6], Sec. 4.7 and Notes to Ch. 4).

Definition 4 (The logic). *Let* **SNG** *be the smallest set of formulas containing all of the above axiom schemata and closed under application of the following rules:*

$$(\text{MODUS PONENS}) \quad \frac{\alpha \rightarrow \beta, \alpha}{\beta} \qquad (K\text{–NECESSITATION}) \quad \frac{\alpha}{K\alpha}$$

$$(\Box\text{–NECESSITATION}) \quad \frac{\alpha}{\Box\alpha} \qquad (\text{A–NECESSITATION}) \quad \frac{\alpha}{\mathsf{A}\alpha}$$

$$(\text{NAME}_{stat}) \quad \frac{j \rightarrow \beta}{\beta} \qquad (\text{NAME}_{sets}) \quad \frac{B \rightarrow \beta}{\beta}$$

$$(K\text{–ENRICHMENT}) \quad \frac{\mathsf{E}\,(i \wedge A \wedge L(j \wedge \alpha)) \rightarrow \beta}{\mathsf{E}(i \wedge A \wedge L\alpha) \rightarrow \beta}$$

$$(\Box\text{–ENRICHMENT}) \quad \frac{\mathsf{E}\,(i \wedge A \wedge \Diamond(B \wedge \alpha)) \rightarrow \beta}{\mathsf{E}(i \wedge A \wedge \Diamond\alpha) \rightarrow \beta}$$

$$(\text{A–ENRICHMENT}) \quad \frac{\mathsf{E}\,(i \wedge A \wedge \mathsf{E}(j \wedge B \wedge \alpha)) \rightarrow \beta}{\mathsf{E}(i \wedge A \wedge \mathsf{E}\alpha) \rightarrow \beta},$$

where $\alpha, \beta \in \text{WFF}$, $i, j \in \mathrm{N}_{stat}$, $A, B \in \mathrm{N}_{sets}$, *and* j, B *are 'new' each time.*

The reader can easily convince himself or herself that **SNG** is sound with respect to the class of all SSNs. But completeness too is yielded for this logic.

Theorem 1 (Completeness). *Every formula* $\alpha \in \text{WFF}$ *that is valid in all SSNs is contained in the logic* **SNG**.

The properties of the canonical model $\mathcal{M}_{\text{SNG}}$ of **SNG** are exploited for the proof of Theorem 1. Since we put the main emphasis on decidability in this paper we explain only the starting point here and sketch the proceeding very roughly.

First, the part of $\mathcal{M}_{\text{SNG}}$ accessible from a distinguished maximal consistent set (which is determined by a given non-derivable formula) has to be named, and all existential demands concerning the modalities have to be realized. To this end, call a maximal consistent set s of formulas

- *named* iff s contains some $i \in \mathrm{N}_{stat}$ and some $A \in \mathrm{N}_{sets}$, and
- *enriched* iff
    1. $\mathsf{E}(i \wedge A \wedge L\alpha) \in s$ implies $\mathsf{E}\,(i \wedge A \wedge L(j \wedge \alpha)) \in s$ for some $j \in \mathrm{N}_{stat}$,
    2. $\mathsf{E}(i \wedge A \wedge \Diamond\alpha) \in s$ implies $\mathsf{E}\,(i \wedge A \wedge \Diamond(B \wedge \alpha)) \in s$ for some $B \in \mathrm{N}_{sets}$, and
    3. $\mathsf{E}(i \wedge A \wedge \mathsf{E}\alpha) \in s$ implies $\mathsf{E}\,(i \wedge A \wedge \mathsf{E}(j \wedge B \wedge \alpha)) \in s$ for some $j \in \mathrm{N}_{stat}$ and $B \in \mathrm{N}_{sets}$.

Let $\widetilde{\mathrm{N}}_{stat}$ and $\widetilde{\mathrm{N}}_{sets}$ be two denumerable sets of *new* symbols, and $\widetilde{\text{WFF}}$ the set of formulas extended accordingly. Then we obtain the following *Modified Lindenbaum Lemma*.

Lemma 1. *Every maximal consistent set* $s \subseteq \text{WFF}$ *can be extended to a named and enriched maximal consistent set* $\tilde{s} \subseteq \widetilde{\text{WFF}}$.

Furthermore, the following *Existence Lemma* can be proved for the structure $\widetilde{\mathcal{M}}$ of which the domain D consists of all named points that are reachable (via the canonical accessibility relations) from $\widetilde{s}$ and the relations are the induced ones.

Lemma 2. *Let $\nabla \in \{L, \Diamond, \mathsf{E}\}$. Assume that $s \in D$ contains the formula $\nabla \alpha$. Then there exists some $t \in D$ which is accessible from s with respect to ∇ and contains α.*

Both lemmata can be proved with the aid of the new rules (among other things).

Now, the axioms of the second group force a set space structure already on the model $\widetilde{\mathcal{M}}$. Moreover, the third group of axioms guarantees that the modalities behave correctly as set space modalities there. Thus we get in fact 'canonical' completeness of the system, via an appropriate *Truth Lemma*.

And we obtain even more. The fact that we are dealing with *named* objects enables us to extend the completeness proof to other classes of set spaces simply by adding a suitable defining correspondent. In this way we get, eg:

Theorem 2 (Extended completeness). *Let $\mathbf{S}$ be the formula schema from Proposition 1.1. Then the system $\mathbf{SNG} + \mathbf{S}$ is sound and complete with respect to the class of all separated set spaces.*

A corresponding result holds for connected frames, and for other classes of set frames, too.

4   Decidability

Subsequently it is shown that $\mathbf{SNG}$ is a decidable set of formulas. As a preparatory step we must consider a class of auxiliary Kripke structures which are a bit involved.

Definition 5 (Cross axiom models with names). *A quintupel*

$$\mathfrak{M} := \left(W, \xrightarrow{L}, \xrightarrow{\Diamond}, \xrightarrow{\mathsf{E}}, V\right)$$

is called a cross axiom model with names *(or, in short, a CMN), iff the following conditions are satisfied:*

- *W is a non-empty set,*
- *the relation $\xrightarrow{L} \subseteq W \times W$ (belonging to K) is an equivalence, the relation $\xrightarrow{\Diamond} \subseteq W \times W$ (belonging to $\Box$) is reflexive and transitive, and the relation $\xrightarrow{\mathsf{E}}$ (belonging to A) is universal,*
- *there is some $w \in W$ such that $W = \{v \mid (w, v) \in (\xrightarrow{L} \cup \xrightarrow{\Diamond})^*\}$,*
- *for all $u, v, w \in W$ such that $u \xrightarrow{\Diamond} v \xrightarrow{L} w$ there exists $t \in W$ such that $u \xrightarrow{L} t \xrightarrow{\Diamond} w$,*

– $V : \mathrm{PROP} \cup \mathrm{N}_{stat} \cup \mathrm{N}_{sets} \longrightarrow \mathcal{P}(W)$ *is a mapping such that*
  • *for all* $c \in \mathrm{PROP} \cup \mathrm{N}_{stat}$ *and* $u, v \in W$ *satisfying* $u \overset{\diamond}{\longrightarrow} v$ *it holds that* $u \in V(c)$ *iff* $v \in V(c)$,
  • *the intersection of* $V(i)$ *with any* $\overset{L}{\longrightarrow}$ *-equivalence class is either empty or a singleton set for every* $i \in \mathrm{N}_{stat}$, *and*
  • *the set* $V(A)$ *equals either* $\emptyset$ *or a unique* $\overset{L}{\longrightarrow}$ *-equivalence class for every* $A \in \mathrm{N}_{sets}$.

The term 'cross axiom model' refers to the fourth item above and originates from [4]. It is shown there that the basic logic of set spaces is sound and complete for interpretations in cross axiom models (without names), and that a corresponding finite model property is satisfied (though that logic lacks the finite model property with respect to set spaces).

We would like to establish the finite model property of **SNG** with respect to CMNs. However, not every CMN validates necessarily all of the above axioms. But, for a start, we get at least:

Proposition 2. *Apart from, possibly, Axioms 15 and 16 each of the above schemata is valid in every CMN.*

Proof. As we are dealing with usual Kripke models presently, names of states and sets are treated like proposition variables. I.e., we have that $\mathfrak{M}, w \models c :$ $\iff w \in V(c)$, for all points w of $\mathfrak{M}$ and $c \in \mathrm{PROP} \cup \mathrm{N}_{stat} \cup \mathrm{N}_{sets}$. Now the verification of the axioms is rather straightforward.

By way of example we prove the validity of Axiom 14. Suppose that there is a point w of $\mathfrak{M}$ and an instance $\mathsf{A}(A \wedge L\alpha \to L\beta) \vee \mathsf{A}(A \wedge L\beta \to L\alpha)$ of that schema such that $\mathfrak{M}, w \not\models \mathsf{A}(A \wedge L\alpha \to L\beta) \vee \mathsf{A}(A \wedge L\beta \to L\alpha)$. Then $\mathfrak{M}, w \models \mathsf{E}(A \wedge L\alpha \wedge \neg L\beta) \wedge \mathsf{E}(A \wedge L\beta \wedge \neg L\alpha)$. Consequently, there exist w_1, w_2 such that $\mathfrak{M}, w_1 \models A \wedge L\alpha \wedge K\neg\beta$ and $\mathfrak{M}, w_2 \models A \wedge L\beta \wedge K\neg\alpha$. As $\mathfrak{M}, w_1 \models A$ and $\mathfrak{M}, w_2 \models A$ we conclude that $w_1 \overset{L}{\longrightarrow} w_2$. Thus $\mathfrak{M}, w_2 \models K\neg\beta$ because $\mathfrak{M}, w_1 \models K\neg\beta$ and $\overset{L}{\longrightarrow}$ is transitive. However, this contradicts the fact that $\mathfrak{M}, w_2 \models L\beta$. Our supposition is, therefore, false; i.e., Axiom 14 is valid in $\mathfrak{M}$.

We call a CMN $\mathfrak{M}$ *good,* iff *all* the axioms are valid in $\mathfrak{M}$. We want to show now that **SNG** satisfies the *strong finite model property* (cf [6], Def. 6.6) with respect to the class of all good CMNs. With that the desired decidability of **SNG** follows, using [6], Th. 6.7.

To establish this finite model property we use the method of *filtration*. So, let $\alpha \in \mathrm{WFF}$ be a consistent formula for which we want to find a model of size at most $f(|\alpha|)$, where f is some computable function and $|\alpha|$ denotes the length of α. Moreover, let $\mathrm{sf}(\alpha)$ be the set of all subformulas of α. We construct an appropriate filter set Σ via the following sets of formulas. We first let

$$\Sigma_0 := \mathrm{sf}(\alpha) \cup \{\Box\neg A \mid A \in \mathrm{N}_{sets} \cap \mathrm{sf}(\alpha)\},$$

and secondly $\Sigma^\neg := \{\neg\beta \mid \beta \in \Sigma_0\}$. Then we take the set Σ' of all finite conjunctions of pairwise distinct elements of $\Sigma_0 \cup \Sigma^\neg$. Finally, we close Σ' under single applications of the operator L. Let Σ be the union of all these intermediate sets of formulas. Note that Σ is finite and subformula closed, and $2^{c \cdot |\alpha|}$ is an upper bound of the cardinality of Σ (for some constant c).

Let $\mathcal{C}$ be the submodel of the canonical model generated by a maximal consistent set containing α. Moreover, let the Kripke model

$$\mathfrak{M} := \left(W, \xrightarrow{L}, \xrightarrow{\Diamond}, \xrightarrow{E}, V \right)$$

be obtained from $\mathcal{C}$ as follows:

- W is the filtration of the carrier set of $\mathcal{C}$ with respect to Σ,
- $\xrightarrow{L}, \xrightarrow{\Diamond}$ and $\xrightarrow{E}$ are the *smallest* filtrations (cf [6], p. 79) of the accessibility relations of $\mathcal{C}$ belonging to the respective modalities with respect to Σ, and
- V is induced by the canonical valuation.

Then, exploiting the structure of the filter set Σ and the minimality of the filtrations we get the following crucial proposition.

Proposition 3. *The just defined model $\mathfrak{M}$ constitutes a filtration of $\mathcal{C}$. Moreover, adjusting the valuation V suitably, $\mathfrak{M}$ can be turned into a good CMN $\mathfrak{M}'$ which is semantically equivalent to $\mathfrak{M}$ with respect to α.*

Proof. The first assertion of the proposition is quite obvious. However, we introduce some notations for later purposes. As is well-known, the domain W of $\mathfrak{M}$ consists of certain equivalence classes. Let us make explicit the corresponding equivalence relation. For all $s, t \in C$, where C is the domain of $\mathcal{C}$, we define $s \sim t :\iff s \cap \Gamma = t \cap \Gamma$, and we let $\underline{s}$ denote the $\sim$-equivalence class of s. I.e., $W = \{\underline{s} \mid s \in C\}$. We use $\xrightarrow[\mathcal{C}]{L}, \xrightarrow[\mathcal{C}]{\Diamond}, \xrightarrow[\mathcal{C}]{E}$, and $V_{\mathcal{C}}$, to denote the accessibility relations and the valuation of $\mathcal{C}$, respectively. Then,

$$\underline{s} \xrightarrow{L} \underline{t} \iff \exists s' \in \underline{s}, t' \in \underline{t} : s' \xrightarrow[\mathcal{C}]{L} t'$$

holds by definition (correspondingly for the other modalities). Moreover, V satisfies

$$\underline{s} \in V(\sigma) \iff s \in V_{\mathcal{C}}(\sigma), \text{ for all } s \in C \text{ and } \sigma \in (\text{PROP} \cup N_{stat} \cup N_{sets}) \cap \Gamma.$$

This explains how the valuation V of $\mathfrak{M}$ is 'induced' by $V_{\mathcal{C}}$.

Now we prove the second assertion. To this end we modify V in the following way. We let

$$V'(\sigma) := \begin{cases} V(\sigma) & \text{if } \sigma \in \Gamma \\ \emptyset & \text{otherwise,} \end{cases}$$

for all $\sigma \in \text{PROP} \cup N_{stat} \cup N_{sets}$. Then we define

$$\mathfrak{M}' := \left(W, \xrightarrow{L}, \xrightarrow{\Diamond}, \xrightarrow{E}, V' \right).$$

With that we go successively through the items of Definition 5. It follows from [4], 2.10, that $\xrightarrow{L}$ is an equivalence, $\xrightarrow{\Diamond}$ is reflexive and transitive and the *cross property* holds[5]; in addition, the rigidity of the proposition variables (as stated in the first condition of the last item of Definition 5) is satisfied for cross axiom models without names already.

Since $\mathcal{C}$ is generated we conclude that $\xrightarrow[\mathcal{C}]{\mathsf{E}}$ is universal (by using Axioms 18 – 21). Thus $\xrightarrow{\mathsf{E}}$ is universal as well. Moreover, Axiom 22 gives us the validity of item 3 for $\mathcal{C}$, hence also for $\mathfrak{M}'$.

The rigidity of the names of states follows in the same manner as the rigidity of the proposition variables (with the aid of Axiom 11 instead of Axiom 6).

Suppose now that $i \in \mathrm{N}_{stat}$, $\underline{s}, \underline{t} \in V'(i)$ and $\underline{s} \xrightarrow{L} \underline{t}$. From this assumption we conclude that $i \in \Gamma$. Thus $s, t \in V_{\mathcal{C}}(i)$, and there are $s' \in \underline{s}$ and $t' \in \underline{t}$ such that $s' \xrightarrow[\mathcal{C}]{L} t'$. Furthermore, $i \in s' \cap t'$. With the aid of Axiom 12 we get that $s' = t'$. Therefore, $\underline{s} = \underline{t}$, as we wanted to show.

Finally, let $\underline{s}, \underline{t} \in V'(A)$ for some $A \in \mathrm{N}_{sets}$. Then, an argument similar to the one just given shows that $\underline{s} \xrightarrow{L} \underline{t}$. (Axiom 14 has to be used for that, actually.) The denotation of A is, therefore, contained in a single $\xrightarrow{L}$–equivalence class, say the class of $\underline{s}$. But since s then contains A, we infer from this that $\mathfrak{M}', \underline{t} \models A$ holds whenever $\underline{s} \xrightarrow{L} \underline{t}$ is valid (by utilizing Axiom 13). Thus the denotation of A equals the class of $\underline{s}$, as desired.

As we have that $\mathfrak{M}, \underline{s} \models \alpha \iff \mathfrak{M}', \underline{s} \models \alpha$ for all $\underline{s} \in W$ (since both structures are filtrations of the same model), it remains to be proved that $\mathfrak{M}'$ is good.

First we show that Axiom 16 is valid in $\mathfrak{M}'$. So, let $w \in W$ and asssume that $\mathfrak{M}', w \models K\Diamond A$. Then, for all $u \in W$ such that $w \xrightarrow{L} u$ there exists $v_u \in W$ satisfying $u \xrightarrow{\Diamond} v_u$ and $\mathfrak{M}', v_u \models A$, i.e., $v_u \in V'(A)$. In particular, v_w is contained in $V'(A)$. Thus $V'(A) \neq \emptyset$. This implies $A \in \Gamma$.

Let $s \in C'$ be a representative of w, i.e. $w = \underline{s}$, and let $t \in C'$ satisfy $s \xrightarrow[\mathcal{C}]{L} t$. Then $w \xrightarrow{L} \underline{t}$, i.e., $\mathfrak{M}', \underline{t} \models \Diamond A$. Since $\Diamond A$ is contained in Γ due to the definition of this set, we infer $\Diamond A \in t$ from that. As we have chosen an arbitrary $t \in C'$ such that $s \xrightarrow[\mathcal{C}]{L} t$ we conclude that $K\Diamond A \in s$. From Axiom 16 we obtain then $A \in s$. This implies $\mathfrak{M}', w \models A$. Therefore, Axiom 16 is valid in $\mathfrak{M}'$.

Finally, the validity of Axiom 15 is to be established. To this end, assume that $\mathfrak{M}', w \models K(\Diamond B \to \Diamond A) \land L\Diamond B$. From the second conjunct we can infer $V'(B) \neq \emptyset$ in a similar way as above. It follows that $B \in \Gamma$. But the first conjunct implies then $A \in \Gamma$.

Suppose towards a contradiction that $\mathfrak{M}', w \not\models \Box(A \to L\Diamond B)$. Then there exists $\underline{t} \in W$ such that $w \xrightarrow{\Diamond} \underline{t}$ and $\mathfrak{M}', \underline{t} \models A \land \neg L\Diamond B$. Due to the definition of the smallest filtration there are $s' \in w = \underline{s}$ and $t' \in \underline{t}$ which are $\xrightarrow[\mathcal{C}]{\Diamond}$–connected: $s' \xrightarrow[\mathcal{C}]{\Diamond} t'$. Moreover, since both A and $L\Diamond B$ are contained in Γ we have that

---

[5] A simplified proof of these facts was done in [14].

$A \wedge \neg L \Diamond B \in t'$, because of the well-known properties of maximal consistent sets. This means that $\Diamond(A \wedge \neg L \Diamond B) \in s'$.

If we could prove that $K(\Diamond B \to \Diamond A) \wedge L \Diamond B \in s'$, then we will be done because all instances of Axiom 15 are contained in s'. Now, the second conjunct of this formula is clearly contained in s' since it is contained in Γ. As to the first conjunct, let u be an arbitrary element of C' such that $s' \xrightarrow[C]{L} u$. It suffices to show that $\neg \Diamond B \vee \Diamond A \in u$, i.e., $\neg \Diamond B \in u$ or $\Diamond A \in u$. But both formulas are contained in Γ, and $\mathfrak{M}', \underline{u} \models \neg \Diamond B$ or $\mathfrak{M}', \underline{u} \models \Diamond A$ is valid according to our assumption. Thus $\neg \Diamond B \vee \Diamond A \in u$ really holds. This proves the validity of Axiom 15 in $\mathfrak{M}'$.

All in all, we have proved Proposition 3 by that.

Accordingly, α is satisfiable in a finite model of the axioms, of which the size is in $O\left(2^{c \cdot |\alpha|}\right)$. The main result of the present paper can be inferred from this easily now.

Theorem 3 (Decidability). *The logic* **SNG** *is decidable.*

As to the complexity of **SNG**, the proof of Theorem 3 yields obviously NEX-PTIME as an upper bound of the corresponding satisfiability problem. We believe that it is possible to force this bound down to EXPTIME. On the other hand, [15], Theorem 4.5, gives us some reasons for suspecting that this problem is EXPTIME–hard. However, we do not know the exact complexity up to now[6].

5   Further Results and Concluding Remarks

In this section we compare our approach with other combined hybrid logics of knowledge or topology. Moreover, we solve an open decidability problem for one of these systems. Finally, a short summary of the present paper is given.

Though *topologic* supports a modal treatment of knowledge for as diverse contexts as, eg, *topological spaces* and *branching time structures,* cf [16] and [17], the underlying language lacks expressive power in other areas of application. Eg, *linear time structures* cannot be dealt with accordingly. Thus the language has to be strengthened. It turns out that *hybrid* means can help a lot here.

In usual hybrid logic one has *nominals* acting as names of states to hand, among other things. Furthermore, a one-place *satisfaction operator* $@_i$ comes along with every nominal i. This operator enables one to evaluate a formula at the denotation of i. As stated above already, right after Definition 2, it is true that satisfaction operators accompanying names of *states* cannot be added consistently to the language of *topologic*. However, by *naming neighbourhood situations* this works well, actually, and, eg, linear time structures can then be captured; cf [18] and [8].

However, we were not content with this because of our further-reaching topological requirements. Note that still none of the fundamental notions introduced

---

[6] Note that we do not know the complexity of basic *topologic* either.

in Definition 3 can be specified by means of the language naming neighbourhood situations. Thus we gave up that 'strict' hybrid approach and turned to the one presented here, which follows in a sense the *sorting strategy* from [5], Sec. 7. The new system proves to be very flexible, reflects clearly the dependence of properties on points and sets, respectively, and comprises the previous one with regard to expressiveness. Furthermore, by a suitable reduction argument we can prove that the earlier system is decidable.

Theorem 4 (Decidability of orthodox hybrid logic). *Letting nominals denote neighbourhood situations, the corresponding hybrid logic of set spaces is decidable.*

Proof. First we revisit briefly the hybrid logic of set spaces considered in the papers cited above. All notations which are not new are kept from the previous sections of the present paper.

The syntax of that 'orthodox' hybrid language is defined by the rule

$$\alpha ::= p \mid \mathfrak{n} \mid \neg\alpha \mid \alpha \wedge \beta \mid K\alpha \mid \Box\alpha \mid @_{\mathfrak{n}}\alpha,$$

where $p \in \mathrm{PROP}$ and $\mathfrak{n} \in \mathrm{NNS}$; the elements of the latter set of symbols are called *names of neighbourhood situations,* and PROP and NNS are assumed to be disjoint. Moreover, $@_{\mathfrak{n}}$ denotes the satisfaction operator belonging to the name $\mathfrak{n} \in \mathrm{NNS}$. Finally, a special name $\mathfrak{n}_0 \in \mathrm{NNS}$ is distinguished in advance.

Hybrid set spaces are based on set frames $(X, \mathcal{O})$ in the usual way, i.e., by means of a *hybrid valuation* $V : \mathrm{PROP} \cup \mathrm{NNS} \longrightarrow \mathcal{P}(X) \cup \mathcal{N}$. We have, in particular, that $V(\mathfrak{n}) \in \mathcal{N}$ for all $\mathfrak{n} \in \mathrm{NNS}$, and $V(\mathfrak{n}_0) = x, X$ for some $x \in X$.

Due to this restricted interpretation of $\mathfrak{n}_0$ we are compelled to change both the language underlying **SNG** and this logic slightly. A special set name Y has to be distinguished, which always denotes the domain X. Moreover, Axiom 22 has to be replaced with $\mathsf{E}Y \wedge (Y \wedge \mathsf{E}\alpha \to L\Diamond\alpha)$. The reader can easily check that Theorems 1 – 3 hold for the resulting variant, **SNG'**, too.

Let **SH** be the set of all formulas that are valid in every hybrid set space. By reducing the **SH**–satisfiability problem to the **SNG'**–satisfiability problem we prove now that **SH** is decidable. In fact, this follows easily from Theorem 3 after a successful reduction.

By induction on the structure of formulas we define a function T translating every hybrid formula into a formula of the comparison language. To this end we assign injectively a pair $i_{\mathfrak{n}} \in \mathrm{N}_{stat}$ and $A_{\mathfrak{n}} \in \mathrm{N}_{sets}$ to every $\mathfrak{n} \in \mathrm{NNS}$; in particular, $A_{\mathfrak{n}_0} = Y$. Then we let

$$\begin{aligned}
T(p) &:= p \\
T(\mathfrak{n}) &:= i_{\mathfrak{n}} \wedge A_{\mathfrak{n}} \\
T(\neg\alpha) &:= \neg T(\alpha) \\
T(\alpha \wedge \beta) &:= T(\alpha) \wedge T(\beta) \\
T(K\alpha) &:= KT(\alpha) \\
T(\Box\alpha) &:= \Box T(\alpha) \\
T(@_{\mathfrak{n}}\alpha) &:= \mathsf{E}(i_{\mathfrak{n}} \wedge A_{\mathfrak{n}} \wedge T(\alpha)),
\end{aligned}$$

for all $p \in \text{PROP}$, $\mathfrak{n} \in \text{NNS}$ and formulas α, β of the source language. T is obviously a computable function.

Now we take any hybrid formula α. Let $\mathfrak{n}_1, \ldots, \mathfrak{n}_n$ be the names of neighbourhood situations occurring in α. Then we have that

$$(*) \quad \alpha \text{ is satisfiable} \iff T(\alpha) \wedge \bigwedge_{i=1,\ldots,n} \mathsf{E}\,(i_\mathfrak{n} \wedge A_\mathfrak{n}) \text{ is satisfiable}$$

with regard to the respective class of models.

In fact, an SSN $\mathcal{M}_{\overline{\mathcal{M}}}$ satisfying the second conjunct on the right-hand side of $(*)$ can easily be assigned to any hybrid set space $\overline{\mathcal{M}}$ based on the same frame, and vice versa, so that each of the following assertions can be proved by induction on α :

$$x, U \models_{\overline{\mathcal{M}}} \alpha \iff x, U \models_{\mathcal{M}_{\overline{\mathcal{M}}}} T(\alpha) \text{ and } x, U \models_{\mathcal{M}} T(\alpha) \iff x, U \models_{\overline{\mathcal{M}}_{\mathcal{M}}} \alpha,$$

where x, U is any neighbourhood situation of the underlying frame. In this way, $(*)$ is yielded.

This concludes the proof of Theorem 4.

Note that the decidability results contained in [18] and [19] concern only special classes of hybrid set spaces. The general case stated in Theorem 4 could not be established in a more direct way up to now.

It should be remarked that a (usual) hybrid extension of the *classical* topological semantics of modal logic (going back to [20]) was briefly considered in [21] (with regard to expressiveness).

Just to sum up, in the present paper we developed the fundamental matters of a logic, **SNG**, for set spaces with names. It turned out that this system is sufficiently expressive as well as computationally well-behaved to serve as an adequate basis for both reasoning about knowledge and specifying topological properties. Our results (concerning expressive power, completeness and, mainly, decidability) and a comparison to other approaches showed that **SNG** is a rather satisfactory system of that kind. However, some open problems for other interesting classes of set spaces seemingly cannot be solved within the framework of **SNG**, eg, the question of decidability for the logic of *directed spaces* (which model 'converging' knowledge acquisition procedures); cf [22].

References

1. Fagin, R., Halpern, J.Y., Moses, Y., Vardi, M.Y.: Reasoning about Knowledge. MIT Press, Cambridge, MA (1995)
2. Meyer, J.J.C., van der Hoek, W.: Epistemic Logic for AI and Computer Science. Volume 41 of Cambridge Tracts in Theoretical Computer Science. Cambridge University Press, Cambridge (1995)
3. Moss, L.S., Parikh, R.: Topological reasoning and the logic of knowledge. In Moses, Y., ed.: Theoretical Aspects of Reasoning about Knowledge (TARK 1992), San Francisco, CA, Morgan Kaufmann (1992) 95–105

4. Dabrowski, A., Moss, L.S., Parikh, R.: Topological reasoning and the logic of knowledge. Annals of Pure and Applied Logic **78** (1996) 73–110
5. Blackburn, P.: Representation, reasoning, and relational structures: a hybrid logic manifesto. Logic Journal of the IGPL **8** (2000) 339–365
6. Blackburn, P., de Rijke, M., Venema, Y.: Modal Logic. Volume 53 of Cambridge Tracts in Theoretical Computer Science. Cambridge University Press, Cambridge (2001)
7. Areces, C., Blackburn, P., Marx, M.: The computational complexity of hybrid temporal logics. Logic Journal of the IGPL **8** (2000) 653–679
8. Heinemann, B.: Knowledge over dense flows of time (from a hybrid point of view). In Agrawal, M., Seth, A., eds.: FST TCS 2002: Foundations of Software Technology and Theoretical Computer Science. Volume 2556 of Lecture Notes in Computer Science., Berlin, Springer (2002) 194–205
9. Heinemann, B.: Separating sets by modal formulas. In Haeberer, A.M., ed.: Algebraic Methodology and Software Technology, AMAST'98. Volume 1548 of Lecture Notes in Computer Science., Springer (1999) 140–153
10. Heinemann, B.: Generalizing the modal and temporal logic of linear time. In Rus, T., ed.: Algebraic Methodology and Software Technology, AMAST 2000. Volume 1816 of Lecture Notes in Computer Science., Berlin, Springer (2000) 41–56
11. Weihrauch, K.: Computable Analysis. Springer, Berlin (2000)
12. Bourbaki, N.: General Topology, Part 1. Hermann, Paris (1966)
13. Heinemann, B.: Extended canonicity of certain topological properties of set spaces. In Vardi, M., Voronkov, A., eds.: Logic for Programming, Artificial Intelligence, and Reasoning. Volume 2850 of Lecture Notes in Artificial Intelligence., Berlin, Springer (2003) 135–149
14. Krommes, G.: A new proof of decidability for the modal logic of subset spaces. In ten Cate, B., ed.: Proceedings of the Eighth ESSLLI Student Session, Vienna, Austria (2003) 137–147
15. Blackburn, P., Spaan, E.: A modal perspective on the computational complexity of attribute value grammar. Journal of Logic, Language and Information **2** (1993) 129–169
16. Georgatos, K.: Knowledge theoretic properties of topological spaces. In Masuch, M., Pólos, L., eds.: Knowledge Representation and Uncertainty, Logic at Work. Volume 808 of Lecture Notes in Artificial Intelligence., Springer (1994) 147–159
17. Georgatos, K.: Knowledge on treelike spaces. Studia Logica **59** (1997) 271–301
18. Heinemann, B.: A hybrid treatment of evolutionary sets. In Coello, C.A.C., de Albornoz, A., Sucar, L.E., Battistutti, O.C., eds.: MICAI'2002: Advances in Artificial Intelligence. Volume 2313 of Lecture Notes in Artificial Intelligence., Berlin, Springer (2002) 204–213
19. Heinemann, B.: An application of monodic first-order temporal logic to reasoning about knowledge. In Reynolds, M., Sattar, A., eds.: TIME-ICTL 2003, Proceedings, Los Alamitos, Ca., IEEE Computer Society Press (2003) 10–16
20. McKinsey, J.C.C.: A solution to the decision problem for the Lewis systems S2 and S4, with an application to topology. Journal of Symbolic Logic **6** (1941) 117–141
21. Gabelaia, D.: Modal definability in topology. Master's thesis, ILLC, Universiteit van Amsterdam (2001)
22. Weiss, M.A., Parikh, R.: Completeness of certain bimodal logics for subset spaces. Studia Logica **71** (2002) 1–30

A Language
for Configuring Multi-level Specifications

Gillian Hill[1] and Steven Vickers[2]

[1] Department of Computer Science, City University,
Northampton Square, London, EC1V OHB and
Department of Computing, Imperial College, London SW7 2AZ
[2] School of Computer Science, The University of Birmingham,
Birmingham, B15 2TT

Abstract. This paper shows how systems can be built from their component parts with specified sharing. Its principle contribution is a modular language for configuring systems. A configuration is a description in the new language of how a system is constructed hierarchically from specifications of its component parts. Category theory has been used to represent the composition of specifications that share a component part by constructing colimits of diagrams. We reformulated this application of category theory to view both configured specifications and their diagrams as algebraic presentations of presheaves. The framework of presheaves leads naturally to a configuration language that expresses structuring from instances of specifications, and also incorporates a new notion of instance reduction to extract the component instances from a particular configuration. The language now expresses the hierarchical structuring of multi-level configured specifications. The syntax is simple because it is independent of any specification language; structuring a diagram to represent a configuration is simple because there is no need to calculate a colimit; and combining specifications is simple because structuring is by configuration morphisms with no need to flatten either specifications or their diagrams to calculate colimits.

1   Introduction

Large complex systems are put together, or configured, from smaller parts, some of which have already been put together from even smaller parts. This paper presents a modular language that expresses the hierarchical structuring of a system from specifications of the component parts. We review briefly the mathematical framework for configuration in order to focus on the constructs of the language. Systems configuration involves specifying each of the components of the system as well as the relationship of sharing between these components. The structure of the system is therefore expressed directly and mathematically by the syntax of the configuration language, while the history of system construction is kept at a second level of mathematical structure by the accumulation of many levels of configured specifications as configuration proceeds. We propose a new and simple concept of 'instance' of a specification to manage the complexity of large systems which may require many instances of their component parts.

C. Rattray et al. (Eds.): AMAST 2004, LNCS 3116, pp. 196–210, 2004.
© Springer-Verlag Berlin Heidelberg 2004

1.1   The Development of the Work

The motivation for our work has been to contribute to research into the modularization of systems. Our aim has been to design a language for configuring systems that is easy to use and involves concepts that should seem natural to software engineers. The language is simple because no assumptions are made about the underlying logic for specification. In earlier work we used the term 'module' to mean a 'uniquely named instance of a specification'. We now use the term 'instance', in order to avoid confusion with the use of 'module' to mean a 'composite structure wrapped up to form a single unit'. This latter use of 'module' is closer to the meaning of a configured specification.

Mathematically we were influenced by Burstall and Goguen, who gave a categorical semantics for their specification language Clear, in [2,3]. Categorical *colimits* were used for building complex specifications in [3,12]. We followed Oriat [9] in using colimits to express configuration in a way that was independent of any particular specification language. Oriat compared two approaches, one using diagrams and the other using a calculus of pushouts. Both in effect described the finite cocompletion of a category $\mathcal{C}$ of primitive (unconfigured) specifications.

In [13] we used instead *finitely presented presheaves*. This is a mathematically equivalent way of making a cocompletion, but leads to a different notation that very naturally describes how a configuration specifies *instances* of the component specifications, brought together with specified sharing of subcomponents. In flavour it is not unlike object-oriented languages, with the relationship between instances and specifications being analogous to that between objects and classes [8,1] (though [13] points out some respects in which the analogy cannot be pushed too far).

As a simple example of our notation we describe, in this paper, a shop in which there are two counters sharing a single queue in which customers wait for whichever counter becomes available. We also discuss how the abstract presheaf structure is a means for describing what 'subcomponents' are, with a categorical morphism from one specification, S, to another, T, representing a means by which each instance of T may be found to bring with it an instance of S — for example, how each shop counter has a queue associated with it.

However, the approach of [13] was entirely 'flat', in that each configuration was described in terms of its primitive components. A more modular style of configuration, developed in [6], allows multi-level configuration of either primitive or previously configured components. The structure of the categorical framework is simply a hierarchy of categories, in which each configuration belongs to a level and is represented by a structured categorical diagram. Morphisms, as simple implementations between configured specifications, are allowed to cross the levels of the hierarchy. There is a notion of assignment between the instances of specifications, and in addition proof obligations are discharged. A case study, of configuration up to four levels, illustrates the expressiveness of the language. The category theory becomes somewhat deeper, with the interesting possibility of incorporating recursively defined configurations, and is still to be worked out in detail. However, the configuration language is subject to only two simple modifications, and it is the aim of this paper to describe them.

1.2   The Structure of the Paper

In Sect. 2 the key idea of 'composites as presheaves' is introduced as an alternative to the established work on 'composites as colimits'. Presheaves provide a firm mathematical basis for the configuration language: presheaf presentations correspond to the components of a configuration and the relationship of sharing a common component; presheaf homomorphisms correspond to morphisms between configurations. In Sect. 3 we review the configuration language of [13]. Mathematically, it is formally equivalent to presenting presheaves by generators and relations, and that provides a well defined abstract semantics. Specification-ally, however, one should read each configuration as specifying components and sharing. In Sect. 4 it is extended to a modular language for multi-level configuration, with two new language constructions ('basic up' morphisms, and 'indirect' morphisms). We present the case study briefly in Sect. 5, and in Sect. 6 we draw conclusions.

2   Composite Specifications as Presheaves

We gave the theoretical framework chosen for configuration in "Presheaves as Configured Specifications", [13]. Most of the technical details of the paper are due to Steven Vickers. Configuration builds composite specifications as presheaves because they express colimits in category theory. Previous research has viewed composite specifications as colimits; the approaches have varied, however, in the choice of a category with appropriate colimits. For example, the pioneering work by Burstall and Goguen on expressing the structuring of specifications by constructing the colimits of diagrams, in [2, 3], was continued in the algebraic approach to specification [5, 4, 10] and also in proof-theoretic approaches [7, 11]. All these research methods depended on the different specification logics that were used, because they constructed colimits over some cocomplete category of specifications.

A contrasting aim of configuration is to separate the specification logic of the primitive (unconfigured) specifications from their configuration. Colimits are expressed in a category of configurations which is a free cocompletion of the category of primitive specifications. There are no assumptions about the underlying logic. This more general approach allows the category of primitive specifications to be incomplete.

We followed Oriat [9] in working more generally. She models the composition of specifications by working within an equiv-category of diagrams, which is finitely cocomplete. Her equiv-category of base specifications need not be complete, however. Oriat's constructions on diagrams are shown in [13] to be mathematically equivalent to the construction of presheaves in configuration.

2.1   Presheaves

The mathematical theory of presheaves provides an alternative construction to Oriat's cocomplete category of diagrams for modelling the composition of diagrams. Formally, the category $\mathbf{Set}^{\mathcal{C}^{op}}$ is the category of presheaves over a small category, $\mathcal{C}$. It follows that a *presheaf*, as an object in the category, is a functor from $\mathcal{C}^{op}$ to $\mathbf{Set}$, and a *presheaf morphism* is a natural transformation from one presheaf to another. The category $\mathbf{Set}^{\mathcal{C}^{op}}$ is a free cocompletion of $\mathcal{C}$. The theory is difficult, and it is understandable that its suitability for the practical application of building specifications might be questioned. There are, however, three main reasons why presheaves express configurations precisely: when presented algebraically, a presheaf expresses the structure of a configuration; a presheaf over $\mathcal{C}$ is formally a colimit of a diagram in $\mathcal{C}$; for each morphism in $\mathcal{C}$, a presheaf presentation provides a contravariant operator from which instance reduction is defined between configurations.

The fact that $\mathbf{Set}^{\mathcal{C}^{op}}$ is cocomplete means it has all small colimits. Intuitively, the fact that it is freely cocomplete means that it contains all the colimit objects and the morphisms to the colimit objects, but no more. Although expressing colimits by presheaves is more complicated theoretically than by just using diagrams, presenting presheaves algebraically simplifies the theory so that it is appropriate for configuration.

2.2   Presheaves Presented Algebraically

The key idea is that using generators and relations algebraically to present a presheaf corresponds directly to specifying components and the sharing of subcomponents in a composite system. This correspondence gives a direct physical interpretation to the configuration language.

Presheaves are presented, in detail in [13], as algebras for a many-sorted algebraic theory PreSh($\mathcal{C}$). The sorts of the theory are the objects of $\mathcal{C}$, and for each morphism $u : Y \rightarrow X$ in $\mathcal{C}$, there is a unary operator $\omega_u : X \rightarrow Y$.

The definition of an *algebra* P for PreSh($\mathcal{C}$) gives:

- for each object X of $\mathcal{C}$, a set $P(X)$, the *carrier* at X;
- for each morphism $u : Y \rightarrow X$, an operation $P(u) : P(X) \rightarrow P(Y)$ (written $x \mapsto ux$).

Algebras and homomorphisms for PreSh($\mathcal{C}$) are equivalent to presheaves and presheaf morphisms. The correspondence with configurations becomes apparent when presheaves are presented, as algebras of the algebraic theory PreSh($\mathcal{C}$), by generators and relations. We give only the main points of the correspondence:

- A *set of generators* (with respect to PreSh($\mathcal{C}$)) is a set G equipped with a function $D : G \rightarrow$ ob $\mathcal{C}$, assigning a sort to each generator in G. In configuration the generators stand for instances of specifications. Instead of denoting the sort of a generator by $D(g) = X$, writing $g : X$ is more suggestive of declaring an instance of the specification X.

- If G is a set of generators, then a *relation* over G is a pair (e_1, e_2) (written as an equation $e_1 = e_2$) where e_1 and e_2 are two expressions of the same sort, X, say. In configuration, the expressions will describe instances of the same specification. Expressions are built out of G by applying a unary operation that corresponds to a morphism. Relations can be reduced to the form $ug_1 = ug_2$.
- A *presentation* is a pair (G, R) where G is a set of generators and R is a set of relations over G. The presheaf that is presented by (G, R) is denoted $\mathrm{PreSh}\langle G \mid R\rangle$. Presheaf presentations correspond to configurations.

Example 1. Suppose $\mathcal{C}$ is the category with two objects, X and Y, and one morphism $u : X \to Y$ (and two identity morphisms). A presheaf P over $\mathcal{C}$ is a pair of sets $P(X)$ and $P(Y)$ equipped with a function, the u operation from $P(Y)$ to $P(X)$. Suppose P is presented by generators g_1 and g_2 (both of sort Y) subject to $ug_1 = ug_2$. This is denoted by:

$$P = \mathrm{PreSh}\langle g_1, g_2 : Y \mid ug_1 = ug_2\rangle$$

Then $P(Y) = \{g_1, g_2\}$, and $P(X)$ has a single element to which u maps both g_1 and g_2. In configuration this single element is the reduction by u of g_1 and g_2.

An advantage of the correspondence with presheaves for configuration is that instead of describing an entire presheaf, by objects and morphisms, enough elements are presented to generate the rest algebraically. Although diagrams provide a simpler way of describing colimits than presheaves, the presentation by generators and relations is more natural than diagrams for expressing the configuration of components (by generators) and the sharing of components (by shared reducts).

2.3   Primitive Specifications

Configuration is over an *arbitrary* base category $\mathcal{C}$. The objects of $\mathcal{C}$ are primitive (unconfigured) specifications that, for instance, may be named after the theory presentations in the category Thpr, but are without their logical properties. For example, a theory presentation for a queue could be named as a primitive specification *Queue* in $\mathcal{C}$. The morphisms in $\mathcal{C}$ are named after the interpretations between theory presentations in mor Thpr. The category $\mathcal{C}$ is the working category for configuration: its objects are those specifications that represent the basic components of the particular system to be configured. The structure of $\mathcal{C}$ is not restricted by making it cocomplete; colimits are constructed as presheaves over $\mathcal{C}$ in a free cocompletion. This means that presheaves express configuration from primitive specifications without referring to their logical properties. Already configuration is shown to contrast with other approaches, such as [11], that work with a category of specifications over some chosen logic; presheaves are colimits whereas other approaches construct colimits of diagrams.

3   The Language for Flat Configurations

This section presents the language of [13], expressing the flat configuration of a system from primitive component parts. It assumes some fixed small category $\mathcal{C}$, whose objects stand for the primitive specifications, and constructs a category $\text{Config}(\mathcal{C})$ whose objects stand for the configured specifications.

It is also important to understand the role of the morphisms. If $f : S \rightarrow T$ is a morphism (in $\mathcal{C}$ or in $\text{Config}(\mathcal{C})$), then it is intended to be interpreted as showing a way by which each instance of the specification T can be 'reduced to' an instance of S. If IT is a T instance, then we write f IT for the correspondingly reduced S instance. A typical example of what 'reduced to' means is when each instance of T — that is to say, each thing satisfying the specification T — already contains within it (as a subcomponent) an instance of S. There may be different modes of reduction. For example, if each T instance contains *two* S instances in it, then there must be two morphisms $S \rightarrow T$.

3.1   Flat Configurations

The configured specification, S, structured from instances of primitive specifications, could be expressed by:

spec S **is**
   components
      $\text{IS}_1 : S_1$;
      $\vdots$
      $\text{IS}_i : S_i$;
      $\vdots$
      $\text{IS}_n : S_n$
   equations
      e1: $f \text{ IS}_i = g \text{ IS}_j$
      $\vdots$
endspec

The relation $e1$ states the equality between the two reducts, instances of the primitive specification T that is the common source of the morphisms f to S_i and g to S_j. The specification S_i, an object in $\mathcal{C}$, only becomes a specification in the flat world $\text{Config}(\mathcal{C})$ when it is configured as $conf\_S_i$ and declares a formal name for a single instance of S_i:

spec $conf\_S_i$ **is**
   components
      $\text{I}_{S_i} : S_i$
endspec

Intuitively, $conf\_S_i$ puts a wrapper round the named instance I_{S_i} of S_i.

Example 2. A system of counters in a post office has queues of people waiting to be served. Let *Counter* and *Queue* be specifications whose instances are actual counters and actual queuing lines. Each counter has a queue, and this instance reduction from *Counter* instances to *Queue* instances is to be represented by a morphism $i : Queue \rightarrow Counter$. The configured specification that expresses the sharing of that queue by two counters in a post office is presented as:

spec *SharingOfQueue* **is**
   components
      $C_1 : Counter$;
      $C_2 : Counter$;
   equations
      e1: $i\ C_1 = i\ C_2$
endspec

Although the instance of the shared queue is not declared in this general form, the expressions $i\ C_1$ and $i\ C_2$ of $e1$ each describe the instance reduct for the specification *Queue*. The specification *conf_Counter* could be configured in Config($\mathcal{C}$) by 'wrapping it up' as:

spec *conf_Counter* **is**
   components
      IC : *Counter*
endspec

3.2   Morphisms between Flat Configurations

A morphism from one configuration, S, to another, T, is again going to represent instance reduction, showing how any instance of T can be reduced to an instance of S. We shall view this as *implementation*. Any T instance must contain all the components of S, with the correct sharing, and so provide an implementation of the specification S. The implementation is expressed by interpreting the individual components of S in T according to the assignments I $\mapsto f$ J, for I, a component of S, and J, a component of T. In addition a proof must also be given that the assignments respect the equations in S. The syntax for a configuration morphism as an implementation must therefore include both assignment of components and proof that equations hold. That proof, that is fundamental to the formal building of a system from its components, is made in the syntax of the configuration language using equations in T in a forwards or backwards direction.

Example 3. (from Ex. 2) We define two morphisms, f and g, from the configuration *conf_Counter* to *SharingOfQueue*, and a morphism, h, from *SharingOfQueue* to *conf_Counter*. f and g pick out the two counters C_1 and C_2 of *SharingOfQueue*, thus showing two ways by which a *SharingOfQueue* instance can be reduced to a *conf_Counter* instance. h describes a degenerate way in which single *conf_Counter* instance can be used to provide a *SharingOfQueue* instance, with the single counter doing all the work for two counters.

implementation f: *conf_Counter*
$\qquad\qquad\qquad \to SharingOfQueue$
IC $\mapsto$ $id\,_{Counter}$ C_1;
endimp

implementation g: *conf_Counter*
$\qquad\qquad\qquad \to SharingOfQueue$
IC $\mapsto$ $id\,_{Counter}$ C_2;
endimp

implementation h: *SharingOfQueue*
$\qquad\qquad\qquad \to conf\_Counter$
$C_1 \mapsto id\,_{Counter}$ IC;
$C_2 \mapsto id\,_{Counter}$ IC;
To check $e1$ *of SharingOfQueue*:
$i\ C_1 \mapsto i$; $id\,_{Counter}$ IC
$\qquad \hookleftarrow i\ C_2$
endimp

The composition of morphisms is expressed by the notation ; . The proof that
the equation $e1 : i\ C_1 = i\ C_2$ in *SharingOfQueue* is respected by the assignment
of instances to *conf_Counter* is simple. The symbol $\mapsto$ denotes the assignment
from the instance on the left hand side of $e1$ of *SharingOfQueue* to the instance
of *conf_Counter*. Finally the symbol $\hookleftarrow$ denotes the assignment from $i\ C_2$ on the
right hand side of $e1$ in *SharingOfQueue* to i ; $id\,_{Counter}$ IC in *conf_Counter*.
The morphism h makes the point that the mathematics of colimits as used
for specification can specify equalities but not inequalities.

4   The Language for Multi-level Configurations

The aim of this section is to extend the configuration language by modularity to
express the hierarchical structuring of multi-level configurations, independently
of any logic. The syntax of the modular configuration language directly expresses
the structure of a system, so that the user of the configuration language is able
to record the history of configuration in easily understood amounts.

Configuration offers a semantics for the structuring of specifications which is
new in two respects. The first is that flattening can be avoided because configurations are isomorphic to their flattened form. The second respect is that the
manipulations do not rely on a flattened form even existing. The language allows
morphisms to be defined with 'relative' flattening down a few levels in the hierarchical configuration but without necessarily reaching a primitive level. To match
this, [6] does not construct the mathematical workspace inductively, starting
with the primitive level and working up, but instead offers an axiomatic approach that identifies the structure needed to interpret the language constructs.
Potentially then, the workspace can contain configurations of infinite depth and
give meaning to recursively defined configurations.

4.1   The Objects and Morphisms in the Configuration Workspace

Providing a new mathematical semantics for structuring multi-level specifications in a categorical workspace leads to a new engineering style of manipulation
for the specifications. The primitive and configured specifications are collected
together in a single category and configuration becomes a construction that can
be applied with arbitrary objects and morphisms. Since S and $conf\_S$ are now
objects in the same category they are assumed to be isomorphic, and this isomorphism leads to the extra syntactic features of *basic up* and *indirect* morphisms
in the multi-level language.

Objects are either primitive or configured.

Primitive objects are drawn from a category $\mathcal{C}$.

Configured objects use the keywords **spec** and **endspec** as before to put together components with sharing. However, now their component specifications
may themselves be either primitive or configured, possibly with some of each.

Morphisms may be defined between any objects in the workspace, and are
needed to construct new objects or to prove that objects are equivalent. Again,
they represent a contravariant notion of instance reduction, that gets an instance
of the source specification from an instance of the target.

Primitive morphisms from $\mathcal{C}$ are between primitive specifications.

Configuration morphisms are defined as in Sect. 3.

However, new morphisms are needed to make any configuration S isomorphic
to the configured specification $conf\_S$ that declares an instance of S.

4.2   Basic Up Morphisms

These morphisms arise from the need for a morphism from $S \rightarrow conf\_S$. Suppose
I_S is declared as the component in $conf\_S$. Our syntactic device is to use that
instance name also as the name of the morphism, $I_S : S \rightarrow conf\_S$. If IS: S is a
component in a configuration T, then as in Sect. 3, we can define a configured
morphism

implementation h: $conf\_S \rightarrow T$
$I_S \mapsto id_S$ IS
endimp

The morphism h can be composed with the isomorphism $S \rightarrow conf\_S$ to get
a morphism f from S to T. Again we apply the device of using the instance
name IS as the name of this composite morphism, $IS : S \rightarrow T$, and this is the
most general form of what we shall call a *basic up* morphism. Note that S may
be either primitive or configured.

4.3   Indirect Morphisms

These arise from the morphism $conf\_S \rightarrow S$ and are defined as *indirect implementations* that use the keyword **given**. This syntax provides a formal name for
an instance in the target specification of the morphism:

implementation $f: T \to S$
given instance IS: S

$\vdots$

endimp

Here the middle, omitted, part is just the usual format (as before) for the body of a configuration morphism. The instance name provided can be taken as defining an anonymous configuration which is isomorphic to $conf\_S$:

spec - - - is
  components
    IS : S
endspec

The indirect definition of f supplies the data for a morphism from T to this anonymous configuration. This is then composed with the isomorphism $conf\_S \to S$ to give the indirect morphism $f : T \to S$. Again indirect morphisms arise from the need to have every S isomorphic to $conf\_S$. The isomorphism $conf\_S \to S$ can itself be denoted using the 'given' notation.

4.4 Morphisms between Multi-level Configurations

We have defined morphisms from configured specifications to primitives. We also need to define them between configured specifications.

Example 4. (from Ex. 2) Second level and first level configurations illustrate two ways of making a post office with three counters and one shared queue:

spec *ExtendedShop* **is**
  components
    C_1QC_2 : *SharingOfQueue* ;
    C_3 : *Counter* ;
  equations
    e1: i $C_3 = i$; C_1 C_1QC_2
endspec

The morphism C_1 is a basic up morphism.

spec *NewShop* **is**
  components
    C_1 : *Counter* ;
    C_2 : *Counter* ;
    C_3 : *Counter* ;
  equations
    e1: i $C_1 = i$ C_2 ;
    e2: i $C_1 = i$ C_3
endspec

These configurations are isomorphic, but the isomorphism g: *ExtendedShop* → *NewShop* cannot be defined except indirectly, with **given**. The syntax of the indirect implementation, g, also uses a keyword **where** to introduce a locally defined morphism, f: *SharingOfQueue* → *NewShop*.

implementation g: *ExtendedShop*
                                → *NewShop*
given instance INS: *NewShop*
$C_1 Q C_2 \mapsto f$ INS ;
$C_3 \mapsto C_3$ INS ;
where

        implementation f: *SharingOfQueue*
                   → *NewShop*
    $C_1 \mapsto C_1$;
    $C_2 \mapsto C_2$;
    To check e1 of SharingOfQueue:
    $i\ C_1 \mapsto i\ C_1$
        $= i\ C_2$ *by e1 of NewShop*
        $\hookleftarrow i\ C_2$
    endimp
To check e1 of ExtendedShop:
$i\ C_3 \mapsto i\ ;\ C_3$ INS
    $= i\ ;\ C_1$ INS *by e2 of NewShop*
    $= i\ ;\ C_1\ ;\ f$ INS
    $\hookleftarrow i\ ;\ C_1\ C_1 Q C_2$
endimp

The proof for equation $e1$ of *ExtendedShop* uses the fact that C_1 INS $= C_1\ ;\ f$ INS. This comes directly out of the definition of f, from $C_1 \mapsto C_1$.

5   A Case Study

We use the new configuration language in a case study, based on an example of Oriat's [9], to express alternative configurations for the theory of rings. In [6] the aim of the case study is to compare Oriat's method of composing specifications, by constructing the pushouts of diagrams, with the method of configuration. Since in configuration both specifications and their diagrams express algebraic presentations of presheaves, and finitely presented presheaves express colimits, the need to construct pushout diagrams is bypassed. Since equivalence between configurations can be proved textually, Oriat's need to flatten diagrams (to construct their colimits) and to complete diagrams before normalizing them can also be bypassed.

5.1   Building Flat Configurations from Primitive Specifications

The theory presentations and theory morphisms that underly the primitive specifications for the components used to configure a ring are expressed in the style of Z schemas. As in Sect. 2.3 we use the name of each theory presentation, forgetting its logical properties, to identify a primitive specification. The simplest component of the mathematical structure of a ring expresses a single sort s.

$$\boxed{\;\underline{A\,sort[s]}}$$

The schema $Bin\text{-}op$ specifies a sort, also called s, and a binary operator op:

$$
\begin{array}{|l}
\underline{Bin\text{-}op[s]} \\
op : s \times s \to s \\
\end{array}
$$

The theory morphism $s : A\,sort \to Bin\text{-}op$ maps the sort of $A\,sort$ to the sort of $Bin\text{-}op$. The schema for the structure of a monoid is:

$$
\begin{array}{|l}
\underline{Monoid[s]} \\
\times : s \times s \to s \\
1 :\to s \\
\hline
\forall\, x, y, z : s \,.\, (x \times y) \times z = x \times (y \times z) \\
\forall\, x : s \,.\, 1 \times x = x \\
\forall\, x : s \,.\, x \times 1 = x \\
\end{array}
$$

The theory morphism $b : Bin\text{-}op \to Monoid$ maps the sort of $Bin\text{-}op$ to the sort of the monoid, and the operator op of $Bin\text{-}op$ to the operator $\times$ in the monoid. The theory presentation for an Abelian group is formed from $Monoid$ by adding an inverse function and the property of commutativity for the binary operator, $+$. The theory morphism m maps the operator $\times$ of $Monoid$ to the operator $+$ of $Abel\text{-}group$ and the constant 1 of $Monoid$ to the constant 0 of $Abel\text{-}group$.

$$
\begin{array}{|l}
\underline{Abel\text{-}group[s]} \\
+ : s, s \to s \\
0 :\to s \\
inv : s \to s \\
\hline
\forall\, x, y, z : s \,.\, (x + y) + z = x + (y + z) \\
\forall\, x : s \,.\, 0 + x = x \\
\forall\, x : s \,.\, inv(x) + x = 0 \\
\forall\, x, y : s \,.\, x + y = y + x \\
\end{array}
$$

Finally the schema $Distributive$ specifies two binary operators that are related by the property of distributivity. There are two morphisms from $Bin\text{-}op$ to $Distributive$: the morphism m_+ maps op to $+$; the morphism $m_\times$ maps op to $\times$. The axioms for the distributive structure express both left and right distributivity for $\times$ over $+$.

$$\boxed{\begin{array}{l} \underline{\textit{Distributive}[s]} \\ \quad + : s, s \rightarrow s \\ \quad \times : s, s \rightarrow s \\ \hline \quad \forall\, x, y, z : s \,.\, x \times (y + z) = (x \times y) + (x \times z) \\ \quad \forall\, x, y, z : s \,.\, (y + z) \times x = (y \times x) + (z \times x) \end{array}}$$

In the text of the configured specifications we use abbreviations for the instance names. Of four equivalent specifications for the flat configuration of a ring the following is the most compact:

spec Ring1 **is**
  components
    M : Monoid ;
    A : Abel-group ;
    D : Distributive ;
  equations
    e1: b ; m A $= m_+$ D ;
    e2: b M $= m_\times$ D
endspec

The specification *Ring1* describes the sharing of the boolean operators explicitly. The instance reduct b ; m A gives the binary operator for addition, derived by reduction from the instance A of *Abel-group*. The instance reduct b M is the operator for multiplication, derived by reduction from the instance M of *Monoid*. That is, *e1* describes the sharing of the addition instance of *Bin-op*, and *e2* describes the sharing of the multiplication instance of *Bin-op*.

5.2 Natural Uses of Modularization

In Oriat's language of terms, all colimits of representative diagrams are pushouts. In the configuration language, modularization is only used if required specificationally: it is not imposed by pushout terms. Configurations that correspond to Oriat's modular constructions of a ring are built in [6]. Two of these are more natural because, although they are built by adding distributivity to a pseudo-ring, neither requires the construction of an extra configuration for the pair of binary operators. Together with the flat *Ring1*, we select these modularized configurations as the ideal configurations for a ring. *Ring4*, a fourth-level specification, illustrates the flexibility of our language by expressing the sharing of each instance of the binary operator in an equation. The history of the configuration is presented first in the three lower-level configurations.

spec Pair_Bin-op_and_Asort **is**
  components
    a : Bin-op ;
    m : Bin-op ;
  equations
    e1: s a $= s$ m *sharing the instance* s
endspec

spec Pair_Bin-op_Asort_and_Monoid **is**
  components
    M : Monoid ;
    ams : Pair_Bin-op_and_Asort;
  equations
  e1: b M $=$ m ams
endspec

spec Pair_Bin-op_Asort_Monoid_and_Abel-group **is**
  components
    amsM : Pair_Bin-op_Asort_and_Monoid;
    A : Abel-group ;
  equations
   e1: a ; ams amsM $=$ b ; m A
endspec

spec Ring4 **is**
  components
    D : Distributive ;
    amsMA : Pair_Bin-op_Asort_Monoid_and_Abel-group ;
  equations
   e1: $m_\times$ D $=$ m ; ams ; $amsM$ amsMA ; *sharing the instance* m
   e2: m_+ D $=$ a ; ams ; $amsM$ amsMA *sharing the instance* a
endspec

6   Conclusions

We thank the reviewers for inspiring us to improve the paper. Our goal has been to introduce, independently of specification language, a modular configuration language that expresses the construction of large complex systems from their component parts, with specified sharing. We have already presented in [13] a configuration language based on components and sharing that is independent of specification language. It has an abstract semantics using presheaves that is mathematically equivalent to the diagrammatic approach of [9]. However, it is limited to flat configurations: it has no modularity and is unable to express any further structuring to multi-level configurations. The modularity here, avoiding the need to flatten structured specifications, has been achieved categorically in [6] by having explicit isomorphisms between unflattened configurations that would become equivalent when flattened. Linguistically it works by the use of two new constructions, the basic ups and the indirect configuration morphisms, whose interpretation provides those isomorphisms. Although the configuration language has been presented with a detailed case study in [6], more work is required on the semantics of the language. The need to avoid the absolute flattening of configured specifications to a primitive level suggests that a hierarchical workspace of infinite depth should be constructed with the potential to deal with recursively defined configurations.

References

1. Booch, G.: Object-Oriented Analysis and Design. The Benjamin Cummings Publishing Company, Inc. second edition (1994)
2. Burstall, R. M., Goguen, J. A.: Putting Theories Together to Make Specifications. Proc.of the 5th. International Joint Conference on Artificial Intelligence, Cambridge, Mass. (1977) 1045–1058
3. Burstall, R. M., Goguen, J. A.: The Semantics of Clear, A Specification Language. Abstract Software Specifications, LNCS 86 (1979)
4. Ehrig, H., Fey, W., Hansen, H., Lowe, M., Papisi-Presicce, F.: Algebraic Theory of Modular Specification Development. Technical report Technical University of Berlin (1987)
5. Ehrig, H., Mahr, B.: Fundamentals of Algebraic Specification 1: Equations and Initial Semantics. Springer-Verlag (1985)
6. Hill, G.: A Language for Configuring Systems. PhD Thesis, Department of Computing, Imperial College, University of London (2002)
7. Maibaum, T. S. E., Sadler, M. R., Veloso, P. A. S.: Logical Specification and Implementation. Foundations of Software Technology and Theoretical Computer Science, LNCS **181** Springer-Verlag (1984)
8. Meyer, B.: Reusability. IEEE Software, (1987) 50–63.
9. Oriat, C.: Detecting Equivalence of Modular Specifications with Categorical Diagrams. Theoretical Computer Science, 247 (2000) 141–190
10. Sannella, D., Tarlecki, A.: Toward Formal Development of Programs from Algebraic Specifications: Implementations Revisited. Acta Informatica 25 (1988) **3** 233–281
11. Srinivas, Y. V., Jellig, R.: Formal Support for Composing Software. Proc. of Conference on the Mathematics of Program Construction, Kloster Irsee, Germany July (1995) Kestrel Institute Technical Report KES.U.94.5
12. Veloso, P. A. S., Fiadeiro, J., Veloso, S. R. M.: On local modularity and interpolation in entailment systems. Inform. Proc. Lett. 82(4)(2002) 203 – 211
13. Vickers, S., Hill, G.: Presheaves as Configured Specifications. Formal Aspects of Computing 13 (2001) 32–49

Flexible Proof Reuse for Software Verification

Chris Hunter, Peter Robinson, and Paul Strooper

School of Information Technology and Electrical Engineering
The University of Queensland, Brisbane, Australia
{chris,pjr,pstroop}@itee.uq.edu.au

Abstract. Proof reuse, or analogical reasoning, involves reusing the proof of a source theorem in the proof of a target conjecture. We have developed a method for proof reuse that is based on the generalisation – replay paradigm described in the literature, in which a generalisation of the source proof is replayed to construct the target proof. In this paper, we describe the novel aspects of our method, which include a technique for producing more accurate source proof generalisations (using knowledge of the target goal), as well as a flexible replay strategy that allows the user to set various parameters to control the size and the shape of the search space. Finally, we report on the results of applying this method to a case study from the realm of software verification.

1 Introduction

Formal methods provide the promise of software that is provably free of faults. To have confidence a method has been applied correctly, proofs associated with the method must be mechanically verified. For most applications, however, such mechanical verification is infeasible due to the amount of user interaction required for even quite simple proofs.

One way to reduce the cost of mechanical verification is to reuse an existing proof, called the *source* proof, to help prove a *target* conjecture. Methods described in the literature work either by

- verifying axioms used in the source proof for the domain of the target conjecture (e.g. Kolbe and Walther's Plagiator system [10]), or
- using the structure of the source proof to construct a proof of the target conjecture, either at the object level (e.g. the work of Melis and Schairer [6]) or at the meta level (e.g. the Abalone system of Melis and Whittle [7]).

In this paper we describe a method for proof reuse that makes use of the source proof structure at the object level. It is based on the two-phase paradigm [6,7] that involves *generalising* a source proof to produce a proof schema that can then be *replayed* on a target conjecture.

The paper is organised as follows. In Sect. 2 we introduce our case study – verifying the correctness of an algorithm for solving the Dutch National Flag problem – which we will use to demonstrate and evaluate our method. In Sect. 3 we provide a brief overview of our basic generalisation – replay approach to reuse.

C. Rattray et al. (Eds.): AMAST 2004, LNCS 3116, pp. 211–225, 2004.

In Sect. 4 we build on these ideas by describing a new approach for dealing with situations in which the source proof schema does not match the target goal. Next, in Sect. 5 we outline a strategy that we have developed and implemented for replaying a source proof. The strategy is based on a mix of heuristics and practical experience with replaying actual proofs. Unlike the replay strategy of [6], we allow a general search at each step of the replay, rather than simply applying (after appropriate preparation) the tactic used at the analogous point in the search proof. This gives our strategy added flexibility but means we must control the size of the search space. In Sect. 6 we report on the results of applying our method to the Dutch National Flag problem. Finally in Sect. 7 we present our conclusions.

2   A Case Study – The Dutch National Flag

The specific software verification domain for which we have developed our reuse method involves proving the correctness of imperative programs using a Hoare-style logic [2]. One of the examples we have used, and the one to which we will refer throughout the rest of this paper, is Dijkstra's *Dutch National Flag Problem* [1], in which coloured array elements must be partitioned in accordance with the Dutch national flag (red followed by white followed by blue). A standard solution described by Kaldewaij [5], shown in Fig. 1, uses a single loop to maintain three (initially empty) non-overlapping segments, each containing elements of one colour. The loop gradually increases the size of these segments until all the elements have been processed.

```
|[    con N : int{N ≥ 0};
     var h : array[0..N) of [red, white, blue];  r, w, b :int
         r, w, b := 0, 0, N;
         do w ≠ b →
             Invariant ≡ {
                 (∀ i : 0 ≤ i < r : h.i = red) ∧ (∀ i : r ≤ i < w : h.i = white) ∧
                 (∀ i : b ≤ i < N : h.i = blue) ∧ 0 ≤ r ≤ w ≤ b ≤ n}
             if h.w = red → swap.h.w.r;  r, w := r + 1, w + 1
             [] h.w = white → w := w + 1
             [] h.w = blue → swap.h.w.(b − 1);  b := b − 1
             fi
         od
     {(∀ i : 0 ≤ i < r : h.i = red) ∧ (∀ i : r ≤ i < w : h.i = white) ∧
     (∀ i : w ≤ i < N : h.i = blue)}
]|
```

Fig. 1. A program written in the guarded command language for solving the *Dutch National Flag problem*

The proof of correctness of this program consists of a number of parts, some of which are quite easy to show, e.g. showing the assignments are type correct. However, in this paper we will focus on the longest (accounting for over half of the total user interactions) and most difficult part of the proof, which is showing the invariant predicate (*Invariant*) is maintained by each iteration of the loop. This requires showing that the combination of the invariant and the loop guard holds across each branch of the conditional statement, e.g. for the third branch (the *Blue* branch), this requires the following to be shown (we use the notation $A \longrightarrow B$ to mean that B holds under the assumptions A – our work is equally applicable to both natural deduction and sequent calculus theoreom provers):

$$(\forall i : 0 \leq i < r : h.i = red) \wedge (\forall i : r \leq i < w : h.i = white) \wedge$$
$$(\forall i : b \leq i < n : h.i = blue) \wedge w \neq b \wedge h.w = blue \wedge 0 \leq r \leq w \leq b \leq n$$
$$\longrightarrow$$
$$(\forall i : 0 \leq i < r : h(w, b-1 : h.(b-1), h.w).i = red) \wedge$$
$$(\forall i : r \leq i < w : h(w, b-1 : h.(b-1), h.w).i = white) \wedge$$
$$(\forall i : b-1 \leq i < n : h(w, b-1 : h.(b-1), h.w).i = blue)$$
$$0 \leq r \leq w \leq b-1 \leq n$$

To prove this, we need to show each conjunct of the right hand side holds given the assumptions on the left. For our purposes we will ignore the final conjunct (whose proof is not interesting) and concentrate on the first three, which we will subsequently refer to as *Blue*.1, *Blue*.2 and *Blue*.3 (and similarly for the *Red* branch). We will also ignore the *White* branch since its proof is trivial.

3   Proof Reuse – An Overview

In this section we briefly outline our basic two-step generalisation – replay model of proof reuse. In the *generalisation* phase, a proof schema of the source proof is constructed. The schema construction method involves reconstructing the source proof using a variable in place of the original source goal, and using generalised inference rules in which function symbols have been replaced with function variables. Starting with a variable instead of the original goal allows us to generalise subterms in the goal that are not relevant to the proof – the variable is instantiated as part of the schema construction as necessary by the application of the generalised inference rules. Given a proof P, we write $Schema(P)$ to denote the proof schema constructed from P.

As an example, consider the following proof fragment $P.1$, taken from the proof of *Blue*.1 (the proof is meant to be read down the page in a goal-directed fashion, child nodes are indented, multiple assumptions on the left of the $\longrightarrow$ are separated by commas, and assumptions used in the application of a rule are underlined):

$$(x < y \wedge y < z) \Rightarrow x \neq z - 1 \qquad \text{(imp\_intro)}$$
$$\underline{x < y \wedge y < z} \longrightarrow x \neq z - 1 \qquad \text{(lt\_lt\_trans)}$$
$$x < y \wedge y < z, \underline{x < z - 1} \longrightarrow x \neq z - 1 \qquad \text{(lt\_imp\_neq)}$$

Its corresponding proof schema $Schema(P.1)$ is:

$$\frac{\dfrac{(X\ f_1\ Y \wedge Y\ f_1\ Z) \Rightarrow X\ f_2\ (Z\ f_3\ W)}{X\ f_1\ Y \wedge Y\ f_1\ Z \longrightarrow X\ f_2\ (Z\ f_3\ W)}}{X\ f_1\ Y \wedge Y\ f_1\ Z, X\ f_1\ (Z\ f_3\ W) \longrightarrow X\ f_2\ (Z\ f_3\ W)}$$

For clarity the logical connectives $\wedge$ and $\Rightarrow$ have not been generalised to function variables – in practice we treat them the same as any other function symbols. Note that unlike the work of [4], the schema is not a proof of a valid schematic theorem – it is simply used to guide the subsequent replay phase.

   In the *replay* phase, a schema is used to direct the proof of a target goal. This is achieved by replaying the source proof but with the single rule application at each step replaced by a search. The aim of the search is to produce a set of subgoals that match the corresponding subgoals of the schema; unifying each subgoal with its analog instantiates the schema.

   For example, suppose we want to use the schema in the previous example to prove the following:

$$(x > y \wedge y > z) \Rightarrow x - 1 \neq z \tag{L.1}$$

Because $L.1$ is not an instance of the schema goal, the first step is to use a rule to transform $x - 1 \neq z$ to $x \neq z + 1$. Instantiating the schema goal with the result gives the following:

$$\frac{\dfrac{(x > y \wedge y > z) \Rightarrow x \neq z + 1}{x > y \wedge y > z \longrightarrow x \neq z + 1}}{x > y \wedge y > z, \underline{x > z + 1} \longrightarrow x \neq z + 1}$$

Note that although this is a total instantiation of the schema, the remainder of the replay must still be completed to justify the proof steps suggested by the instantiated schema. This is a trivial process for our simple example since each of the required steps corresponds to an existing inference rule.

4   Goal-Directed Schema Generalisation

In the previous section we assumed the target conjecture was simply unified with the schema root (possibly after an initial transformation). This will not always be possible, however. For example, suppose we try to use $Schema(P.1)$ to help prove the goal:

$$(x < y \wedge y \leq z) \Rightarrow x \neq z \tag{L.2}$$

Here, f_1 cannot be instantiated to both $<$ and $\leq$, and the term $(Z\ f_3\ W)$ does not match the analogous term z in the target goal[1]. Using higher-order unification [3]

---

[1] Note that for this simple example, the analogical replay may, depending on the search depth, find that $L.2$ could be transformed to $x < y \wedge y < (z + 1) \Rightarrow x \neq (z + 1) - 1$ (which *is* an instance of the schema root). In general, such a transformation may not exist.

(the approach many accounts of proof reuse in the literature take) would allow the generalisation of the $(Z \; f_3 \; W)$ term, however no term can be globally substituted for f_1 to produce a sufficiently general schema.

In this section we describe an algorithm that uses both a more fine-grained version of unification together with anti-unification to generalise a schema based on the form of the target goal (hence the term goal-directed). To motivate the design of the algorithm, the exposition follows the calculation of the above generalised schema.

4.1   An Algorithm for Schema Generalisation

The first step of our schema generalisation algorithm relies on anti-unification [8], an operation that calculates the *least general generalisation (LGG)* of a set of terms. Given two terms t_1 and t_2 we say that t_1 is more general than t_2 if we can find a substitution θ such that $t_1\theta = t_2$. We say that g is a generalisation of a set of terms E provided it is more general than each term in E, and define it to be the LGG of E provided every generalisation of E is more general than g. A set of terms has at most one LGG (up to variable renaming).

The standard anti-unification operation fails for terms built from either different function symbols, or for function symbols with different arities, however for our purposes we require this condition to be relaxed (since we will always want the anti-unification to succeed). Specifically, if the function symbols have different arities, the terms are generalised to a variable, while if the arities are the same but the function symbols are different, the generalisation consists of a term with the same arity but with a variable in place of the function symbol (the subterms are recursively anti-unified as normal).

The procedure we use for schema generalisation is a variant of the original schema construction technique that starts not with a variable, but with the LGG of the target goal and the root of the original schema. For our example, consider the target goal $L.2$ and the root of $Schema(P.1)$:

$$(x < y \wedge y \leq z) \Rightarrow x \neq z \qquad \text{and} \qquad (X \; f_1 \; Y \wedge Y \; f_1 \; Z) \Rightarrow X \; f_2 \; (Z \; f_3 \; W)$$

The LGG of these terms is (where we have used variable names from the root of $Schema(P.1)$ wherever possible):

$$(X \; f_1 \; Y \wedge Y \; f_4 \; Z) \Rightarrow X \; f_2 \; A$$

By reconstructing the schema using the generalised target goal as a base, the idea is that the original schema will be generalised in *exactly* those locations that are required to accommodate the target goal, i.e., a kind of least generalisation of the schema. For this to happen, however, we must ensure that we only allow a (partial) one-sided unification when the generalised rule is applied – specifically, the generalised target goal should be frozen so the schema construction does not erase the results of performing the original anti-unification (freezing a term prevents variables occurring in that term from being instantiated).

As a consequence of beginning the schema construction with a (frozen) term that does not unify with the goal of the original schema, a schema construction step is now more complicated than just a simple unification of an open node in the schema with the goal of a generalised rule (otherwise at least one such step will fail). Specifically, at each step a structural mapping between terms is constructed and used to partially unify the terms, producing a more fine-grained unification that always succeeds.

To see how this works, we show how the schema in our previous example can be calculated. We start with the frozen generalised target goal:

$$(X \ f_1 \ Y \wedge Y \ f_4 \ Z) \Rightarrow X \ f_2 \ A$$

The first step of the schema construction (applying the generalised form of *imp_intro*) can be carried out using normal unification. The schema afterwards is:

$$(X \ f_1 \ Y \wedge Y \ f_4 \ Z) \Rightarrow X \ f_2 \ A$$
$$(X \ f_1 \ Y \wedge Y \ f_4 \ Z) \longrightarrow X \ f_2 \ A$$

Now consider the rule *lt_lt_trans* (left) and its generalised form (right) that we wish to apply:

$$\frac{\Gamma, H < J - 1 \longrightarrow G}{\Gamma \longrightarrow G} \qquad\qquad \frac{\Gamma, H \ f_5 \ (J \ f_6 \ K) \longrightarrow G}{\Gamma \longrightarrow G}$$
$$(\texttt{provided } H < I \wedge I < J) \qquad\qquad (\texttt{provided } H \ f_5 \ I \wedge I \ f_5 \ J)$$

The `provided` clause is a constraint on the applicability of *lt_lt_trans*; more precisely, the application of the rule will cause the $(H < I \wedge I < J)$ term to be unified with an element of the context Γ, e.g. with $(x < y \wedge y < z)$ in the original proof. Such a unification is not possible in the construction of the schema, however, since the $(X \ f_1 \ Y \wedge Y \ f_4 \ Z)$ term is frozen. To proceed with the application of this rule in the construction of the generalised schema, we consider a structural subterm mapping between the open schema node and the generalised conclusion of *lt_lt_trans* (including the `provided` clause), which we represent as a set of unification problems *PSet*:

$$PSet = \{\underline{X} \sim H, \underline{f_1} \sim f_5, \underline{Y} \sim I, \underline{f_4} \sim f_5, \underline{Z} \sim J, (\underline{X \ f_2 \ A}) \sim G\}$$

Here, frozen variables have been underlined. Now we are faced with a problem – not all of these unification problems can be solved together. In particular, if $\underline{f_1} \sim f_5$ is solved by instantiating f_5 to $\underline{f_1}$, we cannot solve $\underline{f_4} \sim f_5$, but if we solve $\underline{f_4} \sim f_5$ first we cannot solve $\underline{f_1} \sim f_5$. Informally then, we want to perform those unifications that have no effect on the success of other unification problems. We call the set of such unifications the **Unification-Independent Subset** (*UIS*) – it is defined as follows.

Definition 1. *We say a set of unification problems R can be **solved** provided each of the constituent unification problems can be solved (the order in which the problems are attempted is irrelevant). A set of unification problems R that* *cannot be solved is said to be a* **minimal failure set** *iff for every problem $r \in R$, $R \backslash \{r\}$ can be solved. The UIS of a set of unification problems R (UIS(R)) is then the set of problems that don't occur in any of the minimal failure sets of R.*

We can calculate the *UIS* with the algorithm of Fig. 2. The algorithm in the worst case has to analyse 2^n sets (summing the first n rows of Pascal's triangle). Importantly, in our domain, because n is small (< 10 for all the inference rules used in the Dutch National Flag case study), this ensures the algorithm executes (virtually) instantaneously.

$UIS(R) \mathrel{\hat=}$
     $n := 1$, $FS := \{\}$, *working* $:= \{\{q\} \mid q \in R\}$
     while $\#\{s \mid s \in working \wedge \#s = n\} > 0$ **do**
         for $s \in \{s \mid s \in working \wedge \#s = n\}$ **do**
             if s cannot be solved **and not** $\exists f : FS \bullet f \subseteq s$ **then**
                 $FS := FS \cup \{s\}$
             elseif s can be solved **then**
                 working $:= working \cup \bigcup \{\{r\} \cup s \mid r \in R \backslash s\}$
         $n := n + 1$
     return $R \backslash \bigcup FS$

Fig. 2. An algorithm for calculating the Unification-Independent Subset of a set of unification problems R

Returning to our example schema generalisation, solving all the unification problems in *UIS(PSet)* gives the following:

$$\frac{(X \; f_1 \; Y \wedge Y \; f_4 \; Z) \Rightarrow X \; f_2 \; A \qquad (X \; f_1 \; Y \wedge Y \; f_4 \; Z) \longrightarrow X \; f_2 \; A}{(X \; f_1 \; Y \wedge Y \; f_4 \; Z), X \; f_5 \; (Z \; f_6 \; K) \longrightarrow X \; f_2 \; A}$$

To finish the schema construction, we want to apply the generalised form (right) of *lt_imp_neq* (left):

$$\frac{}{L < M \longrightarrow L \neq M} \qquad \frac{}{L \; f_8 \; M \longrightarrow L \; f_9 \; M}$$

The structural mapping for this step is:

$$\{X \sim L, f_5 \sim f_8, (Z \; f_6 \; K) \sim M, A \sim M, f_2 \sim f_9\}$$

Performing all of the unifications except for $(Z \; f_6 \; K) \sim M$ and $A \sim M$ gives:

$$\frac{(X \; f_1 \; Y \wedge Y \; f_4 \; Z) \Rightarrow X \; f_2 \; A \qquad (X \; f_1 \; Y \wedge Y \; f_4 \; Z) \longrightarrow X \; f_2 \; A}{X \; f_1 \; Y \wedge Y \; f_4 \; Z, X \; f_8 \; (Z \; f_6 \; K) \longrightarrow X \; f_2 \; A}$$

While the resulting schema is more general than the original schema, it is not quite what we want – intuitively, every occurrence of $(Z\ f_6\ K)$ should be replaced by a variable. If we reexamine the structural mapping we can see why – the same range variable M is related to both the frozen term A (a term from the goal) and the compound term $(Z\ f_6\ K)$ whose functor (f_6) is *not* frozen (i.e. a term introduced by the proof). Hence we replace all occurrences of $(Z\ f_6\ K)$ in the schema with A. This gives the fully generalised schema:

$$\cfrac{\cfrac{(X\ f_1\ Y \wedge Y\ f_4\ Z) \Rightarrow X\ f_2\ A}{X\ f_1\ Y \wedge Y\ f_4\ Z \longrightarrow X\ f_2\ A}}{X\ f_1\ Y \wedge Y\ f_4\ Z, \underline{X\ f_8\ A} \longrightarrow X\ f_2\ A}$$

This schema can be used to produce the following analogous proof:

$$\cfrac{\cfrac{(x < y \wedge y \leq z) \Rightarrow x \neq z}{x < y \wedge y \leq z \longrightarrow x \neq z}}{x < y \wedge y \leq z, \underline{x < z} \longrightarrow x \neq z} \quad \begin{matrix} \text{(imp\_intro)} \\ \text{(lt\_leq\_trans)} \\ \text{(lt\_imp\_neq)} \end{matrix}$$

It must be noted that the development of the algorithm (and the generalised schemas it produces) was motivated and guided to an extent by experience with actual examples. It will always be possible to choose a combination of proof schema and target goal for which the algorithm arguably does not produce the ideal schema, possibly because the schema and target goal make no sense as a reuse combination. The results of Sect. 6 provide an indication of the usefulness of the algorithm in practice. In addition, by avoiding the use of higher-order unification, we avoid having to heuristically choose between the multitude of solutions to a particular unification problem – this becomes an efficiency issue for a flexible replay strategy such as the one we propose in the following section.

5   A Replay Strategy

In this section we describe an algorithm that addresses the following problem: given a single source proof, how can this proof be used to guide the proof of a target conjecture within a reasonable amount of time?

Because the algorithm essentially replays the source proof (but with each step of the source proof replaced by a search), we need a mechanism for evaluating how closely potential continuations in the target proof match the corresponding point in the generalised schema. For this purpose we use the sum of two metrics – the straightforward structural similarity metric of Fig. 3 and the generalisation metric of Fig. 4. The development of these metrics was guided by heuristics and experience.

The structural metric is a straightforward comparison of the structural similarity of two terms. If, at the outermost level the two terms are the same, i.e. quantified expressions, compound terms with the same arity or atoms/variables, the measure is one plus the sum of the recursively calculated measures of the components of the terms (where applicable). Otherwise, the measure is equal to zero.

$struct\_sim(M, N) \mathrel{\hat{=}}$
    if $(var(M) \vee atomic(M))$ **and** $(var(N) \vee atomic(N))$ **then**
        return 1
    elseif $M = (M_q \ x \bullet M_b)$ **and** $N = (N_q \ y \bullet N_b)$ **then**
        return $1 + struct\_sim(M_b, N_b)$
    elseif $M = M_f(MA_1, ..., MA_n)$ **and** $N = N_f(NA_1, ..., NA_n)$ **then**
        return $1 + \sum_{i=1}^{n} struct\_sim(MA_i, NA_i)$
    else
        return 0

Fig. 3. Metric for measuring the structural similarity of two terms – $(M_q \ x \bullet M_b)$ denotes the quantified term where M_b is quantified by M_q over x

The generalisation metric measures, for two terms t_1 and t_2, how much t_1 has to be generalised (via anti-unification) to accommodate t_2 (in practice t_1 is a node in the schema and t_2 is a proof node). For example, consider the terms $t_1 = A(B(C), D(E, E, E))$ and $t_2 = I(J(K, L), M(O, Q, Q))$. The anti-unification AU_{tt} of t_1 and t_2 is $W(U, X(V, Y, Y))$. Writing this term as $A(U, D(V, E, E))$, where variable names from t_1 have been used where possible, highlights what we mean by "the amount t_1 has to be generalised to accommodate t_2" – the new variables U and V have replaced the $B(C)$ term and the first occurrence of E respectively. We can measure the amount of generalisation by analysing the output of a structural mapping function applied to t_1 and AU_{tt} that has been modified to return a bag of mappings (that may contain repeated elements):

$$[\![A \mapsto W, B(C) \mapsto U, D \mapsto X, E \mapsto V, E \mapsto Y, E \mapsto Y]\!]$$

Because AU_{tt} is a generalisation of t_1, this bag can only contain two types of elements: mappings from non-variable terms to variables and mappings from variables to variables. For the non-variable mapping case, each (non-variable) domain element contributes its term weight to the measure. For the variable-variable case, we consider the set of subbags that have common domain elements. For our example this is:

$$\{ [\![A \mapsto W]\!], [\![D \mapsto X]\!], [\![E \mapsto V, E \mapsto Y, E \mapsto Y]\!] \}$$
$$=$$
$$\{ [\![(A \mapsto W) \mapsto 1]\!], [\![(D \mapsto X) \mapsto 1]\!], [\![(E \mapsto V) \mapsto 1, (E \mapsto Y) \mapsto 2]\!] \}$$

Then each subbag contributes the number of elements in that bag minus the frequency of the element that occurs the most. In this case, the only subbag that contributes to the measure is $[\![(E \mapsto V) \mapsto 1, (E \mapsto Y) \mapsto 2]\!]$ which contributes 1.

We combine the metrics as follows.

Definition 2. *For proof node p_n of the form $\Gamma_p \longrightarrow G_p$, and schema node q_n of the form $\Gamma_q \longrightarrow G_q$, we define the **measure** of p_n and q_n, $measure(p_n, q_n)$ as*

$gen\_required(t_1, t_2)$
    $total := 0$
    $AU_{tt} := anti\_unify(t_1, t_2)$
    $ML := structural\_mapping\_bag(t_1, AU_{tt})$
    for $e \in$ dom ML such that $non\_var(e.1)$ **do**
        $total := total + ML\sharp e * term\_weight(e.1)$
    for $e \in$ dom dom ML such that $var(e)$ **do**
        $subbag := \{x : \text{dom } ML \mid x.1 = e\} \lhd ML$
        $total := total + \#subbag - \mathbf{max} \ (\text{ran } subbag)$
    return $total$

structural_mapping_bag is similar to the structural mapping function used in the previous section but returns a bag of mappings. The expression $ML\sharp f$ gives the frequency of f in the bag ML. dom ML gives the set of elements in the bag ML (in this case a set of ordered pairs), dom dom ML gives the set of domain elements of that set. $e.1$ refers to the domain element of the ordered pair e. Finally, $term\_weight(T)$ returns a number representing the size of the term T, and is equivalent to $struct\_sim(T, T)$.

Fig. 4. Required generalisation metric

the maximum of the expression $struct\_sim \ ((\Gamma_{pc}, G_p), (\Gamma_q, G_q)) - (gen\_required$ $((\Gamma_{pc}, G_p), (\Gamma_q, G_q))$, where Γ_{pc} ranges over all the permutations of $\#\Gamma_q$ elements of Γ_p.

In other words, *measure* chooses which context elements to measure against so as to maximise the result. To describe the main replay algorithm itself we need to consider one final definition.

Definition 3. *For two nodes p_n and q_n of the form $\Gamma_p \longrightarrow G_p$ and $\Gamma_q \longrightarrow G_q$, we define the* **node unification** *operation $p_n \sim_n q_n$ to be $G_p \sim G_q$; solve(A), where A is a set consisting of one problem $c_p \sim c_q$ for each element $c_p \in \Gamma_p$, where for each c_q, $c_q \in \Gamma_q$, (solve(A) simply performs the unifications in A).*

Informally, apart from unifying the succeedants, node unification requires that each element of the context of p_n be unified with an element of the context of q_n – this is needed when proof nodes are unified with schema nodes.

Now we can turn to the main replay algorithm itself which we describe in two parts – an initial portion which deals with the initial search and generalisation (*replay*), and a portion that replays each step of the original proof (*replay_step*). These are shown in Fig. 5.

The algorithm is parameterised by the amount of initial search performed (*IDEPTH*), the amount of search performed to find an analogous continuation of each step of the source proof (*SDEPTH*), and the depth of the search used to automatically close any leftover proof nodes (*ADEPTH*).

$replay(p_n, S) \cong$

 $C = generate\_continuations(p_n, IDEPTH)$ (INITIAL SEARCH)

 choose $c \in C$; $o_n \in open\_nodes(c)$ such that the value of

   $measure(o_n, root(S))$ is maximised

  $S_g = construct\_gen\_schema(root(S), o_n)$

  $solve(o_n \sim root(S_g))$

  for each $r_n \in open\_nodes(c)\backslash\{o_n\}$

   $close(r_n, ADEPTH)$ (INITIAL CLOSE)

  $replay\_step(o_n, root(S_g), construct\_sub\_schemas(root(S)))$

$replay\_step(p_n, s_n, SubSchemas) \cong$

 $C = generate\_continuations(p_n, SDEPTH)$ (STEP SEARCH)

 choose $c \in C$; $o_s \subseteq open\_nodes(c)$ such that $can\_solve(A)$,

   where A is a set consisting of one problem $o_{si} \sim_n s_{ci}$ for each

   $o_{si} \in \rho_s$, where $s_{ci} \in children(s_n)$

  for each $(o_{si} \sim_n s_{ci}) \in A$

  $o_{si} \sim_n s_{ci}$

  $replay\_step(o_{si}, s_{ci}, SubSchemas)$

 OR

 choose $S \in SubSchemas$ such that the value of $measure(p_n, root(S))$

   is maximised

  $S_g := construct\_gen\_schema(root(S), p_n)$

  $replay\_step(p_n, root(S_g), \{\})$

 OR

 $close(p_n, ADEPTH)$ (AUTO CLOSE)

$generate\_continuations(p_n, D)$ generates a set of (incomplete) proof fragments
rooted at p_n of depth D, where we use depth to mean the number of
proof steps (either tactics or atomic inference rules) applied to construct
a subproof. $open\_nodes(c)$ returns the open subgoals of the proof fragment
c. $close(o_n, DEPTH)$ attempts to find a proof of the open node o_n in at
most $DEPTH$ proof steps via an exhaustive search. $can\_solve(A)$ holds if
the set of (node) unification problems A can be solved. For a schema S,
$construct\_sub\_schemas(S)$ returns a set of schemas, one constructed at each
node in S (note that this is not the same as simply taking subtrees of S).
$construct\_gen\_schema$ refers to the algorithm described in the previous section.
The (**choose** $c \in S$ such that E) construct chooses an element c of S that makes
the expression E true.

Fig. 5. The main replay algorithm – the four different ways in which steps in the target
proof can be generated have been labelled

$replay$ takes two arguments – the goal p_n to be proved, and a schema S
that will be used in the analogous proof construction. Initially, $replay$ chooses
the continuation from the original target node that contains an open node that
most closely matches the schema root – additionally it must be possible to close
(i.e. prove) the remaining open nodes. To ensure the replay terminates in a

reasonable amount of time, the replay only attempts to proceed from the best initial continuation. The best initial open node is used as the target goal for the purposes of constructing the generalised proof schema. *replay* then hands over to *replay_step* which recursively follows the schema. At each step, it does one of the following:

1. chooses a continuation such that for each open node there is a unifiable child node of the current schema, unifies and recursively replays each of them (note that this is how the analogous proof of Sect. 3 would be constructed), or
2. provided the replay is currently operating on the original schema (to prevent recursive explosion), chooses a schema, closest matching first, constructed from a subproof of the original proof and replays the node with that schema but with no initial search, or
3. closes the node via an automatic brute-force search

For option 2, we make a trade-off – we allow multiple subschemas to be tried however we do not permit any initial search – due to the need to measure many permutations of context for multiple open nodes in many continuations, initial search is a very expensive operation. Option 2 also prevents the recursive application of subschemas – once the replay begins operating on a subschema, it cannot further on in the replay start operating on yet another subschema. Also, for the purposes of this section we are assuming a single source proof as input, however in a situation in which, for example, the entire proof of correctness of the Dutch National Flag problem was being constructed, there would obviously be opportunities to use subschemas from outside of the chosen source proof. Finally, for option 3, to keep things simple we are just using a basic search – obviously this could be replaced with more domain-specific alternatives, e.g. tactics specialising in propositional logic.

Using subschemas gives the replay strategy considerable flexibility by allowing a general rearrangement of subtrees in the original schema, including, for example, the deletion of parts of the original schema (achieved by choosing a subschema from further up the tree). Additionally, it allows a fresh instantiation of the schema at a given point to be made, something which we have found to be occasionally necessary.

5.1   An Example Replay

As an example of how this strategy works in practice, we consider the *Blue*.1 subproof of Fig. 6. This is a high-level view of the proof in that most of the steps shown actually correspond to the application of high-level tactics (and hence the application of multiple inference rules/step) – for the purposes of this exposition, this prevents the essence of the replay from being obscured.

We have implemented our replay strategy in the Ergo theorem prover [9]. Fig. 7 shows the proof of *Blue*.3 that was discovered fully automatically by our implementation (with $IDEPTH = 1$, $ADEPTH = 2$ and $SDEPTH = 1$) using the *Blue*.1 subproof as source. Each node has been labelled with the part of

$[1] \ldots \longrightarrow \forall i \bullet (0 \leqslant i \wedge i < r) \Rightarrow h(w, b-1 : h.(b-1), h.w).i = red$ (not_eq)

$[1.1] \ldots, \forall i \bullet (0 \leqslant i \wedge i < r) \Rightarrow h.i = red \longrightarrow$ (assump)

$$\forall i \bullet (0 \leqslant i \wedge i < r) \Rightarrow h.i = red$$

$[1.2] \ldots, \underline{i \leq r, r \leq w} \longrightarrow i \neq w$ (lt_leq_trans)

$[1.2.1] \ldots, \underline{i < w} \longrightarrow i \neq w$ (lt_imp_neq)

$[1.3] \ldots, \underline{i < r, r \leq w} \longrightarrow i \neq b - 1$ (lt_leq_trans)

$[1.3.1] \ldots, \underline{i < w, w \leq b, w \neq b}, \longrightarrow i \neq b - 1$ (lte_neq_imp_lt)

$[1.3.1.1] \ldots, \underline{i < w, w < b} \longrightarrow i \neq b - 1$ (lt_lt_trans)

$[1.3.1.1.1] \ldots, \underline{i < b - 1} \longrightarrow i \neq b - 1$ (lt_imp_neq)

Fig. 6. Proof of *Blue*.1

the replay strategy that produced it along with the root schema node it was operating on at the time.

Notice the initial step – splitting the $\forall$ quantifier yields a proof node that more closely resembles the root of *Schema*(*Blue*.1). Also of interest is the use of the subschema 1.3.1 to close the node 1.1.2 – this schema was the closest match that allowed node 1.1.2 to be closed.

6   Case Study Results

In this section we report on the results of applying the replay strategy described in the previous section to the Dutch National Flag case study. Fig. 8 shows the outcome of using each subproof of the invariant maintenance proof to prove each of the other subproofs.

Each 3-tuple represents the minimum amount of search required to discover the analogous proof – the first number is the auto-close depth, the second is amount of step search, and the third is the amount of initial search. To give an idea of execution times, on a Pentium IV, schema construction/generalisation took approximately 2 seconds (for subproofs of around 50 proof nodes). For the replay itself, using a tactic/rule base of around 100 rules, the replay strategy took no longer than a minute to terminate.

Even with these limits, the results of Fig. 8 suggest around 70% of user interactions can be saved for this part of the proof. Of course, user interactions is just one measure of proof complexity – another is the amount of time the user spends thinking about the proof. Experience has shown that for this kind of proof, the first subproof is usually the most time-consuming – indeed, in constructing the proof by hand, the first subproof took approximately twice as long as each of the others. Even given this, proof reuse still clearly provides significant savings.

Finally, we would briefly comment on the problems the replay strategy has with finding an analogous proof for *Red*.2 – this is due to inaccurate measures, especially at the initial search stage. In general, we have found that it is important to do as much initial search as possible so that the generalised schema that

INITIAL SEARCH(1)

[1] $..., b \leqslant n \longrightarrow \forall i \bullet (b - 1 \leqslant i \wedge i < n) \Rightarrow$ (split_l)
$h(w, b - 1 : h.(b - 1), h.w).i = blue$

  STEP SEARCH(1)

  [1.1] $... \longrightarrow \forall i \bullet (b \leqslant i \wedge i < n) \Rightarrow$ (not_eq)
  $h(w, b - 1 : h.(b - 1), h.w).i = blue$

    STEP SEARCH(1)

    [1.1.1] $..., \forall i \bullet (b \leqslant i \wedge i < n) \Rightarrow h.i = blue \longrightarrow$ (assump)
    $\forall i \bullet (b \leqslant i \wedge i < n) \Rightarrow h.i = blue$

    STEP SEARCH(1.3.1)

    [1.1.2] $..., w \leqslant b, w \neq b \longrightarrow i \neq w$ (lt_leq_trans)

      AUTO CLOSE(1.3.1)

      [1.1.2.1] $..., w < b, b \leqslant i \longrightarrow i \neq w$ (lte_neq_imp_lt)

        AUTO CLOSE(1.3.1)

        [1.1.2.1.1] $..., w < i \longrightarrow i \neq w$ (neq_flip;lt_imp_neq)

    AUTO CLOSE(1)

    [1.1.3] $..., b \leqslant i \longrightarrow i \neq b - 1$ (leq_imp_minus1)

      AUTO CLOSE(1)

      [1.1.3.1] $..., b - 1 < i, \longrightarrow i \neq b - 1$ (neq_flip;lt_imp_neq)

  INITIAL CLOSE(1)

  [1.2] $... \longrightarrow h(w, b - 1 : h.(b - 1), h.w).(b - 1) = blue$ (ref_eq)

    INITIAL CLOSE(1)

    [1.2.1] $..., h.w = blue \longrightarrow h.w = blue$ (assump)

Fig. 7. The proof of Blue.3

is constructed is as accurate as possible. For the case of *Red*.2, either reattempting the reuse using lower-rated initial continuations (i.e. modifying the replay strategy) or simply increasing the depth of the initial search allows the reuse to succeed.

7 Conclusion

In this paper we have described a method for reusing proofs. We contend that the combination of our goal-directed generalisation technique together with a flexible replay strategy can provide significant savings for the kinds of proofs that arise in the realm of software verification. Recently we have developed a reuse agent based on the reuse technique described in this paper as part of an agent-based distributed interactive theorem proving environment. Early indications of the agent's effectiveness are promising – because the agent is running in the background, the search space of the replay can be increased. Further, the agent

	Blue.1	Blue.2	Blue.3	Red.1	Red.2	Red.3
Blue.1	(0,1,0)	(1,1,0)	(2,1,1)	(2,1,1)	x	(3,1,0)
Blue.2	(3,1,0)	(0,1,0)	(2,1,1)	(2,1,1)	x	(3,1,0)
Blue.3	(2,1,0)	(2,1,0)	(0,1,0)	(3,1,0)	x	(3,1,0)
Red.1	(3,1,0)	(2,1,0)	(3,1,1)	(0,1,0)	(2,1,2)	(4,1,0)
Red.2	(4,1,0)	(3,1,0)	x	(3,1,0)	(0,1,0)	(3,1,0)
Red.3	(2,1,0)	(2,1,0)	(2,1,1)	(2,1,1)	x	(0,1,0)

Fig. 8. Results of using each subproof to prove every other subproof – an 'x' indicates the replay strategy failed to find an analogous proof given maximum search depths of (4,2,2)

can also generically make use of the capabilities of other agents in situations where the replay encounters subgoals that it cannot solve itself.

References

1. E. W. Dijkstra. *A Discipline of Programming.* Prentice-Hall, Englewood Cliffs, New Jersey, 1976.
2. C. A. R. Hoare. An axiomatic basis for computer programming. In *C. A. R. Hoare and C. B. Jones (Ed.), Essays in Computing Science, Prentice Hall.* 1989.
3. G. Huet. A unification algorithm for typed lambda-calculus. *Theoretical Computer Science,* 1:27–57, 1975.
4. E. B. Johnsen and C. Lüth. Abstracting transformations for refinement. *Nordic Journal of Computing,* 10:316–336, 2003.
5. A. Kaldewaij. *Programming: The derivation of algorithms.* Prentice Hall, 1990.
6. E. Melis and A. Schairer. Similarities and reuse of proofs in formal software verification. In *Proceedings of EWCBR-98,* pages 76–87.
7. E. Melis and J. Whittle. Internal analogy in theorem proving. In M. A. McRobbie and J. K. Slaney, editors, *Proceedings of CADE-96,* pages 92–105.
8. J. Reynolds. Transformational systems and algebraic structure of atomic formulas. *Machine Intelligence,* 5:135–152, 1970.
9. M. Utting, P. Robinson, and R. Nickson. Ergo 6: a generic proof engine that uses Prolog proof technology. *LMS Journal of Computation and Mathematics,* 5:194–219, 2002.
10. C. Walther and T. Kolbe. Proving theorems by reuse. *Artificial Intelligence,* 116(1–2):17–66, 2000.

Deductive Verification
of Distributed Groupware Systems

Abdessamad Imine, Pascal Molli, Gérald Oster, and Michaël Rusinowitch

LORIA, INRIA - Lorraine
Campus Scientifique, 54506 Vandœuvre-Lès-Nancy Cedex, France
{imine,molli,oster,rusi}@loria.fr

Abstract. Distributed groupware systems consist of a group of users manipulating a shared object (like a text document, a filesystem, etc). Operational Transformation (OT) algorithms are applied for achieving convergence in these systems. However, the design of such algorithms is a difficult and error-prone activity, since building the correct operations for maintaining good convergence properties of the local copies requires examining a large number of situations. In this paper, we present the modelling and deductive verification of OT algorithms with algebraic specifications. We show that many OT algorithms in the literature do not satisfy convergence properties unlike what was stated by their authors.

1   Introduction

Distributed groupware systems consist of two or more users (sites) that simultaneously manipulate objects (*i.e.* text, image, graphic, etc.) without the need for physical proximity and enables them to synchronously observe each other changes. The shared objects are *replicated* at the local memory of each participating user. Every operation is executed locally first and then *broadcasted* for execution at other sites. So, the operations are applied in different orders at different *replicas* (or copies) of the object. This potentially leads to *divergent* (or different) replicas – an undesirable situation for distributed groupware systems [16].

Operational Transformation is an approach which has been proposed to overcome the divergence problem, especially for building real-time groupware [4, 15]. This approach consists of an algorithm which transforms, *i.e.* to adjust parameters, the remote operation according to local concurrent ones in order to achieve convergence. It has been used in several group editors [4, 13, 15, 14, 18], and more recently it is employed in other distributed systems as the generic synchronizer $S5$ [11]. The advantages of this approach are: (i) it is independent of the replica state and depends only on concurrent operations; (ii) it enables an unconstrained concurrency, *i.e.* no global order on operations; (iii) it ensures a good responsiveness in real-time interaction context. However, if OT algorithms are not correct then the consistency of shared data is not ensured. Accordingly, it is critical to verify such algorithms to avoid the loss of data when broadcasting operations. According to [13], the OT algorithm needs to fulfill two convergence conditions

C. Rattray et al. (Eds.): AMAST 2004, LNCS 3116, pp. 226–240, 2004.

C_1 and C_2 which will be detailed in Section 2. Finding such an OT algorithm and proving that it satisfies C_1 and C_2 is not an easy task. This proof is often difficult – even impossible – to produce by hand and unmanageably complicated.

Our Solution. To overcome this problem, it is necessary to encourage OT algorithm developers to write a formal specification, *i.e.* a description about the replica behaviour, and then verify the correctness of the OT algorithm *w.r.t.* convergence conditions by using a theorem prover. However, effective use of a theorem prover typically requires expertise that is uncommon among software engineers. So, our work is aimed at designing and implementing techniques underlying the development of OT algorithms which meet the following requirements: (i) *Writing formal specifications must be effortless.* (ii) *High degree of automation in the proof process.* The developers should use the theorem prover as a (push-button) probing tool to verify convergence conditions.

Using Observational Semantics, we treat a replica object as a black box [6]. We specify interactions between a replica object and a user. Operations which observe the replica states are called *attributes*, and operations which change the states are called *methods*. We can only recognize the current state by observing states modified by methods through attributes. So, we consider method sequences followed by an attribute as observation tools. We have implemented our approach as a tool which enables a developer to define all replica operations (methods and attributes) and the OT algorithm. From this description, our tool generates an algebraic specification described in terms of conditional equations. As verification back-end we use SPIKE, a first-order implicit induction prover, which is suitable for reasoning about conditional equations [2,3].

The main contribution of this paper is to show that a lightweight use of formal verification techniques is feasible (i) to write easily a formal specification of a replica object, and (ii) to have its OT algorithm checked *w.r.t.* convergence conditions so as to increase confidence in the correctness of the OT algorithm. Moreover, using our theorem-proving approach we have obtained surprising results. Indeed, we have detected bugs in several distributed groupware systems designed by specialists from the domain [7,8].

Related Work. To our best knowledge, there is no other work on formal verification of OT algorithms. In [8], we represented the replica as an abstract data type, but the proof effort became costlier when the data is complex (*e.g.* an XML tree). Indeed, a property proof involving the data could call for numerous sub-proofs of properties about its logical structure. In [10,9], we used the situation calculus for hiding the internal state of replica but this formalism turned out to be inappropriate to accurately model the notion of state. In this work, we found that the observational semantics is natural enough to abstract away from the internal replica structure. It describes the behaviour of a replica object as viewed by an external user. Since OT algorithms rely on replica methods, then the process verification becomes effortless.

Structure of the Paper. This paper is organized as follows: Section 2 summarizes the basic concepts of OT approach. The ingredients of our formalization for specifying replica object and OT algorithm are given in Section 3. In Section 4, we present how to express convergence conditions in our algebraic framework. Section 5 briefly describes our tool and numerous bugs that we have found in some distributed groupware systems. Finally, we give conclusions and present future work.

2   Operational Transformation Approach

Distributed groupware systems typically maintain shared state of some sort (*i.e.* text document, XML tree, etc.). Shared states are usually *replicated* among a group of sites. Updates are performed first at local sites and then propagated to remote sites. When the replicated state is being modified, all the replicas must be modified in a consistent way. In the following, we consider a distributed groupware system as a Group of Sites (GS) where every site has its own replica. Notation $[op_1 \, op_2 \, \ldots \, op_n]$ represents an operation sequence. We denote $X \bullet st = st'$ when an operation (or an operation sequence) X is executed on a replica state st and produces a replica state st'.

Definition 1. (Convergence Property). *The convergence property states that all replicas are identical at all sites after all generated operations have been executed at all sites.*

Example 1. Consider the following group text editor scenario (see Figure 1): there are two sites working on a shared document represented by a string of characters. Initially, all the copies hold the string "efect". The document is modified with the operation $Ins(p, c)$ for inserting a character c at position p. Users 1 and 2 generate two concurrent operations: $op_1 = Ins(2, \text{"f"})$ and $op_2 = Ins(6, \text{"s"})$ respectively. When op_1 is received and executed on site 2, it produces the expected string "effects". But, when op_2 is received on site 1, it does not take into account the fact that op_1 has been executed before it. Consequently, we obtain a *divergence* problem between sites 1 and 2.

As a solution to divergence problem, we can use an OT algorithm which has been proposed in [4]. It takes two *concurrent* operations op_1 and op_2 defined on the *same state* and returns op_1' which is equivalent to op_1 but is defined on a state where op_2 has been applied. We denote this algorithm by a function T.

Example 2. In Figure 2, we illustrate the effect of T on the previous example. When op_2 is received on site 1, op_2 needs to be transformed according to op_1 as follows: $T((Ins(6, \text{"s"}), Ins(2, \text{"f"})) = Ins(7, \text{"s"})$. The insertion position of op_2 is incremented because op_1 has inserted a character at position 2, which is before the character inserted by op_2. Next, op_2' is executed on site 1. In the same way, when op_1 is received on site 2, it is transformed as follows: $T(Ins(2, \text{"f"}), Ins(6, \text{"s"})) = Ins(2, \text{"f"})$; op_1 remains the same because "f" is inserted before "s".

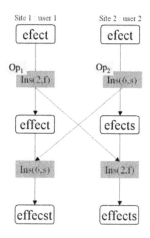

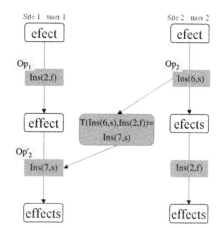

Fig. 1. Incorrect integration.   **Fig. 2.** Integration with transformation.

Intuitively we can write the transformation T as follows:

```
T(Ins(p1,c1),Ins(p2,c2)) = if (p1 < p2) return Ins(p1,c1)
                           else return Ins(p1+1,c1)
                           endif;
```

However, using an OT algorithm requires to satisfy two conditions [13]. Given two operations op_1 and op_2, let $op'_2 = T(op_2, op_1)$ and $op'_1 = T(op_1, op_2)$, the conditions are as follows:

- **Condition** C_1: $[op_1\, op'_2] \bullet st = [op_2\, op'_1] \bullet st$, for every state st.
- **Condition** C_2: $T(T(op, op_1), op'_2) = T(T(op, op_2), op'_1)$.

C_1 defines a *state identity* and ensures that if op_1 and op_2 are concurrent, the effect of executing op_1 before op_2 is the same as executing op_2 before op_1. This condition is necessary but not sufficient when the number of concurrent operations is greater than two. As for C_2, it ensures that transforming op along equivalent and different operation sequences will give the same result. In [14], the authors have proved that conditions C_1 and C_2 are sufficient to ensure the convergence property for *any number* of concurrent operations which can be executed in *arbitrary order*.

Proving the correctness of OT algorithms, *w.r.t* C_1 and C_2 is very complex and error prone even on a simple string object. Consequently, to be able to develop the transformational approach and to safely use it in other replication-based distributed systems with simple or more complex objects, proving conditions on OT algorithms must be assisted by an automatic theorem prover.

3   Formal Specification

We present in this section the theoretical background of our framework. We first briefly review the basics of algebraic specification. Then, we give the ingredients of our formalization for specifying and reasoning on OT algorithms.

3.1   Algebraic Preliminaries

We assume that the reader is familiar with the basic concepts of algebraic specifications [19], term rewriting and equational reasoning [17]. Let S be a set (of sorts). An S-sorted set is a family of sets $X = \{X_s\}_{s \in S}$ indexed by S. A many-sorted signature Σ is a triplet (S, F, X) where S is a set (of sorts), F is a $S^* \times S$-sorted set (of function symbols) and X is a family of S-sorted variables. $\natural(\omega, s)$ denotes the number of occurrences of the sort s in the sequence ω. We assume that we have a partition of F in two subsets: the first one C contains the constructor symbols and the second one D is the set of defined symbols, such that C and D are disjoint. Let $T(F, X)$ be the set of sorted terms. When a term does not contain variables, it is called ground term. The set of all ground terms is $T(F)$. A substitution η assigns terms of appropriate sorts to variables. If t is a term, then $t\theta$ denotes the application of substitution θ to t. If η applies every variable to ground term, then η is a ground substitution. We denote by $\equiv$ the syntactic equivalence between objects. An equation is a formula of the form $l = r$. A conditional equation is a formula of the following form: $\bigwedge_{i=1}^{n} a_i = b_i \implies l = r$. It will be written $\bigwedge_{i=1}^{n} a_i = b_i \implies l \to r$ and called a conditional rewrite rule when using an order on terms. The precondition of rule $\bigwedge_{i=1}^{n} a_i = b_i \implies l \to r$ is $\bigwedge_{i=1}^{n} a_i = b_i$. The term l is the left-hand side of the rule. A set of conditional rewrite rules is called a rewrite system. A constructor is free if it is not the root of a left-hand side of a rule. A term is strongly irreducible if none of its non-variable subterms matches a left-hand side of a rule in a rewrite system. An algebraic specification is a pair $(\Sigma, \mathcal{A})$ where Σ is a many-sorted signature and $\mathcal{A}$ is a rewrite system (called the axioms of $(\Sigma, \mathcal{A})$). A clause is an expression of the form: $\bigwedge_{i=1}^{n} a_i = b_i \implies \bigvee_{j=1}^{m} a'_j = b'_j$. The clause C is a Horn clause if $m \leqslant 1$. The clause C is a logical consequence of $\mathcal{A}$ if C is valid in any model of $\mathcal{A}$, denoted by $\mathcal{A} \models C$. C is said inductively valid in $\mathcal{A}$ (denoted $\mathcal{A} \models_{Ind} C$), if for any ground substitution σ, (for all i, $\mathcal{E} \models a_i \sigma = b_i \sigma$) implies (there exists j, $\mathcal{A} \models a'_j \sigma = b'_j \sigma$). An Observational signature is a many-sorted signature $\Sigma = (S, S_{obs}, F, X)$ where $S_{obs} \subseteq S$ is the set of observable sorts. An Observational specification is a pair $(\Sigma, \mathcal{A})$ where Σ is an observational signature and $\mathcal{A}$ is a set of axioms.

3.2   Replica Specification

The main component in distributed groupware system is the replica. Every replica has a set of operations. The methods are operations which modify the replica state. The attributes are operations which extract informations from this state. In some cases, the replica state can be small, like a text document. In other cases, it can be large like a database, an XML tree or a filesystem. In this work, we use an observational technique which conceals the internal state of the replica by extracting informations from the method sequence executed on it.

We use the sort State for representing the space of replica state. This sort has two constructor functions: (i) the constant constructor S_0 (the initial state), and (ii) a constructor Do which given a method and a state gives the resulting state provided that the execution of this method is possible. The sort Meth represents the set of methods. Every method type has its own constructor. These constructors are free since methods are assumed to be distinct. For every method, we should indicate conditions under which it is enabled. For this we use a boolean function $Poss$ defined by a set of conditional equations. We define also an *idle* method which has null effect on the replica state, *i.e.* constructor Nop. As to attributes, we define them as function symbols which are monadic on the State sort. These attribute functions serve as observers and are inductively expressed upon the State sort. The OT algorithm is defined by the function symbol T which takes two methods as arguments and produces another method. We then formally define a replica specification:

Definition 2. (Replica Specification). *Given S the set of all sorts, $S_{bs} = \{\text{State}, \text{Meth}\}$ the set of* basic sorts *and $S_{is} = S \setminus S_{bs}$ the set of* individual sorts *(data sorts). A replica specification π is an observational specification $OSP^\pi = (\Sigma^\pi, \mathcal{A}^\pi)$ such that:*

- *Σ^π is an observational signature where the set of non-observable sorts is a singleton, i.e. $S \setminus S_{obs} = \{\text{State}\}$. F is defined as $C \cup D$, such that: (i) $C_{\epsilon,\text{State}} = \{S_0\}$, $C_{\text{Meth State},\text{State}} = \{Do\}$ and $C_{w,s} = \emptyset$ if s is State or w contains an element of S_{bs}. (ii) $D_{\text{Meth Meth},\text{Meth}} = \{T\}$, $D_{\text{Meth State},\text{Bool}} = \{Poss\}$ and $D_{w,s} = \emptyset$ if either s is State, w contains Meth, or $\natural(w, \text{State}) > 1$.*
- *$\mathcal{A}^\pi = \mathcal{D}_P \cup \mathcal{D}_A \cup \mathcal{D}_T$ is the set of axioms (written as conditional equations) such that: (i) $\mathcal{D}_P$ is the set of method precondition axioms (Poss); (ii) $\mathcal{D}_A$ is the set of axioms for every attribute; (iii) $\mathcal{D}_T$ contains axioms defining the transformation function T.*

Example 3. Consider the group text editor designed by Ellis and Gibbs [4] who are the pioneers of the OT approach. The string replica has two methods: $Ins(p, c, pr)$ and $Del(p, pr)$ [4]. In this example, Ins and Del are extended with a new parameter pr. This one represents the priority to solve conflict when transforming two Ins operations and it is based on the site identifier where operations have been generated. The replica string has two attributes: *Length* for extracting the length of the string and *Car* for giving the character of the string at given position and state. The replica specification is given in Figure 3. The set $C_{w,\text{Meth}}$ ($w \in S_{is}^*$) contains all constructor methods which represents the method types of a replica. All the necessary conditions for executing a method are given by $\mathcal{D}_P$ (lines $1-2$). The set $D_{w \text{ State},s}$ contains all replica attributes where $w \in S_{is}^*$ and $s \in S_{is}$. $\mathcal{D}_A$ is illustrated in lines $3-9$. Lines $10-25$ gives the definition of T. *true* and *false* are boolean constants.

Sorts: State, Meth, bool, nat, char

Constructors
$S0$: $\rightarrow$ State
Do : Meth $\times$ State $\rightarrow$ State
Ins : nat $\times$ char $\times$ nat $\rightarrow$ Meth
Del : nat $\times$ nat $\rightarrow$ Meth
Nop : $\rightarrow$ Meth

Defined Operations
$Poss$: Meth $\times$ State $\rightarrow$ bool
$Length$: State $\rightarrow$ nat
Car : nat $\times$ State $\rightarrow$ char
T : Meth $\times$ Meth $\rightarrow$ Meth

Variables
$x, p1, p2, pr1, pr2$: nat;
st : State;
m : Meth;
$c1, c2$: char;

Axioms
1. $Poss(Ins(p1, c1, pr1), st) = (p1 \leq Length(st))$;
2. $Poss(Del(p1, pr1), st) = (p1 < Length(st))$;

3. $Length(Do(Ins(p1, c1, pr1), st)) = Length(st) + 1$;
4. $Length(Do(Del(p1, pr1), st)) = Length(st) - 1$;
5. $x = p1 \quad\Rightarrow Car(x, Do(Ins(p1, c1, pr1), st)) = c1$;
6. $(x > p1) = true \Rightarrow Car(x, Do(Ins(p1, c1, pr1), st)) = Car(x - 1, st)$;
7. $(x < p1) = true \Rightarrow Car(x, Do(Ins(p1, c1, pr1), st)) = Car(x, st)$;
8. $(x \geq p1) = true \Rightarrow Car(x, Do(Del(p1), st)) = Car(x + 1, st)$;
9. $(x < p1)) = true \Rightarrow Car(x, Do(Del(p1), st)) = Car(x, st)$;

10. $T(Nop, m) = Nop$;
11. $T(m, Nop) = m$;
12. $(p1 < p2) = true \quad\Rightarrow T(Ins(p1, c1, pr1), Ins(p2, c2, pr2)) = Ins(p1, c1, pr1)$;
13. $(p1 > p2) = true \quad\Rightarrow T(Ins(p1, c1, pr1), Ins(p2, c2, pr2)) = Ins(p1 + 1, c1, pr1)$;
14. $p1 = p2 \wedge c1 = c2 \Rightarrow T(Ins(p1, c1, pr1), Ins(p2, c2, pr2)) = Nop$;
15. $p1 = p2 \wedge c1 \neq c2 \wedge (pr1 > pr2) = true \Rightarrow$
16. $\qquad\qquad T(Ins(p1, c1, pr1), Ins(p2, c2, pr2)) = Ins(p1 + 1, c1, pr1)$;
17. $p1 = p2 \wedge c1 \neq c2 \wedge (pr1 < pr2) = true \Rightarrow$
18. $\qquad\qquad T(Ins(p1, c1, pr1), Ins(p2, c2, pr2)) = Ins(p1, c1, pr1)$;
19. $(p1 < p2) = true \Rightarrow T(Ins(p1, c1, pr1), Del(p2, pr2)) = Ins(p1, c1, pr1)$;
20. $(p1 \geq p2) = true \Rightarrow T(Ins(p1, c1, pr1), Ins(p2, c2, pr2)) = Ins(p1 - 1, c1, pr1)$;
21. $(p1 < p2) = true \Rightarrow T(Del(p1, pr1), Del(p2, pr2)) = Del(p1, pr1)$;
22. $(p1 > p2) = true \Rightarrow T(Del(p1, pr1), Del(p2, pr2)) = Del(p1 - 1, pr1)$;
23. $p1 = p2 \quad\Rightarrow T(Del(p1, pr1), Del(p2, pr2)) = Nop$;
24. $(p1 < p2) = true \Rightarrow T(Del(p1, pr1), Ins(p2, c2, pr2)) = Del(p1, pr1)$;
25. $(p1 \geq p2) = true \Rightarrow T(Del(p1, pr1), Ins(p2, c2, pr2)) = Del(p1 + 1, pr1)$;

Fig. 3. Replica specification for text editor.

We use an observational semantics which is based on weakening the satisfaction relation [3, 1, 6, 5]. Informally speaking, the replica objects which cannot be distinguished by *experiments* are considered as observationally equal. When using algebraic specifications, such experiments can be formally defined by *contexts* of observable sorts and operators over the signature of the specification.

Definition 3. (Context). *Let π be a replica specification and $T_\pi(F, X)$ its term algebra.*

1. *A Σ_π-context of sort* State *is a non-ground term $c \in T_\pi(F, X)$ with a distinguished linear variable z_{State} of sort* State. *This variable is called the context variable of c. To indicate the context variable occurring in c, we often write $c[z_{State}]$.*
2. *A Σ_π-context c is called an observable Σ_π-context if the sort of c is in S_{is}, and a State Σ_π-context if the sort of c is* State.
3. *A State Σ_π-context can be regarded as a sequence of methods. An observable Σ_π-context is the sequence formed by an attribute and a State Σ_π-context.*
4. *A Σ_π-context c is appropriate for a term $t \in T_\pi(F, X)$ iff the sort of t matches that of z_{State}. $c[t]$ defines the replacement of z_{State} by t in $c[z_{State}]$.*
5. *$ObsCt_{\Sigma_\pi}$ denotes the set of all observable Σ_π-contexts.*

Example 4. Consider the replica specification in Figure 3. There are infinitely many observable Σ_π-contexts: $Length(z_{State})$, $Car(x, z_{State})$, $Car(x, Do(Ins(p, c, pr), z_{State}))$, ..., $Length(Do^n(Del(p), z_{State})$.

Our notion of observational validity is based on the idea that two replica objects in given algebra are observationally equal if they cannot be distinguished by computation with observable results.

Definition 4. (Observational Validity). *Two terms t_1 and t_2 are observational equal if for all $c \in ObsCt_{\Sigma_\pi}$ $\mathcal{A}^\pi \models c[t_1] = c[t_2]$. We denote it by $\mathcal{A}^\pi \models_{obs} t_1 = t_2$ or simply $t_1 =_{obs} t_2$.*

The observational validity induces a congruence on $T(F)$ [3].

Definition 5. (State Property). *Let $P \equiv \bigwedge_{i=1}^{n} a_i = b_i \implies t_1 = t_2$. We say that P is a state property (or observational valid) and we denote it by $\mathcal{A}^\pi \models_{obs} P$ if: for all ground substitutions σ, $(\forall i \in [1..n]\ a_i\sigma =_{obs} b_i\sigma)$ implies $t_1\sigma =_{obs} t_2\sigma$.*

Our purpose is to propose a technique allowing to prove and disprove (or refute) state properties. Note that our state properties are Horn clauses and therefore in the scope of observational properties mentionned in [3]. In this work, the authors introduced the concept of *critical contexts*. They allow to prove observational theorems by reasoning on the ground irreducible observable contexts rather than on the whole set of observable contexts. In the following, we denote by $\mathcal{R}$ a conditional rewrite system which is obtained by orienting axioms of $\mathcal{A}^\pi$.

Definition 6. (Inconsistent State Property). *We say that the state property $P \equiv \bigwedge_{i=1}^{n} a_i = b_i \implies t_1 = t_2$ is provably inconsistent iff there exists a substitution σ and a critical context c such that: (i) $\forall i \in [1..n]$, $a_i\sigma = b_i\sigma$ is an inductive theorem w.r.t. $\mathcal{R}$. (ii) $c[t_1 = t_2]$ is strongly irreducible by $\mathcal{R}$.*

Provably inconsistent state properties are not observationally valid.

The computation of critical contexts requires that axioms are sufficiently complete [19]. Moreover, the refutation of any theorem imposes that $\mathcal{R}$ is ground convergent. For more details on how to compute critical context and refute observational theorems can be found in [3]. We denote the inference system of [3] $\text{PROOF}(E)$, which takes as argument a set of conjectures to be proved and returns a set of lemmas remaining to be proved in order to show E is observationally valid. If $\text{PROOF}(E)$ returns an empty set then E is observationally valid.

4   Proving Convergence Properties

Before stating the properties that a replica object has to satisfy for ensuring convergence, we give some notations. Let $m_1, m_2, \ldots, m_n$ and st be terms of sorts Meth and State respectively. We define (the empty sequence is denoted by $[]$):

1. $[](st) = st$ and $[m_1 \, m_2 \, \ldots \, m_n](st) = Do(m_n, \ldots, Do(m_2, Do(m_1, st))\ldots)$ is the application of the method sequence $m_1, m_2, \ldots, m_n$ on the state st.
2. $Legal([], st) = true$ and $Legal([m_1 \, m_2 \, \ldots m_n], st) = Poss(m_1, st) \wedge Poss(m_2, [m_1](st)) \wedge \ldots \wedge Poss(m_n, [m_1 \, m_2 \, \ldots \, m_{n-1}](st))$.
3. $T^*(m, []) = m$ and $T^*(m, [m_1 \, m_2 \, \ldots \, m_n]) = T^*(T(m, m_1), [m_2 \, \ldots \, m_n])$. $T^*(m, seq)$ denotes the operation resulting from transforming m along the sequence seq.

Definition 7. (Sequence Equivalence). *Given two method sequences seq_1 and seq_2. We say that seq_1 and seq_2 are equivalent iff $seq_1(st) =_{obs} seq_2(st)$ for every state st.*

In the following, we present how to express the satisfaction of conditions C_1 and C_2 as properties to be checked in our algebraic framework. Let $OSP^\pi = (\Sigma^\pi, \mathcal{A}^\pi)$ and $\mathcal{M}^\pi$ be a replica specification and the method set respectively, corresponding to a replica object π.

4.1   Condition C_1

As mentionned before, we use an observational approach for comparing two replica states. Accordingly, we define the condition C_1 by the following state property (the variable st is universally quantified):

$$\Phi_1(m_1, m_2) \equiv (Legal([m_1 \, T(m_2, m_1)], st) = true \wedge$$
$$Legal([m_2 \, T(m_1, m_2)], st) = true) \tag{1}$$
$$\implies [m_1 \, T(m_2, m_1)](st) =_{obs} [m_2 \, T(m_1, m_2)](st)$$

The first convergence property is formulated as a conjecture to be proved from the replica specification. It means: for all methods m_1 and m_2 and for every state st, such that m_1 and m_2 are enabled on st, then the states $[m_1 \, T(m_2, m_1)](st)$ and $[m_2 \, T(m_1, m_2)](st)$ are observationally equal. This conjecture is defined as follows:

```
Input   : A replica specification π.
Output : S a set of CP1-scenarios.

S ← ∅;
foreach method M₁ in Mᵀ
    foreach method M₂ in Mᵀ
        E ← {Φ₁(M₁, M₂)};
        E ← PROOF(E);
        if E ≠ ∅ then S ← S ∪ {(M₁, M₂, E)};
    endfor
endfor
```

Fig. 4. Checking Algorithm of Convergence Property $CP1$.

Conjecture 1 (Convergence Property $CP1$). *A replica object* π *satisfies the condition* C_1 *iff* $\Phi_1(m_1, m_2)$ *is a state property for all* m_1 *and* m_2.

According to Definition 7, the convergence property $CP1$ means that the sequences $[m_1 \, T(m_2, m_1)]$ and $[m_2 \, T(m_1, m_2)]$ are equivalent.

Definition 8. (CP1-scenario). *A $CP1$-scenario is a triple $(M_1, M_2, \mathcal{E})$ where M_1 and M_2 are two methods and $\mathcal{E}$ is the set of conjectures obtained by the function* PROOF($\{\Phi_1(M_1, M_2)\}$).

In Figure 4, we present an algorithm allowing us to verify the convergence property $CP1$ by detecting all $CP1$-scenarios which violate this property. The $CP1$-scenarios simply consist of methods and conditions over argument methods which may lead to potential divergence situations.

Example 5. Consider the group editor of Example 3. When applying our algorithm to replica specification of Figure 3, we have detected that convergence property $CP1$ is violated by giving the $CP1$-scenario depicted in Figure 5. From this scenario, we can extract the following informations: (i) the methods $(Ins(u1, u2, u3)$ and $Del(u4, u5))$ that cause divergence problem; (ii) the observation (the attribute Car) that distinguishes the resulting states, and; (iii) the conditions over method arguments (Preconditions) which lead to divergence situation. The counter-example is simple (as illustrated in Figure 6; for clarity we have omitted priority parameter): (i) $user_1$ inserts x in position 2 (op_1) while $user_2$ concurrently deletes the character at the same position (op_2). (ii) When op_2 is received by site 1, op_2 must be transformed according to op_1. So $T(Del(2), Ins(2, x))$ is called and $Del(3)$ is returned. (iii) In the same way, op_1 is received on site 2 and must be transformed according to op_2. $T(Ins(2, x), Del(2))$ is called and returns $Ins(3, x)$. Condition C_1 is violated. Accordingly, the final results on both sites are different.

The error comes from the definition of $T(Ins(p1, c1, pr1), Del(p2, pr2))$. The condition $p_1 < p_2$ should be rewritten $p_1 \le p_2$. Other bugs have been detected in other string-based group editors [13, 15]. More details may be found in [8].

```
Scenario 1:
-----------

op1 : Ins(u1,u2,u3)
op2 : Del(u4,u5)

S1 [op1;T(op2,op1)]:
   [Ins(u1,u2,u3);Del(u1+1,u5)]

S2 [op2;T(op1,op2)]:
   [Del(u1,u5);Ins(u1-1,u2,u3)]

Instance: Car(u1,S1) = Car(u1,S2)

Preconditions:
(u1 <= Length(u5))=true /\
(u4 < Length(u5))=true /\
u1 = u4;
```

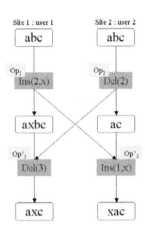

Fig. 5. Output of our algorithm.   **Fig. 6.** Scenario violating $CP1$.

4.2   Condition C_2

C_2 stipulates a *method identity* between two equivalent sequences. Given three methods m_1, m_2 and m_3, transforming m_3 with respect to two sequences $[m_1\,T(m_2,m_1)]$ and $[m_2\,T(m_1,m_2)]$ must give the same method. We define C_2 by the following property:

$$\Phi_2(m_1,m_2,m_3) \equiv T^*(m_3,[m_1\,T(m_2,m_1)]) = T^*(m_3,[m_2\,T(m_1,m_2)])$$

The second convergence property is formulated as a conjecture to prove from the replica specification.

Conjecture 2 (Convergence Property $CP2$). *A replica object π satisfies the condition C_2 iff: $\mathcal{A}^\pi \models_{obs} \Phi_1(m_1,m_2,m_3)$ for all methods m_1, m_2 and m_3.*

Definition 9. ($CP2$-scenarios). *A $CP2$−scenario is represented by a quadruple $(M_1,M_2,M_3,\mathcal{E})$ where M_1, M_2 and M_3 are tree methods and $\mathcal{E}$ is the set of conjectures obtained by the function $\mathrm{PROOF}(\Phi_2(M_1,M_2,M_3))$.*

A $CP2$-scenario simply gives methods and conditions that may lead to potential divergence situations. In Figure 7, we present an algorithm for checking the convergence property $CP2$.

Example 6. Consider the replica specification of Figure 3 with the modification regarding T for satisfying the convergence property $CP1$ (see Example 5). Using our algorithm, we have detected that convergence property $CP2$ is not satisfied. In Figure 8 we give one of the $CP2$-scenarios output by our algorithm. When analyzing this scenario, we notice that transforming *op1* along sequences $S1$ and

> **Input**   : A replica specification π.
> **Output** : $\mathcal{S}$ a set of $CP2$-scenarios.
>
> $\mathcal{S} \leftarrow \emptyset$;
> **foreach** method M_1 in $\mathcal{M}^\pi$
>   **foreach** method M_2 in $\mathcal{M}^\pi$
>   **foreach** method M_3 in $\mathcal{M}^\pi$
> $E \leftarrow \{\Phi_2(M_1, M_2, M_3)\}$;
> $E \leftarrow \text{PROOF}(E)$;
>   **if** $E \neq \emptyset$ **then** $\mathcal{S} \leftarrow \mathcal{S} \cup \{(M_1, M_2, M_3, E)\}$;
>   **endfor**
>   **endfor**
> **endfor**

Fig. 7. Checking Algorithm of Convergence Property $CP2$.

$S2$ produces different methods (i.e. $Ins(u1+1, u2, u3) \neq Ins(u1, u2, u3)$). There is a divergence problem caused by the triple (Ins, Del, Ins). Consider for instance in Figure 9, three sites 1, 2, 3 start from the same initial state "abc". They generate operations $op_1 = Ins(3, y, 1)$, $op_2 = Del(2, 2)$ and $op_3 = Ins(2, x, 3)$ concurrently, which change their states to "abyc", "ac" and "axbc" respectively. At site 1, when op_2 is received, it is transformed against op_1 resulting in $op_2' = Del(2, 2)$. After executing op_2' the state becomes "ayc". When op_3 arrives, it is transformed against $[op_1\, op_2']$ resulting in $op''_3 = Ins(2, x, 3)$ whose execution leads to state "axyc".

At site 2, op_1 arrives first and is transformed against op_2 resulting in $op_1' = Ins(2, y, 1)$. After op_1' is executed, the state becomes "ayc". And when op_3 arrives it is transformed first against op_2 resulting in $op'^2_3 = Ins(2, x, 3)$. Then op'^2_3 is transformed against op_1'. Since the priority of op'^2_3 is greater than that of op_1', it is shifted and we obtain $op''^1_3 = Ins(3, x, 3)$. After executing $op''^2_3 = Ins(3, x, 3)$, the state of site 2 becomes "ayxc" which is not identical to the state ("axyc") of site 1. Consequently, this OT algorithm does not verify convergence property $CP2$.

5   Implementation

We have updated our tool VOTE (Validation of Operational Transformation Environment) [9] by implementing our observational approach-based technique. This tool is designed to automatically check convergence properties $CP1$ and $CP2$. It builds an algebraic specification described in terms of conditional equations. As a verification back-end (implementation of PROOF function) we use SPIKE [3], an automated induction-based theorem prover. When SPIKE is called, either the convergence properties proof succeed and OT algorithm is validated, or the SPIKE's proof-trace is used for extracting all scenarios which may lead to potential divergence situations. There are two possible scenarios: the first one is unnecessary because contains valid conjectures (resulting from correct scenario)

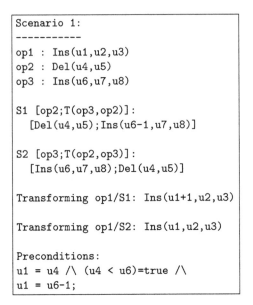

```
Scenario 1:
-----------
op1 : Ins(u1,u2,u3)
op2 : Del(u4,u5)
op3 : Ins(u6,u7,u8)

S1 [op2;T(op3,op2)]:
  [Del(u4,u5);Ins(u6-1,u7,u8)]

S2 [op3;T(op2,op3)]:
  [Ins(u6,u7,u8);Del(u4,u5)]

Transforming op1/S1: Ins(u1+1,u2,u3)

Transforming op1/S2: Ins(u1,u2,u3)

Preconditions:
u1 = u4 /\ (u4 < u6)=true /\
u1 = u6-1;
```

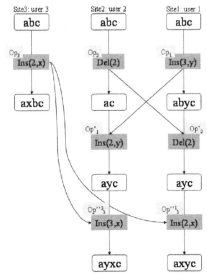

Fig. 8. Output of our algorithm.   **Fig. 9.** Scenario violating $CP2$.

Table 1. Case studies.

Distributed Systems	C_1	C_2
GROVE	violated	violated
Joint Emacs	violated	violated
REDUCE	correct	violated
SAMS	correct	violated
$S5$	correct	violated

that SPIKE cannot reduce[1]. Such cases can be overcame by only introducing lemmas. The second one concerns cases violating convergence properties. VOTE gives all necessary informations (methods and conditions) to understand the divergence origin. Consequently, these informations help developer to correct its OT algorithm.

We have detected a lot of bugs in well-known group editors such that GROVE [4], Joint Emacs [13], REDUCE[2] [15] and SAMS[3] [12] which are based on transformational approach for maintaining consistency of shared data. The results of our experiments are reported in Table 1. GROVE, Joint Emacs and REDUCE are group text editors whereas SAMS is XML document-based group editor. $S5$[4] is a file synchronizer which uses an OT algorithm for synchronizing many file system replicas [11].

---

[1] like $Car((p+1) - 1, st) = Car((p-1) + 1, st)$.
[2] http://www.cit.gu.edu.au/~scz/projects/reduce
[3] http://woinville.loria.fr/sams
[4] http://woinville.loria.fr/ls

6   Conclusion

We have presented our formal approach which is intended to automatically detect copies divergence in distributed groupware systems. To meet convergence requirement, the OT algorithm of these systems must be checked *w.r.t.* the convergence conditions C_1 and C_2. This task is difficult – even impossible – to carry out by hand due to the numerous cases to test. To overcome this problem, we have proposed an algebraic framework to assist the development of correct OT algorithms. Thanks to our framework, we have detected bugs in many well-known systems. So, we think that our approach is very valuable because: (i) it can help significantly to increase confidence in an OT algorithm; (ii) having the theorem prover ensures that all cases are considered and quickly produces counterexample scenarios; (iii) formalization is very easy and effortless. A drawback of this framework is that the user have to identify which set of characteristics gives a complete observation of the replica object. However, this can also be viewed as an advantage because the complexity of the proof is highly reduced.

Future Work. Many features are planned to deal effective and large systems. We plan to ensure the correct composition of many OT algorithms for handling composed objects. Finally, we intend to integrate in our framework the generation of Java classes from correct OT algorithms.

References

1. Michel Bidoit, Rolf Hennicker, and Martin Wirsing, *Behavioural and Abstractor Specifications*, Science of Computer Programming **25** (1995), no. 2-3, 149–186.
2. A. Bouhoula, E. Kounalis, and M. Rusinowitch, *Automated Mathematical Induction*, Journal of Logic and Computation **5(5)** (1995), 631–668.
3. Adel Bouhoula and Michael Rusinowitch, *Observational Proofs by Rewriting*, Theoretical Computer Science **275** (2002), no. 1–2, 675–698.
4. Clarence A. Ellis and Simon J. Gibbs, *Concurrency Control in Groupware Systems*, SIGMOD Conference, vol. 18, 1989, pp. 399–407.
5. Joseph Goguen, Kai Lin, and Grigore Roşu, *Circular Coinductive Rewriting*, Proceedings, 15th International Conference on Automated Software Engineering (ASE'00), Institute of Electrical and Electronics Engineers Computer Society, 2000, Grenoble, France, 11-15 September 2000.
6. Joseph Goguen and Grant Malcolm, *A Hidden Agenda*, Theoretical Computer Science **245** (2000), no. 1, 55–101.
7. Abdessamad Imine, Pascal Molli, Gérald Oster, and Michaël Rusinowitch, *Development of Transformation Functions Assisted by a Theorem Prover*, Fourth International Workshop on Collaborative Editing (ACM CSCW'02), Collaborative Computing in IEEE Distributed Systems Online (2002).
8. ———, *Proving Correctness of Transformation Functions in Real-Time Groupware*, in 8th European Conference of Computer-supported Cooperative Work (Helsinki, Finland), 14.-18. September 2003.
9. Abdessamad Imine, Pascal Molli, Gérald Oster, and Pascal Urso, *Vote: Group Editors Analyzing Tool*, Electronic Notes in Theoretical Computer Science (Ingo Dahn and Laurent Vigneron, eds.), vol. 86, Elsevier, 2003.

10. Abdessamad Imine and Pascal Urso, *Automatic Detection of Copies Divergence in Collaborative Editing Systems*, Electronic Notes in Theoretical Computer Science (Thomas Arts and Wan Fokkink, eds.), vol. 80, Elsevier, 2003.

11. Pascal Molli, Gérald Oster, Hala Skaf-Molli, and Abdessamad Imine, *Using the Transformational Approach to Build a Safe and Generic Data Synchronizer*, Proceedings of the 2003 international ACM SIGGROUP conference on Supporting group work, ACM Press, 2003, pp. 212–220.

12. Pascal Molli, Hala Skaf-Molli, Gérald Oster, and Sébastien Jourdain, *SAMS: Synchronous, Asynchronous, Multi-Synchronous Environments*, The Seventh International Conference on CSCW in Design (Rio de Janeiro, Brazil), September 2002.

13. Matthias Ressel, Doris Nitsche-Ruhland, and Rul Gunzenhauser, *An Integrating, Transformation-Oriented Approach to Concurrency Control and Undo in Group Editors*, Proceedings of the ACM Conference on Computer Supported Cooperative Work (CSCW'96) (Boston, Massachusetts, USA), November 1996, pp. 288–297.

14. Maher Suleiman, Michèle Cart, and Jean Ferrié, *Concurrent Operations in a Distributed and Mobile Collaborative Environment*, Proceedings of the Fourteenth International Conference on Data Engineering, February 23-27, 1998, Orlando, Florida, USA, IEEE Computer Society, 1998, pp. 36–45.

15. Chengzheng Sun, Xiaohua Jia, Yanchun Zhang, Yun Yang, and David Chen, *Achieving Convergence, Causality-preservation and Intention-preservation in real-time Cooperative Editing Systems*, ACM Transactions on Computer-Human Interaction (TOCHI) 5 (1998), no. 1, 63–108.

16. Andrew S. Tanenbaum, *Distributed Operating Systems*, Prentice-Hall, Inc., 2002.

17. Terese, *Term Rewriting Systems*, Cambridge University Press, 2003.

18. Nicolas Vidot, Michèle Cart, Jean Ferrié, and Maher Suleiman, *Copies Convergence in a Distributed Real-Time Collaborative Environment*, Proceedings of the ACM Conference on Computer Supported Cooperative Work (CSCW'00) (Philadelphia, Pennsylvania, USA), December 2000.

19. Martin Wirsing, *Algebraic Specification*, Handbook of theoretical computer science (vol. B): formal models and semantics (1990), 675–788.

Formal Verification of a Commercial Smart Card Applet with Multiple Tools[*]

Bart Jacobs[1], Claude Marché[2], and Nicole Rauch[3]

[1] University of Nijmegen, The Netherlands
[2] PCRI, LRI (CNRS UMR 8623), INRIA Futurs, Université Paris-Sud, France
[3] University of Kaiserslautern, Germany

Abstract. This paper presents a major Java source code verification case study that was carried out within the European VerifiCard project. It involves a realistic smart card applet from the company Schlumber-gerSema that has been verified with several tools in parallel, in order to assess the state of the art in formal verification. The paper describes part of the verification – using the static checker ESC/Java2 and the verifiers Jive, Loop and Krakatoa– and reports on the experiences and outlook.

1 Introduction

The European project VerifiCard has been running for almost three years from 2001 to 2004. Its aim was to apply formal techniques in the area of Java-based smart cards, both at the byte code and source code level, in order to provide the smart card industry with tools and techniques for higher levels of certification – typically within the Common Criteria framework. The VerifiCard project did not focus on the use of one particular tool, but involved the parallel use of several tools in order to stimulate coordination and comparison. At the start of the project it was felt that the area was too young and immature to restrict to one particular technique or tool. This approach is reflected in the current paper.

Within this framework the VerifiCard participant SchlumbergerSema[1] has provided the project with source code of one of its applets, together with a set of security properties that it wants certainty about. It is a commercially developed applet that has been sold to customers, and therefore, several aspects of it (and of the required security properties) have to remain confidential. We thus refer to the applet simply as the "SLB applet".

For the specification of Java source code properties the language JML [2] is developing into a (*de facto*) standard – in part as a result of the activities within the VerifiCard project. This paper describes our experiences in applying tools for JML-based source code verification, namely Jive [14], Loop [9] and Kraka-toa [13], as well as the static checking tool ESC/Java [12] (and its successor

---

[*] Funded by EU IST project IST-2000-26328-VERIFICARD, www.verificard.org
[1] Since jan. 1 2004 SchlumbergerSema's smart card group has become part of Axalto. Here we shall use the old name SchlumbergerSema that was used within the project.

C. Rattray et al. (Eds.): AMAST 2004, LNCS 3116, pp. 241–257, 2004.

ESC/JAVA2 [3]). The paper presents a short overview of the methodologies that
the different verification tools are based upon, demonstrates how the tools have
been applied to the case study, and what their respective achievements in the
verification of the applet are – namely the detection of some bugs.

The paper is organised as follows. The SLB applet and its specification are
introduced in Section 2. The subsequent five sections present the various tools
that have been applied to the applet and the results achieved with them. Finally,
Section 8 draws some conclusions.

2   Outline of the SLB Applet

This section presents some background information about the applet under investigation. The confidential nature of the applet imposes some restrictions on
what can be said. However, these restrictions do not really affect the explanations
of the scientific aspects of the investigation.

The applet has been provided by SchlumbergerSema "as is", that is without
any additional documentation. The code itself does contain some rudimentary
comments, but they are mostly concerned with specific details, and not with the
big picture. We had to reconstruct that ourselves. This first step of the work, done
by Hans Meijer and Bart Jacobs at Nijmegen, was to write a JML specification
for the SLB applet. The specification was based purely on code inspection. The
unclarities that emerged were discussed with SchlumbergerSema. At this stage
the specification was only typechecked, and not verified. It was then distributed
to the four verification teams, who adapted it to their own needs.

2.1   Structure of the Code

The main part of the SLB applet is a 2-dimensional array of bytes, representing
a rudimentary file system. One immediate problem is that Java Card does not
allow multi-dimensional arrays. The trick used in the applet is to define an array:

```
Object [] o_Records;
```

Within the applet, each object in o_Records will then store a byte array, called
a *record*. This intention can be made explicit in a JML class invariant, as:

```
  o_Records[i] instanceof byte[] &&
2 ((byte[]) o_Records[i]).length == record_length && ...
```

where i lies in an appropriate range. As stated, these records all have the same
length, described here as a JML model field record_length. The first entry
o_Records[i][0], if non-zero, is used to indicate the length of the data in the
record o_Records[i].

All this makes the class invariant an essential part of the specification: it
expresses the basic ideas that the applet developers had in mind. These ideas
are not explicit in the code, nor in the documentation (that we have seen). At
the same time, the class invariant is one of the most complicated parts of the
specification, and is sometimes needed to establish additional safety properties.
The class invariant complicates the verification, because for each method it must

be shown that the invariant is maintained. But in the different verifications described below, only the LOOP approach actually used the entire invariant. A detailed description of the class invariant is therefore postponed to Section 5.

The JML specification developed at this stage involved not only a class invariant, but also specifications for all methods in the applet class. A central method (as in any Java Card applet) is the **process** method: based on a case analysis using the APDU's class and instruction entries, it passes the card's incoming byte sequence (called APDU for application data unit) to one of the private methods **processSelectCmd**, **processReadRecord**, **processDeleteRecord**, **processPut-Data**, or **processAppendRecord**. Additionally, the **process** method involves a debug option. The private methods called by **process** typically start with low-level byte operations on the APDU's byte array, to perform appropriate checks or extract the relevant data. The associated JML specifications incorporate the corresponding checks, and are thus also low-level and rather verbose.

The following actual fields of the SLB applet are most relevant:

```
  static final short  OFF_DATA_IN_RECORD  = (short)01;
2 static final byte   LEN_RECORD_LEN_BYTE = (byte)01;
  static final short MAX_LEN_OPTIONAL_DATA = (short)10;
4 static final short MAX_LEN_OPTIONAL_DATA_AND_HEADER
       = MAX_LEN_OPTIONAL_DATA + (short)3;
6 static final short SIZE_OPTIONAL_DATA_BUFFER
       = (short)3 * MAX_LEN_OPTIONAL_DATA_AND_HEADER;
8 // === Data entries
  byte by_NbRecords;       // number of existing data entries
10 byte by_MaxNbRecord;      // max number of data entries
  byte by_MaxSizeRecord;   // max size of a record
12 byte[] bya_OptionalData;
  Object[] o_Records;
```

We introduced the following additional JML *model fields*, which are specification-only variables [10, 11], typically used as convenient abbreviations or to provide a higher level of abstraction with respect to the actual fields.

```
  /*@ public model final short  OFF_DATA_IN_APDU;
2 @    represents  OFF_DATA_IN_APDU
              <- (short)ISO7816.OFFSET_CDATA;
4 @ public model final short  aid_off_in_data;
  @    represents  aid_off_in_data
6              <- (short)(OFF_DATA_IN_RECORD + 5);
  @ public model final short  aid_off_in_apdu;
8 @    represents  aid_off_in_apdu
              <- (short)(OFF_DATA_IN_APDU + 5);
10 @ public model short  record_count;
  @    represents  record_count <- (short)(by_NbRecords & 0xFF);
12 @ public model short  max_record_count;
  @    represents  max_record_count
14              <- (short)(by_MaxNbRecord & 0xFF);
  @ public model short  record_length;
16 @ // with first byte describing length of subsequent records
  @    represents  record_length
18 @    <- (short)(LEN_RECORD_LEN_BYTE+(by_MaxSizeRecord&0xFF));
  @*/
```

Various methods will be discussed below in separate sections.

2.2   Verification Aims

SchlumbergerSema provided us with a list of security properties to be checked. In this paper, we focus on the property describing error prediction. It is requested that "No exception other than an ISOException should be thrown at toplevel as a result of invoking an applet entry point". The following sections present the approaches of the different tools towards securing this property.

2.3   Production and Evaluation

Even though SchlumbergerSema is remarkably open in providing us access to actual production code, it is (understandably) secretive about many details, in order to protect its commercial interests. What we understand is that the applet under consideration has gone through the internal test and evaluation procedures within SchlumbergerSema before being sold to customers. We have no information about the nature and depth of these procedures, but we understand that there is no formal specification and verification involved. Hence, the verification methods described in this paper extend the internal approach at Schlumberger-Sema, and their results are therefore of interest in order to possibly improve the current practice.

3   ESC/Java2 Approach

ESC/JAVA2 stands for *Extended Static Checker for Java Version 2*, a static checker which has originally been developed at Compaq SRC [12] and which is now being extended at the University of Nijmegen [3]. It has been applied to this case study in order to relate its checking approach to the verification approach followed by the other tools. ESC/JAVA2 is fully automatic (push-button). A drawback is that it is neither sound nor complete. Still, it provides valuable results which are demonstrated below.

3.1   Checking Technique

One approach of using ESC/JAVA2 is to add a JML specification to an applet and to apply ESC/JAVA2 to the specified applet. An alternative approach is to develop a specification interactively by invoking ESC/JAVA2 on an unspecified applet and by removing the warnings step by step by adding appropriate annotations. Remaining warnings can be confirmed or refuted by a verification system. This produces a lightweight specification which captures the applet's requirements to run without errors. In a subsequent step, the user can develop a functional specification. The approach presented here is not intended to produce a functional specification automatically.

```
   private void processDeleteRecord(APDU oApdu) {
2     byte[] byaApdu = checkIncomingData(oApdu, true, false);
      byte byMode = MODE_DELETE_UNSET ;
4     switch(byaApdu[ISO7816.OFFSET_P2]&0x07) {
          case 0x00: // -- modes delete by AID or refresh
6             switch(byaApdu[ISO7816.OFFSET_P1]) {
                  case 0x00:
8                     byMode = MODE_DELETE_BY_AID; break;
                  case 0x01:
10                    byMode = MODE_DELETE_BY_REFRESH; break;
                  default:
12                    ISOException.throwIt(ISO7816.SW_INCORRECT_P1P2);
              } break;
14        case 0x04: // -- mode delete by record number
              byMode = MODE_DELETE_BY_NUMBER; break;
16        default:
              ISOException.throwIt(ISO7816.SW_INCORRECT_P1P2);
18    }
      if((byaApdu[ISO7816.OFFSET_P2]>>3) != 1)
20        ISOException.throwIt(ISO7816.SW_FILE_NOT_FOUND);
      byte[] byaRecordData = null;
22    if (byMode == MODE_DELETE_BY_NUMBER) { /* ... */ }
      else if (byMode == MODE_DELETE_BY_AID) {
24        for(byte i = 0; i < by_NbRecords; i++) {
              byaRecordData = (byte[])(o_Records[i]);
26            if (Util.arrayCompare(byaApdu, ISO7816.OFFSET_CDATA,
                      byaRecordData, (short)(OFF_DATA_IN_RECORD+6),
28                    byaRecordData[OFF_DATA_IN_RECORD+5]) == 0)
              {   /* ... */    }
30        }
      } else if (byMode == MODE_DELETE_BY_REFRESH) { /* ... */ }
32 }
```

Fig. 1. SLB applet, method `processDeleteRecord` (slightly shortened)

```
   requires oApdu != null && oApdu._APDU_state == 1;
2  signals (ArrayIndexOutOfBoundsException u)
     (\exists short i; (0 <= i && i < by_NbRecords) ==>
4    ((short)(ISO7816.OFFSET_CDATA +
     ((byte[])(o_Records[i]))[OFF_DATA_IN_RECORD+5])
6    > oApdu.buffer.length));
   signals (APDUException a) true;
8  signals (ISOException i)  true;
```

Fig. 2. Lightweight ESC/JAVA2 specification of `processDeleteRecord` method

3.2   The `processDeleteRecord` Method

This section introduces the method `processDeleteRecord` of the SLB applet. Figure 1 shows part of the method's source code. During the project, ESC/ JAVA2 has been applied to the unspecified SLB applet in order to check the security property described in Section 2.2. We interactively generated a lightweight

```
  /*@ requires  src != null && srcOff+length <= src.length &&
2 @              dst != null && dstOff+length <= dst.length &&
  @              srcOff >= 0 && dstOff >= 0 && length >= 0;
4 @ signals (NullPointerException) src==null || dst==null;
  @ signals (ArrayIndexOutOfBoundsException) length < 0
6 @          || srcOff < 0 || srcOff+length > src.length
  @          || dstOff < 0 || dstOff+length > dst.length;
8 @*/
  public static final /*@ pure @*/ byte arrayCompare(byte[] src,
10           short srcOff, byte[] dst, short dstOff, short length)
```

Fig. 3. Method `Util.arrayCompare()`: Excerpt of JML specification

specification (see Fig. 2) which captures all conditions that are required by the method to run without exceptions. It is especially interesting that ESC/JAVA2 also points out some possible exceptions: `APDUExceptions` and `ISOExceptions`, which can be triggered from outside the applet. For example, the method `APDU.setIncomingAndReceive`, which is invoked in `checkIncoming Data` (see Fig. 1, line 2), throws an `APDUException` if an I/O error occurred. These exceptions cannot be avoided by setting up requirements to the method, which is clear from the generated specification: it states that these exceptions are thrown unconditionally (see Fig. 2, lines 7 – 8). They must explicitly be caught in the applet or allowed at the top level.

Another reported exception is an `ArrayIndexOutOfBoundsException` which can occur in the call to Java Card's `Util.arrayCompare` method (see Fig. 1, lines 26 – 28). This method's last parameter accepts the length of the interval in which the two arrays are compared. If this length is negative or too large, or if one of the offsets passed as arguments are too large, an `ArrayIndexOutOfBoundsException` is thrown (see Figure 3). Section 4 treats the verification of this exception.

4   Jive Approach

JIVE is a verification system developed at the Universities of Hagen and Kaiserslautern. It is an interactive Hoare logic theorem prover with a special graphical user interface that uses an associated general purpose theorem prover (the user can choose between Isabelle/HOL [17] and PVS [16]). Internally, JIVE operates on Diet Java Card, a desugared subset of Java Card which is very close to the original language. For the verification, a Hoare-style programming logic is used. The pre- and postconditions of the Hoare triples are given in a predicate logic notation. The ability to read JML specifications is currently being added.

Proofs are performed by applying the Hoare rules – either manually or via so-called strategies which are Java programs that automatically apply several Hoare rules. Predicate logical statements over the pre- and postconditions (usually implications that stem from strengthening or weakening steps performed in the Hoare logic) are proven interactively in the associated prover.

One of JIVE's main strengths is its dedicated user interface. At all times, the user has full visual control over the whole proof process. A description of JIVE's architecture is given in [14].

4.1   Verification Technique

In order to prove the given security property with JIVE, one needs to provide a specification that describes the absence of any exception other than an ISOException or APDUException in the poststate of the process method.

Two techniques can be used to prove such a specification. The "standard" approach is to perform a *backward proof* by starting at the method implementation and the given specification and by developing a proof by applying the Hoare-logic rules in backward direction, e.g. by using a weak(est) precondition generation strategy [18]. This approach is useful if there is nothing known about the correctness of the code. But if one already suspects an error in the code, another approach comes in handy: the *assertion-supported forward proof*. In this approach an assertion is inserted into the code which can be pushed through the code [19], again e.g. by means of a weak(est) precondition mechanism. This technique will be explained in more detail below.

4.2   The processDeleteRecord Method

To verify the exception in the method processDeleteRecord which was indicated by ESC/JAVA2 (see Sect. 3), we use the *assertion-supported forward proof* that was briefly sketched above. This strategy allows us to insert an assertion just before a statement s which is suspected to throw an exception. This assertion is then propagated to the beginning of the method body by means of weak(est) precondition generation. There are two possibilities for the assertion: One can either insert a formula P_{ass} which is fulfilled by all states that do not raise an exception at s, or one can insert a formula $P_{ass}^{\neg}$ which is fulfilled by all states that definitely throw an exception at s. These formulae are propagated to preconditions P_{calc} and $P_{calc}^{\neg}$, respectively. How do these generated preconditions relate to the precondition P_{spec} which is part of the method specification?

Let us regard the first case. We know that if the precondition P_{calc} holds, then no exception can occur at s. If P_{calc} is a weakest precondition, i.e. if there are neither loops nor method invocations in the code before s, then the implication $P_{spec} \Rightarrow P_{calc}$ must hold, otherwise we know that an exception can occur at s. If P_{calc} is not weakest, we can still try to prove the implication $P_{spec} \Rightarrow P_{calc}$. If it holds, we know again that no exception can occur, but if it does not hold, we do not know anything. We can only assume that P_{spec} may be too weak and should be strengthened (e.g. by adding P_{calc}). Thus, if we suspect that an exception occurs at s, and if there are loops or method invocations in the code before s, this may not be a good strategy because we might not get a usable result. Therefore, let us take a look at the second case. We know that each state which fulfills the formula $P_{calc}^{\neg}$ is guaranteed to raise an exception at s. Therefore, we need to find a state Σ that fulfills both the given precondition P_{spec} and the generated precondition $P_{calc}^\neg$, in other words, the formula $P_{spec}(\Sigma) \wedge P_{calc}^\neg(\Sigma)$ must be valid. The state Σ is called a *counterexample* to the assumption that no exception is thrown at s.

Regarding the `processDeleteRecord` method and the `ArrayIndexOutOf-BoundsException` that ESC/JAVA2 reported, this means that the assertion $P_{ass}^\neg$, which guarantees for all fulfilling states that an exception is thrown in the subsequent statement, contains the following:

```
    ISO7816.OFFSET_CDATA + byaRecordData[OFF_DATA_IN_RECORD+5]
2                > byaApdu.length
```

Here, it is easy to construct a counterexample as the array `byaRecordData` can contain arbitrary data (see Section 3); therefore, we use the second strategy to verify the possible occurrence of an exception. A more detailed description of the methodology can be found in [19].

5   LOOP Approach

After the initial phase described in Section 2 in which the specification was written, the verification with the LOOP tool started[2]. This resulted in a number of changes in the specification: First, only during actual verification is the specification really tested. A small number of subtle mismatches was found, leading to improvements in the specification. Second, the specification was optimised in its formulation to make it more suitable for verification. For instance, method calls were removed from specifications and replaced by explicit descriptions. This lowers the level of abstraction, but facilitates the verification with the LOOP tool.

In Section 2 we have already seen part of the classwide JML specification, namely the model fields. This part will be extended first with the invariant. Then, Subsection 5.2 will describe the specification and verification of one method in greater detail. The class invariant that we formulated is as follows.

```
/*@ invariant
2    @    bya_FCI != null  && bya_FCI.length == 23  &&
     @    LEN_RECORD_LEN_BYTE + DEF_MAX_ENTRY_SIZE
4    @       >= aid_off_in_data + 1 &&
     @    LEN_RECORD_LEN_BYTE <= OFF_DATA_IN_RECORD &&
6    @    0 <= by_NbRecords &&  0 <= by_MaxNbRecord  &&
     @    0 <= record_count && record_count <= max_record_count &&
8    @    0 <= record_length && record_length > aid_off_in_data &&
     @    bya_OptionalData != null  && o_Records != null  &&
10   @    bya_OptionalData.length == SIZE_OPTIONAL_DATA_BUFFER  &&
     @    o_Records.length == max_record_count  &&
12   @    (\forall short i; 0 <= i && i < max_record_count ==> (
     @       (i >= record_count && o_Records[i] == null)  ||
14   @       (i <  record_count && o_Records[i] != null &&
     @          o_Records[i] instanceof byte[]  &&
```

---

[2] At Nijmegen the (original) ESC/JAVA tool was not applied to the applet, because the invariant (involving a complicated universal quantification) was too difficult to be handled by ESC/JAVA.

```
16   @              ((byte[]) o_Records[i]).length == record_length &&
     @              ((byte[]) o_Records[i])[aid_off_in_data] >= 0 &&
18   @              ((byte[]) o_Records[i])[aid_off_in_data] <
     @                   record_length - aid_off_in_data &&
20   @              (((byte[]) o_Records[i])[0] != 0 ==>
     @                   OFF_DATA_IN_RECORD +
22   @                        (((byte[]) o_Records[i])[0] & 0xFF)
     @                   <= record_length ))));
24   @*/
```

The part of this invariant before the forall-quantification makes the bounds of the various (model) fields explicit. The forall-assertion describes the directory o_Records as an appropriate 2-dimensional array of bytes. It contains information that is not explicit in the code nor in the associated comments (provided by SchlumbergerSema), but was crucial for the proper understanding of the applet.

5.1   Verification Technique

The LOOP tool is a compiler that takes Java+JML input and translates it to logical theories for the PVS [16] theorem prover. The generated PVS-theories incorporate the resulting proof obligations: each JML method specification becomes a PVS predicate that must be proven for the associated translated method. See [9] for a recent overview and further references. These translated method specifications incorporate the class invariants in their pre- and postconditions.

The actual verification happens in the backend theorem prover PVS. A user interacts with PVS by typing appropriate proof commands to guide the theorem prover. High level strategies are available, providing much automation.

Typical of the LOOP tool is the use of a shallow embedding: Java programs become actual functions in PVS, on the basis of an underlying semantics. These functions can be evaluated symbolically, corresponding to execution in Java. The LOOP verifications can take place in different ways.

- Symbolic evaluation (aka. semantical proof). This can be very efficient in proofs, but only works for simple Java program fragments, not involving iterations (such as while and for loops).
- Hoare logic [8]. In this way one can step through the Java code, at each point proving appropriate assertions. This provides more abstraction, but has the disadvantage that the user has to explicitly provide the intermediate assertions. However, this works well for larger method bodies, because they can be broken up into smaller manageable parts. The intermediate assertions can be written directly in JML, and are also translated by the LOOP tool.
- Weakest precondition reasoning [7]. It provides several (forward and backward) strategies to automatically prove assertions for code fragments. This works well for relatively small pieces of code, and can thus be combined efficiently with Hoare logic.

The fact that the entire Java+JML input is represented in PVS – and not broken up by the translation tool – has both advantages and disadvantages: an advantage of breaking up the proof goals inside PVS is that the rules for doing so (coming from Hoare and WP logic) are provably sound; a disadvantage is that size becomes an issue. Indeed, in the applet verification we found that the bottleneck in the verification was the theorem prover PVS. It has problems dealing with the large proof goals resulting from the larger methods (esp. `processAppendRecord` and `processDeleteRecord`). Currently, these scalability problems are being discussed and solved with the PVS development team.

One recent addition to the LOOP tool is the bitvector semantics [6] to handle Java Card's bounded arithmetic (`byte`, `short`) in a precise manner. This semantics is heavily used in the SLB applet case study, with its many low-level operations on bytes (such as masking and shifting).

5.2   The `processAppendRecord` Method

The `processAppendRecord` method has the following header, with explanation as provided by SchlumbergerSema.

```
  /** Add a new record to the directory. The directory
2  * is configured for a maximum number of directory entries
   * (set in the install command)
4  * Entries once added can be removed, corresponding space can
   * be re-used again. ...
6  */
private void processAppendRecord (APDU oApdu) { ... }
```

The first step of the method is to put the byte array contained in the parameter APDU into a local variable `byte[] byaApdu`. The contents of this byte array are then examined, to see if they contain the right values to append a record (also contained in this byte array) to the directory (`o_Records`). If such an examination does not provide a positive result, an `ISOException` is thrown. These exceptions show up in the method's JML method specification, as `signal` clauses.

As part of these examinations an AID lookup is performed, of the form:

```
  if (JCSystem.lookupAID (byaApdu, ADF_OFFSET,
2                          byaApdu[ADF_LEN_OFFSET]) == null)
    ISOException.throwIt (ISO7816.SW_CONDITIONS_NOT_SATISFIED);
```

In the verification of this code fragment the method specification of `lookupAID` from the API class `JCSystem` is used. This specification looks as follows.

```
   /*@ public behavior
2  @    requires true;
   @    assignable \nothing;
4  @    ensures true;
   @    signals (NullPointerException) buffer == null;
6  @    signals (IndexOutOfBoundsException)
   @        offset < 0 || length < 0 ||
8  @        offset + length > buffer.length;
   @*/
10 public static AID lookupAID ( byte[] buffer, short offset,
                                  byte length )
```

We see that an `IndexOutOfBoundsException` is thrown in case the parameter `length` is negative. During the verification of the `processAppendRecord` method we were not able to prove that this requirement holds for the byte value `byaApdu[ADF_LEN_OFFSET]`. Indeed, this requirement is not checked during the examinations at the beginning of the method. This is a bug: an APDU sent to the card can generate an exception that is not caught, and thus cause the application to halt, with status word "no precise diagnostic"[3]. However, this is not a security breach, because the applet remains in a consistent state, since the invariant still holds when such an exception appears.

A similar bug shows up later on in the code:

```
   for(byte i=0 ; i<by_NbRecords ; i++) {
2     byaRecordData = (byte[])(o_Records[i]) ;
      if ((byaRecordData[0] != 0x00)  // if not previously erased
4        && Util.arrayCompare(byaApdu, ADF_OFFSET, byaRecordData,
           (short)(OFF_DATA_IN_RECORD+6),byaApdu[ADF_LEN_OFFSET])
6        == 0)
         ISOException.throwIt(ISO7816.SW_CONDITIONS_NOT_SATISFIED);
```

We see that the `arrayCompare` method from Java Card's `Util` class is called. Its last parameter describes the length of the comparison interval (in two byte arrays). As we have seen in Section 3, if this length is negative or too large, an `IndexOutOfBoundsException` is thrown. This may actually happen in the applet because there are no appropriate checks on the value `byaApdu[ADF_LEN_OFFSET]` of the incoming APDU.

In conclusion, these two bugs appear in the JML specification of the method `processAppendRecord` in the fragment:

```
   @  signals (IndexOutOfBoundsException) // This is a real bug!
2  @      (oApdu.buffer[ADF_LEN_OFFSET] < 0    // for lookupAID
   @      || oApdu.buffer[ADF_LEN_OFFSET] >=
4  @         record_length − aid_off_in_data) // for ArrayCompare
```

These bugs were discovered because the proof could not proceed at the above mentioned program points.

6   Krakatoa Approach

6.1   Verification Technique

The KRAKATOA tool [13] proceeds roughly as follows: from the JML-annotated source code of the API, and the program in consideration, and some method in that program, source scripts for the COQ proof assistant [20] are generated automatically. They provide a modeling of the Java memory heap for the program (depending on the collection of classes and fields); COQ definitions corresponding to the specifications of all constructors and methods of the program; a set of proof obligations, which are not in fact generated directly

---

[3] We have been able to recreate this exception on an actual smart card running this applet. It was the first bug found in the course of this case study.

but by using an external tool called WHY [4, 5]; and a so-called *validation* of the method considered, that is a COQ program equivalent to it, of the form $\forall args \,\forall heap_{in}, pre \Rightarrow \exists result \,\exists heap_{out}, post$.

The application of the Krakatoa approach to the SLB applet case study took some benefit from the achievements of the LOOP team. We actually used JML specifications they wrote both for the Java Card API and the applet, performing only a few minor changes to optimise the specifications for the KRAKATOA tool as well as to avoid some *ad hoc* additions related to the LOOP tool. We focus here on the `processPutData` method.

6.2   The `processPutData` Method

An excerpt of the annotated source of this method is shown in Figure 4. The changes (w.r.t. LOOP specifications) we made are marked by appropriate comments. The first addition is in the precondition, line 2 of the listing. It appears to be needed for performing the proof of the normal postconditions, because otherwise `oApdu.buffer` could be modified, and everything would go wrong. After discussion with the LOOP team, it appears they did not need this assumption because they were able to derive it from the JML class invariant of the `APDU` class: they put a very large minimal bound for the length of APDU buffers, so that these arrays can be provably different just because their lengths are different.

The second addition is in the modifiable clause on line 4 of the listing: mentioning `ISOException.systemInstance.theSw[0]` is needed because this location is modified when an `ISOException` is thrown. This was not required by LOOP, because, as before, they simplified the JML specifications of the `ISOException` class, to get rid of such annoying and essentially irrelevant details.

The third change, line 7 of the listing, is the use of the pure method `getIndex` to replace the expression `>>4` used in LOOP, which is OK only because of the specific values of the `TAG_*` constants. For LOOP it is more convenient to have executable specifications because of its symbolic evaluation feature, but with KRAKATOA it is better to use more abstract specifications: we introduced a JML *pure* method `getIndex` to capture the logical behavior of the switch statement.

The last change, lines 17–21 of the listing, is an annotation for the switch statement, which is not standard JML but an extension allowed in KRAKATOA. This is because we met scalability issues: the amount of generated COQ source is quite large; and much effort has been made to reduce such problems. For instance, the annotation for the `switch` statement above reduced the size of the proofs with a factor four, because in some sense it factorizes all four cases of the `switch`. This clearly suggests that such an extension of JML is useful for theorem proving (indeed this extension is now under consideration).

7   Results and Experiences

In this case study, several exceptions were detected by the LOOP tool and ESC/JAVA2. These are `SystemExceptions`, `ISOExceptions`, `APDUExceptions`,

```
   /*@ requires  oApdu != null
 2   @             && byaOptionalData == bya_OptionalData
     @             && byaOptionalData != oApdu.buffer;  //added
 4   @ modifiable byaOptionalData[*],
     @             ISOException.systemInstance.theSw[0];  // added
 6   @ ensures   (\forall short i; 0 <= i &&
     @             i < sh(oApdu.buffer[ISO7816.OFFSET_LC]) + 3;
 8   @        byaOptionalData[getIndex // instead of >>4 with LOOP
     @          (Util.getShort(oApdu.buffer,ISO7816.OFFSET_P1))
10   @             * MAX_LEN_OPTIONAL_DATA_AND_HEADER + i]
     @             == oApdu.buffer[ISO7816.OFFSET_P1+i]);
12   @ signals (ISOException e) true;
     @ signals (APDUException e) true;
14   @*/
   private void processPutData(APDU oApdu, byte[] byaOptionalData)
16 { ... (some verification of input data skipped)
   byte byIndex = (byte)0;
18   /*@ assignable ISOException.systemInstance.theSw[0];
     @ ensures 0 <= byIndex && byIndex <= 2 &&
20   @        byIndex == getIndex(Util.getShort(oApdu.buffer,
     @                            ISO7816.OFFSET_P1));
22   @ signals (ISOException e) true;
     @*/
24   switch (Util.getShort( byApdu, ISO7816.OFFSET_P1 )) {
       case TAG_FCI_ISSUER_DISCRETIONARY_DATA :
26          byIndex = 0 ; break ;
       case TAG_ISSUER_CODE_TABLE_INDEX :
28          byIndex = 1 ; break ;
       case TAG_LANGUAGE_PREFERENCE :
30          byIndex = 2 ; break ;
       default :
32          ISOException.throwIt(ISO7816.SW_INCORRECT_P1P2);
   }
34   Util.arrayCopy( byaApdu, ISO7816.OFFSET_P1, byaOptionalData,
          (short)(byIndex * MAX_LEN_OPTIONAL_DATA_AND_HEADER),
36          (short)(shLC+(short)3));  // +3 for TAG(2) and LEN(1)
   }
38
   /*@ ensures 0 <= \result && \result <= 2
40   @  && (tag == TAG_FCI_ISSUER_DISCRETIONARY_DATA
     @         ==> \result == 0)
42   @  && (tag == TAG_ISSUER_CODE_TABLE_INDEX ==> \result == 1)
     @  && (tag == TAG_LANGUAGE_PREFERENCE ==> \result == 2);
44   @*/
   private static /*@ pure @*/ byte getIndex(short tag);
```

Fig. 4. Excerpt of `processPutData` method, annotated

`TransactionExceptions` and `ArrayIndexOutOfBoundsExceptions`. Schlumber-gerSema's security policy only accepts `ISOExceptions` at the top level. But `SystemExceptions`, `APDUExceptions` and `TransactionExceptions` are inavoid-able in the applet because they can be initiated externally (see Sect. 3.2 for an example). Here, the two tools clearly pointed out that SchlumbergerSema's security policy is too strict. But of course, the `ArrayIndexOutOfBoundsExceptions` are the most interesting exceptions. The LOOP and JIVE tools verified that these exceptions may occur in the `processAppendRecord` and `processDeleteRecord` methods because the applet uses data which has been passed from the card accepting device (CAD) to the applet in an APDU to index an array, and which has not been checked to be within the array bounds. Usually, this does not pose a problem, but if the applet is being attacked, malformed APDUs will be likely to be sent to it, which should not cause an exception to occur. To our knowledge, this exception cannot be used to exploit the applet nor to permanently damage the contained data. But the applet programmers would nevertheless be well advised to remove this cause of abrupt termination.

The `processPutData` method, as it is annotated in Figure 4, has been fully verified by the KRAKATOA tool. It was shown that no exception can be thrown other than `ISOException` or `APDUException`, and that the method satisfies its normal postcondition and preserves a class invariant.

Writing the specifications was a substantial part of the effort. Ideally, the specifications are already there, written by the programmers themselves. However, just writing specifications without checking them may lead to mismatches. Part of this case study's success is due to the support of JML by all tools except JIVE; only small changes had to be performed to adapt to KRAKATOA and ESC/JAVA2.

This case study has shown that the LOOP approach is capable of analyzing a non-trivial amount of Java Card source code (in the order of hundred's of lines), and to detect errors that had previously gone unnoticed. KRAKATOA also demonstrated its ability to prove non-trivial properties of Java Card applets. This is clearly a success.

For KRAKATOA and LOOP, performance and scalability bottlenecks occurred within the backend theorem provers Coq and PVS, respectively. They slowed down the effort considerably. Improvements are currently under investigation.

The LOOP and KRAKATOA approaches require user interaction with PVS and Coq provers respectively, and knowledge of both the back-end prover and the Java modeling introduced by the respective tool is mandatory. But for an experienced user, proving the obligations generated from one method is not so hard: indeed, much of our time has been spent on improving the specification, the modeling, etc. to make proofs easier. Nevertheless, more automation of proofs is required via dedicated strategies/tactics in the back-end prover. Also, the option of further integration of handling of proof obligations by either an automated prover like Simplify (used by ESC/JAVA2) or an interactive prover (like Coq or PVS) is currently under study.

The validation of suspected errors (e.g. those detected by ESC/JAVA2) is not as costly as a full verification. Although a partial verification as e.g. performed by the JIVE tool does not guarantee that the areas that ESC/JAVA2 reports as correct are indeed error-free, it still is a recommendable approach to software verification because the tradeoff between cost and correctness that comes with it seems to be fair enough.

It is difficult to give exact measurements of time because this case study was used to optimize the tools, so that it is hard to tell what the pure verification time is. Additionally, the JML specifications had to be written in the beginning and improved during verification. All we can say that if no major problems arise, then the verification of individual methods is a matter of days.

8   Conclusions

The SLB applet case study described in this paper has pushed the development of the Java verifiers (JIVE, LOOP, KRAKATOA) to make them powerful enough to handle non-trivial source code. (The development from ESC/JAVA to ESC/-JAVA2 is independent.) Moreover, it has led to a unification in the area and a common focus on JML as specification language.

The semantical complexity of languages like Java and JML means that code verification is still very complicated and requires a substantial amount of user interaction. Nevertheless, the current tools are capable of detecting non-trivial bugs that were not spotted with conventional techniques (testing and code inspection). The case study has made it clear where the complexities are, and where improvements are needed. Here is a brief analysis.

Static checking with ESC/JAVA2 has become a very powerful push-button technique that is so simple to use that good Java developers should be able to use it. This means that they can write their own JML specifications together with the implementation, and check it while they are developing code. The theorem-prover based verification techniques are still complicated and labor-intensive, so that they should be reserved for the cases that ESC/JAVA2 cannot handle, probably by people in research departments or outsiders. The positive feedback of SchlumbergerSema on this case study envisages an approach along these lines.

As an aside, separate from verification: SchlumbergerSema (and other companies like Gemplus) are enthusiastic about the use of JML. They see as one of the next challenges to relate their high-level security requirements to the relatively low-level specifications in JML.

It is in the nature of Java Card applets that they communicate with the outside world via byte sequences (in APDUs). As a result, much of the implementation and specification involves low-level details that are hard to write and easy to get wrong. A more abstract approach is badly needed, for instance based on Java Card RMI (see also [15]), as has recently been added in Java Card 2.2.

When we compare the verification tools used in this paper[4] we see two fundamental differences: First, LOOP and KRAKATOA are front-ends to theorem provers in which the interactive verification takes place on the translated program code. In contrast, the user of the JIVE tool spends most of the time interacting with the original program code. Second, verifications with the KRAKATOA

---

[4] Besides the three verification approaches used in this paper there is also the deep embedding of Java into Isabelle/HOL [17] developed by the VerifiCard partner in Munich. Such a deep embedding is ideal for proving meta-properties (like type-safety), but too complicated for the verification of concrete programs.

and JIVE tools decompose the proof goal into many separate verification conditions. The LOOP tool generates one single proof obligation, which is decomposed (using provably sound rules) within the back-end theorem prover. The first approach leads to very many small obligations, and the latter to one big obligation. Both approaches may lead to too much data to manage. Among these alternative approaches there is (maybe unfortunately) not a single one that emerges as "best". An interesting perspective is to be able to make these tools, including ESC/JAVA2, cooperate.

In conclusion, we see a bright future for Java source code verification (based on JML specifications) in a 2-step approach, where theorem-prover based verification should be smoothly integrated with static checking, and be used to handle the difficult left-overs (like in [1]): Especially methods with loops or recursion or sources of remaining warnings are likely to still contain errors and are therefore good candidates for full verification.

Acknowledgements

Many people contributed to the design of the JML specifications and the proofs of obligations: we gratefully thank Cees-Bart Breunesse, Marek Gawkowski, Hans Meijer, Martijn Oostdijk, Christine Paulin, Erik Poll, Anne Schultz, Xavier Urbain, and Martijn Warnier. Further, we like to thank Boutheïna Chetali and Olivier Ly from SchlumbergerSema for their help and feedback.

References

1. C.-B. Breunesse, N. Cataño, M. Huisman, and B. Jacobs. Formal methods for smart cards: an experience report. Technical Report NIII-R0316, University of Nijmegen, 2003.
2. L. Burdy, Y. Cheon, D. Cok, M. Ernst, J. Kiniry, G. Leavens, K. Leino, and E. Poll. An overview of JML tools and applications. In T. Arts and W. Fokkink, editors, *Formal Methods for Industrial Critical Systems (FMICS'03)*, number 80 in Electr. Notes in Theoretical Computer Science. Elsevier, Amsterdam, 2003. www.elsevier.nl/locate/entcs/volume80.html.
3. ESC/Java2. Open source extended static checking for java version 2 (esc/java 2) project. Security of Systems Group, Univ. of Nijmegen www.cs.kun.nl/ita/research/projects/sos/projects/escjava.html.
4. J.-C. Filliâtre. The Why certification tool. http://why.lri.fr/.
5. J.-C. Filliâtre. Verification of non-functional programs using interpretations in type theory. *Journal of Functional Programming*, 13(4), 2003.
6. B. Jacobs. Java's integral types in PVS. In *FMOODS 2003 Proceedings*, Lecture Notes in Computer Science. Springer, 2003.
7. B. Jacobs. Weakest precondition reasoning for Java programs with JML annotations. *Journal of Logic and Algebraic Programming*, 2004.
8. B. Jacobs and E. Poll. A logic for the Java Modeling Language JML. In *Proc. FASE*, LNCS 2029. Springer, 2001.

9. B. Jacobs and E. Poll. Java program verification at Nijmegen: Developments and perspective. Techn. Rep. NIII-R0318, Comput. Sci. Inst., Univ. of Nijmegen. To appear in the LNCS proceedings of the International Symposium of Software Security (ISSS), Tokyo, 2003.

10. G. Leavens, A. Baker, and C. Ruby. JML: A notation for detailed design. In *Behavioral Specifications of Business and Systems*. Kluwer, 1999.

11. G. Leavens, E. Poll, C. Clifton, Y. Cheon, and C. Ruby. JML reference manual (draft). 2003.

12. K. Leino, G. Nelson, and J. Saxe. ESC/Java's User's Manual. Technical Report 2000-002, Compaq Systems Research Center, 2000.

13. C. Marché, C. Paulin-Mohring, and X. Urbain. The KRAKATOA tool for certification of JAVA/JAVACARD programs annotated in JML. *Journal of Logic and Algebraic Programming*, 58(1–2), 2004.

14. J. Meyer and A. Poetzsch-Heffter. An architecture for interactive program provers. In *TACAS00*, LNCS 276, 2000.

15. M. Oostdijk and M. Warnier. On the combination of Java Card Remote Method Invocation and JML. Technical Report NIII-R0321, University of Nijmegen, 2003.

16. S. Owre, J. Rushby, N. Shankar, and F. von Henke. Formal verification for fault-tolerant architectures: Prolegomena to the design of PVS. *IEEE Trans. on Softw. Eng.*, 21(2):107–125, 1995.

17. L. C. Paulson. *Isabelle: A Generic Theorem Prover*. LNCS 828. Springer, 1994.

18. N. Rauch and A. Poetzsch-Heffter. Predicate transformation as a proof strategy. In *Proc. 4th ECOOP Workshop: FTfJP, Tech. Rep. NIII-R0204*. Computing Science Department, University of Nijmegen, 2002.

19. A. Schultz. Verification of Java Card-applets. Master's thesis, Universität Kaiserslautern, 2003.

20. The Coq Development Team. *The Coq Proof Assistant Reference Manual – Version V8.0*, Jan. 2004. http://coq.inria.fr.

Abstracting Call-Stacks for Interprocedural Verification of Imperative Programs

Bertrand Jeannet[1] and Wendelin Serwe[2]

[1] IRISA, 35042 Rennes, France
Bertrand.Jeannet@inria.fr
[2] INRIA Rhône-Alpes, 38334 St Ismier, France
Wendelin.Serwe@inria.fr

Abstract. We propose a new approach to interprocedural analysis and verification, consisting of deriving an interprocedural analysis method by abstract interpretation of the standard operational semantics of programs. The advantages of this approach are twofold. From a methodological point of view, it provides a direct connection between the concrete semantics of the program and the effective analysis, which facilitates implementation and correctness proofs. This method also integrates two main, distinct methods for interprocedural analysis, namely the call-string and the functional approaches introduced by Sharir and Pnueli. This enables strictly more precise analyses and additional flexibility in the tradeoff between efficiency and precision of the analysis.

1 Introduction

We consider the interprocedural verification of invariance properties of imperative programs. The applications we have in mind are automated debugging [4, 14] or automatic test selection [25], which may require precise and complex analyses. These are *flow-sensitive* (the analysis needs to take conditionals into account accurately), *attribute-dependent* (attributes (or properties) of variables are inter-related), and may require the use of *infinite* or even *infinite-height* lattices, in particular for the analysis of the properties numerical variables.

Our ambition is to design a method which is able to infer precise facts on the possible *call-stacks* of programs (and not only on the possible environments lying on top of the call-stack), and which can reuse existing abstract interpretation techniques and implementations available for intraprocedural analysis. For debugging applications, we would like to be able to answer questions like:

Being in procedure P with environment ϵ_P, and assuming being called successively by R and Q with $x = y$, is it possible to enter procedure S?

Formally, this amounts to ask whether an execution path of the form

$$\langle c_R, ? \rangle \cdot \langle c_Q, x = y \rangle \cdot \langle c_P, \epsilon_P \rangle \rightarrow^* \Gamma \cdot \langle c_S, ? \rangle$$

exists. Such an *automatic state reaching* feature has already been implemented in a debugger for reactive programs [14].

C. Rattray et al. (Eds.): AMAST 2004, LNCS 3116, pp. 258–273, 2004.

We first present our method, before discussing existing approaches to interprocedural analysis and some of their limitations for the applications we have in mind.

Our method. We start from the standard (small-step) operational semantics of imperative programs and derive in two distinct abstraction steps an implementable analysis. Since we aim the verification of invariance properties, the property lattice L is the powerset of states $\wp(S)$. The difficulty in interprocedural analysis is that a state is an unbounded stack of *activation records*[1], i.e. pairs of control points and local environments, so that the state-space has the following form: $S = Act^+$ with $E^+ = \cup_{i \geq 1} E^i$ denoting the Kleene operator on a set E. This means that there are two sources of infinity in the lattice L: the unbounded length of stacks, and the cardinality of the domain Act, considered as infinite as soon as there are numerical variables or recursive data structures.

The first abstraction abstracts the collecting semantics – which manipulates (co-)reachable sets of stacks of activation records – to a semantics manipulating sets of extended activation records. This *stack abstraction*, which deals with the first source of infinity, is independent from a second, more classical *data abstraction*, which abstracts sets of extended activation records in one of the many computable abstract lattices that are available for intraprocedural analysis, in order to deal with the second source of infinity, *cf.* Fig. 1.

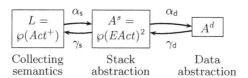

Fig. 1. Analysis scheme

The stack abstraction defines the interprocedural analysis method itself but does not lead directly to an effective analysis. Rather, it just simplifies the concrete lattice of sets of stacks into sets of simpler objects, for which computable abstract lattices already exists. Those need just to be equipped with a correct abstraction of procedure calls and returns operations in the stack abstraction lattice A^s, in order to obtain an implementable interprocedural analysis.

Existing approaches and some of their limitations for our applications. Interprocedural analysis methods have been widely studied and many of them can be classified as one of two classical approaches. We review in section 6 methods that do not fit in this classification. The *functional approach* [6, 28, 19] uses a denotational semantics of the analysed program and proceeds in two steps. The first step computes the predicate transformers associated with the procedures of the program, and the second step uses them to propagate an input predicate along the execution paths, to obtain the predicates holding at each control point. Some drawbacks of this approach are the following:

---

[1] We assume in this paper that there are no global variables.

1. Predicates holding at control points are replaced by predicate transformers, which are more complex objects. This new viewpoint forbids the direct reuse of intraprocedural analyses (which use predicates).
2. The call-stack of the original program disappears, so no property can be specified on it.

The *operational approach*, generalising the "call-string approach" of [28], consists of adopting an operational semantics for programs. Fixpoint computation proceeds as in intraprocedural analysis, i.e. by propagating a predicate along execution paths. However, (an abstraction of) the call-stack has to be taken into account. The general abstraction scheme for call-stacks consists of using "tokens" to abstract stack-tails, and to label current activation records by them. These tokens represent the calling context of the current activation record [28, 18]. We bring a solution to the main limitations of this approach, namely:

- Tokens are used as labels. They are essentially enumerated and are not given any structure (as for instance a lattice structure).
- A notable consequence is that the set of tokens should be *finite* in practice, which means that the abstraction of stack-tails is very rough.

From a more synthetic point of view, both approaches lead to *context-sensitive* analyses. The functional approach takes into account the data part of the calling context of a procedure, whereas the operational approach focuses mainly on the control part. From this point of view, our stack abstraction enables an effective analysis which is both *data* and *control* context-sensitive.

Contributions. Using the principles of abstract interpretation, we unify two main interprocedural analysis methods and give correctness and optimality proofs with a minimal number of concepts. We provide a method which is both data and control context-sensitive, thus strictly more precise. Our approach enables backward analysis and its intersection with forward analysis. Finally, it facilitates implementation issues by clearly separating stack and data abstractions.

Outline. Section 2 presents the considered program model and its semantics. In Section 3 we revisit the functional approach with a first stack abstraction. We show the similarity of the two methods and discuss some advantages of our approach. In Section 4 we refine the previous abstraction by using call-strings, in order to subsume the call-string and the functional approaches. This allows us to accurately specify the above-mentioned debugging problem. Section 5 discusses the data abstraction step, in order to derive an effective analysis from a stack abstraction. Related work is described in Section 6 and Section 7 concludes. Due to the lack of space, we have omitted the proofs. They can be found in [16].

2   Program Model and Standard Semantics

We consider a simple imperative programming language with non-nested procedures and value parameter passing (as in JAVA or ML). We suppose that each

Table 1. Syntactic domains

Var : Variables: Var $= \bigcup_i LVar_i$
$LVar_i$, $\mathbf{loc}_i$: Local variables of procedure P_i
In_i, $\mathbf{fpi}_i$: Formal input parameters of P_i
Out_i, $\mathbf{fpo}_i$: Formal output parameters of P_i
$G_i = \langle Ctrl_i, I_i \rangle$: Flow graph of P_i
$\mathbf{s}_i, \mathbf{e}_i \in Ctrl_i$: Entry and exit points of P_i
$G = \langle Ctrl, I \rangle$: Flow graph of the program

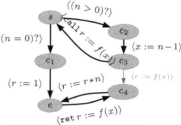

CFG for the Factorial Program

Table 2. Semantic domains

v	$\in Value$	: values
ϵ_i	$\in LEnv_i = LVar_i \rightarrow Value$	: local environments for P_i
ϵ	$\in LEnv = \bigcup_i LEnv_i$	: local environments
$\langle c, \epsilon \rangle$	$\in Act = Ctrl \times LEnv$	: activation record
Γ	$\in State = Act^+$	: stacks/program states

procedure has its own fixed set of variables, and do not consider global variables, which can be passed as additional procedure parameters. Similarly, programs are not restricted to programs manipulating scalar values: pointers and a memory heap can be modelled by adding to all procedures a special parameter modelling the memory heap. We require that formal input parameters *are not modified*. This is crucial to compute a relation between environments at the entry and any other point of a procedure (as in [28, 19]), and this is not restrictive either, as they can always be copied into additional local variables. The main restrictions are thus the absence of exceptions or non-local jumps, variable aliasing on the stack (as it happens with reference parameter passing), pointers to procedures and procedural parameters.

Program Syntax. The syntactic domains are summarised in Table 1.

A *program* is defined by a set $(P_i)_{0 \leq i \leq p}$ of procedures. Since we specify the initial states of an analysis separately, there is no particular "main" procedure.

Each *procedure* $P_i = \langle LVar_i, In_i, Out_i, G_i \rangle$ is defined by its intraprocedural control-flow graph G_i and its sets of local variables $LVar_i$, formal input parameters $In_i \subseteq LVar_i$ and formal output parameters $Out_i \subseteq LVar_i$. We note $\mathbf{x} = \langle \mathbf{x}^{(1)}, \ldots, \mathbf{x}^{(n)} \rangle$ vectors of variables. As vectors, the above-mentioned sets are written $\mathbf{loc}_i$, $\mathbf{fpi}_i$ and $\mathbf{fpo}_i$.

An *intraprocedural control-flow graph* is a graph $G_i = \langle Ctrl_i, I_i \rangle$ where $Ctrl_i$ is the set of *control points* of P_i, containing unique entry and exit control points $\mathbf{s}_i$ and $\mathbf{e}_i$. $I_i : Ctrl_i \times Ctrl_i \rightarrow \mathsf{Inst}$ labels edges between control points with two kinds of instructions: intraprocedural instructions $\langle R \rangle$ and procedure calls $\langle \mathbf{y} := P_j(\mathbf{x}) \rangle$, where $\mathbf{x}$ and $\mathbf{y}$ are the vectors of actual input and output parameters. Intraprocedural instructions are specified as a relation $R \subseteq LEnv^2$ describing the transformation of the top environment. We require the G_i to be deterministic for procedure calls, i.e. if $I_i(c, c')$ is a call then there exists no c'' such that $I_i(c, c'')$ or $I_i(c'', c')$ is a call.

$\dfrac{I(c,c') = \langle R \rangle \quad R(\epsilon, \epsilon')}{\Gamma \cdot \langle c, \epsilon \rangle \rightarrow \Gamma \cdot \langle c', \epsilon' \rangle}$ (Intra)	$\dfrac{I(c, \mathsf{s}_j) = \langle \mathtt{call}\ \mathbf{y} := P_j(\mathbf{x}) \rangle \quad \forall k : \epsilon_j(\mathbf{fpi}_j^{(k)}) = \epsilon(\mathbf{x}^{(k)})}{\Gamma \cdot \langle c, \epsilon \rangle \rightarrow \Gamma \cdot \langle c, \epsilon \rangle \cdot \langle \mathsf{s}_j, \epsilon_j \rangle}$ (Call)
	$\dfrac{I(\mathsf{e}_j, c) = \langle \mathtt{ret}\ \mathbf{y} := P_j(\mathbf{x}) \rangle \quad \epsilon' = \epsilon[\mathbf{y}^{(k)} \mapsto \epsilon_j(\mathbf{fpo}_j^{(k)})]}{\Gamma \cdot \langle \mathsf{call}(c), \epsilon \rangle \cdot \langle \mathsf{e}_j, \epsilon_j \rangle \rightarrow \Gamma \cdot \langle c, \epsilon' \rangle}$ (Return)

Fig. 2. SOS rules defining $\rightarrow$

The *interprocedural control-flow graph* (CFG) $G = \langle Ctrl, I \rangle$ is constructed from the set of intraprocedural ones. $Ctrl$ is defined as $Ctrl = \bigcup_{0 \leq i \leq p} Ctrl_i$ and I is defined as the "union" $\bigcup_i I_i$, where all procedure-call-edges are removed and replaced by two edges usually called *call-to-start* and *exit-to-return* edges:

$$\frac{I_i(c,c') \neq \langle \mathbf{y} := P_j(\mathbf{x}) \rangle}{I(c,c') = I_i(c,c')} \qquad \frac{I_i(c,c') = \langle \mathbf{y} := P_j(\mathbf{x}) \rangle}{\begin{array}{l} I(c,\mathsf{s}_j) = \langle \mathtt{call}\ \mathbf{y} := P_j(\mathbf{x}) \rangle \\ I(\mathsf{e}_j, c') = \langle \mathtt{ret}\ \mathbf{y} := P_j(\mathbf{x}) \rangle \end{array}}$$

Thus there are three kinds of instructions labelling edges of interprocedural CFGs: intraprocedural instructions $\langle R \rangle$, procedure calls $\langle \mathtt{call}\ \mathbf{y} := P_j(\mathbf{x}) \rangle$ and procedure returns $\langle \mathtt{ret}\ \mathbf{y} := P_j(\mathbf{x}) \rangle$, see the factorial program beside Table 1.

In the sequel, we use the following notations. For $c \in Ctrl$, c is a *call-site* to P_j if $\exists c' : I(c,c') = \langle \mathtt{call}\ \mathbf{y} := P_j(\mathbf{x}) \rangle$. $\mathsf{proc}(c)$ denotes j such that $c \in Ctrl_j$. For any edge $c \xrightarrow{\langle \mathbf{y} := P_j(\mathbf{x}) \rangle} c'$, we define $\mathsf{ret}(c) = c'$ and $\mathsf{call}(c') = c$.

Operational Semantics. The semantic domains are summarised in Table 2.

The operational semantics is given by a *transition system* $(State, \rightarrow)$. *States* are *stacks* $\Gamma = \langle c_0, \epsilon_0 \rangle \cdot \ldots \cdot \langle c_n, \epsilon_n \rangle$ of *activation records* (i.e. pairs of a control point c_i and an environment ϵ_i). $\langle c_n, \epsilon_n \rangle$ is the *current activation record* or *top* of Γ; the *tail* of Γ is Γ without its top, i.e. $\langle c_0, \epsilon_0 \rangle \cdot \ldots \cdot \langle c_{n-1}, \epsilon_{n-1} \rangle$. Environments map variables to values; their update is written $\epsilon[x \mapsto v]$. The *transition relation* $\rightarrow \subseteq State \times State$ is defined (in SOS-style) by the rules in Fig. 2. As long as a variable is not initialised, it holds nondeterministically any value in its domain, *cf.* rule (Call). As usual, $\rightarrow^*$ denotes the reflexive-transitive closure of $\rightarrow$.

Standard Collecting Semantics. The forward collecting semantics describes the set of reachable states of a program. It is the natural choice for expressing and verifying invariance properties and is derived from the operational semantics by collecting the states belonging to executions of the program. We define the function $reach : \wp(State) \rightarrow \wp(State)$ computing the states reachable from a set of *initial states* X_0 as:

$$reach(X_0) \stackrel{\text{def}}{=} \{ q \mid \exists q_0 \in X_0 ,\ q_0 \rightarrow^* q \}$$

It is also the least fix-point solution of $X = X_0 \cup post(X)$ where the *forward transfer function* $post$ is defined by $post(X) = \{ q' \mid \exists q \in X : q \rightarrow q' \}$. Table 3 gives the decomposition of $post$ according to the transitions of the CFG,

Table 3. Forward transfer function *post*

$$post(c \xrightarrow{\langle R \rangle} c')(X) = \{\Gamma \cdot \langle c', \epsilon' \rangle \mid \Gamma \cdot \langle c, \epsilon \rangle \in X \wedge R(\epsilon, \epsilon')\}$$

$$post(c \xrightarrow{\langle \text{call } y := P_j(\mathbf{x}) \rangle} s_j)(X) = \{\Gamma \cdot \langle c, \epsilon \rangle \cdot \langle s_j, \epsilon_j \rangle \mid \Gamma \cdot \langle c, \epsilon \rangle \in X \wedge \epsilon_j(\mathbf{fpi}_j^{(k)}) = \epsilon(\mathbf{x}^{(k)})\}$$

$$post(e_j \xrightarrow{\langle \text{ret } y := P_j(\mathbf{x}) \rangle} c)(X) = \left\{ \Gamma \cdot \langle c, \epsilon' \rangle \ \middle| \ \begin{array}{l} \Gamma \cdot \langle \text{call}(c), \epsilon \rangle \cdot \langle e_j, \epsilon_j \rangle \in X \\ \epsilon' = \epsilon[\mathbf{y}^{(k)} \mapsto \epsilon_j(\mathbf{fpo}_j^{(k)})] \end{array} \right\}$$

i.e. $post(X) = \bigcup_{(c,c') \in Ctrl \times Ctrl} post(c \xrightarrow{I(c,c')} c')(X)$. For $X_0, X \subseteq S$, we define $F[X_0](X) \stackrel{\text{def}}{=} X_0 \cup post(X)$. Since $F[X_0]$ is monotone and continuous, Kleene's fix-point theorem allows us to compute the forward collecting semantics by iterated application of $F[X_0]$ starting from $\emptyset$:

$$reach(X_0) \ = \ \text{lfp}(F[X_0]) \ = \ \bigcup_{n \geq 0} (F[X_0])^n(\emptyset)$$

The collecting semantics can also be considered backward, yielding the set of states X from which a given set of *final states* X_0 is reachable. In this case we call X the set of *coreachable states* of X_0. We get the following definitions:

$$pre(X) = \{q \mid \exists q' \in X : q \to q'\}$$
$$coreach(X_0) = \text{lfp}(G[X_0]) \ \text{with} \ G[X_0](X) = X_0 \cup pre(X)$$

Properties of the Stacks in the Standard Semantics. The assumption that formal input parameters are read-only variables induces strong properties on stacks which are the basis of our stack abstractions. A necessary condition for $q = \Gamma \cdot \langle c, \epsilon \rangle$ to lead to $q' = \Gamma \cdot \langle c, \epsilon \rangle \cdot \langle c', \epsilon' \rangle$ (where c is a call site to $P_{proc(c')}$) is that the values of actual input parameters in ϵ have to match those of the formal input parameters in ϵ'. This is formalised by the following definition.

Definition 1. $\langle c, \epsilon \rangle$ *is a* valid calling activation record *(or* valid*) for* $\langle c', \epsilon' \rangle$ *if*
(i) *c is a call site for procedure* P_j: $\exists j : c' = s_j \ \wedge \ I(c, s_j) = \langle \text{call } y := P_j(\mathbf{x}) \rangle$;
(ii) *actual and formal parameters are equal:* $\forall k : \epsilon(\mathbf{x}^{(k)}) = \epsilon'(\mathbf{fpi}_j^{(k)})$.

Extending Definition 1, we call a stack $\langle c_0, \epsilon_0 \rangle \ldots \langle c_n, \epsilon_n \rangle$ *consistent* if $\langle c_i, \epsilon_i \rangle$ is valid for $\langle c_{i+1}, \epsilon_{i+1} \rangle$ $(\forall 0 \leq i < n)$. From now on, we focus on consistent states and restrict *State* to its consistent subset.

3   Revisiting the Functional Approach

We present a first stack abstraction, which allows analysis results at least as accurate as the classical functional approach for forward analysis, but in addition:

- It enables a more direct reuse of standard data abstractions (we manipulate environments and not functions from environments to environments).
- Stacks can be rebuild by concretisation of abstract values.
- Defining backward analysis is straightforward.

3.1   Stack Abstraction

The main idea is to forget all about sequences of activation records in the stack, keeping information only on the activation records themselves. As already observed in a different context [19], we can nevertheless get accurate results.

An abstract state $Y = \langle Y_{hd}, Y_{tl} \rangle$ is composed of two sets of activation records. Y_{hd} represents top activation records, whereas Y_{tl} is the collapse of all activation records in stack-tails. This leads to the following abstract domain:

$$A^f \overset{\text{def}}{=} \wp(Act) \times \wp(Act) \tag{1}$$

which comes equipped with the standard lattice structure $A^f(\sqsubseteq, \sqcup, \sqcap, \top, \bot)$ of a (smashed) Cartesian product of lattices.

We define the Galois connection $\wp(State) \overset{\gamma^f}{\underset{\alpha^f}{\rightleftarrows}} A^f$, with the *abstraction function* $\alpha^f = \langle \alpha_{hd}^f, \alpha_{tl}^f \rangle$ and *concretisation function* γ^f defined by:

$$\alpha^f: \qquad X \qquad \longmapsto \bigsqcup_{q \in X} \alpha^f(\{q\})$$
$$\{\langle c_0, \epsilon_0 \rangle \dots \langle c_n, \epsilon_n \rangle\} \longmapsto \langle \, \{\langle c_n, \epsilon_n \rangle\} \,, \, \{\langle c_i, \epsilon_i \rangle \mid 0 \le i < n\} \, \rangle$$

$$\gamma^f: \quad Y = \langle Y_{hd}, Y_{tl} \rangle \quad \longmapsto \left\{ q = \langle c_0, \epsilon_0 \rangle \dots \langle c_n, \epsilon_n \rangle \, \middle| \, \begin{array}{l} \langle c_n, \epsilon_n \rangle \in Y_{hd} \\ \forall 0 \le i < n : \langle c_i, \epsilon_i \rangle \in Y_{tl} \\ q \text{ is a consistent stack} \end{array} \right\}$$

α_{hd}^f gathers the top activation records, whereas α_{tl}^f collects all the other activation records. To rebuild a stack from an abstract state, γ^f uses the notion of consistent stacks. Notice that $\langle c, \epsilon \rangle \in Y_{tl}$ implies that c is a call site.

The extensive function $\gamma^f \circ \alpha^f$ describes the information lost by A^f. If $\langle c_0, \epsilon_0 \rangle$ is valid for $\langle c_1, \epsilon_1' \rangle$ and $\epsilon_1 \ne \epsilon_1'$, we might have:

$$\langle c_0, \epsilon_0 \rangle \cdot \langle c_1, \epsilon_1' \rangle \in (\gamma^f \circ \alpha^f) \left(\left\{ \langle c_0, \epsilon_0 \rangle \cdot \langle c_1, \epsilon_1 \rangle, \langle c_1, \epsilon_1' \rangle \right\} \right)$$
$$\notin \quad \left\{ \langle c_0, \epsilon_0 \rangle \cdot \langle c_1, \epsilon_1 \rangle, \langle c_1, \epsilon_1' \rangle \right\} \tag{2}$$

However, $\alpha_{hd}^f(X)$ keeps *exact information* on *top* activation records, which is the only information computed by functional approaches.

The main reason why we separate top activation records from those below in the stack is for defining properly an abstract backward analysis, and also for being able to perform a forward analysis from an non empty stack.

3.2   Forward Analysis

Abstract Postcondition $post^f$. To compute reachable states in the abstract domain, we need an abstract postcondition operator $post^f$. Table 4 specifies $post^f$, using the decomposition already used for $post$. We prove in [16] that $post^f$ is a correct approximation of $post$, i.e. $post^f \sqsupseteq \alpha^f \circ post \circ \gamma^f$.

Only procedure returns need comment, since the stack contents before the call has to be recovered. [28, 18, 19] use for this purpose a function *combine* to

Table 4. Equations defining $post^f$

$$post_{hd}^f(c \xrightarrow{\langle R \rangle} c')(Y) = \{\langle c', \epsilon' \rangle | \langle c, \epsilon \rangle \in Y_{hd} \ \wedge \ R(\epsilon, \epsilon')\} \tag{3a}$$

$$post_{hd}^f(c \xrightarrow{\langle \mathtt{call} \ \mathbf{y} := P_j(\mathbf{x}) \rangle} \mathsf{s}_j)(Y) = \{\langle \mathsf{s}_j, \epsilon_j \rangle | \langle c, \epsilon \rangle \in Y_{hd} \ \wedge \ \epsilon_j(\mathbf{fpi}_j^{(k)}) = \epsilon(\mathbf{x}^{(k)})\} \tag{3b}$$

$$post_{hd}^f(e_j \xrightarrow{\langle \mathtt{ret} \ \mathbf{y} := P_j(\mathbf{x}) \rangle} c)(Y) = \left\{ \langle c, \epsilon' \rangle \ \middle| \ \begin{array}{l} \langle \mathtt{call}(c), \epsilon \rangle \in Y_{tl} \ \wedge \ \epsilon(\mathbf{x}^{(k)}) = \epsilon_j(\mathbf{fpi}_j^{(k)}) \\ \langle e_j, \epsilon_j \rangle \in Y_{hd} \ \wedge \ \epsilon' = \epsilon[\mathbf{y}^{(k)} \mapsto \epsilon_j(\mathbf{fpo}_j^{(k)})] \end{array} \right\} \tag{3c}$$

$$post_{tl}^f(c \xrightarrow{\langle \mathtt{call} \ \mathbf{y} := P_j(\mathbf{x}) \rangle} \mathsf{s}_j)(Y) = Y_{tl} \ \cup \ \{\langle c, \epsilon \rangle \in Y_{hd}\} \tag{3d}$$

$$post_{tl}^f(c \xrightarrow{i} c')(Y) = Y_{tl} \quad \text{otherwise } (i \text{ is an instruction}) \tag{3e}$$

combine the environment of the caller at the call site with the environment of the callee at its exit point. In Section 2, we noticed that $\Gamma \cdot \langle \mathtt{call}(c), \epsilon \rangle \cdot \langle e_j, \epsilon_j \rangle$ can be a successor of $\Gamma \cdot \langle \mathtt{call}(c), \epsilon \rangle$ only if $\langle \mathtt{call}(c), \epsilon \rangle$ is valid for $\langle e_j, \epsilon_j \rangle$. So, for $\langle e_j, \epsilon_j \rangle \in Y_{hd}$, (3c) selects the activation records $\langle \mathtt{call}(c), \epsilon \rangle \in Y_{tl}$ valid for it. Then, the reachable activation record(s) at c are obtained by assigning return parameters to $\mathbf{y}$. This is our combine operation, defined by abstraction of the concrete procedure return operation.

Because the input parameters are frozen, at the exit point e_j of a procedure P_j, the top of the stack (in Y_{hd}) contains a relation $\phi_j(\mathbf{fpi}_j, \mathbf{fpo}_j)$ between *reachable* call and return parameters, defined by $\phi_j = \{(\mathbf{x}, \mathbf{y}) \mid \exists \langle e_j, \epsilon \rangle \in Y_{hd} : \mathbf{x} = \epsilon(\mathbf{fpi}_j) \wedge \mathbf{y} = \epsilon(\mathbf{fpo}_j)\}$. Since ϕ_j is the predicate transformer of P_j specialised on the reachable inputs, our handling of procedure returns can be seen as applying the predicate transformer of the callee to the valid calling activation records that are reachable in Y_{tl} at the call site.

Correctness and Optimality. Transposing the notion of reachable states into the abstract lattice A^f, we define $F^f[Y_0](Y) \stackrel{\text{def}}{=} Y_0 \sqcup post^f(Y)$ and $reach^f(Y_0) \stackrel{\text{def}}{=} \mathrm{lfp}(F^f[Y_0]) \ (\forall Y_0 \in A^f)$. Since $post^f$ correctly approximates $post$, we deduce from abstract interpretation theory that we correctly approximate reachable states:

Theorem 1 (Correctness). *For any $Y_0 \in A^f$, $reach^f(Y_0) \sqsupseteq \alpha^f \circ reach \circ \gamma^f(Y_0)$*

The fact that we incrementally build the predicate transformer of a procedure at its exit point together with [28] suggests that we could improve and get the best we can hope for with a Galois connection, that is

$$reach^f \circ \alpha^f = \alpha^f \circ reach$$

This means it doesn't matter if we compute the fixpoint in the concrete lattice and then abstract the result, or directly compute in the abstract lattice.

Since Theorem 1 implies $reach^f \circ \alpha^f \sqsupseteq \alpha^f \circ reach$, we just have to prove the inverse inclusion. We show that $Y = \alpha^f \circ reach(X_0)$ is a post-fixpoint of

Table 5. Equations Defining pre^f

$$pre^f_{hd}(c \xrightarrow{\langle R \rangle} c')(Y) = \{\langle c, \epsilon \rangle \mid \langle c', \epsilon' \rangle \in Y_{hd} \wedge R(\epsilon, \epsilon')\} \tag{4a}$$

$$pre^f_{hd}(c \xrightarrow{\langle \mathtt{call\ y}:=P_j(\mathbf{x})\rangle} \mathsf{s}_j)(Y) = \left\{\langle c, \epsilon \rangle \left| \begin{array}{l} \langle \mathsf{s}_j, \epsilon_j \rangle \in Y_{hd} \wedge \langle c, \epsilon \rangle \in Y_{tl} \\ \epsilon(\mathbf{x}^{(k)}) = \epsilon_j(\mathbf{fpi}^{(k)}) \end{array} \right.\right\} \tag{4b}$$

$$pre^f_{hd}(\mathsf{e}_j \xrightarrow{\langle \mathtt{ret\ y}:=P_j(\mathbf{x})\rangle} c)(Y) = \left\{\langle \mathsf{e}_j, \epsilon_j \rangle \left| \begin{array}{l} \langle c, \epsilon \rangle \in Y_{hd} \wedge \epsilon_j(\mathbf{fpo}_j^{(k)}) = \epsilon(\mathbf{y}^{(k)}) \\ \forall \mathbf{x}^{(k)} \notin \mathbf{y} : \epsilon_j(\mathbf{fpi}_j^{(k)}) = \epsilon(\mathbf{x}^{(k)}) \end{array} \right.\right\} \tag{4c}$$

$$pre^f_{tl}(\mathsf{e}_j \xrightarrow{\langle \mathtt{ret\ y}:=P_j(\mathbf{x})\rangle} c)(Y) = Y_{tl} \cup \left\{\langle \mathsf{call}(c), \epsilon \rangle \left| \begin{array}{l} \langle c, \epsilon' \rangle \in Y_{hd} \\ \forall z \notin \mathbf{y} : \epsilon(z) = \epsilon'(z) \end{array} \right.\right\} \tag{4d}$$

$$pre^f_{tl}(c \xrightarrow{i} c')(Y) = Y_{tl} \quad \text{otherwise } (i \text{ is an instruction}) \tag{4e}$$

$post^f$, i.e. that $post^f(Y) \sqsubseteq Y$, under additional assumptions on X_0. First, we require initial states to be one-element stacks, guaranteeing exactness of their abstraction. Second, we require that initial states/activation records belong to procedures that are never called; otherwise the abstraction might allow procedure returns even if there is no other activation record on the (concrete) stack.

Under these conditions, we get:

Theorem 2 (Optimality). *Let $X_0 \in \wp(State)$ such that $q \in X_0$ implies that $q = \langle c, \epsilon \rangle$ and that $\mathsf{proc}(c)$ is never called by any procedure. Then $reach^f \circ \alpha^f(X_0) = \alpha^f \circ reach(X_0)$. This implies that*

$$reach^f_{hd} \circ \alpha^f(X_0) = \{\langle c, \epsilon \rangle \mid \exists \Gamma : \Gamma \cdot \langle c, \epsilon \rangle \in reach(X_0)\}$$
$$reach^f_{tl} \circ \alpha^f(X_0) = \{\langle c, \epsilon \rangle \mid \exists \Gamma, \Gamma' : \Gamma \cdot \langle c, \epsilon \rangle \cdot \Gamma' \in reach(X_0)\}$$

Whereas our analysis loses information on stack contents, we get *exact results* if we are interested only in the values held by variables at some control points (which is the case for many applications).

As we relate our abstract semantics directly to the standard semantics, we get by Theorem 2 the *Meet Over all Valid Paths* property defined in [28, 19], without having to introduce the notion of interprocedural valid paths. Theorems 1 and 2 are thus very similar to the optimality results in those papers.

3.3 Backward Analysis

Backward analysis is implemented in a similar fashion as forward analysis. Table 5 gives the definition of a correct approximation of pre in A^f. By defining, for $Y_0, Y \in A$, $G^f[Y_0](Y) = Y_0 \sqcup pre^f(Y)$ and $coreach^f(Y_0) = \mathrm{lfp}(G^f[Y_0])$ we get the correctness of the analysis: $coreach^f \sqsupseteq \alpha^f \circ coreach \circ \gamma^f$. Here we do not have any optimality result, as formal output parameters are not frozen during a backward execution. But we could use a similar mechanism by duality.

3.4   Comparison with the Functional Approach

Re-expressing the functional approach (*cf.* introduction) by abstracting the stacks of the concrete semantics presents some advantages. From a conceptual point of view, we manipulate environments instead of functions from environments to environments. The relational character of the stack abstraction is inherited from the relational character of the concrete semantics, induced by the freezing of the formal input parameters.

From an algorithmical point of view, we merge the two fixpoint computations of the traditional functional approach to a unique fixpoint computation. Thus,

- Our method is less compositional: we have to perform a new analysis for different initial states, whereas in the functional approach only the second fixpoint has to be computed again, as the predicate transformers associated with procedures are defined by the first fixpoint and do not depend on the initial states.
- It may however be more accurate: indeed, we compute at the exit point of the procedures their predicate transformers *specialised on their reachable inputs*. As the functions $post^f$ and $F^f[Y_0]$ are *not* distributive, applying them on smaller abstract values may prevent some loss of information. In addition, the data abstraction that follows the stack abstraction is often not distributive either (e.g., [20, 8]), which may increase the gain in precision.

Moreover, backward analysis can be easily defined with a stack abstraction. Backward analysis is especially useful when combined with forward analysis, when one is interested in the set of states reachable from some initial states and leading possibly to some final states, as for the debugging problem of the introduction or test selection. In this case, forward analysis from the initial states returns a set *reach* of reachable states. Then we perform a backward analysis from the final states intersected with *reach* ; practically, $pre^f(Y)$ is replaced by $pre^f(Y) \sqcap reach$ in the definition of $G^f[Y_0]$. The intersection *during the fixpoint iteration* allows a more accurate analysis, as the transfer functions are not distributive. This scheme can be iterated, each fixpoint computation being performed on a restricted state space. Such a scheme has been successfully applied to the verification of reactive systems [15], and we are interested in extending it to recursive programs.

4   Subsuming Functional and Call-String Approaches

Refined Stack Abstraction. In this section we combine the previous abstraction A^f with the call-string approach. We replace activation records $\langle c_n, \epsilon_n \rangle$ used in A^f by objects of the form $\langle c_0 \ldots c_n, \epsilon_n \rangle$, called *extended activation records*. This corresponds to replacing single control points c_n labelling environments by sequences $c_0 \ldots c_n$, called *call strings* in [28]. The abstract semantics proposed in this section can be seen as a synthesis of the two distinct methods described in [28].

Let us note $EAct \stackrel{\text{def}}{=} Ctrl^+ \times LEnv$ the set of extended activation records. We extend the notion of valid calling activation record (see Definition 1) to extended activation records. $\langle \omega, \epsilon \rangle$ is valid for $\langle \omega', \epsilon' \rangle$ if $\omega = \omega_0 \cdot c$, $\omega' = \omega_0 \cdot c \cdot c'$, and $\langle c, \epsilon \rangle$ is valid for $\langle c', \epsilon' \rangle$. This additional condition increases the precision of the analysis.

We extend the domain A^f defined in Section 3.1, (1) by replacing activation records by extended activation records, yielding the abstract domain:

$$A^s \stackrel{\text{def}}{=} \wp(EAct) \times \wp(EAct) \tag{5}$$

A^s is connected to $\wp(State)$ by the Galois connection $\wp(State) \xleftrightarrow[\alpha^s]{\gamma^s} A^s$, where abstraction α^s and concretisation γ^s are defined by:

$$\alpha^s : \quad X \quad \longmapsto \quad \bigsqcup_{q \in X} \alpha^s(\{q\})$$
$$\{\langle c_0, \epsilon_0 \rangle \ldots \langle c_n, \epsilon_n \rangle\} \longmapsto \Big\langle \ \{\langle c_0 \ldots c_n, \epsilon_n \rangle\} \ , \ \{\langle c_0 \ldots c_i, \epsilon_i \rangle \mid 0 \le i < n\} \ \Big\rangle$$

$$\gamma^s : \quad Y = \langle Y_{hd}, Y_{tl} \rangle \quad \longmapsto \left\{ \underbrace{\langle c_0, \epsilon_0 \rangle \cdots \langle c_n, \epsilon_n \rangle}_{=q} \ \middle| \ \begin{array}{l} \langle c_0 \ldots c_n, \epsilon_n \rangle \in Y_{hd} \\ \forall 0 \le i < n : \ \langle c_0 \ldots c_i, \epsilon_i \rangle \in Y_{tl} \\ q \text{ is a consistent stack} \end{array} \right\}$$

A^s loses less information than A^f: thanks to call-strings, we cannot have any more the problem of (2). However we have a subtler phenomenon: if $\langle c_0, \epsilon_0 \rangle$ is a valid calling activation record for $\langle c_1, \epsilon_1' \rangle$ and $\epsilon_1 \ne \epsilon_1'$ we might have

$$\langle c_0, \epsilon_0 \rangle \cdot \langle c_1, \epsilon_1' \rangle \in \gamma^s \circ \alpha^s \Big(\{\langle c_0, \epsilon_0 \rangle \cdot \langle c_1, \epsilon_1 \rangle, \ \langle c_0, \epsilon_0' \rangle \cdot \langle c_1, \epsilon_1' \rangle\} \Big) \tag{6}$$

Hence, while the abstraction keeps the stack length exact, it can still induce some "cross-over" of activation records belonging to different concrete stacks.

Forward Analysis. The transfer function $post^s$, fully defined in [16], is very similar to $post^f$ (*cf.* Tab. 4) but call-strings allow a more accurate "matching" of possible calling contexts with top contexts. For instance,

$$post^s_{hd}(e_j \xrightarrow{\langle \text{ret } y := P_j(x) \rangle} c)(Y) = \left\{ \langle \omega \cdot c, \epsilon' \rangle \ \middle| \ \begin{array}{l} \langle \omega \cdot \text{call}(c) \cdot e_j, \epsilon_j \rangle \in Y_{hd} \\ \langle \omega \cdot \text{call}(c), \epsilon \rangle \in Y_{tl} \\ \epsilon(\mathbf{x}^{(k)}) = \epsilon_j(\mathbf{fpi}_j^{(k)}) \\ \epsilon' = \epsilon[\mathbf{y}^{(k)} \mapsto \epsilon_j(\mathbf{fpo}_j^{(k)})] \end{array} \right\} \tag{7}$$

which is to be compared to (3c).

As in Section 3 and using similar definitions, the forward analysis is not only correct but also optimal. Moreover, the second condition of Theorem 2 is no longer needed, because we know that an extended activation record can return to a caller only when its call-string component is of length at least 2.

Theorem 3 (Optimality). *Let $X_0 \in \wp(State)$ be a set one-element stacks. Then we have $reach^s \circ \alpha^s(X_0) = \alpha^s \circ reach(X_0)$. This implies that*

$$reach^s_{hd} \circ \alpha^f(X_0) = \{\langle c_0 \ldots c_n, \epsilon_n \rangle \mid \langle c_0, \epsilon_0 \rangle \ldots \langle c_n, \epsilon_n \rangle \in reach(X_0)\}$$
$$reach^s_{tl} \circ \alpha^f(X_0) = \{\langle c_0 \ldots c_n, \epsilon_n \rangle \mid \exists \Gamma \in Act^+ : \langle c_0, \epsilon_0 \rangle \ldots \langle c_n, \epsilon_n \rangle \cdot \Gamma \in reach(X_0)\}$$

Discussion. The lattice A^s encompasses the two methods proposed in [28]: abstracted to A^f, it corresponds to the functional approach of that paper, whereas A^s, further abstracted with a data abstraction *which does not relate* the values of input parameters and output parameters, would give the call-string approach. As told in the introduction, A^s leads to an interprocedural analysis which is both control and data context-sensitive. Here we used the call-strings of [28], but this could be generalised to the tokens of [18]. Backward analysis can of course be defined in the lattice A^s similarly as in Section 3.

5   Data Abstractions

Our stack abstractions in some way define each an interprocedural analysis method, but not an effective analysis, due to infinite data values or unbounded size of call strings. We show here which existing abstract lattices are suitable for these methods. [16] shows how to extend them with procedure call and return operations. We will consider only A^s, since A^f is a further abstraction of A^s.

We can use the isomorphism $A^s \simeq Ctrl \rightarrow \left(\wp(Ctrl^* \times LEnv) \right)^2$ to associate a set of values to each control point (as usual when analysing imperative programs). $Ctrl$ being finite, we just need an abstract lattice for $\wp(Ctrl^* \times LEnv)$ in order to get an abstract lattice for A^s. A natural way to achieve this is to build an abstract lattice for $\wp(Ctrl^* \times LEnv)$ by composing abstract domains available for $\wp(Ctrl^*)$ and $\wp(LEnv)$. So we need to abstract $\wp(Ctrl^*)$ and $\wp(LEnv)$ by some lattices A_{Ctrl} and A_{LEnv}, as well as a method for combining them.

Abstractions for call-strings. Several abstract domains are possible for A_{Ctrl}:

1. $\wp(Ctrl^p)$, for some fixed $p \geq 0$: the top p elements of stacks are exactly represented, the others are completely forgotten,
2. Reg($Ctrl$), the set of regular languages over the finite alphabet $Ctrl$, together with a suitable widening operator, as the one suggested by [13] or
3. some subsets of regular languages: simple regular languages, star-free regular languages, etc., with widening operators if necessary.

Notice that apart the first one, these abstractions for call strings are more general than the one suggested in [28], where only finite partitioning of $\wp(Ctrl^*)$ is considered. Here we do not restrict abstract lattices for $\wp(Ctrl^*)$ neither to finite lattices nor to finite-height lattices, thanks to the use of widening.

Abstractions for environments. Any abstract lattice A_{LEnv} available for intraprocedural analysis can be chosen here [8, 20, 10, 27, 17]. However *it should be relational* in order to be effectively able to represent relations between input and output parameters at the exit point of procedures. In particular equality constraints should be possible in A_{LEnv} for implementing an accurate procedure return operation. The typical counter-example is the lattice of intervals for numerical variables, where no relationships between variables can be represented, only a conjunction of invariants for each variable.

Combining the two abstract lattices. To avoid the inherent difficulties of the design of an abstraction combining different data-types (in our case $Ctrl^*$ and $LEnv$), we suggest to combine the lattices abstracting each datatype. We could use the tensor product $A_{Ctrl} \otimes A_{LEnv}$ [21]. However, this product is not finitely representable as soon as either A_{Ctrl} or A_{LEnv} is infinite. Instead, we suggest the simple solution of [15], which is to take the Cartesian product $A = A_{Ctrl} \times A_{LEnv}$. In this lattice, relationships between call strings and data values cannot be directly represented: an abstract value is just the conjunction of an invariant on call strings and an invariant on data values. However, partitioning on A can be used to establish relationships between the two components. This technique has been used for obtaining relational invariants on Boolean and numerical variables in [15], and can be considered as a particular instance of the disjunctive or down-set completion method discussed in [7], where the size of disjuncts is bounded by the size of the partition. For instance, if we partition $A_{Ctrl} \times A_{LEnv}$ according to the k last control points in call-strings, we get a lattice isomorphic to $Ctrl^k \to A_{Ctrl} \times A_{LEnv}$.

Example. Fig. 3 gives a program computing MacCarthy's 91-function and the analysis results, with as initial states $\langle s, n \in \mathbb{Z} \rangle$. The exact denotational semantics of this function is $\lambda n.(if\ n > 101\ then\ n - 10\ else\ 91)$. We use the lattice $A = (Ctrl \cup Ctrl^2) \to (\mathsf{Reg}(Ctrl) \times \mathsf{Pol}(\mathbb{Z}))^2$, i.e. we partition A w.r.t. the two top control points in the call-strings, and we use convex polyhedra for numerical variables. As widening operator on $\mathsf{Reg}(Ctrl)$, we use the bisimulation on automata of order δ, $\sim_\delta$, as in [13], with $\delta = 1$. Starting analysis with a one-element stack, we have $Y_{tl} = Y_{hd}$ for all call-sites and $Y_{tl} = \emptyset$ elsewhere. Thus Fig. 3 shows only Y_{hd}. Observe that the result at point e is both control and data context-sensitive: the partitioning on call-strings induces a differentiation of the possible values of the input n. We do not find the exact result at point e, because of the convex hull on polyhedra performed on the results at points c_1 and c_5, which are *exact*. In particular, although the result at point c_5 depends on the approximate result at point e, it is exact thanks to the context-sensitiveness of the analysis. Clearly, partitioning A w.r.t. only the top control point would have lead to much less precise information, because of the more frequent use of convex hull.

6   Related Work

As explained in the introduction, one can distinguish two main approaches to interprocedural static analysis, namely the *functional* and the *operational*. The functional approach of [6] has been used for instance to analyse the access to arrays in interprocedural Fortran programs [9], using the abstract domain of convex polyhedra [8]. [22] can be seen as an algorithmical implementation of [19] using graph reachability techniques, which can be used with finite lattices and distributive data flow functions. This technique can be applied to all *bit-vector* analyses and the BEBOP tool [1] is based on it. An extension [26] allows to tackle some finite-height infinite lattices, like (linear) constant propagation.

```
proc MC(n: int): (r: int) is
  t1, t2: int;
begin s:{ ⟨s, ?⟩, ⟨ωc₃s, n ≤ 111⟩, ⟨ωc₄s, 91 ≤ n ≤ 101⟩}
  if n > 100 then c₀:{ ⟨c₀, n ≥ 101⟩, ⟨ωc₃c₀, 101 ≤ n ≤ 111⟩, ⟨ωc₄c₀, n = 101⟩}
    r := n - 10  c₁:{ ⟨c₁, r = n − 10 ≥ 91⟩, ⟨ωc₃c₁, 101 ≤ n ≤ 111 ∧ r = n − 10⟩, ⟨ωc₄c₁, r = n − 10 = 91⟩}
  else c₂:{ ⟨c₂ + ωc₃c₂, n ≤ 100⟩}, ⟨ωc₄c₂, 91 ≤ n ≤ 100⟩}
    t1 := n + 11; c₃:{ ⟨c₃ + ωc₃c₃, n ≤ 100 ∧ t1 = n + 11⟩, ⟨ωc₄c₃, 91 ≤ n ≤ 100 ∧ t1 = n + 11⟩}
    t2 := MC(t1); c₄:{ ⟨c₄ + ωc₃c₄, n ≤ 100 ∧ t1 = n + 11 ∧ 91 ≤ t2 ≤ 101 ∧ t2 ≥ n + 1⟩,
                       ⟨ωc₄c₄, 91 ≤ n ≤ 100 ∧ t1 = n + 11 ∧ 91 ≤ t2 ≤ 101 ∧ t2 ≥ n + 1⟩ }
    r := MC(t2); c₅:{ ⟨c₅ + ωc₃c₅, n ≤ 100 ∧ t1 = n + 11 ∧ 91 ≤ t2 ≤ 101 ∧ t2 ≥ n + 1 ∧ r = 91⟩,
                      ⟨ωc₄c₅, 91 ≤ n ≤ 100 ∧ t1 = n + 11 ∧ 91 ≤ t2 ≤ 101 ∧ t2 ≥ n + 1 ∧ r = 91⟩ }
  endif
end MC e:{ ⟨e, r ≥ 91⟩, ⟨ωc₃e, n ≤ 111 ∧ 91 ≤ r ≤ 101 ∧ r ≥ n − 10⟩, ⟨ωc₄e, 91 ≤ n ≤ 101 ∧ r = 91⟩}
```

Fig. 3. MacCarthy's 91-function (where $\omega \overset{\text{def}}{=} (c_3 + c_4)^*$)

In the operational approach, [3] considers more complex Pascal programs with reference parameter passing, which introduces aliasing on the stack (i.e. several variables may refer to the same location in the stack), and nested procedure definitions. Unsurprisingly, the devised solution is quite complex. It has been implemented using the interval domain for integers [5]. Stacks are collapsed more severely than in our model. The proposal of [11], implemented in MOPED [12] and applied to concurrent programs in [2], relies on the result that the set of reachable stacks of a pushdown automata is a regular language, that can be represented by finite-state automata. The analysed program is converted to a pushdown automaton, and is thus restricted to programs manipulating finite-state variables/properties, or requires the finite abstraction of data/properties prior the analysis. A recent extension allows the use of some infinite finite-height lattices [23] and represents a very interesting mix of the two approaches: pushdown automata are here extended by associating transformers to transition rules. This allows to encode the control part of the program and properties belonging to a finite lattice in the pushdown automata, whereas properties belonging to a finite-height lattice can be handled by the transformers attached to the transitions. Finally, [24] directly represents the stack as a linked list of activation records, using the shape analysis of [27].

7 Conclusion

In this paper, we presented an approach to the verification of imperative programs with recursive procedures and variables over possibly infinite domains. We showed that by relying solely on the principles of abstract interpretation, one can derive from the standard semantics of a program an interprocedural analysis method. This is done by abstracting in a proper way sets of call-stacks of the programs. Such an interprocedural analysis method can then be implemented after a second data abstraction. We defined two stack abstractions, the optimality of which suggests that they are good starting points for the following data abstraction. The first one is equivalent to the functional approach, but offers in addition the specialisation of predicate transformers and clarifies the intersection between forward and backward analysis.

The second stack abstraction, which is both data and control context sensitive, integrates the two approaches distinguished in [28]. It allows to specify complex constraints on the stack yielding an analysis which contains strictly more information. This is particularly useful for starting an analysis from initial or final states with a non-empty call-stack and also makes the combination of forward and backward analysis more efficient, as more information can be used for intersecting the two analyses.

It could be argued that our assumptions on the analysed programs are too restrictive. Non-local jumps could be easily added, as in [18], although this would suppress our current optimality results. Allowing reference parameter passing and handling aliasing on the stack would be very useful to tackle C programs. However, it should be noted that all the general approaches, with the notable exception of [3], assume like ourselves that intraprocedural instructions modify only the top activation record in the stack. Thus they cannot tackle directly reference parameter passing. The simplest way for adding this feature in our case would be to add aliasing information in activation records and to use it to properly update variables passed by reference upon procedure returns.

An implementation of our analysis is under work, targeted to programs with enumerated or numeric variables, following the implementation guidelines described in [16]. The next step would be its application to programs manipulating pointers to dynamically allocated objects, using abstract domains such as the one of [27].

References

1. T. Ball and S. K. Rajamani. Bebop: A symbolic model checker for boolean programs. In *SPIN 2000 Workshop*, LNCS 1885, pages 113–130. Springer, 2000.
2. A. Bouajjani, J. Esparza, and T. Touili. A generic approach to the static analysis of concurrent programs with procedures. In *POPL'03*, pages 62–73. ACM, 2003.
3. F. Bourdoncle. Interprocedural abstract interpretation of block structured languages with nested procedures, aliasing and recursivity. In *PLILP'90*, LNCS 456, pages 307–323. Springer, 1990.
4. F. Bourdoncle. Assertion-based debugging of imperative programs by abstract interpretation. In *ESEC'93*, LNCS 717, pages 501–516. Springer, 1993.
5. P. Cousot and R. Cousot. Abstract interpretation: A unified lattice model for static analysis of programs by construction or approximation of fixpoints. In *POPL'77*, pages 238–252. ACM, 1977.
6. P. Cousot and R. Cousot. Static determination of dynamic properties of recursive procedures. In *Formal Description of Programming Concepts*, pages 237–277. North Holland, 1977.
7. P. Cousot and R. Cousot. Abstract interpretation and application to logic programs. *Journal of Logic Programming*, 13(2–3), 1992.
8. P. Cousot and N. Halbwachs. Automatic discovery of linear restraints among variables of a program. In *POPL'78*, pages 84–97. ACM, 1978.

9. B. Creusillet and F. Irigoin. Interprocedural array region analyses. *International Journal of Parallel Programming*, 24(6), 1996.
10. A. Deutsch. Interprocedural may-alias analysis for pointers: Beyond k-limiting. In *PLDI'94*, pages 230–241. ACM, 1994.
11. J. Esparza and J. Knoop. An automata-theoretic approach to interprocedural data-flow analysis. In *FoSSaCS'99*, LNCS 1578, pages 14–30. Springer, 1999.
12. J. Esparza and S. Schwoon. A BDD-based model checker for recursive programs. In *CAV'01*, LNCS 2102, pages 324–336. Springer, 2001.
13. J. Feret. Abstract interpretation-based static analysis of mobile ambients. In *SAS'01*, LNCS 2126, pages 412–430. Springer, 2001.
14. F. Gaucher, E. Jahier, B. Jeannet, and F. Maraninchi. Automatic state reaching for debugging reactive programs. In *AADEBUG'03*, 2003.
15. B. Jeannet. Dynamic partitioning in linear relation analysis. Application to the verification of reactive systems. *Formal Methods in System Design*, 23(1), 2003.
16. B. Jeannet and W. Serwe. Abstracting call-stacks for interprocedural verification of imperative programs. Research Report 4904, INRIA, July 2003.
17. T. Jensen and F. Spoto. Class analysis of object-oriented programs through abstract interpretation. In *FoSSaCS'01*, LNCS 2030, pages 261–275. Springer, 2001.
18. N. D. Jones and S. S. Muchnick. A flexible approach to interprocedural data flow analysis and programs with recursive data structures. In *POPL '82*. ACM, 1982.
19. J. Knoop and B. Steffen. The interprocedural coincidence theorem. In *CC'92*, LNCS 641, pages 125–140. Springer, 1992.
20. A. Miné. The octagon abstract domain. In *AST 2001 in WCRE 2001*, pages 310–319. IEEE, 2001.
21. F. Nielson. Tensor products generalize the relational data flow analysis method. In *4th Hungarian Computer Science Conference*, 1985.
22. T. Reps, S. Horwitz, and M. Sagiv. Precise interprocedural dataflow analysis via graph reachability. In *POPL'95*, pages 49–61. ACM, 1995.
23. T. Reps, S. Schwoon, and S. Jha. Weighted pushdown systems and their application to interprocedural dataflow analysis. In *SAS'03*, LNCS 2694. Springer, 2003.
24. N. Rinetzky and M. Sagiv. Interprocedural shape analysis for recursive programs. In *CC'01*, LNCS 2027, pages 133–149. Springer, 2001.
25. V. Rusu, L. du Bousquet, and T. Jéron. An approach to symbolic test generation. In *IFM'00*, LNCS 1945, pages 338–357. Springer, 2000.
26. M. Sagiv, T. Reps, and S. Horwitz. Precise interprocedural dataflow analysis with applications to constant propagation. *TCS*, 167(1–2):131–170, 1996.
27. M. Sagiv, T. Reps, and R. Wilhelm. Parametric shape analysis via 3-valued logic. *ACM ToPLaS*, 24(3), 2002.
28. M. Sharir and A. Pnueli. Semantic foundations of program analysis. In S. Muchnick and N. Jones, editors, *Program Flow Analysis: Theory and Applications*, chapter 7. Prentice Hall, 1981.

Refining Mobile UML State Machines

Alexander Knapp[1], Stephan Merz[2], and Martin Wirsing[1]

[1] Institut für Informatik, Ludwig-Maximilians-Universität München
{knapp,wirsing}@informatik.uni-muenchen.de
[2] INRIA Lorraine, LORIA, Nancy
Stephan.Merz@loria.fr

Abstract. We study the semantics and refinement of mobile objects, considering an extension of core UML state machines by primitives that designate the location of objects and their moves within a network. Our contribution is twofold: first, we formalize the semantics of state machines in MTLA, an extension of Lamport's Temporal Logic of Actions with spatial modalities. Second, we study refinement concepts for state machines that are semantically justified in MTLA.

1 Introduction

Software development for mobile computing and mobile computations requires appropriate extensions of the traditional methods and concepts for more traditional system models. Moreover, the correctness and security of implementations of systems based on mobile code presents a major concern, as mobile agents may roam the network and must be guaranteed to work reliably in different locations and in different environments.

In this paper, we attempt to combine semi-formal modeling techniques for mobile systems with formal semantics and refinement. For modeling, we consider an extension of state machines in the "Unified Modeling Language" (UML [14]) for mobility. We first formalize the semantics of mobile state machines in MTLA [11], an extension of Lamport's Temporal Logic of Actions [9] with spatial modalities. Building on this logical semantics, we study refinement concepts for mobile state machines. In particular, we consider two notions of spatial refinement: the first one provides for an object to be split into a hierarchy of cooperating objects. The second one can be used to justify implementations of some high-level object by a set of objects that need not reside at the same location.

There has been much interest in formalizing concepts of UML as well as in semantic foundations for mobile computations, and we mention only the most closely related work. Deiß [6] suggested an encoding of (Harel) Statecharts in TLA, without considering either mobility or refinement. Several formal models of mobile computation have been proposed, either in the form of calculi as in [5, 12] or of state machine models as in [8], and sometimes accompanied by logics to describe system behavior [4, 13], but we are not aware of refinement notions for mobile computation. Our definitions of refinement of state machines are partly inspired by [15, 16]; a related notion has been elaborated in [17].

C. Rattray et al. (Eds.): AMAST 2004, LNCS 3116, pp. 274–288, 2004.
© Springer-Verlag Berlin Heidelberg 2004

Fig. 1. Prefix of a run.

1.1 Mobile UML

Mobile UML [2, 3, 10] extends UML [14] by concepts for modeling mobile computation. The extension is described in terms of the UML itself, using stereotypes and tagged values as meta-modeling tools. Most importantly, instances of classes distinguished by the stereotype «location» denote *locations* where other objects may reside. Mobile objects are instances of classes with the stereotype «mobile» and may change their locations over life-time. An actual movement of a mobile object is performed by a move action that takes the target location as its parameter.

1.2 MTLA

The logic MTLA [11] is an extension of Lamport's Temporal Logic of Actions [9] intended for the specification of systems that rely on mobility of code. Due to space restrictions, we refer to [11] for precise definitions of its syntax and semantics and only recall the basic intuitions and notations.

Following the intuition of the Ambient calculus [5] due to Cardelli and Gordon, we represent a configuration of a mobile system as a finite tree of nested locations. In this view, mobility is reflected by modifications of the location hierarchy, as agents move in and out of nested domains. Unlike in the Ambient calculus, we assume that locations carry unique ("physical") names. Moreover, instead of endowing each node of a configuration tree with a process, MTLA associates a local state with every node. A run is modeled as an ω-sequence of configuration trees. For example, Fig. 1 shows three configurations of a system run. The transition from the first to the second configuration models a local action that changes the value of the local attribute ctl associated with location ag. The second transition represents a move of ag from the location n_0 to the location n_1.

The logic MTLA contains both temporal and spatial modalities. Its formulas are evaluated over runs, at a given location. Temporal modalities refer to the truth value of formulas at suffixes of a run. For example $\Box F$ asserts that F holds of all suffixes of the run, at the current location.

Similarly, spatial modalities shift the spatial focus of evaluation, referring to locations below the current one. For example, the formula $m[F]$ asserts that F is true of the current run when evaluated at location m, provided such a location occurs (strictly and at arbitrary depth) below the current location, otherwise $m[F]$ is trivially satisfied. The dual formula $m\langle F\rangle$ asserts that the location m occurs beneath the current location, and that F holds there. For example, the run of Fig. 1 satisfies the formula $ag[ctl = \mathsf{Idle}]$

at the root location. We frequently use a more convenient dot notation to refer to local attributes at a given location and write, e.g., $ag.ctl = \mathsf{Idle}$.

As in TLA, we use formulas to describe systems as well as their properties. State transitions are specified using transition formulas that contain primed symbols, as in $ag.ctl = \mathsf{Idle} \wedge ag.ctl' = \mathsf{Shopping}$. When P is a state formula (i.e., without primed symbols), we write P' for the transition formula obtained by replacing all flexible symbols by their primed counterparts; intuitively, this formula asserts that P holds of the successor state. MTLA adds a transition formula $\alpha.n \gg \beta.n$ where n is a name and α and β are sequences of names. This formula asserts that the subtree rooted at name n within the tree indicated by α moves below the path β. The next-state relation of a system is specified by the temporal formula $\Box[A]_v$ asserting that every transition that modifies the expression v must satisfy the action formula A. Similarly, $\Box[A]_{\alpha.n}$, where n is a name and α is a sequence of names stipulates that every transition that removes or introduces location n below the subtree indicated by α must satisfy A.

Hiding of state components can be expressed in MTLA using existential quantification. For example, $\exists\, ag.ctl : F$ holds if one can assign some value to the attribute ctl of location ag at every state such that F holds of the resulting run. As in TLA, the precise definition is somewhat more complicated in order to preserve invariance under stuttering. One may also quantify over names and write $\exists\, n : F$; this hides the name as well as all its attributes. These quantifiers observe standard proof rules. In particular, we have the introduction axioms

$$(\exists\text{-ref}) \quad F\{t/n, t_1/n.a_1, \ldots, t_k/n.a_k\} \Rightarrow \exists\, n : F$$
$$(\exists\text{-sub}) \quad m\langle\mathbf{true}\rangle \Rightarrow \exists\, n : m.n\langle\mathbf{true}\rangle \qquad (m \not\equiv n)$$

The axiom ($\exists$-ref) asserts that $\exists\, n : F$ can be derived by finding a "spatial refinement mapping" that substitutes witnesses for the hidden name n as well as for its attributes. The axiom ($\exists$-sub) allows us to introduce a new sublocation n of an existing location m.

2   Statecharts and Their MTLA Semantics

We introduce state machines for mobile objects and provide them with a formal semantics based on MTLA. Our concepts are illustrated by means of the "shopper" example: A mobile shopping agent is sent out to gather offers for some item in several shops; when returning to its home base, the shopping agent presents the offers that it has found.

2.1   State Machines for Mobility

UML state machines, an object-oriented variant of Statecharts as defined by Harel [7], are an expressive and feature-rich class of state transition systems with a complex semantics [18]. In this paper, we consider a restricted class of state machines, but extended by a special move action. In particular, we consider neither hierarchical nor pseudo-states, with the exception of a single initial state per state machine. We consider only events triggered by asynchronous signals (excluding call, time, and change events) and

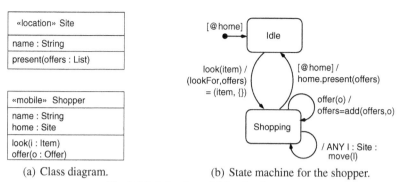

(a) Class diagram.   (b) State machine for the shopper.

Fig. 2. High-level model for the shopper.

ignore deferred events. Although our encoding could be extended to encompass all features of UML state machines, the simplifications we impose let us concentrate on the problems of mobility and refinement that are our primary concern.

Transitions of state machines carry labels of the form $trig[grd]/act$, any and all of which can be absent. The trigger $trig$ denotes a signal receptions of the form $\mathsf{op}(par)$ where op is the name of an operation declared in the class and par is a list of parameters. The guard grd is a Boolean expression over the attributes of the class and the parameters that appear in the trigger clause. In addition, we allow for guards $e_1 \prec e_2$ that refer to the hierarchy of objects; such a clause is true if (the object denoted by) e_1 is currently located beneath e_2. The most common form is $\mathsf{self} \prec e$, requiring the current object to be located below e, which we abbreviate to $@\,e$. The action act denotes the response of an object, beyond the state transition. For simplicity, we assume that all actions are of the form $\mathsf{ANY}\ x : P : upd; send; move$ where each of the constituents may be absent. Herein, P is a predicate over location objects, and $\mathsf{ANY}\ x : P$ functions as a binder that chooses some location object x satisfying P which can be used in the remainder of the action. The upd part is a simultaneous assignment $(a_1, \ldots, a_k) = (e_1, \ldots, e_k)$ of expressions e_i to attributes a_i. The $send$ part is of the form $e.\mathsf{op}(par)$ and denotes the emission of a signal op with parameters par to receiver object e. Finally, the $move$ part consists of a single $\mathsf{move}(e)$ action that indicates that the object should move to the location object whose identity is denoted by e. We require that all free variables in the action are among the attributes of the class, the parameters introduced by the trigger, and the location x bound by ANY. Figure 2(b) shows a first state machine for our shopping agent, based on the class diagram of Fig. 2(a). For the subsequent refinements, we will not explicitly indicate the class diagrams, as they can be inferred from the elements that appear in the state machines.

Our interpretation of transitions deviates in certain ways from the UML standard. First, the UML standard prioritizes triggerless transitions (so-called "completion transitions") over transitions that require an explicit triggering event. In contrast, we consider that completion transitions may be delayed; this less deterministic interpretation is more appropriate for descriptions at higher levels of abstraction. As a second, minor deviation, we allow guards to appear in transitions leaving a state machine's initial state.

2.2   MTLA Semantics of State Machines

We formalize systems of interacting, mobile state machines in MTLA. The formalization enables us to prove properties about systems specified in UML. We will also use it to justify correctness-preserving refinement transformations.

In MTLA, every object is represented by a MTLA location whose local state includes a unique, unmodifiable identifier $self$. We denote by Obj the set of all MTLA locations that represent objects of a given object system. The subset Loc denotes the set of MTLA locations that represent UML Location objects (including Mobile Locations), and the formalization of a system of state machines at a given level of abstraction is with respect to these sets Obj and Loc. An object configuration is represented as a tree of names as described in Sect. 1.2.

The local state at each node represents the attributes of the corresponding object, including $self$. In addition, we use the attributes ctl to hold the current control state of the object (i.e., the active state of the corresponding state machine) and $evts$ to represent the list of events that are waiting to be processed by the object. Objects interact asynchronously by sending and receiving messages. In the MTLA formalization, the communication network is represented explicitly by an attribute $msgs$ located at the root node of the configuration tree.

Every transition of an object is translated into an MTLA action formula that takes a parameter o denoting the location corresponding to the object. For lack of space, we do not give a precise, inductive definition of the translation, but only indicate its general form. In the following, if φ is an MTLA expression (a term or a formula), we write φ^x and φ_o, respectively, for the expressions obtained by replacing x by $x.self$ and by replacing all attributes a of o by $o.a$.

The action formula representing a transition is a conjunction built from the translations of its trigger, guard, and action components. The automaton transition from states src to dest is reflected by a conjunct $o.ctl = \mathsf{src} \wedge o.ctl' = \mathsf{dest}$.

A trigger $\mathsf{op}(par)$ contributes to the definition of the action formula in two ways: first, the parameters par are added to the formal parameters of the action definition. Second, we add the conjunct

$$\neg empty(o.evts) \wedge head(o.evts) = \langle \mathsf{op}, par \rangle \wedge o.evts' = tail(o.evts)$$

asserting that the transition can only be taken if the trigger is actually present in the event queue and that it is removed from the queue upon execution of the transition. For transitions without an explicit trigger we add the conjunct UNCHANGED $o.evts$ to indicate that the event queue is unmodified.

A Boolean guard g over the object's attributes is represented by a formula g_o, indicating that g is true at location o. A constraint $e_1 \prec e_2$ on the hierarchy of objects is represented by a conjunct of the form

$$\bigvee_{o_1, o_2 \in Obj} o_1.self = (e_1)_o \wedge o_2.self = (e_2)_o \wedge o_2.o_1 \langle \mathbf{true} \rangle$$

The representation of an action consists of action formulae for multiple assignment, message sending, and moving. If an action shows an ANY $x : P$ quantifier the conjunction $acts$ of these formulae are bound by a disjunction $\bigvee_{x \in Loc} P_o^x \wedge acts^x$. In more detail, a multiple assignment to attributes is represented by a formula

$$o.a_1' = (e_1)_o^x \wedge \ldots \wedge o.a_k' = (e_k)_o^x \wedge \text{UNCHANGED} \; \langle o.a_{k+1}, \ldots, o.a_n \rangle$$

where $a_{k+1}, \ldots, a_n$ are the attributes of o that are not modified by the assignment and where x is the variable bound by ANY. Sending a message $e.\mathsf{op}(par)$ is modeled by adding a tuple of the form $\langle e_o^x, \mathsf{op}, par_o^x \rangle$ to the network $msgs$. For actions that do not send a message we add the conjunct $msgs' = msgs$. If the action contains a clause $\mathsf{move}(e)$, we add a conjunct

$$\bigvee_{l \in Loc} l.self = e_o^x \wedge \varepsilon.o \gg l.o$$

that asserts that o will move to (the location with identity) e_o. Otherwise we add the conjunct $\bigwedge_{l \in Loc} [\textbf{false}]_{l.o}$, which abbreviates $\bigwedge_{l \in Loc} (l.o \langle \textbf{true} \rangle \Leftrightarrow \bigcirc l.o \langle \textbf{true} \rangle)$, to indicate that the object does not enter or leave any location in Loc.

To model the reception of new events by the object, we add an action $RcvEvt(o, e)$ that removes an event e addressed to o from the network and appends it to the queue $evts$ of unprocessed events while leaving all other attributes unchanged. We also add an action $DiscEvt(o)$ that discards events that do not have associated transitions from the current control state. The entire next-state relation $Next(o)$ of object o is represented as a disjunction of all actions defined from the transitions and the implicit actions $RcvEvt$ and $DiscEvt$, existentially quantifying over all parameters that have been introduced in the translation.

A state predicate $Init(o)$ defining the initial conditions of object o is similarly obtained from the transition from the initial state of the state machine. Finally, the overall specification of the behavior of an object o of class C is given by the MTLA formulas

$$IC(o) \equiv \wedge \; Init(o) \wedge o.evts = \langle \rangle \wedge \square [Next(o)]_{attr(o)} \wedge \square [\textbf{false}]_{o.self} \tag{1}$$
$$\wedge \bigwedge_{l \in Loc} \square [Next(o)]_{l.o}$$

$$C(o) \equiv \exists \, o.ctl, o.evts : IC(o) \tag{2}$$

The "internal" specification $IC(o)$ asserts that the initial state must satisfy the initial condition, that all modifications of attributes of o and all moves of o (entering or leaving any location of Loc) are accounted for by the next-state relation, and that the object identity is immutable. Here, $attr(o)$ denotes the tuple consisting of the explicitly declared attributes and the implicit attributes ctl and $evts$. For example, the formula $IShopper(ag)$ shown in Fig. 3 defines the behavior of an object ag of class Shopper introduced in Fig. 2(b). The "external" specification $C(o)$ is obtained from $IC(o)$ by hiding the implicit attributes ctl and $evts$.

The specification of a finite system of objects consists of the conjunction of the specifications of the individual objects. Moreover, we add conjuncts that describe the hierarchy of locations and objects and that constrain the network. For our shopper example, we might assume a typical system configuration being given by the object diagram in Fig. 4. This configuration can be translated into the formula

$$Sys \equiv \exists \, msgs : \wedge \bigwedge_{i=1}^{N} sh_i \langle self = \mathsf{shop}\text{-}i \wedge joe[\textbf{false}] \wedge \bigwedge_{j=1}^{N} sh_j[\textbf{false}] \rangle \wedge Site(sh_i)$$
$$\wedge \, joe \langle self = \mathsf{joe} \wedge \bigwedge_{i=1}^{N} sh_i[\textbf{false}] \rangle \wedge Site(joe)$$
$$\wedge \, joe.ag \langle self = \mathsf{shopper} \rangle \wedge Shopper(ag)$$
$$\wedge \bigwedge_{l \in Loc} \square [\textbf{false}]_{l.sh_1, \ldots, l.sh_N, l.joe}$$
$$\wedge \, msgs = \langle \rangle \wedge \square \left[\bigvee_{o \in Obj} Next(o) \right]_{msgs}$$

$$
\begin{aligned}
Init(ag) &\equiv ag.ctl = \mathsf{Idle} \land \bigvee_{l \in Loc}(l.ag\langle \mathbf{true}\rangle \land ag.home = l.self) \\
Stationary(ag) &\equiv \bigwedge_{l \in Loc}[\mathbf{false}]_{l.ag} \\
Deq(ag, msg) &\equiv \neg empty(ag.evts) \land head(ag.evts) = msg \land ag.evts' = tail(ag.evts) \\
Look(ag, item) &\equiv \land\ ag.ctl = \mathsf{Idle} \land ag.ctl' = \mathsf{Shopping} \land Deq(ag, \langle \mathsf{look}, item\rangle) \\
&\quad \land\ ag.lookFor' = item \land ag.offers' = \{\} \land \text{UNCHANGED}\ \ ag.home \\
&\quad \land\ msgs' = msgs \land Stationary(ag) \\
Offer(ag, o) &\equiv \land\ ag.ctl = \mathsf{Shopping} \land ag.ctl' = \mathsf{Shopping} \land Deq(ag, \langle \mathsf{offer}, o\rangle) \\
&\quad \land\ ag.offers' = add(ag.offers, o) \land \text{UNCHANGED}\ \langle ag.lookFor, ag.home\rangle \\
&\quad \land\ msgs' = msgs \land Stationary(ag) \\
Present(ag) &\equiv \land\ \bigvee_{l \in Obj} ag.home = l.self \land l.ag\langle\mathbf{true}\rangle \\
&\quad \land\ ag.ctl = \mathsf{Shopping} \land ag.ctl' = \mathsf{Idle} \\
&\quad \land\ \text{UNCHANGED}\ \langle ag.lookFor, ag.offers, ag.home, ag.evts\rangle \\
&\quad \land\ msgs' = msgs \cup \{\langle ag.home, \mathsf{present}, ag.offers\rangle\} \\
&\quad \land\ Stationary(ag) \\
Move(ag) &\equiv \bigvee_{l \in Loc} \land\ l.self \in Site \\
&\quad \land\ ag.ctl = \mathsf{Shopping} \land ag.ctl' = \mathsf{Shopping} \\
&\quad \land\ \text{UNCHANGED}\ \langle ag.lookFor, ag.offers, ag.home, ag.evts\rangle \\
&\quad \land\ msgs' = msgs \land \varepsilon.ag \gg l.ag \\
RcvEvt(ag, e) &\equiv \land\ \langle ag.self, e\rangle \in msgs \land msgs' = msgs \setminus \langle ag.self, e\rangle \\
&\quad \land\ ag.evts' = append(ag.evts, e) \\
&\quad \land\ \text{UNCHANGED}\ \langle ag.ctl, ag.lookFor, ag.offers, ag.home\rangle \\
&\quad \land\ Stationary(ag) \\
DiscEvt(ag) &\equiv \land\ \neg empty(ag.evts) \land ag.evts' = tail(ag.evts) \\
&\quad \land\ \neg\exists i : head(ag.evts) = \langle \mathsf{look}, i\rangle \lor ag.ctl \neq \mathsf{Idle} \\
&\quad \land\ \neg\exists o : head(ag.evts) = \langle \mathsf{offer}, o\rangle \lor ag.ctl \neq \mathsf{Shopping} \\
&\quad \land\ \text{UNCHANGED}\ \langle ag.ctl, ag.lookFor, ag.offers, ag.home\rangle \\
&\quad \land\ msgs' = msgs \land Stationary(ag) \\
Next(ag) &\equiv \lor\ (\exists i : Look(ag, i)) \lor (\exists o : Offer(ag, o)) \lor Present(ag) \\
&\quad \lor\ Move(ag) \lor (\exists e : RcvEvt(ag, e)) \lor DiscEvt(ag) \\
attr(ag) &\equiv \langle ag.ctl, ag.lookFor, ag.offers, ag.home, ag.evts\rangle \\
IShopper(ag) &\equiv \land\ Init(ag) \land ag.evts = \langle\rangle \land \Box[Next(ag)]_{attr(ag)} \land \Box[\mathbf{false}]_{ag.self} \\
&\quad \land\ \bigwedge_{l \in Loc} \Box[Next(ag)]_{l.ag}
\end{aligned}
$$

Fig. 3. MTLA specification of the shopper behavior (see Fig. 2(b)).

Fig. 4. Object diagram for the shopper example.

The formula in the scope of the existential quantifier asserts that the configuration contains the $N + 1$ sites $sh_1, \ldots, sh_N$ and joe, and a shopping agent ag. Moreover, joe and the shops are immobile and unnested locations, whereas ag is situated beneath joe. The last conjunct asserts that messages are only sent and received according to the specifications of the participating objects. The external specification is obtained by

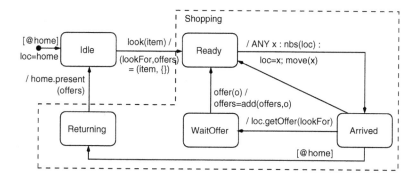

Fig. 5. Refined state machine for the shopper.

hiding, via existential quantification, the set of messages in transit, which is implicit at the UML level.

For this example, Obj is the set $\{sh_1, \ldots, sh_N, joe, ag\}$ and $Loc = Obj \setminus \{ag\}$. Moreover, we define a set $Site$ containing the identities of the elements of Loc, i.e. $Site = \{\text{shop-}1, \ldots, \text{shop-}N, \text{joe}\}$.

One purpose of our formalization is to prove properties about a system of objects. For the shopper example, we can deduce that the shopping agent is always located at its home agent or at one of the shops, expressed by the formula

$$\Box\left(\bigvee_{l \in Loc} l.ag\langle\textbf{true}\rangle \right) \tag{3}$$

3   Refinement of State Machines

In an approach based on refinement, interesting correctness properties of systems can already be established for models expressed at a high level of abstraction. Subsequent models introduce more detail, but ensure that all properties are preserved. In this paper, we focus on the refinement of state machines, and we add a "spatial" dimension to refinement that allows a designer to introduce more structure in the object hierarchy. In particular, a single high-level object can be refined into a tree of sub-objects. Throughout, we assume that the public interface of a refining class contains that of the refined one, and that the sets Obj and Loc of objects and Location objects of the refining model are supersets of those of the refined model.

3.1   Interface Preserving Refinement

Usually, early system models afford a high degree of non-determinism, which is reduced during system design. For example, consider the state machine for the shopping agent shown in Fig. 5, which imposes a number of constraints with respect to the state machine shown in Fig. 2(b). After arriving at a new shop location (whose identity is recorded in the additional attribute loc), the agent may now either query for offers by sending a new message getOffer or it may immediately move on to another neighbor location. In the former case, the agent waits until the offers are received, adds them to its local memory, and then moves on. When the agent arrives at its home location, it may quit the cycle, presenting the collected offers and returning to the Idle state.

Intuitively, the state machine of Fig. 5 is a refinement of the one shown in Fig. 2(b) because the states of the refined state machine can be mapped to those of the high-level state machine such that every transition of the lower-level machine either is explicitly allowed or is invisible at the higher level. In particular, the states Ready, Arrived, Wait-Offer, and Returning can all be mapped to the high-level state Shopping, as indicated by the dashed line enclosing these states. Assuming that the set $nbs(s)$ contains only identities in $Site$, for all $s \in Site$, each transition of the refined model either corresponds to a transition of the abstract model or to a stuttering transition. For example, the transition from Arrived to WaitOffer is invisible at the level of abstraction of the model shown in Fig. 2(b).

We now formalize this intuition by defining the notion of a state machine R refining another state machine M for a class C. Semantically, refinement is represented in linear-time formalisms by trace inclusion or, logically, by validity of implication. However, we will be a little more precise about the context in which M and R are supposed to be embedded. Both machines are specified with respect to attribute and method signatures Σ^R and Σ^M that include all method names that appear in transition labels (either received or sent), and we assume that Σ^R extends Σ^M. Similarly, we assume that the sets Obj^R and Loc^R of MTLA names for the objects and the Location objects at the level of the refinement are supersets of the corresponding sets Obj^M and Loc^M at the abstract level. Finally, the refinement may be subject to global hypotheses about the refined system, such as the hierarchy of names, that are formally asserted by an MTLA state predicate H. Thus, we say that the class R with associated state machine formalized by the MTLA formula C^R refines class M whose state machine is described by C^M under hypothesis H if for all system specifications Sys^M and Sys^R where Sys^R results from Sys^M by replacing all occurrences of $C^M(o)$ by $C^R(o)$ and by conjoining some formulas such that Sys^R implies $\Box H$, the implication $Sys^R \Rightarrow Sys^M$ is valid.

In order to prove that R refines M, we relate the machines by a mapping η that associates with every state s of R a pair $\eta(s) = (Inv(s), Abs(s))$ where $Inv(s)$ is a set of MTLA state predicates, possibly containing spatial operators, and where $Abs(s)$ is a state of M. With such a mapping we associate certain proof obligations: the invariants must be inductive for R, and the (MTLA formalizations of the) transitions of the machine R must imply some transition allowed at the corresponding state of M, or leave unchanged the state of M.

Theorem 1. *Assume that M and R are two state machines for classes C^M and C^R such that the attribute and method signature Σ^R of C^R extends the signature Σ^M of C^M, and that η is a mapping associating with every state s of R a set $Inv(s)$ of MTLA state predicates and a state $Abs(s)$ of M. If all of the following conditions hold then R refines M under hypothesis H. We write $\overline{\varphi}$ for*

$$\varphi\{Abs(o.ctl)/o.ctl, o.evts\rceil_{\Sigma^M} / o.evts, msgs\rceil_{\Sigma^M} / msgs\}$$

where $e\rceil_\Sigma$ denotes the subsequence of elements e whose first component is in Σ.

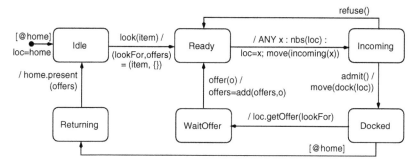

Fig. 6. Spatial refinement of the network sites.

1. $Abs(s_0^R) = s_0^M$ where s_0^M and s_0^R denote the initial states of M and R. Moreover,

$$\models H \wedge Init^R(o) \Rightarrow o[Inv(s_0^R)] \wedge \overline{Init^M}(o)$$

holds for the initial conditions $Init^R$ and $Init^M$ of M and R.

2. For every transition of R with source and target states s and t formalized by the MTLA action formula $A(o, par)$:

$$\models H \wedge H' \wedge o[Inv(s)] \wedge A(o, par) \Rightarrow o[Inv(t)']$$

3. For every state s of R and every outgoing transition of s formalized by formula $A(o, par)$, let $Abs(s)$ denote the corresponding state of M, let $B_1(o, par_1)$, ..., $B_m(o, par_m)$ be the MTLA formulas for the outgoing transitions of $Abs(s)$, let $attr^M(o)$ be the tuple of attributes defined for M and Loc^M the set of locations for M. Then:

$$\models H \wedge H' \wedge o[Inv(s)] \wedge A(o, par) \Rightarrow$$
$$\vee \bigvee_{i=1}^{m} (\exists par_i : \overline{B_i}(o, par_i))$$
$$\vee \text{ UNCHANGED } \langle attr^M(o), msgs\uparrow_{\Sigma M} \rangle \wedge \bigwedge_{l \in Loc_M} [\textbf{false}]_{l.o}$$

Theorem 1 ensures that R can replace M, subject to hypotheses H. In particular, all properties expressed by MTLA formulas that have been established for the high-level system will be preserved by the implementation.

In order to prove that the state machine of Fig. 5 refines that of Fig. 2(b) (with respect to $H \equiv \forall s \in Site : \text{nbs}(s) \in Site$) we must define the mapping η. We have already indicated the definition of the state abstraction mapping Abs. For the mapping Inv, we associate (the MTLA encoding of) @$home$ with state Returning and $ag.loc \in Site$ with all other states. It is then easy to verify the conditions of Theorem 1. In particular, the transitions leaving state Arrived do not modify the shopping agent's attributes, and they do not send messages contained in the original signature. They are therefore allowed by condition (3) of Theorem 1.

Theorem 1 can also be used to justify refinements that modify the spatial hierarchy of locations. Consider the state machine shown in Fig. 6. It is based on the idea that prior to interacting with an object, incoming agents are first placed in a special sublocation for security checking. Instead of a simple, atomic move from one shop to another

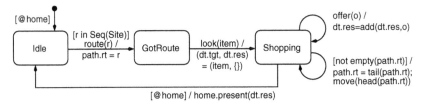

Fig. 7. Spatial refinement of the shopper.

as in Figs. 2(b) and 5, this version moves the shopping agent first to the "incoming" sublocation of the target location. If the agent is accepted by the host, as modeled by the reception of an **admit** signal, it transfers to the "dock" sublocation where the real processing takes place. Otherwise, the host will send a **refuse** signal, and the shopping agent moves on to another neighbor host. Here we assume that every location $l \in Loc$ contains sublocations l_in and l_dock. Moreover, we assume functions incoming and dock that look up the id's of the corresponding sub-locations for a given network site.

Formally, Theorem 1 can again be used to show that the "docked" shopper of Fig. 6 is a refinement of that shown in Fig. 5 with respect to the hypothesis

$$H \equiv \bigwedge_{l \in Loc^M} \begin{array}{l} \wedge \; l.l\_in\langle \mathbf{true} \rangle \wedge l.l\_dock\langle \mathbf{true} \rangle \\ \wedge \; \mathsf{incoming}(l.self) = l\_in.self \wedge \mathsf{dock}(l.self) = l\_dock.self \end{array}$$

The states Incoming and Docked are mapped to the single high-level state Arrived, and the invariant mapping associates (the MTLA encoding of) @ loc with the location Incoming and $ag.loc \in Site$ with all states. Indeed, the move action labeling the transition from the Ready to the Incoming state will be formalized by an MTLA action formula $\bigvee_{l \in Loc^R} \varepsilon.ag \gg l\_in.ag$, which implies the corresponding formula $\bigvee_{l \in Loc^M} \varepsilon.ag \gg l.ag$ formalizing the move between the high-level states Ready and Arrived, using the hypothesis H. Similarly, H and the invariant establish that the move between the Incoming and Docked states maps to a stuttering action: Clearly, the local attributes and the message queue are left unchanged. Moreover, the invariant associated with state Incoming asserts that the agent is located beneath the site (with identity) loc. Therefore, a move to the "dock" sublocation of that same site is invisible with respect to the locations in Loc^M: the action implies $[\mathbf{false}]_{l.ag}$, for all $l \in Loc^M$.

For these kinds of refinement to be admissible, it is essential that the spatial operators of MTLA refer to locations at an arbitrary depth instead of just the children of a node and that it is therefore impossible to specify the precise location of the agent. In fact, we consider the concept of "immediate sublocation" to be as dependent on the current level of abstraction as the notion of "immediate successor state", and MTLA allows to express neither.

3.2   Interface Refinement I: Spatial Distribution of State

Frequently, refinements of the spatial hierarchy will be accompanied by a distribution of the high-level attributes over the hierarchy of sublocations of the refined model. For a simple example, departing again from the high-level shopper of Fig. 2(b), consider the state machine shown in Fig. 7. Here we assume that the shopping agent contains two sub-agents path that determines the path to follow through the network and dt that collects the data, and we have replaced the attributes lookFor and offers of the high-level shopper by attributes tgt and res assigned to the dt sub-agent[1]. The transition from Idle to GotRoute determines the route of the agent. It is guarded by the condition $r \in Seq(Site)$, asserting that r is a list of (identities of) network sites.

Spatial distribution of attributes is similar to the concept of data refinement in standard refinement-based formalisms. Intuitively, the refinement of Fig. 7 is admissible provided that the public interface is preserved. We will therefore assume that the attributes item and offers have been marked as private in the class diagram for the abstract shopper, ensuring that no other object relies on their presence.

Formally, we modify slightly the MTLA formalization of state machines, taking into account the visibility (either "private" or "public") of attributes. We redefine the external specification of the behavior of an object o of class C with private attributes $a_1, \ldots, a_k$ as the MTLA formula

$$C(o) \;\equiv\; \exists\, o.a_1, \ldots, o.a_k, o.ctl, o.evts : IC(o) \tag{4}$$

where $IC(o)$ is defined as before by formula (1). Since the specification of an object system is based on the external object specification, private attributes are invisible at the system level, and the definition of refinement modulo a hypothesis remains as before.

The verification of refinement relies on conditions generalizing those of Theorem 1, provided that the private attributes of the high-level object can be computed from those of the implementation via a refinement mapping [1]. The relation between the two diagrams R and M is therefore given by the mapping η as before, complemented by terms $t_1, \ldots, t_k$ that represent the values of the private high-level attributes $a_1, \ldots, a_k$. These terms have then to be substituted for the attributes in the formulas concerning the high-level state machine M.

Theorem 2. *Extending the context of Theorem 1 by terms* $t_1, \ldots, t_k$, *we now write* $\overline{\varphi}$ *for*

$$\varphi\{Abs(o.ctl)/o.ctl, o.evts\!\restriction_{\Sigma M} /o.evts, msgs\!\restriction_{\Sigma M} /msgs, t_1/o.a_1, \ldots, t_k/o.a_k\}$$

If the set of public attributes of R is a superset of those of M then R refines M under hypothesis H up to hiding of attributes $o.a_1, \ldots o.a_k$ if the conditions of Theorem 1 hold for this new interpretation of substitution.

For the example shown in Fig. 7, the hypothesis is

$$H \equiv ag.path\langle \mathbf{true} \rangle \wedge ag.dt\langle \mathbf{true} \rangle$$

The implementation states Idle and GotRoute are both mapped to the abstract state Idle. The invariant mapping assigns the state formula $ag.path.rt \in Seq(Site)$ to the states GotRoute and Shopping. Finally, the refinement mapping is defined by substituting $ag.dt.res$ and $ag.dt.tgt$ for $ag.offers$ and $ag.lookFor$, respectively. All proof obligations of Theorem 2 are then easily verified.

---

[1] The renaming of the attributes is not necessary, but will make it clear in the following to which model we are referring.

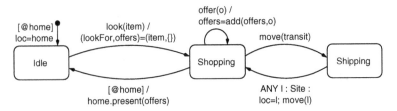

Fig. 8. State machine for the "slow shopper".

3.3   Interface Refinement II: Virtualisation of Locations

Whereas the notions of spatial refinement that we have considered so far have intro-duced new (sub-)objects, we have taken care to preserve the hierarchy of the objects present at the abstract levels. Together with the choice of modalities of MTLA, which cannot express the precise location of an object, we have thus been able to represent refinement as implication and to preserve all MTLA properties. However, it can occa-sionally be desirable to allow for refinements that do not at all times preserve the spatial relationships imposed by the original specification.

For example, the previous specifications of the shopping agent have all assumed that moves between locations happen atomically. Figure 8 presents a variation of the original state machine of Fig. 2(b) where the agent moves to an intermediate transit location, which is not included in $Site$, before moving to the next site. (A subsequent refinement could add more structure to the transit location, modeling the transport of the agent across the network.) We cannot use Theorems 1 or 2 to prove that this model refines the original one because the move to the transit location cannot be mapped to any high-level action. In fact, the MTLA formula representing the "slow shopper" does not imply the formula encoding the original specification, and the invariant formula (3) asserting that the shopping agent is always located at some location that represents a network site does not hold of the slow shopper.

Such relationships can be formalized by considering a weaker notion of refinement, abstracting from some of the names that occur in the original specification. In our run-ning example, the name of the shopping agent should not actually be part of the inter-face: the purpose of the system is that the agent's home site learns about offers made by other network sites; the use of a mobile agent is an implementation detail. We say that an object system formalized by an MTLA formula $Impl$ refines another system formalized by $Spec$ up to hiding of name n if the implication $Impl \Rightarrow \exists\, n : Spec$ holds. In general, the behavior required of object n at the abstract level may be implemented by several implementation objects, hence it does not appear useful to give a "local" rule, similar to Theorems 1 and 2, that attempts to prove refinement by considering a single state machine at a time. Instead, the strategy in proving such a refinement is to define a "spa-tial refinement mapping", using the rules given in Sect. 1.2. For the slow shopper, we first use rule ($\exists$-sub) to introduce a new sublocation, say $l\_virtual$, for every high-level location l and then define a refinement mapping that returns the implementation-level agent as long as it is not at the transit location, and otherwise the location $l\_virtual$ as-sociated with the previous site visited as stored in the attribute loc. The local attributes of the high-level shopper are simply obtained from those of the implementation-level agent. Observe in particular that the invariant (3) cannot be proven of the specification $\exists\, ag : Sys$ because ag is no longer free in that formula.

Refinement up to hiding of names allows for implementations that differ more radically in structure. For example, the single shopping agent of the initial specification could be implemented by a number of shopping agents that roam the network in parallel, cooperating to establish the shopping list. On the other hand, a correct implementation could also be based on a client-server solution instead of using mobile agents.

4   Conclusion

We have studied the applicability of the logic MTLA proposed in [11] in view of formalizing Mobile UML State Machines [3] and of establishing refinement relationships between models described in this language. A configuration of a mobile system is represented as a tree of names, and mobility is reflected by changes to the name hierarchy. MTLA accomodates local attributes at every node in the tree, simplifying the formalization of state-based notations such as UML state machines. The operators of MTLA have been designed to support system refinement; in particular, all spatial operators refer to nodes arbitrarily deep beneath the current node and not just its children as in other spatial logics, e.g. [4].

We have assumed some simplifications and restrictions for our formalization of Mobile UML state machines. In particular, we assume that spatial relationships are specified using constraints $e_1 \prec e_2$, comparing the relative positions of two objects at the current level of abstraction. This assumption has been essential to obtain a sound and elegant representation of refinement as implication of specifications for mobile systems.

Our main objective has been the study of three fundamental refinement principles, focusing on refinements of the spatial hierarchy. We have indicated sufficient conditions for verifying refinement. However, these conditions are incomplete: in particular, it is well known that refinement mappings need to be complemented by devices such as history and prophecy variables in order to obtain completeness [1]. We have also ignored liveness and fairness properties in this paper, and we have mostly restricted ourselves to proving refinement "object by object". We intend to study adequate composition and decomposition concepts in future work.

References

1. M. Abadi and L. Lamport. The existence of refinement mappings. *Theor. Comp. Sci.*, 81(2):253–284, May 1991.
2. H. Baumeister, N. Koch, P. Kosiuczenko, P. Stevens, and M. Wirsing. UML for global computing. In C. Priami, editor, *Global Computing. Programming Environments, Languages, Security, and Analysis of Systems*, volume 2874 of *Lect. Notes in Comp. Sci.*, pages 1–24, Rovereto, Italy, 2003. Springer-Verlag.
3. H. Baumeister, N. Koch, P. Kosiuczenko, and M. Wirsing. Extending activity diagrams to model mobile systems. In M. Aksit, M. Mezini, and R. Unland, editors, *Objects, Components, Architectures, Services, and Applications for a Networked World*, volume 2591 of *Lect. Notes in Comp. Sci.*, pages 278–293, Erfurt, Germany, 2003. Springer-Verlag.

4. L. Caires and L. Cardelli. A spatial logic for concurrency (part I). *Inf. and Comp.*, 186(2):194–235, Nov. 2003.

5. L. Cardelli and A. Gordon. Mobile ambients. *Theor. Comp. Sci.*, 240:177–213, 2000.

6. T. Deiß. An approach to the combination of formal description techniques: Statecharts and TLA. In K. Araki, A. Galloway, and K. Taguchi, editors, *Integrated Formal Methods (IFM 1999)*, pages 231–250, York, UK, 1999. Springer-Verlag.

7. D. Harel. Statecharts: A Visual Formalism for Complex Systems. *Science of Computer Programming*, 8(3):231–274, 1987.

8. T. A. Kuhn and D. v. Oheimb. Interacting state machines for mobility. In K. Araki, S. Gnesi, and D. Mandrioli, editors, *Proc. 12th Intl. FME Symposium (FM2003)*, volume 2805 of *Lect. Notes in Comp. Sci.*, pages 698–718, Pisa, Italy, Sept. 2003. Springer-Verlag.

9. L. Lamport. The Temporal Logic of Actions. *ACM Trans. Prog. Lang. Syst.*, 16(3):872–923, May 1994.

10. D. Latella, M. Massink, H. Baumeister, and M. Wirsing. Mobile UML statecharts with localities. Technical report 37, CNR ISTI, Pisa, Italy, 2003.

11. S. Merz, J. Zappe, and M. Wirsing. A spatio-temporal logic for the specification and refinement of mobile systems. In M. Pezzè, editor, *Fundamental Approaches to Software Engineering (FASE 2003)*, volume 2621 of *Lect. Notes in Comp. Sci.*, pages 87–101, Warsaw, Poland, April 2003. Springer-Verlag.

12. R. D. Nicola, G. Ferrari, and R. Pugliese. Klaim: a kernel language for agents interaction and mobility. *IEEE Trans. Software Eng.*, 24(5):315–330, 1998.

13. R. D. Nicola and M. Loreti. A modal logic for Klaim. In T. Rus, editor, *Algebraic Methodology and Software Technology (AMAST 2000)*, volume 1816 of *Lect. Notes in Comp. Sci.*, pages 339–354, Iowa, 2000. Springer-Verlag.

14. Object Management Group. Unified Modeling Language Specification, Version 1.5. Specification, OMG, 2003. http://cgi.omg.org/cgi-bin/doc?formal/03-03-01.

15. B. Paech and B. Rumpe. A new concept of refinement used for behaviour modelling with automata. In *Formal Methods Europe (FME'94)*, volume 873 of *Lect. Notes in Comp. Sci.*, pages 154–174, Barcelona, Spain, 1994. Springer-Verlag.

16. P. Scholz. A refinement calculus for Statecharts. In *Fundamental Approaches to Software Engineering (FASE'98)*, volume 1382 of *Lect. Notes in Comp. Sci.*, pages 285–301, Lisbon, Portugal, 1998. Springer-Verlag.

17. M. Schrefl and M. Stumptner. Behavior-consistent specialization of object life cycles. *ACM Trans. Software Eng. Meth.*, 11(1):92–148, 2002.

18. M. von der Beeck. Formalization of UML-statecharts. In M. Gogolla and C. Kobryn, editors, *Proc. 4th Int. Conf. UML (UML 2001)*, volume 2185 of *Lect. Notes in Comp. Sci.*, pages 406–421. Springer, 2001.

Verifying Invariants of Component-Based Systems through Refinement

Olga Kouchnarenko* and Arnaud Lanoix

Laboratoire dInformatique de l'Université de Franche-Comté
FRE CNRS 2661
16, route de Gray, 25030 Besançon Cedex France
Ph: (33) 3 81 66 65 24, Fax: (33) 3 81 66 64 50
{kouchna,lanoix}@lifc.univ-fcomte.fr
http://lifc.univ-fcomte.fr/~synco

Abstract. In areas like manufacturing, communications, transportation or aerospace, the increasing size and complexity of reactive systems make their verification difficult to handle. Compositional reasoning is a way to master this problem. In this paper, we propose an approach based on a constraint synchronized product to specify and to verify such systems. This approach supports a compositional refinement for both labelled transition systems and their composition. In this framework, we show how to verify local and global invariance properties during a refinement verification. Thus, these properties are preserved through refinement.
The different aspects of our work are illustrated on the example of a communication protocol between an integrated chip card and a reader interface device.

Keywords: Invariance properties, refinement, compositional verification, synchronized product, preservation.

1 Introduction

Invariance properties (also called invariants) are fundamental in specification and verification of complex reactive systems. Size and complexity of specification and development continue to grow making these systems more and more difficult to understand and to handle.

Compositional reasoning is a way to master this complexity and to postpone the state explosion problem while verifying these systems by enumeration. Our compositional method describes the system's components and their interactions instead of building the whole system. For that, we define a specific composition operator, a kind of constraint synchronized product, that models synchronous and asynchronous behaviours. One of its advantages is that this composition operator supports a refinement of transition systems. It allows us to master the complexity of specifications with a step by step development process. In [13]

---

* This work was partially supported by the LIFC-LORIA/INRIA Research Group CASSIS.

C. Rattray et al. (Eds.): AMAST 2004, LNCS 3116, pp. 289–303, 2004.

the refinement of the whole system is linked with the weak refinement of its expanded components. In addition, this component-based refinement guarantees preservation of the abstract system reachability properties, and the deadlock freedom, for the refined ones. It seems natural to answer the question about invariants'verification in this framework.

In this paper, we focus on two invariance properties: local and global invariants. Local invariants relate to the allowed behaviours of one of the system components whereas global invariants express the allowed whole system's behaviours.

We show that both kinds of invariants can be checked compositionally when systems are specified by the constraint synchronized product. This verification of invariants can be done while verifying our compositional refinement without increasing this algorithm's complexity. Moreover these invariants are preserved by refinement (see Fig. 1).

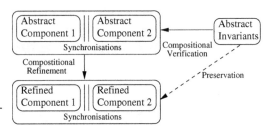

Fig. 1. Compositional verification

This paper is organised as follows. After giving preliminary notions, we define, in Section 2, the behavioural semantics of synchronized component-based systems. In Section 3, we show how this semantics is well-adapted to compositional invariance property verification. Then, in Section 4, we recall a refinement relation and give a compositional refinement verification method preserving invariance properties. Throughout this paper, our work is illustrated on an industrial example of a communication protocol between an integrated chip card and a reader interface device. We end by a tool presentation and some perspectives.

2   Compositional Specification

In this section, some preliminaries are introduced to specify and to model component's behaviours. Then a constraint synchronized product is introduced to specify the synchronized behaviours of components. Finally, we define an expanded component and we establish a link between the constraint synchronized product and the expanded components.

2.1   Preliminaries

We introduce interpreted labelled transition system to specify component's behaviours and properties. Let $V = \{X_1, \ldots, X_n\}$ be a finite set of variables with their respective domains $\mathbb{D}_1, \ldots, \mathbb{D}_n$. Let $AP_V \stackrel{\text{def}}{=} \{X_i = v | X_i \in V \wedge v \in \mathbb{D}_i\}$ be a set of atomic propositions over V.

Definition 1 (Labelled Transition System (LTS)). *A LTS S over V is a tuple $\langle Q, Q_0, E, T, l \rangle$ where*

– Q is a set of states,
– $Q_0 \subseteq Q$ is a set of initial states,
– E is a finite set of transition labels (actions),
– $T \subseteq Q \times E \times Q$ is a labelled transition relation,
– $l : Q \rightarrow 2^{AP_V}$ is a function that labels each state with the set of atomic propositions true in that state; we call l the interpretation.

Let $\mathcal{Q}$ be a state universe. The set of states Q of S is the subset of the reachable states of $\mathcal{Q}$ from Q_0, using T (we denote $q \xrightarrow{e} q'$ an element of T). The set $SP_V \stackrel{\text{def}}{=} \{sp, sp_1, sp_2, \ldots\}$ of state propositions over V is defined by the following grammar: $sp_1, sp_2 ::= ap \mid sp_1 \vee sp_2 \mid \neg sp_1$, where $ap \in AP_V$.

Definition 2 (sp holds on q). *We define a state $q \in Q$ satisfies a state proposition $sp \in SP_V$, written $q \models sp$, as follows.*
– $q \models ap$ iff $ap \in l(q)$,
– $q \models \neg sp$ iff it is not true that $q \models sp$,
– $q \models sp_1 \vee sp_2$ iff $q \models sp_1$ or $q \models sp_2$.

We call invariant *a proposition $sp \in SP_V$ which holds on every state of S, i.e. $\forall q.\ (q \in Q \Rightarrow q \models sp)$, notice $S \models sp$.*

Fig. 2. Protocol T=1 components $Card_A$ and $Device_A$

T=1 is a half duplex block transmission protocol between an Integrated Chip Card (ICC) and a Reader Interface Device (IFD) [12]. Figure 2 presents two LTSs, $Card_A$ and $Device_A$, for the ICC and the IFD. The whole system, $Protocol_A$, can be obtained from both LTSs $Card_A$ and $Device_A$.

The variables cs and cp of the LTS $Card_A$ specify respectively the Card Status (*unknown, sender* or *reader*) and the Card Position (*input* or *output*). The card may be inserted or ejected (actions insertC and ejectC). The card can alternatively receive or send blocks of messages (actions receiveC and sendC). The variables ds and de of the LTS $Device_A$ specify respectively the Device Status and the Device Exchange status (*ready* or *progress*). The IFD can be initialised or aborted (actions initD and abortD). It can alternatively receive or send blocks of messages (actions receiveD and sendD), and communicate with other devices (actions exchD and endD).

2.2   Constraint Synchronized Product

In order to specify interactions between components, we define a *synchronization set*. This set contains tuples of labels with feasibility conditions constraining activations of transitions the labels of which are in the synchronization set. Let '−' denote the fictive action "skip".

Definition 3 (Synchronization Set). *Let* $S_1 = \langle\, Q_1, Q_{01}, E_1, T_1, l_1\,\rangle$ *over* V_1 *and* $S_2 = \langle\, Q_2, Q_{02}, E_2, T_2, l_2\,\rangle$ *over* V_2 *be two LTSs. A synchronization set syn is defined as a subset of* $\{((e_1, e_2) \textbf{ when } sp)|\ e_1 \in E_1 \cup \{-\} \wedge e_2 \in E_2 \cup \{-\} \wedge sp \in SP_{V_1 \cup V_2}\}.$

The whole system is a rearrangement of its separate parts, i.e. the components and their interactions. This arrangement is specified by a constraint synchronized product. To do this, we need to extend Definition 2 to pairs of states.

Definition 4 (*sp* holds on (q_1, q_2)). *We define a state* $(q_1, q_2) \in Q_1 \times Q_2$ *satisfies a state proposition* $sp \in SP_{V_1 \cup V_2}$, *written* $(q_1, q_2) \models sp$, *as follows.*
- $(q_1, q_2) \models sp$ *iff* $sp \in SP_{V_1}$ *and* $q_1 \models sp$,
- $(q_1, q_2) \models sp$ *iff* $sp \in SP_{V_2}$ *and* $q_2 \models sp$,
- $(q_1, q_2) \models \neg sp$ *iff it is not true that* $(q_1, q_2) \models sp$,
- $(q_1, q_2) \models sp_1 \vee sp_2$ *iff either* $(q_1, q_2) \models sp_1$ *or* $(q_1, q_2) \models sp_2$.

Definition 5 (Constraint Synchronized Product). *Let* S_1 *and* S_2 *be LTSs. Let syn be their synchronization set. The constraint synchronized product of* S_1 *and* S_2, *written* $S_1 \times_{syn} S_2$ *is the tuple* $\langle\, Q, Q_0, E, T, l\,\rangle$ *where* $Q \subseteq Q_1 \times Q_2$, $Q_0 \subseteq Q_{01} \times Q_{02}$, $E \subseteq (E_1 \cup \{-\}) \times (E_2 \cup \{-\})$, $l((q_1, q_2)) = l_1(q_1) \cup l_2(q_2)$ *and* $T \subseteq Q \times E \times Q$ *obeys the following rules.*

$$[5.1] \quad \frac{q_1 \overset{e_1}{\to} q_1' \in T_1, (q_1, q_2) \models sp}{(q_1, q_2) \overset{(e_1, -)}{\longrightarrow} (q_1', q_2) \in T} \quad \text{if } ((e_1, -) \textbf{ when } sp) \in syn$$

$$[5.2] \quad \frac{q_2 \overset{e_2}{\to} q_2' \in T_2, (q_1, q_2) \models sp}{(q_1, q_2) \overset{(-, e_2)}{\longrightarrow} (q_1, q_2') \in T} \quad \text{if } ((-, e_2) \textbf{ when } sp) \in syn$$

$$[5.3] \quad \frac{q_1 \overset{e_1}{\to} q_1' \in T_1, q_2 \overset{e_2}{\to} q_2' \in T_2, (q_1, q_2) \models sp}{(q_1, q_2) \overset{(e_1, e_2)}{\longrightarrow} (q_1', q_2') \in T} \quad \text{if } ((e_1, e_2) \textbf{ when } sp) \in syn$$

This definition of composition is more expressive than the classical synchronized product defined by Arnold and Nivat [6, 5] because of feasibility conditions. Our composition models synchronous and asynchronous behaviours. Indeed, each transition of the product can involve either joint transitions of two components or single transition of one component with respect to the synchronization set.

The synchronization set syn_A given in Fig. 3, specifies the authorized behaviours of the whole system $Protocol_A = Card_A \times_{syn_A} Device_A$ (see Fig. 4) that can be built by Definition 5. The first action is the initialisation of the IFD before the card insertion. Then, components can communicate.

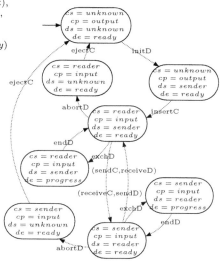

$(-,\text{initD})$ **when** $(cs = unknown) \wedge (cp = output)$,
$(\text{insertC},-)$ **when** $(ds = sender) \wedge (de = ready)$,
$(\text{receiveC},\text{sendD})$ **when** $(cs \neq ds)$,
$(\text{sendC},\text{receiveD})$ **when** $(cs \neq ds)$,
$(\text{ejectC},-)$ **when** $(ds = unknown) \wedge (de = ready)$
$(-,\text{exchD})$ **when** $(cs \neq ds) \wedge (cp = input)$,
$(-,\text{endD})$ **when** $(cs \neq ds) \wedge (cp = input)$,
$(-,\text{abortD})$ **when** $(cs \neq ds) \wedge (cp = input)$,

Fig. 3. Protocol T=1 synchronization set

Alternation of messages is specified by synchronization between action receiveC of $Card_A$ and action sendD of $Device_A$, and by synchronization between action sendC of $Card_A$ and action receiveD of $Device_A$. The IFD can exchange with other devices only if the card is inserted. The IFD must abort before the ICC is ejected.

Fig. 4. Protocol T=1 product

2.3 Expanded Components

Each component is a context-free component. However, there is a need to take its environment into account. For that, we define an *expanded* component, that is a component in the context of the other components under a synchronization set.

Definition 6 (Expanded Component). *Let* S_1 *and* S_2 *be two LTSs. Let* syn *be their synchronization set. The expanded component* S_1^c *corresponding to* S_1 *is defined by the tuple* $\langle Q_1^c, Q_{01}^c, E_1^c, T_1^c, l_1^c \rangle$ *where*

- $Q_1^c \subseteq Q_1 \times Q_2$,
- $Q_{01}^c \subseteq Q_{01} \times Q_{02}$,
- $E_1^c = \{(e_1, e_2) \mid ((e_1, e_2) \text{ when } sp) \in syn \wedge e_1 \in E_1 \wedge e_2 \in E_2 \cup \{-\}\}$,
- $l_1^c((q_1, q_2)) = l_1(q_1) \cup l_2(q_2)$,
- $T_1^c \subseteq Q_1^c \times E_1^c \times Q_1^c$ *is obtained by Rules [5.1] and [5.3].*

The expanded component S_2^c is similarly defined. Notice that both expanded components are over the same set of variables $V_1 \cup V_2$. We define a sum of two LTSs in the same context, i.e. over the same set of variables V.

Definition 7 (Sum of Two LTSs). *Let* $S_1 = \langle Q_1, Q_{01}, E_1, T_1, l_1 \rangle$ *and* $S_2 = \langle Q_2, Q_{02}, E_2, T_2, l_2 \rangle$ *be two LTSs over* V. *The sum of* S_1 *and* S_2, *written* $S_1 \uplus S_2$, *is the tuple* $\langle Q_1 \cup Q_2, Q_{01} \cup Q_{02}, E_1 \cup E_2, T_1 \cup T_2, l \rangle$ *where* l *is defined by*

$$l(q) = \begin{cases} l_1(q) & \text{if } q \in Q_1 \\ l_2(q) & \text{if } q \in Q_2 \end{cases}, \text{ and } \forall q_1 \in Q_1 \forall q_2 \in Q_2.(q_1 = q_2 \Leftrightarrow l_1(q_1) = l_2(q_2)).$$

We have shown in [13] that Definition 7 allows us to link components and expanded components by:

$$S_1 \times_{syn} S_2 = S_1^c \uplus S_2^c \tag{1}$$

For example, we can build expanded components $Card_A^c$ and $Device_A^c$ for the protocol T=1 (see Fig. 5 and 6). We have $Protocol_A = Card_A^c \uplus Device_A^c$ thanks to Equality 1.

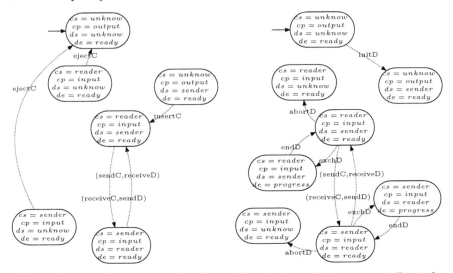

Fig. 5. Expanded component $Card_A^c$ **Fig. 6.** Expanded component $Device_A^c$

Formal framework of expanded components above can be compared with the *assume-guarantee* paradigm (see for instance [9, 18, 19, 10]). In both cases, one wants to take the component's environment into account. In the assume-guarantee paradigm, it is expressed by hypotheses on the component's environment. In our approach, it is done by increasing a component model to obtain an expanded component. Notice that it is not the designer's task to build the expanded components nor their sum. As they can be automatically generated by Definitions 6 and 7, the designer only needs to specify both components and their interactions, i.e. their synchronization set.

3   Verifying Compositional Invariants

In this section, we show that the synchronized component-based systems model is well-adapted to a compositional verification of invariance properties, called *local* and *global* invariants. These invariants are expressed in the propositional logic as state propositions (see Def. 2). Consider two components S_1 over V_1 and S_2 over V_2, and their constraint synchronized product $S_1 \times_{syn} S_2$. Roughly speaking, a *local* invariant I concerns only variables in V_1 (resp. V_2) and relates to states of S_1 (resp. S_2). A *global* invariant I' is defined over $V_1 \cup V_2$ and expresses the authorized states of $S_1 \times_{syn} S_2$.

3.1   Composing Local Invariants

In this section, we present some simple compositional results which are quite standard and trivially follow from the definition of the constraint synchronized product.

Theorem 1 states that a local invariant related to component's states, is preserved on the whole system thanks to the constraint synchronized product. The idea is the following. Local invariant concerns only variables assigned by the component. The only manner to modify the variables'values – in the component or in the whole system – is to take the component's transitions into account. This way, we can state a theorem about the preservation of local invariants for the whole system.

Theorem 1 (Preserving Local Invariant). *Let S_1 over V_1 and S_2 over V_2 be two components. Let syn be their synchronization set. Let $I_1 \in SP_{V_1}$ be a S_1 local invariant. Then we have*

$$\frac{S_1 \models I_1}{S_1 \times_{syn} S_2 \models I_1}$$

Proof by contradiction using Definitions 2, 4 and 5, and the predicate calculus.

Generally when several invariants are verified on the whole system, they cannot be composed: if $S_1 \models I_1$ and $S_2 \models I_2$, we can only conclude that $S_1 \| S_2 \models I_1 \vee I_2$. In the case of our constraint synchronized product, S_1 (resp. S_2) assigns values to V_1 (resp. $V2$) variables and the local invariant I_1 (resp. I_2) relates to only these variables. This is why the stronger predicate $I_1 \wedge I_2$ is an invariant for $S_1 \times_{syn} S_2$.

Theorem 2 (Composing Local Invariants). *Let S_1 over V_1 and S_2 over V_2 be two components. Let syn be their synchronization set. Let $I_1 \in SP_{V_1}$ and $I_2 \in SP_{V_2}$ be respectively S_1 and S_2 local invariants. Then we have*

$$\frac{S_1 \models I_1, \; S_2 \models I_2}{S_1 \times_{syn} S_2 \models I_1 \wedge I_2}$$

Proof by Theorem 1 and the $\wedge$-introduction rule of the sequent calculus.

Theorems 1 and 2 provide a method to verify local invariants in a compositional manner without building the whole system. The constraint synchronized product of n components being defined in [13], both theorems can be given for n components.

For our running example, the invariant $I_{Card} \stackrel{\text{def}}{=} (cs = unknown \Rightarrow cp = output)$ expresses that if the card status is unknown, then the card is ejected, whereas the invariant $I_{Device} \stackrel{\text{def}}{=} (de = progress \Rightarrow ds \in \{sender, reader\})$ expresses that the device does not abort when it communicates with other devices. It is easy to check that I_{Card} holds on $Card_A$, and I_{Device} holds on $Device_A$. By Theorem 2 $I_{Card} \wedge I_{Device}$ holds on $Protocol_A$.

3.2   Verifying Global Invariants

We show in this section that global invariants can be checked compositionally too. A global invariance property holding on all expanded components of a system, holds on the whole system.

Theorem 3 (Verifying Global Invariants). *Let S_1 over V_1 and S_2 over V_2 be two components. Let syn be their synchronization set. Let $I \in SP_{V_1 \cup V_2}$ be a global invariant over $V_1 \cup V_2$. Then we have*

$$\frac{S_1^c \models I, \ S_2^c \models I}{S_1^c \uplus S_2^c \models I}$$

Proof by contradiction using Equality 1, Definition 2 and Definition 7.

Theorem 3 allows us to verify a global invariant of the protocol T=1. $I_{Proto1} \stackrel{\text{def}}{=}$ $(cp = output \Rightarrow ds \in \{unknown, sender\})$ expresses that if ICC is ejected, then IFD is aborted or reinitialised. $I_{Proto2} \stackrel{\text{def}}{=} (de = progress \Rightarrow cp = input)$ expresses that ICC cannot be ejected when IFD communicates with other devices, and $I_{Proto3} \stackrel{\text{def}}{=} (ds \neq unknown \Rightarrow ((ds = sender \wedge cs = reader) \vee (ds = reader \wedge cs = sender)))$ expresses that there is an alternation of messages (in fact, an alternation of status) between ICC and IFD. These global invariants hold on $Card_A^c$ and on $Device_A^c$, so they are valid for $Protocol_A$.

In [4], a compositional verification approach of some CTL properties using a model like expanded components, is proposed. These properties are viewed as universal or existential properties and rules for their compositional verification are given. Notice that the composition operator in [4] and our constraint synchronized product are different because of synchronized transitions. Moreover, in our approach checking global invariants can be done during a model exploration of all the expanded components to verify the refinement. This is the matter of the next section.

4   Invariants Verification through Refinement

In this paper, we deal with specification and verification of component-based finite state systems supporting a top-down refinement paradigm. The refinement relation we consider in this paper, is inspired by the syntactic and semantic concepts of the B refinement method [1, 2]. More, our refinement paradigm guarantees preservation of the abstract system $PLTL$ properties for the refined system as shown in [11].

4.1   Basic Transition Systems Refinement

In [8], the refinement semantics has been expressed as a relation between transition systems. In this section, we give some basic definitions about LTSs refinement.

Let $SA = \langle\, Q_A, Q_{0A}, E_A, T_A, l_A\,\rangle$ be an abstract LTS over V_A and $SR = \langle\, Q_R, Q_{0R}, E_R, T_R, l_R\,\rangle$ a concrete LTS over V_R. The syntactic features of the refinement are as follows. First, $E_A \subseteq E_R$. Second, $V_A \cap V_R = \varnothing$. Third, we have to define a gluing predicate gp expressing how the variables of both LTSs are linked. Let $AP_{V_A \cup V_R}$ be the set of atomic propositions over V_A and V_R, and let $SP_{V_A \cup V_R}$ be the set of state propositions over V_A and V_R. In this setting, gp is a propositional calculus formula over $SP_{V_A \cup V_R}$. We define a binary relation $\mu \subseteq Q_R \times Q_A$ expressing the relation between states of two LTSs[1].

Definition 8 (Glued states). *Let $gp \in SP_{V_A \cup V_R}$ be a gluing predicate. The state $q_R \in Q_R$ is glued to $q_A \in Q_A$ w.r.t. gp, written $q_R\,\mu\,q_A$, iff $(q_A, q_R) \models gp$.*

The semantic features of the refinement are the following.

1. The transitions of S_R, the labels of which are in E_A (i.e. labelled by the "old" labels) are kept. New transitions introduced during the refinement design (i.e. labelled in $E_R \setminus E_A$) are considered as being non-observable; they are labelled by τ and called τ-transitions. Each portion of path containing τ-transitions must end by a transition labelled in E_A (see Fig. 7).

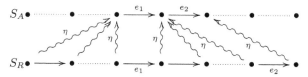

Fig. 7. Path refinement

2. In order to avoid new live-locks, new transitions should not take control forever. So, paths containing infinite sequences of τ-transitions are forbidden.
3. Moreover, new transitions should not introduce deadlocks.

Definition 9 (Strict Refinement Relation [8]). *Let SA and SR be LTSs, and $e \in E_A$. We define the refinement relation η as the greatest binary relation included in μ and satisfying the following conditions:*
1) strict transition refinement
$(q_R\,\eta\,q_A \;\wedge\; q_R \xrightarrow{e} q'_R \in T_R) \Rightarrow \exists q'_A.\,(q_A \xrightarrow{e} q'_A \in T_A \;\wedge\; q'_R\,\eta\,q'_A),$
2) stuttering transition refinement
$(q_R\,\eta\,q_A \;\wedge\; q_R \xrightarrow{\tau} q'_R \in T_R) \Rightarrow (q'_R\,\eta\,q_A),$
3) lack of τ-divergence
$q_R\,\eta\,q_A \Rightarrow \neg\,(q_R \xrightarrow{\tau} q'_R \xrightarrow{\tau} q''_R \xrightarrow{\tau} \ldots \xrightarrow{\tau} \ldots),$
4) lack of new deadlocks
$(q_R\,\eta\,q_A \;\wedge\; q_R \nrightarrow) \Rightarrow (q_A \nrightarrow)$ [2].

---

[1] In [8], we have defined μ to be a function by requiring the invariant $l_2(s_2) \wedge I_2 \Rightarrow l_1(s_1)$. Then, μ^{-1} gives a partition of the state space of refined systems allowing a verification by parts [17].

[2] We note $q \nrightarrow$ when $\forall q', e.\,(q' \in Q \wedge e \in E \Rightarrow q \xrightarrow{e} q' \notin T).$

We say that SR *refines* SA, written $SR \sqsubseteq SA$, when $\forall q_R. \, (q_R \in Q_R \Rightarrow \exists q_A. \, (q_A \in Q_A \wedge q_R \, \eta \, q_A))$.

It has been shown in [8] that η is a kind of τ-simulation. It is well-known that a simulation can be computed iteratively for finite state systems. We have an algorithm based on a depth-first search enumeration of the reachability graph of the refined system. Its complexity order is $O(|SR|)$ where $|SR| = |Q_R| + |T_R|$. However, it is well-known that the algorithmic verification quickly meets its limits when applied to huge systems. Therefore, we have to face the problem of combinatorial explosion during refinement verification. Details introduced by refinement tend to drastically increase the number of states of the system. Compositional approaches partially postpone this problem.

4.2   Compositional Component-Based Systems Refinement

In [13] we conciliate the synchronized component-based specification with the refinement verification. In this section we briefly recall a refinement relation for synchronized component-based systems.

Definition 10 (Weak Refinement Relation). *Let SA and SR be two LTSs, and $e \in E_A$. Let $D \subseteq Q_R$ (initially $D = \varnothing$) be the set of new deadlocks. We define the weak refinement relation η_f as the greatest binary relation included in μ and satisfying conditions 1), 2) and 3) of Definition 9 and the following condition:*
4') old or new deadlocks
$$(q_R \, \eta_f \, q_A \wedge q_R \nrightarrow) \Rightarrow ((q_A \nrightarrow) \vee (q_A \xrightarrow{e} q'_A \in T_A \Rightarrow q_R \in D)).$$

We say that SR *weakly refines* SA, written $SR \sqsubseteq_D SA$, when $\forall q_R.(q_R \in Q_R \Rightarrow \exists q_A. \, (q_A \in Q_A \wedge q_R \, \eta_f \, q_A))$. This relation is weaker than the refinement relation defined in [8]. It is easy to see that

$$(SR \sqsubseteq SA) \Leftrightarrow (SR \sqsubseteq_D SA \wedge D = \varnothing) \tag{2}$$

Moreover, in [13], we give a theorem ensuring refinement of the whole system from weak refinements of its expanded components, and vice versa. In this paper, we express this result using a deadlock reduction. The idea is the following. Some deadlocks result from the omitted details when building the expanded components. A state inducing a new deadlock in an expanded component does not induce a deadlock in the whole system if there is an expanded component where this state is not a deadlock state.

Definition 11 (Reduced Deadlocks). *Let SA_1, SR_2, SR_1 and SR_2 be respectively two abstract and two refined components. Let syn and syn' be two respective synchronization sets. We have $SR_1^c \sqsubseteq_{D_1} SA_1^c$. The reduced deadlock set RD_1 corresponding to D_1 is defined by:*
$$RD_1 \overset{def}{=} D_1 \smallsetminus \{q \mid q \in D_1 \wedge \exists e_2. \, (e_2 \in E_2 \wedge ((-, e_2) \text{ when } p) \in syn' \wedge q \models p)\}$$

Notice that this deadlock reduction is independent of the other expanded components. The reduced deadlock set RD_2 corresponding to D_2 can be computed similarly. If RD_1 and RD_2 are empty, then $RD_1 \cup RD_2$ is also empty and, by Equality 2, the strict refinement of the whole system is ensured.

Theorem 4 (Compositional Refinement). *Let SA_1, SA_2, and SR_1, SR_2 be respectively two abstract and two refined components. Let syn and syn' be two respective synchronization sets. Then we have*

$$\frac{SR_1^c \sqsubseteq_{D_1} SA_1^c, \ RD_1 = \varnothing, \ SR_2^c \sqsubseteq_{D_2} SA_2^c, \ RD_2 = \varnothing}{SR_1 \times_{syn'} SR_2 \sqsubseteq SA_1 \times_{syn} SA_2}$$

This theorem can be given for n components. It provides a compositional refinement verification algorithm based on the computation, for each refined expanded component SR_i^c, of the relation η_f. The complexity order of this refinement verification algorithm is $O(|SR_1^c| + |SR_2^c| + \cdots + |SR_n^c|)$. However, the greatest memory space used by the computation is $max(|SR_1^c|, |SR_2^c|, \ldots, |SR_n^c|)$, at most. Indeed, the building of an expanded component, the weak refinement verification and the deadlock reduction can be done sequentially for each component. Thus, this memory space is, in the worst case, the one of the constraint synchronized product.

4.3 Invariance Properties Verification

In general, behavioural properties established for an abstract system are not preserved when the system is refined to a richer level of details. It is not the case in our approach. Indeed, we show in [11] that our refinement relation preserves the abstract system's $PLTL$ properties. This way, an algorithmic verification of the refinement by model exploration can be associated with a $PLTL$ properties verification by model-checking. Invariance properties can be expressed by $PLTL$ properties like $\square sp$, where sp is a state proposition. Our component-based refinement ensures preservation of this kind of abstract properties for the refined system (see Fig. 1).

Definition 12 (sp is preserved). *Let $q_A \in Q_A$ and $q_R \in Q_R$ be two states. Let $gp \in SP_{V_A \cup V_R}$ be a gluing predicate. We define a state proposition $sp \in SP_{V_A}$ to be preserved for q_R, written $q_R \models_{gp} sp$, iff $q_A \models sp$ and $q_R \mu q_A$.*

It's easy to see that the following holds.

Proposition 1 (Abstract Invariant Preservation). *Let SA and SR be two LTSs. Let $gp \in SP_{V_A \cup V_R}$ be their gluing predicate. Let I_A be from SP_{V_A}. If $SR \sqsubseteq SA$ and $SA \models I_A$ then $SR \models_{gp} I_A$.*

Invariants checking can be done during the refinement verification step without increasing complexity of the corresponding algorithm. Actually, the ideas are as follows.

– Local invariant can be checked on-the-fly while building the corresponding expanded component. By Theorems 2, 4 and Proposition 1 we also have the following preservation result

$$\frac{SR_1 \times_{syn_R} SR_2 \sqsubseteq SA_1 \times_{syn_A} SA_2,\ SA_1 \models I_1,\ SA_2 \models I_2}{SR_1 \times_{syn_R} SR_2 \models_{gp} I_1 \wedge I_2}$$

– Global invariant can be checked during the model exploration step of the weak refinement verification for each expanded component. By Theorems 3, 4 and Proposition 1, we obtain

$$\frac{SR_1 \times_{syn_R} SR_2 \sqsubseteq SA_1 \times_{syn_A} SA_2,\ SA_1^c \models I,\ SA_2^c \models I}{SR_1 \times_{syn} SR_2 \models_{gp} I}$$

To summarise, abstract system invariance properties can be used as hypotheses to ensure their preservation for the refined system. Moreover, they can be expressed on V_R while refining these properties like in [7].

4.4 T=1 Protocol Refinement

This section illustrates the refinement by mean of the protocol T=1. Suppose that different blocks are exchanged between the ICC and the IFD: I-blocks for each piece of Information, R-blocks for Ready Acknowledgement blocks, and S-blocks for Supervisory Transmission control blocks. We assume that information to be sent is divided into 64 I-blocks (LEN). For every I-block, its reception is acknowledged by a R-block sent as a response. The S-block is sent when all I-blocks have been sent. The IFD sends all 64 I-blocks before the ICC sends its I-blocks.

The ICC is refined by the $Card_R$ component. Variables cs and cp are refined, and two new variables cir and cis are introduced. Here cir is a Card I-block Received counter whereas cis is a Card I-block Sent counter. We observe the begin of insertion or ejection of ICC (stInsertC and stEjectC). Four new transitions (receiveIBC, receiveRBC, sendIBC and sendRBC) are introduced to receive or to send I-blocks and R-blocks. Figure 9 shows a part of $Card_R$, and the refinement relation $relRef$ between $Card_R$

	$\|Q\|$	$\|T\|$	$\|Var\|$
$Card_A$	3	5	2
$Device_A$	5	9	2
$Card_A^c$	8	5	4
$Device_A^c$	8	9	4
$Protocol_A$	8	12	4
$Card_R$	774	1032	4
$Device_R$	645	1161	4
$Card_R^c$	774	1032	8
$Device_R^c$	1032	1161	8
$Protocol_R$	1032	1548	8

Fig. 8. Protocol T=1

and $Card_A$. The entire system $Card_R$ cannot be drawn here because of its size (see Fig 8).

In the same manner, IFD is refined to obtain LTS $Device_R$. Then, a refined synchronization set syn_R and two gluing predicates gp_{Card} and gp_{Device} are given to build the refined expanded components $Card_R^c$ and $Device_R^c$. We have $Protocol_R = Card_R \times_{syn_R} Device_R = Card_R^c \uplus Device_R^c$. The weak refinement relation is computed between $Card_R^c$ and $Card_A^c$, and between $Device_R^c$ and $Device_A^c$. Deadlock sets are reduced. Finally, we conclude using Theorem 4 that $Protocol_R$ refines $Protocol_A$.

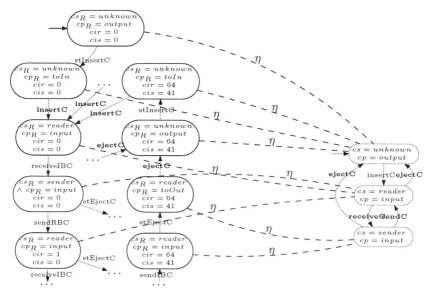

Fig. 9. Protocol T=1 component $Card_R \sqsubseteq Card_A$

All invariance properties which have been verified compositionally on the abstract system $Protocol_A$ (see Section 3), are preserved on the refined system $Protocol_R$. Since $Protocol_R \sqsubseteq Protocol_A$, local invariants I_{Card} and I_{Device} and global invariants I_{Proto1}, I_{Proto2} and I_{Proto3} are automatically verified by preservation on $Protocol_R$. Moreover, the invariant $I_{Card_R} \equiv (cs_R = unknown \Rightarrow cp_R \in \{output, toIn\})$ can be proved trivially using I_{Card} and gp_{Card} as hypotheses.

5   Conclusion and Perspectives

Compositional design and refinement paradigms are very convenient to specify and to verify large embedded reactive systems such as automatic trains, car driving assistance systems or communication protocols.

On one hand, a compositional specification method allows us to ensure refinement of a whole system from weak refinements of its expanded components, and vice versa. Although expanded components seem comparable with the whole system, they are generally smaller than the whole system (see Fig. 8, Fig. 10 and Fig. 11). Notice that the size of expanded components depends on both synchronous and asynchronous transitions from the synchronization set. On the other hand, we want to postpone the model-checking blow-up by verifying invariance properties in a compositional way.

The refinement relation provides a formal framework for verifying reactive systems. When the refinement holds, most of the properties that have been verified on the abstract model are preserved by the refined one (see Fig. 1).

In this paper, we propose a technique to take advantages of the constraint synchronized product to verify invariance properties. We give three theorems allowing us to verify local invariants or global invariants, in a compositional way. Our refinement paradigm is close to the B event systems one [1, 2]. In the B method, refinement is verified by proof. Safety properties, expressed by an invariant, are also verified by proof.

One of the advantages of our approach is that the compositional refinement relation can be verified for finite state systems using an algorithmic method. Therefore, this allows verifying both refinement and invariants by model exploration. Another advantage of our method is the invariance properties preservation through refinement.

	$|Q|$	$|T|$	$|Var|$
$Clip_A$	2	2	1
$Lift_A$	3	4	1
$Handle_A$	2	2	1
$Clip_A^c$	9	7	3
$Lift_A^c$	4	5	3
$Handle_A^c$	9	7	3
$Robot_A$	9	15	3
$Clip_R$	4	3	1
$Lift_R$	7	8	1
$Handle_R$	4	4	1
$Clip_R^c$	19	21	3
$Lift_R^c$	11	12	3
$Handle_R^c$	19	20	3
$Robot_R$	24	39	3

Fig. 10. Robot

An analysis tool related to our compositional refinement verification has been implemented [15][3]. This tool, called *SynCo*, allows us to measure the efficiency of our approach. Component specifications are given in the syntax of Fair Transition Systems (FTS[4]). Our tool builds the expanded components and verifies sequentially their weak refinement. During components and expanded components building, local and global invariants are checked without increasing the algorithm's complexity. When the refinement is not verified, our tool finds the origin of the error. As far as we know, there are other compositional tools implementing reachability compositional verification [16] or invariant and trace refinement checking [3]. However, they are designed for a very different purpose than *SynCo* and, thus, do not combine compositional refinement with compositional invariant verification and preservation.

	$|Q|$	$|T|$	$|Var|$
$Control_A$	3	6	1
$Left_A$	2	2	1
$Right_A$	2	2	1
$Sensor_A$	4	10	1
$Control_A^c$	5	10	4
$Left_A^c$	7	6	4
$Right_A^c$	7	6	4
$Sensor_A^c$	5	14	4
$Wipers_A$	8	22	4
$Control_R$	4	9	2
$Left_R$	5	6	2
$Right_R$	5	6	2
$Sensor_R$	6	16	1
$Control_R^c$	8	17	7
$Left_R^c$	19	20	7
$Right_R^c$	19	20	7
$Sensor_R^c$	10	25	7
$Wipers_R$	20	47	7

Fig. 11. Wipers

We want to extend our compositional framework to take some *PLTL* properties'patterns into account. It could be done in the same manner than in [4]. More generally, we intend to exploit works on a verification by parts [17] and dynamic property refinement [7] in the framework of the component-based system development presented in this paper. We hope to extend the class of properties that can be verified in a compositional way.

References

1. J.-R. Abrial. *The B Book*. Cambridge University Press - ISBN 0521-496195, 1996.
2. J.-R. Abrial and L. Mussat. Introducing dynamic constraints in B. In *Second Conference on the B method, France*, volume 1393 of *LNCS*, pages 83–128. Springer Verlag, April 1998.
3. R. Alur, L. de Alfaro, R. Grosu, T.A. Henzinger, M. Kang, R. Majumdar, F. Mang, C.M. Kirsch, and B.Y. Wang. Mocha: A model checker that exploits design structure. In *23rd International Conference on Software Engineering (ICSE'01)*, may 2001.

---

[3] See for more informations http://lifc.univ-fcomte.fr/~synco.

[4] FTS is one of the input language of the Standford Temporal Prover (STeP).

4. H. A. Andrade and B. Sanders. An approach to compositional model checking. In *International Parallel and Distributed Processing Symposium (IPDPS'02) Workshops: FMPPTA'2002*, Fort Lauderdale, Florida, April 2002. IEEE.
5. A. Arnold. *Systèmes de transitions finis et sémantique des processus communicants.* Collection Etudes et Recherches en Informatiques. Masson, Paris, 1992.
6. A. Arnold and M. Nivat. Comportements de processus. In *Actes du Colloque AFCET - Les Mathématiques de l'Informatique*, pages 35–68, 1982.
7. F. Bellegarde, C. Darlot, J. Julliand, and O. Kouchnarenko. Reformulation: a way to combine dynamic properties and B refinement. In *In Proc. Int. Conf. Formal Method Europe'01, Berlin, Germany*, volume 2021 of *LNCS*, pages 2–19. Springer Verlag, March 2001.
8. F. Bellegarde, J. Julliand, and O. Kouchnarenko. Ready-simulation is not ready to express a modular refinement relation. In *Proc. Int. Conf. on Fundamental Aspects of Software Engineering, FASE'2000*, volume 1783 of *LNCS*, pages 266–283, April 2000.
9. Edmund M. Clarke, Orna Grumberg, and Doron A. Peled. *Model Checking.* The MIT Press, 2000.
10. J.-M. Cobleigh, D. Giannakopoulou, and C. Pasareanu. Learning assumptions for compositional verification. In *9th International Conference on Tools and Algorithms for the Construction and Analysis of Systems (TACAS 2003)*, volume 2619 of *LNCS*, Warsaw, Poland, April 2003. Springer-Verlag.
11. C. Darlot, J. Julliand, and O. Kouchnarenko. Refinement preserves PLTL properties. In D. Bert, J. P. Bowen, S. C. King, and M. Walden, editors, *ZB'2003: Formal Specification and Development in Z and B*, volume 2651 of *LNCS*, Turku, Finland, June 2003.
12. European Normalisation Committee. En27816-3. European standard - identification cards - integrated circuit(s) card with contacts - electronic signal and transmission protocols. Technical Report ISO/CEI 7816-3, 1992.
13. O. Kouchnarenko and A. Lanoix. Refinement and verification of synchronized component-based systems. In K. Araki, S. Gnesi, and Mandrioli D., editors, *Formal Methods 2003 (FM'03)*, volume 2805 of *LNCS*, pages 341–358, Pisa, Italy, September 2003. Springer-Verlag. extended version published as Research Report [14].
14. O. Kouchnarenko and A. Lanoix. Refinement and verification of synchronized component-based systems. INRIA Research Report 4862, June 2003.
15. O. Kouchnarenko and A. Lanoix. SynCo: a refinement analysis tool for synchronized component-based systems. In T. Margaria, editor, *FM'03 Tool Exhibition Notes*, pages 47–51, Pisa, Italy, September 2003.
16. J. Lind-Nielsen, H. R. Andersen, H. Hulgaard, G. Behrmann, K. Kristoffersen, and K. G. Larsen. Verification of large state/event systems using compositionality and dependency analysis. *Formal Methods in System Design*, 18(1):5–23, January 2001.
17. P.-A. Masson, H. Mountassir, and J. Julliand. Modular verification for a class of PLTL properties. In T. Santen W. Grieskamp and B. Stoddart, editors, *2nd international conference on Integrated Formal Methods, IFM 2000*, volume 1945 of *LNCS*, pages 398–419, November 2000.
18. K. L. McMillan. A methodology for hardware verification using compositional model-checking. *Science of Computer Programming*, (37):279–309, 2000.
19. Y.-K. Tsay. Compositional verification in linear-time temporal logic. In J. Tiuryn, editor, *Proc. 3rd Int. Conf. on Foundations of Software Science and Computation Structures (FOSSACS 2000)*, volume 1784 of *LNCS*, pages 344–358. Springer Verlag, 2000.

Modelling Concurrent Interactions

Juliana Küster-Filipe*

Laboratory for Foundations of Computer Science
School of Informatics, University of Edinburgh
King's Buildings, Edinburgh EH9 3JZ, UK
jkfilipe@inf.ed.ac.uk

Abstract. In UML 2.0 sequence diagrams have been considerably extended but their expressiveness and semantics remains problematic in several ways. In other work we have shown how sequence diagrams combined with an OCL liveness template gives us a much richer language for inter-object behaviour specification. In this paper, we give a semantics of these enriched diagrams using labelled event structures. Further, we show how sequence diagrams can be embedded into a true-concurrent two-level logic interpreted over labelled event structures. The top level logic, called *communication* logic, is used to describe inter-object specification, whereas the lower level logic, called *home logic*, describes intra-object behaviour. An interesting consequence of using this logic relates to how state-based behaviour can be synthesised from inter-object specifications. Plans of extending the Edinburgh Concurrency Workbench in this context are discussed.

1   Introduction

One of the major changes made to UML 2.0 with respect to its previous versions concerns sequence diagrams which have been extended to include a number of features borrowed from Message Sequence Charts (MSCs)[11] and, to a limited extent, Live Sequence Charts (LSCs)[4]. As a consequence, UML's sequence diagrams are now more expressive and fundamentally better structured. However, there are still several problems with their informal description in the UML 2.0 specification [8].

A major change in sequence diagrams is that interactions can be structured using so-called interaction fragments. There are several possible fragments, for example, **alt** (alternative behaviour), **par** (parallel behaviour), **neg** (forbidden behaviour), **assert** (mandatory behaviour - though we will mention some ambiguities in the specification concerning this fragment), and so on. Compared to LSCs, sequence diagrams in UML 2.0 can still not adequately distinguish between mandatory and possible behaviour. For instance, it is still not possible to distinguish between a message that if sent *may* or *must* be received, or to enforce progress of an instance along its lifeline. To address this limitation we have proposed in [2] to enrich a sequence diagram with liveness constraints expressed

---

* Work reported here was supported by the EPSRC grant GR/R16891.

C. Rattray et al. (Eds.): AMAST 2004, LNCS 3116, pp. 304–318, 2004.
© Springer-Verlag Berlin Heidelberg 2004

in UML's Object Constraint Language (OCL) using an OCL template defined in [1]. The template is written in the context of a Classifier or Type (as OCL constraints generally are) and consists essentially of an **after:** clause followed by an **eventually:** clause. The idea is that whenever the **after** clause holds, at some point in the future the **eventually** clause *must* hold.

The contribution of the present paper is twofold: on the one side we provide a semantics to UML 2.0 sequence diagrams (as well as liveness enriched sequence diagrams), and on the other side provide a means for reasoning about the specified inter-object behaviour. We also envisage verification of scenario-based inter-object behavioural models with respect to state-based behavioural models. For this purpose we are currently extending the Edinburgh Concurrency Workbench (CWB)[1]. We discuss the extension at the end of the paper.

In this paper, we give a semantics to sequence diagrams using labelled event structures [10]. We show how to construct such a model from a sequence diagram. Event structures allow us to describe distributed computations as event occurrences together with relations for expressing causal dependency, called *causality*, and nondeterminism, called *conflict*. In addition, from these relations a further relation for denoting *concurrency* is derived (events not related by causality or conflict are necessarily concurrent). The causality relation implies a (partial) order among event occurrences, while the conflict relation expresses how the occurrence of certain events excludes the occurrence of others. Essentially, event structures constitute a simple and very natural model to capture the behaviour specified in a sequence diagram. Further information from a sequence diagram is attached to the formal model in the form of labels (messages, state invariants, etc.).

Further, we also show how a sequence diagram can be specified as a collection of formulae in a true-concurrent two-level logic interpreted over labelled event structures. The top level logic, called *communication* logic, is used to describe inter-object specification. It can be understood as a way to model an observer of the interaction, for example that whenever a message is sent it is always eventually received; or that certain interactions are happening concurrently. By contrast, the lower level logic, called *home logic*, describes intra-object behaviour. It can be used to capture local state invariants, interaction constraints, and the interaction from a local perspective. Additionally, OCL liveness constraints are translated into our communication/home logic depending on whether they correspond to a global (observer viewpoint) or a local (instance viewpoint) constraint.

The paper is structured as follows. In Section 2, we give an overview of sequence diagrams in UML 2.0. In Section 3 we introduce our underlying semantic model, namely labelled event structures, and show how to build a model for a given sequence diagram. In Section 4, we describe a simple distributed concurrent logic for reasoning about the specified inter-object behaviour and imposing further interaction constraints. The paper finishes with some concluding remarks and ideas for future work.

---

[1] For information on the tool see http://www.lfcs.inf.ed.ac.uk/cwb/

2   Sequence Diagrams in UML 2.0

Graphically, a sequence diagram has two dimensions: the horizontal dimension represents the instances participating in the scenario; the vertical dimension represents time. Objects have a vertical dashed line called *lifeline*. The lifeline represents the existence of the instance at a particular time; the order of events along a lifeline is significant denoting the order in which these events will occur.

A *message* is a communication between two instances that conveys information with the expectation that action will ensue. A message will cause an operation to be invoked, a signal to be raised, or an instance to be created or destroyed. Messages are shown as a horizontal arrows from the lifeline of one instance to the lifeline of another instance. A message specifies not only the kind of communication between instances, but also the sender and receiver event occurrences associated to it.

UML 2.0 sequence diagrams may contain sub-interactions called *interaction fragments* which can be structured and combined using *interaction operators*. The semantics of the resulting combined fragment depends upon the operator and is described informally in the UML 2.0 superstructure specification [8]. Below we give the meaning of some operators used in this paper as defined in [8]:

alt designates that the fragment represents a choice of behaviour. At most one of the operands will execute. The operand that executes must have a guard expression that evaluates to true at this point in the interaction.

par designates that the fragment represents a parallel merge between the behaviours of the operands. The event occurrences of the different operands can be interleaved in any way as long as the ordering imposed by each operand as such is preserved.

seq designates that the fragment represents a weak sequencing between the behaviours of the operands, i.e. the ordering of event occurrences within each of the operands are maintained whereas event occurrences on different lifelines from different operands may come in any order, and event occurrences on the same lifeline from different operands are ordered such that an event occurrence of the first operand comes before that of the second operand.

neg designates that the fragment represents traces that are defined to be invalid. All interaction fragments that are different from negative are considered positive meaning that they describe traces that are valid and should be possible.

assert designates that the fragment represents an assertion. The sequences of the operand of the assertion are the only valid continuations.

We borrow some concepts introduced in LSCs that are important to derive a semantic model, but are currently missing in sequence diagrams. The lifeline of an instance consists of several points called *locations* corresponding to the occurrence of events. All instances have at least three locations: an initial location, a location corresponding to the beginning of the main chart (the whole diagram for sequence diagrams), and a final location. Locations are also associated with the sending and receiving of messages, the beginning and the end of subcharts (interaction fragments for sequence diagrams), and the evaluation of a condition or an assignment. The locations along a single lifeline are ordered top-down; therefore, a *partial order* is induced among locations determining the order of execution.

Another notion important for LSCs is *temperature*. Every element in an LSC has a temperature which can be either hot or cold. This is used to distinguish between possible (cold) and mandatory (hot) elements and behaviour. An element can be a location, a message and a subchart. Consequently, if a location is hot/cold it must/may be reached; if a message is hot/cold it must/may be received after it has been sent; and if a subchart is hot/cold it describes a collection of interactions that must/may happen.

Sequence diagrams can only express the possibility that a certain scenario occurs. That is, sequence diagrams model behaviour in the form of *possible* interactions, i.e. communication patterns that *may* occur between a set of instances. Furthermore, sequence diagrams, in their current setting, seem to be able to express *necessity* only to a very limited extent. In particular, it is not clear whether the intention of the new operator called **assert** is to specify mandatory behaviour. The superstructure specification is ambiguous in the definition of this operator, and it is not obvious from the text whether this operator enforces a sequence of messages to happen, or they are "expected" to happen (see [8], pages 412, 442). However, even if the former case were the intended one, it would still only solve the problem of expressing necessity in sequence diagrams at the interaction level, but not at the local level of a single message or location. Messages and locations in sequence diagrams are implicitly cold. If sent a message *may* be received, but it does not have to be. Similarly, any location in a sequence diagram *may* be reached but it does not have to be. This reflects that an instance is not actually forced to progress along its lifeline.

To address the important dichotomy between *must* and *may* we have shown in [2] how to achieve necessity, both at the global and the local level, using an extension of OCL for liveness. The idea is that by default a sequence diagram only reflects possible behaviour (except for the **assert** operator) or possible but forbidden behaviour (given by the **neg** operator). To impose additionally that a location must be reached or a message must be received, we have to enrich the model with corresponding liveness constraints written in an appropriate OCL template. A more powerful version of the template can also be used to express global liveness. For instance, that after a sequence of interactions has occurred, another sequence of interactions must occur. We omit further details on the OCL constraints in this paper as they are not essential. It suffices to understand that the OCL liveness constraints change the temperature of associated locations from cold to hot.

UML 2.0 provides different kinds of conditions in sequence diagrams.

1. An *interaction constraint* is a boolean expression shown in square brackets covering the lifeline where the first event will occur, positioned above that event inside an interaction operand.

2. A *state invariant* is a constraint on the state of an instance, for example on the values of its attributes. The constraint is assumed to be evaluated during run time immediately prior to the execution of the next event occurrence. If the constraint is true the trace is a valid trace; otherwise, the trace is invalid. Notationally, state invariants are shown as a constraint inside a state symbol or in curly brackets, and are placed on a lifeline.

Finally, in previous versions of UML it was not possible to express that, at any time, a specific scenario should not occur. However, as mentioned at the beginning of this section, a new operator called **neg** has been introduced for this purpose in UML 2.0. Therefore, to model what is called an anti-scenario in LSCs, we simply place it inside an interaction fragment within the scope of a **neg**. Our semantics is only defined for sequence diagrams which do not contain a **neg** interaction fragment. The reason for this is that there is no real need to use such an interaction fragment to indicate forbidden behaviour. For example, as done in LSCs, it suffices to use a false state invariant to indicate that a particular scenario is not allowed. We show how to capture undesired behaviour as logical formulae.

3   The Model

3.1   Basic Definitions

We recall some basic notions on the model we use, namely *labelled prime event structures* [10].

Prime event structures, or event structures for short, allow the description of distributed computations as event occurrences together with relations for expressing causal dependency and nondeterminism. The first relation is designated *causality*, and the second *conflict*. The causality relation implies a (partial) order among event occurrences, while the conflict relation expresses how the occurrence of certain events excludes the occurrence of others. Consider the following definition of event structures as in [3].

Event Structure. A *event structure* is a triple $E = (Ev, \rightarrow^*, \#)$ where Ev is a set of events and $\rightarrow^*, \# \subseteq Ev \times Ev$ are binary relations called *causality* and *conflict*, respectively. Causality $\rightarrow^*$ is a partial order. Conflict $\#$ is symmetric and irreflexive, and propagates over causality, i.e., $e \# e' \rightarrow^* e'' \Rightarrow e \# e''$ for all $e, e', e'' \in Ev$. Two events $e, e' \in Ev$ are *concurrent*, $e \ co \ e'$ iff $\neg(e \rightarrow^* e' \vee e' \rightarrow^* e \vee e \# e')$.

From the two relations defined on the set of events, a further relation is derived, namely the *concurrency* relation co. As stated, two events are concurrent iff they are completely unrelated, i.e., neither related by causality nor by conflict. Moreover, an event structure is called *sequential* if the concurrency relation co is empty.

In our approach to inter-object behaviour specification, we will consider a restriction of event structures sometimes referred to as *discrete* event structures. An event structure is said to be *discrete* if the set of previous occurrences of an event is finite.

Discrete Event Structure. Let $E = (Ev, \rightarrow^*, \#)$ be an event structure. E is a *discrete event structure* iff for each event $e \in Ev$, the *local configuration* of e given by $\downarrow e = \{e' \mid e' \rightarrow^* e\}$ is finite.

The finiteness assumption of the so-called local configuration is motivated by the fact that system's computations always have a starting point, which means that any event in a computation can only have finitely many previous occurrences.

Consequently, we are able to talk about immediate causality in such structures. Two events are related by *immediate* causality if there are no other event occurrences in between. Formally, if $\forall_{e'' \in Ev} (e \rightarrow^* e'' \rightarrow^* e' \Rightarrow (e'' = e \vee e'' = e'))$ holds. If $e \rightarrow^* e'$ are related by immediate causality then e is said to be an *immediate predecessor* of e' and e' is said to be an *immediate successor* of e. We may write $e \rightarrow e'$ instead of $e \rightarrow^* e'$ to denote immediate causality. Furthermore, we also use the notation $e \rightarrow^+ e'$ whenever $e \rightarrow^* e'$ and $e \neq e'$.

Hereafter, discrete event structures are designated *event structures* for short.

Configuration. Let $E = (Ev, \rightarrow^*, \#)$ be an event structure and $C \subseteq Ev$. C is a *configuration* in E iff it is both (1) conflict free: for all $e, e' \in C$, $\neg(e \# e')$, and (2) downwards closed: for any $e \in C$ and $e' \in Ev$, if $e' \rightarrow^* e$ then $e' \in C$.

A maximal configuration denotes a run. A run is sometimes called *life cycle*.

Finally, in order to use event structures to provide a denotational semantics to languages, it is necessary to link the event structures to the language they are supposed to describe. This is achieved by attaching a labelling function to the set of events. A generic labelling function is as defined next.

Labelling Function. Let $E = (Ev, \rightarrow^*, \#)$ be an event structure, and L be an arbitrary set. A *labelling function* for E is a total function $l : Ev \rightarrow L$ mapping each event into an element of the set L.

An event structure together with a labelling function defines a so-called labelled event structure.

Labelled Event Structure. Let $E = (Ev, \rightarrow^*, \#)$ be an event structure, L be a set of labels, and $l : Ev \rightarrow L$ be a labelling function for E. A *labelled event structure* is a pair $(E, l : Ev \rightarrow L)$.

Usually, events model the occurrence of actions, and a possible labelling function maps each event into an action symbol or a set of action symbols. We see next how to use event structures for sequence diagrams in UML 2.0 and what labelling function we need in this case.

3.2   Event Structures for Sequence Diagrams

Consider the sequence diagram in **Fig. 1** used here to illustrate our semantics for sequence diagrams.

We define the signature of a sequence diagram in UML 2.0.

Sequence Diagram. A sequence diagram is given by a tuple
$SD = (I, Loc, Loc_{ini}, Mes, E, Path, X_I)$ where

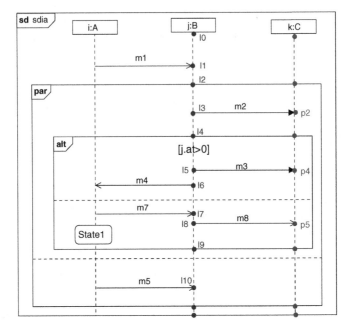

Fig. 1. An example of a sequence diagram in UML 2.0.

- I is a set of *instance identifiers* corresponding to the objects participating in the interaction described by the diagram;
- Loc is a set of *locations*;
- $Loc_{ini} \subseteq Loc$ is a set of *initial locations*;
- Mes is a set of *message labels*;
- $E \subseteq Loc \times Mes \times Loc$ is a set of *edges* where an edge (l_1, m, l_2) represents a message m sent from location l_1 to location l_2. E is such that:
  (i)$\forall_{e=(l_1,m,l_2)\in E} l_1 \neq l_2$ and
  (ii)$\forall_{e=(l_1,m,l_2)\in E} \neg \exists_{e'=(l'_1,m',l'_2)\neq e\in E} l'_1 = l_1 \vee l'_2 = l_2$;
- $Path$ is a given set of well-formed path terms for the diagram.
- $\{X_i\}_{i\in I}$ is a family of I-indexed sets of constraint symbols where $X = X_{Int} \cup X_{st}$.

The following table contains additional functions defined over SD and associated conditions.

$loc : I \rightarrow 2^{Loc}$	(1) $\forall_{i,j\in I, i\neq j} loc(i) \cap loc(j) = \emptyset$
	(2) $\forall_{i\in I} loc(i) \cap Loc_{ini} \neq \emptyset$
	(3) $\forall_{i\in I, l_1, l_2 \in loc(i)\cap Loc_{ini}} l_1 = l_2$
$time : Loc \rightarrow \mathbb{N}_0$	(4) $\forall_{l\in Loc_{ini}} time(l) = 0$
	(5) $\forall_{i\in I} \forall_{l_1, l_2 \in loc(i), l_1 \neq l_2} time(l_1) \neq time(l_2)$
$loc\_const : loc(i) \rightarrow 2^{\Phi(X_i)}$	
$scope : Loc \rightarrow Path$	(6) $\forall_{e=(l_1,m,l_2)\in E} scope(l_1) = scope(l_2)$
$temp : E \rightarrow Boolean$	
$comm\_synch : E \rightarrow Boolean$	

A few explanations regarding this definition. The conditions on edges state that (i) an edge cannot start and end at the same location, and (ii) a location can only be the source or target of at most one edge. The function loc associates to each instance a set of locations. According to condition (1), $loc(i)$ gives the locations along the lifeline of i and these are unique for i. Each instance in a diagram has at least one initial location in Loc_{ini} (condition (2)) and with condition (3) we further assume that each instance has a unique initial location corresponding to the start point of its lifeline[2].

The function $time$ associates to each location a natural number (according to its position along a lifeline in the diagram) and is assumed given. Initial locations have associated time value 0 (condition (4)). Further, all locations of a particular instance have necessarily different time values (condition (5)), but locations of different instances can still have the same time value. Notice that time here does not necessarily mean *occurrence time* (though within a **seq** interaction fragment it does) but an implicit *visual time* value according to the diagram. Such locations can still be concurrent (if they belong to different operands in a **par** fragment) or in conflict (if they belong to different operands in a **alt** fragment).

As we have mentioned, in a sequence diagram we may find simple constraints associated to locations, namely interaction constraints or state invariants. Further, these constraints are always local to a particular instance. Consequently, we introduced X_i as a set of constraint symbols local to $i \in I$ to be able to refer to such constraints (typically these constraint symbols will correspond to integer-typed attributes $X_{Int i}$ or state names $X_{st i}$). We assume that for a set of constraint symbols X_i, constraints $\varphi \in \Phi(X_i)$ are of the form

$$\varphi := x \leq c \mid c \leq x \mid x = c \mid x < c \mid c < x \mid y \mid \varphi \wedge \varphi$$

where $x \in X_{Int i}$, c is an integer constant, $y \in X_{st i}$ is a state name. The function loc_const associates to each location a(n ordered) set of constraints over X_i (if defined it is as given in the diagram). In the case of a location marking the beginning of an interaction fragment **alt** one constraint is implicitly associated to each possible operand. For example, location l_4 in **Fig. 1** has an associated set of constraints given by $loc\_const(l_4) = \{j.at > 0, j.at \leq 0\}$ where $j.at \in X_{Int j}$.

Further, we assume a function $scope$ that associates to each location in a diagram a path term. We do not define here the grammar for generating $Path$ terms. It suffices to understand that path terms are encoded in such a way that it is possible to distinguish between a location that is:

1. marking the beginning of an interaction fragment. Here a path term has the form $\alpha.par(n)$ for an interaction fragment **par** with $n \in \mathbb{N}$ operands. For example, for locations l_2, l_4 in **Fig. 1** we have $scope(l_2) = sdia.par(2)$, and $scope(l_4) = sdia.par(2)\#1.alt(2)$.

---

[2] This is not an essential condition, and could be dropped for describing agent/role interactions.

2. inside an operand of an interaction fragment. Here a path term has the form $\alpha.par(n)\#k$ where $1 \leq k \leq n$ indicates that the location is within the k-th operand of the interaction fragment. For example, $scope(l_3) = sdia.par(2)\#1$ and $scope(l_5) = sdia.par(2)\#1.alt(2)\#1$.

3. marking the end of an interaction fragment. Here a path term has the form $\overline{\alpha.par(n)}$. For example, for l_9, l_{11} we have $scope(l_9) = sdia.par(2)\#1.\overline{alt(2)}$ and $scope(l_{11}) = \overline{sdia.par(2)}$.

Further, we assume that $Path$ contains the $shortest$ path terms of a diagram only, that is, a path term $sdia.par(2).\overline{par(2)}.alt(2)$ is not allowed and instead the expected path term should be simply $sdia.alt(2)$.

Condition (6) states an important property on edges, namely that the locations involved must belong to the same scope. This due to the fact that by definition, messages in a sequence diagram cannot cross borders of operands or interaction fragments (cf. [8]).

On edges and locations we define a function $temp$ which returns true if the edge (and associated message) or location is hot and false if it is $cold$. A further function on edges is $comm\_synch$ which returns true if the communication is synchronous and false otherwise.

Consider an additional auxiliary function $alt\_occ : Loc \times I \to \mathbb{N}_0$ that given a location and an instance gives a natural number corresponding to the total number of operands of all **alt** interaction fragments that appear in the diagram above the given location and are complete (the end location of the fragment is also above the given location).

$$alt\_occ(l,i) = \begin{cases} 0 & \Leftarrow l \in loc(i) \cap Loc_{ini} \\ m + alt\_occ(l',i) & \Leftarrow l, l' \in loc(i), time(l) = time(l') + 1 \\ & \qquad \text{and } scope(l) = \alpha.\overline{alt(m)} \\ alt\_occ(l',i) & \Leftarrow l, l' \in loc(i), time(l) = time(l') + 1 \\ & \qquad \text{and } scope(l) \neq \alpha.\overline{alt(m)} \\ \bot & \Leftarrow \text{otherwise} \end{cases}$$

The function is recursive and only defined for a pair (l,i) where l is a location on the lifeline of i. For example, take location l_9 of **Fig. 1**, $alt\_occ(l_9,j) = 2$. In fact, $alt\_occ(l,j) = 0$ for all $l \in loc(j)$ with $time(l) < time(l_9)$, and $alt\_occ(l,j) = 2$ for all $l \in loc(j)$ with $time(l_9) \leq time(l)$.

We now define a model associated to an instance participating in the interaction described by a sequence diagram.

Instance Model. Let $SD = (I, Loc, Loc_{ini}, Mes, E, Path, X_I)$ be a sequence diagram and i be an arbitrary instance in I. A model for i is a labelled event structure $M_i = (ES_i, \mu_i)$ where $ES_i = (Ev_i, \to_i^*, \#_i)$ with Ev_i such that there is an injective function $ev\_map : loc(i) \to 2^{Ev_i}$ such that

$$ev\_map(l) = \begin{cases} \{e_{l_1}, \ldots, e_{l_m}\} & \Leftarrow m = alt\_occ(l,i) \neq 0 \\ \{e_l\} & \Leftarrow m = alt\_occ(l,i) = 0 \end{cases}$$

For arbitrary $e_1 \neq e_2 \in Ev_i$, $e_1 \#_i e_2$ iff either

- $\exists_{l \in loc(i)} e_1, e_2 \in ev\_map(l)$, or
- $\exists_{l_1 \neq l_2 \in loc(i)} e_1 \in ev\_map(l_1) \wedge e_2 \in ev\_map(l_2) \wedge scope(l_1) = \alpha.alt(m)\#k \wedge$
  $scope(l_2) = \alpha.alt(m)\#n \wedge n \neq k \wedge alt\_occ(l_1, i) = alt\_occ(l_2, i)$

The local causality relation $\rightarrow_i^*$ is such that for arbitrary $l_1, l_2 \in loc(i)$ with $time(l_1) \leq time(l_2)$ and $(scope(l_1) \neq \alpha.alt(m)\#k$ or $scope(l_2) \neq \alpha.alt(m)\#n)$ and $(scope(l_1) \neq \alpha.par(m)\#k$ or $scope(l_2) \neq \alpha.par(m)\#n)$, the following holds

$$\forall_{e_1 \in ev\_map(l_1)} ! \exists_{e_2 \in ev\_map(l_2)} e_1 \rightarrow_i^* e_2$$

Additionally, all events must satisfy the following

$$\forall_{e_1, e_2, e_3 \in Ev_i} e_1 \#_i e_2 \rightarrow_i^* e_3 \Rightarrow \neg(e_1 \rightarrow_i^* e_3)$$

The labelling function $\mu_i : Ev_i \rightarrow Mes \times \{s, r\} \cup 2^{\Phi(X_i)}$ is a function defined as follows

$$\mu_i(e) = \begin{cases} (m, s) \Leftarrow \exists_{w=(l,m,l_1) \in E} \text{ for some } l \in loc(i), e \in ev\_map(l) \\ (m, r) \Leftarrow \exists_{w=(l_1,m,l) \in E} \text{ for some } l \in loc(i), e \in ev\_map(l) \\ \Gamma \Leftarrow \exists_{\Gamma \subset \Phi(X_i)} loc\_const(l) = \Gamma \text{ for some } l \in loc(i), e \in ev\_map(l) \\ \perp \Leftarrow otherwise \end{cases}$$

Some explanations for this definition are in order. A location in a diagram does not correspond to a unique event in the event structure. The reason for this is that according to the definition of (prime) event structures the conflict relation propagates over causality, that is, two events that are in conflict have all their subsequent events also in conflict. In order to respect this, locations must have associated a set of events (one or more). Take **Fig. 1**, locations l_5, l_7 are within an **alt** interaction fragment. This means that they are mutually exclusive (captured by the $\#_j$ event relation). Exiting the interaction fragment means that location l_9 is reached, but this location is reached in either the case that l_5 or l_7 happened, so l_9 has to have two associated events, one for each of the possible operands chosen before (the number of events needed for a location is in fact given by the auxiliary function alt_occ). Naturally, all events associated to one location are in conflict. Locations that are ordered with respect to $time$ have (some of) their associated events related by causality as long as the locations are not within the *same* **par** or **alt** interaction fragment. Take **Fig. 1** again. No events associated to locations l_3 and l_{10} are related by causality because they are within the same **par** interaction fragment. In fact they are also not in conflict by definition, so they have to be concurrent. Locations l_5 and l_9 are ordered by $time$, and satisfy the conditions for the causality relation. Consequently, for all events associated to l_5 (which is only one, say e_5, because $alt\_conc(l_5, j) = 0$) there is a unique event within the set $ev\_map(l_9)$ (let this event be e_{91}) such that $e_5 \rightarrow_j^* e_{91}$.

Finally, the labelling function is defined for an event e if this event is associated to one of the following kinds of locations: a *send location* (the label is (m, s)

where m is the message being sent); a *receive location* (the label is (m, r) where m is the message being received); a *condition location*, that is, a location which has associated an interaction constraint or a state invariant.

The above defined model for an instance is a well-defined labelled event structure. We omit the proof here for space reasons.

From the models of the instances involved in the interaction we can obtain a model for the entire interaction diagram given in the following definition.

Diagram Model. Let $SD = (I, Loc, Loc_{ini}, Mes, E, Path, X_I)$ be a sequence diagram, and $M_i = (ES_i, \mu_i)$ be a model for $i \in I$. A model of the diagram is given by $M = (ES, \mu)$ where $ES = (Ev, \rightarrow^*, \#)$ and Ev is defined as follows

$$Ev = \{e \in Ev_i \mid e \in ev\_map(l) \text{ for some } l \in loc(i) \text{ such that}$$
$$\neg \exists_{w \in E}(w = (l_1, m, l) \vee w = (l, m, l_1)) \wedge comm\_synch(w) = true$$
$$\text{for some } l_1 \in Loc \text{ and } m \in Mes\}$$

$$\bigcup$$

$$\{(e_1, e_2) \in Ev_i \times Ev_j \mid \text{for each } e_1 \in ev\_map(l_1) \text{ there is a unique}$$
$$e_2 \in ev\_map(l_2) \text{ for some } l_1 \in loc(i), l_2 \in loc(j)$$
$$\text{iff } \exists_{w \in E} w = (l_1, m, l_2) \text{ and } comm\_synch(w) = true$$
$$\text{for some } m \in Mes\}$$

For arbitrary $e \neq e' \in Ev$, $e \# e'$ iff one of the following cases holds

1. $e, e' \in Ev_i$ for some $i \in I$ and $e \#_i e'$
2. $e \in Ev_i$, $e' \in Ev_j$ for some $i \neq j \in I$ and $e \in ev\_map(l_1)$, $e' \in ev\_map(l_2)$, $scope(l_1) = \alpha.alt(m)\#k$, $scope(l_2) = \alpha.alt(m)\#n$ for $n \neq k$ and further $alt\_occ(l_1, i) = alt\_occ(l_2, j)$.
3. $e \in Ev_i$ for some $i \in I$ and $e' = (e_1, e_2)$ with $e_1 \in Ev_j, e_2 \in Ev_k$ for some $j \neq k \in I$ and either:
   (a) $(i = j \wedge e \#_i e_1)$ or $(i = k \wedge e \#_i e_2)$
   (b) e and e_1 satisfy condition 2. above.
4. $e = (e_1, e_2)$ and $e' = (e_1', e_2')$ with $e_1 \in Ev_i, e_2 \in Ev_j, e_1' \in Ev_p, e_2' \in Ev_q$ for $i \neq j \in I$ and $p \neq q \in I$. e_1 and e_1' satisfy the above condition 1. or 2.

For arbitrary $e, e' \in Ev$, $e \rightarrow^* e'$ iff one of the following cases holds

1. $e, e' \in Ev_i$ for some $i \in I$ and $e \rightarrow_i^* e'$
2. $e \in Ev_i$ for some $i \in I$ and $e' = (e_1, e_2)$ with $e_1 \in Ev_j, e_2 \in Ev_k$ for some $j \neq k \in I$. Either: $(i = j \wedge e \rightarrow_i^* e_1)$ or $(i = k \wedge e \rightarrow_i^* e_2)$
3. $e = (e_1, e_2)$ and $e' = (e_1', e_2')$ with $e_1 \in Ev_i, e_2 \in Ev_j, e_1' \in Ev_p, e_2' \in Ev_q$ for $i \neq j \in I$ and $p \neq q \in I$. One of the following holds: (a) $i = p$ and $e_1 \rightarrow_i^* e_1'$; (b) $j = p$ and $e_2 \rightarrow_j^* e_1'$; (c) $i = q$ and $e_1 \rightarrow_i^* e_2'$ or (d) $j = q$ and $e_2 \rightarrow_j^* e_2'$.
4. there is a $w \in E$ such that $w = (l_1, m, l_2)$ for some $m \in Mes$ such that $comm\_synch(w) = false$, $e \in ev\_map(l_1), e' \in ev\_map(l_2)$ and $\neg(e \# e')$.

Additionally, for all $w = (l_1, m, l_2) \in E$ with $comm\_synch(w) = false$ and $temp(w) = true$ the following holds:

$$\forall_{e_1 \in ev\_map(l_1)} ! \exists_{e_2 \in ev\_map(l_2)} e_1 \rightarrow^* e_2$$

The labelling function $\mu : Ev \rightarrow Mes \times \{s, r\} \cup 2^{\Phi(X_I)}$ is defined as follows:

$$\mu(e) = \begin{cases} m & \Leftarrow e = (e_1, e_2), e_1 \in Ev_i, e_2 \in Ev_j, \mu_i(e_1) = (m, s), \mu_j(e_2) = (m, r) \\ \mu_i(e) & \Leftarrow e \in Ev_i \end{cases}$$

To summarise, the above set of events Ev is such that it contains all the local events of the instance models except for those that belong to locations participating in a synchronous communication - for which pairs of events are built (the first argument for the send event, the second for the receive event). Concerning the relations, local relations propagate to the global level. Events are in conflict at the diagram level if they are associated to locations that fall within the same scope of an **alt** interaction fragment. New events are added to the causality relation, namely those that are associated to an asynchronous message. Here, a distinction is made between hot asynchronous messages and other messages, namely, for hot messages we require that a message sent *must* be received, that is, in event terms there must exist a unique receive event causally related to the send event. Finally, the labelling function is such that a synchronous communication pair of events is labelled by the message exchanged, and all other events are labelled with the local label (an asynchronous message send/receive, a set of interaction constraints or state invariants).

The above defined model for a sequence diagram is a well-defined labelled event structure. We omit the proof here for space reasons.

Consider **Fig. 2**, it shows an event-based model for part of the interaction from the diagram of **Fig. 1**. It shows the two possible traces in the interaction. Only some labels are included for increased readability: instances j and k synchronise at event e_1 with message $m2$, and at event e_2 with message $m3$. Event e_{l_8} denotes the sending of message $m8$, and event e_{p_5} denotes the corresponding message receive.

4   A Concurrent Communication Logic

We describe briefly the main idea of our logic, a distributed temporal logic called MDTL, and how it can be used to specify interactions. More details on the logic, including the semantics, can be found in [7, 6].

An instance involved in the interaction described by a sequence diagram SD has a *home* logic to describe internal properties (for example, state invariants and interaction constraints) or describe interactions from a local point of view. Further, a *communication* logic describes interactions between several instances. To some extent, the communication logic describes an *observer* of the interaction. The abstract syntax of MDTL where i and k are instances, is given as follows (in a simplified variant for our purposes):

MDTL ::= $\{i.H_i\}_{i \in I} \mid C$
H_i ::= ATOM$_i \mid \neg H_i \mid H_i \Rightarrow H_i \mid H_i \, \mathcal{U}_\forall \, H_i \mid H_i \, \mathcal{U}_\exists \, H_i \mid \Delta H_i$
C ::= $i.Mes! \leftrightarrow k.Mes? \mid i.Mes! \rightarrow k.Mes? \mid H_{obs}$
ATOM$_i$ ::= $true \mid \Phi(X_i) \mid Mes! \mid Mes?$

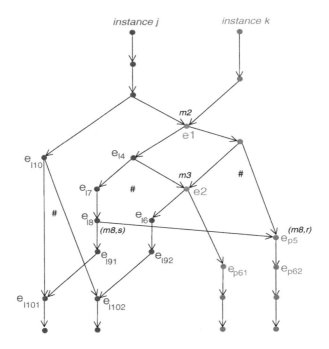

Fig. 2. An event structure for the example interaction between j and k.

The home logic H_i is basically an extension of CTL (notice that $\mathcal{U}_\forall$ corresponds to all paths, whilst $\mathcal{U}_\exists$ corresponds to there is a path) with a concurrency operator Δ. The intuition of a formula $i.(\varphi_1 \wedge \Delta\varphi_2)$ is that from the point of view of instance i, φ_1 holds and φ_2 holds concurrently. Mes denotes a message term where $Mes!$ is used to denote the sending of a message and $Mes?$ to denote the receipt of a message. In the communication logic C, $\leftrightarrow$ is used for synchronous communication and $\rightarrow$ and for asynchronous communication, and in both cases for denoting *hot* messages, that is, a message that if sent must be received. Notice that the communication logic can refer to H_{obs} where obs is used to denote an interaction observer. $\Phi(X_i)$ is the constraint logic introduced earlier in Section 3.2.

We can use MDTL to describe the complete interaction specified in a sequence diagram. Furthermore, we can impose constraints on the interaction, describe the *forbidden* behaviour associated to an interaction or that a certain collection of communications is *mandatory* (which is captured to some extent by an **assert** interaction fragment). For space reasons we just give some examples.

Imagine that we want to state that an asynchronous hot message if sent must always be received. Take the sequence diagram of **Fig. 1** and assume we have an OCL liveness constraint attached to the diagram stating that *if sent message m1 must always be received*. The corresponding communication logic formula is written as follows:

$$i.m1! \rightarrow j.m1?$$

After m1 has been sent and until m2 has been received no other messages are sent or received is given by the next communication formula.

$$m1! \wedge (\neg(m! \vee m?) \, \mathcal{U}_\forall \, m2?)$$

Consider now, a few local statements to illustrate the home logic. The formula below states that

$$j.(m2! \wedge \Delta m5?)$$

from the point of view of j, sending message m2 and receiving message m5 happens concurrently (in either order or at the same time).

After i has sent m7, i eventually reaches State1, which can be used to denote internal liveness, that is, that an instance must progress along its lifeline (which cannot be stated directly in UML 2.0 sequence diagrams).

$$i.(m7! \wedge (true \, \mathcal{U}_\forall \, State1))$$

5   Conclusions

We have given a semantics to sequence diagrams in UML 2.0 based on labelled event structures. The presented semantics given is not complete as we have not considered all interaction fragments permitted in UML (for example **strict** and **loop**). Extending the presented model with such fragments is straightforward.

We have presented a simple concurrent distributed temporal logic and showed how interactions and various constraints can be described in this logic. There are essentially two ways in which we can use this logic. Firstly, to capture some interaction properties (e.g., forbidden behaviour, liveness properties, state invariants, etc). In this case we can check whether the inter-object behavioural model (a labelled event structure) satisfies the properties. Secondly, to capture the entire interaction of a sequence diagram as a set of formulae. An interesting consequence of this case is that we can verify the sequence diagram against the state-based behavioural model directly through model checking.

We are currently working on the integration of our concurrent logic into the Edinburgh Concurrency Workbench (CWB). Ultimately, our aim is to establish a connection between CWB and UML 2.0 enabling the verification of temporal formulae over behavioural models. The temporal formulae to be verified can be given either directly in our simple temporal extension of OCL2.0 [1, 2] or indirectly through a sequence diagram in UML2.0, both of which translate into formulae in our concurrent logic. Depending on what is being verified, the behaviour model itself can be given by sequence or state diagrams in UML 2.0.

There are many variants of MSCs and high-level MSCs as well as work on providing a formal semantics to them. Some MSC extensions consider liveness properties as well (e.g., [5]), or are given a semantics based on Petri nets with unfoldings defined as event structures [9]. Additionally, there is a lot of recent work using LSCs which have a well defined operational semantics [4]. By contrast, our approach is based on the most recent version of the standard UML 2.0

using OCL 2.0 for additional liveness properties if required. We do not know of other work which addresses the latest extension of sequence diagrams in UML 2.0 or explores the powerful combination of UML and OCL. We also believe that the underlying embedding onto a true-concurrent logic as ours is novel and offers interesting perspectives concerning synthesis and verification which we will explore in the future.

References

1. J. Bradfield, J. Küster-Filipe, and P. Stevens. Enriching OCL using observational mu-calculus. In *Proceedings of the 5th International Conference on Fundamental Approaches to Software Engineering, LNCS* 2306, pages 203–217. Springer, 2002.
2. A. Cavarra and J. Küster-Filipe. Combining sequence diagrams and OCL for liveness. In *Proceedings of the Semantic Foundations of Engineering Design Languages (SFEDL), ETAPS 2004. Barcelona, Spain*, 2004. To appear in ENTCS.
3. H.-D. Ehrich and A. Sernadas. Local Specification of Distributed Families of Sequential Objects. In *Recent Trends in Data Types Specification*, pages 219–235. Springer, LNCS 906, 1995.
4. D. Harel and R. Marelly. *Come, Let's Play: Scenario-based Programming Using LSCs and the Play-Engine.* Springer, 2003.
5. I. Krueger. Capturing overlapping, triggered, and preemptive collaborations using MSCs. In *Proceedings of the 6th International Conference on Fundamental Approaches to Software Engineering, LNCS* 2621, pages 387–402. Springer, 2003.
6. J. Küster-Filipe. *Foundations of a Module Concept for Distributed Object Systems.* PhD thesis, Technische Universität Braunschweig, Germany, September 2000.
7. J. Küster-Filipe. Fundamentals of a module logic for distributed object systems. *Journal of Functional and Logic Programming*, 2000(3), March 2000.
8. OMG. *UML 2.0 Superstructure Draft Adopted Specification.* OMG document ptc/03-08-02, available from www.uml.org, August 2003.
9. A. Roychoudhury and P.S. Thiagarajan. Communicating transaction processes. In *Proceedings of the Third International Conference on Application of Concurrency to System Design*, pages 157–166. IEEE Computer Society Press, 2003.
10. G. Winskel and M. Nielsen. Models for Concurrency. In S. Abramsky, D.M. Gabbay, and T.S.E. Maibaum, editors, *Handbook of Logic in Computer Science, Vol. 4, Semantic Modelling*, pages 1–148. Oxford Science Publications, 1995.
11. ITU-TS Recommendation Z.120. *Message Sequence Chart (MSC).* ITU-TS, Geneva, 1996.

Proof Support for RAISE
by a Reuse Approach Based on Institutions

Morten P. Lindegaard and Anne E. Haxthausen

Informatics and Mathematical Modelling, Technical University of Denmark,
DK-2800 Kgs. Lyngby, Denmark
{mpl,ah}@imm.dtu.dk

Abstract. This paper explains how proof support for the RAISE Specification Language (RSL) can be obtained by reusing the Isabelle/HOL theorem prover. An institution for a subset of RSL is defined together with a light institution comorphism from this institution to an existing institution for higher-order logic (HOL). A translation of RSL specifications to Isabelle/HOL theories is derived from the light institution comorphism and proved sound wrt. semantic entailment. Finally, the results of some case studies are reported.

Keywords: Institutions, RSL, HOL, algebraic semantics, proof support.

1 Introduction

The RAISE Specification Language (RSL) [14] is a wide-spectrum specification language associated with a development method [15] based on stepwise refinement. Proving properties of specifications is important to the method and proof tools [4] are available. However, a higher degree of automation is desired.

The goal of this paper is to outline how increased proof support for RAISE can be obtained. An approach is to use an existing theorem prover having the desired properties. A good candidate for that is the theorem prover Isabelle [13] which offers support for higher-order logic (HOL). The idea is to translate RSL proof-obligations to HOL and then prove the translated proof obligations using Isabelle. The translation must be sound in the sense that properties proved about the translated specifications in HOL also hold for the original specifications in RSL.

An approach for defining the translation in a formal way, so that it can be proved sound, is to define a morphism from an institution for RSL to an institution for HOL. The concept of *institutions* [16] formalizes the informal notion of logical systems with signatures, sentences, models, and a satisfaction relation. Many variants of morphisms (defining translations of signatures, sentences, and models) between institutions exist, cf. [8]. However, we define a new, more simple kind, *light institution comorphisms*, that is sufficient for our purpose. An institution for HOL already exists, so our task is to define an institution for RSL and a light institution comorphism from the institution of RSL to that of HOL, and to prove that the light institution comorphism satisfies certain requirements that ensure the above mentioned soundness property.

C. Rattray et al. (Eds.): AMAST 2004, LNCS 3116, pp. 319–333, 2004.

1.1   Outline of the Paper

First, Sect. 2 introduces the concepts of institutions and light institution comorphisms. Then, in Sect. 3 and in Sect. 4, we define institutions for a subset of RSL and for HOL, respectively. Next, in Sect. 5, we present a light institution comorphism from the RSL institution to that of HOL, prove that it is sound wrt. semantic entailment, and explain how it induces a translation from RSL to Isabelle/HOL. Finally, achievements, related work, and future work is summarized in Sect. 6.

2   Formal Background

In this section, we define the notions of institutions and light institution comorphisms. Moreover, we explain how theorem provers may be reused via light institution comorphisms.

2.1   Institutions

The concept of institutions [16] formalizes the informal notion of a logical system.

Definition 1. *An* institution I *is a quadruple* $\langle Sign, Sen, Mod, \models \rangle$ *where*

- *Sign is a category of signatures*
- *Sen : Sign $\longrightarrow$ Set is a functor that maps each signature to the set of sentences over the signature*
- *Mod : Sign $\longrightarrow$ Cat^{op} is a functor that maps each signature to the category of models over the signature*
- *$\models_\Sigma \subseteq |Mod(\Sigma)| \times Sen(\Sigma)$ is a satisfaction relation for each $\Sigma \in |Sign|$*

so that

$$m' \models_{\Sigma'} Sen(\sigma)(e) \qquad \textit{iff} \qquad Mod(\sigma)(m') \models_\Sigma e$$

for each $\sigma : \Sigma \longrightarrow \Sigma'$ *in Sign, $m' \in |Mod(\Sigma')|$, and $e \in Sen(\Sigma)$. This requirement is known as* the satisfaction condition.

For a category C, its collection of objects is denoted by $|C|$. The notation $\sigma(e)$ is often used for $Sen(\sigma)(e)$, and the notation $m|_\sigma$ is often used for $Mod(\sigma)(m)$.

2.2   Light Institution Comorphisms

Many different kinds of mappings between institutions and names for these have been suggested, cf. e.g. [8] which gives a survey of this. In this section we introduce a new kind, *light institution comorphisms*, that are weak versions of simple institution representations [11].

In the following, let $I = \langle Sign, Sen, Mod, \models \rangle$, $I' = \langle Sign', Sen', Mod', \models' \rangle$ be institutions and $Pres'$ be the category of I'-presentations such that an object of $Pres'$ is of the form $\langle \Sigma', E' \rangle$ where $\Sigma' \in |Sign'|$ and $E' \subseteq Sen'(\Sigma')$. The notation $Sig(P)$ is used for the signature of a presentation P. Furthermore, let Mod' denote the functor from $Pres'$ into Cat^{op}, for which $Mod'(\langle \Sigma', E' \rangle)$ is the category of the models over the signature Σ' that satisfy the set of sentences E'.

Definition 2. *A* light institution comorphism $\varrho : I \longrightarrow I'$ *is a triple* $\langle \varrho, \alpha, \beta \rangle$ *where*

- $\varrho : |Sign| \longrightarrow |Pres'|$ *is a function that maps I-signatures to I'-presentations*
- $\alpha = (\alpha_\Sigma : Sen(\Sigma) \longrightarrow Sen'(Sig(\varrho(\Sigma))))_{\Sigma \in |Sign|}$ *is a family of functions*
- $\beta = (\beta_\Sigma : |Mod'(\varrho(\Sigma))| \longrightarrow |Mod(\Sigma)|)_{\Sigma \in |Sign|}$ *is a family of partial functions*

so that

$$ m' \models'_{Sig(\varrho(\Sigma))} \alpha_\Sigma(e) \qquad iff \qquad \beta_\Sigma(m') \models_\Sigma e $$

for each $\Sigma \in |Sign|$, $e \in Sen(\Sigma)$, and $m' \in |Mod'(\varrho(\Sigma))|$ for which $\beta_\Sigma(m')$ is defined. This requirement is known as the representation condition.

The partiality of β_Σ reflects that the institution I' may have a larger collection of models than the institution I.

Light institution comorphisms are weak versions of *simple institution representations*[11]: ϱ is a function and not a functor. Consequently, ϱ is not concerned with mapping signature morphisms. As a consequence, it has no meaning to require α and β to be natural. Furthermore, β is a family of functions and not a family of functors. Finally, β_Σ may be partial and the representation condition is only required to hold when $\beta_\Sigma(m')$ is defined.

Light institution comorphisms are weaker than simple institution representations since they do not enable reuse of theorem provers for structured specifications built using signature morphisms. Moreover, the partiality of β_Σ causes the below Theorem 1 to be one-directional.

The reason for using light institution comorphisms is that they enable reuse of theorem provers for flat specifications without requiring a mapping of signature morphisms and naturality of α and β.

The notion of *institution comorphisms* [8] has previously been weakened: *Semi-natural* institution comorphisms [8] relax the requirements of naturality, and *simulations* [2] allow mappings of models to be partial (but they must be surjective).

2.3   The Reusing Theorem
In this section, we describe how one can prove semantic consequences of a specification in one institution by using a theorem prover for another institution.

For light institution comorphisms, like for some other kinds of mappings between institutions (e.g. for *maps of institutions* [5], i.e. theoroidal institution comorphisms satisfying certain requirements), there is a reusing theorem:

Theorem 1. *Let* $\langle \varrho, \alpha, \beta \rangle : I \longrightarrow I'$ *be a light institution comorphism for which β_Σ is surjective for each $\Sigma \in |Sign|$. Then, for any $\Sigma \in |Sign|$*

$$ \langle \Sigma, E \rangle \models_\Sigma \varphi \qquad if \qquad \langle \Sigma', \alpha_\Sigma(E) \cup E' \rangle \models'_{\Sigma'} \alpha_\Sigma(\varphi) $$

where $\varrho(\Sigma) = \langle \Sigma', E' \rangle$.

---

[1] *Simple institution representations* are like *simple theoroidal institution comorphisms* [8] except that signatures are mapped to presentations instead of theories.

Proof. See [10].

If β_Σ is total for each $\Sigma \in |Sign|$, the theorem holds in both directions, ensuring a sense of completeness.

The theorem suggests the following method for proving semantic consequences for an institution I:

1. Find another institution I' that has a sound proof system (with tool support), i.e.,
$$SP' \vdash'_{\Sigma'} e' \text{ implies } SP' \models'_{\Sigma'} e'.$$

2. Define a light institution comorphism $\langle \varrho, \alpha, \beta \rangle : I \longrightarrow I'$ such that β_Σ is surjective for each Σ.

3. To prove φ to be a semantic consequence of a specification $\langle \Sigma, E \rangle$, translate the specification and the proof obligation φ along the light institution comorphism into I' and use the proof system (theorem prover tool) of I' to prove that the translated proof obligation is a consequence of the translated specification.

3   An Institution for a Subset of RSL

RSL [14] is a wide-spectrum language which encompasses and integrates different specification styles (algebraic, model-oriented, applicative, imperative, and concurrent) in a common conceptual framework. However, for the purpose of providing proof support, we only consider an applicative subset, as, according to the RAISE method, most proofs are made in the early development steps where specifications are applicative.

In this section, we first describe the chosen subset, *mRSL*, and then we define an institution for this subset. See [10] for details.

3.1   An Applicative Subset of RSL

An mRSL *specification* is a named class expression consisting of declarations of types, values, and axioms.

A *type declaration* is either a sort declaration or an abbreviation type declaration. A *sort declaration*, as known from algebraic specification, introduces a name for a new type without specifying it any further, while an *abbreviation type declaration* declares a name to be an abbreviation for a type expression that is constructed in a model-oriented way from declared type names, type literals like **Int**, **Real**, and **Bool**, and type operators like $\times$ (for Cartesian products) and $\stackrel{\sim}{\to}$ (for partial functions).

A *value declaration* consists of a name and a type expression.

An *axiom declaration* consists of a name and an *equivalence expression* of the form $ve_1 \equiv ve_2$, where ve_1 and ve_2 are value expressions. Value expressions include value names, value literals for the built-in types, basic expressions (e.g., for denoting non-termination), product expressions, λ-abstractions, function applications, infix expressions, prefix expressions, quantified expressions, if-expressions, and equivalence expressions.

Example. As an example of an (m)RSL specification, we show an algebraic specification of a gas burner:

> **scheme** GAS =
>   **class**
>   **type** State
>   **value** gas : State $\overset{\sim}{\to}$ **Bool**
>   **value** flame : State $\overset{\sim}{\to}$ **Bool**
>   **value** leak : State $\overset{\sim}{\to}$ **Bool**
>   **axiom** [leak_def] leak $\equiv$ (λ s : State $\bullet$ gas(s) $\wedge \sim$flame(s))
>   **axiom** [gas_total] ($\forall$ s : State $\bullet \exists$ b : **Bool** $\bullet$ (gas(s) = b)) $\equiv$ **true**
>   **axiom** [flame_total] ($\forall$ s : State $\bullet \exists$ b : **Bool** $\bullet$ (flame(s) = b)) $\equiv$ **true**
>   **end**

The axioms state that there is a leak when the gas is on and there is no flame and that gas and flame are total functions.

Semantics. A (well-formed) mRSL specification determines a signature, Σ, and a set of Σ-sentences, E, of the underlying institution (that will be described in next subsection). The signature is derivable from the type declarations and value declarations, whereas the set of sentences is derivable from the axiom declarations in an obvious way.

The *semantics* of the specification is the loose semantics of the presentation $\langle \Sigma, E \rangle$, i.e., the class of Σ-models that satisfy all the sentences in E.

3.2 An Institution for the Subset of RSL

An institution I_{mRSL} for the mRSL subset is defined in the following.

Signatures. $Sign_{mRSL}$ is the category of signatures.

For a set X of identifiers, $T(X)$ denotes the set of type expressions that can be generated using identifiers in X, mRSL type literals, and type operators.

A *signature* Σ is a quadruple $\langle S, A, \Psi, V \rangle$, where S is a set of sort names, A is a set of abbreviation type names, $\Psi : A \to T(S \cup A)$ is a function that maps abbreviation type names to type expressions they are abbreviations for, and V is a $T(S)$-sorted set of value names; the sets S and A must be disjoint, type abbreviations must not be cyclic and the sets of V are disjoint (disallowing overloading).

Let $\Sigma = \langle S, A, \Psi, V \rangle$ and $\Sigma' = \langle S', A', \Psi', V' \rangle$ be signatures. A *signature morphism* $\sigma : \Sigma \longrightarrow \Sigma'$ is a triple $\sigma = \langle \sigma_S, \sigma_A, \sigma_V \rangle$ where

- $\sigma_S : S \longrightarrow S' \cup A'$ is a mapping of sort names
- $\sigma_A : A \longrightarrow A'$ is a mapping of abbreviation type names
- $\sigma_V : V \longrightarrow V'$ is a mapping of value names.

such that

- the mapping σ_V is $T(S)$-sorted and respects σ_S:

$$\sigma_V = (\sigma_{V_t} : V_t \longrightarrow V'_{\Psi'^*(\sigma_S(t))})_{t \in T(S)}$$

where σ_S is lifted to work on type expressions in $T(S)$ by replacing occurrences of sort names as prescribed by σ_S, and where Ψ^* is the function that takes a type expression to its canonical form by recursively expanding abbreviation type names as prescribed by Ψ, and
- the following diagram commutes:

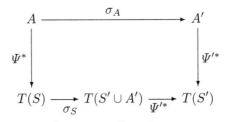

Note that signature morphisms are allowed to map sort names to sort names as well as abbreviation names, but abbreviation names are only allowed to be mapped to abbreviation names. This reflects the fact that the RAISE method allows sorts to be refined to concrete types, but not vice versa.

Sentences. The functor $Sen_{mRSL} : Sign_{mRSL} \longrightarrow Set$ maps each signature to the set of sentences over the signature, and lifts each signature morphism to work on sentences.

A Σ-*sentence* is an equivalence expression (as defined by the mRSL grammar and informally described in Sect. 3.1) which is well-formed wrt. Σ.

Well-formedness of value expressions (and thereby also of equivalence expressions) wrt. a signature $\Sigma = \langle S, A, \Psi, V \rangle$ is defined in the usual way by a set of deduction rules defining type assertions of the form $\Sigma \vdash ve : t$ where Σ is a signature, ve is a value expression, and t is a canonical type expression. Informally a value expression is well-formed if it is type correct and any value name is in V or an enclosing typing and any type name is in $S \cup A$.

Signature morphisms $\sigma = \langle \sigma_S, \sigma_A, \sigma_V \rangle$ are lifted to work on sentences by replacing occurrences of sort names, abbreviation type names, and value names as prescribed by σ_S, σ_A, and σ_V, respectively.

Models. The contravariant functor $Mod_{mRSL} : Sign_{mRSL} \longrightarrow Cat^{op}$ maps each signature to the discrete category of models over the signature, and maps each signature morphism to the reduct functor between the categories of models.

A Σ-*model* interprets each type name in Σ as a type and each value name in Σ as a value. Before giving the precise definition, we first define the semantic domains of types and values.

For each canonical type expression t ($\in T(\emptyset)$), there is a value domain $Value_t$ which is a set of values. Inheriting terminology from full RSL, a value expression denotes a "process". For each type expression t, there is a process domain extending the value domain for t with the diverging process $\perp$:

$$Procs_t = Value_t \cup \{\bot\}.$$

Examples of value domains: $Value_{\textbf{Int}}$ is the set of all integers, and $Value_{t_1 \overset{\sim}{\to} t_2}$ is the set of all functions from $Value_{t_1}$ to $Procs_{t_2}$. The process domains (rather than value domains) are used to give semantics to value expressions, since these may not terminate.

The domain of types can now be defined:

$$Types = \{type \mid type \subseteq Value_t \wedge type \neq \emptyset \wedge t \in T(\emptyset)\},$$

i.e., a type is a set of values.

Let $\Sigma = \langle S, A, \Psi, V \rangle$ be a signature. A Σ-model is a triple $\langle m_S, m_A, m_V \rangle$ where

- $m_S : S \longrightarrow Types$
- $m_A : A \longrightarrow Types$
- $m_V = (m_{V_t} : V_t \longrightarrow m_{T(S)}(t))_{t \in T(S)}.$

m_S interprets each sort name in S as a set in $Types$. Type literals and type operators have fixed interpretations (according to $Value_t$) which are used to lift m_S to a map $m_{T(S)}$ that interprets type expressions that may contain sorts in S.

m_A interprets each abbreviation type name a in A as a type in $Types$ such that $m_A(a) = m_{T(S)}(\Psi^*(a))$.

m_V interprets each value name in V_t as a value in the type denoted by its type expression t.

Let $\Sigma = \langle S, A, \Psi, V \rangle$, $\Sigma' = \langle S', A', \Psi', V' \rangle$ be signatures, let $\sigma = \langle \sigma_S, \sigma_A, \sigma_V \rangle : \Sigma \longrightarrow \Sigma'$ be a signature morphism, and let $m' = \langle m'_{S'}, m'_{A'}, m'_{V'} \rangle$ be a Σ'-model. Then the σ-reduct $m'|_\sigma$ of m' is a Σ-model $m = \langle m_S, m_A, m_V \rangle$, where

- $m_S(s) = m'_{T(S')}(\Psi'^*(\sigma_S(s)))$ for all $s \in S$
- $m_A(a) = m'_{A'}(\sigma_A(a))$ for all $a \in A$
- $m_V = (m'_{V'_{\Psi'^*(\sigma_S(t))}} \circ \sigma_{V_t} : V_t \longrightarrow m'_{T(S')}(\Psi'^*(\sigma_S(t))))_{t \in T(S)}.$

The Satisfaction Relation. The satisfaction relation $\models$ is expressed in terms of the dynamic denotational semantics M:

$$m \models_\Sigma value\_expr_1 \equiv value\_expr_2$$

if and only if

$$M_\Sigma(m)(value\_expr_1) = M_\Sigma(m)(value\_expr_2).$$

The meaning function M is indexed by a signature Σ so that its type is $M_\Sigma : Mod(\Sigma) \longrightarrow ValExpr(\Sigma) \longrightarrow Procs$ where $ValExpr(\Sigma)$ denotes value expressions over Σ and $Procs$ is the union of process domains $Procs_t$, $t \in T(\emptyset)$.

$M_\Sigma(m)$ essentially evaluates a value expression, interpreting value names as in the model m. The result is a process. As an example, if m interprets the value c as 2, then $M_\Sigma(m)(c + 2)$ denotes a process that terminates with the result 4.

The Satisfaction Condition. Truth is invariant under change of notation:

$$m' \models_{\Sigma'} \sigma(e) \qquad \text{iff} \qquad m'|_\sigma \models_\Sigma e$$

for all signatures Σ and Σ', signature morphisms $\sigma : \Sigma \longrightarrow \Sigma'$, Σ'-models m', and Σ-sentences e.

Proof. By induction on the structure of e.

Example. Consider the specification of a gas burner presented in Sect. 3.1. Its signature is $\Sigma = \langle S, A, \Psi, V \rangle$, where $S = \{\text{State}\}$, $A = \emptyset$, Ψ is the empty function, $V_{\text{State} \overset{\sim}{\to} \textbf{Bool}} = \{\text{gas}, \text{flame}, \text{leak}\}$, and $V_t = \emptyset$ for types t different from State $\overset{\sim}{\to}$ **Bool**. Its set of sentences contains the three axioms.

4 HOL

In this section, we outline parts of the institution for higher-order logic from [3], and we briefly touch upon Isabelle/HOL.

Higher-order logic, presented in [1], is based on Church's theory of simple types [6]. The technical details of the institutional formulation are based on [9].

4.1 Institution for HOL

We now outline those parts of the institution I_{HOL} for HOL defined in [3] that are relevant for translating from mRSL to HOL.

Signatures. $Sign_{HOL}$ is the category of signatures as described below.

Let $TyVars$ be an infinite set of *type variables*, and $TyNames$ an infinite set of *names of type constants*. A *type constant* is a pair (ν, n) with the name $\nu \in TyNames$ and the arity n (a natural number).

A *type structure* Ω is a set of type constants (with distinct names). The set $Types_\Omega$ of types over a type structure Ω is generated by the grammar:

$$\tau \quad ::= \quad \gamma \quad | \quad c \quad | \quad (\tau_1, ...\tau_n)\nu \quad | \quad \tau_1 \to \tau_2$$

where $\gamma \in TyVars$, $(c, 0) \in \Omega$, and $(\nu, n) \in \Omega$.

A *signature* Σ is a pair $\langle \Omega, C \rangle$, where Ω is a type structure and C is a set of constants typed by types over Ω (i.e., types in $Types_\Omega$) such that

- Ω contains $(bool, 0)$, $(\iota, 0)$, and $(\times, 2)$, and
- C contains the primitive constants $=_{\gamma \to \gamma \to bool}$ (equality), $\epsilon_{(\gamma \to bool) \to \gamma}$ (choice), $pair_{\gamma \to \delta \to (\gamma, \delta) \times}$ (pair construction), $fst_{(\gamma, \delta) \times \to \gamma}$ (projection), and $snd_{(\gamma, \delta) \times \to \delta}$ (projection).

Sentences. For each signature $\Sigma = \langle \Omega, C \rangle$, the set $Sen_{HOL}(\Sigma)$ of sentences over Σ is the set of closed Σ-formulas, i.e., the set of Σ-terms of type *bool* that do not contain free variables.

Σ-terms are formed by variables typed by types in $Types_{\Omega}$, constants in C, function applications, and λ-abstractions in the usual way described by the following grammar:

$$t_{\tau} ::= x_{\tau} \mid c_{\tau} \mid (t_{\tau' \to \tau} t'_{\tau'})_{\tau} \mid (\lambda x_{\tau_1}.t_{\tau_2})_{\tau_1 \to \tau_2}$$

where the subscripts indicate the types of the terms.

Models. For each signature $\Sigma = \langle \Omega, C \rangle$, the category $Mod_{HOL}(\Sigma)$ is the discrete category of standard Σ-models as described below.

The syntax of higher-order logic comprises types and terms. Types denote sets, and terms denote elements of sets. The semantics is given using a fixed collection of sets $\mathcal{U}$, called the *universe*, with the following properties:

- the elements of the universe $\mathcal{U}$ are non-empty sets,
- if $S \in \mathcal{U}$ then $\mathcal{U}$ contains all non-empty subsets of S and the power set of S,
- $\mathcal{U}$ is closed under Cartesian product,
- $\mathcal{U}$ contains a distinguished infinite set Ind, and
- there is a distinguished element $ch \in \Pi_{S \in \mathcal{U}} S$ which is a function; $ch(S) \in S$ witnesses that the sets $S \in \mathcal{U}$ are non-empty.

The properties imply that $\mathcal{U}$ contains finite Cartesian products and function spaces. Furthermore, particular sets are selected such that $\mathcal{U}$ contains a distinguished singleton set $\mathbf{1} = \{0\}$ and a distinguished two-element set $\mathbf{2} = \{0, 1\}$.

A *model* m_{Ω} of a type structure Ω maps each type constant (ν, n) in Ω to an n-ary function $m_{\Omega}(\nu) : \mathcal{U}^n \longrightarrow \mathcal{U}$ if $n > 0$. If $n = 0$ then $m_{\Omega}(\nu) \in \mathcal{U}$.

A *model* of a signature $\langle \Omega, C \rangle$ consists of a model m_{Ω} of Ω and an interpretation m_C of C. If a type τ contains no type variables, m_C maps each constant c_{τ} in C to an element in the set $[\![\tau]\!]_{m_{\Omega}}$ which is the meaning of τ wrt. m_{Ω}. If τ contains $n > 0$ type variables, $[\![\tau]\!]_{m_{\Omega}}$ is a function $\mathcal{U}^n \longrightarrow \mathcal{U}$, and m_C maps each constant c_{τ} to a function $\mathcal{U}^n \longrightarrow S$ where $S \in \mathcal{U}$ is a type that depends on the argument (in $\mathcal{U}^n$) given to the function.

The model *is standard* if

- m_{Ω} interprets *bool* as the distinguished two-element set $\{0, 1\}$, ι as the distinguished infinite set Ind, and $(\times, 2)$ as Cartesian product, and
- m_C interprets the primitive constants that are required to be in a signature accordingly.

The Satisfaction Relation. A standard Σ-model m satisfies a Σ-sentence φ, written $m \models_{\Sigma} \varphi$, if and only if the meaning of φ with respect to m is 1 for all interpretations of type variables.

The meaning wrt. a model m of a closed term without type variables is an element of a set in $\mathcal{U}$ and depends on the interpretation m_C of constants in the term. The meaning of a term with free variables and/or type variables is a function that takes arguments corresponding to the interpretation of the free variables and type variables and returns an element of a set in $\mathcal{U}$.

4.2   Isabelle/HOL

Isabelle is a generic theorem prover, i.e., logics such as HOL can be expressed in the meta-logic of Isabelle. The implementation of higher-order logic in Isabelle (Isabelle/HOL [12]) is based on the description of HOL in [9]. A module or specification in Isabelle is called a *theory*, and users of Isabelle/HOL write theories that extend a theory containing definitions and declarations of Isabelle/HOL.

Considering Isabelle/HOL, it is the entailment relation (i.e., the proof rules) that alone defines the logic. Isabelle/HOL has eight primitive inference rules and a number of derived rules. Isabelle/HOL follows the standard higher-order logic approach of taking a few connectives as primitives and then defining the rest. We refer to [12] for a detailed account of Isabelle/HOL.

5   A Light Institution Comorphism from mRSL to HOL

To reuse the theorem prover Isabelle, we must translate mRSL specifications into Isabelle/HOL theories. In this section, we present a light institution comorphism from mRSL to HOL and describe how it provides a translation. For further details, see [10].

5.1   A Light Institution Comorphism from mRSL to HOL

We now describe a light institution comorphism, $\langle \varrho, \alpha, \beta \rangle : I_{mRSL} \longrightarrow I_{HOL}$.

The idea is to encode the semantics of mRSL in HOL.

Mapping Signatures. ϱ maps an mRSL signature $\Sigma = \langle S, A, \Psi, V \rangle$ to a HOL presentation $\langle \langle \Omega_{mRSL} \cup \Omega_{\Sigma}, C_{mRSL} \cup C_{\Sigma} \rangle, E_{mRSL} \rangle$ where

- Ω_{mRSL} is a type structure containing type constants:
  - $(t, 0)$ for all mRSL type literals t
  - $(P, 1)$ for lifting value domains to process domains
- C_{mRSL} is a set of typed constants for expressing the mRSL semantics in HOL
- E_{mRSL} is a set of sentences specifying the constants in Ω_{mRSL} and C_{mRSL}

and

- Ω_{Σ} is a type structure containing type constants $(s, 0)$ for all sort types $s \in S$
- C_{Σ} is a set containing a typed constant, v_{τ}, for each value v in V_t, $t \in T(S)$, where $\tau = A_{\Sigma}^{te}(t)$ and $A_{\Sigma}^{te} : T(S) \to Types_{\Omega_{mRSL} \cup \Omega_{\Sigma}}$ is a function translating mRSL type expressions to HOL types (e.g., $A_{\Sigma}^{te}(t_1 \overset{\sim}{\to} t_2) = A_{\Sigma}^{te}(t_1) \to (A_{\Sigma}^{te}(t_2))P$ to encode partial mRSL functions as total HOL functions).

Mapping Sentences. For a given mRSL signature Σ, α_Σ maps mRSL Σ-sentences (equivalence expressions) into HOL $Sig(\varrho(\Sigma))$-sentences as follows:

$$\alpha_\Sigma(ve_1 \equiv ve_2) \doteq A_\Sigma(ve_1) = A_\Sigma(ve_2)$$

where A_Σ is a function mapping mRSL Σ-value expressions into HOL $Sig(\varrho(\Sigma))$-terms that represent their semantics, i.e., an mRSL value expression is mapped to a HOL term that represents a simplified process.

Mapping Models. For a given mRSL signature Σ, β_Σ maps HOL $Sig(\varrho(\Sigma))$-models into mRSL Σ-models.

The definition of β_Σ is based on a function B that maps mRSL types (value domains) to sets in $\mathcal{U}$ and a function D that maps values to elements of sets in $\mathcal{U}$, i.e.,

$$B : Types \longrightarrow \mathcal{U} \quad \text{and} \quad D : type \longrightarrow B(type) \quad \text{for } type \in Types.$$

The functions B and D are *injective* and define a *standard encoding* of mRSL types in HOL. B^{-1} and D^{-1} denotes the *left inverses* of B and D, respectively.

Let $\Sigma = \langle S, A, \Psi, V \rangle$ be an mRSL signature, and let $m = \langle m_\Omega, m_C \rangle$ be a $Sig(\varrho(\Sigma))$-model with standard encoding of mRSL types so that (1) for all $s \in S$ there exists a type t such that $m_\Omega(s, 0) = B(t)$ and (2) for all $v \in V_{t'}$ there exists an element dv of the type denoted by t' such that $m_C(v_{A_\Sigma^{te}(t')}) = D(dv)$. Then $\beta_\Sigma(m)$ is defined as $\langle m_S, m_A, m_V \rangle$ where

- $m_S(s) = B^{-1}(m_\Omega(s, 0))$ for $s \in S$
- m_A is defined by m_S and Ψ, as usual
- $m_V = (m_{V_t} : V_t \longrightarrow m_{T(S)}(t))_{t \in T(S)}$, where $m_{V_t}(c) = D^{-1}(m_C(c_\tau))$ for $c \in V_t$, $t \in T(S)$, and $\tau = A_\Sigma^{te}(t)$.

Proposition 1. *For all mRSL signatures Σ, mRSL Σ-value expressions ve, and HOL $Sig(\varrho(\Sigma))$-models m, if $\beta_\Sigma(m)$ is defined, then*

$$D(M_\Sigma(\beta_\Sigma(m))(ve)) = [\![A_\Sigma(ve)]\!]_m.$$

Proof. By induction on the structure of ve. See [10] for details.

Proposition 2. $\langle \varrho, \alpha, \beta \rangle : I_{mRSL} \longrightarrow I_{HOL}$ *is a light institution comorphism.*

Proof. It should be shown that

$$m \models_{Sig(\varrho(\Sigma))} \alpha_\Sigma(\varphi) \quad \text{iff} \quad \beta_\Sigma(m) \models_\Sigma \varphi$$

for all mRSL signatures Σ, mRSL Σ-sentences φ, and HOL $\varrho(\Sigma)$-models m for which $\beta_\Sigma(m)$ is defined. This follows from the definition of $\models_{Sig(\varrho(\Sigma))}$, Prop. 1, the definition of α_Σ, injectivity of the function D, and the definition of $\models_\Sigma$.

Proposition 3. β_Σ *is surjective for $\Sigma \in |Sign_{mRSL}|$.*

Proof. For each signature $\Sigma = \langle S, A, \Psi, V \rangle$ and mRSL Σ-model $m = \langle m_S, m_A, m_V \rangle$ there is a HOL $Sig(\varrho(\Sigma))$-model $m' = \langle m'_\Omega, m'_C \rangle$ such that $m'_\Omega(s, 0) = B(m_S(s))$ for $s \in S$ and $m'_C(c_\tau) = D(m_V(c))$ for c in V. Since $\beta_\Sigma(m') = m$, the mapping β_Σ is surjective.

By Prop. 3, Theorem 1, and soundness of Isabelle/HOL, we can use the light institution comorphism to translate mRSL specifications and proof obligations to HOL and reuse Isabelle/HOL as illustrated below:

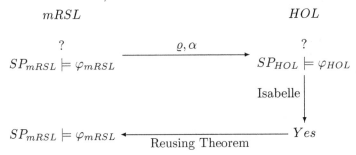

5.2 Translating mRSL Specifications

We now describe how the light institution comorphism $\langle \varrho, \alpha, \beta \rangle$ provides a translation of mRSL specifications to Isabelle/HOL theories.

An Isabelle/HOL theory, mRSL, is constructed to represent the fixed contribution $\langle \langle \Omega_{mRSL}, C_{mRSL} \rangle, E_{mRSL} \rangle$ of definitions of data types and constants needed for expressing the mRSL semantics in HOL.

Consider an mRSL specification SP representing an mRSL signature Σ and a set E of mRSL sentences, and let $\langle \langle \Omega_{mRSL} \cup \Omega_\Sigma, C_{mRSL} \cup C_\Sigma \rangle, E_{mRSL} \rangle = \varrho(\Sigma)$ and let $E' = \alpha_\Sigma(E)$. The specification SP is translated to an Isabelle/HOL theory which extends mRSL with (1) declarations that represent Ω_Σ and C_Σ and (2) the axioms in E'.

More precisely:

– sort declarations **type** s are translated to `typedecl s`,
– abbreviation type declarations **type** $T = type\_expr$ are translated to
  `type s = τ`, where $\tau = A_\Sigma^{te}(type\_expr)$,
– value declarations **value** $v : type\_expr$ are translated to
  `consts v :: "τ"`, where $\tau = A_\Sigma^{te}(type\_expr)$, and
– axiom declarations **axiom** $[id]$ $ve_1 \equiv ve_2$ are translated to
  `axioms id : "ve₁ = ve₂"`, where $ve_i = A_\Sigma(ve_i)$

The translation may be described as *a shallow embedding encoding the semantics*. It is shallow since we use data types that are part of Isabelle/HOL rather than axiomatizing the data types from mRSL. Moreover, it encodes the semantics since the translation of sentences (from axioms) is based on the denotational semantics.

Example. As an example, we here show the result of translating the specification from Sec. 3.1 to an Isabelle/HOL theory:

```
theory GAS = mRSL :
typedecl State
consts gas :: "State => (bool P)"
       flame :: "State => (bool P)"
       leak :: "State => (bool P)"
axioms leak_def :
 "((Proc leak)) = (v2p (% s :: State . ((appl (Proc gas) (Proc s))
 AND (NOT (appl (Proc flame) (Proc s))))))"
axioms gas_total :
 "(v2p (ALL s . ((v2p (EX b . ((appl (Proc gas) (Proc s) =rsl
 (Proc b)) = true))) = true))) = true"
axioms flame_total :
 "(v2p (ALL s . ((v2p (EX b . ((appl (Proc flame) (Proc s) =rsl
 (Proc b)) = true))) = true))) = true"
end
```

where v2p lifts values to processes, appl expresses function application, and AND
and NOT express the connectives $\wedge$ and $\sim$.

We can prove that the specification has the following property:

$$(\forall\ s : \text{State} \bullet \text{flame}(s) \Rightarrow \sim\!\text{leak}(s))$$

expressed by the following lemma in Isabelle/HOL:

```
lemma flame_no_leak :
 "(v2p (ALL s :: State . ((appl (Proc flame) (Proc s))
 IMP (NOT (appl (Proc leak) (Proc s))))) = true)) = true"
```

5.3 Case Studies

Case studies concerning use of the light institution comorphism to provide increased proof support are described in [10]; they include logic circuits, a generalized railway crossing, and an encoding of Duration Calculus. The translation
and reuse of Isabelle provided increased proof support in most cases.

Also a series of RSL proof rules has been translated to Isabelle/HOL and
proved as theorems using Isabelle. This not only provides helpful theorems but
also formally proves the RSL proof rules sound with respect to the denotational
semantics. So far, only a simplified semantics for a subset of RSL is considered,
but a formal proof relating the proof rules and semantics is novel work.

6   Conclusion

In this section we summarize achievements, related work and future work.

6.1 Achievements

The major contribution of the work reported in this paper is the provision of
proof support for an applicative subset, mRSL, of RSL in the form of a translation from this subset into Isabelle/HOL. We choose an applicative subset, since

according to the RAISE method most proofs are done for applicative specifications. Case studies have shown that this proof support has a higher degree of automation than the original RAISE proof tools.

In order to make the translation sound (so that Isabelle/HOL proofs made for translated mRSL proof obligations are sound wrt. the mRSL semantics) we took the following approach: First we formulated an institution for the considered subset of RSL and gave an institution-independent semantics of the subset. Then we defined a light institution comorphism from the mRSL institution into an institution for HOL and proved that this fulfills a condition that ensures the desired soundness properties.

The presented institution and semantics for mRSL not only provide the foundations for providing proof support, but are also new, interesting results on their own. Institutions have typically been defined for algebraic specification languages, while our institution is defined for a language that combines features from algebraic specification and model-oriented specification. Compared with the original semantics of RSL, our new semantics is more elegant and easy to understand as simpler semantic domains are used (this is possible as we only consider a subset of the full language) and a clear structure is provided by the institutional setting. Moreover, an advantage of defining the semantics in an institutional way is the institution-independent results that then come for free.

6.2   Related Work

The RAISE tool suite developed at UNU/IIST [7] provides a translator from another applicative subset of RSL to PVS so that the PVS theorem prover can be reused. This subset does not include partial functions, but on the other hand provides more type constructors. The translation is not based on institutions and has not been proved sound.

Other kinds of institution comorphisms have been used to provide proof support for other languages than RSL, e.g., the algebraic specification language CASL [3, 11].

6.3   Future Work

We plan to extend the subset with more type constructors and subtypes to make the applicative subset more complete. This can easily be done within the framework we have set up. It should also be considered to which extent the module structuring operations can be included.

Acknowledgments

The authors would like to thank Till Mossakowski for discussing early parts of our work. The diagrams in this paper were created using Paul Taylor's package diagrams.tex for drawing diagrams in TeX and LaTeX.

References

1. Andrews, P.B.: *An Introduction to Mathematical Logic and Type Theory: To Truth through Proof.* Academic Press, Inc. 1986.
2. Astesiano, E., Cerioli, M.: Relationships between Logical Frameworks. In Bidoit, M., Choppy, C. (eds.): *Recent Trends in Data Type Specifications. 8th Workshop on Specification of Abstract Data Types.* LNCS 655, pp. 126-143. Springer, 1993.
3. Borzyszkowski, T.: Higher-Order Logic and Theorem Proving for Structured Specifications. In Choppy, C., Bert, D., Mosses, P. (eds.): *Recent Trends in Algebraic Development Techniques, Selected Papers, 14th International Workshop WADT'99.* LNCS 1827, pp. 401-418. Springer, 1999.
4. Bruun, P.M., Dandanell, B., Gørtz, J., Haff, P., Heilmann, S., Prehn, S., Haastrup, P., Reher, J., Snog, H., Zierau, E.: *RAISE Tools Reference Manual.* LACOS/CRI/DOC13/0/V2. 1993.
5. Cerioli, M., Meseguer, J.: May I Borrow Your Logic? In A.M. Borzyszkowski, S. Sokolowski (eds.): *Mathematical Foundation of Computer Science '93*, LNCS 711, pp. 342-351. Springer, 1993.
6. Church, A.: A Formulation of the Simple Theory of Types. *Journal of Symbolic Logic.* Vol. 5, issue 2, pp. 56-68. 1940.
7. George, C.: *RAISE Tools User Guide.* Technical Report 227, UNU/IIST, 2001.
8. Goguen, J., Roşu, G.: Institution Morphisms. *Formal Aspects of Computing*, 13:274-307, 2002.
9. Gordon, M.J.C., Melham, T.F. (eds.): *Introduction to HOL – A theorem proving environment for higher order logic.* Cambridge University Press, 1993.
10. Lindegaard, M.P.: *Proof Support for RAISE – by a Reuse Approach based on Institutions.* Ph.D. thesis, no. 132, IMM, Technical University of Denmark, 2004.
11. Mossakowski, T.: *Relating CASL with Other Specification Languages: the Institution Level.* Theoretical Computer Science 286, pp. 367-475. 2002.
12. Nipkow, T., Paulson, L.C., Wenzel, M.: *Isabelle/HOL – A Proof Assistant for Higher-Order Logic.* LNCS 2283. Springer, 2002.
13. Paulson, L.C.: *Isabelle – A Generic Theorem Prover.* LNCS 828. Springer, 1994.
14. The RAISE Language Group: *The RAISE Specification Language.* The BCS Practitioners Series. Prentice Hall Int., 1992.
15. The RAISE Method Group: *The RAISE Development Method.* The BCS Practitioners Series. Prentice Hall Int., 1995.
16. Tarlecki, A.: Institutions: An Abstract Framework for Formal Specifications. In Astesiano, E., Kreowski, H.-J., Krieg-Brückner, B. (eds.): *Algebraic Foundations of Systems Specification.* IFIP state-of-the-art report. Springer, 1999.
17. Tarlecki, A.: Moving between logical systems. In Haveraaen, M., Owe, O., Dahl, O.-J. (eds.): *Recent Trends in Data Type Specifications. 11th Workshop on Specification of Abstract Data Types.* LNCS 1130, pp. 478-502. Springer, 1996.

Separate Compositional Analysis of Class-Based Object-Oriented Languages

Francesco Logozzo

STIX - Ecole Polytechnique
F-91128 Palaiseau, France
Francesco.Logozzo@polytechnique.fr

Abstract. We present a separate compositional analysis for object-oriented languages. We show how a generic static analysis of a context that uses an object can be split into two separate semantic functions involving respectively only the context and the object. The fundamental idea is to use a regular expressions for approximating the interactions between the context and the object. Then, we introduce an iterative schema for composing the two semantic functions. A first advantage is that the analysis can be parallelized, with a consequent gain in memory and time. Furthermore, the iteration process returns at each step an upper approximation of the concrete semantics, so that the iterations can be stopped as soon as the desired degree of precision is reached. Finally, we instantiate our approach to a core object-oriented language with aliasing.

1 Introduction

One important facet of object-oriented design is encapsulation. Encapsulation hides the objects' inner details from the outside world and allows a hierarchical structuring of code. As a consequence, a program written in the object-oriented style has often the structure of $C[o]$, where $C[\cdot]$ is a context which interacts with an encapsulated object o. The interaction between the context and the object can be of two kinds. The first, direct, one is through method invocations. The context invokes a method of the object which may return some value and modify its internal state. In particular, the value returned by the method can be a pointer to the value of a field. Thus, the object may expose a part of its internal state to the context, which can arbitrarily change the value of the field. We call this second kind of interaction an indirect one.

We are interested in an analysis that exploits the encapsulation features of the object-oriented languages, so that the context and the object can be analyzed separately. In fact, most available analyses are not separated e.g. [7], or they are imprecise as they assume the worst case for the calling context, e.g. [8]. A separate analysis presents several advantages. First, it may significantly reduce the overall analysis cost both in time and space, as e.g. different computers can be used for the analysis of the context and the object. Second, as the total memory consumption is reduced, very precise analyses can be used for the context and/or

C. Rattray et al. (Eds.): AMAST 2004, LNCS 3116, pp. 334–348, 2004.

the object. Third, it allows a form of modular analysis: if o is replaced by another object o′ then the analysis of C[o′] requires just the analysis of o′. For instance, this is the case when C[·] is a function and o and o′ are actual parameters, or when o′ is a refinement of o, e.g. o′ is a sub-object of o.

We present a compositional separate analysis of class-based object oriented languages. We illustrate and prove our results for a core object-oriented language with aliasing. In particular in the considered language the identity of an object is given by the memory address where its environment is stored. This implies that we handle objects aliasing and that objects are semantic rather than syntactic entities. This is in line with mainstream object-oriented languages, so the presented framework can be easily extended to cope with realistic languages.

In Sect. 2, we define the syntax and the concrete semantics for our language and in Sect. 3 we present a generic *monolithic* static analysis of the context and the object $[\![C[o]]\!]^a$, parameterized by an abstract domain D^a. In Sect. 4, we show how it can be split into two semantic functions, Γ and Θ, corresponding respectively to the analysis of the context and the object. The fundamental idea is the use of regular expressions for approximating the interactions between the context and the object. Therefore, we refine the abstract domain D^a with a domain of regular expressions. We have that:

- The object analysis Θ is a function that takes as input a map from objects to regular expressions. It returns a map from objects to their approximations.
- The context analysis Γ is a function that takes as input the approximation of the semantics of the objects. It returns an abstract value and a map from objects to regular expressions.

The functions Θ and Γ are mutually recursive. Thus, we handle this situation with the usual iterative approach. In particular, we begin by assuming the worst-case for the objects approximations and the contexts. Then, we show that the iterations form a chain of increasing precision, each step being a sound upper-approximation of $[\![C[o]]\!]^a$. This implies that the iterations can be stopped as soon as the desired degree of precision is reached, enabling a trade-off between precision and cost.

2   Concrete Semantics

We begin by defining the syntax and the semantics of a minimal Java-like language. We make some simplifying assumptions. First, in order to simplify the notation we assume the existence of just one class. The generalization of the results to the case of an arbitrary number of classes is straightforward. Second, we distinguish between a context, for which we will give the detailed syntax, and a class, for which we give just the interface, i.e. the fields and the methods, but not the definition of the methods body. This is not restrictive, as the notion of context is relative. For example, a context that accesses an object o can be the body of a method of an object o′. Such an o′ can be accessed by further context, so that the contexts can be encapsulated.

2.1 Syntax

The class syntax can be abstractly modeled as a triplet $\langle \mathtt{init}, \mathtt{F}, \mathtt{M} \rangle$ where $\mathtt{init}$ is the class constructor, $\mathtt{F}$ is a set of variables and $\mathtt{M}$ is a set of function definitions. We do not require to have typed fields or methods and without any loss of generality we assume that a class has just a single constructor and each access to a field $\mathtt{f}$ is done through $\mathtt{set\_f}/\mathtt{get\_f}$.

The syntax of a context is quite standard, except that we distinguish three kinds of assignments: the assignment of the value of a side-effects free expression to a variable, the assignment of the address pointed by a variable to another one and the assignment of the return value of a method call to a variable. So, let $\mathtt{x}$, $\mathtt{x_1}$, $\mathtt{x_2}$ and $\mathtt{o}$ be variables, let $\mathtt{E}$ be an expression and let $\mathtt{b}$ be a boolean expression. Then the language of contexts is generated by the following grammar:

$$\mathtt{C} ::= \ \mathtt{A \ o} \ = \ \mathtt{new \ A(E)} \mid \mathtt{C_1 ; C_2} \mid \mathtt{skip} \mid \mathtt{x} = \mathtt{E} \mid \mathtt{x_1} = \mathtt{x_2}$$
$$\mid \mathtt{x} = \mathtt{o.m(E)} \mid \mathtt{if \ b \ then \ C_1 \ else \ C_2} \mid \mathtt{while \ b \ do \ C}.$$

$\mathtt{C}$ denotes an arbitrary context, $\mathtt{C}[\cdot]$ denotes a context that may contain one or more objects and $\mathtt{C}[\mathtt{o}]$ denotes a context that uses an object $\mathtt{o}$. However, as we allow aliasing of objects, we cannot give the formal definition of $\mathtt{C}[\mathtt{o}]$ on a strictly syntactic basis. Therefore, such a definition is postponed to the next section.

2.2 Semantic Domains

The first step for the specification of the concrete semantics is the definition of the concrete domain. In our case, we are interested in a domain that models the fact that an object has its own identity and environment. Moreover, we need to express object aliasing. In order to fulfill the above requirements, we consider a domain whose elements are pairs of environments and stores. An environment is a map from variables to memory addresses, $\mathsf{Env} = [\mathsf{Var} \to \mathsf{Addr}]$, and a store is a map from addresses to memory elements, $\mathsf{Store} = [\mathsf{Addr} \to \mathsf{Val}]$. A memory element can be a primitive value as well as an environment or an address, i.e. $\mathsf{Env} \subseteq \mathsf{Val}$ and $\mathsf{Addr} \subseteq \mathsf{Val}$. In such a setting, the identity of an object is the memory address where its environment is stored. Therefore, two distinct variables are aliases for an object if they reference the same memory address.

2.3 Object Semantics

We consider an input/output semantics for the class constructor and the methods. The semantics of the constructor is a function $\mathcal{I}[\![\mathtt{init}]\!] \in [\mathsf{Val} \times \mathsf{Store} \to \mathsf{Env} \times \mathsf{Store}]$ which takes as input the constructor's actual parameter and a store. It returns the (initialized) object environment and the (possibly modified) store. It is worth noting that the constructor does not return any value to the context.

The semantics of a method $\mathtt{m}$ is a function $\mathcal{M}[\![\mathtt{m}]\!] \in [\mathsf{Val} \times \mathsf{Env} \times \mathsf{Store} \to \mathsf{Val} \times \mathsf{Env} \times \mathsf{Store}]$. It takes as input the method's actual parameter, the object

$$\mathcal{C}[\![A \text{ o } = \text{ new } A(E)]\!] = \lambda e, s. let \ v = \mathcal{E}[\![E]\!](e, s), a = \text{alloc}(s),$$
$$(e_0, s_0) = \mathcal{I}[\![\text{init}]\!](v, s),$$
$$in \ (e[o \mapsto a], s_0[a \mapsto e_0])$$
$$\mathcal{C}[\![C_1; C_2]\!] = \lambda e, s.\mathcal{C}[\![C_2]\!](\mathcal{C}[\![C_1]\!](e, s)) \qquad \mathcal{C}[\![\text{skip}]\!] = \lambda e, s.(e, s)$$
$$\mathcal{C}[\![x = E]\!] = \lambda e, s.(e, s[e(x) \mapsto \mathcal{E}[\![E]\!](e, s)])$$
$$\mathcal{C}[\![x_1 = x_2]\!] = \lambda e, s.(e, s[e(x_1) \mapsto e(x_2)])$$
$$\mathcal{C}[\![x = \text{o.m}(E)]\!] = \lambda e, s. let \ v = \mathcal{E}[\![E]\!](e, s), (v_0, e_0, s_0) = \mathcal{M}[\![m]\!](v, s(e(o)), s),$$
$$in \ (e, s_0[e(x) \mapsto v_0, e(o) \mapsto e_0])$$
$$\mathcal{C}[\![\text{if b then } C_1 \text{ else } C_2]\!] = \lambda e, s. if \ \mathcal{B}[\![b]\!](e, s) = \text{tt } then \ \mathcal{C}[\![C_1]\!](e, s) \ else \ \mathcal{C}[\![C_2]\!](e, s)$$
$$\mathcal{C}[\![\text{while b do } C]\!] = \text{lfp} \lambda \phi. \lambda e, s. if \ \mathcal{B}[\![b]\!](e, s) = \text{tt } then \ \phi(\mathcal{C}[\![C]\!](e, s)) \ else \ (e, s)$$

Fig. 1. Semantics of the context

environment and the store. It returns a (possibly void) value, the new object environment and the modified store. It is worth noting that as $\text{Addr} \subseteq \text{Val}$ the method may expose a part of the object's internal state to the context.

2.4   Context Semantics

We define the context semantics in denotational style, by induction on the syntax. The semantics of expressions and that of the boolean expressions are assumed to be side-effect free, such that $\mathcal{E}[\![e]\!] \in [\text{Env} \times \text{Store} \rightarrow \text{Val}]$ and $\mathcal{B}[\![b]\!] \in [\text{Env} \times \text{Store} \rightarrow \{\text{tt}, \text{ff}\}]$. A function $\text{alloc} \in [\text{Store} \rightarrow \text{Addr}]$ returns a fresh memory address. The semantics of a context, $\mathcal{C}[\![C]\!] \in [\text{Env} \times \text{Store} \rightarrow \text{Env} \times \text{Store}]$ is given in Fig. 1.

Some comments on the context semantics. When a class A is instantiated, the initial value is evaluated, and the class constructor is invoked with that value and the store. The class constructor returns the environment e_0 of the new object and the modified store s_0. Then the environment and the store change, so that o points to the memory allocated for storing e_0. When a method of the object o is invoked, its environment is fetched from the memory and passed to the method. This implies that the method has no access to the caller environment, but only to that of the object it belongs to. In other words, the context has the burden of setting the right environment for a method call, so that the handling of this is somehow transparent to the callee. For the rest, the semantics in Fig. 1 is a quite standard denotational semantics. In particular, the loop semantics is handled by the least fixpoint operator on the flat Scott-domain $\text{Env} \times \text{Store} \cup \{\bot\}$.

Using the context semantics, we can formally define the writing $C[o]$, i.e. the context $C[\cdot]$ that uses an object o. Let $(e_0, s_0) \in \text{Env} \times \text{Store}$, such that $\mathcal{C}[\![C]\!](e_0, s_0) = (e, s)$. Then a context C uses an object o if $\exists x \in \text{Var}.e(x) = o \wedge s(e(x)) \in \text{Env}$.

2.5   Collecting Semantics

A semantic property is a set of possible semantics of a program. The set of seman-
tic properties $\mathcal{P}(\mathsf{Env} \times \mathsf{Store})$ is a complete boolean lattice $\langle \mathcal{P}(\mathsf{Env} \times \mathsf{Store}), \subseteq, \emptyset,$
$\mathsf{Env} \times \mathsf{Store}, \cup, \cap \rangle$ for subset inclusion, that is logical implication. The standard
collecting semantics of a program, $[\![\mathsf{C}]\!](In) = \{\mathsf{C}[\![\mathsf{C}]\!](\mathsf{e},\mathsf{s}) \mid (\mathsf{e},\mathsf{s}) \in In\}$, is the
strongest program property. The goal of a static analysis is to find a computable
approximation of $[\![\mathsf{C}]\!]$.

3   Monolithic Abstract Semantics

We proceed to the definition of a generic abstract semantics for the language
presented in the previous section. First we consider the abstract semantic domains. Afterward, we present the abstract semantics for the class constructor
and methods, and for the context.

3.1   Abstract Semantic Domains

The values in $\mathcal{P}(\mathsf{Val})$ are approximated by an abstract domain $\mathsf{Val}^{\mathsf{a}}$. The correspondence between the two domains is given by the Galois connection [2]:

$$\langle \mathcal{P}(\mathsf{Val}), \subseteq, \emptyset, \mathsf{Val}, \cup, \cap \rangle \xrightleftharpoons[\alpha_v]{\gamma_v} \langle \mathsf{Val}^{\mathsf{a}}, \sqsubseteq^{\mathsf{a}}_v, \bot^{\mathsf{a}}_v, \top^{\mathsf{a}}_v, \sqcup^{\mathsf{a}}_v, \sqcap^{\mathsf{a}}_v \rangle.$$

The set of abstract addresses is $\mathsf{Addr}^{\mathsf{a}} \subseteq \mathsf{Val}^{\mathsf{a}}$. We assume $\mathsf{Addr}^{\mathsf{a}}$ to be a sublattice
of $\mathsf{Val}^{\mathsf{a}}$. If $\mathsf{o} \in \mathsf{Addr}$ denotes an object in the concrete, then $\vartheta = \alpha_v(\{\mathsf{o}\})$ is
the corresponding abstract address. On the other hand, ϑ stands for the set
of concrete addresses $\gamma_v(\vartheta)$, which may contain several objects. Therefore, ϑ
approximates all the objects in $\gamma_v(\vartheta)$. We call ϑ an abstract object.

The domain D^{a} abstracts the domain of concrete properties $\mathcal{P}(\mathsf{Env} \times \mathsf{Store})$
by means of a Galois connection:

$$\langle \mathcal{P}(\mathsf{Env} \times \mathsf{Store}), \subseteq, \emptyset, \mathsf{Env} \times \mathsf{Store}, \cup, \cap \rangle \xrightleftharpoons[\alpha]{\gamma} \langle \mathsf{D}^{\mathsf{a}}, \sqsubseteq^{\mathsf{a}}, \bot^{\mathsf{a}}, \top^{\mathsf{a}}, \sqcup^{\mathsf{a}}, \sqcap^{\mathsf{a}} \rangle.$$

We call an element of D^{a} an abstract state. In general, the domain D^{a} is a
relational abstraction of $\mathcal{P}(\mathsf{Env} \times \mathsf{Store})$. We consider two projections such that
for each $\mathsf{d}^{\mathsf{a}} \in \mathsf{D}^{\mathsf{a}}$, $\mathsf{d}^{\mathsf{a}} \downharpoonright_{\mathsf{e}}$ and $\mathsf{d}^{\mathsf{a}} \downharpoonright_{\mathsf{s}}$ are, respectively, the projections of d^{a} on
the environment and the store. We use the brackets $\lfloor \cdot \rfloor$ to denote the inverse
operation of the projection, i.e. given an abstraction for the environment and
the store it returns the abstract state. Moreover, some operations are defined
on D^{a}: $\mathsf{alloc}^{\mathsf{a}}$, $\mathsf{assign}^{\mathsf{a}}$, $\mathsf{true}^{\mathsf{a}}$ and $\mathsf{false}^{\mathsf{a}}$. The first one, $\mathsf{alloc}^{\mathsf{a}} \in [\mathsf{D}^{\mathsf{a}} \to \mathsf{Addr}^{\mathsf{a}}]$, is
the abstract counterpart for memory allocation. It takes an approximation of
the state and it returns an abstract address where the object environment can
be stored. It satisfies the soundness requirement: $\forall \mathsf{d}^{\mathsf{a}} \in \mathsf{D}^{\mathsf{a}}.\{\mathsf{alloc}(\mathsf{s}) \mid (\mathsf{e},\mathsf{s}) \in$
$\gamma(\mathsf{d}^{\mathsf{a}})\} \subseteq \gamma_v(\mathsf{alloc}^{\mathsf{a}}(\mathsf{d}^{\mathsf{a}}))$.

The function $\mathsf{assign}^{\mathsf{a}} \in [\mathsf{D}^{\mathsf{a}} \times (\mathsf{Var} \times \mathsf{D}^{\mathsf{a}})^{+} \to \mathsf{D}^{\mathsf{a}}]$ handles the assignment in the
abstract domain. It takes as input an abstract state and a non-empty list of bind-
ings from variables to values. It returns the new abstract state. With an abuse

of notation, we sometimes write $\mathsf{assign}^{\mathsf{a}}(\mathsf{d}^{\mathsf{a}}, \mathsf{d}^{\mathsf{a}} \downarrow_s \mapsto \mathsf{d}^{\mathsf{a}}_0 \downarrow_s)$ to denote that the abstract store $\mathsf{d}^{\mathsf{a}} \downarrow_s$ is updated by $\mathsf{d}^{\mathsf{a}}_0 \downarrow_s$. Moreover, $\mathsf{true}^{\mathsf{a}}, \mathsf{false}^{\mathsf{a}} \in [\mathrm{BExp} \times \mathsf{D}^{\mathsf{a}} \rightarrow \mathsf{D}^{\mathsf{a}}]$ are the functions that given a boolean expression and an abstract element d^{a} return an abstraction of the pairs $(e, s) \in \gamma(\mathsf{d}^{\mathsf{a}})$ that make the condition respectively true or false. For instance $\mathsf{true}^{\mathsf{a}}$ is such that:

$$\forall b \in \mathrm{BExp}. \forall \mathsf{d}^{\mathsf{a}} \in \mathsf{D}^{\mathsf{a}}. \{(e, s) \mid \mathcal{B}[\![b]\!](e, s) = \mathsf{tt}\} \cap \gamma(\mathsf{d}^{\mathsf{a}}) \subseteq \mathsf{true}^{\mathsf{a}}(b, \mathsf{d}^{\mathsf{a}}).$$

3.2 Abstract Object Semantics

The abstract semantics for the class constructor and methods mimics the concrete one. Therefore, the abstract counterpart for the constructor semantics is a function $\mathcal{J}[\![\mathsf{init}]\!]^{\mathsf{a}} \in [\mathsf{Val}^{\mathsf{a}} \times \mathsf{D}^{\mathsf{a}} \rightarrow \mathsf{D}^{\mathsf{a}}]$, which takes an abstract value and an abstract state and returns an abstract environment, that of the new object, and an abstract store. The abstract semantics for methods is a function $\mathcal{M}[\![m]\!]^{\mathsf{a}} \in [\mathsf{Val}^{\mathsf{a}} \times \mathsf{D}^{\mathsf{a}} \rightarrow \mathsf{Val}^{\mathsf{a}} \times \mathsf{D}^{\mathsf{a}}]$. The input is an abstract value and an abstract state, and the output is an abstraction of the return value and a modified abstract state.

3.3 Monolithic Abstract Context Semantics

The abstract semantics for contexts is defined on the top of the abstract semantics for the expressions and the basic operations of the abstract domain D^{a}. In particular, the abstract semantics of expressions is $\mathcal{E}[\![e]\!]^{\mathsf{a}} \in [\mathsf{D}^{\mathsf{a}} \rightarrow \mathsf{Val}^{\mathsf{a}}]$. It must satisfy the soundness requirement: $\forall (e, s) \in \mathsf{Env} \times \mathsf{Store}. \, \mathcal{E}[\![e]\!](e, s) \in \gamma_v \circ \mathcal{E}[\![e]\!]^{\mathsf{a}} \circ \alpha(\{(e, s)\})$.

The generic abstract semantics mimics the concrete semantics. In particular, when a method m is invoked, the corresponding abstract function $\mathcal{M}[\![m]\!]^{\mathsf{a}}$ is used. In practice, this means that the body of a method m is analyzed from scratch at each invocation. Therefore the encapsulation of the object w.r.t. context is not exploited in the analysis. We call such an abstract semantics a *monolithic* abstract semantics in order to differentiate it from the separate compositional abstract semantics that we will introduce in the next section.

Finally, the monolithic abstract context semantics $[\![C]\!]^{\mathsf{a}} \in [\mathsf{D}^{\mathsf{a}} \rightarrow \mathsf{D}^{\mathsf{a}}]$ is defined in Fig. 2. The semantics in Fig. 2 is quite similar to the concrete one in Fig. 1. It is worth noting that the burden of handling the assignment is left to the underlying abstract domain D^{a}, and in particular to the function $\mathsf{assign}^{\mathsf{a}}$. We use the notation $\mathsf{lfp}^{\sqsubseteq}_{\mathsf{d}} \lambda x. F(x, z)$ to denote the least fixpoint w.r.t. the order $\sqsubseteq$, greater than d of the equation $F(x, z) = (x, y)$, for some z and y. Nevertheless, in general the abstract domains $\mathsf{Val}^{\mathsf{a}}$ and D^{a} may not respect the Ascending Chain Condition (ACC), so that the convergence of the analysis is enforced through the widening operators $\nabla^{\mathsf{a}}_v \in [\mathsf{Val}^{\mathsf{a}} \times \mathsf{Val}^{\mathsf{a}} \rightarrow \mathsf{Val}^{\mathsf{a}}]$ and $\nabla^{\mathsf{a}} \in [\mathsf{D}^{\mathsf{a}} \times \mathsf{D}^{\mathsf{a}} \rightarrow \mathsf{D}^{\mathsf{a}}]$. The soundness of the above semantics is a consequence of the definitions of this section:

Theorem 1 (Soundness of $[\![C]\!]^{\mathsf{a}}$). *The monolithic context abstract semantics is a sound approximation of the concrete semantics:* $\forall In \in \mathcal{P}(\mathsf{Env} \times \mathsf{Store})$. $[\![C]\!](In) \subseteq \gamma \circ [\![C]\!]^{\mathsf{a}} \circ \alpha(In)$.

$$[\![\texttt{A o = new A(E)}]\!]^a = \lambda d^a.let\ v^a = \mathcal{E}[\![E]\!]^a(d^a), \vartheta = alloc^a(d^a),$$
$$d_0{}^a = \mathcal{J}[\![\texttt{init}]\!]^a(v^a, d^a)$$
$$in\ assign^a(d^a, \vartheta \mapsto d_0{}^a \downharpoonright_e, d^a \downharpoonright_s \mapsto d_0{}^a \downharpoonright_s)$$

$$[\![\texttt{C}_1; \texttt{C}_2]\!]^a = \lambda d^a.[\![C_2]\!]^a([\![C_1]\!]^a(d^a)) \qquad [\![\texttt{skip}]\!]^a = \lambda d^a.d^a$$

$$[\![\texttt{x = E}]\!]^a = \lambda d^a.assign^a(d^a, x \mapsto \mathcal{E}[\![E]\!]^a(d^a))$$

$$[\![\texttt{x}_1 = \texttt{x}_2]\!]^a = \lambda d^a.assign^a(d^a, x_1 \mapsto d^a \downharpoonright_e(x_2))$$

$$[\![\texttt{x = o.m(E)}]\!]^a = \lambda d^a.let\ v^a = \mathcal{E}[\![E]\!]^a(d^a), \vartheta = d^a \downharpoonright_e(o),$$
$$(v_0{}^a, d_0{}^a) = \mathcal{M}[\![m]\!]^a(v^a, \lfloor d^a \downharpoonright_s(\vartheta), d^a \downharpoonright_s \rfloor),$$
$$in\ assign^a(d^a, x \mapsto v_0{}^a, \vartheta \mapsto d_0{}^a \downharpoonright_e, d^a \downharpoonright_s \mapsto d_0{}^a \downharpoonright_s)$$

$$[\![\texttt{if b then C}_1 \texttt{ else C}_2]\!]^a = \lambda d^a.[\![C_1]\!]^a(true^a(b, d^a)) \sqcup^a [\![C_2]\!]^a(false^a(b, d^a))$$

$$[\![\texttt{while b do C}]\!]^a = \lambda d^a.false^a(b, lfp_{d^a}^{\subseteq} \lambda x.[\![C]\!]^a(true^a(b, x)))$$

Fig. 2. Monolithic abstract semantics

4   Separate Abstract Semantics

The abstract semantics $[\![\cdot]\!]^a$ defined in the previous section does not take into account the encapsulation features of object-oriented languages, so that, for instance each time a method of an object is invoked, its body must be analyzed. In this section we show how to split $[\![\cdot]\!]^a$ into two parts. The first part analyzes the context using an approximation of the object. The latter analyzes the object using an approximation of the context.

4.1   Regular Expressions Domain

The main idea for the separate analysis is to refine the abstract domain D^a with the abstract domain R of regular expressions over the infinite alphabet $(\{\texttt{init}\} \cup \mathcal{P}(\mathsf{M})) \times \mathsf{Val}^a \times \mathsf{D}^a$. Given an object, the intuition behind the refinement is to use a regular expression to abstract the method's invocations performed by the context. In particular, each *letter* in the alphabet represents a set of methods that can be invoked, an approximation of their input values and an approximation of the state. Such a regular expression is built during the analysis of the context. Then it is used for the analysis of the object.

The definition of the regular expressions in R is given by structural induction. The base cases are the *null* string ε and the letters l of the alphabet $(\{\texttt{init}\} \cup \mathcal{P}(\mathsf{M})) \times \mathsf{Val}^a \times \mathsf{D}^a$. Then, if r_1 and r_2 are regular expressions so are the concatenation $r_1 \cdot r_2$, the union $r_1 + r_2$ and the Kleene-closure r_1^*.

The language generated by a regular expression r is defined by structural induction:

$$\mathcal{L}(\langle ms, v^a, s^a \rangle) = \{\langle m, v, s \rangle \mid m \in ms, v \in \gamma_v(v^a), s \in \gamma(s^a)\} \qquad \mathcal{L}(\varepsilon) = \emptyset$$
$$\mathcal{L}(r_1 \cdot r_2) = \{s_1 \cdot s_2 \mid s_1 \in \mathcal{L}(r_1), s_2 \in \mathcal{L}(r_2)\} \qquad \mathcal{L}(r_1 + r_2) = \mathcal{L}(r_1) \cup \mathcal{L}(r_2)$$
$$\mathcal{L}(r^*) = lfp_{\emptyset}^{\subseteq} \lambda X.\mathcal{L}(r) \cup \{s_1 \cdot s_2 \mid s_1 \in X, s_2 \in \mathcal{L}(r)\}.$$

$$\top_r \nabla_r x = x \nabla_r \top_r = \top_r \qquad\qquad x \nabla_r \varepsilon = \varepsilon \nabla_r x = x$$
$$\langle \mathsf{m}, \mathsf{v}^{\mathsf{a}}, \mathsf{s}^{\mathsf{a}} \rangle \nabla_r \langle \mathsf{m}_1, \mathsf{v}_1^{\mathsf{a}}, \mathsf{s}_1^{\mathsf{a}} \rangle = \langle \mathsf{m} \cup \mathsf{m}_1, \mathsf{v}^{\mathsf{a}} \nabla_v^{\mathsf{a}} \mathsf{v}_1^{\mathsf{a}}, \mathsf{s}^{\mathsf{a}} \nabla^{\mathsf{a}} \mathsf{s}_1^{\mathsf{a}} \rangle \qquad (r_1 \cdot r_2) \nabla_r n = (r_1 \nabla_r n) \cdot r_2$$
$$(r_1 + r_2) \nabla_r n = (r_1 \nabla_r n) + (r_2 \nabla_r n) \qquad\qquad r^* \nabla_r n = (r \nabla_r n)^*$$
$$(r_1 \cdot r_2) \nabla_r (r_1' \cdot r_2') = (r_1 \nabla_r r_1') \cdot (r_2 \nabla_r r_2') \qquad\qquad r_1^* \nabla_r r_2^* = (r_1 \nabla_r r_2)^*$$
$$(r_1 + r_2) \nabla_r (r_1' + r_2') = (r_1 \nabla_r r_1') + (r_2 \nabla_r r_2')$$
$$x \nabla_r y = \top_r \qquad \text{in all the other cases}$$

Fig. 3. Widening on regular expressions

The order on regular expressions is a direct consequence of the above definition: $\forall r_1, r_2 \in \mathsf{R}.r_1 \sqsubseteq_r r_2 \Longleftrightarrow \mathcal{L}(r_1) \subseteq \mathcal{L}(r_2)$. So, two expressions are equivalent if they generate the same language: $r_1 \equiv r_2 \Longleftrightarrow \mathcal{L}(r_1) = \mathcal{L}(r_2)$. From now on, we consider all the operations and definitions on regular expressions modulo the equivalence $\equiv$. The expression $\top_r = \langle \{\mathtt{init}\} \cup \mathsf{M}, \top_v^{\mathsf{a}}, \top^{\mathsf{a}} \rangle^* \in \mathsf{R}$ stands for a context that may invoke any method, with any input value and with any memory configuration for a non-specified number of times. So, it gives no information. Thus, it is the largest element of $\langle \mathsf{R}, \sqsubseteq_r \rangle$. The join of two regular expressions is simply their union: $\forall r_1, r_2 \in \mathsf{R}.r_1 \sqcup_r r_2 = r_1 + r_2$. Similarly, the meet operator $\sqcap_r$ can be defined, so that $\langle \mathsf{R}, \sqsubseteq_r, \varepsilon, \top_r, \sqcup_r, \sqcap_r \rangle$ is a complete lattice.

The domain R does not satisfy the ACC, so we define the widening operator of Fig.3 to deal with strictly increasing chains of regular expressions. There are two intuitions behind the operator in Fig.3. The first one is to preserve the syntactic structure of the regular expressions between two successive iterations, so that the number of $\{\cdot, +,^*\}$ does not increase. The second one is to propagate the ∇_r inside the regular expressions in order to use the widenings on $\mathsf{Val}^{\mathsf{a}}$ and D^{a}. Convergence is assured as M is a finite set, and ∇_v^{a} and ∇^{a} are widenings on the respective domains.

For the purpose of our analysis, we need to associate with each abstract address, i.e. a set of concrete objects, a regular expression that denotes the interaction of the context on it. As a consequence we consider the functional lifting[1] $\dot{\mathsf{R}} = [\mathsf{Addr}^{\mathsf{a}} \to \mathsf{R}]$. The order $\dot{\sqsubseteq}_r$ is defined pointwise: $\forall \dot{r}_1, \dot{r}_2 \in \dot{\mathsf{R}}.\dot{r}_1 \dot{\sqsubseteq}_r \dot{r}_2 \Leftrightarrow \forall \vartheta \in \mathsf{Addr}^{\mathsf{a}}.\dot{r}_1(\vartheta) \sqsubseteq_r \dot{r}_2(\vartheta)$. In a similar way, the join and the meet are defined point-wise, so that $\langle \dot{\mathsf{R}}, \dot{\sqsubseteq}_r, \lambda \vartheta.\varepsilon, \lambda \vartheta.\top_r, \dot{\sqcup}_r, \dot{\sqcap}_r \rangle$ is a complete lattice. We call an element $\dot{r} \in \dot{\mathsf{R}}$ an interaction history.

4.2   Separate Object Analysis

The goal of the separate object analysis is to infer an object invariant and the method postconditions when the instantiation context is approximated by a regular expression. Thus, the input of the abstract semantics $\mathcal{O}[\![\vartheta]\!]^{\mathsf{a}}$ is a regular

---

[1] We use the notation that given a domain D^{a}, $\dot{\mathsf{D}}^{\mathsf{a}}$ stands for the domain of functions $[\mathsf{Addr}^{\mathsf{a}} \to \mathsf{D}^{\mathsf{a}}]$. The operations on $\dot{\mathsf{D}}^{\mathsf{a}}$ are the pointwise extension of that of D^{a}: given an operation $\diamond$, then $\forall \dot{d}_1^{\mathsf{a}}, \dot{d}_2^{\mathsf{a}} \in \dot{\mathsf{D}}^{\mathsf{a}}. \dot{d}_1^{\mathsf{a}} \dot{\diamond} \dot{d}_2^{\mathsf{a}} = \lambda \vartheta.\dot{d}_1^{\mathsf{a}}(\vartheta) \diamond \dot{d}_2^{\mathsf{a}}(\vartheta)$.

$$\mathcal{O}[\![\vartheta]\!]^{\mathsf{a}}(\varepsilon, \langle \mathsf{i}^{\mathsf{a}}, \mathsf{p}^{\mathsf{a}} \rangle) = \langle \mathsf{i}^{\mathsf{a}}, \mathsf{p}^{\mathsf{a}} \rangle$$

$$\mathcal{O}[\![\vartheta]\!]^{\mathsf{a}}(\langle \{\mathtt{init}\}, \mathsf{v}^{\mathsf{a}}, \mathsf{s}^{\mathsf{a}} \rangle, \langle \mathsf{i}^{\mathsf{a}}, \mathsf{p}^{\mathsf{a}} \rangle) = let \ \langle \mathsf{e}_0^{\mathsf{a}}, \mathsf{s}_0^{\mathsf{a}} \rangle = \mathcal{I}[\![\mathtt{init}]\!]^{\mathsf{a}}(\mathsf{v}^{\mathsf{a}}, \mathsf{s}^{\mathsf{a}} \sqcup^{\mathsf{a}} \mathsf{i}^{\mathsf{a}})$$
$$in \ \langle \mathsf{i}^{\mathsf{a}} \sqcup^{\mathsf{a}} \mathsf{s}_0^{\mathsf{a}}, \mathsf{p}^{\mathsf{a}}[\mathtt{init} \mapsto \langle \perp_v^{\mathsf{a}}, \mathsf{e}_0^{\mathsf{a}} \rangle] \rangle$$

$$\mathcal{O}[\![\vartheta]\!]^{\mathsf{a}}(\langle \mathsf{ms}, \mathsf{v}^{\mathsf{a}}, \mathsf{s}^{\mathsf{a}} \rangle, \langle \mathsf{i}^{\mathsf{a}}, \mathsf{p}^{\mathsf{a}} \rangle) = let \ \forall \mathsf{m}_i \in \mathsf{ms}. \langle \mathsf{v}_i^{\mathsf{a}}, \mathsf{s}_i^{\mathsf{a}} \rangle = \mathcal{M}[\![\mathsf{m}_i]\!]^{\mathsf{a}}(\mathsf{v}^{\mathsf{a}}, \mathsf{s}^{\mathsf{a}} \sqcup^{\mathsf{a}} \mathsf{i}^{\mathsf{a}}), \ \langle \mathsf{w}_i^{\mathsf{a}}, \mathsf{q}_i^{\mathsf{a}} \rangle = \mathsf{p}^{\mathsf{a}}(\mathsf{m}_i)$$
$$in \ (\mathsf{i}^{\mathsf{a}} \sqcup^{\mathsf{a}} \bigsqcup{}^{\mathsf{a}} \mathsf{s}_i^{\mathsf{a}}, \mathsf{p}^{\mathsf{a}}[\mathsf{m}_i \mapsto \langle \mathsf{w}_i^{\mathsf{a}} \sqcup_v^{\mathsf{a}} \mathsf{v}_i^{\mathsf{a}}, \mathsf{q}_i^{\mathsf{a}} \sqcup^{\mathsf{a}} \mathsf{s}_i^{\mathsf{a}} \rangle])$$

$$\mathcal{O}[\![\vartheta]\!]^{\mathsf{a}}(r_1 \cdot r_2, \langle \mathsf{i}^{\mathsf{a}}, \mathsf{p}^{\mathsf{a}} \rangle) = let \ (\mathsf{i}_1^{\mathsf{a}}, \mathsf{p}_1^{\mathsf{a}}) = \mathcal{O}[\![\vartheta]\!]^{\mathsf{a}}(r_1, \langle \mathsf{i}^{\mathsf{a}}, \mathsf{p}^{\mathsf{a}} \rangle), (\mathsf{i}_2^{\mathsf{a}}, \mathsf{p}_2^{\mathsf{a}}) = \mathcal{O}[\![\vartheta]\!]^{\mathsf{a}}(r_2, (\mathsf{i}_1^{\mathsf{a}}, \mathsf{p}_1^{\mathsf{a}}))$$
$$in \ (\mathsf{i}^{\mathsf{a}}, \mathsf{p}^{\mathsf{a}}) \sqcup_o^{\mathsf{a}}(\mathsf{i}_1^{\mathsf{a}}, \mathsf{p}_1^{\mathsf{a}}) \sqcup_o^{\mathsf{a}}(\mathsf{i}_2^{\mathsf{a}}, \mathsf{p}_2^{\mathsf{a}})$$

$$\mathcal{O}[\![\vartheta]\!]^{\mathsf{a}}(r_1 + r_2, \langle \mathsf{i}^{\mathsf{a}}, \mathsf{p}^{\mathsf{a}} \rangle) = let \ (\mathsf{i}_1^{\mathsf{a}}, \mathsf{p}_1^{\mathsf{a}}) = \mathcal{O}[\![\vartheta]\!]^{\mathsf{a}}(r_1, \langle \mathsf{i}^{\mathsf{a}}, \mathsf{p}^{\mathsf{a}} \rangle), (\mathsf{i}_2^{\mathsf{a}}, \mathsf{p}_2^{\mathsf{a}}) = \mathcal{O}[\![\vartheta]\!]^{\mathsf{a}}(r_2, \langle \mathsf{i}^{\mathsf{a}}, \mathsf{p}^{\mathsf{a}} \rangle)$$
$$in \ (\mathsf{i}^{\mathsf{a}}, \mathsf{p}^{\mathsf{a}}) \sqcup_o^{\mathsf{a}}(\mathsf{i}_1^{\mathsf{a}}, \mathsf{p}_1^{\mathsf{a}}) \sqcup_o^{\mathsf{a}}(\mathsf{i}_2^{\mathsf{a}}, \mathsf{p}_2^{\mathsf{a}})$$

$$\mathcal{O}[\![\vartheta]\!]^{\mathsf{a}}(r^*, \langle \mathsf{i}^{\mathsf{a}}, \mathsf{p}^{\mathsf{a}} \rangle) = \mathrm{lfp}_{\langle \mathsf{i}^{\mathsf{a}}, \mathsf{p}^{\mathsf{a}} \rangle}^{\sqsubseteq_o^{\mathsf{a}}} \ \lambda x, y. \mathcal{O}[\![\vartheta]\!]^{\mathsf{a}}(r, (x, y))$$

Fig. 4. Separate object abstract semantics

expression r and an initial abstract value for the object fields and the method preconditions. The output is an invariant for the object fields and the method postconditions, under the context represented by r. A postcondition is a pair consisting of an approximation of the return value and an abstract state. Thus, the result is an element of the abstract domain $\mathsf{O}^{\mathsf{a}} = \mathsf{D}^{\mathsf{a}} \times [\mathsf{M} \to \mathsf{Val}^{\mathsf{a}} \times \mathsf{D}^{\mathsf{a}}]$. From basic domain theory, the orders on D^{a} and $\mathsf{Val}^{\mathsf{a}}$ induce the order on O^{a}. So, the order is $\sqsubseteq_o^{\mathsf{a}} = \sqsubseteq^{\mathsf{a}} \times (\sqsubseteq_v^{\mathsf{a}} \times \sqsubseteq^{\mathsf{a}})$, the least element is $\perp_o^{\mathsf{a}} = \langle \perp^{\mathsf{a}}, \lambda m. \langle \perp_v^{\mathsf{a}}, \perp^{\mathsf{a}} \rangle \rangle$ and the largest $\top_o^{\mathsf{a}} = \langle \top^{\mathsf{a}}, \lambda m. \langle \top_v^{\mathsf{a}}, \top^{\mathsf{a}} \rangle \rangle$. The meet, the join and the widening can be defined in a similar fashion, so that $\langle \mathsf{O}^{\mathsf{a}}, \sqsubseteq_o^{\mathsf{a}}, \perp_o^{\mathsf{a}}, \top_o^{\mathsf{a}}, \sqcup_o^{\mathsf{a}}, \sqcap_o^{\mathsf{a}} \rangle$ is a complete lattice. Finally, the separate object abstract semantics, $\mathcal{O}[\![\vartheta]\!]^{\mathsf{a}} \in [\mathsf{R} \times \mathsf{O}^{\mathsf{a}} \to \mathsf{O}^{\mathsf{a}}]$, is defined in Fig. 4. Its definition is by structural induction on the regular expression r.

Some comments on the separate object semantics. The base cases are the empty expression ε and the letters $\langle \mathsf{ms}, \mathsf{v}^{\mathsf{a}}, \mathsf{s}^{\mathsf{a}} \rangle$ and $\langle \{\mathtt{init}\}, \mathsf{v}^{\mathsf{a}}, \mathsf{s}^{\mathsf{a}} \rangle$. In the first case the context does not perform any action, so that the state of the object does not change at all. In the latter, the context may invoke any method $\mathsf{m}_i \in \mathsf{ms}$. The abstract value $\bigsqcup{}^{\mathsf{a}} \mathsf{s}_i^{\mathsf{a}}$ approximates the object field values after calling the method m_1 or m_2 or ... or m_n. As a consequence, $\mathsf{i}^{\mathsf{a}} \sqcup^{\mathsf{a}} \bigsqcup{}^{\mathsf{a}} \mathsf{s}_i^{\mathsf{a}}$ approximates the object fields before and after executing any method in ms. Hence, it is an object invariant. On the other hand, if $\langle \mathsf{w}_i^{\mathsf{a}}, \mathsf{q}_i^{\mathsf{a}} \rangle$ is the initial approximation of the return values and the states reached after the execution of a method $\mathsf{m}_i \in \mathsf{ms}$, then $\langle \mathsf{w}_i^{\mathsf{a}} \sqcup_v^{\mathsf{a}} \mathsf{v}_i^{\mathsf{a}}, \mathsf{q}_i^{\mathsf{a}} \sqcup^{\mathsf{a}} \mathsf{s}_i^{\mathsf{a}} \rangle$ is the postcondition of m_i after its execution. The case of the constructor $\mathtt{init}$ is quite similar.

As for the inductive cases are concerned, the rules for concatenation and union formalize respectively that *"the context first performs r_1 and then r_2"* and *"the context can perform either r_1 or r_2"*. Finally, the rule for the Kleene-closure is a little bit more tricky. In fact the intuitive meaning of r^* is that, starting from an initial abstract value $\langle \mathsf{i}^{\mathsf{a}}, \mathsf{p}^{\mathsf{a}} \rangle$ the context performs the interaction encoded by r an unspecified number of times. We handle this case by considering the least fixpoint greater than $\langle \mathsf{i}^{\mathsf{a}}, \mathsf{p}^{\mathsf{a}} \rangle$ according to the order $\sqsubseteq_o^{\mathsf{a}}$ on O^{a}. If the abstract domains D^{a} and $\mathsf{Val}^{\mathsf{a}}$ do not respect the ACC then the convergence of the iterations must be enforced using the following pointwise widening operator:

$$\lambda(\mathsf{i}^{\mathsf{a}}, \mathsf{p}^{\mathsf{a}}), (\mathsf{i}'^{\mathsf{a}}, \mathsf{p}'^{\mathsf{a}}).(\mathsf{i}^{\mathsf{a}} \nabla^{\mathsf{a}} \mathsf{i}'^{\mathsf{a}}, \lambda\mathsf{m}.\mathsf{p}^{\mathsf{a}}(\mathsf{m}) \ (\nabla_v^{\mathsf{a}} \times \nabla^{\mathsf{a}}) \ \mathsf{p}'^{\mathsf{a}}(\mathsf{m})).$$

The regular expression $r_\top = \langle\{\mathtt{init}\}, \top_v^{\mathsf{a}}, \top^{\mathsf{a}}\rangle \cdot \top_r$ stands for a context that calls at first the class constructor with an unknown value and then may invoke any object method, with any possible value, an unspecified number of times. Thus the abstract value $\langle \mathsf{i}^{\mathsf{a}}, \mathsf{p}^{\mathsf{a}}\rangle = \mathcal{O}[\![\vartheta]\!]^{\mathsf{a}}(r_\top, \perp_o^{\mathsf{a}})$ is such that i^{a} is a property of the object fields valid for all the object instances, in any context. So it is a class invariant in the sense of [5, 4]. In the following we refer to it as $[\![\mathtt{A}]\!]^{\mathsf{a}}$.

4.3   Separate Context Analysis

The separate context analysis $\mathcal{C}[\![\mathsf{C}[\cdot]]\!]^{\mathsf{a}}$ has two goals. The first goal is to analyze $\mathsf{C}[\mathsf{o}]$ without referring to the o code, but just to a pre-computed approximation of its semantics. The second goal is to infer, for each object o a regular expression r that describes the interaction of the context with o. This r can then be used to refine the approximation of the object semantics. In general, a context creates several objects and it interacts with each of them in a different way. As a consequence, in the definition of the abstract context semantics $\mathcal{C}[\![\cdot]\!]^{\mathsf{a}}$ we use a domain $\dot{\mathsf{O}}^{\mathsf{a}} = [\mathsf{Addr}^{\mathsf{a}} \rightarrow \mathsf{O}^{\mathsf{a}}]$, whose elements are maps from abstract objects to their approximations. The definition of $\mathcal{C}[\![\mathsf{C}]\!]^{\mathsf{a}} \in [\mathsf{D}^{\mathsf{a}} \times \dot{\mathsf{O}}^{\mathsf{a}} \times \dot{\mathsf{R}} \rightarrow \mathsf{D}^{\mathsf{a}} \times \dot{\mathsf{R}}]$ is given by structural induction on C in Fig. 5.

Some comments on the separate context semantics. The semantics takes three parameters: an abstract state, an approximation of the semantics of the objects and the invocation history. When a class is instantiated, the semantics $\mathcal{C}[\![\cdot]\!]^{\mathsf{a}}$ (abstractly) evaluates the value to pass to the constructor $\mathtt{init}$ and it obtains an address ϑ for the new object. Then, it uses the object abstraction $\dot{\vartheta}(\vartheta)$ to get the constructor postcondition $\mathsf{p}^{\mathsf{a}}(\mathtt{init})$ and it updates the invocation history. In general, the abstract address ϑ identifies a set $\gamma_v(\vartheta)$ of concrete objects. So, the semantics adds an entry to the ϑ history corresponding to the invocation of $\mathtt{init}$, with an input v^{a} and an abstract state d^{a}. Eventually, the result is the new abstract state, obtained considering the store after the execution of the constructor, and the updated invocation history. The sequence, the $\mathtt{skip}$ and the two assignments do not interact with objects so, in these cases, $\mathcal{C}[\![\cdot]\!]^{\mathsf{a}}$ is very close to the corresponding semantics of Fig. 2. The definition of $\mathcal{C}[\![\cdot]\!]^{\mathsf{a}}$ for method invocation is similar to the constructor's one: it fetches the (abstract) address corresponding to o and the corresponding invariant. Then, it updates the abstract state, using the m postcondition, and the invocation history. The definition of the conditional merges the abstract states and the invocation histories originating from the two branches. Eventually, the loop is handled by the least fixpoint operator on the abstract domain $\mathsf{D}^{\mathsf{a}} \times \dot{\mathsf{R}}$. In particular we consider the least fixpoint greater than $(\mathsf{d}^{\mathsf{a}}, \lambda\vartheta. \varepsilon)$ as we need to compute an invocation history that is valid for all the iterations of the loop body. The history for the whole $\mathtt{while}$ command is the concatenation of the input history with

$$\mathcal{C}[\![A\ o\ =\ \mathtt{new}\ A(e)]\!]^a = \lambda d^a, \dot{\vartheta}, \dot{r}.let\ v^a = \mathcal{E}[\![e]\!]^a d^a, \vartheta = \mathtt{alloc}^a(d^a),$$

$$\langle i^a, p^a \rangle = \dot{\vartheta}(\vartheta), \langle \bot_v^a, d_0^a \rangle = p^a(\mathtt{init}),$$

$$\dot{r}' = \dot{r}[\vartheta \mapsto \langle \{\mathtt{init}\}, v^a, d^a \rangle \sqcup_r \dot{r}(\vartheta)]$$

$$in\ (\mathtt{assign}^a(d^a, \vartheta \mapsto d_0^a \mid_e, d^a \mid_s \mapsto d_0^a \mid_s), \dot{r}')$$

$$\mathcal{C}[\![C_1; C_2]\!]^a = \lambda d^a, \dot{\vartheta}, \dot{r}.let\ (d_1^a, \dot{r}_1) = \mathcal{C}[\![C_1]\!]^a(d^a, \dot{\vartheta}, \dot{r})$$

$$in\ \mathcal{C}[\![C_2]\!]^a(d_1^a, \dot{\vartheta}, \dot{r}_1)$$

$$\mathcal{C}[\![\mathtt{skip}]\!]^a = \lambda d^a, \dot{\vartheta}, \dot{r}.(d^a, \dot{r})$$

$$\mathcal{C}[\![x = e]\!]^a = \lambda d^a, \dot{\vartheta}, \dot{r}.(\mathtt{assign}^a(d^a, x \mapsto \mathcal{E}[\![e]\!]^a(d^a)), \dot{r})$$

$$\mathcal{C}[\![x_1 = x_2]\!]^a = \lambda d^a, \dot{\vartheta}, \dot{r}.(\mathtt{assign}^a(d^a, x_1 \mapsto d^a \mid_e(x_2)), \dot{r})$$

$$\mathcal{C}[\![x = o.m(e)]\!]^a = \lambda d^a, \dot{\vartheta}, \dot{r}.let\ v^a = \mathcal{E}[\![e]\!]^a(d^a), \vartheta = d^a \mid_e(o),$$

$$\langle i^a, p^a \rangle = \dot{\vartheta}(\vartheta), \langle v_m^a, q_m^a \rangle = p^a(m)$$

$$d'^a = \mathtt{assign}^a(d^a, x \mapsto v_m^a, \vartheta \mapsto q_m^a \mid_e, d^a \mid_s \mapsto q_m^a \mid_s),$$

$$in\ (d'^a, \dot{r}[\vartheta \mapsto \dot{r}(\vartheta) \cdot \langle m, v^a, \lfloor d^a \mid_s(\vartheta), d^a \mid_s \rfloor \rangle])$$

$$\mathcal{C}[\![\mathtt{if}\ b\ \mathtt{then}\ C_1\ \mathtt{else}\ C_2]\!]^a = \lambda d^a, \dot{\vartheta}, \dot{r}.let\ (d_1^a, \dot{r}_1) = \mathcal{C}[\![C_1]\!]^a(\mathtt{true}^a(b, d^a), \dot{\vartheta}, \dot{r}),$$

$$(d_2^a, \dot{r}_2) = \mathcal{C}[\![C_2]\!]^a(\mathtt{false}^a(b, d^a), \dot{\vartheta}, \dot{r})$$

$$in\ (d_1^a \sqcup d_2^a, \dot{r}_1 \sqcup_r \dot{r}_2)$$

$$\mathcal{C}[\![\mathtt{while}\ b\ \mathtt{do}\ C]\!]^a = \lambda d^a, \dot{\vartheta}, \dot{r}.let\ (d'^a, \dot{r}') = \mathtt{lfp}_{(d^a, \lambda\dot{\vartheta}.\varepsilon)}^{\sqsubseteq^a \times \sqsubseteq_r} \lambda(x, y).\mathcal{C}[\![C]\!]^a(\mathtt{true}^a(b, x), \dot{\vartheta}, y)$$

$$in\ (\mathtt{false}^a(b, d'^a), \lambda\vartheta.\dot{r}(\vartheta) \cdot (\dot{r}'(\vartheta))^*)$$

Fig. 5. Separate context semantics

the body one, repeated an unspecified number of times. As usual, the convergence of the analysis can be forced through the use of the widening operator $\lambda(d_1^a, \dot{r}_1).(d_2^a, \dot{r}_2), (d_1^a \nabla^a d_2^a, \dot{r}_1 \dot{\nabla}_r \dot{r}_2)$.

We conclude this section with two soundness lemmata. The proof for both can be found in [6]. The first one states that for each initial value and object approximation, all the history traces computed by $\mathcal{C}[\![\cdot]\!]^a$ are of the form of $\langle \{\mathtt{init}\}, v^a, s^a \rangle \cdot r$, for some $v^a \in \mathsf{Val}^a, s^a \in D^a$ and regular expression r. Intuitively, it means that the first interaction of the context with an object is the invocation of $\mathtt{init}$ with some value and store configuration. This fact can be used to show that $[\![A]\!]^a$, as defined in the previous section, overapproximates the semantics of all the objects. Thus, that it is a sound class invariant.

Lemma 1 (Soundness of the class invariant). *Let $d_0^a \in D^a$, $\dot{\vartheta} \in \dot{O}^a$ and $\mathcal{C}[\![C]\!]^a(d_0^a, \dot{\vartheta}, \lambda\vartheta.\varepsilon) = (d^a, \dot{r})$. Then for all the abstract objects ϑ such that $\dot{r}(\vartheta) \neq \varepsilon$:*

(i) $\dot{r}(\vartheta) = \langle \{\mathtt{init}\}, v^a, s^a \rangle \cdot r$, *for some* $v^a \in \mathsf{Val}^a, s^a \in D^a$ *and* $r \in R$;
(ii) $O[\![\vartheta]\!]^a(\dot{r}(\vartheta), \dot{\bot}_o) \sqsubseteq_o^a [\![A]\!]^a$.

The next lemma shows that the history traces computed by $\mathcal{C}[\![\cdot]\!]^a$ are an overapproximation of the history traces computed by $[\![\cdot]\!]^a$. Thus, the soundness of $[\![\cdot]\!]^a$ implies that the history traces are a sound approximation of the context.

Lemma 2 (Soundness of the history traces). *Let* $[\![C[o]]\!]^a(\bot^a) = d^a$, $\alpha_v(\{o\})$
$= \vartheta$ *and* $t = \langle \text{init}, v^a, s^a \rangle \cdot \langle m_1, v_1^a, s_1^a \rangle \ldots \langle m_n, v_n^a, s_n^a \rangle$ *a sequence of method invocations of* ϑ *when the rules of Fig. 2 are used to derive* d^a. *Then* $\mathcal{C}[\![C[o]]\!]^a(\bot^a,$
$\lambda\vartheta.[\![A]\!]^a, \lambda\vartheta.\varepsilon) = (d'^a, \dot{r}')$ *is such that* $d^a \sqsubseteq^a d'^a$ *and* $\mathcal{L}(t) \subseteq \mathcal{L}(\dot{r}'(\vartheta))$.

4.4   Putting It All Together

In this section we show how to combine the two abstract semantic functions
$\mathcal{O}[\![\cdot]\!]^a$ and $\mathcal{C}[\![\cdot]\!]^a$ in order to obtain a separate compositional analysis of $C[o]$. The
functions $\mathcal{O}[\![\cdot]\!]^a$ and $\mathcal{C}[\![\cdot]\!]^a$ are mutually related. The first one takes as input an
approximation of the context and it returns an approximation of the object
semantics. The second one takes as input an approximation of the objects. It
returns an abstract state and, for each abstract object ϑ, an approximation
of the context that interacts with ϑ. Then it is natural to handle this mutual
dependence with a fixpoint operator.

Nevertheless, before doing it we need to define formally the function $\Theta \in$
$[\dot{R} \rightarrow \dot{O}^a]$, that maps an interaction history $\dot{r}$ to a function $\dot{\vartheta}$ from abstract objects
to their approximation. First we consider the set of the abstract objects that
interact with the context, i.e. the abstract addresses whom interaction history is
non-empty: $I = \{\vartheta \mid \dot{r}(\vartheta) \neq \varepsilon\}$. Next, we define a function that maps elements
of I to their abstract semantics and the others to the class invariant $[\![A]\!]^a$:

$$\dot{\vartheta}_{\dot{r}} = \lambda\vartheta. \begin{cases} \mathcal{O}[\![\vartheta]\!]^a(\dot{r}(\vartheta), \bot_o^a) & \text{if } \vartheta \in I \\ [\![A]\!]^a & \text{otherwise.} \end{cases} \tag{1}$$

Moreover, we require that the more precise the abstract object, the more precise
its abstract semantics. Therefore we perform the downward closure of $\dot{\vartheta}_{\dot{r}}$, to
make it monotonic. Finally, the object abstractions function in a context $\dot{r}$,
$\Theta \in [\dot{R} \rightarrow \dot{O}^a]$, is defined as $\Theta(\dot{r}) = \lambda\vartheta. \sqcap_o^a \{\dot{\vartheta}_{\dot{r}}(\vartheta') \mid \vartheta' \in \text{Addr}^a \text{ and } \vartheta \sqsubseteq_v^a \vartheta'\}$. The
function Θ is well-defined as Addr^a is a sublattice of Val^a and the monotonicity
of $\Theta(\dot{r})$ is a direct consequence of the definition.

Using the above definition and defining $\Gamma(\dot{\vartheta}) = \mathcal{C}[\![C]\!]^a(\bot^a, \dot{\vartheta}, \lambda\vartheta.\varepsilon)$, it is now
possible to formally state the interdependence between the context and the objects semantics as follows:

$$\begin{aligned} \dot{\vartheta} &= \Theta(\dot{r}) \\ (d^a, \dot{r}) &= \Gamma(\dot{\vartheta}). \end{aligned} \tag{2}$$

A solution to the recursive equation (2) can be found with the standard iterative
techniques. Nevertheless, our goal is to parallelize the iterative computation
of Θ and Γ, in order to speed up the whole analysis. Therefore, we start the
iterations by considering a worst-case approximation for $\dot{r}$ and $\dot{\vartheta}$: $\dot{r}_0 = \lambda\vartheta.r_\top$ and
$\dot{\vartheta}_0 = \lambda\vartheta.\top_o^a$. In other words, we assume an unknown context when computing
the abstract object semantics and an unknown semantics when analyzing the
context. Then we obtain $\dot{\vartheta}_1 = \Theta(\dot{r}_0)$ and $(d_1^a, \dot{r}_1) = \Gamma(\dot{\vartheta}_0)$.

As we consider the worst-case approximation for the objects semantics, the abstract state d_1^a is an upper approximation of $[\![C]\!]^a(\bot^a)$. Furthermore, it is easy to see that $\dot{r}_1 \dot{\sqsubseteq}_r \dot{r}_0$ and $\dot{\vartheta}_1 \dot{\sqsubseteq}_o \dot{\vartheta}_0$. Roughly speaking, this means that after one iteration we have a better approximation of the context and the object semantics. As a consequence, if we compute $\dot{\vartheta}_2 = \Theta(\dot{r}_1)$ and $(d_2^a, \dot{r}_2) = \Gamma(\dot{\vartheta}_1)$, we obtain a better approximation for the abstract state, the semantics of the objects and that of the context. This process can be iterated, so that at step $i + 1$ we have:

$$\dot{\vartheta}_{i+1} = \Theta(\dot{r}_i)$$
$$(d_{i+1}^a, \dot{r}_{i+1}) = \Gamma(\dot{\vartheta}_i). \tag{3}$$

The next theorem synthesizes what has been said so far. It states that the iterations of (3) form a decreasing chain and that at each iteration step d_{i+1}^a is an sound approximation of the monolithic abstract semantics. Hence, of the concrete semantics:

Theorem 2 (Soundness). *Let* C *be a context. Then* $\forall i \geq 0$.

(i) $d_{i+1}^a \sqsubseteq^a d_i^a$, $\dot{r}_{i+1} \dot{\sqsubseteq}_r \dot{r}_i$ *and* $\dot{\vartheta}_{i+1} \dot{\sqsubseteq}_o \dot{\vartheta}_i$.

(ii) $[\![C]\!]^a(\bot^a) \sqsubseteq^a d_i^a$.

Roughly speaking the first point of the theorem states that the more the iterations the more precise the result of the analysis. On the other hand, the second point states that the abstract states are all above the result of the monolithic abstract semantics. As a consequence it is possible to stop the iterations at a step $\underline{i}$, the resulting abstract state $d_{\underline{i}}^a$ being a sound approximation of the concrete semantics.

An analysis based on (3) has several advantages. First, it is possible to use the asynchronous iterations with memory [1] in order to parallelize the analysis of the context and the objects. Intuitively, this is a consequence of the fact that at each iteration, the result of Θ and Γ depends just on the result of the previous iteration. Furthermore, Θ computes the abstract semantics for several, independent, abstract objects (cf. (1)). Therefore, even the effective implementation of Θ may take advantage of a further parallelization. Finally the fact that each iteration is a sound approximation allows a fine tuning of the trade-off precision/cost. In particular, we can stop the iterations as soon as the desired degree of precision is reached.

Example 1. As an example, we can consider the context and the class A in Fig. 6, where Prop is the property: $(o_1.a + o_1.b) - (o_2.a + o_2.b) + (o_1.y + o_2.y) \geq 0$.

We are interested in proving that the assert condition is never violated. In order to do it, we instantiate the abstract domain D^a with Polyhedra [3], and we consider the two abstract objects ϑ_1 and ϑ_2 corresponding respectively to o_1 and o_2. According to the iteration schema (3), the first step approximates the objects semantics with the class invariant: $\Theta(\dot{r}_0) = \lambda \vartheta.\langle i^a, \lambda m.p^a(m)\rangle$. The object

o₁ = new A(5, 10);
o₂ = new A(3, 10);
while ... do
   if o₁.get_y() + o₂.get_y() ≥ 0 then
     o₁.addA(5); o₁.addB(3);
   else
     o₂.addA(7); o₂.addA(1);
  { assert(Prop) }

(a) The context

$F : \{a, b, y\}$

$init(a_0, c_0) : a = a_0; b = c_0 - a_0; y = 0$

$addA(x) : a = a + x; b = b - x; y = y + 1$
$addB(x) : a = a - x; b = b + x; y = y - 1$
$get\_y() : \texttt{return } y$

(b) The class A

Fig. 6. Example of a context and a class

fields invariant is $i^a = \{a + b = c_0\}$ and the method postconditions are:

$$p^a = \begin{cases} \texttt{init} \mapsto \langle \perp_v^a, i^a \cup \{y = 0\} \rangle \\ \texttt{addA} \mapsto \langle \perp_v^a, i^a \cup \{y = y + 1\} \rangle \\ \texttt{addB} \mapsto \langle \perp_v^a, i^a \cup \{y = y - 1\} \rangle \\ \texttt{get\_y} \mapsto \langle \top_v^a, i^a \rangle. \end{cases}$$

On the other hand, as far as the context analysis is concerned, we have $(\emptyset, \dot{r}_1) = \Gamma(\dot{\vartheta}_0)$, where $\dot{r}_1$ is the interaction history below. For lack of space we simplify the structure of the interaction history by omitting the abstract state. Nevertheless, in the example this is not problematic, as the objects do not expose the internal state.

$$\dot{r}_1 = \begin{cases} \vartheta_1 \mapsto \langle \{\texttt{init}\}, (5, 10) \rangle \cdot (\langle \{\texttt{get\_y}\}, \emptyset \rangle \cdot \langle \{\texttt{addA}\}, 5 \rangle \cdot \langle \{\texttt{addB}\}, 3 \rangle)^* \\ \vartheta_2 \mapsto \langle \{\texttt{init}\}, (7, 10) \rangle \cdot (\langle \{\texttt{get\_y}\}, \emptyset \rangle \cdot \langle \{\texttt{addA}\}, 7 \rangle \cdot \langle \{\texttt{addA}\}, 1 \rangle)^*. \end{cases}$$

The result of the next iteration, $\Gamma(\dot{\vartheta}_1)$, is still too imprecise for verifying the assertion, as the object fields invariant i^a implies that $(o_1.a + o_1.b) - (o_2.a + o_2.b) = 0$, but nothing can be said about $o_1.y + o_2.y$. Nevertheless, the analysis of the object semantics under the context $\dot{r}_1$ results in a more precise approximation of the objects semantics. In particular, we obtain for the first object the field invariant $i_1^a = i^a \cup \{0 \leq y \leq 1\}$ and for the latter $i_2^a = i^a \cup \{y \geq 0\}$. As a consequence, a further iteration is enough to infer that the condition Prop is verified. From Th. 2 it follows that the result is sound, even if it is not the most precise one. In fact it is easy to see that a further iteration gives a more precise result, proving that the else branch in the conditional is never taken. Therefore that $o_2.y$ is identically equal to zero.

5   Conclusions and Future Work

In this work we introduced a separate compositional analysis and we proved it correct for a small yet realistic object-oriented language. In particular we presented an iteration schema for the computation of the abstract semantics that approximates it from above. The central idea for the parallelization is the use of a domain of regular expressions to encode the interactions between the context and the objects.

In future work we plan to study the practical effectiveness of the presented technique, for example with regard to memory consumption. Moreover, it would be interesting to study how many iterations are needed in order to reach an acceptable degree of precisions. As far as the theoretical point of view is concerned, a straightforward extension of this work is a direct handling of inheritance. Nevertheless, in our opinion the combination of the present work with modular techniques for the handling of inheritance presents some more challenges that deserve to be explored [5].

Acknowledgments

We would like to thank R. Cousot, J. Feret, C. Hymans, A. Miné and X. Rival for their comments.

References

1. P. Cousot. Asynchronous iterative methods for solving a fixed point system of monotone equations in a complete lattice. Technical Report R.R. 88, Laboratoire IMAG, Université scientifique et médicale de Grenoble, 1977.
2. P. Cousot and R. Cousot. Abstract interpretation: a unified lattice model for static analysis of programs by construction or approximation of fixpoints. In *POPL '77*. ACM Press, 1977.
3. P. Cousot and N. Halbwachs. Automatic discovery of linear restraints among variables of a program. In *POPL '78*. ACM Press, 1978.
4. F. Logozzo. Class-level modular analysis for object oriented languages. In *SAS'03*, volume 2694 of *LNCS*. Springer-Verlag, 2003.
5. F. Logozzo. Automatic inference of class invariants. In *VMCAI'04*, volume 2937 of *LNCS*. Springer-Verlag, 2004.
6. F. Logozzo. *Modular Static Analysis of Object Oriented Languages*. PhD thesis, École Polytechnique, France, 2004. To appear.
7. I. Pollet, B. Le Charlier, and A. Cortesi. Distinctness and sharing domains for static analysis of Java programs. In *ECOOP'01*, volume 2072 of *LNCS*. Springer-Verlag, 2001.
8. A. Rountev, A. Milanova, and B.G. Ryder. Fragment class analysis for testing of polymorphism in Java software. In *ICSE'03*. IEEE Press, 2003.

Abstract Domains for Property Checking Driven Analysis of Temporal Properties

Damien Massé

École Normale Supérieure & École Polytechnique, France
damien.masse@polytechnique.fr
http://www.stix.polytechnique.fr/~dmasse/

Abstract. Abstract interpretation-based static analysis infers properties from the source code of a program. When the goal is to check a temporal specification on the program, we need the analysis to be as precise as possible to avoid false negatives. In previous work [9], we suggested a method called "property checking driven analysis" to automatically use the specification to check during the analysis in order to refine it. However, this approach requires to abstract domains of lower closure operators, something which was not developed. In this paper, we describe some abstractions on lower closure operators developed for a small analyzer of temporal properties. We examine the need for weak relational abstractions, and show that using our new approach can give more precise results than using a traditional abstract interpretation-based analysis with expensive abstract domains.

1 Introduction

The objective of static program analysis is to automatically infer run-time properties on programs at compile-time. Since the properties are often undecidable, static program analysis uses approximations. The acceptable level of approximation depends on the application of the static analyzer: for example, when the goal is to remove unnecessary run-time checks [4], missing a few possible optimizations is acceptable. However, when the goal is to prove some specifications about the program (e.g. the absence of run-time errors), the analysis must be very precise (to avoid any false negative), and efficient enough to check large programs. To obtain this result, an approach is to design a special-purpose static program analyzer adapted to a restricted class of programs and a restricted class of specifications. This method was proposed and successfully applied in [1], for the verification of real-time embedded software. However, the analyzer so designed handled only one specification (the absence of run-time error).

It may be possible to extend the class of specifications (e.g. to a larger class of temporal properties) if the specification itself is used to refine the analysis. In previous work [9], we proposed to use the specification in a different, "reverse" analysis (e.g. if the abstract semantics is computed with a backward analysis, the "reverse" analysis is a forward one) which computes a "guide" for the computation of the abstract semantics. This approach, which we called "property

C. Rattray et al. (Eds.): AMAST 2004, LNCS 3116, pp. 349–363, 2004.

checking driven analysis", uses domains of lower closure operators as the concrete description of the "guide" for the main analysis. Thus, implementing this method requires to design efficient and precise abstractions of such domains.

This paper extends [9] by showing how to construct such an abstract domain, and presents a prototype implementation. We show different abstractions of domains of lower closure operators, based on the structure of the underlying domains. Since a non-relational domain appears insufficient to keep the precision of the analysis, we develop a *weak relational domain* for abstract environments, which uses a notion of local dependence between variables related to the temporal specification. Finally we present the results of the analyzer on a few examples, and we discuss the merits of our approach compared with analyses using traditional abstract domains.

2   Concrete Semantics

In this section, we introduce briefly the language and the kind of temporal specifications we intend to analyse.

2.1   Language and States

Since the inspiration for the analyzer is Cousot's Marktoberdorf generic analyzer [3], we analyse roughly the same language. It is a very simple imperative language, with only integers and without functions. Integer values range from `min_int` to `max_int`, and there is an arithmetic error Ω. We denote by $\mathbb{I}$ the interval $[\texttt{min\_int}, \texttt{max\_int}]$, and by $\mathbb{I}_\Omega$ the set $\mathbb{I} \cup \{\Omega\}$.

The language uses arithmetic expression Aexp, boolean expression Bexp, command Com and list of commands Seq. The syntax is defined as:

$$A \in \text{Aexp} ::= n \mid \texttt{x} \mid ?\,\texttt{in}\,[A_1, A_2] \mid un\,A \mid A_1\,bin\,A_2$$
$$B \in \text{Bexp} ::= \texttt{true} \mid \texttt{false} \mid A_1 = A_2 \mid A_1 < A_2$$
$$\mid B_1\,\texttt{and}\,B_2 \mid B_1\,\texttt{or}\,B_2$$
$$C \in \text{Com} ::= \texttt{skip} \mid \texttt{x} := A \mid \texttt{if}\,B\,\texttt{then}\,S_1\,\texttt{else}\,S_2\,\texttt{fi}$$
$$\mid \texttt{while}\,B\,\texttt{do}\,S\,\texttt{od}$$
$$S \in \text{Seq} ::= C \mid C\,;\,S$$

Here, n are integers, $x \in \mathbb{V}$ variables, $un \in \{+, -\}$ unary operators and $bin \in \{+, -, *, /, \text{mod}\}$ binary operators. The difference with the analyzer developed in [3] is the non-deterministic operation ? in $[A_1, A_2]$, which returns an integer between the value v_1 of A_1 and the value v_2 of A_2 if $v_1 \le v_2$, and Ω if $v_1 > v_2$. This operation intends to model all non-deterministic aspects of the program, e.g. user inputs, sensors or configurations. The exact *meaning* of each non-deterministic operation (i.e. how it relates to the specification) is determined in the temporal specification.

Hence, a state $\sigma \in \Sigma$ is defined as a pair of a program point p (in a set of program points Lab) and an environement in $\mathbb{V} \to \mathbb{I}_\Omega$:

$$\Sigma = \text{Lab} \times (\mathbb{V} \to \mathbb{I}_\Omega)$$

2.2   Temporal Specification and Semantics

To simplify the abstraction, we choose to analyse only simple specifications which make the distinction between non-deterministic operations as "internal" and "external" non-determinism. These properties are expressed with μ-calculus formulas of the form (we identify atomic predicates and the states they represent):

$$I \models {}^{\mu}_{\nu}X.(A \vee (B \wedge \Diamond X) \vee (C \wedge \Box X))$$

with ${}^{\mu}_{\nu} \in \{\mu, \nu\}$. We can note that this class of temporal specification is stable by negation (i.e., if ϕ belongs to this class, so does $\neg\phi$).

Here, I are initial states, A final states, B states with "internal" non-determinism (the result of non-deterministic operations there can be chosen in order to satisfy the specification), and C states with "external" non-determinism (the specification must be satisfied for any result of the non-deterministic operations there). States which are not in $A \vee B \vee C$ are error states. For the sake of simplicity we assume that B and C are disjoint. Infinite computations are taken into account by the difference between μ and ν. This class of specifications includes all the **CTL** operators, and some kind of game specifications as well.

In a framework using states, we use the predicate transformers pre ($y \in preX$ iff at least one successor of y is in X) and $\widetilde{pre}$ ($y \in \widetilde{pre}X$ iff all successors of y are in X). Then the specification is (with lgfp being either lfp or gfp):

$$I \subseteq \text{lgfp } X.(A \cup (B \cap preX) \cup (C \cap \widetilde{pre}X)).$$

Since the class of specifications is stable by negation, we can choose to express the negation of the specification. Then it is expressed by the formula:

$$I \cap \text{lgfp } X.(A \cup (B \cap preX) \cup (C \cap \widetilde{pre}X)) = \emptyset \tag{1}$$

Then, the goal of the analyzer is to over-approximate $\mathcal{S} = I \cap \text{lgfp } X.(A \cup (B \cap preX) \cup (C \cap \widetilde{pre}X))$. If we get $\emptyset$, we prove the property. Otherwise, we get initial states which may not satisfy the specification.

2.3   Property Checking Driven Analysis

The goal of our analyzer is to illustrate the technique of property checking driven analysis developed in [9]. In this section we give a summary of this technique. Starting from a concrete semantics $\mathcal{S} = I \cap \text{lgfp } \phi$, this method shows how to construct a lower closure operator[1] ρ such that:

$$I \cap \text{lgfp } (\rho \circ \phi) = I \cap \text{lgfp } \phi.$$

In this formula, ρ reduces the value obtained at each iteration of the fixpoint, while ensuring that the fixpoint will be greater than or equal to $\mathcal{S}$. Hence, ρ focuses the fixpoint computation on parts useful for the computation of $\mathcal{S}$.

---

[1] Lower closure operators are monotonic, reductive and idempotent. Basic results on lower closure operators are recalled in [9].

In practice, the fixpoints are transfered into the abstract domain using abstract interpretation results, and an over-approximation ρ' of ρ is used. When ρ' is included in the abstract fixpoint computation, it leads to more precise (and still sound) results for $\mathcal{S}$. Thus, the lower closure operator "drives" the analysis towards the verification of the specification. We can notice that this process does not change the abstraction itself, which is defined in the fixpoint transfer: it is not an abstraction refinement, but an analysis refinement.

Some notation is necessary to exhibit the formulas expressing ρ. The lattice of lower closure operators on a domain D is written $(lco\,(D)\,,\sqsubseteq)$. A lower closure operator ρ is characterized by the set of its fixpoints $\rho(D)$, which is an *upper Moore family*: $\rho(D) = \mathcal{M}(\rho(D)) = \{\cup A \mid A \subseteq \rho(D)\}$. As usual, ρ will denote either the operator or the set of its fixpoints $\rho(D)$, depending on the context.

With $\phi = \lambda X.(A \cup (B \cap preX) \cup (C \cap \widetilde{pre}X))$, applying the results of [9] gives:

$$\rho_0 = \lambda X.X \cap I$$
$$\rho\, = \mathrm{lfp}\ \eta.(\rho_0 \sqcup F_\phi(\eta))$$

where F_ϕ is a complex function defined in [9].

Here, ρ is constructed iteratively from ρ_0 (the fixpoints of which are the subsets of I, i.e. all the possible values for $\mathcal{S}$). Intuitively, at each iteration, F_ϕ add as new fixpoints the minimum sets which can lead, by an application of ϕ, to a superset of an existing fixpoint X of ρ. All the fixpoints of ρ will then form a sub-lattice of $\wp(\Sigma)$ on which we can restrict the computation of lgfp ϕ, without modifying the intersection of the fixpoint with I.

To create a minimum set Y leading to a superset of X, we first test if X is included in $A \cup B \cup C$. If this is not the case, then we cannot define any Y since $\phi(\Sigma) \not\supseteq X$. But if $X \subseteq A \cup B \cup C$, we construct a possible Y by choosing one successor for each element of $X \setminus B$, and all the successors of $X \setminus C$. Then these sets are added as new fixpoints.

As the approximation of ρ involves only looking for successors of states, we can consider it as our "forward" analysis (it starts from I and goes forward) which will "drive" the "backward" analysis lgfp ϕ. We show in [9] that this process can be iterated (as the result of the backward analysis can be used to get more precise guide) until a fixpoint is obtained. In practice, as was observed previously in other abstract interpretation-based analyzers which use backward-forward combination (e.g. in [2]), this fixpoint is reached after a few iterations.

3   Construction of the Abstract Domain

Of course, ρ is not computable, we use abstract interpretation to approximate it. Hence, we need to find abstractions on the domain of lower closure operators[2]. Identifying ρ by the set of its fixpoints, an over-approximation of ρ is simply a superset of ρ.

---
[2] Note that this abstraction is orthogonal to the abstraction of $\wp(\Sigma)$ used in the computation of the abstract semantics.

Given a set of states X, we can see the application of ρ on X as the removal of states in X. A state σ may be removed for two reasons:

- The first case is that $\sigma \notin \rho(\Sigma)$. Then, σ is not in any fixpoint of ρ, and σ is totally useless for the computation of $I \cap \text{lgfp } \rho \circ \phi$. This happens if σ never appears in the computation of the "successors" of I.
- There are some fixpoints of ρ which contain σ, but none is included in X. In this case, σ is removed because it is useless *without* other states. Such a state may be removed at some point during the computation of lgfp $\rho \circ \phi$, and reintroduced later in the computation along with other states such that it is not removed. The construction of ρ presented in the previous section guarantees that any "useful" state for the computation of $I \cap \text{lgfp } \phi$ will eventually appear in the computation of lgfp $\rho \circ \phi$.

This distinction gives a starting point for the development of approximations for lower closures. The states appearing in our first case can be obtained through an approximation of $\rho(\Sigma)$. As it is a subset of Σ, we can do that with any existing abstraction of Σ. To remove other states, we need some kind of dependence relations between useful states. Hence, a main issue of this work is to express the dependence (related to the temporal specification) between states.

3.1 "Interval" Abstraction

The "interval" abstraction is the only one described in [9]. From an abstraction of P to $P^\sharp$ we derive an abstraction of $lco\,(P)$ where each lower closure ρ is represented by two elements of $P^\sharp$: an "upper" one which abstracts the greatest fixpoint of ρ, and a "lower" one which gives a lower bound on the non-empty fixpoints of ρ.

Formally, from a Galois connection $P \xrightleftharpoons[\alpha]{\gamma} (P^\sharp, \sqsubseteq^\sharp)$, we construct a new Galois connection $lco\,(P) \xrightleftharpoons[\alpha^\bullet]{\gamma^\bullet} (P^\sharp, \sqsupseteq^\sharp) \times (P^\sharp, \sqsubseteq^\sharp)$ defined as:

$$\alpha^\bullet(\rho) = (\sqcap^\sharp \{\alpha(Y) \mid Y \in \rho \setminus \{\emptyset\}\}, \alpha(\rho(\Sigma)))$$
$$\gamma^\bullet(l, u) = \{X \mid l \sqsubseteq^\sharp \alpha(X) \sqsubseteq^\sharp u\}$$

With the functional view of lower closures, $\gamma^\bullet$ becomes:

$$\gamma^\bullet(l, u) = \lambda X. \begin{cases} X \cap \gamma(u) & \text{if } \alpha(X) \sqsupseteq^\sharp l \\ \emptyset & \text{otherwise} \end{cases}$$

Hence the lower bound gives the relations between the elements of P. The precision of these relations relies on the precision of the abstract intersection: $\{X \in P \mid \alpha(X) \sqsubseteq^\sharp \sqcap^\sharp E\}$ should be the smallest possible for a given E.

Example 1. Applying this construction directly on a whole existing abstraction of $\wp(\Sigma)$ is difficult and gives imprecise results (because the abstract intersection is not precise enough). Rather, we use it on abstractions of basic elements, and we use the result as a base for the construction of our abstract domain.

An example is given in [9] for integers, using the interval domain. We use the same approach to abstract $lco\,(\wp\,(\mathbb{I}))$: the initial abstract domain is $(\mathbb{I}\times\mathbb{I},\sqsubseteq)$, with $(a_1,b_1)\sqsubseteq(a_2,b_2)\Longleftrightarrow(a_2\leq a_1)\wedge(b_1\leq b_2)$ (this is an extension of the interval domain, as we do not restrict a_i to be less than b_i), and $\alpha(X)=\min X,\max X$.

Then a lower closure ρ is abstracted by two intervals $((\mathrm{Mm},\mathrm{mM}),(\mathrm{mm},$ $\mathrm{MM}))$ [3], such that:

$$\forall X\subseteq\mathbb{I},\ \rho(X)\subseteq\begin{cases}\emptyset & \text{if } \min X>\mathrm{Mm} \text{ or } \max X<\mathrm{mM}\\ X\cap[\mathrm{mm},\mathrm{MM}] & \text{otherwise}\end{cases}$$

3.2   Abstraction of $lco\,(\wp\,(\mathbb{I}_\Omega))$

The abstraction of $lco\,(\wp\,(\mathbb{I}_\Omega))$ can be deduced from an abstraction $lco\,(\wp\,(\mathbb{I}))$, using the fact that $\mathbb{I}_\Omega=\mathbb{I}\cup\{\Omega\}$. More generally, we study possible abstractions of $lco\,(\wp\,(A\cup B))$ (with A and B disjoint) using $lco\,(\wp\,(A))$ and $lco\,(\wp\,(B))$. A very simple abstraction is the non-relational abstraction:

Proposition 1. *Let* A,B *be two sets with* $A\cap B=\emptyset$. *Then* $lco\,(\wp\,(A\cup B))\xrightleftharpoons[\alpha]{\gamma}$ $lco\,(\wp\,(A))\times lco\,(\wp\,(B))$ *defined as:*

$$\alpha(\rho)=(\lambda X.\rho(X\cup B)\cap A),(\lambda Y.\rho(A\cup Y)\cap B)$$
$$\gamma(\rho_A,\rho_B)=(\lambda Z.\rho_A(Z\cap A)\cup\rho_B(Z\cap B))$$

is a Galois connection.

This abstraction is non-relational because it does not keep any constraints between the two sets. Applied for our domain of values, it would abstract $lco\,(\wp\,(\mathbb{I}_\Omega))$ to $lco\,(\wp\,(\mathbb{I}))\times\{\rho_\emptyset,\rho_\Omega\}$, such that $\rho_\emptyset=\lambda x.\emptyset$ means that no error appears in any fixpoint, whereas $\rho_\Omega=\lambda x.x$ means that some fixpoints have arithmetic errors.

However, it may be useful to know facts like "all non-empty fixpoints have arithmetic errors" (in this case, we can stop the analysis), so we will use three "possible error values" instead of two. To get this abstraction, we define the set $T=\{\mathrm{INI},\mathrm{ERR},\mathrm{TOP}\}$, with the following intuitive meaning:

- INI means that no error is possible. The lower closure associated ρ satisfies $\rho(X)=\rho(X\setminus\{\Omega\})$ for all $X\subseteq\mathbb{I}_\Omega$.
- ERR means that all elements of ρ, except $\emptyset$, contains Ω. Hence this is the "error" case.
- TOP is the "do not know" answer.

Then we construct an abstraction α from $lco\,(\wp\,(\mathbb{I}_\Omega))$ to $lco\,(\wp\,(\mathbb{I}))\times T$. Noting $\alpha_1(\rho),\alpha_2(\rho)$ the two components of the abstraction, we define:

$$\alpha_1(\rho)=\lambda X.\rho(X\cup\{\Omega\})\cap\mathbb{I}$$
$$\alpha_2(\rho)=\begin{cases}\mathrm{INI} & \text{if } \forall X\subseteq\mathbb{I}_\Omega,\rho(X)=\rho(X\setminus\{\Omega\})\\ \mathrm{ERR} & \text{if } \rho\neq\lambda X.\emptyset\ \wedge\forall X\subseteq\mathbb{I},\rho(X)=\emptyset\\ \mathrm{TOP} & \text{otherwise}\end{cases}$$

---

[3] Which stands for $((\max\min,\min\max),(\min\min,\max\max))$, as these are in fact the bounds of the bounds of the non-empty fixpoints of ρ.

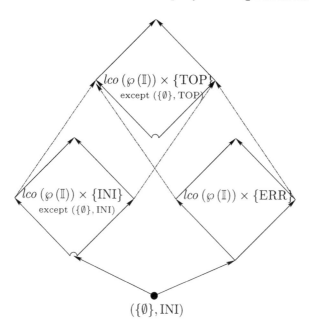

Fig. 1. Lattice $lco\,(\wp\,(\mathbb{I})) \times T$, divided in three parts, one for each value of T. Note that $(\{\emptyset\}, \mathrm{INI})$ is the global bottom, and that $(\{\emptyset\}, \mathrm{TOP})$ is collapsed with $(\{\emptyset\}, \mathrm{ERR})$.

Then $lco\,(\wp\,(\mathbb{I})) \times T$ can be defined as a lattice with the structure defined Fig. 1, and α is the abstraction of a Galois connection.

Our abstract domain of values is then $\mathbb{I}_{\mathrm{int}} \times T$. This loses the relation between the Ω and the non-error values in a same set in ρ. However, we keep the option "error everywhere", which enables to know whether the error is not avoidable.

3.3   Abstract Environment

To abstract an environment, we need to abstract lower closures on powerset of Cartesian products. There is an intuitive non-relational abstraction (i.e. which abstracts each variable separately), but we will see that it is not sufficient in general, as it loses dependency information between source of non-determinism. Hence we will describe a *weak relational abstraction* which expresses the dependence between the possible values of several variables.

Proposition 2 (Non-relational abstraction). *Let A, B be two sets, and $\pi_A :$ $\wp\,(A \times B) \to \wp\,(A)$, $\pi_B : \wp\,(A \times B) \to \wp\,(B)$ the projections of subsets of $A \times B$ on their components. Then $lco\,(\wp\,(A \times B)) \xrightleftharpoons[\alpha]{\gamma} lco\,(\wp\,(A)) \times lco\,(\wp\,(B))$ defined as:*

$$\alpha(\rho) = (\lambda X.\pi_A(\rho(X \times B))), (\lambda Y.\pi_B(\rho(A \times Y)))$$
$$\gamma(\rho_A, \rho_B) = (\lambda Z.\rho_A(\pi_A(Z)) \times \rho_B(\pi_B(Z)))$$

is a Galois connection.

Hence, we can produce a non-relational abstraction of the whole environment, but this abstraction is not sufficient. Let us look at the following program:

```
x := ? in [0,1] ;
y := ? in [0,1] ;
z := x + y
```

If we want to prove that, by choosing the value for x and y, we can satisfy $z = 2$ at the end of the program, we must know that x and y are independent before computing z (e.g. we do not have $x = 1 - y$). However, with a completely non-relational abstraction, we know that x and y can satisfy $x = 1$ and $y = 1$, but we do not know that these properties can be true simultaneously. The forward analysis would give the same result if we replace the second line by $y := 1 - x$.

This limitation makes the analysis much less useful. Thus, we need to keep a kind of independence relation between variables, which is used to know that the constraints expressed on each variable are effective simultaneously. We do this by a weak relational abstraction, keeping only a relation between each variables.

Weak relational abstraction: two variables case. First, we give a weak abstraction of $lco\,(\wp\,(\mathbb{I}_\Omega)) \times lco\,(\wp\,(\mathbb{I}_\Omega))$ (like with two variables). Like the non-relational abstraction, we keep an abstract value for each component, but we add a boolean expressing the dependence between the components (*true* means that the components may depend on each other). Saying that x and y are independent in a lower closure ρ means that ρ is a Moore family generated by Cartesian products of sets of values (i.e., all the fixpoints of ρ are union of the Cartesian products generated by the fixpoints of the abstract values of each components).

Hence, the abstract domain is $lco\,(\wp\,(\mathbb{I}_\Omega)) \times lco\,(\wp\,(\mathbb{I}_\Omega)) \times \mathbb{B}$. The concretization of $(\rho_x, \rho_y, false)$, expressed as a Moore family, is $\rho = \mathcal{M}(X \times Y \mid X \in \rho_x \wedge Y \in \rho_y)$: ρ is generated by Cartesian products. The concretization of $(\rho_x, \rho_y, true)$ is $\rho = \{Z \in \wp\,(\mathbb{I}_\Omega) \times \wp\,(\mathbb{I}_\Omega) \mid \pi_x(Z) \in \rho_x \wedge \pi_y(Z) \in \rho_y\}$ with $\pi_x(Z)$ (resp. $\pi_y(Z)$) the projection of Z to its first (resp. second) component: this is the non-relational concretization of (ρ_x, ρ_y).

Example 2. To illustrate our abstraction, we restrict the values to $S = \{0,1\}$. We want to study an abstraction of $lco\,(\wp\,(S \times S))$ to $lco\,(\wp\,(S)) \times lco\,(\wp\,(S)) \times \mathbb{B}$, or, more precisely, we want to describe the meaning of the abstract values. To simplify, we suppose that the first component of the abstract value is $\{\emptyset, \{0,1\}\}$, and that the second satisfies $\rho(\{0,1\}) = \{0,1\}$. Still, we have eight possible cases (four values for the second lower closure, and two for the boolean), all described in Fig. 2, with the order between them. When the boolean is *false*, the values are independent, which means that all the generators of the lower closure are Cartesian products, whereas when the boolean is *true* we can keep any generator.

The relation is very weak, and its meaning is restricted to the "lower" part of the abstraction, but it is sufficient to keep the independence of the random generators, which was our goal. In particular, the abstract functions will be easier to compute than with stronger relations like octagons[10].

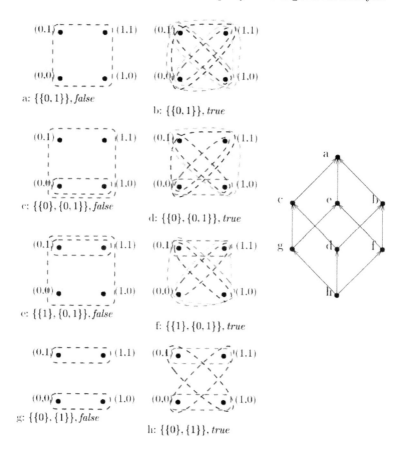

Fig. 2. The eight cases of example 2. Each case is described by the value of the boolean, and the generators of the second lower closure operator in the abstract value. For each case, we give the generators of the concretized lower closure. We give also the order between them. We can see that adding the independence boolean (case *false*) restrict greatly the concrete lower closure.

Weak relational abstraction: general case. Our first generic abstract domain is $(\mathbb{V} \to lco\,(\wp\,(\mathbb{I}_\Omega))) \times \wp\,(\wp\,(\mathbb{V}))$, with the concretization:

$$\gamma(f, B) = \{Z \mid \forall x \in \mathbb{V}, \pi_x(Z) \in f(x) \land \forall V \notin B, \pi_V(Z) = \times_{x \in V} \pi_x(Z)\}$$

The principle is to associate a boolean to any subset V of $\mathbb{V}$, describing the idea that these variables are "globally" dependent. Note that three variables may be "independent" when taken pairwise without being "globally" independent (a set of points in dimension 3 may look like a Cartesian product when projected in any direction, but not be itself a Cartesian product, cf. Fig 3 for an example). Thus we cannot deduce exactly the "global" dependence function from the dependence relations between pairs of variables.

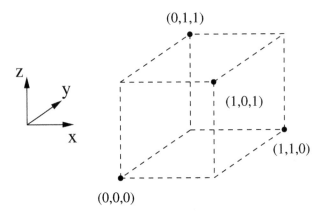

Fig. 3. An example of a set of 3-dimensional points which gives Cartesian products when projected in every direction, but is not a Cartesian product.

However, keeping an element of $\wp(\wp(\mathbb{V})))$ is too costly and, in practice, it is not useful[4]. Rather, we keep only a symmetric relation between variables (i.e. a subset of $\wp(\mathbb{V} \times \mathbb{V})$) expressing the dependence of the variables, but such that all sets of variables completely unrelated must be globally independent (thus, this is a subset of the real dependence relations between each couple of variables, but from it we can reconstruct the whole set of dependencies). This construction can be viewed as an abstraction (as we construct a sound, but not complete, "global" dependence function), where the concretization function $\gamma_{\mathcal{V}}$ between $\wp(\mathbb{V} \times \mathbb{V})$ and $\wp(\wp(\mathbb{V})))$ is:

$$\gamma_{\mathcal{V}}(R) = \{\{v_i\}_{i \in \Delta} \mid \exists (i,j) \in \Delta^2, (v_i, v_j) \in R\}.$$

Hence, the domain of abstract environments is $(\mathbb{V} \to \mathbb{I}_{\text{int}} \times T) \times \wp(\mathbb{V} \times \mathbb{V})$. Due to this relational abstraction, we do not have a best abstraction function α, but we still have the concretization function γ. Though we cannot use the Galois connection framework, we can use the frameworks developed in [5].

3.4   Abstract Domain

With the abstract domain constructed above, we can use the non-relational abstraction developed for powerset of unions:

$$lco\left(\wp\left(\text{Lab} \times (\mathbb{V} \to \mathbb{I}_\Omega)\right)\right) \xrightleftharpoons[\alpha]{\gamma} \text{Lab} \to lco\left(\wp\left(\mathbb{V} \to \mathbb{I}_\Omega\right)\right)$$

This abstraction forgets information on conditionals and loops. This may be a problem, for example, with the program:

```
x:= ? in [0,1] ;
if (x=0) then x:=0 else x:=1 ;
```

---

[4] It appears to be difficult to infer the dependencies with this level of precision.

(1)   Initial states: I

(2)   x:= ? in {0,1}

(3)   y:= ? in {0,1}

(4)   x:= x+y   Final states: $F : x = 1$

1	{ x:[[ini: (-oo,+oo,-oo,+oo)]]; y:[[ini: (-oo,+oo,-oo,+oo)],×] }	1	{ BOT }
2	{ x:[[ini: (0,1,0,1)]]; y:[[ini: (-oo,+oo,-oo,+oo) },×] }	2	{ BOT }
3	{ x:[[ini: (0,1,0,1)]]; y:[[ini: (0,0,1,1)]] }	3	{ BOT }
4	{ x:[[ini: (0,1,1,2)]]; y:[[ini: (0,0,1,1) },×] }	4	{ x:[[ini: (1,1,1,1)]]; y:[[ini: (0,0,1,1) },×] }

Result after a forward analysis.   Result after a forward analysis
                                          followed by a backward analysis.

Fig. 4. Example described in section 4.1: program and results.

Before the test, there is only one non-empty fixpoint for x ($\{0, 1\}$), but after the test, we get two fixpoints ($\{0\}, \{1\}$). Hence we lose information. In practice, we think that it is possible to deal with this issue without modifying the abstract domain, by a good implementation of the abstract test during the backward analysis. Here, it means that we should be able to see that both branches of the test are taken. Hence, if one branch can not satisfy the specification, the backward analysis must propagate that the specification is not satisfied.

4   Examples of Analyses

We wrote a prototype analyzer in OCaml with a graphical interface, from Cousot's Marktoberdorf generic analyzer [3]. The abstract domain for $\wp(\Sigma)$ is the domain of intervals, and the abstract domain for lower closures is the domain defined in the previous section. As a simple prototype, we did not try to optimize the abstract operations. Thus, its complexity is exponential in the depth of the nested loops, cubic in the number of variables and linear in the size of the program.

4.1   First Example

We analyse the very small program presented, with its results, in Fig. 4.

The first random generator (for **x**) is "internal", whereas the second (for **y**) is "external". The initial states are the non-error states of program point (1), and the final states are the states of program point (4) satisfying $x = 1$. The specification is that it is not possible to reach the final state when only the value of **x** is chosen[5]. Using the notations of the formula (1), it means that $A = F$, C are the states at program point (2), and B are the other states.

---

[5] This example was given in [8] as a verification which does not work with interval analysis.

For each program point and each variable, we give:

1. The error status (ini, err or top), corresponding to the values of T.
2. Four integers (mm, Mm, mM, MM) describing the possible values of the variables. Informally, it means that for each generator, the variable has a value in $[mm, Mm]$ and in $[mM, MM]$. We changed the order of the integers (compared with section 3.1) so that this informal definition is more readable.
3. The list of variables which are dependent of this variable. Since the relation of dependence is symmetric, we give it only once.

For example, in program point (3), with only the forward analysis, we get x:[ini: (0,1,0,1)]; y:[ini: (0,0,1,1)], which means informally that x is in $\{0,1\}$, y is in $\{0,1\}$ and takes all values in $\{0,1\}$ whatever the value of x (as they are independent). If the computations were exact, the generators of the concrete lower closure operators would be $\{\{(x:0,y:0),(x:0,y:1)\},\{(x:1,y:0),(x:1,y:1)\}\}$, which are exactly the concretization of the abstract environment. Here we do not lose information.

In program point (4), the result means that x is in $[0,2]$ and takes at least one value in $[0,1]$ and one value in $[1,2]$, y takes all the values of $[0,1]$, and the two variables are dependent. Note that the real generators would be $\{\{(x:0,y:0),(x:1,y:1)\},\{(x:1,y:0),(x:2,y:1)\}$. We lose much information, but thanks to the following backward analysis this is not a problem.

With a backward analysis following the forward analysis, we get $x=1$ in program point (4), and y takes all the values in $[0,1]$, which gives the elements $\{(x:1,y:0),(x:1,y:1)\}$, Hence, in program point (3), it yields $\{(x:1,y:0),(x:0,y:1)\}$ which is not possible given the fact that x and y must be independent. Thus, the application of the lower closure operator gives BOT (which represents $\perp$), which proves our specification.

4.2   Second Example

The program we analyse is described in Fig. 5. Here again we have two random operations. The first one (for the test) is internal, whereas the second (for n) is external. We want to be sure that by controlling the test, we can prevent $x=0$ after the loop. Hence, with the notation of the equation 1, $A = F$, C are the states at program point (2), and B are the other states. The result of the analysis is given Fig. 5. Since we get $\perp$ for the initial point after the backward analysis, the analyzer proved our specification.

The difficult point in this analysis is the backward computation at program point (2). The analyzer detects that both branches of the analyzer are taken, independently of the variables x and n, using the weak relational analysis. Thus, it computes an intersection of the environments of program points (3) and (5).

5   Discussion

In this section we examine the improvement obtained by property-checking driven analyses, with respect to the precision of the analysis.

```
(0)                    Initial states: F :  x = 1
(1)
(2)        while (n>0) do {
(3)                if (? in [0,1]=0) then
(4)                        x = x * (n-1);
(5)                else
(6)                        x = x * n;
(7)                    fi
(8)            n = n - (? in [0,1]);
(9)        }
                       Final states: F :  x = 0
```

0	{ n:[[ini: (-00,+00,-00,+00)]]; ×:[[ini: (1,1,1,1)],n] }	0	{ BOT }
1	{ n:[[ini: (-00,+00,-00,+00)]]; ×:[[ini: (0,+00,0,+00)],n] }	1	{ n:[[ini: (-00,+00,-00,+00)]]; ×:[[ini: (0,0,0,0)],n] }
2	{ n:[[ini: (1,+00,1,+00)]]; ×:[[ini: (0,+00,0,+00)],n] }	2	{ n:[[ini: (1,+00,1,+00)]]; ×:[[ini: (0,0,0,0)],n] }
3	{ n:[[ini: (1,+00,1,+00)]]; ×:[[ini: (0,+00,0,+00)],n] }	3	{ n:[[ini: (1,+00,1,+00)]]; ×:[[ini: (0,+00,0,+00)],n] }
4	{ n:[[ini: (1,+00,1,+00)]]; ×:[[ini: (0,+00,0,+00)],n] }	4	{ n:[[ini: (1,+00,1,+00)]]; ×:[[ini: (0,0,0,0)],n] }
5	{ n:[[ini: (1,+00,1,+00)]]; ×:[[ini: (0,+00,0,+00)],n] }	5	{ n:[[ini: (1,+00,1,+00)]]; ×:[[ini: (0,0,0,0)],n] }
6	{ n:[[ini: (1,+00,1,+00)]]; ×:[[ini: (0,+00,0,+00)],n] }	6	{ n:[[ini: (1,+00,1,+00)]]; ×:[[ini: (0,0,0,0)],n] }
7	{ n:[[ini: (1,+00,1,+00)]]; ×:[[ini: (0,+00,0,+00)],n] }	7	{ n:[[ini: (1,+00,1,+00)]]; ×:[[ini: (0,0,0,0)],n] }
8	{ n:[[ini: (0,+00,0,+00)]]; ×:[[ini: (0,+00,0,+00)],n] }	8	{ n:[[ini: (0,+00,0,+00)]]; ×:[[ini: (0,0,0,0)],n] }
9	{ n:[[ini: (-00,0,-00,0)]]; ×:[[ini: (0,+00,0,+00)],n] }	9	{ n:[[ini: (-00,0,-00,0)]]; ×:[[ini: (0,0,0,0)],n] }

Result after a forward analysis. | Result after a forward analysis followed by a backward analysis.

Fig. 5. Example of section 4.2: program and results.

Starting from the definition of S (as in equation (1)), the first idea to overapproximate S with abstract interpretation-based analyses is to choose an abstract domain, develop in this abstract domain approximations for pre and $\widetilde{pre}$, and directly approximate the fixpoint using the fixpoint transfer theorem. A more precise method, still easier than property-checking driven analysis, is to combine the previous analysis with a reachability analysis starting from I, a method proposed in [7]. This method gives a backward-forward analysis quite similar to ours, but which does not use the lower closure framework. Thus, comparing the two analyses is a good method to see what is gained with this framework.

The first thing we can see is that when the specification does not use $\widetilde{pre}$ (e.g. we just want to prove that some states are unreachable), our technique is useless. Then the analysis is at worst as imprecise as a classical interval analysis (though the complexity is worse).

To show that our analysis can be useful, let us examine a very small example:

$$I : y \in] - \infty, +\infty[$$
$$x := ? \text{ in } [0,2];$$
$$y := f(x,y) ;$$
$$A : y \in [0,8]$$

Here, $f(x,y)$ should be read as an arbitrary expression depending on x and y, e.g. $x+y$ or $x*y$. We want y to be in $[0,8]$ at the end of the program, whatever the result of the non-deterministic operator.

A "classical" analyzer will know that x is in $[0, 2]$ when $f(x, y)$ is computed. In the backward analysis, it must be able to carry a relation between x and y, sufficient to deduce the correct constraints on y when it analyses the command x:= ? in [0,2]. The relation the analyzer can carry is closely related to the abstract domain, and may not be suitable for the function f.

On the other hand, our analyzer can deduce precise constraints on y directly during the backward analysis of $y := f(x, y)$, knowing that x must take the values 0 and 2. Hence it does not need to transmit complex relations.

Let us consider this example when $f(x, y) = x * y$. Even with an expensive abstract domain like polyhedra, we cannot express the good relation between x and y, and the result of the analysis remain imprecise (i.e. it finds that y is in $[-\infty, +\infty]$ at the beginning of the program).Whereas our analyzer finds that y must take values in $[0, 4]$, which is the optimal solution. This example shows that our approach can give better results than the traditional approach, even with expensive abstract domains.

6   Conclusion and Future Work

We presented in this paper the construction of an abstract domain used in a small prototype analyzer for automatic program verification of temporal properties. This analyzer uses a technique based on lower closure operators to exploit the information given by the specification in the analysis. We showed how to abstract the domain of lower closure operators, from the classical interval abstraction, to keep relational information about the states of the program. Even with these weak abstractions, the method may give more precise results than classical abstract analyses with expensive relational domains. Though we analyse only a small class of specifications, the same method can be used for larger classes by including temporal formulas in the concrete domain of the semantics (as was done in [8]).

Our analyzer uses similar abstractions for both analyses (the direct one, on sets of states, and the "reverse" one, on lower closures). Future work can stem from the idea that these abstractions are, in fact, orthogonal. Thus, we can use our abstractions on lower closures along with polyhedra on sets of states, improving the precision of the analysis (without designing a complex abstraction on lower closures based on polyhedra). In general, we believe that property-checking driven analysis can be applied to other abstraction-related analyses. Especially, it would be interesting to examine exactly how it can be compared with abstraction refinements and domain completions [6] and how we can use both technique efficiently.

Acknowledgments

This work was partly done while the author was a visitor in the Cambridge University Computer Laboratory. I wish to thank the anonymous referees, as well as Alan Mycroft and Radhia Cousot for their comments and suggestions.

References

1. B. Blanchet, P. Cousot, R. Cousot, J. Feret, L. Mauborgne, A. Miné, D. Monniaux, and X. Rival. Design and implementation of a special-purpose static program analyzer for safety-critical real-time embedded software, invited chapter. In T. Mogensen, D.A. Schmidt, and I.H. Sudborough, editors, *The Essence of Computation: Complexity, Analysis, Transformation. Essays Dedicated to Neil D. Jones*, LNCS 2566, pages 85–108. Springer-Verlag, October 2002.
2. F. Bourdoncle. Abstract debugging of higher-order imperative languages. In *Proceedings of SIGPLAN '93 Conference on Programming Language Design and Implementation*, pages 46–55, 1993.
3. P. Cousot. The calculational design of a generic abstract interpreter. In M. Broy and R. Steinbrüggen, editors, *Calculational System Design*. NATO ASI Series F. IOS Press, Amsterdam, 1999. Generic Abstract Interpreter available on http://www.di.ens.fr/~cousot/Marktoberdorf98.shtml.
4. P. Cousot and R. Cousot. Static determination of dynamic properties of programs. In *Proceedings of the Second International Symposium on Programming*, pages 106–130. Dunod, Paris, France, 1976.
5. P. Cousot and R. Cousot. Abstract interpretation frameworks. *Journal of Logic and Computation*, 2(4):511–547, August 1992.
6. R. Giacobazzi, F. Ranzato, and F. Scozzari. Making abstract interpretations complete. *Journal of the ACM*, 47(2):361–416, 2000.
7. D. Massé. Combining backward and forward analyses of temporal properties. In O. Danvy and A. Filinski, editors, *Proceedings of the Second Symposium PADO'2001, Programs as Data Objects*, volume 2053 of *Lecture Notes in Computer Sciences*, pages 155–172, Århus, Denmark, 21 – 23 May 2001. Springer-Verlag, Berlin, Germany.
8. D. Massé. Semantics for abstract interpretation-based static analyzes of temporal properties. In M. Hermenegildo, editor, *Proceedings of the Ninth Static Analysis Symposium SAS'02*, volume 2477 of *Lecture Notes in Computer Sciences*, pages 428 – 443, Madrid, Spain, 17 – 20 September 2002. Springer-Verlag, Berlin, Germany.
9. D. Massé. Property checking driven abstract interpretation-based static analysis. In *Proceedings of the Fourth International Conference on Verification, Model Checking and Abstract Interpretation (VMCAI'03)*, volume 2575 of *Lecture Notes in Computer Sciences*, pages 56–69, New York, USA, January 9–11 2003. Springer-Verlag, Berlin, Germany. Available on http://www.stix.polytechnique.fr/~dmasse/papiers/VMCAI2003.pdf.
10. A. Miné. The octagon abstract domain. In *AST 2001 in WCRE 2001*, IEEE, pages 310–319. IEEE CS Press, October 2001. http://www.di.ens.fr/~mine/publi/article-mine-ast01.pdf.

Modular Rewriting Semantics
of Programming Languages

José Meseguer[1] and Christiano Braga[2]

[1] University of Illinois at Urbana-Champaign, USA
[2] Universidade Federal Fluminense, Niterói, Brazil

Abstract. We present a general method to achieve modularity of semantic definitions of programming languages specified as rewrite theories. This provides modularity for a language specification method that combines and extends the best features of both SOS and algebraic semantics. The relationship to Mosses' modular operational semantics (MSOS) is explored in detail, yielding a semantics-preserving translation that could support execution and analysis of MSOS specifications in Maude.

1 Introduction

This work presents a general method to achieve modularity of semantic definitions of programming languages specified as theories in rewriting logic [16, 6], a logical framework which can naturally represent many different logics, languages, operational semantics formalisms, and models of computation [16, 14, 15]. Since equational logic is a sublogic of rewriting logic, this language specification style generalizes the well-known *algebraic semantics* of programming languages (see, e.g., [13] for an early paper, [24] for the relationship with action semantics, and [12] for a recent textbook). The point of this generalization is that equational logic is well suited for specifying *deterministic* languages, but ill suited for concurrent language specification. In rewriting logic, deterministic features are described by *equations*, but concurrent ones are instead described by *rewrite rules* with a concurrent transition semantics. Our modularity techniques yield also new modularity techniques for algebraic semantics as a special case.

It has also been clear from the early stages [16, 20, 14] that there is a natural semantic mapping of structural operational semantics (SOS) definitions [27] into rewriting logic. In essence, an SOS rule is mapped to a *conditional* rewrite rule. In a sense, we can view rewriting logic semantics as a synthesis of algebraic semantics and SOS, that adds a crucial distinction between equations and rules (determinism vs. concurrency) missing in each of those two formalisms. This distinction is of more than academic interest. The point is that, since rewriting with rules R takes place *modulo* the equations E [16], only the rules R contribute to the size of a system's state space, which can be drastically smaller than if all axioms had been given as rules. This observation, combined with the fact that rewriting logic has several high-performance implementations [1, 11, 8] and associated formal verification tools [9, 15], means that we can use rewriting logic language

C. Rattray et al. (Eds.): AMAST 2004, LNCS 3116, pp. 364–378, 2004.

definitions to obtain practical language analysis tools *for free*. For example, in the JavaFAN formal analysis tool [10] the semantics of the JVM is defined as a rewrite theory in Maude which is then used to perform formal analyses such as symbolic simulation, search, and LTL model checking of Java programs with a performance that compares favorably with that of other Java analysis tools. Indeed, the fact that rewriting logic specifications provide in practice an easy way to develop executable formal definitions of programming languages, which can then be subjected to different tool-supported formal analyses, is by now well established [33, 2, 34, 30, 29, 18, 31, 7, 28, 32, 10].

The new question that this work addresses is: how can rewriting logic specifications of programming languages be made *modular*, so that the semantics of each feature can be given once and for all, instead of having to redefine the semantic axioms in a language extension? In this regard, we have learned much from Mosses' elegant solution to the SOS modularity problem though his modular structural operational semantics (MSOS) [23, 25, 26, 22]. To maximize modularity and extensibility, the techniques we propose make the semantic rules as abstract and as general as possible using two key ideas: *record inheritance* through associative commutative matching (a technique also used in MSOS); and systematic use of *abstract interfaces*. We compare in detail our modularity techniques with those of MSOS. This takes the form of a translation map τ mapping an MSOS specification to an (also modular) rewrite theory. The map τ has very strong semantics-preserving properties, including a bisimilarity result between transition systems for an MSOS specification and its resulting translation. This work further advances a line of joint work with Hermann Haeusler and Peter Mosses on semantics-preserving translations from MSOS to rewriting logic [3, 2, 4]. The translation τ and its properties are discussed in Section 4. As for the rest of the paper, Section 2 summarizes prerequisites, Section 3 presents our modularity techniques, and Section 5 gives some conclusions.

2   Membership Equational Logic and Rewriting Logic

We gather here prerequisites about membership equational logic (MEL) [17] and rewriting logic [16, 6]. Maude 2.0 [8] supports all the logical features described below, with a syntax almost identical to the mathematical notation.

2.1   Membership Equational Logic

A MEL *signature* is a triple (K, Σ, S) (just Σ in the following), with K a set of *kinds*, $\Sigma = \{\Sigma_{w,k}\}_{(w,k)\in K^*\times K}$ a many-kinded signature and $S = \{S_k\}_{k\in K}$ a K-kinded family of disjoint sets of sorts. The kind of a sort s is denoted by $[s]$. A MEL Σ-algebra A contains a set A_k for each kind $k \in K$, a function $A_f: A_{k_1} \times \cdots \times A_{k_n} \to A_k$ for each operator $f \in \Sigma_{k_1\cdots k_n, k}$ and a subset $A_s \subseteq A_k$ for each sort $s \in S_k$, with the meaning that the elements in sorts are well-defined, while elements without a sort are *errors*. We write $\mathbb{T}_{\Sigma,k}$ and $\mathbb{T}_{\Sigma}(X)_k$ to denote respectively the set of ground Σ-terms with kind k and of Σ-terms

$$\frac{t \in \mathbb{T}_\Sigma(X)_k}{(\forall X)\, t \to t}\ \textbf{Reflexivity} \qquad \frac{(\forall X)\, t_1 \to t_2, \qquad (\forall X)\, t_2 \to t_3}{(\forall X)\, t_1 \to t_3}\ \textbf{Transitivity}$$

$$\frac{E \vdash (\forall X)\, t = u, \qquad (\forall X)\, u \to u', \qquad E \vdash (\forall X)\, u' = t'}{(\forall X)\, t \to t'}\ \textbf{Equality}$$

$$\frac{\begin{array}{c} f \in \Sigma_{k_1 \cdots k_n, k}, \qquad t_i, t_i' \in \mathbb{T}_\Sigma(X)_{k_i}\ \text{for}\ i \in \{1,\dots,n\} \\ t_i' = t_i\ \text{for}\ i \in \phi(f), \qquad (\forall X)\, t_j \to t_j'\ \text{for}\ j \in \nu(f) \end{array}}{(\forall X)\, f(t_1,\dots,t_n) \to f(t_1',\dots,t_n')}\ \textbf{Congruence}$$

$$\frac{\begin{array}{c} (\forall X)\, r: t \to t'\ \text{if}\ \bigwedge_{i \in I} p_i = q_i\ \wedge\ \bigwedge_{j \in J} w_j : s_j\ \wedge\ \bigwedge_{l \in L} t_l \to t_l' \in R \\ \theta, \theta' : X \to \mathbb{T}_\Sigma(Y), \qquad \theta(x) = \theta'(x)\ \text{for}\ x \in \phi(t,t') \\ E \vdash (\forall Y)\, \theta(p_i) = \theta(q_i)\ \text{for}\ i \in I, \qquad E \vdash (\forall Y)\, \theta(w_j) : s_j\ \text{for}\ j \in J \\ (\forall Y)\, \theta(t_l) \to \theta(t_l')\ \text{for}\ l \in L, \qquad (\forall Y)\, \theta(x) \to \theta'(x)\ \text{for}\ x \in \nu(t,t') \end{array}}{(\forall Y)\, \theta(t) \to \theta'(t')}\ \begin{array}{l}\textbf{Nested}\\ \textbf{Replacement}\end{array}$$

Fig. 1. Deduction rules for rewriting logic.

with kind k over variables in X, where $X = \{x_1 : k_1,\dots,x_n : k_n\}$ is a set of kinded variables. Given a MEL signature Σ, *atomic formulae* have either the form $t = t'$ (Σ-equation) or $t : s$ (Σ-membership) with $t, t' \in \mathbb{T}_\Sigma(X)_k$ and $s \in S_k$; and Σ-*sentences* are conditional formulae of the form $(\forall X)\, \varphi$ if $\bigwedge_i p_i = q_i\ \wedge\ \bigwedge_j w_j : s_j$, where φ is either a Σ-equation or a Σ-membership, and all the variables in φ, p_i, q_i, and w_j are in X. A MEL theory is a pair (Σ, E) with Σ a MEL signature and E a set of Σ-sentences. We refer to [17] for the detailed presentation of (Σ, E)-algebras, sound and complete deduction rules, and initial and free algebras. Order-sorted notation $s_1 < s_2$ can be used to abbreviate the conditional membership $(\forall x : k)\, x : s_2$ if $x : s_1$. Similarly, an operator declaration $f : s_1 \times \cdots \times s_n \to s$ corresponds to declaring f at the kind level and giving the membership axiom $(\forall x_1 : k_1,\dots,x_n : k_n)\, f(x_1,\dots,x_n) :$ s if $\bigwedge_{1 \le i \le n} x_i : s_i$. We write $(\forall x_1 : s_1,\dots,x_n : s_n)\, t = t'$ in place of $(\forall x_1 : k_1,\dots,x_n : k_n)\, t = t'$ if $\bigwedge_{1 \le i \le n} x_i : s_i$.

2.2   Rewrite Theories and Deduction

We present the general version of rewrite theories over **MEL** theories defined in [6]. A *rewrite theory* is a tuple $\mathcal{R} = (\Sigma, E, \phi, R)$ consisting of: (i) a **MEL** theory (Σ, E); (ii) a function $\phi : \Sigma \to \wp_f(\mathbb{N})$ assigning to each function symbol $f : k_1 \cdots k_n \to k$ in Σ a set $\phi(f) \subseteq \{1,\dots,n\}$ of *frozen argument positions*; (iii) a set R of (universally quantified) labeled conditional rewrite rules r having the general form

$$(\forall X)\ r\colon t \to t' \text{ if } \bigwedge_{i \in I} p_i = q_i\ \wedge\ \bigwedge_{j \in J} w_j : s_j\ \wedge\ \bigwedge_{l \in L} t_l \to t'_l \qquad (1)$$

where, for appropriate kinds k and k_l in K, $t, t' \in \mathbb{T}_\Sigma(X)_k$ and $t_l, t'_l \in \mathbb{T}_\Sigma(X)_{k_l}$ for $l \in L$.

The function ϕ specifies which arguments of a function symbol f *cannot be rewritten*, which are called *frozen positions*. Note that if the ith position of f is frozen, then in $f(t_1, \ldots, t_n)$ any subterm of t_i becomes also frozen. That is, the freezing idea extends to subterms and in particular to variables. Given two terms t, t' we can then define the sets $\phi(t, t')$ and $\nu(t, t')$ of their frozen (resp. unfrozen) variables (see [6]). Given a rewrite theory $\mathcal{R} = (\Sigma, E, \phi, R)$, a *sequent* of $\mathcal{R}$ is a pair of (universally quantified) terms of the same kind t, t', denoted $(\forall X)t \to t'$ with $X = \{x_1 : k_1, \ldots, x_n : k_n\}$ a set of kinded variables and $t, t' \in \mathbb{T}_\Sigma(X)_k$ for some k. We say that $\mathcal{R}$ *entails* the sequent $(\forall X)\ t \to t'$, and write $\mathcal{R} \vdash (\forall X)\ t \to t'$, if the sequent $(\forall X)\ t \to t'$ can be obtained by means of the inference rules in Figure 1. (**Reflexivity**), (**Transitivity**), and (**Equality**) are the usual rules for idle rewrites, concatenation of rewrites, and rewriting modulo the **MEL** theory E. (**Congruence**) allows rewriting the arguments of a generalized operator, subject to the condition that frozen arguments must stay idle. (**Nested Replacement**) characterizes the concurrent application of a rewrite rule in its most general form (1).

3   Modular Language Specifications in Rewriting Logic

Modularity is only meaningful in the context of an *incremental* specification, where syntax and corresponding semantic axioms are introduced for groups of related features. We can describe an *incremental* presentation of the syntax of a programming language $\mathcal{L}$ as an indexed family of syntax definitions $\{\mathcal{L}_i\}_{i \in I}$, where the index set I is a *poset* with a top element $\top$, such that: (i) if $i \leq j$, then $\mathcal{L}_i \subseteq \mathcal{L}_j$, and (ii) $\mathcal{L}_\top = \mathcal{L}$. An *incremental rewriting semantics* for $\mathcal{L}$ is then an indexed family of rewrite theories $\{\mathcal{R}_{\mathcal{L}_i}\}_{i \in I}$, with $\mathcal{R}_{\mathcal{L}_i}$ defining the semantics of the language fragment $\mathcal{L}_i$. Modularity of the incremental rewriting semantics $\{\mathcal{R}_{\mathcal{L}_i} = (\Sigma_i, E_i, \phi_i, R_i)\}_{i \in I}$ means in essence two things. First of all, it should satisfy the following *monotonicity property*: if $i \leq j$, then there is a theory inclusion $\mathcal{R}_{\mathcal{L}_i} \subseteq \mathcal{R}_{\mathcal{L}_j}$. This is not easy to achieve, but one can always achieve it *a posteriori*, by carving out submodules of the top module $\mathcal{R}_{\mathcal{L}_\top}$ for each of the language fragments $\mathcal{L}_i$. This way of "cheating" can be immediately recognized by asking the specifier to tell us what the SOS rules are going to be when new features are added in a further extension. Therefore, besides monotonicity, we need the second requirement of *extensibility*. This, together with monotonicity, means that the semantics of each language feature can be defined *once and for all*, so that we never have to *retract* earlier semantic definitions in a later language extension. One is then interested in *methods* that can be used to develop incremental rewriting semantics definitions of programming languages that are as modular as possible.

The method we propose uses pairs, called *configurations*; the first component is the *program text*, and the second a *record* with the different *semantic entities*

that change as the program is computed. That is, we organize all the semantic
entities associated to a program's computation in a record data structure. For
example, one of the record's fields may be the *store*, another the *environment* of
declarations, and yet another the *traces* left by a concurrent process' execution.
We can specify configurations in Maude with the following membership equational theory (a functional module importing the RECORD module shown later in
protecting mode, that is, adding no more data ("no junk") and no new equalities ("no confusion") to records; the ctor keyword indicates a *constructor*):

```
fmod CONF is
 protecting RECORD .
 sorts Program Conf .
 op <_,_> : Program Record -> Conf [ctor] .
endfm
```

The first key modularity technique is *record inheritance*, which is accomplished through pattern matching *modulo* associativity, commutativity, and identity. Features added later to a language may necessitate adding new semantic
components to the record; but the axioms of older features can be given once
and for all in full generality: they will apply just the same with new components
in the record. Here is the Maude specification of the membership equational
theory of records. Note Maude's convention of identifying kinds with connected
components in the subsort inclusion poset, and naming them as equivalence
classes of sorts in such components. For example, [PreRecord] denotes the kind
determined by the connected component {Field,PreRecord}.

```
fmod RECORD is
 sorts Index Component Field PreRecord Record Truth .
 subsort Field < PreRecord .
 op tt : -> Truth .
 op null : -> PreRecord [ctor] .
 op _,_ : PreRecord PreRecord -> PreRecord [ctor assoc comm id: null] .
 op _:_ : [Index] [Component] -> [Field] [ctor] .
 op {_} : [PreRecord] -> [Record] [ctor] .
 op duplicated : [PreRecord] -> [Truth] .
 var I : Index . vars C C' : Component . var PR : PreRecord .
 eq duplicated((I : C),(I : C'), PR) = tt .
 cmb {PR} : Record if duplicated(PR) =/= tt .
endfm
```

A Field is defined as a pair of an Index and a Component; illegal pairs will
be of kind [Field]. A PreRecord is a possibly empty (null) multiset of fields,
formed with the union operator _,_ which is declared to be *associative* (assoc),
commutative (comm) and to have null as its *identity* (id). Maude will then
apply all equations and rules *modulo* such equational axioms [8]. Note the conditional membership (cmb) defining a Record as an "encapsulated" PreRecord
with no duplicated fields. *Record inheritance* means that we can always consider
a record with more fields as a special case of one with fewer fields. Matching
modulo associativity, commutativity, and identity supports record inheritance,

because we can always use an extra variable PR of sort PreRecord to match *any extra fields the record may have*. For example, a function get-env extracting the environment component can be defined by

```
eq get-env({env : E , PR}) = E .
```

and will apply to records with any extra fields that are matched by PR.

The second key modularity technique is the systematic use of *abstract interfaces*. That is, the sorts specifying key syntactic and semantic entities (e.g., Program, Store, Env) are *abstract sorts* for which we:

- only specify the *abstract functions* manipulating them, that is, a given *signature*, or *interface*, of abstract sorts and functions; *no axioms* are specified about such functions *at the level of abstract sorts*;
- in a language specification no *concrete* syntactic or semantic sorts are ever identified with abstract sorts: they are always either specified as *subsorts* of corresponding abstract sorts, or mapped to abstract sorts by *coercions*; it is *only at the level of such concrete sorts* that *axioms* about abstract functions (e.g., functions manipulating the store) are specified.

This means that we *make no a priori ontological commitments* as to the nature of the syntactic or semantic entities. It also means that, since the only commitments ever made happen always at the level of *concrete sorts*, one remains forever free to introduce new meaning and structure in any language extension.

A third modularity technique regards the form of the rules. We require that the rewrite rules in the rewrite theories $\mathcal{R}_{\mathcal{L}_i}$ are semantic rules

$$\langle f(t_1, \ldots, t_n), u \rangle \longrightarrow \langle t', u' \rangle \ \text{ if } \ C,$$

where f is a language feature, e.g., if-then-else, u and u' are record expressions, and u contains a variable PR of sort PreRecord standing for unspecified additional fields and allowing the rule to match by record inheritance. In addition, the following *information hiding* discipline should be followed in u, u', and in any record expressions appearing in C: besides basic record syntax, only function symbols appearing in the *abstract interfaces* of some of the record's fields can appear in record expressions; any auxiliary functions defined in concrete sorts of those field's components should never be mentioned. This information hiding makes the rules highly extensible, because the concrete representations of the auxiliary semantic entities can be changed or extended without having to change the rules at all. For example, we can change or extend the internal representations of traces, stores, or environments, without the rules being affected in any way: all we have to do is add new equations defining the semantics of the abstract functions on the new concrete data representations. For two interesting applications of this modularity discipline, one in which the semantics of CCS is extended from the strong transition semantics to the weak transition semantics, and another in which the bc sublanguage of C is enriched with annotations for physical units in the style of [7], so that the values stored now need to be pairs of a number and a unit expression, see [5]. Another example illustrating our general methodology is given in [19], Appendices A–B.

The combination of these three techniques can be of great help in making semantic definitions modular and easily extensible. That is, we can develop in this way modular incremental semantic definitions for a language $\mathcal{L}$ as a poset-indexed hierarchy $\{\mathcal{R}_{\mathcal{L}_i} = (\Sigma_i, E_i, \phi_i, R_i)\}_{i \in I}$ of rewrite theory *inclusions*, with the full language definition as the top theory in the hierarchy and with the theory CONF – which contains RECORD – as the bottom of the hierarchy. By following the methods described above, such a modular definition will then be much more easily extensible than if such methods had not been followed. As we explain in more detail in Section 4, the above methods are closely related to Mosses' MSOS methodology; however, besides the fact that we are exploring such methods in a different semantic framework, it appears that our systematic use of abstract interfaces is an aspect considerably less developed in the MSOS approach.

An important variant of our approach is to choose the MEL sublogic of rewriting logic as the *logical framework* in which to define the semantics of a language. As argued in [18], this is a perfectly good possibility for *sequential* or *deterministic* languages, but is typically a bad choice for concurrent languages. Of course, in this case the semantics is no longer given by rewrite rules, but by *conditional equations* of the form,

$$\langle f(t_1, \ldots, t_n), u \rangle = \langle t', u' \rangle \ \text{ if } \ C.$$

This variant provides modularity techniques for a long strand of work in the so-called *algebraic* or *initial algebra semantics* of programming languages (see, e.g., [13, 24, 12]). In fact, as explained in the Introduction, the best approach in practice is to *combine* the equational and the rewriting variants of the modular language specification methodology just described. This is easy in rewriting logic, because of its explicit distinction between equations and rules.

3.1   Controlling Rewrite Steps in Conditions

When relating SOS rules to rewrite rules a few technicalities are needed to obtain an exact correspondence. The key issue is the *number of steps* of rewrites in conditions. In an SOS rule

$$\frac{P_1 \longrightarrow P_1' \quad \cdots \quad P_n \longrightarrow P_n'}{Q \longrightarrow Q'}$$

the rewrites $P_i \longrightarrow P_i'$ in the condition are *one-step rewrites*. By contrast, in a rewrite rule

$$Q \longrightarrow Q' \ \text{ if } \ P_1 \longrightarrow P_1' \wedge \ldots \wedge P_n \longrightarrow P_n',$$

because of the **(Reflexivity)** and **(Transitivity)** inference rules of rewriting logic (see Figure 1) the rewrites $P_i \longrightarrow P_i'$ in the condition are considerably more general: they can have zero, one, or more steps of rewriting. The point is that, by definition, in rewriting logic *all finitary computations are always derivable as sequents*, whereas in SOS they are not, unless special provisions are made such as a big-step style or adding transitivity as an explicit SOS rule. The question we now address is how to accomplish two simultaneous goals: (i) to represent SOS rules in an exact way, with one step rewrites in conditions; and (ii) to get also all finitary computations in the rewriting logic representation, so that we always get a language interpreter, even when the SOS rules do not directly provide one. To accomplish goals (i) and (ii), we extend the module CONF to a system module (rewrite theory):

```
mod RCONF is extending CONF .
 op {_,_} : [Program] [Record] -> [Conf] [ctor] .
 op [_,_] : [Program] [Record] -> [Conf] [ctor] .
 vars P P' : Program .  vars R R' : Record .
 crl [step] : < P , R > => < P' , R' > if { P , R } => [ P' , R' ] .
endm
```

We furthermore require semantic rewrite rules to be of the form,

$$\{t, u\} \longrightarrow [t', u'] \text{ if } \{v_1, w_1\} \longrightarrow [v'_1, w'_1] \wedge \ldots \wedge \{v_n, w_n\} \longrightarrow [v'_n, w'_n] \wedge C, \tag{2}$$

where $n \geq 0$, and C is a (possibly empty) equational condition involving only equations and memberships. Note that a rewrite theory $\mathcal{R}$ containing only RCONF and rules of the form (2) will be such that one-step rewrites correspond to proofs of sequents $\mathcal{R} \vdash \{v, w\} \longrightarrow [v', w']$, whereas finitary computations correspond to proofs of sequents $\mathcal{R} \vdash \langle v, w \rangle \longrightarrow \langle v', w' \rangle$.

4 Relationship to MSOS

Our modular rewriting logic ideas have a close relationship to a new formulation of MSOS initiated in [26] and further developed by Peter Mosses under the name of *definitive semantics* [21, 22]. This allows us to define a quite succinct mapping from MSOS to rewriting logic that is semantics-preserving in a very strong sense. In Mosses' definitive semantics, labeled transitions are of the form $t \xrightarrow{u} t'$, where t, t' are *program expressions* (which can involve values), and u is a *record* which specifies the semantic information *before and after* the transition takes place. This is accomplished by postulating that the indices of the record are classified in three different types:

- *read-only*, where the index, say i, is a name with no additional structure;
- *write-only*, where the index has primed form, that is, is of the form i'; and
- *read-write*, where there are in fact *two* indices in the record: a plain i, and its primed variant i'.

Furthermore, the values of any write only index must range over some *free monoid of lists*, with an associative append operation $\_.\_$ and neutral element *nil*. For example, an environment index *env* will be read-only, a trace index tr' for a concurrent process will be write only, and store indices st, st' will be read-write. As usual, the primed indices indicate the relevant changes *after* the transition takes place. Such records can be naturally viewed as the arrows of a *label category*, so that several-step *computations*[1] can be defined by composing the labels and forgetting the intermediate stages as follows: $(t \xrightarrow{u} t'); (t' \xrightarrow{u'} t'') = (t \xrightarrow{u;u'} t'')$. For the composition $u; u'$ to be defined, we must have (in usual record notation) $u.i = u'.i$ for each read-only index i, and $u.j' = u'.j$ for each read-write index j. The composition $u; u'$ then has $(u; u').i = u.i = u'.i$, $(u; u').j = u.j$, $(u; u').j' = u'.j'$, and $(u; u').k = (u.k).(u'.k)$ for each write-only index k. *Identities* are then records u such that $u.j = u.j'$, and $u.k = nil$. Note that we can easily axiomatize these records and their composition and units in an extension of our RECORD module. In particular, we will have a subsort IRecord < Record of identity records, another subsort IPreRecord < PreRecord of identity prerecords, and a partial record composition operator $\_; \_$. In what follows we will use variables $X, X' \dots$, of sort Record, variables $U, U', \dots$, of sort IRecord, variables $PR, PR' \dots$, of sort PreRecord, and variables $UPR, UPR', \dots$, of sort IPreRecord.

In MSOS a rule defining the semantics of a feature f has the general form,

$$\frac{v_1 \xrightarrow{u_1} v_1' \ \dots \ v_n \xrightarrow{u_n} v_n' \quad cnd}{f(t_1, \dots, t_n) \xrightarrow{u} t'} \tag{3}$$

where $f(t_1, \dots, t_n), t'$, and the v_i, v_i' are *program expressions* (which can involve values), $u, u_1, \dots, u_n$ are record expressions, and *cnd* is a side condition involving equations and perhaps predicates. The key idea is that matching of such MSOS rules uses *record inheritance*. Our concrete notation is similar to a record-pattern notational variant for MSOS mentioned in passing in [22]. For example, rule [10] in [19] Appendix B could be expressed in our notational variant of MSOS as follows:

$$\frac{\sigma' = \sigma[l \mapsto v]}{l := v \xrightarrow{\{st:\sigma, st':\sigma', UPR\}} noop} \tag{4}$$

A definitive MSOS specification $\mathcal{S}$ has rules of the form (3), but must also specify: (i) the *syntax* of programs in the language $\mathcal{L}$ of interest, and (ii) the *semantic entities* used in the different record components. We choose membership equational logic to specify (i) and (ii). Therefore, in what follows we assume that an MSOS specification is in essence equivalent to a triple (Σ, E, R), with (Σ, E) a membership equational theory containing abstract sorts Program and Component, and importing the above-sketched extension of the RECORD specification as a subtheory, and with no other operations in Σ having kinds in the RECORD extension as arguments, except for the [Component] kind. The rules R are then MSOS rules of the general form (3), where $f(t_1, \dots, t_n), t'$, and the v_i, v_i' are terms of sort Program, $u, u_1, \dots, u_n$ are expressions of sort Record, and *cnd* is a conjunction of equations and memberships. Furthermore, we assume that all MSOS rules in R are in the following *normal form*[2]: (i) the side condition

---

[1] A somewhat subtle point explained in what follows is that to organize the computations themselves as a category we need some more structure on the objects.

[2] This assumption does not seem overly restrictive in practice; a detailed characterization of the class of normalizable MSOS rules should be developed.

cnd does not involve any record, field, index, or [Truth] (different from [Bool]) expressions in its terms or subterms; and (ii) a record expression appearing in either the premises or the conclusion is either: (1) a variable of the general form X, or U; or (2) a constructor term of the general form $\{i_1 : w_1, \ldots, i_n : w_n, PR\}$, or $\{i_1 : w_1, \ldots, i_n : w_n, UPR\}$, with $n \geq 0$, some of the indices perhaps primed, and the w_j terms with sorts in the corresponding field components. For example, rule [9] in [19] Appendix B might be expressed as the MSOS rule

$$\frac{U = \{env : \rho,\, UPR\} \qquad v = \rho(x)}{x \xrightarrow{\ U\ } v}$$

which fails to satisfy (i) above, but we have an equivalent normal form

$$\frac{v = \rho(x)}{x \xrightarrow{\ \{env : \rho,\, UPR\}\ } v} \tag{5}$$

The desired translation is a mapping $\tau : (\Sigma, E, R) \mapsto (\Sigma', E', \phi, R')$ where:

1. (Σ', E') is obtained from (Σ, E) by: (i) omitting all the primed indices and their related equations and memberships from the record subspecification (but adding the unprimed version of each write-only index); (ii) defining subsorts ROPreRecord, RWPreRecord, and WOPreRecord (all containing the constant null) of the PreRecord sort, corresponding to those parts of the record involving read-only, read-write, and write-only fields (we use variables $A, A', \ldots$, $B, B', \ldots$, and $C, C', \ldots$, to range over those respective subsorts); (iii) In WOPreRecord we also equationally axiomatize a *prefix predicate* $\sqsubseteq$, where $C \sqsubseteq C'$ means that for each write-only field k the string $C.k$ is a (possibly identical) prefix of the string $C'.k$; and (iv) adding the signature of the module RCONF;
2. ϕ declares all operators in Σ' as unfrozen; and
3. R' contains the step rule in RCONF, as well as for each MSOS rule in R, in the normal form described above, a corresponding rewrite rule of the form,

$$\{f(t_1, \ldots, t_n), u^{pre}\} \longrightarrow [t', u^{post}]$$

$$\text{if } \{v_1, u_1^{pre}\} \longrightarrow [v_1', u_1^{post}] \wedge \ldots \wedge \{v_n, u_n^{pre}\} \longrightarrow [v_n', u_n^{post}] \wedge \ cnd',$$

where for each record expression u in the MSOS rule, u^{pre} and u^{post} are defined as follows. For u a record expression of the general form X or $\{PR\}$, u^{pre} is a record expression of the form $\{A, B, C\}$, and u^{post} has the form $\{A, B', C'\}$. For u a record expression of the general form U or $\{UPR\}$, u^{pre} is of the general form R or $\{PR\}$, and $u^{pre} = u^{post}$. Otherwise, for u of the general form $\{i_1 : w_1, \ldots, i_n : w_n, PR\}$, with $n \geq 1$, u^{pre} and u^{post} are record expressions similar to u where: (i) if a read-only field expression $i : w$ appears in u, then it appears in both u^{pre} and u^{post}; (ii) if a write-only field expression $i' : w$ appears in u, if u labels the conclusion, then u^{pre} contains a field expression of the form $i : l$, with l a new list variable in the corresponding data type, and u^{post} contains a field expression of the form $i : l.w$ (if u labels a condition, u^{pre} contains $i : nil$, and u^{post} contains $i : w$); (iii) if a read-write pair of field expressions $i : w, i' : w'$ appear in u, then u^{pre} contains $i : w$, and u^{post} contains $i : w'$; and (iv) PR is translated in u^{pre} as A, B, C, and in u^{post} as A, B', C'. Finally, if u is of the general form $\{i_1 : w_1, \ldots, i_n : w_n, UPR\}$, with $n \geq 1$. Then u^{pre} and u^{post} are record expressions like u where cases (i)–(iii) as handled as before, and (iv) UPR is translated in both u^{pre} and u^{post} as PR. Furthermore, the condition cnd' is either cnd itself, or is obtained by conjoining to cnd the prefix predicate $C \sqsubseteq C'$ in case subexpressions of the form A, B, C, and A, B', C' were introduced in the terms u^{pre} and u^{post} in the conclusion. We assume throughout some reasonable naming conventions, such as using different translated names for different variables, introducing the same new names for repeated variables, avoiding variable capture, and so on.

For example, the MSOS rule (4) would be translated as the conditional rewrite rule:

$$\{l := v, \{(st : \sigma), PR\}\} \longrightarrow [noop, \{(st : \sigma'), PR\}] \quad \text{if} \quad \sigma' = \sigma[l \mapsto v].$$

The translation τ just defined is *semantics-preserving* in a very strong sense. To make this clear, we need to discuss the *semantic models* associated to (finite) *computations* in both formalisms. We shall focus on transitions involving *ground* terms t, t' of sort **Program**, and with u a ground record expression. First of all, note that an MSOS specification S defines a *category* $\mathbb{C}_S$ of *finite computations*, whose arrows have the form $\langle t, u \rangle \xrightarrow{w} \langle t', u' \rangle$, where $u = dom(w)$, and $u' = cod(w)$ are the source and target identity records of the label w, and for some $n \geq 0$ there are n composable transitions (i.e., their labels are composable) derivable from S,

$$t \xrightarrow{w_1} t_1, \ldots t_{n-1} \xrightarrow{w_n} t'$$

such that $w = w_1; \ldots; w_n$. Note that the objects of $\mathbb{C}_S$ must be pairs $\langle t, u \rangle$, whose identity arrow is u, since without the second component the identities of $\mathbb{C}_S$ would be underdetermined. By considering the labeled subgraph of $\mathbb{C}_S$ determined by the computations corresponding to one-step transitions we obtain a *labeled transition system*[3] that we denote $\mathbb{L}_S$.

In rewriting logic, the simplest model associated to a rewrite theory $\mathcal{R}$ is its *initial reachability model* $\mathbb{T}_{Reach(\mathcal{R})}$ [6], which defines the $\mathcal{R}$-reachability relation $[t] \longrightarrow [t']$ between equivalence classes[4] of ground terms modulo the equations in $\mathcal{R}$ by the equivalence, $[t] \longrightarrow [t'] \Leftrightarrow \mathcal{R} \vdash t \longrightarrow t'$. In particular, for $\mathcal{R} = \tau(S)$, we can restrict this model to the sort **Conf** and, because of the **(Transitivity)** and **(Reflexivity)** inference rules, we then get a *preorder relation* on equivalence classes of **Conf** ground terms modulo the equations E' in $\tau(S)$, that is, a preorder *category* $\mathbb{T}_{Reach(\tau(S))}|_{\text{Conf}}$. As pointed out in Section 3.1, we will have a $\tau(S)$-reachability relation $\langle v, w \rangle \longrightarrow \langle v', w' \rangle$ iff for some $n \geq 0$, there is a sequence

---

[3] Note that this is a more detailed labeled transition system than the one associated to S in [22], since the states are pairs $\langle t, u \rangle$.

[4] To keep the notation lighter, in what follows we will often leave implicit the equivalence class notation, and will denote reachability using representative terms.

of rewrite steps, each obtained by application of the (**Nested Replacement**) inference rule to the `step` rule, and by (**Equality**). Any such application of the `step` rule exactly mimics a one-step rewrite with a rule $r' \in R'$, which is the translation of an MSOS rule $r \in R$ in S. The semantics-preserving nature of τ takes the form of a *functor* $\pi : \mathbb{T}_{Reach(\tau(S))}|_{\texttt{Conf}} \longrightarrow \mathbb{C}_S$ surjective on objects and arrows and defined on arrows by:

$$\pi : (\langle t, w \rangle \longrightarrow \langle t', w' \rangle) \mapsto (\langle t, \rho(w) \rangle \overset{w \mapsto w'}{\longrightarrow} \langle t', \rho(w') \rangle)$$

where the function ρ maps each record w without primed indices in (Σ', E') to an identity record $\rho(w)$ in (Σ, E) by leaving all read-only fields untouched, adding a primed copy of each read-write index (with same value), and making all write only fields primed and all with value nil; and where the label $w \mapsto w'$ is defined as follows. For a read-only index i, w, w', and $w \mapsto w'$ all agree on the field $i : x$; for a write-only index i, if w contains the field $i : l$, then w' will contain a field of the form $i : l.l'$, and $w \mapsto w'$ contains the field $i' : l'$; for a read-write index i, if w contains the field $i : x$, and w' the field $i : y$, then $w \mapsto w'$ contains the fields $i : x$ and $i' : y$. The well-definedness of this functor follows by induction on the length of the rewrites/computations from the strong bisimulation result for one-step rewrites stated below. Furthermore, it also requires showing the well-definedness of the operation $w \mapsto w'$. This follows easily by induction on the length of the rewrites from the following lemma, which is itself an easy consequence of the definition of u^{pre} and u^{post} in the translation τ:

Lemma 1. *In any one-step rewrite $\langle t, w \rangle \longrightarrow \langle t', w' \rangle$, the record w' has the same read-only fields as w, and the value of any write-only field in w is a prefix of the corresponding value in w'.*

We can associate to $\tau(S)$ the labeled transition system $\mathbb{L}^{\pi}_{\mathbb{T}_{Reach(\tau(S))}}$, whose transitions are of the form $\langle t, w \rangle \overset{w \mapsto w'}{\longrightarrow} \langle t', w' \rangle$, where $\langle t, w \rangle \longrightarrow \langle t', w' \rangle$ is a *one-step* $\tau(S)$-rewrite. The semantics-preserving nature of τ can be further expressed by the following theorem (a proof sketch is given in [19], Appendix C):

Theorem 1. *(Strong Bisimulation). The projection function $\pi : \langle t, w \rangle \mapsto \langle t, \rho(w) \rangle$, together with its inverse relation π^{-1}, define a strong bisimulation of labeled transition systems $\pi : \mathbb{L}^{\pi}_{\mathbb{T}_{Reach(\tau(S))}} \longrightarrow \mathbb{L}_S$.*

5   Concluding Remarks

We have presented a general method to make semantic definitions of programming languages both modular and executable in rewriting logic. It would be natural to extend these techniques in the direction of "true concurrency." The point is that SOS provides an *interleaving semantics*, based on labeled transition systems, whereas rewriting logic provides a "true concurrency" semantics, in which many rewrites can happen concurrently. This is one of the reasons for a gradual shift towards so-called *reduction semantics* for concurrent languages, which is a special case of rewriting semantics. Although an interleaving semantics remains a possible choice, the SOS idea of executing a *single program* becomes less natural when languages are concurrent or even mobile.

We have also discussed the relationship to MSOS. Once implemented, our translation from MSOS to rewriting logic would make available in a sound and simple way the possibility of executing MSOS specifications in Maude. As already done for a previous translation [2], this can be the basis of a Maude-based tool to execute MSOS specifications. Since Maude has breadth-first search, an efficient LTL model checker [8], and several other formal tools [9, 15], MSOS specifications can then be formally analyzed in various ways. Besides the example in [19], we have developed several variants of a modular rewriting semantics for CCS and for the bc language in [5]. More work on the MSOS translation and on experimentation remains ahead.

Acknowledgments

This research has been supported by ONR Grant N00014-02-1-0715, NSF Grant CCR-0234524, and by CNPq under processes 552192/2002-3 and 300294/2003-4. We have benefitted much from our collaboration with Hermann Haeusler and Peter Mosses on the MSOS-rewriting logic connection; and from Peter Mosses' very helpful comments that have facilitated a precise comparison between his work and ours. We also thank the referees, Narciso Martí-Oliet, and Grigore Roşu for their detailed comments on an earlier draft.

References

1. P. Borovanský, C. Kirchner, H. Kirchner, and P.-E. Moreau. ELAN from a rewriting logic point of view. *Theoretical Computer Science*, 285:155–185, 2002.
2. C. Braga. *Rewriting Logic as a Semantic Framework for Modular Structural Operational Semantics*. PhD thesis, Departamento de Informática, Pontificia Universidade Católica de Rio de Janeiro, Brasil, 2001.
3. C. Braga, E. H. Haeusler, J. Meseguer, and P. D. Mosses. Maude Action Tool: Using reflection to map action semantics to rewriting logic. In T. Rus, editor, *Algebraic Methodology and Software Technology, 8th International Conference, AMAST 2000, Iowa City, Iowa, USA, May 20–27, 2000, Proceedings*, volume 1816 of *Springer LNCS*, pages 407–421, 2000.
4. C. Braga, E. H. Haeusler, J. Meseguer, and P. D. Mosses. Mapping modular SOS to rewriting logic. In M. Leuschel, editor, *12th International Workshop, LOPSTR 2002, Madrid, Spain*, volume 2664 of *Springer LNCS*, pages 262–277, 2002.
5. C. Braga and J. Meseguer. Modular rewriting semantics in practice. in Proc. *WRLA'04*, ENTCS.
6. R. Bruni and J. Meseguer. Generalized rewrite theories. In J. Baeten, J. Lenstra, J. Parrow, and G. Woeginger, editors, *Proceedings of ICALP 2003, 30th International Colloquium on Automata, Languages and Programming*, volume 2719 of *Springer LNCS*, pages 252–266, 2003.

7. F. Chen, G. Roşu, and R. P. Venkatesan. Rule-based analysis of dimensional safety. In *Rewriting Techniques and Applications (RTA'03)*, volume 2706 of *Springer LNCS*, pages 197–207, 2003.

8. M. Clavel, F. Durán, S. Eker, P. Lincoln, N. Martí-Oliet, J. Meseguer, and C. Talcott. Maude 2.0 Manual. June 2003, http://maude.cs.uiuc.edu.

9. M. Clavel, F. Durán, S. Eker, and J. Meseguer. Building equational proving tools by reflection in rewriting logic. In *CAFE: An Industrial-Strength Algebraic Formal Method*. Elsevier, 2000. http://maude.cs.uiuc.edu.

10. A. Farzan, J. Meseguer, and G. Roşu. Formal JVM code analysis in JavaFAN. This volume.

11. K. Futatsugi and R. Diaconescu. *CafeOBJ Report*. World Scientific, AMAST Series, 1998.

12. J. A. Goguen and G. Malcolm. *Algebraic Semantics of Imperative Programs*. MIT Press, 1996.

13. J. A. Goguen and K. Parsaye-Ghomi. Algebraic denotational semantics using parameterized abstract modules. In J. Diaz and I. Ramos, editors, *Formalizing Programming Concepts*, pages 292–309. Springer-Verlag, 1981. LNCS, Volume 107.

14. N. Martí-Oliet and J. Meseguer. Rewriting logic as a logical and semantic framework. In D. Gabbay and F. Guenthner, editors, *Handbook of Philosophical Logic, 2nd. Edition*, pages 1–87. Kluwer Academic Publishers, 2002. First published as SRI Tech. Report SRI-CSL-93-05, August 1993.

15. N. Martí-Oliet and J. Meseguer. Rewriting logic: roadmap and bibliography. *Theoretical Computer Science*, 285:121–154, 2002.

16. J. Meseguer. Conditional rewriting logic as a unified model of concurrency. *Theoretical Computer Science*, 96(1):73–155, 1992.

17. J. Meseguer. Membership algebra as a logical framework for equational specification. In F. Parisi-Presicce, editor, *Proc. WADT'97*, pages 18–61. Springer LNCS 1376, 1998.

18. J. Meseguer. Software specification and verification in rewriting logic. In M. Broy and M. Pizka, editors, *Models, Algebras, and Logic of Engineering Software, NATO Advanced Study Institute, Marktoberdorf, Germany, July 30 – August 11, 2002*, pages 133–193. IOS Press, 2003.

19. J. Meseguer and C. Braga. Modular rewriting semantics of programming languages. January 2004, http://maude.cs.uiuc.edu.

20. J. Meseguer, K. Futatsugi, and T. Winkler. Using rewriting logic to specify, program, integrate, and reuse open concurrent systems of cooperating agents. In *Proceedings of the 1992 International Symposium on New Models for Software Architecture, Tokyo, Japan, November 1992*, pages 61–106. Research Institute of Software Engineering, 1992.

21. P. D. Mosses. Definitive semantics. Version 0.2, May 31, 2003, http://www.mimuw.edu.pl/~mosses/DS-03.

22. P. D. Mosses. Modular structural operational semantics. Manuscript, September 2003, to appear in *J. Logic and Algebraic Programming*.

23. P. D. Mosses. Semantics, modularity, and rewriting logic. *Proc. 2nd Intl. Workshop on Rewriting Logic and its Applications*, ENTCS, Vol. 15, North Holland, 1998.

24. P. D. Mosses. Unified algebras and action semantics. In *Proc. Symp. on Theoretical Aspects of Computer Science, STACS'89*. Springer LNCS 349, 1989.

25. P. D. Mosses. Foundations of modular SOS. In *Proceedings of MFCS'99, 24th International Symposium on Mathematical Foundations of Computer Science*, pages 70–80. Springer LNCS 1672, 1999.

26. P. D. Mosses. Pragmatics of modular SOS. In *Proceedings of AMAST'02, 9th Intl. Conf. on Algebraic Methodology and Software Technology*, pages 21–40. Springer LNCS 2422, 2002.

27. G. D. Plotkin. A structural approach to operational semantics. Technical Report DAIMI FN-19, Computer Science Dept., Aarhus University, 1981.

28. G. Roşu, R. P. Venkatesan, J. Whittle, and L. Leustean. Certifying optimality of state estimation programs. In *Computer Aided Verification (CAV'03)*, pages 301–314. Springer, 2003. LNCS 2725.

29. M.-O. Stehr and C. Talcott. Plan in Maude: Specifying an active network programming language. In F. Gadducci and U. Montanari, editors, *Proc. 4th. Intl. Workshop on Rewriting Logic and its Applications*. ENTCS, Elsevier, 2002.

30. P. Thati, K. Sen, and N. Martí-Oliet. An executable specification of asynchronous Pi-Calculus semantics and may testing in Maude 2.0. In F. Gadducci and U. Montanari, editors, *Proc. 4th. Intl. Workshop on Rewriting Logic and its Applications*. ENTCS, Elsevier, 2002.

31. A. Verdejo. *Maude como marco semántico ejecutable*. PhD thesis, Facultad de Informática, Universidad Complutense, Madrid, Spain, 2003.

32. A. Verdejo and N. Martí-Oliet. Executable structural operational semantics in Maude. Manuscript, Dto. Sistemas Informáticos y Programación, Universidad Complutense, Madrid, August 2003.

33. A. Verdejo and N. Martí-Oliet. Executing and verifying CCS in Maude. Technical Report 99-00, Dto. Sistemas Informáticos y Programación, Universidad Complutense, Madrid; also, `http://maude.cs.uiuc.edu`.

34. A. Verdejo and N. Martí-Oliet. Implementing CCS in Maude 2. In F. Gadducci and U. Montanari, editors, *Proc. 4th. Intl. Workshop on Rewriting Logic and its Applications*. ENTCS, Elsevier, 2002.

Modal Kleene Algebra and Partial Correctness

Bernhard Möller and Georg Struth*

Institut für Informatik, Universität Augsburg
Universitätsstr. 14, D-86135 Augsburg, Germany
{moeller,struth}@informatik.uni-augsburg.de

Abstract. Modal Kleene algebra is Kleene algebra enriched by forward and backward box and diamond operators. We formalize the symmetries of these operators as Galois connections and dualities. We study their properties in the associated semirings of operators. Modal Kleene algebra provides a unifying semantics for various program calculi and enhances efficient cross-theory reasoning in this class, often in a very concise state-free style. This claim is supported by novel algebraic soundness and completeness proofs for Hoare logic.

1 Introduction

Complex hardware and software development usually depends on many different models and formalisms. This calls for a unifying semantics and for calculi that enhance safe cross-theory reasoning. During the last decade, variants of Kleene algebra (KA) have emerged as foundational structures with widespread applications in computer science ranging from program and protocol analysis [3, 12, 22], program development [2, 18] and compiler optimization [14] to rewriting theory [20] and concurrency control [3]. The development has been initialized by two seminal papers by Kozen, the first one providing a particularly useful and elegant axiomatization of KA as the algebra of regular events [11], the second one extending KA to Kleene algebra with tests (KAT) for modeling the usual constructs of sequential programming [12]. But although KAT subsumes propositional Hoare logic (PHL) [13], it seems not appropriate as a unifying core calculus, since it does not admit an explicit definition of modalities as they occur in many popular methods.

KAT has recently been enriched by simple equational axioms for abstract domain and codomain operations [4]. This Kleene algebra with domain (KAD) is more expressive than KAT. It does not only allow relational reasoning about hardware and software [4], it also subsumes propositional dynamic logic and supplies it with a natural algebraic semantics [7].

This motivates the following question: Is KAD suitable as a calculus for cross-theory reasoning and as a unifying semantics? Answering this question, however, requires further consideration of the modal aspects of KAD in general and its semantical impact for Hoare logic in particular[1].

---

* Supported by DFG Project InopSys (Interoperability of System Calculi).

[1] The relation between KAD and temporal logics will be the subject of another paper.

C. Rattray et al. (Eds.): AMAST 2004, LNCS 3116, pp. 379–393, 2004.

Our Contributions. First, we use the abstract image and preimage operations of KAD for defining forward and backward box and diamond operators as modal operators à la Jónsson and Tarski [10]. We show that these operators are related by two fundamental symmetries: Galois connections and dualities. The former serve as theorem generators, yielding a number of modal properties for free. The latter serve as theorem transformers, passing properties of one modal operator automatically to its relatives. We also develop further natural and interesting algebraic properties, including continuity of domain and codomain. Most of them immediately transfer to predicate transformer algebras.

Second, we study the algebra of modal operators over KAD, which under suitable conditions is again a Kleene algebra. This abstraction supports even more concise state-free modal reasoning and leads to further structural insight.

Third, we apply modal Kleene algebra by giving purely calculational algebraic proofs of soundness and relative completeness for PHL. We use this formalism both for a faithful encoding of Hoare's syntax and for modeling the standard weakest liberal precondition semantics. Our encoding and soundness proof – all inference rules of PHL are theorems in KAD – is more direct and concise than previous KAT-based ones [13]. In particular, when abstracted to the algebra of modal operators, the Hoare rules immediately reflect natural properties. Our novel algebraic proof of relative completeness is much shorter and more abstract, thus applicable to more models, than the standard ones (e.g. [1]). It exploits a Galois connection between forward boxes and backward diamonds that is beyond the expressiveness of most related modal formalisms.

These technical results support our claim that KAD may serve both as a calculus for cross-theory reasoning with various calculi for imperative programs and state transition systems and as a unifying semantics for modal, relational and further algebraic approaches. The economy of concepts in Kleene algebra imposes a discipline of thought which usually leads to simpler and more perspicuous proofs and to a larger class of application models than with alternative approaches, for instance relational algebra (cf. [19]) or temporal algebra [21], where some of our issues have also been treated. This is also interesting from a pedagogical point of view, since taxonomic knowledge about various structures and complex axiomatizations can be replaced by systematic knowledge about a few simple operations together with symmetries and abstraction techniques, a particular advantage of the algebraic approach. Finally, our results are of independent interest for the foundations of modalities.

In this extended abstract, we can only describe the main ideas of our approach. See [17] for a full technical treatment and [5] for a synopsis of related results on modal Kleene algebra and for further support for our claims.

Outline. The remainder is organized as follows: Section 2 introduces KAD and its basic properties. Section 3 introduces modal operators and the associated algebras of modal operators. Section 4 develops the basic calculus of modal operators. The syntax and semantics of Hoare logic and its soundness and completeness proofs in KAD are the subject of Section 5, Section 6 and Section 7. Section 8 contains a summary, a discussion of further results and an outlook.

2   Kleene Algebra with Domain

A *Kleene algebra* [11] is a structure $(K, +, \cdot, *, 0, 1)$ such that $(K, +, \cdot, 0, 1)$ is an (additively) idempotent semiring (an i-semiring) and $*$ is a unary operation axiomatized by the identities and quasi-identities

$$1 + aa^* \leq a^*, \quad (*\text{-}1) \qquad\qquad b + ac \leq c \Rightarrow a^*b \leq c, \quad (*\text{-}3)$$
$$1 + a^*a \leq a^*, \quad (*\text{-}2) \qquad\qquad b + ca \leq c \Rightarrow ba^* \leq c, \quad (*\text{-}4)$$

for all $a, b, c \in K$ (the operation $\cdot$ is omitted here and in the sequel). If the structure satisfies $(*\text{-}1)$, $(*\text{-}2)$ and $(*\text{-}3)$, but not necessarily $(*\text{-}4)$, we call it a *left Kleene algebra*. It is called a *right Kleene algebra*, if $(*\text{-}1)$, $(*\text{-}2)$ and $(*\text{-}4)$, but not necessarily $(*\text{-}3)$ holds. The natural ordering $\leq$ on K is defined by $a \leq b$ iff $a + b = b$. Models of Kleene algebra are relations under set union, relational composition and reflexive transitive closure, sets of regular languages (regular events) over some finite alphabet under the regular operations or programs under non-deterministic choice, sequential composition and finite iteration.

A *Boolean algebra* is a complemented distributive lattice. By overloading, we usually write $+$ and $\cdot$ also for the Boolean join and meet operation and use 0 and 1 for the least and greatest elements of the lattice. The symbol $\neg$ denotes the operation of complementation. We will consistently use the letters $a, b, c \ldots$ for Kleenean elements and $p, q, r, \ldots$ for Boolean elements.

A *Kleene algebra with tests* [12] is a two-sorted structure (K, B), where K is a Kleene algebra and $B \subseteq K$ is a Boolean algebra such that the B operations coincide with the restrictions of the K operations to B. In particular, $p \leq 1$ for all $p \in B$. In general, B is only a subalgebra of the subalgebra of all elements below 1 in K, since elements of the latter need not be multiplicatively idempotent. We call elements of B *tests* and write $\text{test}(K)$ instead of B. All $p \in \text{test}(K)$ satisfy $p^* = 1$. The class of Kleene algebras with tests is denoted by KAT.

When a Kleenean element a describes an action or abstract program and a test p a proposition or assertion, the product pa describes a restricted program that executes a when the starting state satisfies assertion p and aborts otherwise. Dually, ap describes a restriction of a in its possible result states. We now introduce an abstract domain operator that assigns to a the test that describes precisely its enabling states.

A *Kleene algebra with domain* [4] is a structure (K, δ), where $K \in$ KAT and the *domain operation* $\delta : K \to \text{test}(K)$ satisfies for all $a, b \in K$ and $p \in \text{test}(K)$

$$a \leq \delta(a)a, \quad (\text{d1}) \qquad \delta(pa) \leq p, \quad (\text{d2}) \qquad \delta(a\delta(b)) \leq \delta(ab). \quad (\text{d3})$$

KAD denotes the class of Kleene algebras with domain.

Let us explain these axioms. Since $\delta(a) \leq 1$ by $\delta(a) \in \text{test}(K)$, isotonicity of multiplication shows that (d1) can be strengthened to an equality expressing that restriction to the full domain is no restriction at all. Axiom (d1) means that after restriction the remaining domain must satisfy the restricting test. (d3) states that the domain of ab is not determined by the inner structure of b or its codomain; information about $\delta(b)$ in interaction with a suffices. It also ensures that the modal operators introduced below distribute through multiplication.

Moreover, (d1) is equivalent to one implication in each of the statements

$$\delta(a) \leq p \Leftrightarow a \leq pa, \quad \text{(llp)} \qquad \delta(a) \leq p \Leftrightarrow \neg pa \leq 0, \quad \text{(gla)}$$

that constitute elimination laws for δ, while (d2) is equivalent to the other implications. (llp) says that $\delta(a)$ is the least left preserver of a. (gla) says that $\neg\delta(a)$ is the greatest left annihilator of a.

All domain axioms hold in the relational model, but (d1) and (d2) suffice for many applications, such as, for instance, proving soundness of propositional Hoare logic. Our completeness proof, however, depends on (d3). We will always explicitly mention where (d3) has to be used.

Because of (llp), domain is uniquely characterised by the two domain axioms. Moreover, if $\mathsf{test}(K)$ is complete then a domain operation always exists. If $\mathsf{test}(K)$ is not complete, this need not be the case.

Many natural properties follow from the axioms. Domain is fully strict ($\delta(a) = 0 \Leftrightarrow a = 0$), stable on tests ($\delta(p) = p$) and satisfies the import/export law ($\delta(pa) = p\,\delta(a)$). See [4] for further information.

Moreover, the Galois-like characterization (llp) implies that the domain operation satisfies a continuity property.

Proposition 2.1. *Domain commutes with all existing suprema in* KAD*; in particular, it is additive* ($\delta(a+b) = \delta(a) + \delta(b)$) *and isotone* ($a \leq b \Rightarrow \delta(a) \leq \delta(b)$).

Proof. Let $b = \sup(a : a \in A)$ exist for some set $A \subseteq K$. We must show that $\delta(b) = \sup(\delta(a) : a \in A)$. First, by isotonicity of domain, $\delta(b)$ is an upper bound of the set $\delta(A) = \{\delta(a) : a \in A\}$, since b is an upper bound of A.

To show that $\delta(b)$ is the least upper bound of $\delta(A)$, let p be an arbitrary upper bound of $\delta(A)$. Then for all $a \in A$, $\delta(a) \leq p \Leftrightarrow a \leq pa \Rightarrow a \leq pb$, by (llp). Hence pb is an upper bound of A and therefore $b \leq pb$. But by (llp) this is equivalent to $\delta(b) \leq p$. $\square$

A codomain operation ρ can easily be axiomatized as a domain operation on the opposite semiring. As usual in algebra, opposition just swaps the order of multiplication. An alternative definition uses the operation of converse, which can be axiomatized for $K \in \mathsf{KA}$ as follows. For all $a, b, p \in K$ with $p \leq 1$,

$$a^{\circ\circ} = a, \quad (a+b)^\circ = a^\circ + b^\circ, \quad (ab)^\circ = b^\circ a^\circ, \quad (a^*)^\circ = (a^\circ)^*, \quad p^\circ \leq p.$$

Consequently, $p^\circ = p$ and $a \leq b \Leftrightarrow a^\circ \leq b^\circ$. Codomain is defined by $\rho(a) = \delta(a^\circ)$.

3   Modalities

We now define various modal operators in KAD. Their names are justified, since they induce mappings on test algebras that form Boolean algebras with operators in the sense of Jónsson and Tarski. They can also be interpreted, respectively, as disjunctive or conjunctive predicate transformers. This links KAD with the syntax and semantics of Hoare logic.

The first definition introduces forward and backward diamond operators in the standard way via abstract preimage and image.

$$|a\rangle p = \delta(ap), \quad (1) \qquad\qquad \langle a|p = \rho(pa). \quad (2)$$

Conversely therefore, $\delta(a) = |a\rangle 1$ and $\rho(a) = \langle a|1$. Forward and backward diamonds are duals with respect to converse.

$$|a\rangle p = \langle a^\circ|p, \qquad \langle a|p = |a^\circ\rangle p. \tag{3}$$

They are also related by an *exchange law*.

Lemma 3.1. *Let $K \in$ KAD. For all $a \in K$ and $p, q \in \text{test}(K)$,*

$$|a\rangle p \leq \neg q \Leftrightarrow \langle a|q \leq \neg p. \tag{4}$$

Proof. Expanding the definitions of forward and backward diamonds and using (gla) we calculate $|a\rangle p \leq \neg q \Leftrightarrow qap \leq 0 \Leftrightarrow \langle a|q \leq \neg p$. □

Therefore, even in absence of converse, forward and backward diamond are interdefinable. Moreover, both operators are unique. Duality with respect to complementation transforms diamonds into boxes:

$$|a]p = \neg|a\rangle \neg p, \qquad [a|p = \neg\langle a|\neg p. \tag{5}$$

By (4) and (5), this symmetry can also be expressed by Galois connections.

Lemma 3.2. *Let $K \in$ KAD. For all $a \in K$, the operators $|a\rangle$, $\langle a|$ and $\langle a|$, $|a]$ are lower and upper adjoints of Galois connections. For all $p, q \in \text{test}(K)$,*

$$|a\rangle p \leq q \Leftrightarrow p \leq [a|q, \qquad \langle a|p \leq q \Leftrightarrow p \leq |a]q. \tag{6}$$

Exploiting the symmetries further yields the dualities $|a]p = [a^\circ|p$ and $[a|p = |a^\circ]p$ and the exchange law $|a]p \leq \neg q \Leftrightarrow [a|q \leq \neg p$. In later sections, we will use these Galois connections as theorem generators and the dualities as theorem transformers. We write $\langle a\rangle p$ and and $[a]p$ if the direction does not matter.

Many modal properties can be expressed and calculated more succinctly in a point-free style in the operator semirings induced by the modal operators. While such structure-preserving abstractions are standard in algebra, they have no immediate logical analogues. See [17] for more information.

Proposition 3.3. *Let $\langle K \rangle$ be the set of all mappings $\lambda x.\langle a\rangle x$ on some $K \in$ KAD, where $a \in K$. Defining addition and multiplication on $\langle K \rangle$ by*

$$(\langle a\rangle + \langle b\rangle)(p) = \langle a\rangle p + \langle b\rangle p, \quad (7) \qquad (\langle a\rangle \cdot \langle b\rangle)(p) = \langle a\rangle(\langle b\rangle p), \quad (8)$$

the structure $(\langle K \rangle, +, \cdot, \langle 0 \rangle, \langle 1 \rangle)$ is an i-semiring. Depending on whether $\langle . \rangle$ is $|.\rangle$ or $\langle .|$, we call it the forward diamond semiring *or* backward diamond semiring.

The natural ordering on $\langle K \rangle$ is defined by point-wise lifting as

$$\langle a \rangle \leq \langle b \rangle \Leftrightarrow \forall p. \langle a \rangle p \leq \langle b \rangle p. \tag{9}$$

By duality with respect to complementation, also the structures $([K], \sqcap, \cdot, [0], [1])$ are i-semirings, the *forward* and *backward box semiring*, respectively. Here, $\sqcap$ is the lower bound operation on box operators defined by

$$([a] \sqcap [b])(p) = ([a]p)([b]p); \tag{10}$$

the natural ordering is lifted as for diamonds. This yields an interesting correspondence with disjunctive and conjunctive predicate transformer algebras.

Using the point-wise lifting we can write formulas like $\langle a \rangle + \langle b \rangle = \langle a + b \rangle$ and $([a] \sqcap [b]) = \langle a + b \rangle$ in a point-free style. We will strongly use point-free resoning in the following sections. This will yield shorter specifications and simpler and more concise proofs.

4   The Algebra of Modalities

We now develop the basic laws of an algebra of modal operators in KAD. We further investigate their symmetries in terms of Galois connections and of duality in order to derive further properties. But since our modal operators are not completely characterized by the symmetries, we also present properties that are based directly on domain and codomain. See [17] for a more technical discussion.

Expanding the definitions, we can show the following simple properties of the units of the operator semirings.

Lemma 4.1. *Let* $K \in \mathsf{KAD}$ *and* $p \in \mathsf{test}(K)$. *Then* $\langle 0 \rangle p = 0 = \neg[0]p$ *and* $[1] = \langle 1 \rangle$.

The Galois connections (6) give us the following two theorems for free.

Lemma 4.2. *Let* $K \in \mathsf{KAD}$. *For all* $a \in K$, *we have the* cancellation laws

$$|a\rangle[a| \leq \langle 1 \rangle \leq [a||a\rangle, \qquad \langle a||a] \leq \langle 1 \rangle \leq |a]\langle a|. \tag{11}$$

Proposition 4.3. *Let* $K \in \mathsf{KAD}$ *and* $a \in K$. *Then* $\langle a \rangle$ *and* $[a]$ *commute with all existing suprema and infima, respectively. If* $\mathsf{test}(K)$ *is a complete Boolean lattice then* $\langle a \rangle$ *is universally disjunctive and* $[a]$ *is universally conjunctive, that is, the operators commute with all suprema and infima, respectively.*

Proof. By Lemma 3.2, boxes and diamonds of KAD are upper and lower adjoints of a Galois connection. Then the results follow from general properties.   □

As special cases we obtain, for all $a \in K$ and $p, q \in \mathsf{test}(K)$,

$$\langle a \rangle 0 = 0, \qquad \langle a \rangle (p + q) = \langle a \rangle p + \langle a \rangle q,$$
$$|a]1 = 1 \qquad |a](pq) = (|a]p)(|a]q).$$

Consequently, $(\mathsf{test}(K), \{\langle a \rangle : a \in K\})$ and $(\mathsf{test}(K), \{[a] : a \in K\})$ are *Boolean algebras with operators* in the sense of Jónsson and Tarski [10]. This justifies calling our boxes and diamonds *modal operators*.

We now collect some further natural algebraic properties of modal operators. We restrict our attention to diamonds. Corresponding statements for boxes can immediately be inferred by duality.

Lemma 4.4. *Let $K \in \mathsf{KAD}$. For all $a, b \in K$ and $p, q \in \mathsf{test}(K)$,*

$$\langle a + b \rangle = \langle a \rangle + \langle b \rangle, \quad (12) \qquad\qquad a \le b \Rightarrow \langle a \rangle \le \langle b \rangle, \quad (15)$$

$$|ab\rangle \le |a\rangle|b\rangle, \quad (13) \qquad\qquad |paq\rangle = |p\rangle|a\rangle|q\rangle, \quad (16)$$

$$\langle ab| \le \langle b|\langle a|, \quad (14) \qquad\qquad \langle paq| = \langle q|\langle a|\langle p|. \quad (17)$$

The properties (13) and (14) can be proved using (d1) and (d2) only; since we additionally have (d3), they even become equalities. Spelling out (12) for box yields, for instance, $[a + b] = [a] \sqcap [b]$, while (13) yields $|a\rangle|b\rangle \le |ab\rangle$. Moreover, boxes are antitonic: $a \le b$ implies $[b] \le [a]$.

The following statements show that a star operation can be defined on a semiring of modal operators.

Proposition 4.5. *Let $|K\rangle$ be the forward diamond semiring over $K \in \mathsf{KAD}$. Defining a star on $|K\rangle$ by*

$$|a\rangle^*(p) = |a^*\rangle p, \quad (18)$$

for all $a \in K$ and $p \in \mathsf{test}(K)$ turns $|K\rangle$ into a left Kleene algebra.

To see that (*-1)-(*-3) hold in the forward diamond semiring $|K\rangle$, we use that that the identities $|1\rangle + |aa^*\rangle = |a^*\rangle$ and $|1\rangle + |a\rangle|a^*\rangle \ge |a^*\rangle$ have been shown in [4], whereas $|1\rangle + |a\rangle|a^*\rangle = |a^*\rangle$ follows using (d3). Moreover, we have the quasi-identity $p + |a\rangle q \le q \Rightarrow |a^*\rangle p \le q$ (see again [4]) and therefore also

$$|b\rangle + |a\rangle|c\rangle \le |c\rangle \Rightarrow |a^*\rangle|b\rangle \le |c\rangle. \quad (19)$$

For $\langle K|$, we obtain a right Kleene algebra by similar arguments.

In case of a complete test algebra, we obtain a full Kleene algebra.

Lemma 4.6. *Let $K \in \mathsf{KAT}$. Then $\lambda x.p + x$ on $\mathsf{test}(K)$ commutes with all existing suprema and $\lambda x.px$ on $\mathsf{test}(K)$ commutes with all existing infima.*

Proposition 4.7. *Let $K \in \mathsf{KAD}$ and let $\mathsf{test}(K)$ be a complete Boolean lattice. Then for all $a \in K$, the operators $\langle a \rangle^*$ and $[a]^*$ exist. Moreover,*

$$\langle a \rangle^* = \sup(\langle a \rangle^i : i \ge 0), \qquad [a]^* = \inf([a]^i : i \ge 0).$$

This follows from Proposition 4.3, Lemma 4.6 and Kleene's fixed-point theorem.

Proposition 4.8. *Let $K \in \mathsf{KAD}$ with $\mathsf{test}(K)$ a complete Boolean lattice. Then the i-semiring $|K\rangle$ can uniquely be extended to a Kleene algebra.*

Instead of calculating with domain and modal operator laws, we can therefore calculate many modal properties simply in Kleene algebra at this higher level of abstraction (see below).

5   Hoare Logic

We now apply our results to obtain completely calculational algebraic soundness and completeness proofs for propositional Hoare logic. We first present the syntax and semantics of Hoare logic.

Let Φ be a set of *propositions* built from a set Π with the usual Boolean connectives. Let Σ be a set of *statements* defined by the following grammar from a set Γ of atomic commands.

$$\Sigma ::= \text{abort} \mid \text{skip} \mid \Gamma \mid \Sigma \, ; \, \Sigma \mid \text{ if } \Phi \text{ then } \Sigma \text{ else } \Sigma \mid \text{ while } \Phi \text{ do } \Sigma \, .$$

The basic formulas of Hoare logic are *partial correctness assertions* (PCAs) of the form $\{\phi\} \, \alpha \, \{\psi\}$, with $\phi, \psi \in \Phi$ (the *pre-* and *postcondition*) and $\alpha \in \Sigma$.

To define a semantics with respect to KAD, let $K \in$ KAD. We assign to each propositional variable $\pi \in \Pi$ a test $[\![\pi]\!] \in \text{test}(K)$ and to each atomic command $\gamma \in \Gamma$ a Kleenean element $[\![\gamma]\!] \in K$. Moreover, we assign 0 to $[\![\text{abort}]\!]$ and 1 to $[\![\text{skip}]\!]$. The remainder is the usual homomorphic extension.

$$[\![\phi \wedge \psi]\!] = [\![\phi]\!][\![\psi]\!], \tag{20}$$

$$[\![\neg\phi]\!] = \neg[\![\phi]\!], \tag{21}$$

$$[\![\alpha \, ; \, \beta]\!] = [\![\alpha]\!][\![\beta]\!], \tag{22}$$

$$[\![\text{ if } \phi \text{ then } \alpha \text{ else } \beta]\!] = [\![\phi]\!][\![\alpha]\!] + \neg[\![\phi]\!][\![\beta]\!], \tag{23}$$

$$[\![\text{ while } \phi \text{ do } \alpha]\!] = ([\![\phi]\!][\![\alpha]\!])^{*}\neg[\![\phi]\!]. \tag{24}$$

We follow [13] in defining validity of formulas and PCAs. $\models \phi \Leftrightarrow [\![\phi]\!] = 1$, for all $\phi \in \Phi$. In particular, $\models \phi \rightarrow \psi \Leftrightarrow [\![\phi]\!] \leq [\![\psi]\!]$. Moreover,

$$\models \{\phi\} \, \alpha \, \{\psi\} \Leftrightarrow [\![\phi]\!][\![\alpha]\!]\neg[\![\psi]\!] \leq 0.$$

Using (gla) and Boolean algebra, we rewrite this definition more intuitively as

$$\models \{\phi\} \, \alpha \, \{\psi\} \Leftrightarrow \langle [\![\alpha]\!] | [\![\phi]\!] \leq [\![\psi]\!].$$

In the relational model of KAD, the expression $\langle [\![\alpha]\!] | [\![\phi]\!]$ denotes the set of all states that can be reached from states in $[\![\phi]\!]$ through $[\![\alpha]\!]$. Therefore, the formula $\langle [\![\phi]\!] | [\![\alpha]\!] \leq [\![\psi]\!]$ is indeed a faithful translation of $\{\phi\} \, \alpha \, \{\psi\}$ that, by the exchange law of Lemma 3.1, is consistent with the standard wlp-semantics (see also Section 7 for further details).

To shorten notation, we will henceforth confuse syntax and semantics and use Kleene algebra notation everywhere. Thus we express validity of a PCA as

$$\models \{p\} \, a \, \{q\} \Leftrightarrow \langle a | p \leq q. \tag{25}$$

The Hoare calculus for partial correctness of deterministic sequential programs consists of the following inference rules.

(Abort) $\qquad\qquad\qquad\qquad\qquad\qquad\qquad$ $\{p\}$ abort $\{q\}$,

(Skip) $\qquad\qquad\qquad\qquad\qquad\qquad\qquad\quad$ $\{p\}$ skip $\{p\}$,

(Assignment) $\qquad\qquad\qquad\qquad\qquad\qquad$ $\{p[e/x]\}\ x := e\ \{p\}$,

(Composition) $\qquad\qquad\qquad\qquad$ $\dfrac{\{p\}\ a\ \{q\}\quad \{q\}\ b\ \{r\}}{\{p\}\ a\,;\,b\ \{r\}}$,

(Conditional) $\qquad\qquad\qquad$ $\dfrac{\{p\wedge q\}\ a\ \{r\}\quad \{\neg p\wedge q\}\ b\ \{r\}}{\{q\}\ \text{if } p \text{ then } a \text{ else } b\ \{r\}}$,

(While) $\qquad\qquad\qquad\qquad\quad$ $\dfrac{\{p\wedge q\}\ a\ \{q\}}{\{q\}\ \text{while } p \text{ do } a\ \{\neg p\wedge q\}}$,

(Weakening) $\qquad\qquad\qquad$ $\dfrac{p_1\to p\quad \{p\}\ a\ \{q\}\quad q\to q_1}{\{p_1\}\ a\ \{q_1\}}$.

A rule with premises $P_1,\cdots,P_n$ and conclusion P is *sound* if $P_1,\ldots,P_n \models P$. Derivations are defined in the standard way.

(Assignment) is a non-propositional inference rule that deals with the internal structure of states. We therefore do not encode it directly into our framework, but instead use the set Γ of atomic commands as a parameter in our approach. The requirement of sufficient expressiveness on Γ that ensures completeness of the calculus will be discussed in Section 7. Following [13], we call this abstract form of Hoare logic *propositional Hoare logic* (PHL).

6   Soundness of Propositional Hoare Logic

We now prove soundness of PHL with respect to the KAD-semantics. More precisely, we show that the encoded inference rules of PHL are theorems of KAD. This subsumption is a popular exercise for many logics and algebras of programs, among them propositional dynamic logic [8] and KAT [13], which are both subsumed by KAD. However our result is interesting for two reasons, a syntactic and a semantic one. First, our encoding of PHL is more simple, abstract and direct, and Hoare-style reasoning in KAD is more flexible than in previous approaches. However we do not sacrifice algorithmic power. Second, the properties of our modal operators defined in terms of abstract image and preimage operations reflect precisely those of the standard partial correctness semantics [1, 15] and show that KAD provides a natural abstract algebraic semantics for PHL.

A first point-wise encoding of the soundness conditions for the Hoare rules is rather straightforward from (25). (Composition), for instance, becomes

$$\langle a|p \le q \wedge \langle b|q \le r \Rightarrow \langle ab|p \le r.$$

This is a theorem of KAD, since

$$\langle ab|p \leq \langle b|\langle a|p \leq \langle b|q \leq r$$

by (decomposition). As a second example, (While) becomes

$$\langle a|(pq) \leq q \Rightarrow \langle (pa)^* \neg p|q \leq \neg pq.$$

This is also a theorem of KAD. Using (induction), we calculate

$$\langle a|(pq) \leq q \Rightarrow \langle (pa)^*|q \leq q \Rightarrow \neg p(\langle (pa)^*|q) \leq \neg pq \Leftrightarrow \langle (pa)^* \neg p|q \leq \neg pq.$$

Point-wise encodings and proofs for the remaining PHL-rules are similar. Consequently, soundness of PHL can be proved literally in one line per inference rule from natural properties of KAD. In KAT, (Composition), for instance, must be encoded quite indirectly as

$$pa \leq aq \wedge qb \leq br \Rightarrow pab \leq abr$$

and the proof of theoremhood is based on rather syntactic commutation properties (cf. [13]). We can obtain this encoding also in KAD, using (llp). More generally, (llp) and (gla) provide translations of all PHL-rules into KAT and, using a result from [9], connect validity with respect to PHL with PSPACE automata-theoretic decision procedures. See [17] for a deeper discussion.

Compared with standard textbooks (cf. [1, 15]), our proof is about ten times shorter. In addition, the textbook proofs are only semi-formal, since many logical and set-theoretic assumptions are left implicit. A complete formalization would produce further overhead.

We now give another point-free soundness proof of PHL in KAD that is even more abstract and concise. In particular, the properties expressed by the Hoare rules now correspond to natural algebraic properties of the algebra of modal operators.

Proposition 6.1. *Let $K \in$ KAD. Then the soundness conditions for the inference rules of PHL can be encoded as follows. For all $a, b \in K$ and $p \in \mathsf{test}(K)$,*

(Abort)	$\langle 0| \leq \langle q|,$
(Skip)	$\langle 1| \leq \langle 1|,$
(Composition)	$\langle ab| \leq \langle b|\langle a|,$
(Conditional)	$\langle pa + \neg pb| \leq \langle a|\langle p| + \langle b|\langle \neg p|,$
(While)	$\langle a|\langle p| \leq \langle 1| \Rightarrow \langle \neg p|\langle (pa)^*| \leq \langle \neg p|,$
(Weakening)	$\langle p_1| \leq \langle p| \wedge \langle p|\langle a| \leq \langle q| \wedge \langle q| \leq \langle q_1| \Rightarrow \langle q_1|\langle a| \leq \langle q_1|.$

The point-free encoding is derived from the point-wise one using the *principle of indirect inequality*: $p \leq q$ iff $q \leq r$ implies $p \leq r$ for all r.

(Skip) and (Abort) now reflect natural or even trivial semiring properties. (Conditional) expresses (additivity) and (import/export) of the operator semiring, (While) expresses a variant of (induction). (Composition) expresses (decomposition); it becomes an equality when (d3) is assumed on the underlying KAD. (Weakening) is the only rule where at first sight, nothing has be gained by the lifting. However, its correctness proof can now be based entirely on semiring properties, instead of expanding to properties of domain. These facts are immediately reflected by the following subsumption result.

Theorem 6.2. *The point-free encodings of the* PHL-*rules are theorems in* KAD.

Proof. The point-free variants of (Abort) and (Skip) are trivial consequences of Lemma 4.1. The point-free variant of (Composition) is nothing but (14). The point-free variant of (Conditional) is evident from (12) and (14). (While) follows immediately from (19) and isotonicity. (Weakening) holds by isotonicity of multiplication in i-semirings.   □

Theorem 6.3. PHL *is sound with respect to the* KAD *semantics.*

Proof. By induction on the structure of PHL derivations, using Theorem 6.2.   □

As observed in [13], all Horn clauses built from PCAs in PHL that are valid with respect to the standard semantics are theorems of KAT; whence a fortiori of KAD. PHL is too weak to derive all such formulas. Consequently, KAT and KAD have not only the derivable, but also the admissible rules of PHL as theorems.

7   Completeness of Propositional Hoare Logic

In this section we provide a novel algebraic completeness proof for the inference rules of PHL, using modal Kleene algebra as a semantics. Conventional completeness proofs use the *weakest liberal precondition* semantics. For a set S of program states, a relational program $P \subseteq S \times S$ and set $T \subseteq S$ of target states one defines

$$\mathsf{wlp}(P, T) = \{s \in S : P(s) \subseteq T\}, \tag{26}$$

where $P(s)$ is the image of s under P. Equivalently, $\mathsf{wlp}(P, T)$ is the largest subset $U \subseteq S$ such that $P(U) \subseteq T$. In a modal setting the wlp-operator can then of course be identified with the forward box operator. Confusing again syntax and semantics, the Galois connection (6) and (25) immediately imply that

$$\models \{p\}\, \alpha\, \{q\} \ \Leftrightarrow\ p \le |a]q. \tag{27}$$

On the one hand, this Galois connection connects PHL syntax and semantics in a very concise way. One the other hand, we get the entire wlp-calculus for free by dualizing our results from Section 4.

For the standard completeness proofs (see e.g. [1]) it is crucial that the underlying assertion language is sufficiently expressive. This implies that for all statements $\alpha \in \Sigma$ and all postconditions $\psi \in \Phi$ there is an assertion $\phi \in \Phi$ that expresses the weakest liberal precondition for ψ under α, i.e.,

$$\llbracket \phi \rrbracket = \mathsf{wlp}(\llbracket \alpha \rrbracket, \llbracket \psi \rrbracket). \tag{28}$$

Using (28) we can continue working semantically in KAD. We extend the original calculus so that all predicates are denoted by propositional variables. Completeness of this extension will then imply completeness of the former calculus.

For every atomic command $\gamma \in \Gamma$ and test q we add an axiom

$$\{|g]q\} \ g \ \{q\}, \tag{29}$$

where $g = \llbracket \gamma \rrbracket$. (Assignment) has precisely this form.

Before the completeness proof, we state some technical properties of boxes in connection with conditionals and loops. Logical variants appear in [1].

Proposition 7.1. *Let $K \in$ KAD. Let $a, b, c, w \in K$ and $p, q \in \mathsf{test}(K)$.*

(i) For $c = $ if p then a else b,

$$p\,(|c]q) = p\,(|a]q), \quad (30) \qquad\qquad \neg p\,(|c]q) = \neg p\,(|b]q). \quad (31)$$

(ii) For $w = $ while q do a,

$$p\,(|w]q) = p|a]\,(|w]q), \quad (32) \qquad\qquad \neg p\,(|w]q) \le q. \quad (33)$$

The proofs need a few lines of calculus using the properties from Section 4. Now we can proceed, as for instance in [1].

Lemma 7.2. *Let $K \in$ KAD. For all $a \in K$ that are denotable by PHL commands and all $q \in \mathsf{test}(K)$, the PCA $\{|a]q\} \ a \ \{q\}$ is derivable in PHL.*

Proof. Let $\vdash \{p\} \ a \ \{q\}$ denote that $\{p\} \ a \ \{q\}$ is derivable in PHL. The proof is by induction on the structure of command a.

(i) a is either skip or abort or denotes an atomic command. Then the claim is trivial, since PHL contains the respective PCA as an axiom.

(ii) Let $a = b$ and $c = bc$. By the induction hypothesis,

$$\vdash \{|b]\,(|c]q)\} \ b \ \{|c]q\}, \qquad \vdash \{|c]q\} \ c \ \{q\}.$$

Now (Composition) shows $\vdash \{|b]\,(|c]q)\} \ bc \ \{q\}$, which by the additional assumption of (d3) and the dual of (13) is equivalent to $\vdash \{|bc]q\} \ bc \ \{q\}$. Note that this is the only part of the proof where (d3) is used.

(iii) Let $a = $ if p then b else c. By the induction hypothesis,

$$\vdash \{|b]q\} \ b \ \{q\}, \qquad \vdash \{|c]q\} \ c \ \{q\}.$$

Hence, by (Weakening), also

$$\vdash \{p\,(|b]q)\} \ b \ \{q\}, \qquad \vdash \{\neg p\,(|b]q)\} \ b \ \{q\}.$$

By (30) and (31) these statements are equivalent to

$$\vdash \{p(|a]q)\}\ b\ \{q\}, \qquad \vdash \{\neg p(|a]q)\}\ c\ \{q\},$$

so that (Conditional) shows the claim.

(iv) Let $a = $ while p do b. Let $c = |a]q$. By the induction hypothesis,

$$\vdash \{|a]c\}\ b\ \{c\}.$$

By (32) this is equivalent to $\vdash \{pc\}\ b\ \{c\}$. (While) shows that $\vdash \{c\}\ a\ \{\neg pc\}$ and (33) and (Weakening) yield $\vdash \{|a]q\}\ a\ \{q\}$, as required, $\qquad\qquad\square$

We are now prepared for the main theorem of this section.

Theorem 7.3. PHL *is relatively complete for the partial correctness semantics of deterministic programs in* KAD.

Proof. We must show that $\models \{p\}\ a\ \{q\}$ implies $\vdash \{p\}\ a\ \{q\}$. This follows from (27), Lemma 7.2 and (Weakening). $\qquad\qquad\square$

Alternatively, we could also use our encodings of PCAs in KAD in the completeness proof. We could write $\langle a|_\vdash p \leq q$ instead of $\vdash \{p\}\ a\ \{q\}$ to further stress the fact that our proof is entirely in Kleene algebra and to denote that only the encodings of PHL-rules are allowed for transforming the indexed diamonds. Using this encoding, the statement $\langle a|_\vdash (|a]p) \leq p$, or even $\langle a|_\vdash |a] \leq |1]$, looks very much like a cancellation property of a Galois connection. This fact certainly deserves further consideration.

8   Conclusion and Outlook

We have investigated Kleene algebra with domain as a modal Kleene algebra. Modal operators have been defined as abstractions of relational image and preimage operations. Their symmetries have been formalized in terms of Galois connections and dualities. We have also studied the semirings induced by the modal operators. This additional level of abstraction yields very concise state-free specifications and proofs of modal properties.

Our results show the usefulness of modal Kleene algebra both as a calculus for cross-theoretic reasoning with various calculi for imperative programs and state transition systems, and as a unifying semantics for modal, relational and further algebraic approaches. While an analogous claim has already been verified for relational approaches [4] and for propositional dynamic logic [7], we provide algebraic soundness and completeness proofs for Hoare logic that use modal Kleene algebra both at the syntactic and at the semantic side. In particular the state-free soundness proof and the completeness proof exhibit very nicely the natural algebraic properties that are implicit in the partial correctness assertions and Hoare rules.

Compared with other formalisms, modal Kleene algebra is also very flexible. E.g., in [17], we show that several inference rules that are derivable in PHL are theorems of modal Kleene algebra. There, it is not always preferable to reason entirely using the modalities. Especially when the rules encode commutativity conditions, the subtheory KAT may provide more direct proofs.

It is also interesting to investigate in how far modalities can be eliminated from KAD formulas. In combination with hypothesis elimination techniques, we obtain a linear translation of certain KAD-expression into identities over KAT, whose validity can be decided by automata in PSPACE [17].

Modal Kleene algebra also subsumes Hoare logic for programs with bounded nondeterminism. Guarded commands, for instance, can be encoded as

$$\text{if } p_1 \to a_1 \, [] \cdots [] \, p_n \to a_n \text{ fi } = \sup(p_i a_i : 1 \le i \le n),$$
$$\text{do } p_1 \to a_1 \, [] \cdots [] \, p_n \to a_n \text{ od} = (\sup(p_i a_i : 1 \le i \le n))^* \inf(\neg p_i : 1 \le i \le n).$$

Program termination can also be modelled in modal Kleene algebra [4, 6]. This suggests extending our approach to Hoare logics for total correctness. Moreover, since modal Kleene algebra allows the specification of syntax and relational semantics of modal calculi in one single formalism, one can use it to develop a calculational modal correspondence theory; see [4, 5, 18] for first results. To further establish modal Kleene algebra as a unifying framework, we also plan to consider temporal logics like LTL or CTL; for LTL an account of this along the lines of [21] is contained in [5]. Recently, the modal operators have also been incorporated into *Lazy Kleene Algebra* [16], a framework that extends the work of Cohen [3] and von Wright [22] and is designed to deal with both terminating and non-terminating computations and hence also with reactive systems. It is a challenging task to apply the framework of modal Kleene algebra to other problems and structures for further extending its practical relevance.

Acknowledgment

We would like to thank Jules Desharnais, Thorsten Ehm and Joakim von Wright for valuable discussions and comments.

References

1. K.-R. Apt and E.-R. Olderog. *Verification of Sequential and Concurrent Programs.* Springer, 2nd edition, 1997.
2. K. Clenaghan. Calculational graph algorithmics: Reconciling two approaches with dynamic algebra. Technical Report CS-R9518, CWI, Amsterdam, 1994.
3. E. Cohen. Separation and reduction. In R. Backhouse and J. N. Oliveira, editors, *Proc. of Mathematics of Program Construction, 5th International Conference, MPC 2000*, volume 1837 of *LNCS*, pages 45–59. Springer, 2000.
4. J. Desharnais, B. Möller, and G. Struth. Kleene algebra with domain. Technical Report 2003-07, Universität Augsburg, Institut für Informatik, 2003.

5. J. Desharnais, B. Möller, and G. Struth. Applications of modal Kleene algebra – a survey. Technical Report DIUL-RR-0401, Département d'informatique et de génie logiciel, Université Laval, Québec, 2004.
6. J. Desharnais, B. Möller, and G. Struth. Termination in modal Kleene algebra. Technical Report 2004-04, Universität Augsburg, Institut für Informatik, 2004.
7. T. Ehm, B. Möller, and G. Struth. Kleene modules. In R. Berghammer and B. Möller, editors, *Proc. 7th Seminar on Relational Methods in Computer Science and 2nd International Workshop on Applications of Kleene Algebra*, volume 3051 of *LNCS*. Springer, 2004. (to appear).
8. J. M. Fischer and R. F. Ladner. Propositional dynamic logic of regular programs. *J. Comput. System Sci.*, 18(2):194–211, 1979.
9. C. Hardin and D. Kozen. On the elimination of hypotheses in Kleene algebra with tests. Technical Report 2002-1879, Computer Science Department, Cornell University, October 2002.
10. B. Jónsson and A. Tarski. Boolean algebras with operators, Part I. *American Journal of Mathematics*, 73:891–939, 1951.
11. D. Kozen. A completeness theorem for Kleene algebras and the algebra of regular events. *Information and Computation*, 110(2):366–390, 1994.
12. D. Kozen. Kleene algebra with tests. *Trans. Programming Languages and Systems*, 19(3):427–443, 1997.
13. D. Kozen. On Hoare logic and Kleene algebra with tests. *Trans. Computational Logic*, 1(1):60–76, 2001.
14. D. Kozen and M.-C. Patron. Certification of compiler optimizations using Kleene algebra with tests. In J. Lloyd, editor, *1st International Conference on Computational Logic*, volume 1861 of *LNCS*, pages 568–582. Springer, 2000.
15. J. Loeckx and K. Sieber. *The Foundations of Program Verification*. Wiley Teubner, 2nd edition, 1987.
16. B. Möller. Lazy Kleene algebra. In D. Kozen, editor, *Proc. of Mathematics of Program Construction, 7th International Conference, MPC 2004*, LNCS. Springer, 2004. (to appear). Preliminary version: Report No. 2003-17, Institut für Informatik, Universität Augsburg, December 2003.
17. B. Möller and G. Struth. Modal Kleene algebra and partial correctness. Technical Report 2003-08, Universität Augsburg, Institut für Informatik, 2003.
18. B. Möller and G. Struth. Greedy-like algorithms in modal Kleene algebra. In R. Berghammer and B. Möller, editors, *Proc. 7th Seminar on Relational Methods in Computer Science and 2nd International Workshop on Applications of Kleene Algebra*, volume 3051 of *LNCS*. Springer, 2004. (to appear).
19. G. W. Schmidt and T. Ströhlein. *Relations and Graphs: Discrete Mathematics for Computer Scientists*. EATCS Monographs on Theoretical Computer Science. Springer, 1993.
20. G. Struth. Calculating Church-Rosser proofs in Kleene algebra. In H.C.M. de Swart, editor, *Relational Methods in Computer Science, 6th International Conference*, volume 2561 of *LNCS*, pages 276–290. Springer, 2002.
21. B. von Karger. Temporal algebra. *Mathematical Structures in Computer Science*, 8(3):277–320, 1998.
22. J. von Wright. From Kleene algebra to refinement algebra. In B. Möller and E. Boiten, editors, *Mathematics of Program Construction, 6th International Conference, MPC 2002*, volume 2386 of *LNCS*, pages 233–262. Springer, 2002.

Modularity and the Rule of Adaptation

Cees Pierik[1] and Frank S. de Boer[1,2,3]

[1] Institute of Information and Computing Sciences, Utrecht University,
The Netherlands
{cees,frankb}@cs.uu.nl
[2] CWI, Amsterdam, The Netherlands
[3] LIACS, Leiden University, The Netherlands

Abstract. This paper presents a new rule for reasoning about method calls in object-oriented programs. It is an adaptation of Hoare's rule of adaptation to the object-oriented paradigm, which takes both the write effects and the creational effects of a method into account. The new rule contributes in various ways to the modularity of the specification. We also argue that our rule of adaptation is the missing link between Hoare logics and proof outlines for object-oriented programs.

1   Introduction

It is often argued that encapsulation is one of the strong features of the object-oriented paradigm. Encapsulation is obtained by forbidding direct access to the internal state of other objects. Thus one can only query or alter the state of an object by invoking a method. Therefore method calls are the main computational mechanism in object-oriented programs.

Different calls to a particular method may occur in completely different contexts. Preferably, a method has one fixed pre/post specification that acts as a contract between the designer of the method and its clients. This contract should enable a client to infer the behavior of a call to this method. Essential for this scenario is that the reasoning rule for method calls is able to adapt the method specification to the required specification of the call. This can be achieved for procedural languages with global variables by means of the rule of adaptation. This rule was introduced by Hoare [5], and improved by Olderog [11].

The above mentioned rules suffice in a context where only simple global variables occur. However, the set of variables of modern object-oriented programs is more complex since each existing object has its own internal state. Moreover, the state can be *extended* during the execution of a method by object creations. A rule of adaptation for object-oriented programs has to take these state extensions into account.

The main contribution of this paper is a rule of adaptation for object-oriented programs. The rule enables in various ways a modular specification of object-oriented programs. For example, it results in an important separation of concerns for local variables in proof outlines of object-oriented programs.

The rule that we present in this paper integrates write effects [4], and creational effects of a method in a natural way in the programming logic. By the

C. Rattray et al. (Eds.): AMAST 2004, LNCS 3116, pp. 394–408, 2004.

creational effects of a method we mean the types of the objects that a method creates. These optimizations of the rule also contribute to the modularity of the required specification. The integration of write effects ensures that a method specification only needs to specify its effect on the fields that actually occur in the implementation of the method. Thus one does not have to revise the specification whenever a field is added to a class. The integration of creational effects permits method specifications only to report the creation of new objects if their implementation actually contains statements that create objects. They need not explicitly express their absence.

Our adaptation rule has been integrated in a tool that computes the verification conditions of proof outlines for a sequential subset of Java. The verification conditions are passed to a theorem prover. The tool is a successor of the tool described in earlier work [1]. The adaptation rule is in particular suitable for proof outline logics since it describes the verification condition that results from the specification of a call and the corresponding method. Thus it enables a shift from Hoare logics (and the embedding and manipulation of Hoare triples in a theorem prover) to proof outlines for object-oriented languages.

2   Related Work

The past decade has seen a large interest in proof methods for object-oriented programming. An enumeration of the results in this area will therefore likely be incomplete. Several Hoare logics for sequential object-oriented languages have been proposed [14, 10, 6, 15, 13].

Kleymann [7] has introduced a new rule of consequence for sequential programs. It enables one to adapt the values of logical variables in specifications. The outline of the rule resembles the outline of the rule that we propose. However, it solves none of the typical object-oriented problems like the state extensions that are due to object creation.

One particular advantage of the adaptation rule we present in this paper is its treatment of local variables. The values of the local variables of the caller are not changed during a method invocation. A common technique that is used to reflect this fact is to allow the use of a substitution rule (see, for example, [14, 15, 13]). Such a rule replaces a logical variable by a local variable in both the precondition and the postcondition of a method call:

$$\frac{\{P\} \text{ call } \{Q\}}{\{P[u/z]\} \text{ call } \{Q[u/z]\}} \ .$$

In this paper we use the convention that u always denotes a local variable. In the above rule z is a logical variable. Applying this rule is the only way to prove, for example, that the value of a local variable of the caller does not change.

Example 1. Let $m()\{$ skip $\}$ be the declaration of method m with body skip. Obviously, the Hoare triple $\{u = 1\}$ $m()$; $\{u = 1\}$ is valid. To prove this we must first derive $\{z = 1\}$ skip $\{z = 1\}$. A rule for reasoning about method calls then typically lets us infer $\{z = 1\}$ $m()$; $\{z = 1\}$. An application of the substitution rule with $[u/z]$ then yields the desired specification. Observe that the introduction of the logical variable z is necessary because the local variable u is out of scope in the body of the method.

The above example reveals an important drawback of this technique. It turns out that to prove a property of the local variables of the caller one has to express this property in the proof outline of the procedure in terms of logical variables. This clearly violates the modularity of the method specification. Our rule of adaptation provides a means to prove such facts by expressing it in the proof outline of the caller only.

3   Object-Oriented Programs

The rule of adaptation that we propose in this paper is suited for reasoning about method calls in object-oriented programs. An adaptation rule has to take into account all changes to the program state that can result from execution of a method. For this reason it cannot be stated independent of the rest of the programming language. Therefore we outline in this section a sequential object-oriented Java-like language to which our adaptation rule will be tailored. The chosen language illustrates the main features of object-oriented languages. At the same time it leaves out some features of Java that would merely complicate the definitions.

The syntax of the programming language is summarized in Fig. 1. A program π consists of a set of classes. The declaration of a class specifies the fields (or instance variables) of the class (denoted by the sequence $\bar{x}$), and a set of methods. A clause extends c' indicates that the class is an extension of another class c'. In that case, the class is said to be a subclass of class c'. We assume that a class extends the root class Object if the clause is omitted. A class inherits all fields and methods of its superclass. We impose no restrictions concerning the fields or methods that a class declares in relation to the fields and methods that it inherits. That is, we allow both field shadowing and method overriding.

A method m specifies a sequence of formal parameters $u_1, \ldots, u_n$, a sequence of additional local variables $\bar{v}$, a statement S and the return value e. For brevity, we leave the return type of the method implicit. The formal parameters and the sequence $\bar{v}$ make up the local variables of the method. The scope of a local variable is the method to which is belongs. We use u as a typical element of the set of local variables of a method. It denotes either a formal parameter or an additional local variable from $\bar{v}$.

Assignments are divided in two kinds. Assignments to local variables have the form $u := e$, where e denotes an expression that has no side effects. This kind of assignment has no permanent effect on the program state (the values of local variables are discarded when the method terminates). Assignments to instance variables have the form $e.x := e'$. Execution of such a statement results in the assignment of the value of e' to field x of object e.

op ∈ Op   an operator on elements of a primitive type
$y \in \text{Loc} ::= u \mid e.x$
$e \in \text{Expr} ::= \textbf{null} \mid \textbf{this} \mid y \mid e_1 = e_2 \mid (c)e \mid e \textbf{ instanceof } c \mid e_1 \,?\, e_2 : e_3$
$\qquad \mid \text{op}(e_1, \ldots, e_n)$
$S \in \text{Stat} ::= y := e \mid S \,;\, S \mid u := \textbf{new } c() \mid u := e_0.m(e_1, \ldots, e_n)$
$\qquad \mid \textbf{if } (e) \,\{\, S \,\} \textbf{ else } \{\, S \,\} \mid \textbf{while } (e) \,\{\, S \,\}$
$meth \in \text{Meth} ::= m(u_1, \ldots, u_n) \,\{\, \bar{v} \; S \textbf{ return } e \,\}$
$class \in \text{Class} ::= \textbf{class } c \,(\epsilon \mid \textbf{extends } c) \,\{\, \bar{x} \; meth^* \,\}$
$\pi \in \text{Prog} ::= class^*$

Fig. 1. The syntax of the programming language

A statement $u := \textbf{new } c()$ involves the creation of a new object of class c. A reference to this new object is assigned to the local variable u. Method invocations are denoted by $e_0.m(e_1, \ldots, e_n)$. Here e_0 is the object of which method m is invoked. The expressions $e_1, \ldots, e_n$ are the actual parameters of the invocation.

The expressions listed are a minimal subset that suffices for our present purposes. The **instanceof** operator is used to handle dynamic binding. An expression e **instanceof** c is true if e denotes an instance of (some subclass of) class c. Casts of the form $(c)e$ can be used to cope with field shadowing [13]. They change the type of their argument e to c. An expression $e_1 \,?\, e_2 : e_3$ is a conditional expression. Such expressions are important for reasoning about aliases [13].

We consider only two primitive types in this paper: **int** and **boolean**. Variables in the program may also have a reference type. In that case the type of the variable is one of the classes declared in the program. We will tacitly assume that all programs are well-typed. We refer to the type of an expression e by $\lVert e \rVert$. The variable t ranges over the set of types.

Semantics. A complete description of the formal semantics of the presented language can be found in the full version of this paper [12]. Here we restrict our attention to a formal description of the states of object-oriented programs. This enables us to give a formal semantics to method calls in the final part of this section. Moreover, it should be helpful for a proper understanding of the rest of this paper.

We start our description with some auxiliary definitions for a program π. The program π determines among other things the subclass relation. By $c \preceq c'$ we denote that c is a subclass of c'. This relation is reflexive and transitive. We assume that $F_{\preceq}(c)$ denotes the direct superclass of a class c. It is undefined if c has no superclass. We say that c' is a *proper* subclass of c if $c' \preceq c$ and $c' \neq c$.

Let Var denote a set of variables. We assume that at least all local variables that are used in π and the special-purpose variable **this** are elements of Var. Let IVar_c denote the set of instance variables declared in class c. Due to inheritance an object can have several fields with the same identifier. An expression $e.x$

always corresponds to the first declaration of a instance variable x as found by an upward search that starts in class $\|e\|$. The upward search is formalized by the function origin, which is declared as follows.

$$\text{origin}(c, x) = \begin{cases} c & \text{if } x \in \text{IVar}_c \\ \text{origin}(F_{\prec}(C), x) & \text{otherwise} \end{cases}$$

We represent objects as follows. Each object has its own identity and belongs to a certain class. Let $\mathcal{C}$ be the set of classes declared in π. For each class $c \in \mathcal{C}$ we introduce the infinite set $O^c = \{c\} \times \mathbb{N}$ of objects of class c (here $\mathbb{N}$ denotes the set of natural numbers).

The domain of a variable of type t is denoted by $\text{dom}(t)$. The domains $\text{dom}(\texttt{boolean})$ and $\text{dom}(\texttt{int})$ are the set of boolean and integer values, respectively. Due to subtyping a variable of some reference type c can also refer to an object of some subclass of c. Let $\texttt{subs}(c)$ be the set of all subclasses of class c. Then $\text{dom}(c)$ is the set $(\bigcup_{c' \in \texttt{subs}(c)} O^{c'}) \cup \{\nu\}$. Here ν is the value of $\texttt{null}$.

A state (σ, τ) of a program π consists of a heap σ and an environment τ. An environment $\tau \in \mathrm{T}$ assigns values to the local variables. Formally, T is the set $\prod_{z \in \text{Var}} \text{dom}(\|z\|)$. Note that by $\prod_{a \in A}(P(a))$ we mean a (generalized) cartesian product. An element of this set is a function that assigns to every element $a \in A$ an element of the set $P(a)$.

A heap consists of the existing objects. Each object in turn has its own internal state (the values of its instance variables). The internal state of an object $o \in O^c$ is a total function that maps the instance variables of class c and its superclass to their values. Let $\texttt{supers}(c)$ be the set $\{c' \in \mathcal{C} | c \preceq c'\}$. The internal state of an instance of class c is an element of the set $\text{internal}(c)$, which is the set

$$\prod_{c' \in \texttt{supers}(c)} \prod_{x \in \text{IVar}_{c'}} \text{dom}(\|x\|) \ .$$

A heap σ is a partial function that maps each *existing* object to its internal state. That is, σ is an element of the set $\prod_{c \in \mathcal{C}} \left(\mathbb{N} \rightharpoonup \text{internal}(c) \right)$. Note that $\sigma(c)$ is not defined for objects that do not exist in a particular heap σ. Thus σ specifies the set of existing objects. We will only consider states that are *consistent*. A heap is consistent if all instance variable of existing objects refer to existing objects or ν. Similarly, an environment is consistent with a heap if all variables refer to existing objects or ν.

Expressions are evaluated relative to a program π and a state (σ, τ). The value of an expression e is denoted by $\mathcal{E}(e)(\sigma, \tau)$. We leave the program π implicit. The definition of the function $\mathcal{E}$ can be found in the full paper [12].

Method Calls. Our rule of adaptation is designed for reasoning about method calls in object-oriented languages. In this section we discuss the structural operational semantics of such method calls. By $\langle S, (\sigma, \tau) \rangle \rightarrow (\sigma', \tau')$ we denote that a computation of S that starts in a state (σ, τ) ends in a state (σ', τ').

Recall that a call to method m of object e_0 is of the form $e_0.m(e_1, \ldots, e_n)$. We assume that all methods are public. This implies that we have to deal with dynamic binding if a method is overridden in a subclass. However, we will first describe the case of a method that is not overridden. This implies that we can simply lookup the implementation of method m in class $\|e_0\|$ (we ignore method overloading). Let this implementation be $m(u_1, \ldots, u_n)\{ \bar{v} \; S \; \mathtt{return} \; e \; \}$. The call starts with the context switch in which the value of e_0 is assigned to $\mathtt{this}$ and the values of the actual parameters $\bar{e} = e_1, \ldots, e_n$ are assigned to the formal parameters $\bar{u} = u_1, \ldots, u_n$. The additional local variables $\bar{v}$ initially have their default value. The default value depends on the type of the variable. We assume that $\mathsf{def}(t)$ denotes the default value of type t. After execution of the body the value of e is assigned to u. These steps are described by the following rule.

$$\frac{\begin{array}{c} \langle S, (\sigma, \tau_i) \rangle \to (\sigma', \tau_i') \\ \tau_i = \tau[\mathtt{this}, \bar{u}, \bar{v} \mapsto \mathcal{E}(e_0)(\sigma, \tau), \mathcal{E}(\bar{e})(\sigma, \tau), \mathsf{def}(\|\bar{v}\|)] \\ \tau' = \tau[u \mapsto \mathcal{E}(e)(\sigma', \tau_i')] \end{array}}{\langle u := e_0.m(e_1, \ldots, e_n), (\sigma, \tau) \rangle \to (\sigma', \tau')}$$

Observe that the environment τ' only differs from τ in the value that is assigned to u. All other local variables are not changed by the method call. On the other hand, the fields of the objects in σ may have different values in σ'. Moreover, the heap σ' possibly contains objects that did not exist in σ.

Evaluation of calls to public methods requires an evaluation of e_0 to determine to which implementation the call is bound. Let $\mathcal{E}(e_0)(\sigma, \tau) = (c, n)$. Then this call is bound to the implementation of method m that is found in class c or otherwise the first implementation of this method in a superclass of c. In all other aspects the evaluation of the call is similar to the rule above.

4   The Assertion Language

The proof rule that we propose is tailored to a specific assertion language called AsO (Assertion language for Object structures). We will describe the syntax and semantics of AsO in this section. Moreover, we will explain some of the design decisions behind the language.

Any Hoare logic is tailored to a specific assertion language in which the assertions are expressed. The assertion language decides the abstraction level of the logic. An important design decision behind AsO is to keep its abstraction level as close as possible to the programming language. In other words, we refrain as much as possible from introducing constructs that do not occur in the programming language. This makes it easier for programmers to annotate their programs.

The set of expressions in AsO is obtained by extending the set of program expressions with the following clauses.

$$l \in \mathrm{LExpr} \; ::= \; \ldots \mid z \mid l[l'] \mid l.\mathtt{length}$$

The variable z denotes a logical variable. A logical variable is simply a placeholder for an arbitrary value. Logical variables are, for example, used to refer to the old values of variables in the postcondition of a method.

The only real additions are the two operations on *finite sequences*. We allow logical variables of type t^* for some type t of the programming language. This means that its value is a finite sequence of elements from the domain of t. Finite sequences can be used to specify properties of object structures. For example, they enable us to express that there exists a sequence of objects in which each objects has a pointer to its successor in the sequence. In fact, finite sequences are also essential for our adaptation rule as will become clear in the rest of this paper. We write $l[l']$ to select the element at position l' in the sequence l. The length of a sequence l is denoted by $l.\texttt{length}$.

Formulas in AsO are built from expressions in the usual way.

$$P, Q \in \text{Ass} ::= l_1 = l_2 \mid \neg P \mid P \wedge Q \mid \exists z : t(P)$$

A formula $\exists z : c(P)$ means that P holds for an *existing* object of (a subclass of) class c or $\texttt{null}$. A formula $\exists z : c^*(P)$ indicates that P holds for a sequence of such objects. A formal definition of the semantics of AsO can also be found in the full paper [12]. We sometimes omit the type in $\exists z : t(P)$ if it is clear from the context.

The standard abbreviations like $p \vee q$ for $\neg(\neg p \wedge \neg q)$ and $\forall z(P)$ for $\neg \exists z \neg (P)$ are valid. We also use two other useful abbreviations. A formula $z \in z'$ will stand for $\exists i (0 \leq i < z'.\text{length} \wedge z = z'[i])$, and $z \subseteq z'$ abbreviates the formula $\forall i (0 \leq i < z.\text{length} \rightarrow z[i] \in z')$.

Note that the validity of assertions may also be affected by the creation of new objects. We can, for example, express in the assertion language that there exist no objects of some class c by means of the formula $\forall z : c(z = \texttt{null})$. This formula clearly does not hold in the state that results from executing $u := \texttt{new } c()$.

5   The Effects of a Method

The rule of adaptation that we present in the next section integrates write effects [4] in a natural way in the verification conditions of method calls. Write effects specify what variables are altered by a method. Thus they implicitly contain information about which variables remain unchanged by a method. We approximate the set of variables that is written by a method in the sense that we do not attempt to solve the aliasing problem in write effects. We only analyze which fields are possibly assigned.

Similarly, one can statically collect information about which objects are possibly created by a class. We will call this information the *creational* effects of a method. These effects can be used to guarantee that a method creates no objects of a particular class. The creation of objects can affect the validity of assertions as explained in Sect. 4. Surprisingly, a specification language for Java such as JML [8] seems to have no clause that enables the designer of a method to specify its creational effects. The adaptation rule in this paper reveals that such information is equally important as the well-known frame conditions.

By the *effects* of a method we mean a pair that consists of the creational effects and the write effects of the method. The creational effects of a method is the set of classes of which objects are possibly created by the method. The write effects is the set of fields that are possibly assigned by the method. A field is described by a pair that consists of the class in which the field is declared and its identifier. We denote effects by ψ or by its pair of constituents (cs, fs). In the rest of this section we give a formal definition of the effects of a method.

The effects of a method implementation *meth* are given by $\mathsf{sef}(meth)(\mathsf{ms})$. The second parameter is the set of methods that have already been considered. This parameter prevents a circular definition in case of recursive methods. The effects of a method implementation in some class c are the effects of its body. Let $meth \equiv m(u_1, \ldots, u_n)\{\ \bar{v}\ S\ \mathtt{return}\ e\ \}$. Then

$$\mathsf{sef}(meth)(ms) = \begin{cases} \mathsf{sef}(S)(ms \cup \{(c, m)\}) & \text{if } (c, m) \notin ms \\ (\emptyset, \emptyset) & \text{otherwise} \end{cases}.$$

The following cases list the effects of basic statements.

$$\begin{aligned}
\mathsf{sef}(u := e)(ms) &= (\emptyset, \emptyset) \\
\mathsf{sef}(e.x := e')(ms) &= (\emptyset, \{(\mathsf{origin}(\llbracket e \rrbracket, x), x)\}) \\
\mathsf{sef}(S_1\ ;\ S_2)(ms) &= \mathsf{sef}(S_1)(ms) \cup \mathsf{sef}(S_2)(ms) \\
\mathsf{sef}(u := \mathtt{new}\ c())(ms) &= (\{c\}, \emptyset) \\
\mathsf{sef}(\mathtt{if}\ (e)\ \{\ S_1\ \}\ \mathtt{else}\ \{\ S_2\ \})(ms) &= \mathsf{sef}(S_1)(ms) \cup \mathsf{sef}(S_2)(ms) \\
\mathsf{sef}(\mathtt{while}\ (e)\ \{\ S\ \})(ms) &= \mathsf{sef}(S)(ms)
\end{aligned}$$

The union of two effects is defined as follows:

$$(cs_1, fs_1) \cup (cs_2, fs_2) = (cs_1 \cup cs_2, fs_1 \cup fs_2) .$$

Finally, we will define $\mathsf{sef}(u := e_0.m(e_1, \ldots, e_n))(ms)$. Due to dynamic binding we cannot (in general) statically decide which implementation will be bound to this call. Therefore we consider all possibilities. We denote the set of classes that provide an implementation for this call by $\mathsf{impls}(\llbracket e_0 \rrbracket, m)$. More precisely, $\mathsf{impls}(c, m)$ denotes the set of classes that contains

- class c if it provides an implementation of method m or otherwise the class from which c inherits the implementation of method m;
- All subclasses of class c that provide an implementation of method m.

Let $\mathsf{impl}(c, m)$ denote the implementation of method m in class c. The definition of the final clause is then as follows.

$$\mathsf{sef}(u := e_0.m(e_1, \ldots, e_n))(ms) = \bigcup\nolimits_{c \in \mathsf{impls}(\llbracket e_0 \rrbracket, m)} (\mathsf{sef}(\mathsf{impl}(c, m))(ms))$$

6   A Rule of Adaptation for OO

The Starting Point. The rule of adaptation that we propose in this paper was inspired by the profound analysis by Olderog of several earlier rules of adaptation [11]. Olderog showed that the precondition of Hoare's original proposal was not the weakest possible precondition that fits in the adaptation rule. He described a way to actually *derive* a weaker precondition. His analysis also leads in a similar way to a rule of adaptation that is based on the strongest possible *postcondition*. This particular rule appears also in [16]. It has the following form.

$$
\frac{\{P'\}S\{Q'\}}{P[\bar{y}/\bar{x}] \wedge \forall \bar{z}(P'[\bar{y}/\bar{x}] \rightarrow Q') \rightarrow Q}{\{P\}S\{Q\}} \tag{1}
$$

Here $\bar{y}$ denotes a sequence of fresh logical variables (not occurring free in P, P', Q or Q'). The sequence $\bar{x}$ contains all program variables that occur in the statement S, and $\bar{z}$ is a list of the logical variables that occur free in P' or Q'.

The logical variables in $\bar{y}$ are placeholders of the values of the program variables $\bar{x}$ in the initial state. The antecedent of the verification condition of this rule says that P holds in the initial state. Typically, this implies that P' also holds in the old state for particular values of its logical variables. For those values of the logical variables we then infer that Q' holds in the final state. And Q' must in turn imply Q.

One particular advantage of this adaptation rule is that it considers the old values of the variables (in the state preceding the call) instead of the new values after the call. This is important due to the state extensions in object-oriented programs. In the precondition variant we are forced to reason about objects that do not (yet) exist. But our assertion language only describes properties of *existing* objects. A postcondition variant of the adaptation rule only requires us to consider the objects that existed in the initial state from the perspective of the final state. That boils down to reasoning about subsets of the objects that exist in the state after the call.

Modelling the Old Heap. To design an object-oriented variant of the above adaptation rule we have to analyze which parts of the state are modified by a method call. Recall from Sect. 3 that the environment of the caller (and hence every local variable of the caller) does not change during a method call. Therefore we do not have to distinguish between the old values of local variables of the caller and their new values. The heap however may change in two ways. We have to deal with both heap *modifications* and heap *extensions*.

Observe that the above rule introduces a sequence of logical variables $\bar{y}$ that represent the old values of the program variables. A first challenge for an object-oriented version of the rule of adaptation is to introduce logical variables that model the old heap. The heap comprises the existing objects and the values of their instance variables. We can model the old heap by means of a fresh logical variable μ of type `Object*`. We will assume that this sequence contains the objects that existed in the old heap.

The internal state of an object consists of the values of its instance variables. For each instance variable x of some type t defined in some class c we introduce a fresh logical variable $\mu(x_c)$ of type t^*. The idea is that if an object o is stored at position i in the sequence μ then the value of $o.x$ is $\mu(x_c)[i]$. This straightforward model of the old heap presupposes that we have some way of finding the index of an object in the old heap. For this purpose we introduce a function f that yields the index in the old heap of every object.

This model of the old heap is based on the following assumptions.

- $\forall z : \texttt{Object} \forall i (0 \leq i < \mu.\texttt{length} \wedge z = \mu[i] \rightarrow f(z) = i)$,
  which states that the index function yields the index of each object in μ; It also implies that each object occurs at most once in the old heap;
- $\mu(x_c) \subseteq \mu$,
  for every field x declared in some class c such that $\|x\| \in C$. This formula boils down to consistency of the old heap.

We denote the conjunction of these two assumptions by $\text{heap}(\bar{\mu}, f)$. This formula should be available in the theorem prover as an axiom for the verification conditions.

Bounded Quantification. An important concept in our rule of adaptation is that of bounded quantification. Methods can not only modify the heap but also extend it by creating new objects. Thus a property that holds for all objects in the state before the call may not hold for all objects after the call. Therefore we sometimes have to restrict quantification to the objects that existed before the call. However, we only restrict quantification if the creational effects of the method indicate that an object of this class might be created by the method.

Let μ be the sequence that models the objects in the old heap. Let the effects of the method that is called be ψ. Let $\psi = (cs, fs)$. Then we define the bounded variant $\exists z_\mu^\psi : t(P)$ of an expression $\exists z : t(P)$ as follows.

$$\exists z_\mu^\psi : t(P) \equiv \exists z : t(P) \text{ for } t \in \{\texttt{int}(*), \texttt{boolean}(*)\}$$

$$\exists z_\mu^\psi : c(P) \equiv \begin{cases} \exists z : c(z \in \mu \wedge P) & \text{if } \bigvee_{c' \in cs}(c' \preceq c) \\ \exists z : c(P) & \text{otherwise} \end{cases}$$

$$\exists z_\mu^\psi : c^*(P) \equiv \begin{cases} \exists z : c^*(z \subseteq \mu \wedge P) & \text{if } \bigvee_{c' \in cs}(c' \preceq c) \\ \exists z : c^*(P) & \text{otherwise} \end{cases}$$

Note that quantification becomes bounded if the creational effects list a subclass of the class that we consider. This is the right condition because the quantification domain of a class also contains the objects of subclasses.

Restricting Assertions to the Old Heap. Another challenge is to find a counterpart of the substitution $[\bar{y}/\bar{x}]$. This substitution must do two things. It has to replace references to instance variables by the corresponding logical variables. And it must restrict quantification to the objects that existed in the old heap as argued above. We introduce the substitution $\downarrow_\psi$ for this purpose. Note that this substitution takes the effects ψ of the method that is called into account.

The most interesting case of this substitution is $(l.x) \downarrow_\psi$. It replaces $l.x$ by its value in the old heap as described above. Let $\psi = (cs, fs)$. Assume that origin($\llbracket l \rrbracket, x$) $= c$ (field x is defined in class c). Then we define this case as follows.

$$(l.x)\downarrow_\psi \equiv \begin{cases} \mu(x_c)[\mathsf{f}(l\downarrow_\psi)] & \text{if } (c,x) \in \mathit{fs} \\ (l\downarrow_\psi).x & \text{otherwise} \end{cases}$$

We only substitute the instance variable if the write effects indicate that the method possibly changes its value.

Another interesting case is $(\exists z : t(P))\downarrow_\psi$, which is defined as follows.

$$(\exists z : t(P))\downarrow_\psi \equiv \exists z^\psi_\mu : t(P\downarrow_\psi))$$

Thus the substitution $\downarrow_\psi$ restricts quantification to the objects that existed in the state before the call. The quantification is not bounded if the creational effects of the method guarantee that no objects of a class will be created. All other cases of the substitution $\downarrow_\psi$ correspond to the usual notion of structural substitution.

The Adaptation Rule. We can now state our rule of adaptation. We start with a simplified version of the rule for calls that have but one implementation. This is, for example, the case if a method is private, or if it is not overridden in some subclass. Let the statement $u = e_0.m(e_1, \ldots, e_n)$ involve a call to such a method. Let $\mathit{meth} \equiv m(u_1, \ldots, u_n)\{\ \bar{v}\ S\ \mathtt{return}\ e\ \}$ be the implementation of method m that is bound to the call. Let the effects ψ be $\mathsf{sef}(\mathit{meth})(\emptyset)$. Then we have the following rule of adaptation (2) for such calls.

$$\frac{\{P'\}\ \bar{v}\ S\ \{Q'[e/\mathsf{result}]\}}{\{P\}\ u := e_0.m(e_1, \ldots, e_n)\ \{Q\}} \quad (2)$$
$$\mathsf{locs} \wedge P\downarrow_\psi \wedge \forall \bar{z}^\psi_\mu (P'[e_0, \bar{e}/\mathtt{this}, \bar{u}]\downarrow_\psi \rightarrow \exists \bar{v}'(Q')) \rightarrow Q[\mathsf{result}/u]$$

Observe that the rule has the same outline as the former rule except for the predicate locs. This predicates states that the local variables of the caller refer to objects in the old heap. It is the following formula: $\bigwedge_{u \in U} u \in \mu$, where U is the set of all local variables that occur either in P, Q or e_i, for $i \in \{0 \ldots n\}$.

The list $\bar{z}$ again contains all logical variables that occur free in P' or Q' (except the special-purpose logical variable result). The sequence $\bar{v}'$ is a sequence of all local variables that occur free in Q' including $\mathtt{this}$. We quantify over these local variables of the callee to prevent confusion with local variables of the caller in P or Q. The precondition of the method P' may only mention the formal parameters and $\mathtt{this}$. All other local variables are out of scope in P'.

The simultaneous substitution $[e_0, \bar{e}/\mathtt{this}, \bar{u}]$ models the context switch. It replaces $\mathtt{this}$ by e_0, and the formal parameters $\bar{u} = u_1, \ldots, u_n$ by the actual parameters $\bar{e} = e_1, \ldots, e_n$. To prevent problems with field shadowing these substitutions should preserve the types of the expressions by introducing casts if necessary (see [13] for further details).

The special-purpose logical variable result denotes the result value. It is only allowed in postconditions of methods. The substitution $[e/\mathsf{result}]$ models a virtual assignment of the result value to result. Similarly, the substitution $[\mathsf{result}/u]$ models an assignment of this value to u.

Observe that the second premiss of the rule is the verification condition of a call with annotation $\{P\}\ u = e_0.m(e_1, \ldots, e_n)\ \{Q\}$ in an object-oriented program where the corresponding implementation of method m has precondition P' and postcondition Q'. To be able to express the verification condition we extended the assertion language by allowing the function symbol f in assertions. However, this symbol will only occur in verification conditions. The proof outlines retain their desired abstraction level.

The rule looks rather complex because it has to account for all possible contexts. In concrete contexts the resulting verification condition is often simpler. We give an example in Sect. 7.

Next, we analyze reasoning about method invocations that are dynamically bound to an implementation. In such cases we have to consider all implementations of method m that might possibly be bound to the call. Suppose again that we consider a call $u := e_0.m(e_1, \ldots, e_n)$. Recall that $\mathsf{impls}(\llbracket e_0 \rrbracket, m)$ denotes the set of classes that provide an implementation for this call. This set tells us how many implementations we must consider.

A second consideration concerns the set of classes that inherit a particular implementation in some class $c \in \mathsf{impls}(\llbracket e_0 \rrbracket, m)$. A class inherits the implementation in class c if it is a subclass of class c and it is not a subclass of some class that overrides the implementation in class c. We denote the set of classes that override the implementation of method m in class c by $\mathsf{overrides}(c)(m)$. We have $c' \in \mathsf{overrides}(c)(m)$ if

- c' is a proper subclass of c that provides an implementation of method m, and
- there does not exist another proper subclass c'' of c such that c' is a proper subclass of c'', and c'' also provides an implementation of method m.

With this definition we can formulate in what circumstances the implementation of m in class c is bound to a method call $e_0.m(e_1, \ldots, e_n)$. That occurs when e_0 is an instance of a subclass of c and it is not an instance of a class that overrides this implementation. These conditions are expressed in the assertion language by the following formula: $e_0\ \mathtt{instanceof}\ c \wedge \bigwedge_{c' \in \mathsf{overrides}(c)(m)} \neg(e_0\ \mathtt{instanceof}\ c')$. We denote this formula by $\mathsf{boundto}(e_0, c, m)$.

Let $\mathsf{impls}(\llbracket e_0 \rrbracket, m) = \{c_1, \ldots, c_k\}$. Let $m(u_1^i, \ldots, u_n^i)\{\ \bar{v}_i\ S_i\ \mathtt{return}\ e_i\ \}$ be the implementation of method m in class c_i, for $i = 1, \ldots, k$ with effects ψ_i. We assume that the implementation of method m in class c_i has precondition P_i' and postcondition Q_i'. Let B_i denote $\{P_i'\}\ \bar{v}_i\ S_i\ \{Q_i'[e_i/\mathsf{result}]\}$. That is, B_i describes the specification of the implementation in class c_i. The verification condition V_i for the implementation in class c_i is the implication

$$\mathsf{locs} \wedge P\!\downarrow_{\psi_i} \wedge \mathsf{boundto}(e_0, c_i, m)\!\downarrow_{\psi_i} \wedge$$
$$\forall \bar{z}_{\bar{\mu}}^{\psi_i} (P_i'[e_0, \bar{e}/\mathtt{this}, \bar{u}]\!\downarrow_{\psi_i} \to \exists \bar{v}'(Q_i')) \to Q[\mathsf{result}/u]\ .\quad (V_i)$$

Note that we have strengthened the antecedent with the clause that implies that the receiver is an object of a class that inherits the implementation of class c_i.

```
class c {
  int x;

  requires u = U ∧ this = O
  ensures O.x = U
  public void set(int u) { this.x := u; }

  requires u = U ∧ this.x = X
  ensures this.x = U ∧ result = (X = U)
  public boolean testAndSet(int u) {
    boolean b := (this.x = u);
    /*{ u = U ∧ b = (X = U) }*/
    this.set(u);
    return b; }
}
```

Fig. 2. An example proof outline

The rule of adaptation for dynamically-bound method calls then simply checks if all specifications of implementations that might possibly be bound to the call can be adapted to the specification of the call. The rule results in one verification condition for each implementation.

$$\frac{B_1, \ldots, B_k \quad V_1, \ldots, V_k}{\{P\} \, u = e_0.m(e_1, \ldots, e_n) \, \{Q\}} \tag{3}$$

7 An Example

The rule as given above is certainly more complex that most other well-known Hoare rules. It can be conveniently used in proof outlines however if the verification condition is computed automatically. We will give a small example of a program and the resulting verification condition in this section. The example mainly illustrates the elegant way in which the rule handles local variables.

The example proof outline is listed in Fig. 2. The code defines a simple Java class with two methods. Each method is preceded by a precondition (requires clause) and a postcondition (ensures clause). The capital letters denote logical variables.

It may seem strange that we also introduce a logical variable O in the precondition of the **set** method to refer to the old value of **this** in its postcondition. This is done because the verification condition puts an existential quantifier around occurrences of **this** in the postcondition of the method to prevent confusion with occurrences of **this** in the specification of the caller. A possible refinement of the rule could make these steps superfluous by automatically adding the clause **this** $= z$ to the precondition of the method and replacing **this** in the postcondition by z (where z is a fresh logical variable).

The call **this.set(u)** is preceded by an intermediate assertion that is the precondition of the call. The postcondition of this call is obtained by substituting result in the postcondition of the method by b. We assume that the methods are not overridden in subclasses. Therefore this call has the following verification condition.

$$u = U \wedge b = (X = U) \wedge \texttt{this instanceof}\ c$$
$$\wedge\ \forall O : c(\forall U : \texttt{int}(u = U \wedge \texttt{this} = O \rightarrow O.x = U))$$
$$\rightarrow \texttt{this}.x = U \wedge b = (X = U)$$

The effects of the $\texttt{set}$-method are $(\emptyset, \{(c, x)\})$. That is, it creates no new objects and only modifies field x in class c. That explains why quantification in this verification condition is not bounded. The predicate $\textsf{locs}$ is not needed to prove this particular verification condition and is therefore left out. It is not difficult to see that the above verification condition is valid if one chooses $\texttt{this}$ for O.

Observe that the clause $b = (X = U)$ in the precondition of the call can be used to prove the same clause in the postcondition. The specification of the $\texttt{set}$ method is unrelated to the local variables of the caller.

8   Conclusions

The main result of this paper is a (to the best of our knowledge) first rule of adaptation for the object-oriented paradigm. The rule not only enables us to adapt the values of logical variables but also results in an important separation of concerns for local variables in proof outlines for object-oriented programs. This improves the modularity of the specification. The rule uses a static approximation of the effects of a method to further improve the modularity of the specification. We prove soundness of the adaptation rule in the full version [12] of this paper.

The adaptation rule complements our earlier work on a (relatively) complete Hoare logic [13] for object-oriented programs. It is a well-known fact that the rule of adaptation can replace the substitution rule, the invariance rule, and the elimination rule without compromising completeness [11]. In the full version we show that this is also the case for the rule presented here.

Proof outlines are not only a compact representation of correctness proofs but are also useful for documentation purposes. The rule of adaptation describes the verification condition that results from the specification of a call and the corresponding method. That is precisely what is needed in a proof outline logic. Thus the rule enables a shift from Hoare logics to proof outlines for OO. It is unclear if and how the rules that it replaces can be used in proof outlines.

Future work includes an investigation of the sharpness [2, 9] of the proof rule.

References

1. de Boer, F., Pierik, C.: Computer-aided specification and verification of annotated object-oriented programs. In Jacobs, B., Rensink, A., eds.: FMOODS V, Kluwer Academic Publishers (2002) 163–177

2. Bijlsma, A., Matthews, P., Wiltink, J.: A sharp proof rule for procedures in wp semantics. Acta Informatica **26** (1989) 409–419

3. Cataño, N., Huisman, M.: Chase: a static checker for JML's assignable clause. In: Verification, Model Checking and Abstract Interpretation (VMCAI '03). Volume 2575 of LNCS. (2003) 26–40

4. Greenhouse, A., Boyland, J.: An object-oriented effects system. In Guerraoui, R., ed.: Proc. of ECOOP '99. Volume 1628 of LNCS. (1999) 205–229

5. Hoare, C.: Procedures and parameters:an axiomatic approach. In Engeler, E., ed.: Symp. on Semantics of Algorithmic Languages. Volume 188 of Lecture Notes in Mathematics. (1971) 102–116

6. Huisman, M.: Reasoning about Java programs in higher order logic with PVS and Isabelle. PhD thesis, Katholieke Universiteit Nijmegen (2001)

7. Kleymann, T.: Hoare logic and auxiliary variables. Formal Aspects of Computing **11** (1999) 541–566

8. Leavens, G.T., Baker, A.L., Ruby, C.: Preliminary design of JML: A behavioral interface specification language for Java. Technical Report 98-06u, Department of Computer Science, Iowa State University (2003)

9. Naumann, D.A.: Calculating sharp adaptation rules. Information Processing Letters **7** (2000) 201–208

10. von Oheimb, D.: Hoare logic for Java in Isabelle/HOL. Concurrency and Computation: Practice and Experience **13** (2001) 1173–1214

11. Olderog, E.R.: On the notion of expressiveness and the rule of adaptation. Theoretical Computer Science **24** (1983) 337–347

12. Pierik, C., de Boer, F.S.: A rule of adaptation for OO. Technical Report UU-CS-2003-032, Institute of Information and Computing Sciences, Utrecht University, The Netherlands (2003)

13. Pierik, C., de Boer, F.S.: A syntax-directed Hoare logic for object-oriented programming concepts. In Najm, E., Nestmann, U., Stevens, P., eds.: Formal Methods for Open Object-Based Distributed Systems (FMOODS) VI. Volume 2884 of LNCS. (2003) 64–78

14. Poetzsch-Heffter, A., Müller, P.: A programming logic for sequential Java. In Swierstra, S.D., ed.: ESOP '99. Volume 1576 of LNCS. (1999) 162–176

15. Reus, B., Wirsing, M., Hennicker, R.: A Hoare calculus for verifying Java realizations of OCL-constrained design models. In Hussmann, H., ed.: FASE 2001. Volume 2029 of LNCS. (2001) 300–317

16. Zwiers, J., Hannemann, U., Lakhnech, Y., de Roever, W.P., Stomp, F.: Modular completeness: Integrating the reuse of specified software in top-down program development. In: Formal Methods Europe, FME'96 Symposium. Volume 1051 of LNCS. (1996) 595–608

Modal Abstractions in μCRL⋆

Jaco van de Pol and Miguel Valero Espada

Centrum voor Wiskunde en Informatica,
P.O. Box 94079, 1090 GB Amsterdam, The Netherlands
{Jaco.van.de.Pol,Miguel.Valero.Espada}@cwi.nl

Abstract. We describe a framework to generate modal abstract approximations from process algebraic specifications, written in the language
μCRL. We allow abstraction of state variables and action labels. Moreover, we introduce a new format for process specifications called *Modal
Linear Process Equation* (MLPE). Every transition step may lead to a set
of abstract tates labelled with a set of abstract actions. We use MLPEs
to characterize abstract interpretations of systems and to generate *Modal
Labelled Transition Systems*, in which transitions may have two modalities *may* and *must*. We prove that the abstractions are sound for the
full action-based μ-calculus. Finally, we apply the result to check some
safety and liveness properties for the bounded retransmission protocol.

1 Introduction

The theory of abstract interpretation [4, 13] denotes a classical framework for
program analysis. It extracts program approximations by eliminating uninter-
esting information. Computations over concrete universes of data are performed
over smaller abstract domains. The application of abstract interpretation to the
verification of systems is suitable since it allows to formally transform possibly infinite instances of specifications into smaller and finite ones. By loosing
some information we can compute a desirable view of the analysed system that
preserves some interesting properties of the original.

The achievement of this paper is to enhance existing process algebraic verification tools (e.g. LOTOS, μCRL) with state-of-the-art abstract interpretation
techniques that exist for state-based reactive systems. These techniques are based
on homomorphisms [3] (easier to use) or Galois Connections [16, 5, 12, 9] (more
precise abstractions). The latter are sound for safety as well as liveness properties. A three-valued logic ensures that the theory can be used for proofs and
refutations of temporal properties. We transpose this to a process algebra setting,
allowing abstraction of states and action labels, and treating homomorphisms
and Galois Connections in a uniform way. A preliminary step was already taken
in [7]; those authors show how process algebras can benefit from abstract interpretation *in principle*. To this end they work with a basic LOTOS language and

---

⋆ Partially supported by PROGRESS, the embedded systems research program of the
Dutch organisation for Scientific Research NWO, the Dutch Ministry of Economic
Affairs and the Technology Foundation STW, grant CES.5009.

C. Rattray et al. (Eds.): AMAST 2004, LNCS 3116, pp. 409–425, 2004.

a simple temporal logic; their abstractions preserve linear-time safety properties only.

Semantically, our method is based on *Modal Labelled Transition Systems* [15]. MLTSs are *mixed* transition systems in which transitions are labelled with actions and with two modalities: *may* and *must*. They are appropriate structures to define abstraction/refinement relations between processes. *May* transitions determine the actions that possibly occur in all refinements of the system while *must* transitions denote the ones that necessarily happen. On the one hand, the *may* part corresponds to an over-approximation that preserves *safety* properties of the concrete instance and on the other hand the *must* part under-approximates the model and reflects *liveness* properties. We define approximations and prove that they are sound for all properties in the full (action-based) μ-calculus [14], including negation. We had to extend the existing theory by allowing abstraction and information ordering of action labels, which is already visible in the semantics of μ-calculus formulas. This is treated in Section 2.

This theory is applied to μCRL specifications [10], which (as in LOTOS) consist of an ADT part defining data operations, and a process specification part, specifying an event-based reactive system. Processes are defined using sequential and parallel composition, non-deterministic choice and hiding. Furthermore, atomic actions, conditions and recursion are present, and may depend on data parameters. The μCRL toolset [1, 2] transforms specifications to *linear process equations* (LPE). An LPE is a concise representation of all possible interleavings of a system in which parallel composition and hiding are eliminated. Several tools that manipulate LPEs have been developed; they do, for example, symbolic model checking, state space reduction, confluence analysis, etc... The μCRL language and tool set have been used in numerous verifications of communication and security protocols and standards, distributed algorithms and industrial embedded systems.

We implement abstract interpretation as a transformation of LPEs to MLPEs (modal LPEs). MLPEs capture the extra non-determinism arising from abstract interpretation. They allow a simple transition to lead to a *set* of states with a *set* of action labels. We show that the MLTS generated from an MLPE is a proper abstraction of the LTS generated from the original LPE. This implies soundness for μ-calculus properties. Section 4 is devoted to this part. The next figure shows the different possibilities to extract abstract approximations from a concrete specifications.

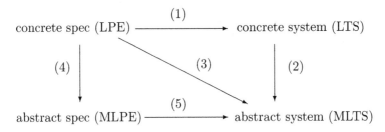

From a concrete system, encoded as an LPE, we can:

- Generate the concrete transition system (1), from which we compute the abstraction (2). This solution is not very useful for verification because the generation of the concrete transition system may be impossible due to the size of the state space.
- Generate directly the abstract *Modal*-LTS (3), by interpreting the concrete specification over the abstract domain. This solution avoids the generation of the concrete transition system.
- First, generate a symbolic abstraction of the concrete system (4), and then extract the abstract transition system (5).

Typically, standard abstract interpretation frameworks implement the second approach, however we believe that the third one is more modular. MLPEs act as intermediate representation that may be subjected to new transformations. Furthermore, MLPEs can be encoded as LPEs thus our method integrates perfectly with the existing transformation and state space generation tools of the μCRL toolset. Also, the three valued model checking problem can be rephrased as the usual model checking problem, along the lines of [9]. This enables the reuse of the model checkers in the CADP toolset [8]. The latter two transformations are not detailed in the current paper.

The main part of the theory mentioned above has been defined and proved correct in an elegant way using the computer assisted theorem prover PVS [18]. The use of the theorem prover gives extra confidence about the correctness of the theory. Furthermore, the definitions and proofs can be reused to easily extend the theory, to prove other transformations, or to apply the same techniques to another specification language. Also, this prover could be used to prove the safety conditions generated for user-specified abstractions.

To improve usability, we have predefined a few standard abstractions. Of course, the user can define specific abstractions, which in general lead to the generation of safety conditions. Finally, thanks to the uniform treatment, the tool can automatically lift a (predefined or user specified) homomorphism to a Galois Connection, thus combining ease with precision.

As running example we use a simple system in which two processes communicate by sending natural numbers through a channel described as a FIFO buffer of size N, see the figure below. The system has two sources of infinity: the size of the buffer and the value of the data, which should be abstracted if we want to apply model checking techniques to its verification. Finally, we demonstrate the method by checking some safety and liveness properties of the bounded retransmission protocol in Section 5.

2   Transition Systems

Part of the results included in this section are well known in the field of abstract interpretation. We adapt classical frameworks for generating safe abstract approximations, by doing a non-trivial extension of them in order to allow the explicit abstraction of action labels which will permit to manipulate infinitely branching systems. First, we start by defining some general concepts and then we continue by introducing the two abstraction techniques.

The semantics of a system can be defined by a *Labelled Transition System*. We assume a non-empty set S of states, together with a non-empty set of transition labels A:

Definition 1 *A transition is a triple $s \xrightarrow{a} s'$ with $a \in A$ and $s,s' \in S$. We define a Labelled Transition System (LTS) as tuple $(S, A, \rightarrow, s_0)$ in which S and A are defined as above and $\rightarrow$ is a possibly infinite set of transitions and s_0 in S is the initial state.*

To model abstractions we are going to use a different structure that allows to represent approximations of the concrete system in a more suitable way. As introduced before, in *Modal Labelled Transition Systems* transitions have two modalities *may* and *must* which denote the possible and necessary steps in the refinements. This concept was introduced by Larsen and Thomsen [15]. Let us see the definition:

Definition 2 *A Modal Labelled Transition System (MLTS) or may/must labelled transition system (may/must-LTS) is a tuple $(S, A, \rightarrow_\diamond, \rightarrow_\square, s_0)$ where S, A and s_0 are as in the previous definition and $\rightarrow_\diamond, \rightarrow_\square$ are possibly infinite sets of (may or must) transitions of the form $s \xrightarrow{a}_x s'$ with $s,s' \in S$, $a \in A$ and $x \in \{\diamond, \square\}$. We require that every must-transition is a may-transition $(\xrightarrow{a}_\square \subseteq \xrightarrow{a}_\diamond)$.*

MLTSs are suitable structures for stepwise refinements and abstractions. A refinement step of a system is done by preserving or extending the existing *must*-transitions and by preserving or removing the *may*-transitions. Abstraction is done the other way around. To every LTS corresponds a trivially equivalent MLTS constructed by labelling all transitions with both modalities; we will call it the corresponding *concrete* MLTS.

Having a set of states S and a set of action labels A with their corresponding abstract sets, denoted by $\widehat{S}$ and $\widehat{A}$, we define a homomorphism H as a pair of total and surjective functions $\langle h_S, h_A \rangle$, where h_S is a mapping from states to abstract states, i.e., $h_S : S \rightarrow \widehat{S}$, and h_A maps action labels to abstract action labels, i.e., $h_A : A \rightarrow \widehat{A}$. The abstract state $\widehat{s}$ corresponds to all the states s for which $h_S(s) = \widehat{s}$, and the abstract action label $\widehat{a}$ corresponds to all the actions a for which $h_A(a) = \widehat{a}$.

Definition 3 *Given a concrete MLTS P $(S, A, \rightarrow_\diamond, \rightarrow_\square, s_0)$ and a mapping H, we say that $\widehat{P}$ defined by $(\widehat{S}, \widehat{A}, \rightarrow_\diamond, \rightarrow_\square, \widehat{s}_0)$ is the minimal may/must$_H$-abstraction (denoted by $\widehat{P} = min_H(P)$) if and only if $h_S(s_0) = \widehat{s}_0$ and the following conditions hold:*

$$- \widehat{s} \xrightarrow{\widehat{a}}_\diamond \widehat{r} \iff \exists s, r, a.\, h_S(s) = \widehat{s} \wedge h_S(r) = \widehat{r} \wedge h_A(a) = \widehat{a} \wedge s \xrightarrow{a}_\diamond r$$
$$- \widehat{s} \xrightarrow{\widehat{a}}_\square \widehat{r} \iff \forall s.h_S(s) = \widehat{s}.\, (\exists r, a.\, h_S(r) = \widehat{r} \wedge h_A(a) = \widehat{a} \wedge s \xrightarrow{a}_\square r)$$

This definition gives the most accurate abstraction of a concrete system by using a homomorphism, in other words the one that preserves most information of the original system. The figure below shows, on the left side, the *concrete* MLTS corresponding to the buffer model[1]. In the middle we see the minimal abstraction of the system by only abstracting states with h_S defined as follows: it maps the initial state to the abstract state e, which means *empty*, the states in which there are N entries in the buffer to f, which represents *full*, and the rest of the states to m, which means something in the *middle*; we can see that the resulting abstract system is infinitely branching, therefore in order to be able to do model checking to the system we need also to abstract the action labels. We define h_A as follows: it maps all the write actions to $\widehat{w}$ and all the read actions to $\widehat{r}$. On the right side, we see the result of applying both abstractions. In the final system, we have removed all the information about the values that are in the buffer and the transferred data, only preserving the information about whether the buffer is empty, full or none of them. This abstraction allows to have a small finite model which keeps some information about the original. The example clearly illustrates the importance of the abstraction of action labels.

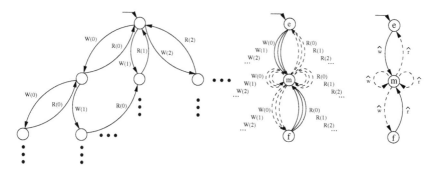

Instead of using mappings between concrete and abstract domains we can define relations. The other classical approach we present is based on Galois Connections between domains, and it was introduced in the late seventies by Cousot and Cousot [4], see also [5] for a good introduction. Two functions α and γ over two partially ordered sets $(P, \subseteq)$ and $(Q, \preccurlyeq)$ such that $\alpha : P \to Q$ and $\gamma : Q \to P$ form a Galois Connection if and only if the following conditions hold:

1. α and γ are total and monotonic.
2. $\forall p : P, p \subseteq \gamma \circ \alpha(p)$.
3. $\forall q : Q, \alpha \circ \gamma(q) \preccurlyeq q$.

---

[1] For clarity we use dashed lines to represent *may* transitions and normal lines to represent *must* transitions. Note that when there is a *must* transition we do not include the corresponding *may* one.

α is the lower adjoint and γ is the upper adjoint of the Galois Connection, and they uniquely determine each other. Galois Connections enjoy suitable properties to perform abstractions. Let $\mathcal{P}(S)$ and $\mathcal{P}(A)$ be partially ordered sets ordered by the set inclusion operator and the abstract $\widehat{S}$ and $\widehat{A}$ both being posets equipped with some order $\preccurlyeq$. The order gives a relation of the precision of the information contained in the elements of the domain. We define a pair G of Galois Connections: $G = \langle(\alpha_S, \gamma_S),(\alpha_A, \gamma_A)\rangle$. α is usually called the abstraction function and γ the concretization function. As in the previous case we define the minimal abstraction, as follows:

Definition 4 *Given two systems P and $\widehat{P}$ defined as in definition 3 and a pair of Galois Connections G, $\widehat{P}$ is the minimal may/must$_G$-abstraction (denoted by $\widehat{P} = min_G(P)$) if and only if $s_0 \in \gamma_S(\widehat{s}_0)$ and the following conditions hold:*

$$- \widehat{s} \xrightarrow{\widehat{a}}_{\diamond} \widehat{r} \iff \exists s \in \gamma_S(\widehat{s}),\, r \in \gamma_S(\widehat{r}),\, a \in \gamma_A(\widehat{a}).\, s \xrightarrow{a}_{\diamond} r$$

$$- \widehat{s} \xrightarrow{\widehat{a}}_{\square} \widehat{r} \iff \forall s \in \gamma_S(\widehat{s}).\, (\exists r \in \gamma_S(\widehat{r}),\, a \in \gamma_A(\widehat{a}).\, s \xrightarrow{a}_{\square} r)$$

The following figure presents a part of the minimal abstraction of the buffer system[2]. On the left side we can see the two abstract lattices, and on the right side we see the transitions corresponding to the abstract write and read actions. The abstract lattices are: $\{\bot, e, m, f, nE, nF, \top\}$ and $\{\bot, \widehat{w}, \widehat{r}, \top\}$, the Galois Connection is trivially defined following the previous example and considering that nE represents the states in which the buffer is not empty and nF in which it is not full.

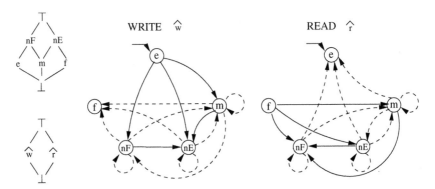

Due to the order over the abstract states and actions, the minimal system defined by the above presented definition is saturated of *may* and *must* transitions, i.e. there are transitions that do not add any extra information. We can easily see in the previous figure that the *must* part is saturated for example, the transition $e \xrightarrow{\widehat{w}}_{\square} nE$ does not add any information because we have $e \xrightarrow{\widehat{w}}_{\square} m$ which is more precise. We can restrict our definition by requiring that the abstract transitions

---

[2] For readability, we separate write transitions: $\xrightarrow{\widehat{w}}$ and read transitions $\xrightarrow{\widehat{r}}$ and we do not include transitions to and from $\top$, or labelled with it.

are performed only between the most precise descriptions of the concrete transitions, as done in [5][3]. In the previous figure, this would remove: $e \xrightarrow{\widehat{w}}_{\diamond,\square} nF$, $e \xrightarrow{\widehat{w}}_{\diamond,\square} nE$, $m \xrightarrow{\widehat{w}}_{\diamond} nF$, $m \xrightarrow{\widehat{r}}_{\diamond} nE$, $f \xrightarrow{\widehat{r}}_{\diamond,\square} nF$, $f \xrightarrow{\widehat{r}}_{\diamond,\square} nE$, $nF \xrightarrow{\widehat{w},\widehat{r}}_{\diamond} nF$ and $nE \xrightarrow{\widehat{w},\widehat{r}}_{\diamond} nE$. We call the system in which all redundant transitions have been eliminated *restricted* and it is denoted by $\widehat{P}\!\downarrow$.

We formalize now the approximation relation between MLTSs:

Definition 5 *Given two MLTSs P $(S, A, \rightarrow_{\diamond}, \rightarrow_{\square}, s_0)$ and Q $(S, A, \rightarrow_{\diamond}, \rightarrow_{\square}, s_0)$ built over the same sets of states $\langle S, \preccurlyeq_S \rangle$ and actions $\langle A, \preccurlyeq_A \rangle$; Q is an abstraction of P, denoted by $P \sqsubseteq_{\preccurlyeq} Q$, if the following conditions hold:*

- $\forall s, a, r, s'.\, s \xrightarrow{a}_{\diamond} r \wedge s \preccurlyeq_S s' \implies \exists a', r'.\, s' \xrightarrow{a'}_{\diamond} r' \wedge r \preccurlyeq_S r' \wedge a \preccurlyeq_A a'$
- $\forall s', a', r', s.\, s' \xrightarrow{a'}_{\square} r' \wedge s \preccurlyeq_S s' \implies \exists a, r.\, s \xrightarrow{a}_{\square} r \wedge r \preccurlyeq_S r' \wedge a \preccurlyeq_A a'$

$P \sqsubseteq_{\preccurlyeq} Q$ means that Q is more abstract than P and it preserves all the information of the *may*-transitions of P and at least all *must* transitions present in Q are reflected in P. The *may* part of Q is an over-approximation of P and the *must* part is an under-approximation. The refinement relation is the dual of the abstraction. Note that for the homomorphism approach there is no order defined between states or actions so we substitute $\preccurlyeq$ by $=$.

3   Logical Characterization

To express properties about systems we are going to adapt the highly expressive temporal logic (action-based) μ-calculus [14] which is defined by the following syntax, where a is an action in A:

$$\varphi ::= T \mid F \mid \neg\varphi \mid \varphi_1 \wedge \varphi_2 \mid \varphi_1 \vee \varphi_2 \mid [a]\varphi \mid \langle a \rangle \varphi \mid Y \mid \mu Y.\varphi \mid \nu Y.\varphi$$

Formulas are assumed to be monotonic. Following [12], a given formula is interpreted dually over an MLTS, i.e. there will be two sets of states that satisfy it. A set of states that necessarily satisfy the formula and a set of states that possibly satisfy it. Thus, the semantics of the formulas is given by $\llbracket \varphi \rrbracket \in 2^S \times 2^S$ and the projections $\llbracket \varphi \rrbracket^{nec}$ and $\llbracket \varphi \rrbracket^{pos}$ give the first and the second component, respectively. We show below the simultaneous recursive definitions of the evaluation of a state formula. In the state formulas, the propositional context $\rho : Y \rightarrow 2^S \times 2^S$ assigns state sets to propositional variables, and the $\oslash$ operator denotes context overriding. Note that the precision order between action labels plays an important role in the definition of the semantics of the modalities.

---

[3] The main difference with Dams' approach is that we preserve the condition $\xrightarrow{a}_{\square} \subseteq \xrightarrow{a}_{\diamond}$.

$$\llbracket F \rrbracket_\rho = \langle \emptyset, \emptyset \rangle$$
$$\llbracket T \rrbracket_\rho = \langle S, S \rangle$$
$$\llbracket \neg\varphi \rrbracket_\rho = \langle S \backslash \llbracket \varphi \rrbracket_\rho^{pos}, S \backslash \llbracket \varphi \rrbracket_\rho^{nec} \rangle \quad \text{(Note the switch of } pos \text{ and } nec)$$
$$\llbracket \varphi_1 \wedge \varphi_2 \rrbracket_\rho = \langle \llbracket \varphi_1 \rrbracket_\rho^{nec} \cap \llbracket \varphi_2 \rrbracket_\rho^{nec}, \llbracket \varphi_1 \rrbracket_\rho^{pos} \cap \llbracket \varphi_2 \rrbracket_\rho^{pos} \rangle$$
$$\llbracket \varphi_1 \vee \varphi_2 \rrbracket_\rho = \langle \llbracket \varphi_1 \rrbracket_\rho^{nec} \cup \llbracket \varphi_2 \rrbracket_\rho^{nec}, \llbracket \varphi_1 \rrbracket_\rho^{pos} \cup \llbracket \varphi_2 \rrbracket_\rho^{pos} \rangle$$
$$\llbracket [a]\varphi \rrbracket_\rho = \langle \{ s \mid \forall r, a'. a \preccurlyeq a' \wedge s \xrightarrow{a'}_\diamond r \Rightarrow r \in \llbracket \varphi \rrbracket_\rho^{nec} \},$$
$$\qquad\qquad \{ s \mid \forall r, a'. a' \preccurlyeq a \wedge s \xrightarrow{a'}_\square r \Rightarrow r \in \llbracket \varphi \rrbracket_\rho^{pos} \} \rangle$$
$$\llbracket \langle a \rangle \varphi \rrbracket_\rho = \langle \{ s \mid \exists r, a'. a' \preccurlyeq a \wedge s \xrightarrow{a'}_\square r \wedge r \in \llbracket \varphi \rrbracket_\rho^{nec} \},$$
$$\qquad\qquad \{ s \mid \exists r, a'. a \preccurlyeq a' \wedge s \xrightarrow{a'}_\diamond r \wedge r \in \llbracket \varphi \rrbracket_\rho^{pos} \} \rangle$$
$$\llbracket Y \rrbracket_\rho = \rho(Y)$$
$$\llbracket \mu Y.\varphi \rrbracket_\rho = \langle \bigcap \{ S' \subseteq S \mid \Phi_\rho^{nec}(S') \subseteq S' \}, \bigcap \{ S' \subseteq S \mid \Phi_\rho^{pos}(S') \subseteq S' \} \rangle$$
$$\llbracket \nu Y.\varphi \rrbracket_\rho = \langle \bigcup \{ S' \subseteq S \mid S' \subseteq \Phi_\rho^{nec}(S') \}, \bigcup \{ S' \subseteq S \mid S' \subseteq \Phi_\rho^{pos}(S') \} \rangle$$
$$\text{where } \Phi_\rho^x : 2^S \to 2^S \text{ with } x \in \{nec, pos\}, \Phi_\rho^x(S') = \llbracket \varphi \rrbracket_{(\rho \oslash [S'/Y])}^x$$

We say that a state s necessarily satisfies a formula φ, denoted by $s \models^{nec} \varphi$, iff $s \in \llbracket \varphi \rrbracket^{nec}$ and dually s possibly satisfies a formula φ, denoted by $s \models^{pos} \varphi$, iff $s \in \llbracket \varphi \rrbracket^{pos}$. We remark that from the semantics of the negation follows:

- s necessarily satisfies $\neg\varphi$ iff s not possibly satisfies φ.
- s possibly satisfies $\neg\varphi$ iff s not necessarily satisfies φ.

It is not difficult to see that if s necessarily satisfies a formula φ then also s possibly satisfies φ. This follows from the fact that every *must*-transition is also a *may*-transition. Without this condition we would be able to prove $s \models^{nec} \varphi$ and $s \models^{nec} \neg\varphi$ for some φ which will lead to an *inconsistent* logic. In fact, it cannot be proved for any φ $s \models^{nec} \varphi$ and also $s \models^{nec} \neg\varphi$, i.e. the necessarily interpretation is consistent and it is always possible to prove $s \models^{pos} \varphi$ or $s \models^{pos} \neg\varphi$ which means that the possibly interpretation is complete. The semantics gives a three valued logic:

- s necessarily satisfies φ.
- s possibly satisfies φ but not necessarily satisfies φ.
- s not possibly satisfies φ.

Since the abstraction of a system preserves some information of the original one, the idea is to prove properties on the abstract and then to infer the result for the original. Since action labels occur in μ-calculus formulas, formulas over A are distinct from formulas over $\widehat{A}$. Therefore, we need to define the meaning of the satisfaction relation of an abstract formula on a concrete state: $\llbracket \widehat{\varphi} \rrbracket_\xi$ where ξ is either h_A or α_A depending on whether we use a homomorphism or a Galois Connection. $\llbracket \widehat{\varphi} \rrbracket_\xi$ gives the set of concrete states that (*necessarily* or *possibly*) satisfy an abstract formula. An extract of the *necessarily* semantics is given below, note that for the rest of the cases (T, F, $\wedge$, $\vee$ and fixpoints) the definition does not change, and the *possibly* semantics are dual:

$$\llbracket [\widehat{a}]\widehat{\varphi}\rrbracket_\xi^{nec} = \{s \mid \forall r, a.\, \widehat{a} \preccurlyeq \xi(\{a\}) \wedge s \xrightarrow{a}_\diamond r \Rightarrow r \in \llbracket\widehat{\varphi}\rrbracket_\xi^{nec}\}$$

$$\llbracket \langle \widehat{a}\rangle\widehat{\varphi}\rrbracket_\xi^{nec} = \{s \mid \exists r, a.\, \xi(\{a\}) \preccurlyeq \widehat{a} \wedge s \xrightarrow{a}_\Box r \wedge r \in \llbracket\widehat{\varphi}\rrbracket_\xi^{nec}\}$$

A concrete state s necessarily satisfies the abstract formula $\widehat{\varphi}$, denoted by $s \models_\xi^{nec} \widehat{\varphi}$, iff $s \in \llbracket\widehat{\varphi}\rrbracket_\xi^{nec}$. And dually, $s \models_\xi^{pos} \widehat{\varphi}$, iff $s \in \llbracket\widehat{\varphi}\rrbracket_\xi^{pos}$. Now we can give the property preservation result:

Theorem 6 Let P be the MLTS $(S, A, \rightarrow_\diamond, \rightarrow_\Box, s_0)$, X be either a homomorphism $H = \langle h_S, h_A\rangle$ between (S, A) and $(\widehat{S}, \widehat{A})$ [in which case ξ stands for h] or a Galois Connection $G = \langle(\alpha_S, \gamma_S), (\alpha_A, \gamma_A)\rangle$ between $(\mathcal{P}(S), \mathcal{P}(A))$ and $(\widehat{S}, \widehat{A})$ [in which case ξ stands for α] and let $\widehat{P}\!\downarrow$ (over $\widehat{S}$ and $\widehat{A}$) be the minimal (restricted) abstraction of P. And finally let $\widehat{\varphi}$ be a formula over $\widehat{A}$, then for all $p \in S$ and $\widehat{p} \in \widehat{S}$ such that $\xi(\{p\}) \preccurlyeq \widehat{p}$:

- $\widehat{p} \models^{nec} \widehat{\varphi} \Rightarrow p \models_\xi^{nec} \widehat{\varphi}$

- $\widehat{p} \not\models^{pos} \widehat{\varphi} \Rightarrow p \not\models_\xi^{pos} \widehat{\varphi}$

The proof follows from the fact that every *may* trace of P is mimicked on $\widehat{P}$ by some related states and, on the other hand, every *must* trace of $\widehat{P}$ is present in P. We refer to the technical report [19] for the proof.

The theorem states that we can infer the satisfaction of a formula on a concrete system if it is *necessarily* satisfied on the abstract. Furthermore, if the formula is not *possibly* satisfied on the abstract it will not hold on the concrete either, so it can be refuted. For example, the two presented abstractions (by homomorphism and by Galois Connection) prove: "*It is possible to write something if the buffer is empty*" expressed as $e \models^{nec} \langle\widehat{w}\rangle T$, which means that in the concrete system s_0 either satisfies $\langle w(0)\rangle T$ or $\langle w(1)\rangle T$ or $\langle w(2)\rangle T$ or ... Furthermore, we can prove on the abstract $f \not\models^{pos} \langle\widehat{w}\rangle T$ which means that in the concrete, all the corresponding concrete states satisfy neither $\langle w(0)\rangle T$ nor $\langle w(1)\rangle T$ nor $\langle w(2)\rangle T$ nor ...

In general, abstractions produced by using Galois Connections preserve more information than the ones generated by homomorphisms, however the state space reduction is stronger in the latter case. For example, the abstraction with the Galois Connection can prove *absence of deadlock* since there is a *must* transition from every state that *may* be reached, however the abstraction done by the homomorphism is not able to prove the property for there is no *must* transition from the state *middle* so we cannot infer the existence of some transition from the related states in the concrete system.

Abstract approximations also preserve properties, this idea is stated in the following lemma:

Lemma 7 *Given two MLTSs P and Q, over the same sets of states and labels S and A, with $P \sqsubseteq_\preccurlyeq Q$ then for all p and q in S such that $p \preccurlyeq q$ and for all formula φ then:*

- $q \models_Q^{nec} \varphi \Rightarrow p \models_P^{nec} \varphi$

- $q \not\models_Q^{pos} \varphi \Rightarrow p \not\models_P^{pos} \varphi$

This result is useful because, by performing symbolic abstractions (see next section) we generate approximations of the minimal abstraction, so the lemma states that we can still infer the satisfaction/refutation of the properties from the approximation to the original.

4   Process Abstraction

μCRL is a combination of process algebra and abstract data types. Data is represented by an *algebraic specification* $\Omega = (\Sigma, E)$, in which Σ denotes a many-sorted *signature* (S, F), where S is the set of sorts and F the set of functions, and E a set of Σ-*equations*, see [10]. All specifications must include the boolean data type with the constants true and false (T and F). From process algebra μCRL inherits the following operators: p.q (perform p and then perform q); p + q (perform arbitrarily either p or q); $\sum_{d:D}$ p(d) (perform $p(d)$ with an arbitrarily chosen d of sort D); p ◁ b ▷ q (if b is true, perform p, otherwise perform q); p || q (run processes p and q in parallel). δ stands for deadlock. Atomic actions may have data parameters. Every μCRL system can be transformed to a special format, called *Linear Process Equation* or *Operator*. An LPE (see definition below) is a single μCRL process which represents the complete system and from which communication and parallel composition operators have been eliminated.

$$X(d : D) = \sum_{i \in I} \sum_{e_i : E_i} a_i(f_i[d, e_i]).X(g_i[d, e_i]) \triangleleft c_i[d, e_i] \triangleright \delta \qquad (1)$$

In the definition, d denotes a vector of parameters d of type D that represents the state of the system at every moment. We use the keyword *init* to declare the initial vector of values of d. The process is composed by a finite number I of summands, every summand i has a list of local variables e_i, of possibly infinite domains, and it is of the following form: a condition $c_i[d, e_i]$, if the evaluation of the condition is true the process executes the action a_i with the parameter $f_i[d, e_i]$ and will move to a new state $g_i[d, e_i]$, which is a vector of terms of type D. $f_i[d, e_i]$, $g_i[d, e_i]$ and $c_i[d, e_i]$ are terms built recursively over variables $x \in [d, e_i]$, applications of function over terms $t = f(t')$ and vectors of terms. To every LPE specification corresponds a labelled transition system. The semantics of the system described by an LPE is given by the following rules:

- $s_0 = init_{lpe}$
- $s \xrightarrow{a} s'$ iff there exist $i \in I$ and $e : E_i$ such that $c_i[s, e] = $ T, $a_i(f_i[s, e]) = a$ and $g_i[s, e] = s'$

Terms are interpreted over the universe of values $\mathcal{D}$. The following μCRL specification corresponds to the LPE of the example, in which the buffer is modeled by a list:

$$X(l : List) = \sum_{d:D} W(d).X(cons(d,l)) \vartriangleleft length(l) < N \vartriangleright \delta +$$

$$R(head(l)).X(tail(l)) \vartriangleleft 0 < length(l) \vartriangleright \delta$$

4.1 Abstract Interpretation of μCRL: Data Part

In order to abstract μCRL specifications we may define the relations between concrete and abstract values by means of a mapping $H : D \rightarrow \widehat{D}$ or a Galois Connection $(\alpha : \mathcal{P}(\mathcal{D}) \rightarrow \widehat{D}, \gamma : \widehat{D} \rightarrow \mathcal{P}(\mathcal{D}))$.

To abstract data terms, first a mapping $\widehat{}$ on (lists of) data sorts must be provided. $\widehat{D}$ represents the sort to which D is abstracted. We require that $\widehat{Bool} = Bool$. Next, for each function symbol f, its abstract counterpart $\widehat{f}$ must be provided. In particular, if $f : D \rightarrow E$ then $\widehat{f} : \mathcal{P}(\widehat{D}) \rightarrow \mathcal{P}(\widehat{E})$. These ingredients allow to extend $\widehat{}$ to data terms, by replacing all symbols by their abstract counterpart. In particular, a data term $t : D$ with variables $x : E$ is abstracted to a data term $\widehat{t} : \mathcal{P}(\widehat{D})$ with variables $\widehat{x} : \mathcal{P}(\widehat{E})$.

We now explain why we lifted all symbols to sets of abstract sorts. For example, in our buffer specification we may have a function S which computes the successor of a natural number. The abstract version of S may be defined as follows:

- For the homomorphism case: $\widehat{S}(empty) = middle$, $\widehat{S}(middle) = middle$ or $\widehat{S}(middle) = full$. It will be undefined for $full$.
- For the Galois Connection approach: $\widehat{S}(empty) = middle$, $\widehat{S}(middle) = nonEmpty$, $\widehat{S}(nonFull) = nonEmpty$, $\widehat{S}(nonEmpty) = nonEmpty$, $\widehat{S}(full) = \top$, $\widehat{S}(\bot) = \bot$ and $\widehat{S}(\top) = \top$.

Abstract interpretation of functions may add non-determinism to the system, for example $\widehat{S}(middle)$ in the homomorphism case may return different values ($middle$ and $full$). Furthermore, not all the sorts of a specification need to be abstracted, for example, a predicate $\widehat{empty?}$ applied to $empty$ will return true, however, applied to $nonFull$ it can be either true or false. To deal with these two considerations, we have lifted abstract functions to return sets of values. So for the homomorphism case, $\widehat{S}(\{middle\}) = \{middle, full\}$. For the Galois case, $\widehat{S}(\{empty\}) = \{nonEmpty\}$ and $\widehat{empty?}(nonFull) = \{T, F\}$.

Not all possible abstract interpretations are correct; in order to generate *safe* abstractions the data terms involved in the specification and their abstract versions have to satisfy a formal requirement, usually called *safety condition*. The condition for the homomorphism function and Galois connections are expressed as follows (note that in the second case t is applied pointwisely to sets).

- Homomorphism: for all d in D. $H(t[d]) \in \widehat{t}[\{H(d)\}]$
- Galois connection: for all $\widehat{d}$ in $\widehat{D}$. $\widehat{t}[\widehat{d}] \succcurlyeq \alpha(t[\gamma(\widehat{d})])$

4.2   Abstract Interpretation of μCRL: Process Part

We will now present a new format, the *Modal Linear Process Equation* (MLPE), and the abstraction mapping from LPEs to MLPEs. An MLPE has the following form:

$$X(d : \mathcal{P}(D)) = \sum_{i \in I} \sum_{e_i : E_i} a_i(F_i[d, e_i]).X(G_i[d, e_i]) \lhd C_i[d, e_i] \rhd \delta \qquad (2)$$

The definition is similar to the one of *Linear Process Equation*, the difference is that the state is represented by a list of powersets of abstract values and for every i: C_i returns a non-empty set of booleans, G_i a non-empty set of states and F_i also a non-empty set of action parameters. Actions are parameterized with sets of values, as well. From an MLPE we can generate a *Modal Labelled Transition System* following these semantic rules:

- $S_0 = \text{init}_{mlpe}$
- $S \xrightarrow{A}_\square S'$ iff there exist $i \in I$ and $e \in E_i$ such that $\mathbf{F} \notin C_i[S, e]$, $A = a(F_i[S, e])$ and $S' = G_i[S, e]$
- $S \xrightarrow{A}_\diamond S'$ iff there exist $i \in I$ and $e \in E_i$ such that $\mathbf{T} \in C_i[S, e]$, $A = a(F_i[S, e])$ and $S' = G_i[S, e]$

To compute an abstract interpretation of a linear process, we define the operator " $\bar{}$ ": $LPE \to MLPE$ that pushes the abstraction through the process operators till the data part:

$p = X(t)$ then $\bar{p} = X(\hat{t})$ where X is a process name
$p = a(t)$ then $\bar{p} = \bar{a}(\hat{t})$ where a is an action label
$p = p_0 + \ldots + p_n$ then $\bar{p} = \bar{p}_0 + \ldots + \bar{p}_n$
$p = p_l \lhd t_c \rhd p_r$ then $\bar{p} = \bar{p}_l \lhd \hat{t}_c \rhd \bar{p}_r$

$p = p_0.p_1$ then $\bar{p} = \bar{p}_0.\bar{p}_1$
$p = \delta$ then $\bar{p} = \delta$
$p = \sum_{e:E} p$ then $\bar{p} = \sum_{\hat{e}:\hat{E}} \bar{p}$

MLPEs allow to capture in a uniform way both approaches: Galois Connection and Homomorphism as well as the combination of both consisting in the lifting of a mapping to a Galois Connection. The lifting is done by considering the abstract domain as a lattice over the powerset of the abstract values ordered by subset inclusion and $\alpha(X)$ will be defined as $\{H(x) \mid x \in X\}$ and $\gamma(\hat{X})$ as $\{x \mid H(x) \in \hat{X}\}$. In the example, the abstract values $nonEmpty$ and $nonFull$ will be captured by $\{middle, full\}$ and $\{empty, middle\}$ respectively. The successor of $nonFull$ will be the union of the successor of $empty$ and $middle$: $\{middle, full\}$. In this example, the lifting of the homomorphism saves the extra effort of defining abstract functions. In case we use a plain homomorphism (without lifting it to a Galois Connection), we restrict the semantics of the MLPE by letting S_0, S, A and S' be only singleton sets. The MLPE below models the buffer example:

$$X(\hat{l} : \mathcal{P}(\widehat{List})) = \sum_{\hat{d}:\hat{D}} \widehat{w}(\hat{d}).X(\widehat{cons}(\hat{d}, \hat{l})) \lhd \widehat{length}(\hat{l}) < \hat{N} \rhd \delta +$$

$$\widehat{r}(\widehat{head}(\hat{l})).X(\widehat{tail}(\hat{l})) \lhd \hat{0} < \widehat{length}(\hat{l}) \rhd \delta$$

Just considering that all functions are pointwisely extended in order to deal with sets of values, the above MLPE can be used equally for any kind of relation between the data domains: homomorphisms, arbitrary Galois Connections and lifted homomorphisms. The following theorem asserts that the abstract interpretation of a linear process produces an abstract approximation of the (restricted) minimal abstraction of the original.

Theorem 8 *Let* lpe *be Linear Process Equation (as defined in (1)),* mlpe *a Modal Linear Process Equation (as defined in (2)) and* X *an abstraction relation between their data domains (where* X *is either a homomorphism or a Galois Connection). Then if:*

1. mlpe *is the abstract interpretation of* lpe (mlpe $= \overline{\text{lpe}}$),
2. mlts *is the* concrete *MLTS generated from* lpe
3. $\widehat{\text{mlts}\!\downarrow}$ *is the (restricted) minimal w.r.t* X *of* mlts
4. *All pairs* (f, F), (g, G) *and* (c, C) *satisfy the safety condition.*

 - *Then, the MLTS (* $\widehat{\text{mlts}}$ *) generated from* mlpe *is an abstraction of* $\widehat{\text{mlts}\!\downarrow}$*, i.e,* $(\widehat{\text{mlts}\!\downarrow} \sqsubseteq_{\preccurlyeq} \widehat{\text{mlts}})$

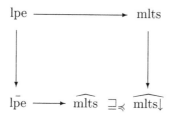

The proof is done by checking that every *may* transition generated by the abstract *Modal Linear Process Equation* has at least one more precise counterpart in the (restricted) minimal abstraction of the concrete system (and the other way around for the *must* transitions). By Theorem 6 and Lemma 7, we can prove (refute) properties for *lpe* by considering *mlpe* directly.

5 The Bounded Retransmission Protocol

The BRP is a simplified variant of a Philips' telecommunication protocol that allows to transfer large files across a lossy channel. Files are divided in packets and are transmitted by a sender through the channel. The receiver acknowledges every delivered data packet. Both data and confirmation messages, may be lost. The sender will attempt to retransmit each packet a limited number of times.

    The protocol has a number of parameters, such as the length of the lists, the maximum number of retransmissions and the contents of the data, that cause the state space of the system to be infinite and limit the application of automatic verification techniques such as model checking. Following the ideas presented along the paper, we abstract the specification by eliminating some uninteresting information. We base our solution on the μCRL model presented in the paper [11], in which Groote and van de Pol proved using algebraic methods that the model is branching bisimilar to the desired external behavior also specified in μCRL. This proof requires a strong and creative human interaction in order to be accomplished. However by computing an abstraction of the original specification we can automatically model check some properties.

The system is composed by a sender that gets a file which consists of a list of elements. It delivers the file frame by frame through a channel. The receiver sends an acknowledgment for each frame, when it receives a packet it delivers it to the external receiver client attaching a positive indication I_{fst}, I_{inc} or I_{ok}. The sender, after each frame, waits for the acknowledgments, if the confirmation message does not arrive, it retransmits the packet. If the transmission was successful, i.e. all the acknowledgments have arrived, then the sender informs the sending client with a positive indication. When the maximum number of retransmissions is exceeded, the transmission is canceled and I_{nok} is sent to the exterior by both participants. If the confirmation of the last frame is lost the sender cannot know whether the receiver has received the complete list, therefore it sends "I don't know" to the sending client, I_{dk}.

We are interested in proving that the external indications delivered by the sender and the receiver are "consistent". For that purpose, we chose an abstraction that abstracts away the data stored in the file and maps the list to three critical values: abs_empty, abs_one, abs_more. The first one denotes the empty list, abs_one when it has only one element, and abs_more when it has more than one. The maximum number of retransmissions is removed (abstracted to a single abstract value abs_N) which makes the sender to non-deterministically choose between resending a lost packet or giving up the transmission.

The next table presents the abstract specification of the data and the abstraction mappings. The concrete specification of the list and the sort Nat are standard. The function indl indicates by a bit whether a list is at its end (it is empty or has only one element) or not.

sort abs_D, abs_Nat, abs_List	$tail(abs\_more) =$
func $abs\_D :\rightarrow abs\_D$	$\quad \{abs\_more, abs\_one\}$
$\quad abs\_N :\rightarrow abs\_Nat$	$last(abs\_empty) = \{t\}$
$\quad abs\_empty :\rightarrow abs\_List$	$last(abs\_one) = \{t\}$
$\quad abs\_one :\rightarrow abs\_List$	$last(abs\_more) = \{f\}$
$\quad abs\_more :\rightarrow abs\_List$	$indl(abs\_empty) = \{e_1\}$
$\quad succ : abs\_Nat \rightarrow \mathcal{P}(abs\_Nat)$	$indl(abs\_one) = \{e_1\}$
$\quad lt : abs\_Nat \times abs\_Nat \rightarrow \mathcal{P}(Bool)$	$indl(abs\_more) = \{e_0\}$
$\quad head : abs\_List \rightarrow \mathcal{P}(abs\_D)$	**func** $H : List \rightarrow abs\_List$
$\quad tail : abs\_List \rightarrow \mathcal{P}(abs\_List)$	$\quad H : D \rightarrow abs\_D$
$\quad last : abs\_List \rightarrow \mathcal{P}(Bool)$	$\quad H : Nat \rightarrow abs\_Nat$
$\quad indl : abs\_List \rightarrow \mathcal{P}(Bit)$	**var** $l : List$
var $\hat{l} :\rightarrow abs\_List$	$\quad d, d' : D$
$\quad \hat{m}, \hat{n} : abs\_Nat$	$\quad m : Nat$
rew $succ(\hat{m}) = \{abs\_N\}$	**rew** $H(empty) = abs\_empty$
$\quad lt(\hat{m}, \hat{n}) = \{t, f\}$	$\quad H(add(d, empty)) = abs\_one$
$\quad head(\hat{l}) = \{abs\_D\}$	$\quad H(add(d, add(d', l))) = abs\_more$
$\quad tail(abs\_empty) = \{abs\_empty\}$	$\quad H(d) = abs\_D$
$\quad tail(abs\_one) = \{abs\_empty\}$	$\quad H(m) = abs\_N$

This abstraction allows to reason about the execution of the final part of the protocol without knowing the exact content of data files or the number of retrials. For example the following *safety* property: *"after a positive notification by the receiver, the sender cannot send a negative one"* is *necessarily* satisfied by the abstract system. We express the property in the Regular Alternation-free μ-calculus, which is a fragment of the modal μ-calculus extended with regular expressions, as follows[4]:

$$[T*.\text{'R}_{may}(.*, \{I_{ok}\})\text{'}.(\neg\text{'S}.*\text{'})*.\text{'S}_{may}(\{I_{nok}\})\text{'}]\,F$$

The following *liveness* property expresses that: *"After a negative notification by the receiver, there exists a path which leads to a negative or* don't know *notification by the sender"*.

$$[T*.\text{'R}_{may}(.*, \{I_{nok}\})\text{'}]\,\langle\text{'}.*_{must}\text{'}*.(\text{'S}_{must}(\{I_{nok}\})\text{'} \vee \text{'S}_{must}(\{I_{dk}\})\text{'})\rangle\,T$$

The next property is stronger than the previous, instead of only requesting that there exists a path it states that the expected sender notification is inevitably achieved:

$$[T*.\text{'R}_{may}(.*, \{I_{nok}\})\text{'}]\,\mu X(\langle\text{'}.*_{must}\text{'}\rangle\,T \wedge [\neg(\text{'S}_{must}(\{I_{nok}\})\text{'} \vee \text{'S}_{must}(\{I_{dk}\})\text{'})]\,X)$$

These three properties are *necessarily* satisfied in the abstract system, therefore we can infer its satisfaction in the original one. However, the following property which states that *"after a positive notification by the receiver there exists a path which leads to a positive or* don't know *notification by the sender"* is not satisfied in the abstract system. The reason is that we have abstracted away the maximum number of retransmissions, therefore if all the acknowledgments are lost the sender can retransmit the frames forever:

$$[T*.\text{'R}_{may}(.*, \{I_{ok}\})\text{'}]\,\langle\text{'}.*_{must}\text{'}*.(\text{'S}_{must}(\{I_{ok}\})\text{'} \vee \text{'S}_{must}(\{I_{dk}\})\text{'})\rangle\,T$$

Other papers have verified some properties of the protocol using abstract interpretation, we refer among others to [17,?]. The approach of Manna et al. is based on automatic predicate abstractions and is limited to the proof of invariants. However Dams and Gerth propose a number of creative abstractions in order to prove the satisfaction of some safety properties about the sequentiality on the delivering of the frames.

6   Conclusion

In order to apply the abstract interpretation framework for reactive systems [5, 12] to μCRL processes, we extended it with the explicit abstraction of action labels. This required non-trivial changes in most definitions. A μCRL specification in LPE-form can be abstracted to a modal LPE; the state space of this MLPE corresponds to a reduced MLTS, approximating the original LTS. This

---

[4] We have used the CADP model checker to prove the properties we describe, therefore we present the formulas with the CADP syntax. The reader not familiar with the logic can safely skip them.

approximation can be used to verify and refute safety as well as liveness formulas for the original system.

The resulting approach incorporates the homomorphism approach (which is easier to understand and use) and the Galois Connection approach (which preserves more properties, especially liveness properties). A user-defined homomorphism can also be lifted to a Galois Connection automatically, combining ease with precision.

We already have a working prototype implementation, which will be described in a separate paper. It is based on a projection of MLPEs to LPEs, in order to reuse existing state space generation tools. We will apply the techniques from [12] in order to translate the three-valued model checking problem to regular model checking, in order to also reuse a model checker for the modal μ-calculus, e.g. CADP [8]. Another interesting question, optimality of abstractions, can in principle be addressed along the lines of [5].

Our theory allows the collection of standard data abstractions with accompanying safety criteria proofs. Such abstraction libraries can be freely used in various protocols, without additional proof obligations. This will enhance automatic verification of protocols in specialized domains.

Model checking becomes more and more crucial for the correctness of software, but in practice additional techniques, such as abstraction, are needed. This may affect the correctness and modularity of the resulting verification methodology and tools. We support modularity by implementing abstraction as an LPE to LPE transformation, which can be composed freely by other existing transformations [1]. We feel that it is important to provide a rigorous basis for verification technology, so we have checked the main part of the results in this paper in the theorem prover PVS.

References

1. S.C.C. Blom, W. Fokkink, J.F. Groote, I.A. van Langevelde, B. Lisser, and J.C. van de Pol. μCRL: A toolset for analysing algebraic specifications. In *CAV*, LNCS 2102, pages 250–254, 2001.
2. S.C.C. Blom, J.F. Groote, I.A. van Langevelde, B. Lisser, and J.C. van de Pol. New developments around the μCRL tool set. In *ENTCS 80*, 2003.
3. E. M. Clarke, O. Grumberg, and D. E. Long. Model checking and abstraction. In *TOPLAS*, ACM 16, pages 1512–1542, 1994.
4. P. Cousot and R. Cousot. Abstract interpretation: A unified lattice model for static analysis of programs by construction of approximation of fixed points. In *POPL*, ACM, pages 238–252, 1977.
5. D. Dams. Abstract Interpretation and Partition Refinement for Model Checking. PhD thesis, Eindhoven University of Technology, 1996.
6. D. Dams and R. Gerth. The bounded retransmission protocol revisited. In *ENTCS 9*, 2000.
7. A. Fantechi, S. Gnesi, and D. Latella. Towards automatic temporal logic verification of value passing process algebra using abstract interpretation. In *CONCUR*, LNCS 1119, pages 562–578, 1996.

8. J.-C. Fernandez, H. Garavel, A. Kerbrat, L. Mounier, R. Mateescu, and M. Sighire-anu. CADP – a protocol validation and verification toolbox. In *CAV*, LNCS 1102, pages 437–440, 1996.

9. P. Godefroid, M. Huth, and R. Jagadeesan. Abstraction-based model checking using modal transition systems. In *CONCUR*, LNCS 2154, pages 426–440, 2001.

10. J. F. Groote and A. Ponse. The syntax and semantics of μCRL. In *ACP*, Workshops in Computing Series, pages 26–62, 1995.

11. J. F. Groote and J.C. van de Pol. A bounded retransmission protocol for large data packets. In *AMAST*, LNCS 1101, pages 536–550, 1996.

12. M. Huth, R. Jagadeesan, and D. Schmidt. Modal transition systems: a foundation for three-valued program analysis. In *ESOP*, LNCS 2028, pages 155–169, 2001.

13. N. D. Jones and F. Nielson. Abstract interpretation: A semantics-based tool for program analysis. In *Handbook of Logic in Computer Science*, pages 527–636. 1995.

14. D. Kozen. Results on the propositional μ-calculus. In *ICALP*, LNCS 140, pages 348–359, 1982.

15. K. G. Larsen and B. Thomsen. A modal process logic. In *LICS*, pages 203–210. IEEE, 1988.

16. C. Loiseaux, S. Graf, J. Sifakis, A. Bouajjani, and S. Bensalem. Property preserving abstractions for the verification of concurrent systems. *Formal Methods in System Design 6*, pages 11–44, 1995.

17. Z. Manna, M. Colon, B. Finkbeiner, H. Sipma, and T. E. Uribe. Abstraction and modular verification of infinite-state reactive systems. In *RTSE*, LNCS 1526, pages 273–292, 1997.

18. S. Owre, S. Rajan, J.M. Rushb, N. Shankar, and M.K. Srivas. PVS: Combining specification, proof checking, and model checking. In *CAV*, LNCS 1102, pages 411–414, 1996.

19. J.C. van de Pol and M. Valero Espada. Modal abstraction in μCRL. Technical Report SEN-R0401, CWI, 2004.

Semantics of Plan Revision in Intelligent Agents

M. Birna van Riemsdijk[1], John-Jules Charles Meyer[1], and Frank S. de Boer[1,2,3]

[1] ICS, Utrecht University, The Netherlands
[2] CWI, Amsterdam, The Netherlands
[3] LIACS, Leiden University, The Netherlands

Abstract. In this paper, we give an operational and denotational semantics for a 3APL meta-language, with which various 3APL interpreters can be programmed. We moreover prove equivalence of these two semantics. Furthermore, we relate this 3APL meta-language to object-level 3APL by providing a specific interpreter, the semantics of which will prove to be equivalent to object-level 3APL.

1 Introduction

An agent is commonly seen as an encapsulated computer system that is situated in some environment and that is capable of flexible, autonomous action in that environment in order to meet its design objectives [19]. Autonomy means that an agent encapsulates its state and makes decisions about what to do based on this state, without the direct intervention of humans or others. Agents are situated in some environment which can change during the execution of the agent. This requires *flexible* problem solving behaviour, i.e. the agent should be able to respond adequately to changes in its environment. Programming flexible computing entities is not a trivial task. Consider for example a standard procedural language. The assumption in these languages is, that the environment does not change while some procedure is executing. If problems do occur during the execution of a procedure, the program might throw an exception and terminate (see also [20]). This works well for many applications, but we need something more if change is the norm and not the exception.

A philosophical view that is well recognized in the AI literature is, that rational behaviour can be explained in terms of the concepts of *beliefs*, *goals* and *plans*[1] [1,13,2]. This view has been taken up within the AI community in the sense that it might be possible to *program* flexible, autonomous agents *using* these concepts. The idea is, that an agent tries to fulfill its goals by selecting appropriate plans, depending on its beliefs about the world. Beliefs should thus represent the world or environment of the agent; the goals represent the state of the world the agent wants to realize and plans are the means to achieve these goals. When programming in terms of these concepts, beliefs can be compared

---

[1] In the literature, also the concepts of desires and intentions are often used, besides or instead of goals and plans, respectively. This is however not important for the current discussion.

C. Rattray et al. (Eds.): AMAST 2004, LNCS 3116, pp. 426–442, 2004.

to the program state, plans can be compared to statements, i.e. plans constitute the procedural part of the agent, and goals can be viewed as the (desired) post-conditions of executing the statement or plan. Through executing a plan, the world and therefore the beliefs reflecting the world will change and this execution should have the desired result, i.e. achievement of goals.

This view has been adopted by the designers of the agent programming language $3APL^2$ [8]. The dynamic parts of a 3APL agent thus consist of a set of beliefs, a plan[3] and a set of goals[4]. A plan can consist of sequences of so-called basic actions and abstract plans. Basic actions change the beliefs[5] if executed and abstract plans can be compared to procedure names. To provide for the possibility of programming flexible behaviour, so-called *plan revision* rules were added to the language. These rules can be compared to procedures in the sense that they have a head (the procedure name) and a body (a plan or statement). The operational meaning of plan revision rules is similar to that of procedures: if the procedure name or head is encountered in a statement or plan, this name or head is replaced by the body of the procedure or rule, respectively (see [4] for the operational semantics of procedure calls). The difference however is, that the head in a plan revision rule can be *any* plan (or statement) and not just a procedure name. In procedural languages it is furthermore usually assumed that procedure names are distinct. In 3APL however, it is possible that multiple rules are applicable at the same time. This provides for very general and flexible plan revision capabilities, which is a distinguishing feature of 3APL compared to other agent programming languages [12, 14, 6].

As argued, we consider these general plan revision capabilities to be an essential part of agenthood. The introduction of these capabilities now gives rise to interesting issues concerning the *semantics of plan execution*, the exploration of which is the topic of this paper.

Semantics of plan execution can be considered on two levels. On the one hand, the semantics of *object-level* 3APL can be studied as a function yielding the result of executing a plan on an initial belief base, where the plan can be revised through plan revision rules during execution. An interesting question is, whether a denotational semantic function can be defined that is compositional in its plan argument.

On the other hand, the semantics of a 3APL *interpreter* language or *meta-language* can be studied, where a plan and a belief base are considered the data on which the interpreter or meta-program operates. This meta-language is the main focus of this paper. To be more specific, we define a meta-language and provide an operational and denotational semantics for it. These will be proven equivalent. We furthermore define a very general interpreter in this language, the semantics of which will prove to be equivalent to the semantics of object-level 3APL.

---

[2] 3APL is to be pronounced as "triple-a-p-l".

[3] In the original version this was a set of plans.

[4] The addition of goals was a recent extension [18].

[5] A change in the environment is a possible "side effect" of the execution of a basic action.

For regular procedural programming languages, studying a specific interpreter language is in general not very interesting. In the context of agent programming languages it however *is*, for several reasons. First of all, 3APL and agent-oriented programming languages in general are non-deterministic by nature. In the case of 3APL for example, it will often occur that several plan revision rules are applicable at the same time. Choosing a rule for application (or choosing whether to execute an action from the plan or to apply a rule if both are possible), is the task of a 3APL interpreter. The choices made, affect the outcome of the execution of the agent. In the context of agents, it is interesting to study various interpreters, as different interpreters will give rise to different *agent types*. An interpreter that for example always executes a rule if possible, thereby deferring action execution, will yield a thoughtful and passive agent. In a similar way, very bold agents can be constructed or agents with characteristics anywhere on this spectrum. These conceptual ideas about various agent types fit well within the agent metaphor and therefore it is worthwhile to study an interpreter language and the interpreters that can be programmed in it (see also [3]).

Secondly, as pointed out by Hindriks [7], differences between various agent languages often mainly come down to differences in their meta-level reasoning cycle or interpreter. To provide for a *comparison* between languages, it is thus important to separate the semantic specification of object-level and meta-level execution.

Finally, and this was the original motivation for this work, we hope that the specification of a denotational semantics for the meta-language might shed some light onto the issue of specifying a denotational semantics for object-level 3APL. It however seems, contrary to what one might think, that the denotational semantics of the meta-language cannot be used to define a denotational semantics for object-level 3APL. We will elaborate on this issue in section 6.2.

2   Syntax

2.1   Object-Level

As stated in the introduction, the latest version of 3APL incorporates beliefs, goals and plans. In this paper, we will however consider a version of 3APL with only beliefs and plans as was defined in [8]. The reason is, that in this paper we focus on the semantics of plan execution, for the treatment of which only beliefs and plans will suffice. The language defined in [8] is a first-order language, a propositional and otherwise slightly simplified version of which we will use in this paper.

In the sequel, a language defined by inclusion shall be the smallest language containing the specified elements.

Definition 1. *(belief bases)* Assume a propositional language $\mathcal{L}$ with typical formula ψ and the connectives $\wedge$ and $\neg$ with the usual meaning. Then the set of possible belief bases Σ with typical element σ is defined to be $\wp(\mathcal{L})$.

Definition 2. *(plans)* Assume that a set BasicAction with typical element a is given, together with a set AbstractPlan. The symbol E denotes the empty plan. Then the set of plans Π with typical element π is defined as follows:

- $\{E\} \cup \mathsf{BasicAction} \cup \mathsf{AbstractPlan} \subseteq \Pi$,
- if $c \in (\{E\} \cup \mathsf{BasicAction} \cup \mathsf{AbstractPlan})$ and $\pi \in \Pi$ then $c \,;\pi \in \Pi$ [6].

A plan $E; \pi$ is identified with the plan π.

For reasons of presentation and technical convenience, we exclude non-deterministic choice and test from plans. This is no fundamental restriction as non-determinism is introduced by plan revision rules (to be introduced below). Furthermore, tests can be modelled as basic actions that do not affect the state if executed (for semantics of basic actions see definition 8).

A plan and a belief base can together constitute the so-called mental state of a 3APL agent. A mental state can be compared to what is usually called a configuration in procedural languages, i.e. a statement-state pair.

Definition 3. *(mental states)* Let Σ be the set of belief bases and let Π be the set of plans. Then $\Pi \times \Sigma$ is the set S of possible mental states of a 3APL agent.

Definition 4. *(plan revision (PR) rules)* A PR rule ρ is a triple $\pi_h \mid \psi \rightsquigarrow \pi_b$ such that $\psi \in \mathcal{L}$, $\pi_h, \pi_b \in \Pi$ and $\pi_h \neq E$.

Definition 5. *(3APL agent)* A 3APL agent $\mathcal{A}$ is a tuple $\langle \pi_0, \sigma_0, \mathsf{BasicAction},$ AbstractPlan, Rule, $\mathcal{T} \rangle$ where $\langle \pi_0, \sigma_0 \rangle$ is the initial mental state, BasicAction, AbstractPlan and Rule are sets of basic actions, abstract plans and PR rules respectively and $\mathcal{T} : (\mathsf{BasicAction} \times \Sigma) \to \Sigma$ is a belief update function.

In the following, when referring to agent $\mathcal{A}$, we will assume this agent to have a set of basic actions BasicAction, a set of abstract plans AbstractPlan, a set of PR rules Rule and a belief update function $\mathcal{T}$.

2.2   Meta-level

In this section, we define the meta-language that can be used to write 3APL interpreters. The programs that can be written in this language will be called *meta-programs*. Like regular imperative programs, these programs are state transformers. The kind of states they transform however do not simply consist of an assignment of values to variables like in regular imperative programming, but the states that are transformed are 3APL mental states. In section 3.1, we will define the operational semantics of our meta-programs. We will do this using the concept of a *meta-configuration*. A meta-configuration consists of a meta-program and a mental state, i.e. the meta-program is the procedural part and the mental state is the "data" on which the meta-program operates.

The basic elements of meta-programs are the *execute* action and the *apply(ρ)* action (called *meta-actions*). The *execute* action is used to specify that a basic

---

[6] For technical convenience, plans are defined to have a list structure.

action from the plan of an agent should be executed. The $apply(\rho)$ action is used to specify that a PR rule ρ should be applied to the plan. Composite meta-programs can be constructed in a standard way.

Below, the meta-programs and meta-configurations for agent $\mathcal{A}$ are defined.

Definition 6. *(meta-programs)* We assume a set $Bexp$ of boolean expressions with typical element b. Let $b \in Bexp$ and $\rho \in \mathsf{Rule}$, then the set $Prog$ of meta-programs with typical element P is defined as follows:

$$P ::= execute \mid apply(\rho) \mid \texttt{while } b \texttt{ do } P \texttt{ od} \mid P_1; P_2 \mid P_1 + P_2.$$

Definition 7. *(meta-configurations)* Let $Prog$ be the set of meta-programs and let S be the set of mental states. Then $Prog \times S$ is the set of possible meta-configurations.

3   Operational Semantics

In [8], the operational semantics of 3APL is defined using transition systems [11]. A transition system for a programming language consists of a set of derivation rules for deriving transitions for this language. A transition is a transformation of one configuration into another and it corresponds to a single computation step. In the following section, we will repeat the transition system for 3APL given in [8] (adapted to fit our simplified language) and we will call it the *object-level* transition system. We will furthermore give a transition system for the meta-programs defined in section 2.2 (the *meta-level* transition system). Then in the last section, we will define the operational semantics of the object- and meta-programs using the defined transition systems.

3.1   Transition Systems

The transition systems defined in the following sections assume 3APL agent $\mathcal{A}$.

Object-Level. The object-level transition system (Trans_o) is defined by the rules given below. The transitions are labeled to denote the kind of transition.

Definition 8. *(action execution)* Let $a \in \mathsf{BasicAction}$.

$$\frac{\mathcal{T}(a, \sigma) = \sigma'}{\langle a; \pi, \sigma \rangle \rightarrow_{execute} \langle \pi, \sigma' \rangle}$$

In the next definition, we use the operator $\bullet$. The statement $\pi_1 \bullet \pi_2$ denotes a plan of which π_1 is the first part and π_2 is the second, i.e. π_1 is the prefix of this plan. We need this operator because plans are defined to have a list structure (see definition 2).

Definition 9. *(rule application)* Let $\rho : \pi_h \mid \psi \rightsquigarrow \pi_b \in \mathsf{Rule}$.

$$\frac{\sigma \models \psi}{\langle \pi_h \bullet \pi, \sigma \rangle \rightarrow_{apply(\rho)} \langle \pi_b \bullet \pi, \sigma \rangle}$$

Meta-level. The meta-level transition system (Trans_m) is defined by the rules below, specifying which transitions from one meta-configuration to another are possible. As for the object-level transition system, the transitions are labelled to denote the kind of transition.

An *execute* meta-action is used to execute a basic action. It can thus only be executed in a mental state, if the first element of the plan in that mental state is a basic action. As in the object-level transition system, the basic action a must be executable and the result of executing a on belief base σ is defined using the function $\mathcal{T}$. After executing the meta-action *execute*, the meta-program is empty and the basic action is gone from the plan. Furthermore, the belief base is changed as defined through $\mathcal{T}$.

Definition 10. *(action execution)* Let $a \in \mathsf{BasicAction}$.

$$\frac{\mathcal{T}(a, \sigma) = \sigma'}{\langle execute, (a; \pi, \sigma)\rangle \to_{execute} \langle E, (\pi, \sigma')\rangle}$$

A meta-action $apply(\rho)$ is used to specify that PR rule ρ should be applied. It can be executed in a mental state if ρ is applicable in that mental state. The execution of the meta-action in a mental state results in the plan of that mental state being changed as specified by the rule.

Definition 11. *(rule application)* Let $\rho : \pi_h \mid \psi \leadsto \pi_b \in \mathsf{Rule}$.

$$\frac{\sigma \models \psi}{\langle apply(\rho), (\pi_h \bullet \pi, \sigma)\rangle \to_{apply(\rho)} \langle E, (\pi_b \bullet \pi, \sigma)\rangle}$$

In order to define the transition rule for the `while` construct, we first need to specify the semantics of boolean expressions $Bexp$.

Definition 12. *(semantics of boolean expressions)* We assume a function $\mathcal{W}$ of type $Bexp \to (S \to W)$ yielding the semantics of boolean expressions, where W is the set of truth values $\{tt, ff\}$ with typical formula β.

The transition for the `while` construct is then defined in a standard way below. The transition is labeled with *idle*, to denote that this is a transition that does not have a counterpart in the object-level transition system.

Definition 13. *(while)*

$$\frac{\mathcal{W}(b)(s)}{\langle \texttt{while } b \texttt{ do } P \texttt{ od}, s\rangle \to_{idle} \langle P; \texttt{while } b \texttt{ do } P \texttt{ od}, s\rangle}$$

$$\frac{\neg\mathcal{W}(b)(s)}{\langle \texttt{while } b \texttt{ do } P \texttt{ od}, s\rangle \to_{idle} \langle E, s\rangle}$$

The transitions for sequential composition and non-deterministic choice are defined as follows in a standard way. The variable x is used to pass on the type of transition through the derivation.

Definition 14. *(sequential composition)* Let $x \in \{execute, apply(\rho), idle \mid \rho \in$ Rule$\}$.

$$\frac{\langle P_1, s \rangle \to_x \langle P_1', s' \rangle}{\langle P_1; P_2, s \rangle \to_x \langle P_1'; P_2, s' \rangle}$$

Definition 15. *(non-deterministic choice)* Let $x \in \{execute, apply(\rho), idle \mid \rho \in$ Rule$\}$.

$$\frac{\langle P_1, s \rangle \to_x \langle P_1', s' \rangle}{\langle P_1 + P_2, s \rangle \to_x \langle P_1', s' \rangle} \qquad \frac{\langle P_2, s \rangle \to_x \langle P_2', s' \rangle}{\langle P_1 + P_2, s \rangle \to_x \langle P_2', s' \rangle}$$

3.2   Operational Semantics

Using the transition systems defined in the previous section, transitions can be derived for 3APL and for the meta-programs. Individual transitions can be put in sequel, yielding so called *computation sequences*. In the following definitions, we define computation sequences and we specify the functions yielding these sequences, for the object- and meta-level transition systems. We also define the function κ, yielding the last element of a computation sequence if this sequence is finite and the special state $\perp$ otherwise. These functions will be used to define the operational semantics.

Definition 16. *(computation sequences)* The sets S^+ and S^∞ of respectively finite and infinite computation sequences are defined as follows:

$$S^+ = \{s_1, \ldots, s_i, \ldots, s_n \mid s_i \in S, 1 \leq i \leq n, n \in \mathbb{N}\},$$
$$S^\infty = \{s_1, \ldots, s_i, \ldots \quad \mid s_i \in S, i \in \mathbb{N}\}.$$

Let $S_\perp = S \cup \{\perp\}$ and $\delta \in S^+ \cup S^\infty$. The function $\kappa : (S^+ \cup S^\infty) \to S_\perp$ is defined by:

$$\kappa(\delta) = \begin{cases} \text{last element of } \delta & \text{if } \delta \in S^+, \\ \perp & \text{otherwise.} \end{cases}$$

The function κ is extended to handle sets of computation sequences as follows:

$$\kappa(\{\delta_i \mid i \in I\}) = \{\kappa(\delta_i) \mid i \in I\}.$$

Definition 17. *(functions for calculating computation sequences)* The functions $\mathcal{C}_o$ and $\mathcal{C}_m$ are respectively of type $S \to \wp(S^+ \cup S^\infty)$ and $Prog \to (S \to \wp(S^+ \cup S^\infty))$.

$$\mathcal{C}_o(s) \quad = \{s_1, \ldots, s_n \quad \in \wp(S^+) \mid s \to_{t_1} s_1 \to_{t_2} \cdots \to_{t_n} \langle E, \sigma_n \rangle$$
$$\text{is a finite sequence of transitions in Trans}_o\} \cup$$
$$\{s_1, \ldots, s_i, \ldots \in \wp(S^\infty) \mid s \to_{t_1} s_1 \to_{t_2} \cdots \to_{t_i} s_i \to_{t_{i+1}} \cdots$$
$$\text{is an infinite sequence of transitions in Trans}_o\}$$
$$\mathcal{C}_m(P)(s) = \{s_1, \ldots, s_n \quad \in \wp(S^+) \mid \langle P, s \rangle \to_{x_1} \langle P_1, s_1 \rangle \to_{x_2} \cdots \to_{x_n} \langle E, s_n \rangle$$
$$\text{is a finite sequence of transitions in Trans}_m\} \cup$$
$$\{s_1, \ldots, s_i, \ldots \in \wp(S^\infty) \mid \langle P, s \rangle \to_{x_1} \langle P_1, s_1 \rangle \to_{x_2} \cdots$$
$$\to_{x_i} \langle P_i, s_i \rangle \to_{x_{i+1}} \cdots$$
$$\text{is an infinite sequence of transitions in Trans}_m\}$$

Note that both $\mathcal{C}_o$ and $\mathcal{C}_m$ return sequences of mental states. $\mathcal{C}_o$ just returns the mental states comprising the sequences of transitions derived in Trans_o, whereas $\mathcal{C}_m$ removes the meta-program component of the meta-configurations of the transition sequences derived in Trans_m. The reason for defining these functions in this way is, that we want to prove equivalence of the object- and meta-level transition systems: both yield the same transition sequences with respect to the mental states (or that is for a certain meta-program, see section 4). Also note that for $\mathcal{C}_o$ as well as for $\mathcal{C}_m$, we only take into account infinite sequences and successfully terminating sequences, i.e. those sequences ending in a mental state or meta-configuration with an empty plan or meta-program respectively.

The operational semantics of object- and meta-level programs are functions $\mathcal{O}_o$ and $\mathcal{O}_m$, yielding, for each mental state s and possibly meta-program P, a set of mental states corresponding to the final states reachable through executing the plan of s or executing the meta-program P respectively. If there is an infinite execution path, the set of mental states will contain the element $\perp$.

Definition 18. *(operational semantics)* Let $s \in S$. The functions $\mathcal{O}_o$ and $\mathcal{O}_m$ are respectively of type $S_\perp \to \wp(S_\perp)$ and $Prog \to (S_\perp \to \wp(S_\perp))$.

$$\begin{aligned}
\mathcal{O}_o(s) &= \kappa(\mathcal{C}_o(s)) \\
\mathcal{O}_m(P)(s) &= \kappa(\mathcal{C}_m(P)(s)) \\
\mathcal{O}_o(\perp) &= \mathcal{O}_m(P)(\perp) = \{\perp\}
\end{aligned}$$

Note that the operational semantic functions can take any state $s \in S_\perp$, including $\perp$, as input. This will turn out to be necessary for giving the equivalence result of section 6.

4   Equivalence of $\mathcal{O}_o$ and $\mathcal{O}_m$

In the previous section, we have defined the operational semantics for 3APL and for meta-programs. Using the meta-language, one can write various 3APL interpreters. Here we will consider an interpreter of which the operational semantics will prove to be equivalent to the object-level operational semantics of 3APL. This interpreter for agent $\mathcal{A}$ is defined by the following meta-program.

Definition 19. *(interpreter)* Let $\bigcup_{i=1}^{n} \rho_i = \mathsf{Rule}$, $s \in S$ and let $notEmptyPlan \in Bexp$ be a boolean expression such that $\mathcal{W}(notEmptyPlan)(s) = tt$ if the plan component of s is not equal to E and $\mathcal{W}(notEmptyPlan)(s) = ff$ otherwise. Then the interpreter can be defined as follows.

$$\texttt{while } notEmptyPlan \texttt{ do } (execute + apply(\rho_1) + \ldots + apply(\rho_n)) \texttt{ od}$$

In the sequel, we will use the keyword interpreter to abbreviate this meta-program.

This interpreter thus iterates the execution of a non-deterministic choice between all basic meta-actions, until the plan component of the mental state is empty.

Intuitively, if there is a possibility for the interpreter to execute some meta-action in mental state s, resulting in a changed state s', it is also possible to go from s to s' in an object-level execution through a corresponding object-level transition. At each iteration, an executable meta-action is non-deterministically chosen for execution. The interpreter thus as it were, non-deterministically chooses a path through the object-level transition tree. The possible transitions defined by this interpreter correspond to the possible transitions in the object-level transition system and therefore the object-level operational semantics is equivalent to the meta-level operational semantics of this meta-program[7].

Theorem 1. *($\mathcal{O}_o = \mathcal{O}_m$(interpreter))*

$$\forall s \in S : \mathcal{O}_o(s) = \mathcal{O}_m(\text{interpreter})(s)$$

Proof. We prove a weak bisimulation (see [17]).

Note that it is easy to show that $\mathcal{O}_o = \mathcal{O}_m(P)$ does not hold for all meta-programs P.

5 Denotational Semantics

In this section, we will define the denotational semantics of meta-programs. The method used is the fixed point approach as can be found in Stoy [15]. The semantics greatly resembles the one in De Bakker ([4], Chapter 7) to which we refer for a detailed explanation of the subject.

A denotational semantics for a programming language in general, is, like an operational semantics, a function taking a statement P and a state s and yielding a state (or set of states in case of a non-deterministic language) resulting from executing P in s. The denotational semantics for meta-programs is thus, like the operational semantics of definition 18, a function taking a meta-program P and mental state s and yielding the set of mental states resulting from executing P in s, i.e. a function of type $Prog \rightarrow (S_\perp \rightarrow \wp(S_\perp))$ [8]. Contrary however to an operational semantic function, a denotational semantic function is not defined using the concept of computation sequences and, in contrast with most operational semantics, it *is* defined compositionally [16, 10, 4].

5.1 Preliminaries

In order to define the denotational semantics of meta-programs, we need some mathematical machinery. Most importantly, the domains used in defining the

---

[7] The result only holds if PR rules of the form $E \mid \psi \rightsquigarrow \pi_b$ are excluded from the set of rules under consideration, as was specified in definition 4. A relaxation of this condition would call for a slightly different interpreter to yield the equivalence result. For reasons of space and clarity, we will however not discuss this possibility here.

[8] The type of the denotational semantic function is actually slightly different as will become clear in the sequel, but that is not important for the current discussion.

semantics of meta-programs are designed as so-called complete partial orders (CPO's). A CPO is a set with an ordering on its elements with certain characteristics. We assume the reader is familiar with the notions of partially ordered sets, least upper bounds and chains, in terms of which the concept of a CPO is defined. For a rigorous treatment of the subject, we refer to De Bakker [4].

Definition 20. *(CPO)* A complete partially ordered set is a set C with a partial order $\sqsubseteq$ which satisfies the following requirements:

1. there is a least element with respect to $\sqsubseteq$, i.e. an element $\bot \in C$ such that $\forall c \in C : \bot \sqsubseteq c$,
2. each chain $\langle c_i \rangle_{i=0}^{\infty}$ in C has a least upper bound $(\bigsqcup_{i=0}^{\infty} c_i) \in C$.

The semantics of meta-programs will be defined using the notion of the least fixed point of a function on a CPO.

Definition 21. *(least fixed point)* Let $(C, \sqsubseteq)$ a CPO, $f : C \to C$ and let $x \in C$.

- x is a fixed point of f iff $f(x) = x$
- x is a least fixed point of f iff x is a fixed point of f and for each fixed point y of f: $x \sqsubseteq y$

The least fixed point of a function f is denoted by μf.

Finally, we will need the following definition and fact.

Definition 22. *(continuity)* Let $(C_1, \sqsubseteq_1), (C_2, \sqsubseteq_2)$ be CPO's. Then a function $f : C_1 \to C_2$ is continuous iff for each chain $\langle c_i \rangle_{i=0}^{\infty}$ in C_1, the following holds:

$$f(\bigsqcup_{i=0}^{\infty} c_i) = \bigsqcup_{i=0}^{\infty} f(c_i).$$

Fact 1. *(fixed point theorem)* Let C be a CPO and let $f : C \to C$. If f is continuous, then the least fixed point μf exists and equals $\bigsqcup_{i=0}^{\infty} f^i(\bot)$, where $f^0(\bot) = \bot$ and $f^{i+1}(\bot) = f(f^i(\bot))$.

For a proof, see for example De Bakker [4].

5.2 Definition

We will now show how the domains used in defining the semantics of meta-programs are designed as CPO's. The reason for designing these as CPO's will become clear in the sequel.

Definition 23. *(domains of interpretation)* Let W be the set of truth values of definition 12 and let S be the set of possible mental states of definition 3. Then the sets $W_\bot$ and $S_\bot$ are defined as CPO's as follows:

$$W_\bot = W \cup \{\bot_{W_\bot}\} \text{ CPO by } \beta_1 \sqsubseteq \beta_2 \text{ iff } \beta_1 = \bot_{W_\bot} \text{ or } \beta_1 = \beta_2,$$
$$S_\bot = S \cup \{\bot\} \quad \text{ CPO analogously.}$$

Note that we use $\perp$ to denote the bottom element of $S_\perp$ and that we use $\perp_C$ for the bottom element of any other set C. As the set of mental states is extended with a bottom element, we extend the semantics of boolean expressions of definition 12 to a strict function, i.e. yielding $\perp_{W_\perp}$ for an input state $\perp$.

In the definition of the denotational semantics, we will use an if-then-else function as defined below.

Definition 24. *(if-then-else)* Let C be a CPO, $c_1, c_2, \perp_C \in C$ and $\beta \in W_\perp$. Then the if-then-else function of type $W_\perp \to C$ is defined as follows.

$$\texttt{if } \beta \texttt{ then } c_1 \texttt{ else } c_2 \texttt{ fi} = \begin{cases} c_1 & \text{if } \beta = tt \\ c_2 & \text{if } \beta = ff \\ \perp_C & \text{if } \beta = \perp_{W_\perp} \end{cases}$$

Because our meta-language is non-deterministic, the denotational semantics is not a function from states to states, but a function from states to *sets of states*. These resulting sets of states can be finite or infinite. In case of bounded non-determinism[9], these infinite sets of states have $\perp$ as one of their members. This property may be explained by viewing the execution of a program as a tree of computations and then using König's lemma which tells us that a finitely-branching tree with infinitely many nodes has at least one infinite path (see [4]). The meta-language is indeed bounded non-deterministic[10] and the result of executing a meta-program P in some state, is thus either a finite set of states or an infinite set of states containing $\perp$. We therefore specify the following domain as the result domain of the denotational semantic function instead of $\wp(S_\perp)$.

Definition 25. *(T)* The set T with typical element τ is defined as follows: $T = \{\tau \in \wp(S_\perp) \mid \tau \text{ finite or } \perp \in \tau\}$.

The advantage of using T instead of $\wp(S_\perp)$ as the result domain, is that T can nicely be designed as a CPO with the following ordering [5].

Definition 26. *(Egli-Milner ordering)* Let $\tau_1, \tau_2 \in T$. $\tau_1 \sqsubseteq \tau_2$ holds iff either $\perp \in \tau_1$ and $\tau_1 \setminus \{\perp\} \subseteq \tau_2$, or $\perp \notin \tau_1$ and $\tau_1 = \tau_2$. Under this ordering, the set $\{\perp\}$ is $\perp_T$.

We are now ready to give the denotational semantics of meta-programs. We will first give the definition and then justify and explain it.

Definition 27. *(denotational semantics of meta-programs)* Let $\phi_1, \phi_2 : S_\perp \to T$. Then we define the following functions.

$$\begin{aligned} \hat{\phi} &: T \to T = \lambda\tau \cdot \bigcup_{s\in\tau} \phi(s) \\ \phi_1 \circ \phi_2 &: S_\perp \to T = \lambda s \cdot \hat{\phi}_1(\phi_2(s)) \end{aligned}$$

---

[9] Bounded non-determinism means that at any state during computation, the number of possible next states is finite.

[10] Only a finite number of rule applications and action executions are possible in any state.

Let $(\pi, \sigma) \in S$. The denotational semantics of meta-programs $\mathcal{M} : Prog \rightarrow (S_\perp \rightarrow T)$ is then defined as follows.

$$\mathcal{M}[\![execute]\!](\pi, \sigma) = \begin{cases} \{(\pi', \sigma')\} & \text{if } \pi = a; \pi' \\ & \text{with } a \in \mathsf{BasicAction} \text{ and } \mathcal{T}(a, \sigma) = \sigma' \\ \emptyset & \text{otherwise} \end{cases}$$

$$\mathcal{M}[\![execute]\!] \perp = \perp_T$$

$$\mathcal{M}[\![apply(\rho)]\!](\pi, \sigma) = \begin{cases} \{(\pi_b \circ \pi', \sigma)\} & \text{if } \sigma \models \psi \text{ and } \pi = \pi_h \circ \pi' \\ & \text{with } \rho : \pi_h \mid \psi \rightsquigarrow \pi_b \in \mathsf{Rule} \\ \emptyset & \text{otherwise} \end{cases}$$

$$\mathcal{M}[\![apply(\rho)]\!] \perp = \perp_T$$
$$\mathcal{M}[\![\mathbf{while}\ b\ \mathbf{do}\ P\ \mathbf{od}]\!] = \mu \Phi$$
$$\mathcal{M}[\![P_1; P_2]\!] = \mathcal{M}[\![P_2]\!] \circ \mathcal{M}[\![P_1]\!]$$
$$\mathcal{M}[\![P_1 + P_2]\!] = \mathcal{M}[\![P_1]\!] \cup \mathcal{M}[\![P_2]\!]$$

The function $\Phi : (S_\perp \rightarrow T) \rightarrow (S_\perp \rightarrow T)$ used above is defined as $\lambda \phi \cdot \lambda s \cdot \mathbf{if}\ \mathcal{W}(b)(s)\ \mathbf{then}\ \hat{\phi}(\mathcal{M}[\![P]\!](s))\ \mathbf{else}\ \{s\}\ \mathbf{fi}$, using definition 24.

Meta-actions. The semantics of meta-actions is straight forward. The result of executing an *execute* meta-action in some mental state s, is a set containing the mental state resulting from executing the basic action of the plan of s. The result is empty if there is no basic action on the plan to execute. The result of executing an $apply(\rho)$ meta-action in state s, is a set containing the mental state resulting from applying ρ in s. If ρ is not applicable, the result is the empty set.

While. The semantics of the $\mathbf{while}$ construct is more involved, but we will only briefly comment on it. For a detailed treatment, we again refer to De Bakker [4].

What we want to do, is define a function specifying the semantics of the $\mathbf{while}$ construct $\mathcal{M}[\![\mathbf{while}\ b\ \mathbf{do}\ P\ \mathbf{od}]\!]$, the type of which should be $S_\perp \rightarrow T$, in accordance with the type of $\mathcal{M}$. The function should be defined compositionally, i.e. it can only use the semantics of the guard and of the body of the $\mathbf{while}$. This is required for $\mathcal{M}$ to be well-defined.

The requirement of compositionality is satisfied, as the semantics is defined to be the least fixed point of the operator Φ, which is defined in terms of the semantics of the guard and body of the $\mathbf{while}$.

The least fixed point of an operator does not always exist. By the fixed point theorem however (fact 1), we now that if the operator is continuous (definition 22), the least fixed point *does* exist and is obtainable within ω steps. By proving that Φ is continuous, we can thus conclude that $\mu \Phi$ exists and therefore that $\mathcal{M}$ is well-defined.

Theorem 2. *(continuity of Φ)* The function Φ as given in definition 27 is continuous.

Proof. For a proof, we refer to [17]. The proof is analogous to continuity proofs given in [4].

Note that in the definition of Φ, the function ϕ is of type $S_\perp \to T$ and $\mathcal{M}[\![P]\!](s) \in T$. This ϕ can thus not be applied directly to this set of states in T, but it must be extended using the $\hat{\ }$ operator to be of type $T \to T$.

Sequential Composition and Non-deterministic Choice. The semantics of the sequential composition and non-deterministic choice operator is as one would expect.

6   Equivalence of Meta-level Operational and Denotational Semantics

In this section, we will state that the operational semantics for meta-programs is equal to the denotational semantics for meta-programs and we will relate this to the equivalence result of section 4. We will furthermore discuss the issue of defining a denotational semantics for object-level 3APL.

6.1   Equivalence Theorem

Theorem 3. $(\mathcal{O}_m = \mathcal{M})$ Let $\mathcal{O}_m : Prog \to (S_\perp \to \wp(S_\perp))$ be the operational semantics of meta-programs (definition 18) and let $\mathcal{M} : Prog \to (S_\perp \to T)$ be the denotational semantics of meta-programs (definition 27). Then, the following equivalence holds for all meta-programs $P \in Prog$ and all mental states $s \in S_\perp$.

$$\mathcal{O}_m(P)(s) = \mathcal{M}(P)(s)$$

Proof. For a proof, we refer to [17]. The proof is constructed using techniques from Kuiper [9].

In section 4, we stated that the object-level operational semantics of 3APL is equal to the meta-level operational semantics of the interpreter we specified in definition 19. Above, we then stated that it holds for any meta-program that its operational semantics is equal to its denotational semantics. This holds in particular for the interpreter of definition 19, i.e. we have the following corollary.

Corollary 1. $(\mathcal{O}_o = \mathcal{M}(\text{interpreter}))$ From theorems 1 and 3 we can conclude that the following holds.

$$\mathcal{O}_o = \mathcal{M}(\text{interpreter})$$

6.2   Denotational Semantics of Object-Level 3APL

Corollary 1 states an equivalence between a denotational semantics and the object-level operational semantics for 3APL. The question is, whether this denotational semantics can be called a denotational semantics for object-level 3APL.

A denotational semantics for object-level 3APL should be a function taking a plan and a belief base and returning the result of executing the plan on this belief base, i.e. a function of type $\Pi \to (\Sigma_\perp \to \wp(\Sigma_\perp))$ or equivalently[11], of type $(\Pi \times \Sigma) \to \wp(\Sigma_\perp)$. The type of $\mathcal{M}(\mathsf{interpreter})$, i.e. $S_\perp \to \wp(S_\perp)$ [12], does not match the desired type. This could however be remedied by defining the following function.

Definition 28. $(\mathcal{N})$ Let snd be a function yielding the second element, i.e. the belief base, of a mental state in S and yielding $\perp_{\Sigma_\perp}$ for input $\perp$. This function is extended to handle sets of mental states through the function $\hat{\ }$, as was done in definition 27. Then $\mathcal{N} : S_\perp \to \wp(\Sigma_\perp)$ is defined as follows.

$$\mathcal{N} = \lambda s \cdot \widehat{snd}(\mathcal{M}[\![\mathsf{interpreter}]\!](s))$$

Disregarding a $\perp$ input, the function $\mathcal{N}$ is of the desired type $(\Pi \times \Sigma) \to \wp(\Sigma_\perp)$. The question now is, whether it is legitimate to characterize the function $\mathcal{N}$ as being a denotational semantics for 3APL. The answer is no, because a denotational semantic function should be compositional in its program argument, which in this case is Π. This is obviously *not* the case for the function $\mathcal{N}$ and therefore this function is not a denotational semantics for 3APL.

So, it seems that the specification of the denotational semantics for meta-programs cannot be used to define a denotational semantics for object-level 3APL. The difficulty of specifying a compositional semantic function is due to the nature of the PR rules: these rules can transform not just atomic statements, but any sequence of statements. The semantics of an atomic statement can thus depend on the statements around it. We will illustrate the problem using an example.

$$a \rightsquigarrow b$$
$$b; c \rightsquigarrow d$$
$$c \rightsquigarrow e$$

Now the question is, how we can define the semantics of $a; c$? Can it be defined in terms of the semantics of a and c? The semantics of a would have to be something involving the semantics of b and the semantics of c something with the semantics of e, taking into account the PR rules given above. The semantics of $a; c$ should however also be defined in terms of the semantics of d, because of the second PR rule: $a; c$ can be rewritten to $b; c$, which can be rewritten to d. Moreover, if b is not a basic action, the third rule cannot be applied and the semantics of e would be irrelevant. So, although we do not have a formal proof, it seems that the semantics of the sequential composition operator[13] of a 3APL plan or program cannot be defined using only the semantics of the parts of which the program is composed.

Another way to look at this issue is the following. In a regular procedural program, computation can be defined using the concept of a program counter. This counter indicates the location in the code, of the statement that is to be

---

[11] For the sake of argument, we for the moment disregard a $\perp_{\Sigma_\perp}$ input.

[12] $\mathcal{M}(\mathsf{interpreter})$ is actually defined to be of type $S_\perp \to T$, but $T \subset \wp(S_\perp)$, so we may extend the result type to $\wp(S_\perp)$.

[13] Or actually of the plan concatenation operator •

executed next or the procedure that is to be called next. If a procedure is called, the program counter jumps to the body of this procedure. Computation of a 3APL program cannot be defined using such a counter. Consider for example the PR rules defined above and assume an initial plan $a; c$. Initially, the program counter would have to be at the start of this initial plan. Then, the first PR rule is "called" and the counter jumps to b, i.e. the body of the first rule. According to the semantics of 3APL, it should be possible to get to the body of the second PR rule, as the statement being executed is $b; c$. There is however no reason for the program counter to jump from the body of the first rule to the body of the second rule.

7   Related Work and Conclusion

The concept of a meta-language for programming 3APL interpreters was first considered by Hindriks [7]. Our meta-language is similar to, but simpler than Hindriks' language. The main difference is that Hindriks includes constructs for explicit selection of a PR rule from a set of applicable ones. These constructs were not needed in this paper. Dastani defines a meta-language for 3APL in [3]. This language is similar to, but more extensive than Hindriks' language. Dastani's main contribution is the definition of constructs for explicit planning. Using these constructs, the possible outcomes of a certain sequence of rule applications and action executions can be calculated in advance, thereby providing the possibility to choose the most beneficial sequence. Contrary to our paper, these papers do not discuss the relation between object-level and meta-level semantics, nor do they give a denotational semantics for the meta-language.

Concluding, we have proven equivalence of an operational and denotational semantics for a 3APL meta-language. We furthermore related this 3APL meta-language to object-level 3APL by proving equivalence between the semantics of a specific interpreter and object-level 3APL. Although these results were obtained for a simplified 3APL language, we conjecture that it will not be fundamentally more difficult to obtain similar results for full first order 3APL[14].

As argued in the introduction, studying interpreter languages of agent programming languages is important. In the context of 3APL and PR rules, it is especially interesting to investigate the possibility of defining a denotational or compositional semantics, for such a compositional semantics could serve as the basis for a (compositional) proof system. It seems, considering the investigations as described in this paper, that it will however be very difficult if not impossible to define a denotational semantics for object-level 3APL. As it *is* possible to define a denotational semantics for the meta-language, an important issue for future research will be to investigate the possibility and usefulness of defining a proof system for the meta-language, using this to prove properties of 3APL agents.

---

[14] The requirement of bounded non-determinism will in particular not be violated.

References

1. M. E. Bratman. *Intention, plans, and practical reason*. Harvard University Press, Massachusetts, 1987.
2. P. R. Cohen and H. J. Levesque. Intention is choice with commitment. *Artificial Intelligence*, 42:213–261, 1990.
3. M. Dastani, F. S. de Boer, F. Dignum, and J.-J. Ch. Meyer. Programming agent deliberation – an approach illustrated using the 3apl language. In *Proceedings of the second international joint conference on autonomous agents and multiagent systems (AAMAS'03)*, pages 97–104, Melbourne, 2003.
4. J. de Bakker. *Mathematical Theory of Program Correctness*. Series in Computer Science. Prentice-Hall International, London, 1980.
5. H. Egli. A mathematical model for nondeterministic computations. Technical report, ETH, Zürich, 1975.
6. G. d. Giacomo, Y. Lespérance, and H. Levesque. *ConGolog*, a Concurrent Programming Language Based on the Situation Calculus. *Artificial Intelligence*, 121(1-2):109–169, 2000.
7. K. Hindriks, F. S. de Boer, W. van der Hoek, and J.-J. C. Meyer. Control structures of rule-based agent languages. In J. Müller, M. P. Singh, and A. S. Rao, editors, *Proceedings of the 5th International Workshop on Intelligent Agents V : Agent Theories, Architectures, and Languages (ATAL-98)*, volume 1555, pages 381–396. Springer-Verlag: Heidelberg, Germany, 1999.
8. K. V. Hindriks, F. S. de Boer, W. van der Hoek, and J.-J. Ch. Meyer. Agent programming in 3APL. *Int. J. of Autonomous Agents and Multi-Agent Systems*, 2(4):357–401, 1999.
9. R. Kuiper. An operational semantics for bounded nondeterminism equivalent to a denotational one. In J. W. de Bakker and J. C. van Vliet, editors, *Proceedings of the International Symposium on Algorithmic Languages*, pages 373–398. North-Holland, 1981.
10. P. D. Mosses. Denotational semantics. In J. van Leeuwen, editor, *Handbook of Theoretical Computer Science*, volume B: Formal Models and Semantics, pages 575–631. Elsevier, Amsterdam, 1990.
11. G. Plotkin. A structural approach to operational semantics. Technical report, Aarhus University, Computer Science Department, 1981.
12. A. S. Rao. AgentSpeak(L): BDI agents speak out in a logical computable language. In W. van der Velde and J. Perram, editors, *Agents Breaking Away (LNAI 1038)*, pages 42–55. Springer-Verlag, 1996.
13. A. S. Rao and M. P. Georgeff. Modeling rational agents within a BDI-architecture. In J. Allen, R. Fikes, and E. Sandewall, editors, *Proceedings of the Second International Conference on Principles of Knowledge Representation and Reasoning (KR'91)*, pages 473–484. Morgan Kaufmann, 1991.
14. Y. Shoham. Agent-oriented programming. *Artificial Intelligence*, 60:51–92, 1993.
15. J. E. Stoy. *Denotational Semantics: The Scott-Strachey Approach to Programming Language Theory*. MIT Press, Cambridge, MA, 1977.
16. R. Tennent. *Semantics of Programming Languages*. Series in Computer Science. Prentice-Hall International, London, 1991.
17. M. B. van Riemsdijk, F. S. de Boer, and J.-J. Ch. Meyer. Semantics of plan revision in intelligent agents. Technical report, Utrecht University, Institute of Information and Computing Sciences, 2003. UU-CS-2004-002.

18. M. B. van Riemsdijk, W. van der Hoek, and J.-J. Ch. Meyer. Agent programming in Dribble: from beliefs to goals using plans. In *Proceedings of the second international joint conference on autonomous agents and multiagent systems (AAMAS'03)*, pages 393–400, Melbourne, 2003.

19. M. Wooldridge. Agent-based software engineering. *IEEE Proceedings Software Engineering*, 144(1):26–37, 1997.

20. M. Wooldridge and P. Ciancarini. Agent-Oriented Software Engineering: The State of the Art. In P. Ciancarini and M. Wooldridge, editors, *First Int. Workshop on Agent-Oriented Software Engineering*, volume 1957, pages 1–28. Springer-Verlag, Berlin, 2000.

Generic Exception Handling
and the Java Monad

Lutz Schröder and Till Mossakowski

BISS, Department of Computer Science, University of Bremen

Abstract. We develop an equational definition of exception monads that characterizes Moggi's exception monad transformer. This axiomatization is then used to define an extension of previously described monad-independent computational logics by abnormal termination. Instantiating this generic formalism with the Java monad used in the LOOP project yields in particular the known Hoare calculi with abnormal termination and JML's method specifications; this opens up the possibility of extending these formalisms by hitherto missing computational features such as I/O and nondeterminism.

Introduction

In the course of efforts to provide proof support for the verification of Java programs, the classical Hoare calculus [4] has been extended to encompass exception handling in Java [5, 7, 8, 26], the main challenge being to deal with poststates of abruptly terminating statements. Exceptions in Java are part of a monad for Java [9], following the paradigm of encapsulation of side effects via monads [12]. Thus, the question arises whether extended Hoare calculi for exceptions can be developed abstractly over any monad with exceptions. We answer this question positively by first characterizing Moggi's exception monad transformer by an equational theory based on a categorical formulation, and then extending our previous work about monad-independent Hoare calculi [21, 23] to calculi for exception monads that take into account both normal and abrupt termination. The advantage of such an approach is that it is not bound to a specific combination of computational effects like the Java monad. Moreover, most of the rules of the calculus come essentially for free by just adapting the normal monad independent Hoare rules [21, 23] using the equational description of exception monads.

As the background formalism for these concepts, we use the logic of HAS-CASL, a higher-order language for functional specification and programming, which is basically the internal language of partial cartesian closed categories (a generalization of toposes). The paper is organized as follows. In Sections 1, 2 and 3, we recall the HASCASL logic, monads and the logic for monads built on top of HASCASL. Section 4 axiomatizes exception monads and proves that they characterize Moggi's exception monad transformer, and Section 5 introduces the Hoare logic for exception monads.

C. Rattray et al. (Eds.): AMAST 2004, LNCS 3116, pp. 443–459, 2004.

1   HasCasl

HasCasl is a higher-order extension of Casl [2], featuring higher-order functions in the style of Moggi's partial λ-calculus [10], type constructors, type classes and constructor classes (for details, see [22, 24]); general recursion is specified on top of this in the style of HOLCF. The semantics of a HasCasl specification is the class of its (set-theoretic) *intensional Henkin models*: a function type need not contain all set-theoretic functions, and two functions that yield the same value on every input need not be equal.

A consequence of the intensional semantics is the presence of an intuitionistic *internal logic* that lives within λ-terms, defined in terms of equality [22]. There is built-in syntactical sugar for the internal logic, invoked by means of the keyword **internal** which signifies that formulas in the following block are to be understood as formulas of the internal logic.

Categorically speaking, HasCasl's Henkin models correspond to models in partial cartesian-closed categories (pccc's) [20], a generalization of toposes [1]. Basic HasCasl can be seen as syntactic sugar over the internal language of a pccc, i.e. essentially intuitionistic higher order logic of partial functions.

2   Monads for Computations

On the basis of Moggi's seminal work [12], monads are being used in both semantics and programming to formalize and encapsulate side effects in an elegant, functional way. Intuitively, a monad associates to each type A a type TA of *computations* of type A; a function with side effects that takes inputs of type A and returns values of type B is, then, just a function of type $A \to TB$, also called a *(B-valued) program*. This approach abstracts away from particular notions of computation such as store, non-determinism, non-termination etc.; a surprisingly large amount of reasoning can in fact be carried out independently of the choice of such a notion.

More formally, a monad on a given category $\mathbf{C}$ can be presented as a *Kleisli triple* $\mathbb{T} = (T, \eta, \_^*)$, where $T : \mathrm{Ob}\,\mathbf{C} \to \mathrm{Ob}\,\mathbf{C}$ is a function, the *unit* η is a family of morphisms $\eta_A : A \to TA$, and $\_^*$ assigns to each morphism $f : A \to TB$ a morphism $f^* : TA \to TB$ such that

$$\eta_A^* = id_{TA}, \quad f^*\eta_A = f, \quad \text{and} \quad g^*f^* = (g^*f)^*.$$

This description is equivalent to the more familiar one via an endofunctor T with unit η and a multiplication μ [1]. A monad defines its *Kleisli category*, which has the same objects as $\mathbf{C}$ and 'functions with side effects' $f : A \to TB$ as morphisms from A to B; the Kleisli composite of two such functions g and f is just g^*f.

In order to support a language with finitary operations and multi-variable contexts (see below), one needs a further technical requirement: a monad is called *strong* if it is equipped with a natural transformation

$$t_{A,B} : A \times TB \to T(A \times B)$$

called *strength*, subject to certain coherence conditions.

Example 1 ([12]). Computationally relevant monads on **Set** (all monads on **Set** are strong) include

- stateful computations with possible non-termination: $TA = (S \to? (A \times S))$, where S is a fixed set of states and $\_ \to? \_$ denotes the partial function type;
- (finite) non-determinism: $TA = \mathcal{P}_{fin}(A)$, where $\mathcal{P}_{fin}$ denotes the finite power set functor;
- exceptions: $TA = A + E$, where E is a fixed set of exceptions;
- interactive input: TA is the least fixed point of $\lambda\gamma. A + (U \to? \gamma)$, where U is a set of input values.
- non-deterministic stateful computations: $TA = S \to \mathcal{P}_{fin}(A \times S)$, where, again, S is a fixed set of states;

Reasoning about a category **C** equipped with a strong monad is greatly facilitated by the fact that proofs can be conducted in an *internal language* [12]. The crucial features of this language are

- A type operator T, where, as above, TA contains computations of type A;
- a polymorphic operator ret : $A \to TA$ corresponding to the unit;
- a binding construct, which we denote in Haskell's do style instead of by let: terms of the form

$$\text{do } x \leftarrow e_1;\ e_2$$

are interpreted by means of the tensorial strength and Kleisli composition [12] — e.g. if the ambient context is $y : A$ and e_1 is B-valued, then the interpretation $[\![\text{do } x \leftarrow e_1;\ e_2]\!]$ is $[\![e_2]\!]^* \circ t_{A,B} \circ [\![(y, e_1)]\!]$. Intuitively, do $x \leftarrow e_1;\ e_2$ computes e_1 and passes the results on to e_2. Nested do expressions like do $x \leftarrow e_1$; do $y \leftarrow e_2$; ... may also be denoted do $x \leftarrow e_1; y \leftarrow e_2$; .... Repeated nestings such as do $x_1 \leftarrow e_1, \ldots, x_n \leftarrow e_n$; e are somewhat inaccurately denoted in the form do $\bar{x} \leftarrow \bar{e}$; e. Sequences of the form $\bar{x} \leftarrow \bar{e}$ are called *program sequences*. Variables x_i that are not used later on may be omitted from the notation.

This language (with further term formation rules and a deduction system) can be used both in order to define morphisms in **C** and in order to prove equalities between them [12]. Indeed, the theory of exception monads presented in Section 4 is formulated in this internal language, over an arbitrary category with equalizers. Only in the context of computational *logic* (e.g. the Hoare calculus introduced in the next section), one needs the framework of a partial cartesian closed category (and its internal language, phrased in HASCASL).

Given a monad, one can generically define control structures such as a while loop (see for example [15]). Such definitions require general recursion, which is realized in HASCASL by means of fixed point recursion on cpos, with the associated fixed point operator on continuous endofunctions denoted by Y [22]. Thus, one has to restrict to monads, called *cpo-monads*, that allow lifting a cpo structure on A to a cpo structure on the type TA of computations in such a way that the monad operations become continuous. The (executable) specification of an iteration construct is shown in Figure 1. The specification imports named specifications BOOL of the booleans and CPOMONAD of cpo-monads; $\_\_ \xrightarrow{c}? \_\_$ and $\_\_ \xrightarrow{c} \_\_$ are the type constructors for the partial and total continuous function space, respectively. The iteration construct behaves like a while loop, except that it additionally passes a result value through the iterations. The while loop is just iteration ignoring the result value.

---

spec ITERATION = CPOMONAD **and** BOOL **then**
    vars $m : CpoMonad$; $a : Cpo$
    op $iter : (a \xrightarrow{c} m\ Bool) \xrightarrow{c} (a \xrightarrow{c}? m\ a) \xrightarrow{c} a \xrightarrow{c}? m\ a$
    program $iter\ test\ f\ x =$
           $do\ b \leftarrow test\ x$
            $if\ b\ then$
              $do\ y \leftarrow f\ x$
                $iter\ test\ f\ y$
              $else\ return\ x$
    op $while(b : m\ Bool)(p : m\ Unit) : m\ Unit = iter\ (\lambda x \bullet b)\ (\lambda x \bullet p)\ ()$

Fig. 1. The iteration control structure

3 Monad-Independent Computational Logic

We now recall notions of side-effect freeness in monads [21, 25] and the recently developed monad-independent computational logics [23, 21]. Throughout, $\mathbb{T}$ will denote a strong monad.

Like traditional Hoare logic, the monad-independent Hoare calculus is concerned with proving *Hoare triples* consisting of a stateful expression together with a pre- and a postcondition. The pre- and postconditions are required to 'leave the state unchanged' in a suitable sense in order to guarantee composability of Hoare triples; they may, however, read the state.

Definition 2. A program p is called *discardable* if

$$(do\ y \leftarrow p;\ ret *) = ret *,$$

where $*$ is the unique element of the unit type.

For example, a program p is discardable in the state monad iff p terminates and does not change the state.

A program p is called *stateless* if it factors through η, i.e. if it is just a value inserted into the monad — otherwise, it is called *stateful*. E.g. in the state monad, p is stateless iff it neither changes nor reads the state. Stateless programs are discardable, but not vice versa.

In order to define the semantics of Hoare triples below, we introduce *global dynamic judgements* of the form $[\bar{x} \leftarrow \bar{p}]\,\phi$, which intuitively state that ϕ holds after execution of the program sequence $\bar{x} \leftarrow \bar{p}$, where ϕ is a stateless formula (i.e. $\phi : \Omega$, where Ω is the type of truth values). Formally, $[\bar{x} \leftarrow \bar{p}]\,\phi$ abbreviates

$$(\text{do } \bar{x} \leftarrow \bar{p};\ \text{ret}(\bar{x}, \phi)) = \text{do } \bar{x} \leftarrow \bar{p};\ \text{ret}(\bar{x}, \top).$$

Definition 3. A program p is *copyable* if

$$(\text{do } x \leftarrow p; y \leftarrow p;\ \text{ret}(x, y)) = \text{do } x \leftarrow p;\ \text{ret}(x, x).$$

A program p *commutes* with a program q if

$$(\text{do } x \leftarrow p; y \leftarrow q;\ \text{ret}(x, y)) = \text{do } y \leftarrow q; x \leftarrow p;\ \text{ret}(x, y).$$

A discardable and copyable program is called *deterministically side-effect free* (dsef) if it commutes with all (equivalently: all Ω-valued) discardable copyable programs. For a type A, the subtype of TA consisting of the dsef computations is denoted DA.

Dsef programs are syntactically treated like stateless values; their properties guarantee that arising ambiguities correspond to actual equalities. A discardable program p is copyable iff it is deterministic in the sense that

$$[x \leftarrow p; y \leftarrow p]\,(x = y).$$

Stateless programs are dsef. In the monads of Example 1, all discardable programs are dsef, with the exception of the monads where non-determinism is involved. In these cases, a discardable program is dsef iff it is deterministic.

Definition 4. A *Hoare triple for partial correctness*, written $\{\phi\}\ \bar{x} \leftarrow \bar{p}\ \{\psi\}$, consists of a program sequence $\bar{x} \leftarrow \bar{p}$, a precondition $\phi : D\Omega$, and a postcondition $\psi : D\Omega$ (which may contain $\bar{x}$). This abbreviates the formula

$$[a \leftarrow \phi; \bar{x} \leftarrow \bar{p}; b \leftarrow \psi]\,a \Rightarrow b.$$

For example, a Hoare triple $\{\phi\}\ x \leftarrow p\ \{\psi\}$ holds in the state monad iff, whenever ϕ holds in a state s, then ψ holds for x after successful execution of p from s with result x. In the non-deterministic state-monad, ψ must be satisfied for all possible pairs of results and post-states for p.

Monad-independent *dynamic logic* is used in order to also capture Hoare triples for *total correctness* [21]. Dynamic logic allows nesting modal operators of the nature 'after execution of p' and the usual connectives of first order logic. This means informally that the state is changed according to the effect of p within the scope of the modal operator, but is 'restored' outside that scope. E.g., in a dynamic logic formula such as

$$[p]\,\phi \implies [q]\,\psi,$$

the subformulas $[p]\,\phi$ and $[q]\,\psi$ are evaluated in the same state, while ϕ and ψ are evaluated in modified states.

Definition 5. A *formula* (of dynamic logic) is a term $\varphi : D\Omega$. The formula ϕ is *valid* if $\varphi = \mathrm{do}\ x \leftarrow \varphi;\ \mathrm{ret}\ \top$.

The usual logical connectives are defined using the do-notation. The question is now whether $D\Omega$ has enough structure to allow also the interpretation of the diamond and box operators $\langle p \rangle$ and $[p]$ of dynamic logic.

Definition 6. $\mathbb{T}$ *admits dynamic logic* if there exist, for each program sequence $\bar{y} \leftarrow \bar{q}$ and each formula $\varphi : D\Omega$, a formula $[\bar{y} \leftarrow \bar{q}]\,\varphi$ such that

$$[\bar{x} \leftarrow \bar{p}]\,(x_i \Rightarrow [\bar{y} \leftarrow \bar{q}]\,\varphi) \iff [\bar{x} \leftarrow \bar{p}; \bar{y} \leftarrow \bar{q}]\,(x_i \Rightarrow \varphi)$$

for each program sequence $\bar{x} \leftarrow \bar{p}$ containing $x_i : \Omega$ and, dually, a formula $\langle \bar{y} \leftarrow \bar{q} \rangle \varphi$ such that

$$[\bar{x} \leftarrow \bar{p}]\,(\langle \bar{y} \leftarrow \bar{q} \rangle \varphi \Rightarrow x_i) \iff [\bar{x} \leftarrow \bar{p}; \bar{y} \leftarrow \bar{q}]\,(\varphi \Rightarrow x_i).$$

(Since the internal logic is intuitionistic, one cannot simply define $\langle \bar{y} \leftarrow \bar{q} \rangle \varphi$ as $\neg[\bar{y} \leftarrow \bar{q}]\,\neg\varphi$.) The formulae $[\bar{y} \leftarrow \bar{q}]\,\varphi$ and $\langle \bar{y} \leftarrow \bar{q} \rangle \varphi$ are uniquely determined. A deduction system is obtained as a collection of lemmas in the internal logic. Most computational monads, with the exception of the continuation monad, admit dynamic logic [21].

Hoare triples for partial correctness $\{\varphi\}\ p\ \{\psi\}$ can be expressed in dynamic logic as $\varphi \Rightarrow [p]\,\psi$, and we now can also give a meaning to Hoare triples for total correctness by interpreting them as partial correctness plus termination:

$$[\varphi]\ p\ [\psi] :\equiv \varphi \Rightarrow (\langle p \rangle \top \wedge [p]\,\psi).$$

4   Exception Monads

We now proceed to develop a general treatment of monads that feature exceptions which can be raised and later caught. To begin, we give an equational definition in categorical terms, which is then translated into the do-notation. In the next section, this definition will be used to formulate Hoare-calculi for exception monads. Throughout, T will denote a strong monad.

There are two essential operations associated to exceptions: an operation *raise* that throws an exception and freezes the state, and a *catch* operation that returns a thrown exception, if any, and unfreezes the state, i.e. resumes normal execution of statements. Obvious variations such as an operation that catches only certain exceptions (e.g. exceptions only of a given class) and lets others pass are easily implemented by combinations of *catch* and *raise*. It will be convenient to give *raise* the polymorphic type $E \rightarrow TA$ (in the most basic example, *raise* is the right injection $E \rightarrow A + E$); of course, the result type A is in fact immaterial since *raise* does not return any results.

There is a certain amount of disagreement in the literature concerning whether raising exceptions should be regarded as a computational feature in its own right, or whether exceptions should just be recorded as part of the global state. In the Java semantics developed in the LOOP project, the former approach is advocated [6, 9], while the latter option is preferred in [14, 26] (but not in [13], motivated by modularity considerations — compare this to Theorem 8 below). In terms of concrete monads, the LOOP project [9] uses the monad

$$J = \lambda A. (S \rightarrow (A \times S + E \times S + 1)),$$

with S the set of states and E the set of exceptions (the parameter A is thought of as the type of results), while

$$K = \lambda A. (S_E \rightarrow (A + 1) \times S_E + 1)$$

is implicitly used in [14, 26], where $S_E = S \times (E + 1)$, and where the use of $A + 1$ accommodates the fact that abnormally terminating statements do not return a value. The monad J is obtained by applying Moggi's exception monad transformer [11] to the usual state monad with non-termination, while K is the state monad with non-termination (and possibly undefined results) for the extended state set S_E. A simultaneous advantage and disadvantage of K is that *catch* is a monadic value, i.e. a statement, while in J, *catch* is a function on monadic values, i.e. a control structure. Thus, K contains monadic values that arbitrarily manipulate the state even though an exception has been thrown, or simultaneously throw an exception and return a value. Consequently, explicit Hoare rules are needed to force normal statements to be skipped in exceptional states [14, 26]; moreover, it becomes necessary to add normality of the state as a precondition to all of the usual Hoare rules. By contrast, in J the state is automatically frozen when an exception is thrown.

We shall model our treatment of generic exception monads along J rather than K, precisely because this allows better control over what monadic values do. Thus, we need to work with a *catch* operation of polymorphic type

$$TA \rightarrow T(A + E),$$

which takes a monadic value x and returns a monadic value that executes x but terminates normally (if at all), returning either a value returned by x or an exception raised by x.

We are now ready to present the announced categorical definition of exception monads. We begin with a short but somewhat mysterious definition and then proceed in opposite direction to the heuristics, explicating the definition stepwise into a number of intuitively convincing equations.

Definition 7. An *exception monad* is a strong monad T, together with a type E of *exceptions* and a natural transformation $catch : T \rightarrow T(\_ + E)$ such that

$$T \xrightarrow{\quad catch \quad} T(\_ + E) \mathrel{\substack{\xrightarrow{\;catch_{(\_+E)}\;} \\ \xrightarrow[\;T\ inl\;]{}}} T(\_ + E + E),$$

is an equalizer diagram of strong monad morphisms.

Recall that, given two strong monads S, T on a category $\mathbf{C}$, a (simplified) *strong monad morphism* is a natural transformation $\sigma : S \to T$, compatible with the remaining data in the sense that $\sigma \eta^S = \eta^T$, $\sigma \mu^S = \mu^T(\sigma * \sigma)$ (where $\sigma * \sigma = T\sigma\sigma_S = \sigma_T S\sigma$), and $\sigma_{(A\times\_)}t_A^S = t_A^T(A \times \sigma)$ (e.g. [11]). Naturality of σ and compatibility with μ are, in terms of the associated Kleisli triples $(T, \eta^T, \_^*)$ and $(S, \eta^S, \_^+)$, equivalent to $\sigma f^* = (\sigma f)^+ \sigma$, i.e. σ is *compatible with binding*.

Next, let us point out which parts of Definition 7 are self-understood. Given a strong monad T, the functor $T(\_ + E)$ is made into a strong monad by taking $\eta^T inl : A \to T(A + E)$ as the unit, and a binding operator that transforms $f : A \to T(B+E)$ into $[f, \eta\, inr]^* : T(A+E) \to T(B+E)$; this is Moggi's monad transformer for exceptions [11] as implemented in the Haskell libraries [15] (of course, this presupposes that $A + E$ exists for all A). It is easy to check that $T\, inl : T \to T\,(\_ + E)$ is a strong monad morphism.

Thus, the actual conditions imposed by the definition are in particular

-   $catch : T \to T(\_ + E)$ is a strong monad morphism. Compatibility with binding amounts to the equation

    $$catch\, f^* = [catch\, f, \eta\, inr]^*\, catch.$$

    We will see that this expresses the fact that in a sequential statement $p; q$, either p raises an exception, which is then passed on, and q is skipped, or p terminates normally and then q is executed. Compatibility of catch with the unit states that pure values do not throw exceptions: $catch\, \eta = \eta\, inl$.
-   $catch_{(\_+E)}\, catch = (T\, inl)\, catch$: this equation states that *catch* does not itself raise exceptions.

Finally, the fact that *catch* not only equalizes $catch_{(\_+E)}$ and $T\, inl$, but is indeed their equalizer, can be captured equationally by means of the *raise* operation, which is conspicuously absent from the basic definition. Indeed we can *construct* this operation: we have a morphism $\eta : \_ + E \to T(\_ + E)$, which equalizes $catch_{(\_+E)}$ and $T\, inl$ because *catch* is a monad morphism. Thus, we obtain a factorization $catch\, f = \eta$, where f is necessarily of the form $[\eta, \cdot]$; the *raise* operator is defined as the second component of this morphism:

$$
\begin{array}{ccc}
T & \xrightarrow{\ catch\ } & T\,(\_ + E) \\
{\scriptstyle [\eta,\, raise]}\big\uparrow & \nearrow{\scriptstyle \eta} & \\
\_ + E & &
\end{array}
$$

— i.e. the defining property of *raise* is the equation

$$catch\, raise = \eta\, inr \qquad\qquad (1)$$

stating that raised exceptions are actually caught. In combination with binding compatibility of *catch*, this implies

$$catch\, [\eta, raise]^* = T\,(\_ + \nabla)\, catch_{(\_+E)}, \qquad\qquad (2)$$

where ∇ denotes the codiagonal $E+E \to E$; note that $T(\_+\nabla) : T(\_+E+E) \to T(\_+E)$ is a strong monad morphism. From this equation, in turn, we can derive, using the fact that *catch* is monic, that

$$[\eta, raise]^* catch = id \qquad (3)$$

— i.e. catching an exception can be undone by re-raising it.

Equations 2 and 3, together with the fact that *catch* equalizes $catch_{(\_+E)}$ and $T\,inl$ and the obvious equation $T(\_+\nabla)\,T\,inl = id$, amount to stating that

$$T \xrightleftharpoons[{[\eta,\,raise]^*}]{catch} T(\_+E) \; \underset{T\,inl}{\overset{catch_{(\_+E)}}{\underset{\longrightarrow}{\rightleftarrows}}}\, T(\_+\nabla)\; T(\_+E+E),$$

is a split equalizer diagram [1] of strong monad morphisms. Thus, we can equivalently describe an exception monad by means of equations 1 and 3 (the latter implies that *catch* is monic), together with the fact that *catch* is a strong monad morphism (since *catch* is monomorphic, it follows easily that $[\eta, raise]^*$ is also a strong monad morphism) and equalizes $catch_{(\_+E)}$ and $T\,inl$.

The arising purely equational presentation of exception monads can be translated into the do-notation as explained in Section 2; the corresponding HASCASL specification is shown in Figure 2. The two imported specifications are the specification MONAD of monads as described in Section 2 and a specification SUMTYPE which provides + as an infix sum type operator, with left and right injections *inl* and *inr*. The axioms (catch-ret) and (catch-seq) state that *catch* is a strong monad morphism (where (catch-seq) covers compatibility with binding as well as with the strength). Axiom (catch-raise) is Equation 1 above, (catch-catch) states that *catch* equalizes $catch_{(\_+E)}$ and $T\,inl$, and (catchN) is Equation 3.

In the notation given in Figure 2, it becomes even more evident that all these axioms are properties that one would intuitively expect of *raise* and *catch*. In fact, as indicated above, the given equations come heuristically *before* Definition 7. Other expected properties follow; e.g., (catchN), (catch-seq), and (catch-raise) imply that 'nothing happens after an exception is raised', i.e. that

$$(do \; x \leftarrow raise \; e; \; p) = raise \; e. \qquad (4)$$

An obvious way to construct exception monads is to use Moggi's exception monad transformer as described above, i.e. to take $T = R(\_+E)$ for a strong monad R, with

$$catch : R(\_+E) \to R((\_+E)+E)$$

being $R\,inl$. Surprisingly, this construction indeed *classifies* exception monads:

Theorem 8. *Let* **C** *be a category with equalizers. Then every exception monad on* **C** *is of the form* $R(\_+E)$ *for some strong monad* R *on* **C***.*

Proof. Let T be an exception monad, and let R be the equalizer of the morphisms

$$catch, T\,inl : T \to T(\_+E).$$

spec EXCEPTION = MONAD **and** SUMTYPE **then**
    types $E: Type, Ex: Monad$
    var $a, b: Type$
    ops $raise: E \to Ex\ Unit;$
           $catch: Ex\ a \to Ex\ (a + E);$
    internal{
    forall $e: E;\ p: Ex\ a;\ q: a \to Ex\ b;$

      • $catch\ (do\ x \leftarrow p;\ q\ x) =$
           $do\ y \leftarrow catch\ p;\ case\ y\ of\ inl\ a \to catch\ q\ a$
                           $|\ inr\ e \to ret\ (inr\ e)$ %(catch-seq)%
      • $catch\ (ret\ x) = ret\ (inl\ x)$ %(catch-ret)%
      • $catch\ (raise\ e) = ret\ (inr\ e)$ %(catch-raise)%
      • $catch\ (catch\ p) = do\ y \leftarrow catch\ p;\ ret\ (inl\ y)$ %(catch-catch)%
      • $p = do\ y \leftarrow catch\ p;\ case\ y\ of\ inl\ x \to ret\ x$
                              $|\ inr\ e \to raise\ e$ %(catchN)%

    }

Fig. 2. Equational specification of exception monads

This equalizer is preserved by the exception monad transformer $\lambda M.\ M\ (\_ + E)$, so that $R\ (\_ + E) \cong T$.

Thus, our definition of exception monads can be regarded as a complete equational characterization of the exception monad transformer. It may appear at first sight that this result is illicitly built into the definition, since the codomain of $catch$ is $T\ (\_ + E)$. However, this is not the case: the result type of $catch$ describes how exceptions are *observed*, and in this sense constitutes rather a minimal expectation. By contrast, the theorem above concerns the entirely different question of how exceptions are *represented* in computations, and the answer is not at all self-understood. One of its implications is that exceptions are never inextricably entwined with other notions of computation — they can always be regarded as added to the remaining computational features in a modular way.

    The classification theorem can also be used to facilitate reasoning about exception monads; e.g. one can prove:

Corollary 9. *If T is an exception monad and $p : TA$ is a dsef computation, then p terminates normally, i.e. $catch\ p = (T\ inl)\ p$.*

(However, discardable computations may raise exceptions!).

Notation. In order to have intermediate results of a program sequence $\bar{x} \leftarrow \bar{p}$ available after a *catch*, the latter must be packaged up and explicitly returned. For this procedure, an abbreviated notation comes in handy: $catch\ \bar{x} \leftarrow \bar{p}$ denotes $catch\ (do\ \bar{x} \leftarrow \bar{p};\ ret\ \bar{x})$.

Remark 10. One subtlety that we have omitted from the development so far is the fact that monad-independent computational logic [21, 23] uses a strengthened version of monads in the sense that binding is required to extend also to partial morphisms, i.e. one has f^* for f partial. The unit laws for partial bind-ing state that $f^*\eta$ agrees with f on the domain of f and behaves like f under binding (i.e. $(f^*\eta)^* = f^*$).

The results obtained so far are adapted to this setting as follows. Slightly sur-prisingly, the correct definition of a morphism σ of monads with partial binding turns out to require, in the notation used above, that the equation $\sigma f^* = (\sigma f)^+ \sigma$ holds strongly, i.e. as an equation between partial morphisms. Then the arising abstract definition of partial exception monads unravels in the same way as above. In particular, the only equation in Figure 2 that actually involves partial binding, (catch-seq), becomes a strong equation (to be read 'one side is defined iff the other is, and then both sides are equal). Moreover, the exception monad transformer indeed transforms monads with partial binding into monads with partial binding, and the analogue of Theorem 8 holds in categories with a notion of partial morphism, i.e. in dominional categories [19] with equalizers.

4   A Generic Hoare Calculus with Abrupt Termination

We now proceed to extend the partial and total generic Hoare calculi for monadic programs introduced in Section 2 to take into account exceptional termination, thus generalizing similar calculi [5, 7, 8]. The reason that the basic version of the generic Hoare calculus is insufficient for purposes of exceptional termination is that, due to Equation 4 above, we have

$$\{\} \, p \, \{\bot\}$$

whenever p raises an exception (indeed, this holds also e.g. in the 'normal' part of the calculi of [5, 7]). Thus, no reasonable statements can be made about what happens when an exception is raised. We remedy this problem by introducing an additional postcondition for exceptional termination, called the *abnormal postcondition* in opposition to the usual postcondition now called the *normal postcondition*.

This might raise the suspicion that the Hoare calculi of [21, 23] are in fact 'insufficiently generic', and that in reality every new computational feature re-quires a whole new Hoare calculus. Besides the examples given in [21, 23], the following heuristic argument supports our claim that this is not the case: the problem '$\{\}$ *raise* e $\{\bot\}$' quoted above is due to the fact that exceptions are constants in the ambient monad, so that exceptional computations of type Ω do not actually contain any truth values. This phenomenon is unique to constant operations, and the only computational interpretation of constants known so far is precisely as exceptions. Thus, it appears that the need for substantially (i.e. other than in terms of additional axioms and rules for monad-specific program constructs) extended Hoare calculi will be limited to the case of exceptions.

We begin by introducing a partial Hoare calculus for abnormal termination in an exception monad T. A corresponding total Hoare calculus, which like the total Hoare calculus for normal termination requires additional assumptions on the monad, will be treated further below. We denote the combination of precondition and normal and abnormal postcondition in the form

$$\{\phi\}\ \bar{x} \leftarrow \bar{p}\ \{\psi\ \|\ S\},$$

where the normal postcondition ψ is a stateful *formula* that may contain the result variables $\bar{x}$, while the abnormal postcondition S is stateful *predicate* on E (i.e. a function $E \rightarrow T\Omega$) and cannot use $\bar{x}$. This restriction reflects the fact that exceptional computations do not have a result; instead, the abnormal postcondition is concerned with a hitherto anonymous exception. The interpretation of such an *extended Hoare assertion* is that, if the program sequence $\bar{x} \leftarrow \bar{p}$ is executed in a state that satisfies ϕ, then if the execution terminates normally, the post-state and the result $\bar{x}$ satisfy ψ, and if the execution terminates abnormally with exception e, the post-state satisfies $S\ e$. Formally, $\{\phi\}\ \bar{x} \leftarrow \bar{p}\ \{\psi\ \|\ S\}$ abbreviates

$$\{\phi\}\ y \leftarrow (catch\ \bar{x} \leftarrow \bar{p})\ \{case\ y\ of\ inl\ \bar{x} \rightarrow \psi\ |\ inr\ e \rightarrow S\ e\}.$$

The associated Hoare calculus subsumes the generic Hoare calculus [23], since one can show that

$$\{\phi\}\ \bar{x} \leftarrow \bar{p}\ \{\psi\} \iff \{\phi\}\ \bar{x} \leftarrow \bar{p}\ \{\psi\ \|\ \top\}.$$

Figure 3 shows a set of proof rules for extended Hoare assertions. Most of the rules except (catch) and (raise) have counterparts in the 'normal' generic Hoare calculus; note in particular the single composition rule (seq), which is similar to the one given in [8]. In the conjunction and weakening rules, notations such as $S_1 \wedge S_2$ stand for the pointwise operations, here: $\lambda e : E.\ (S_1\ e) \wedge (S_2\ e)$. The (Y) rule refers to the fixed-point operator Y (cf. Section 2); this rule applies only to cpo-monads. Application of the Y operator to F requires implicitly that F has the continuous function type $(A \xrightarrow{c} ?TB) \xrightarrow{c} (A \xrightarrow{c} ?TB)$. The square brackets indicate reasoning with local assumptions, discharged by application of the rule. Soundness of the rule is a consequence of the fact that satisfaction of extended Hoare assertions is an admissible predicate. Using this rule, one easily derives rules for particular recursive functions, e.g. the usual while rule or a corresponding rule for the iteration construct described in Section 2,

$$(iter)\ \frac{\{\phi\ x \wedge (b\ x)\}\ y \leftarrow p\ x\ \{\phi\ y\ \|\ S\}}{\{\phi\ e\}\ y \leftarrow iter\ b\ p\ e\ \{\phi\ y \wedge \neg(b\ y)\ \|\ S\}}.$$

(A corresponding rule is listed as a basic rule in the generic Hoare calculus [23], which can indeed be improved by moving to a (Y)-like rule.) The exception-monad specific rules (catch) and (raise) state that *catch* catches a thrown exception but otherwise does not affect the enclosed program, and that *raise* really throws the given exception.

The calculus is sound w.r.t. the coding of Hoare triples as formulas in the internal language:

Theorem 11. *If an extended Hoare assertion is derivable in an exception monad (cpo-monad) by the rules of Figure 3 excluding (including) (Y), then the translated formula is derivable in the internal language.*

The proof uses the definition of extended Hoare assertions via standard Hoare triples, the generic Hoare rules [23], and the equations of Figure 2. Completeness of the calculus is the subject of further research; we conjecture that a completeness proof for the Hoare logic of [23] will lead to completeness of the extended calculus, since each of the equations in Figure 2 is reflected in one of the rules of Figure 3.

$$(\text{seq}) \quad \frac{\{\phi\}\ \bar{x} \leftarrow \bar{p}\ \{\psi \parallel S\} \qquad \{\psi\}\ \bar{y} \leftarrow \bar{q}\ \{\chi \parallel S\}}{\{\phi\}\ \bar{x} \leftarrow \bar{p};\bar{y} \leftarrow \bar{q}\ \{\chi \parallel S\}}$$

$$(\text{wk}) \quad \frac{\{\phi\}\ \bar{x} \leftarrow \bar{p}\ \{\psi \parallel S\} \qquad \phi' \Rightarrow \phi \quad \psi \Rightarrow \psi' \qquad S \Rightarrow S'}{\{\phi'\}\ \bar{x} \leftarrow \bar{p}\ \{\psi' \parallel S'\}}$$

$$(\text{conj}) \quad \frac{\{\phi\}\ \bar{x} \leftarrow \bar{p}\ \{\psi_1 \parallel S_1\} \qquad \{\phi\}\ \bar{x} \leftarrow \bar{p}\ \{\psi_2 \parallel S_2\}}{\{\phi\}\ \bar{x} \leftarrow \bar{p}\ \{\psi_1 \wedge \psi_2 \parallel S_1 \wedge S_2\}}$$

$$(\text{disj}) \quad \frac{\{\phi_1\}\ \bar{x} \leftarrow \bar{p}\ \{\psi \parallel S\} \qquad \{\phi_2\}\ \bar{x} \leftarrow \bar{p}\ \{\psi \parallel S\}}{\{\phi_1 \vee \phi_2\}\ \bar{x} \leftarrow \bar{p}\ \{\psi \parallel S\}}$$

$$(\text{if}) \quad \frac{\{\phi \wedge b\}\ x \leftarrow p\ \{\psi \parallel S\} \qquad \{\phi \wedge \neg b\}\ x \leftarrow q\ \{\psi \parallel S\}}{\{\phi\}\ x \leftarrow \textit{if } b \textit{ then } p \textit{ else } q\ \{\psi \parallel S\}}$$

$$(\text{Y}) \quad \frac{[\{\phi\}\ x \leftarrow p\ y\ \{\psi \parallel S\}] \\ \vdots \\ \{\phi\}\ x \leftarrow (F\ p)\ y\ \{\psi \parallel S\}}{\{\phi\}\ x \leftarrow Y(F)\ y\ \{\psi \parallel S\}}$$

$$(\text{ctr}) \quad \frac{\{\phi\}\ \ldots;x \leftarrow p;y \leftarrow q;\bar{z} \leftarrow \bar{r}\ \{\psi \parallel S\} \qquad x \notin FV(\bar{r}) \cup FV(\psi)}{\{\phi\}\ \ldots;y \leftarrow (\textit{do } x \leftarrow p;\ q);\ldots\ \{\psi \parallel S\}}$$

$$(\text{dsef}) \quad \frac{p\ \text{dsef}}{\{\phi\}\ x \leftarrow p\ \{\phi \wedge x = p \parallel \bot\}}$$

$$(\text{raise}) \quad \frac{}{\{\}\ \textit{raise } e_0\ \{\bot \parallel \lambda e.\, e = e_0\}}$$

$$(\text{stateless}) \quad \frac{}{\{\textit{ret } \phi\}\ q\ \{\textit{ret } \phi \parallel \lambda e.\, \textit{ret } \phi\}}$$

$$(\text{catch}) \quad \frac{\{\phi\}\ \bar{x} \leftarrow \bar{p}\ \{\psi \parallel S\}}{\{\phi\}\ y \leftarrow (\textit{catch } \bar{x} \leftarrow p)\ \{\textit{case } y \textit{ of inl } \bar{x} \rightarrow \psi \mid \textit{inr } e \rightarrow S\ e \parallel \bot\}}$$

Fig. 3. The generic Hoare calculus for partial exception correctness

Total extended Hoare assertions for reasoning about total correctness are, as in the basic case, encoded in generic dynamic logic [21]; this requires additionally that the monad admits dynamic logic. In this respect, exceptions do not cause additional problems:

Theorem 12. *If a strong monad T admits dynamic logic, then so does $T(\_+E)$.*

If an exception monad T admits dynamic logic, then we can define combined total correctness assertions for normal and abnormal termination by

$$[\phi]\ \bar{x} \leftarrow \bar{p}\ [\psi \parallel S] :\equiv \{\phi\}\ \bar{x} \leftarrow \bar{p}\ \{\psi \parallel S\} \wedge \langle catch\ p\rangle\top.$$

A Hoare calculus for such *extended total Hoare assertions* can be derived from the equational axioms for exception monads by means of the proof system for generic dynamic logic [21]. As in the basic case, the rules are essentially the same as for extended partial Hoare assertions, with the exception of the (Y) rule (and, of course, the rules for constructs such as *while* and *iter* derived from it). In fact, there does not seem to be an immediately obvious total analogue of (Y); however, one can easily prove e.g. a total Hoare rule for *iter* (which then specializes to a corresponding total *while* rule (while-total) by taking *while b p = iter b p ()*):

$$
\text{(iter-total)}\quad
\frac{\begin{array}{c} t : A \to DB \\ \_<\_ : B \times B \to \Omega \text{ is well-founded} \\ [\phi\ x \wedge b\ x \wedge (t\ x = z)]\ y \leftarrow p\ x\ [\phi\ y \wedge (t\ y < z) \parallel S] \end{array}}{\{\phi\ e\}\ y \leftarrow iter\ b\ p\ e\ \{\phi\ y \wedge \neg(b\ y) \parallel S\}}
$$

In JML, the effect of a statement is specified by giving clauses for the precondition (`assumes`), the normal postcondition (`ensures`), the abnormal postcondition (`signals`), and a precondition for non-termination (`diverges`) which must hold before execution of a statement in cases where the statement hangs [8]. The specification $\{\texttt{assumes} = \phi, \texttt{ensures} = \psi, \texttt{signals} = S, \texttt{diverges} = \delta\}$ for a statement p can be expressed in the notation above as

$$\{\phi\}\ p\ \{\psi \parallel S\} \quad \text{and} \quad [\phi \wedge \neg\delta]\ p\ [\top \parallel \lambda e.\ \top]$$

(The observation that the `derives` clause can be coded out in this way is made already in [8], and indeed the calculus discussed there forces this coding to be used in proofs. Of course, taken literally, this works only classically, but intuitionistically, one would at any rate prefer a condition that guarantees termination, replacing $\neg\delta$, over one that is entailed by non-termination.) By consequence, also the Hoare calculus of [7] is expressible in our calculus (ignoring, for the time being, class constraints on the involved exceptions; given a formal description of the class mechanism as described e.g. in [6], such constraints can be expressed in the postcondition as carried out in [5]). Explicitly, we can put, e.g.,

$$\{\phi\}\ p\ \{\mathsf{exception}(S)\} :\equiv \{\phi\}\ p\ \{\top \parallel S\} \quad \text{and}$$
$$[\phi]\ p\ [\mathsf{exception}(S)] :\equiv [\phi]\ p\ [\bot \parallel S],$$

thus obtaining conditions stating that, under precondition ϕ,

- if p terminates abnormally with exception e, then the resulting state satisfies $S\,e$ (*partial exception correctness*), and
- p terminates abnormally with exception e in a state that satisfies $S\,e$ (*total exception correctness*), respectively.

In fact, $\{\phi\}\ x \leftarrow p\ \{\psi\ \|\ S\}$ is equivalent to the conjunction of $\{\phi\}\ x \leftarrow p\ \{\psi\}$ and $\{\phi\}\ p\ \{\text{exception}(S)\}$ (while no such simplification holds in general for total extended Hoare assertions).

Example 13. The exceptional Hoare triple $\{\phi\}\ p\ \{\text{exception}(S)\}$ is made explicit in $T\,(\_+E)$ for concrete T as follows:

- T the state monad: if execution of p from a state s satisfying ϕ terminates with exception e, then $S\,e$ holds in the poststate.
- T the non-deterministic state monad: whenever p possibly terminates exceptionally in state s' with exception e, then $S\,e$ holds in s'.
- T the input monad: whenever p throws an exception e after reading a string of inputs, e satisfies $S\,e$.

While there is, due to the fact that the calculus of [5, 7] is tailored specifically to Java, no precise one-to-one correspondence between the rules listed there and our generic rules, the central features of the former do in fact drop out of the generic calculus. In particular, the composition rules of [7] can be derived from rule (seq) in Figure 3 and its total analogue, and the *partial exception while rule* [5] is the projection of the generic while rule (obtained as a specialization of rule (iter) above) to partial exception correctness (i.e. normal correctness is just dropped in the conclusion). More interesting is the derivation of the *total exception while rule*: this rule (reformulated in analogy to the *total break while rule* [7]) translates into our calculus as

$$\frac{\begin{array}{c} [\phi \wedge b]\ p\ [\top\ \|\ \top] \\ \{\phi \wedge b \wedge t = n\}\ p\ \{\phi \wedge b \wedge t < n\ \|\ \top\} \\ \{\phi \wedge b\}\ p\ \{\top\ \|\ S\} \end{array}}{\{\phi \wedge b\}\ while\ b\ p\ \{\bot\ \|\ S\}}$$

(with exceptional Hoare triples already coded out). The premises can be gathered into the single total extended Hoare assertion

$$[\phi \wedge b \wedge t = n]\ p\ [\phi \wedge b \wedge t < n\ \|\ S],$$

and from this we can indeed draw the required conclusion by means of the generic total while rule (while-total) (see above), noting that in the normal postcondition, we get the contradiction $b \wedge \neg b$. Similarly, the Hoare rules for JML method specifications [8] are in direct correspondence with the generic rules of our calculus. All this is by no means intended to disparage JML (which constitutes a much wider effort involving many aspects of a concrete programming language), but rather goes to show that the kernel of known specialized Hoare calculi with exceptions is covered by our generic mechanism.

6   Conclusion and Future Work

We have generalized existing Hoare calculi for the Java monad [5, 7, 8] to a monad-independent Hoare calculus for exception monads, expressed within the specification language HASCASL. To this end, we have extended previous work [21, 23] on monad-independent calculi by adding postconditions for abrupt termination; this was based on an equational characterization of Moggi's exception monad transformer which arose from a rather striking categorical formulation.

This extension of the monad-independent Hoare logic is necessary for two reasons:

– the *catch* operation is not algebraic, but rather acts on top of the monad.
– exceptions are constants, so that the basic monad independent calculus cannot make statements about exceptional post-states

We argue that both these reasons, in particular the latter, apply uniquely to exceptions as a computational feature. By contrast, other computational effects are purely algebraic [18], and moreover based on non-constant algebraic operations. Hence, we expect that the genericity of our approach will allow for an easy integration of further monadic computational effects without any substantial extensions of the overall formalism as in the case of exceptions. Such additional computational effects include in particular input/output [17] and concurrency [3, 16]. While the corresponding monads have been designed for use with the functional language Haskell, we do not anticipate major obstacles in their re-use as extensions of the Java monad [9].

Acknowledgements

This work forms part of the DFG-funded project HasCASL (KR 1191/7-1). The authors wish to thank Christoph Lüth for useful comments and discussions.

References

[1] M. Barr and C. Wells, *Toposes, triples and theories*, Springer, 1984.
[2] M. Bidoit and P. D. Mosses, CASL *user manual*, LNCS, IFIP Series, vol. 2900, Springer, 2003.
[3] K. Claessen, *A poor man's concurrency monad*, J. Funct. Programming **9** (1999), 313–323.
[4] C. A. R. Hoare, *An axiomatic basis for computer programming*, Communications of the ACM **12** (1969), no. 10, 576–580.
[5] M. Huisman, *Java program verification in higher order logic with PVS and Isabelle*, Ph.D. thesis, University of Nijmegen, 2001.
[6] M. Huisman and B. Jacobs, *Inheritance in higher order logic: Modeling and reasoning*, Theorem Proving in Higher Order Logics, LNCS, vol. 1869, Springer, 2000, pp. 301–319.

[7] _____, *Java program verification via a Hoare logic with abrupt termination*, Fundamental Approaches to Software Engineering, LNCS, vol. 1783, Springer, 2000, pp. 284–303.

[8] B. Jacobs and E. Poll, *A logic for the Java Modeling Language JML*, Fundamental Approaches to Software Engineering, LNCS, vol. 2029, Springer, 2001, pp. 284–299.

[9] _____, *Coalgebras and Monads in the Semantics of Java*, Theoret. Comput. Sci. **291** (2003), 329–349.

[10] E. Moggi, *Categories of partial morphisms and the λ_p-calculus*, Category Theory and Computer Programming, LNCS, vol. 240, Springer, 1986, pp. 242–251.

[11] _____, *An abstract view of programming languages*, Tech. Report ECS-LFCS-90-113, Univ. of Edinburgh, 1990.

[12] _____, *Notions of computation and monads*, Inform. and Comput. **93** (1991), 55–92.

[13] T. Nipkow, *Jinja: Towards a comprehensive formal semantics for a Java-like language*, Proc. Marktobderdorf Summer School 2003, IOS Press, to appear.

[14] _____, *Hoare logics in Isabelle/HOL*, Proof and System-Reliability, Kluwer, 2002, pp. 341–367.

[15] S. Peyton-Jones (ed.), *Haskell 98 language and libraries — the revised report*, Cambridge, 2003, also: J. Funct. Programming **13** (2003).

[16] S Peyton Jones, A. Gordon, and S. Finne, *Concurrent Haskell*, Principles of Programming Languages, ACM Press, 1996, pp. 295–308.

[17] S. Peyton Jones and P. Wadler, *Imperative functional programming*, Principles of Programming Languages, ACM Press, 1993, pp. 71–84.

[18] G. Plotkin and J. Power, *Notions of computation determine monads*, Foundations of Software Science and Computation Structures, LNCS, vol. 2303, Springer, 2002, pp. 342–356.

[19] L. Schröder, *Classifying categories for partial equational logic*, Category Theory and Computer Science, ENTCS, vol. 69, 2002.

[20] _____, *Henkin models of the partial λ-calculus*, Computer Science Logic, LNCS, vol. 2803, Springer, 2003, pp. 498–512.

[21] L. Schröder and T. Mossakowski, *Monad-independent dynamic logic in* HASCASL, J. Logic Comput., to appear. Preliminary version in Recent Developments in Algebraic Development Techniques, LNCS 2755 (2003), Springer, pp. 425–441.

[22] L. Schröder and T. Mossakowski, HASCASL*: Towards integrated specification and development of functional programs*, Algebraic Methodology and Software Technology, LNCS, vol. 2422, Springer, 2002, pp. 99–116.

[23] _____, *Monad-independent Hoare logic in* HASCASL, Fundamental Aspects of Software Engineering, LNCS, vol. 2621, 2003, pp. 261–277.

[24] L. Schröder, T. Mossakowski, and C. Maeder, HASCASL *– Integrated functional specification and programming. Language summary*, Available at `http://www.informatik.uni-bremen.de/agbkb/forschung/formal_methods/CoFI/HasCASL`

[25] H. Thielecke, *Categorical structure of continuation passing style*, Ph.D. thesis, University of Edinburgh, 1997.

[26] D. von Oheimb, *Hoare logic for Java in Isabelle/HOL*, Concurrency and Computation: Practice and Experience **13** (2001), 1173–1214.

Expressing Iterative Properties Logically in a Symbolic Setting

Carron Shankland[1], Jeremy Bryans[2], and Lionel Morel[3]

[1] Department of Computing Science and Mathematics, University of Stirling, UK
[2] Centre for Software Reliability, University of Newcastle, UK
[3] Verimag, France

Abstract. We present a logic for reasoning about state transition systems (LOTOS behaviours) which allows properties involving repeated patterns over actions and data to be expressed. The state transition systems are derived from LOTOS behaviours; however, the logic is applicable to any similar formalism. The semantics of the logic is given with respect to *symbolic* transition systems, allowing reasoning about data to be separated from reasoning about flow of control. Several motivational examples are included.

Keywords: LOTOS, formal verification, temporal logics, infinite state systems, symbolic representation.

1 Introduction

When describing a system formally it is often useful to be able to do so at different levels of abstraction. This enhances our confidence in the correctness of our mental model of the system, especially if the different descriptions can be related mathematically to each other. An obvious example is the use of modal and temporal logics to describe abstract properties of a system which can then be validated with respect to a more concrete, implementation focussed, description. For example, the use of PROMELA and CTL in the model checker SPIN [1].

In previous work we have established a symbolic framework for reasoning about Full LOTOS [2] based on *Symbolic Transition Systems* (STS) [3]. This new semantics for LOTOS allows behaviour to be expressed in a more compact, elegant fashion than is possible using the standard semantics [2]. In particular, this is a way of avoiding the state explosion caused by the use of data in LOTOS. A modal logic called FULL has been defined [4], but although this logic has the desirable theoretical property of adequacy with respect to bisimulation [5], it lacks the expressiveness required in many applications.

For example, consider the simple buffer which accepts data at gate in and outputs that data at gate out (in pseudo LOTOS, B = in?x:Nat; out!x; B). An abstract property of this buffer is *"after an in event, an out event occurs"*. Of course, it would be useful to express the repetition of this pattern: *"after the in action, the out action happens, and then repeat"*. We also want to express something about the data involved in the transaction *"if we input the value x*

C. Rattray et al. (Eds.): AMAST 2004, LNCS 3116, pp. 460–474, 2004.
© Springer-Verlag Berlin Heidelberg 2004

then what is output is also x, and never anything else". Again, it is desirable to express this over a number of iterations.

Another simple example is the communication protocol dealing with lossy channels in which an item is sent and resent until an acknowledgement is received *"repeatedly send m until ack m"*. Here the goal is to repeat a single action an indeterminite (but finite) number of times.

These are simple examples, but they capture the principle that such repetitions are central to defining the behaviour of many systems, including communications protocols, games, hardware systems, operating systems, and telecommunications. Moreover, it is useful to be able to express these properties in a high level, abstract language such as temporal logic. Logics of this nature have been defined for LOTOS, most notably in association with the CADP toolset [6–8], and used in, for example, verifications of the reconfiguration protocol [9] and the IEEE 1394 tree identify protocol [10].

Our goal here is to provide a logic which has the expressivity described above, and is based on *symbolic* transition systems, allowing data to be dealt with in a more efficient fashion and avoiding some forms of state explosion. We present a logic, FULL*, which extends FULL [4]. FULL* is designed to be expressive: we want the ability to formulate as many useful properties as possible since we feel this is more important for the software development process than the theoretically desirable adequacy of FULL. The grammar of FULL* draws features familiar from traditional modal logics [11,4], and from XTL [12] which in turn draws on the language of regular expressions. This paper explores the expressivity of the resulting logic via a series of simple examples. In particular, key safety, liveness, response and reachability properties are examined.

2   Reasoning about Infinite State Systems

Over a number of years we have developed a simple and elegant framework in which it is possible to reason about both data and processes (or flow of control). The core of our framework (see Fig. 1) is a symbolic semantics for Full LOTOS [3] using Symbolic Transition Systems (STS) [13].

Although our main interest is in LOTOS, in fact STS may be used to express the semantics of many languages, therefore FULL* is more generally applicable. The reader is assumed to be familiar with state transition systems (see, for example, Fig. 2). Some very small examples with LOTOS syntax are used, but this should not hinder the reader who knows no LOTOS.

The standard semantics [2] are represented in the leftmost portion of Fig. 1, where labelled transition systems give meaning to LOTOS specifications. This is the world in which CADP and XTL are based. The righthand portion of the Figure represents our contribution: namely, the symbolic semantics for LOTOS, an HML-like logic (FULL) [4] and its extension FULL* (described in this paper), and various flavours of bisimulation [3] (not discussed further here). The arrows between the components represent relationships. For example, we have proved that the standard, concrete and symbolic bisimulations are all equivalent

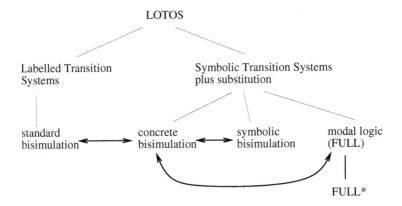

Fig. 1. Symbolic Framework for LOTOS

for closed processes (i.e. those with no free variables), and further, that the same equivalence relation is induced by the modal logic FULL. These results are all essential to showing the strength and self-consistency of our symbolic framework, and its consistency with the standard semantics over closed behaviours. In other words, we have not simply re-defined the LOTOS language in different terms, yielding a different language from that of the standard: we have striven to ensure that the two semantics can be used interchangeably, with ours offering computational advantages for reasoning about processes.

Below we reproduce the definitions associated with *symbolic transition systems* and FULL in order to give a basis for the definition of FULL*.

2.1   Symbolic Transition Systems

Labelled transition systems are commonly used for specification (e.g. process algebra semantics, IO automata). Hennessy and Lin [13] developed a *symbolic* version of transition systems to combat the problems of infinite breadth introduced by transitions over data, and of partial specification introduced through parameterised processes. Related work includes the symbolic simulator of Eertink [14] and the NTIF work of Garavel and Lang [15]. Both of these approaches are quite different from ours, having as a goal a more efficient implementation of simulation, equivalence checking, and/or model checking. Also, the NTIF development is aimed at interpretation of E-LOTOS behaviours.

Here we use a version of the Hennessy and Lin symbolic transition systems customised to fit the philosophy of communication in LOTOS: multiway synchronisation, i.e. more than two processes may communicate, and early acquisition of values, i.e. binding of values to variables at the point of synchronisation.

The main features of the STS are that transitions are decorated with a Boolean, b, and an action, α. The Boolean expresses under which conditions the transition may occur. Actions in LOTOS must be associated with a *gate*, and may additionally offer some *data*. Therefore we split actions into two sets:

SimpleEv and *StructEv*. An action in *SimpleEv* consists only of a gate name, g. An action in *StructEv* consists of a gate name g and a data expression d. We use α to denote an action in $SimpleEv \cup StructEv$.

The data expression may contain free variables. In this case we refer to it as *open*. Alternatively, it has no free variables and is therefore *closed*. Similarly, LOTOS behaviours, or states in a transition system, may be either open or closed. The function $fv()$ can be applied to a behaviour expression to determine the free variables of that expression. We do not repeat the definition of $fv()$ here. It can be found in [3].

Note there is no syntactic distinction between input and output events since this goes against the LOTOS interpretation of events. We can think of data offers such as g!42 as output events, and g?x:Nat as input events; however, it is more accurate to think of these as denoting different sets of values. For "output" events the data expression will evaluate to a single value, while for "input" events the data expression will be a variable name, potentially instantiated by many possible values. In STS the Boolean condition and data expression together encapsulate the set of possible values offered at a gate.

We repeat here the definition given in [3]:

Definition 1. *Symbolic Transition Systems*
An STS is composed of:

- *a non-empty set of states. To each state S, we associate a set of free variables, $fv(S)$. This can be computed from the syntactic behaviour associated with S.*
- *an initial state, S_0.*
- *a set of transitions $S \xrightarrow{b \quad \alpha} S'$, where b is a Boolean expression and $\alpha \in SimpleEv \cup StructEv$, such that $fv(S') \subseteq fv(S) \cup fv(\alpha)$ and $fv(b) \subseteq fv(S) \cup fv(\alpha)$ and $\sharp(fv(\alpha) - fv(S)) \leq 1$.*

That is, any new names in S' must come from the action α, the Boolean condition b may refer to variables in S and any new variable introduced in α, and only one variable may be introduced by α. The last part of the restriction is an artificial imposition by us to make the semantics clearer. In fact, LOTOS syntax allows multiple variables to be introduced and this can be modelled using lists.

Operationally, states are insufficient as a basis for our framework. Consider the simple buffer, repeating the actions in?x:Nat; out!x. To evaluate this process (e.g. with respect to another process for bisimulation, or with respect to a modal formula) a substitution of a value for the variable x is required; however, the substitution must change at every iteration. In order to accommodate this, the framework for reasoning must be based on *terms*.

A term T_σ is a pair (T, σ) where T is a node in the STS being considered and σ is a substitution. σ is a partial function from variables to new variables, or to values, i.e. from Var to $Var \cup Val$. We require $range(\sigma) \subseteq fv(T)$, this means parts of the substitution which are no longer relevant are discarded. In the definition of both FULL and FULL* the substitutions always map variables to values at the time of their first use, therefore for subsequent transitions the data item associated with an action is either a value, or a single variable name, indicating the introduction of that variable.

The notion of transitions between states in a symbolic transition system must be transposed to the notion of transitions between terms:

Definition 2. *Transitions Over Terms*

$$T \xrightarrow{b \quad a} T' \text{ implies } T_\sigma \xrightarrow{b\sigma \quad a} T'_{\sigma'}$$

$$T \xrightarrow{b \quad gE} T' \text{ implies } T_\sigma \xrightarrow{b\sigma \quad gE\sigma} T'_{\sigma'}$$
$$\text{where } fv(E) \subseteq fv(T)$$

$$T \xrightarrow{b \quad gx} T' \text{ implies } T_\sigma \xrightarrow{b\sigma[z/x] \quad gz} T'_{\sigma'[z/x]}$$
$$\text{where } x \notin fv(T) \text{ and } z \notin fv(T_\sigma)$$

In all cases, $\sigma' = fv(T') \lhd \sigma$, that is to say the restriction of σ only containing elements of $fv(T')$.

The third clause above is essential to allow alpha renaming during model checking when matching variables in the transition system to those in properties.

In [4] the logic FULL was defined over transitions on terms. Similarly, the logic FULL* will be presented over transitions on terms.

2.2   Paths

Given that FULL* will include iterative operators we also need to define the notion of a path over terms, denoted π, as a finite sequence of transitions: $\pi = t_1 \xrightarrow{b_1 \quad \alpha_1} t_2 \xrightarrow{b_2 \quad \alpha_2} \cdots t_{n-1} \xrightarrow{b_{n-1} \quad \alpha_{n-1}} t_n$. Given that STS include free variables this means that paths also contain free variables (in the states and in the labels α). This is evident in the rules for matching labels of Sect. 3.

Given π as above, operators on paths include:

$first(\pi)$ the first term of a path: $first(\pi) = t_1$

$last(\pi)$ the last term of a path: $last(\pi) = t_n$

$init(\pi)$ the initial transition on a path: $init(\pi) = t_1 \xrightarrow{b_1 \quad \alpha_1} t_2$

$\pi \cdot \pi_2$ If $\pi_2 = s_1 \xrightarrow{c_1 \quad \beta_1} s_2 \cdots s_{m-1} \xrightarrow{c_{m-1} \quad \beta_{m-1}} s_m$, and $t_n = s_1$, the concatenation operator $\pi \cdot \pi_2$ (or $\pi\pi_2$) is defined as
$$t_1 \xrightarrow{b_1 \quad \alpha_1} t_2 \cdots t_{n-1} \xrightarrow{b_{n-1} \quad \alpha_{n-1}} t_n \xrightarrow{c_1 \quad \beta_1} s_2 \cdots s_{m-1} \xrightarrow{c_{m-1} \quad \beta_{m-1}} s_m$$

Substitutions over paths are dealt with in the same way as substitutions over terms: on taking each transition the substitution is applied if it can be. It disappears as soon as it is no longer required.

2.3   FULL: A Modal Logic for Full LOTOS

FULL was proposed in [4] in order to express properties of STS and to capture the same equivalence as symbolic bisimulation over STS [5]. The form of the logic is inspired by HML (Hennessy-Milner Logic [11]), with the addition of quantifiers over data. FULL has the following syntax:

$$\Phi ::= b \mid \Phi_1 \wedge \Phi_2 \mid \Phi_1 \vee \Phi_2 \mid [a]\Phi \mid \langle a \rangle \Phi \mid [\exists x \ g]\Phi$$
$$\mid [\forall x \ g]\Phi \mid \langle \exists x \ g \rangle \Phi \mid \langle \forall x \ g \rangle \Phi$$

where b is a Boolean expression, $a \in SimpleEv$, $g \in G$ and $x \in Var$.

Formulae are evaluated over terms (defined above), pattern matching on the structure of the logic operators and over the transitions of the system. To illustrate the logic we give the semantics of the different versions of the $\langle\rangle$ operator. The complete semantics of FULL is given in [4].

$$t \models \langle a \rangle \Phi \quad = \exists t', t \xrightarrow{\text{tt} \ a} t' \text{ and } t' \models \Phi$$

$$t \models \langle \exists x \ g \rangle \Phi = \exists v \text{ s.t. either } \exists t', \ t \xrightarrow{\text{tt} \ gv} t' \text{ and } t' \models \Phi_{[v/x]}$$

$$\text{or } \exists t', \ t \xrightarrow{b \ gz} t' \text{ and } b_{[v/z]} \equiv \text{tt and } t'_{[v/z]} \models \Phi_{[v/x]}$$

$$t \models \langle \forall x \ g \rangle \Phi = \forall v \text{either } \exists t', t \xrightarrow{\text{tt} \ gv} t' \text{ and } t' \models \Phi_{[v/x]}$$

$$\text{or } \exists t', \ t \xrightarrow{b \ gz} t' \text{ and } b_{[v/z]} \equiv \text{tt and } t'_{[v/z]} \models \Phi_{[v/x]}$$

where t is a term, z, $x \in Var$ and $v \in Val$.

The features of the logic are that the data and transition quantifiers are tightly tied together, with the data quantifier always coming first. This logic captures all the discriminatory power of symbolic bisimulation, and with it we are able to express certain simple properties capturing ordering of events, and manipulation of data.

3   FULL*: Adding Iteration Operators to FULL

Clearly, a major drawback of FULL is that properties over sequences of actions whose length is unknown cannot be expressed. In turn the essential properties of liveness and safety cannot be expressed, yet FULL has a very desirable theoretical property: it captures exactly symbolic bisimulation over STS.

Here we propose some iteration operators to extend FULL to allow such properties to be expressed. The inspiration for the form of the new logic comes mainly from the work of Mateescu [12] on XTL, which is in turn influenced by the language of regular expressions. The basic innovation of both XTL and FULL* is to allow descriptions of properties of *paths* inside modal operators. A property over a path is expressed as a sequence of actions, possibly involving the repetition operators "+" (1 or more repetitions) and "*" (zero or more repetitions). We believe this is a natural and familiar way to express repetition, making this logic more accessible to LOTOS practitioners. The novel contribution of our work here is adding quantifiers over data and interpreting these operators over symbolic transition systems. In FULL* properties are expressed over paths which may contain free variables, whereas in XTL properties are interpreted over concrete transition systems (with no free variables, and a limited number of values for each data type).

We also take this opportunity to rationalise the semantics of FULL, introducing the $\neg$ operator and separating the definitions of quantification over data from those of quantification over transitions. In other ways the logic becomes more complex. In particular, the effect of the position of quantification has to be carefully considered, since it can completely change the meaning of a property.

Before presenting the formal details of FULL*, we present some examples to illustrate the form of the logic. Recall the simple buffer of the Introduction. At the simplest level we wish to write properties with sequences of events, such as

$$\langle \text{in} \cdot \text{out} \rangle \text{tt}$$

(i.e. one repetition of the actions in and out). We also wish to repeat patterns

$$\langle (\text{in} \cdot \text{out})^* \rangle \text{tt}$$

(i.e. zero or more repetitions of the actions in and out). Adding data yields

$$\langle (\exists x \text{ in } x \cdot \exists y (x = y) \text{ out } y)^* \rangle \text{tt}$$

which we expect to hold. The same property, with $(x = y)$ replaced by $(x \neq y)$ would not be expected to hold since it says the buffer outputs a different value from that input.

3.1   FULL*

Definition 3. *FULL* Grammar*

$$\Phi ::= b \mid \Phi_1 \wedge \Phi_2 \mid \langle R \rangle \Phi \mid \exists x (C) \Gamma \mid \neg \Phi$$
$$\Gamma ::= b \mid \Gamma_1 \wedge \Gamma_2 \mid \langle Q \rangle \Phi \mid \neg \Gamma$$
$$R ::= \alpha \mid \neg R \mid R_1 \wedge R_2 \mid R_1 \cdot R_2 \mid (R)^* \mid (R)^+ \mid \exists x (C) Q$$
$$Q ::= \alpha \mid \neg Q \mid Q_1 \wedge Q_2 \mid Q \cdot R \mid (Q)^* \mid (Q)^+$$

where b and C are Boolean expressions, and $\alpha \in SimpleEv \cup StructEv$.

Given this syntax, $\vee$, $\forall$ and $[\,]$ can all be defined as the duals of $\wedge$, $\exists$ and $\langle \, \rangle$.

The grammar describes modal formulae: Φ and Γ, and patterns (path properties) inside modal operators: R and Q. Both Γ and Q are auxiliary definitions: they are identical to Φ and R except that they may not include quantification as their first operator. This is to make the handling of quantification simpler. By imposing this structure on the grammar we ensure that a quantifier is always followed directly by an instance of the data being bound. For example, we allow

$$\exists x \langle \text{in } x \rangle \exists y \langle \text{out } y \rangle \text{tt}$$

but we do not allow

$$\exists x \exists y \langle \text{in } x \rangle \langle \text{out } y \rangle \text{tt}$$

Although this is an artificial constraint to make the association of quantifiers and variables more straightforward it is not unreasonable since data must always be associated with an action, and the expressivity of the logic is not affected.

Expressions within the logic FULL* are derived by beginning with Φ. We now explain the meaning of these expressions with respect to a Symbolic Transition System, via terms. We split the definition of the semantics into four parts corresponding to Φ, Γ, R and Q. In each part the semantics is given inductively over the structure of the logical formulae. The definitions are mutually recursive.

Definition 4. *FULL* Semantics: Φ*

$$
\begin{aligned}
t &\models b &&= b \\
t &\models \Phi_1 \wedge \Phi_2 &&= t \models \Phi_1 \ \wedge t \models \Phi_2 \\
t &\models \langle R \rangle \Phi &&= \exists \pi \ s.t. \ first(\pi) = t \ \wedge \ \pi \models R \ \wedge last(\pi) \models \Phi \\
t &\models \exists x(C)\Gamma &&= \exists v : Val \ s.t. \ C[v/x] \equiv \mathrm{tt} \ \wedge \ t, v \models \Gamma[v/x] \\
t &\models \neg \Phi &&= t \not\models \Phi
\end{aligned}
$$

Note in particular that the rule for existential quantification requires satisfaction via the rules for Γ expressions. A value has been chosen and must be carried into the next stage of the evaluation so it can be tied to a corresponding datum in the transition system. This binding is not possible in the Φ rules since there may be several transitions (with different actions) from t and only the first part of Γ will determine which particular action is of interest.

Γ expressions are evaluated over a pair (term, value), but note that Γ has already had the appropriate substitution applied $[v/x]$.

Definition 5. *FULL* Semantics: Γ*

$$
\begin{aligned}
t, v &\models b &&= b \\
t, v &\models \Gamma_1 \wedge \Gamma_2 &&= t, v \models \Gamma_1 \ \wedge \ t, v \models \Gamma_2 \\
t, v &\models \langle Q \rangle \Phi &&= \exists \pi \ s.t. \ first(\pi) = t \ \wedge \ \pi_\sigma \models Q \ \wedge \ last(\pi)_\sigma \models \Phi \\
t, v &\models \neg \Gamma &&= t, v \not\models \Gamma
\end{aligned}
$$

where $\sigma = [v/z]$ if $first\_data(\pi) = z$, empty otherwise.

The auxiliary function $first\_data(\pi)$ is defined as:

$$
first\_data(\pi) = \begin{array}{l} \text{if } init(\pi) = t_1 \xrightarrow{b \quad gz} t_2 \text{ then } z \\ \text{if } init(\pi) = t_1 \xrightarrow{b \quad gw} t_2 \text{ then } w \end{array}
$$

As soon as the path is chosen in the rule for $\langle Q \rangle$ it is possible to match v with a corresponding datum in the transition system, which may be a variable (z) or a data value (w). This is done both for the evaluation of Q and for the evaluation of the next section of modal formula Φ. The latter is important, since if there is no substitution here then any matching done in Q is forgotten, and the value v is not propagated. This will be illustrated in Sect. 4. If the datum returned by $first\_data(\pi)$ is a value w, then no substitution is applied in the $\langle Q \rangle \Phi$ rule above. This is because either $v = w$ and the substitution has no effect, or $v \neq w$ and the property will fail when we try to match the first action in Q to the first action of π (see Definition 7). The failure would not happen if a substitution $[v/w]$ had been made.

We now define formulae over paths described by R. We assume that any path chosen in the previous steps is an exact match for the length of R (so we don't have to use $init(\pi)$ for example to extract the first part). This is essential. Consider the case where a long path is extracted to match only a single action R. The rules then ask us to continue evaluating from $last(\pi)$. Clearly this misses out large chunks of behaviour and is undesirable.

Definition 6. *FULL* Semantics: R*

$$
\begin{aligned}
\pi &\models \alpha &&= \pi = t_1 \xrightarrow{tt \ \ \alpha_1} t_2 \wedge \ match(\alpha, \alpha_1) \\
\pi &\models \neg R &&= \pi \not\models R \\
\pi &\models R_1 \wedge R_2 &&= \pi \models R_1 \ \wedge \pi \models R_2 \\
\pi &\models R_1 \cdot R_2 &&= \pi = \pi_1 \pi_2 \ \wedge \ \pi_1 \models R_1 \ \wedge \ \pi_2 \models R_2 \\
\pi &\models (R)^* &&= \pi = \pi_1 \cdots \pi_p, \ p \geq 0 \wedge \forall i.1 \leq i \leq p. \ \pi_i \models R \\
\pi &\models (R)^+ &&= \pi = \pi_1 \cdots \pi_p, \ p > 0 \wedge \forall i.1 \leq i \leq p. \ \pi_i \models R \\
\pi &\models \exists x(C)Q &&= \exists v \ s.t. \ C[v/x] \equiv tt \ \wedge \ \pi_\sigma \models Q[v/x]
\end{aligned}
$$

where p is an integer, and $\sigma = [v/z]$ if first_data$(\pi) = z$, empty otherwise.

The *match* function takes two actions, matching the names of the actions, and data if there is data associated with the first action, ignoring it otherwise. This allows inexact matching of actions. If the logic specifies both gate and data then the transition system must match that exactly, but if only a gate is given in the logic then any data in the transition system is ignored.

Lastly we define formulae over Q patterns. Due to substitutions applied above α here can only be of the form gv if the formula is well formed (there must be data, since the first part of a Q formula has to match a quantifier).

Definition 7. *FULL* Semantics: Q*

$$
\begin{aligned}
\pi &\models \alpha &&= \pi = t_1 \xrightarrow{tt \ \ \alpha} t_2 \\
\pi &\models \neg Q &&= \pi \not\models Q \\
\pi &\models Q_1 \wedge Q_2 &&= \pi \models Q_1 \ \wedge \pi \models Q_2 \\
\pi &\models Q \cdot R &&= \pi = \pi_1 \pi_2 \ \wedge \ \pi_1 \models Q \ \wedge \ \pi_2 \models R \\
\pi &\models (Q)^* &&= \pi = \pi_1 \cdots \pi_p, \ p \geq 0 \wedge \forall i.1 \leq i \leq p. \ \pi_i \models Q \\
\pi &\models (Q)^+ &&= \pi = \pi_1 \cdots \pi_p, \ p > 0 \wedge \forall i.1 \leq i \leq p. \ \pi_i \models Q
\end{aligned}
$$

It can be easily demonstrated that all the properties which can be written with FULL can be expressed in FULL* just by not considering iterative mechanisms and by restricting the position of quantification. Therefore FULL* has at least the expressive power of FULL. More importantly, the addition of iterative operators give the extra flexibility required to express safety and liveness properties, as can be seen in the next section.

4   Examples of the Use of FULL*

The semantics presented above are complex, so we present some examples in detail to illustrate how the operators might be used, and to give an informal explanation of their meaning. We sketch one example unfolding a property using the rules above, but the remainder are omitted due to lack of space.

A Simple Buffer. We return again to the buffer example. Given the transition systems of Fig. 2 we can say

$$
t_1 \models \langle (\exists x \ \mathbf{in} \ x \cdot \mathbf{out} \ x)^* \rangle tt
$$

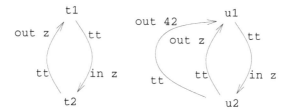

Fig. 2. Buffer processes t and u respectively

where t_1 is the initial state of process t. In fact this is rather weak, since $*$ expressions can be satisfied vacuously.

$$t_1 \models \langle (\exists x \text{ in } x \cdot \text{out } x)^+ \rangle \text{tt}$$

is better, but this property may be satisfied by only a single iteration of the pattern. Really we're trying to express a *liveness property* here, which can be best formulated as

$$t_1 \models \neg \langle \neg (\exists x \text{ in } x \cdot \text{out } x)^+ \rangle \text{tt}$$

i.e. it is not possible that the buffer does not perform a cycle of in and out actions. This property also holds for u_1 which is not the usual sort of buffer, since it sometimes outputs the value 42, regardless of the input, so to more accurately describe the buffer something stronger has to be said:

$$t_1 \models [(\exists x \text{ in } x \cdot \exists y (x \neq y) \text{out } y)^+] \text{ff}$$

This formula says that the pattern inside the box operator is not possible (since it is followed by ff which cannot be satisfied). This is true for t since the pattern says it is possible to output a value different from the one input. *Safety properties* will in general have this form; however, the presence of the box operator means the property could be vacuously true, so it is necessary to always combine this sort of property with ones expressing that some appropriate behaviour is possible (like the liveness property above). Obviously,

$$u_1 \not\models [(\exists x \text{ in } x \cdot \exists y (x \neq y) \text{out } y)^+] \text{ff}$$

since u_1, the initial state of process u of Fig. 2, leads to u_2 which can output 42 at any time.

To focus on the position of quantifiers, and the substitution of values for variables, consider the following

$$t_1 \models [(\exists x \text{ in } x)][\exists y (x = y) \text{out } y] \text{tt}$$

This seems to be a desirable property (similar to that above) but it is meaningless because the value of x is "forgotten" outside its modal quantifier, so the expression $x = y$ cannot be evaluated. So, if properties are to relate different quantified variables, then those variables must be bound inside the same modal operator. Conversely, if one variable is used across a number of modal operators then it must be bound *outside* the modal operators. Consider, for example, the evaluation of the following property:

$$t_1 \models \exists x \langle \texttt{in } x \rangle \langle \texttt{out } x \rangle \texttt{tt}$$

First unfold using the rule for $\exists x(C)\Gamma$ of Definition 4, substituting v for x throughout the formula.

$$\exists v : Val \text{ s.t. } \texttt{tt} \wedge t_1, v \models \langle \texttt{in } v \rangle \langle \texttt{out } v \rangle \texttt{tt}$$

Second, use the rule for $\langle Q \rangle \Phi$ of Definition 5, and σ carries the mapping information $[v/z]$ through the transition system.

$$\exists v : Val \text{ s.t. } \exists \pi \ first(\pi) = t_1 \ \wedge \ \pi_\sigma \models \texttt{in } v \ \wedge \ last(\pi)_\sigma \models \langle \texttt{out } v \rangle \texttt{tt}$$

Otherwise $last(\pi)$ will not be a suitable model for $\langle \texttt{out } v \rangle \texttt{tt}$ since z is not bound to any value. Success of evaluation relies on choosing the correct path π.

Finally, consider the property

$$\exists x \langle (\texttt{in } x \cdot \texttt{out } x)^* \rangle \texttt{tt}$$

This is unsatisfactory because it allows only one value to ever be input (repeatedly). This highlights a feature of the logic: in fact, in repeated patterns the quantifiers are most likely to be *inside* the repetition inside a modal operator, *allowing* new values in each iteration. Also, as was seen above, the scope of a quantifier inside an operator is just that operator, so many properties will consist only of one or two modal operators but with more complex patterns inside. This has implications for *adequacy*, see Sect. 5.

Communications. Consider the classic communications protocol Alternating Bit Protocol (ABP). We do not give the STS here. The idea is that a producer and consumer of data are communicating over lossy channels, therefore to ensure data sent is received some acknowledgements are set up. The protocol itself uses a sender and a receiver, and data is sent with a single bit tag. Each of the sender and receiver maintain a record of the tag, or what the tag ought to be in the case of the receiver, to help detect errors. For example, if the sender sends (d, b) and the receiver gets (d, b'), where b' denotes the inverse of b, then the receiver knows there has been an error and does not send its acknowledgement.

From the outside, the ABP behaves just like the buffer: data items go in and come out reliably, therefore all of the properties described above can be applied if internal behaviour of the protocol is ignored.

Consider first of all a view of the sending and receiving messages. The protocol behaviour says that any particular message is sent repeatedly until correctly received. This *reachability property* is expressed:

$$abp \models \langle \exists m (\texttt{send } m)^+ \cdot \texttt{receive } m \rangle \texttt{tt}$$

From the sender's point of view, the successful **receive** is not apparent until an acknowledgement with the correct tag arrives, so we might require the following property of the sender:

$$sender \models \langle \exists b \exists d \ (\textbf{send} \ (d, b))^+ \cdot \textbf{ack} \ b \rangle tt$$

We take the liberty of using a pair type to represent the data and the tag sent. We can be even more explicit about the behaviour at the sender end:

$$sender \models \langle \exists b \exists d \ (\textbf{send} \ (d, b) \cdot (\textbf{ack} \ err \ \lor \ \textbf{ack} \ b'))^* \cdot \textbf{ack} \ b \rangle tt$$

reflecting the repetition of the pattern that a message is sent, and acknowledgement received (possibly of the error type, possibly with the wrong tag), until a correct acknowledgement is received. Note the use of $*$ here to allow for the possibility that errors may not happen at all.

The sender, once a message is successfully sent, switches to using the tag b'. If the above pattern inside the $\langle \ \rangle$ operator from the $\exists d$ is named p_b and the property with b and b' swapped named p_not_b then

$$sender \models \langle \exists b((p\_b) \cdot (p\_not\_b))^* \rangle tt$$

Now we look more generally at some communication based examples. Consider that the values sent have a particular sequence, in particular, that the value sent increases from each sending to its successor:

$$cp \models [(\forall x(\textbf{send} \ x \cdot \neg(\exists y \ (y < x)\textbf{send} \ y)))^*]tt$$

That is, for all finite paths π beginning at cp, π can be segmented into $\pi_1 \cdots \pi_p$ where $\pi_i \models \forall x(\textbf{send} \ x \cdot \neg(\exists y \ (y < x)\textbf{send} \ y))$. However, if π is grouped into consecutive pairs starting with the first send, then this says nothing about the relationship between the data of the second and third sending. If we call the property above $inc\_pairwise$ then a better formulation may be

$$cp \models inc\_pairwise \land \ [\textbf{send}](inc\_pairwise)$$

effectively staggering the property to consider alternate pairs (2,3), (4,5) and so on, as well as pairs (1,2), (3,4), etc.

Mutual Exclusion. In the classic problem of mutual exclusion there are a number of processes executing which have *critical* and *non-critical* sections. It is required that only one process access its critical section at any one time. A master process therefore controls entry and exit from critical sections. A process i may enter (**InCritical** i) or leave (**OutCritical** i) its critical section. The correctness property for mutual exclusion is then

$$\forall i \ [\textbf{InCritical} \ i] \ \forall j(j \neq i) \ [(\neg \textbf{InCritical} \ j)^* \ \textbf{OutCritical} \ i]tt$$

Assume processes request permission to enter their critical section Ask_Critical i, and then Wait until permission is granted. Processes are guaranteed access to their critical section by:

$$\forall j \ [(\text{Ask\_Critical } j \ \cdot (\text{Wait } j)^* \ \cdot \text{InCritical } j)^*]\text{tt}$$

Note the use of nested occurrences of the * operator.

As with all [ ] formulae care must be taken to also express properties which can do something, to avoid the case that the [ ] formula is satisfied vacuously.

Summary. Some rules of thumb for expressing properties can be given:

- *Safety properties*: express that bad things do not happen, therefore can be described in the form [bad_action]ff.
- *Liveness properties*: express that good things do happen, therefore can be expressed in the form $\neg\langle\neg(\text{good\_action})^+\rangle\text{tt}$.
- *Reachability properties*: express that eventually it is possible to do a good action, ignoring some sequence of other actions which may come first, and can be expressed as $\langle(\text{other\_pattern})^* \cdot \text{good\_action}\rangle\text{tt}$.
- *Response properties*: are similar to reachability properties. These may also use the stronger pattern $[(\text{other\_pattern})^*]\langle\text{good\_action}\rangle\text{tt}$ to indicate that on *all* paths from the current state the desired action is attainable.

5   Comparison with FULL

Clearly, FULL* has more expressive power than FULL. In particular, FULL* can distinguish between processes which are symbolically bisimilar [3].

Given the three bisimilar STS of Fig. 3, A, B and C, with initial states $a1$, $b1$ and $c1$, the following FULL* formulae distinguish them:

$$a1 \models [\forall x \ \text{send } x]\text{tt}$$
$$b1 \not\models [\forall x \ \text{send } x]\text{tt}$$
$$b1 \models [(\exists x(x \le 4) \ \text{send } x]\text{tt}$$
$$c1 \not\models [(\exists x(x \le 4) \ \text{send } x]\text{tt}$$

The crucial point in the formulation of the properties above is that quantification over transitions happens first (which may also constrain data), and then quantification over data occurs. This gives us the opportunity to choose the "wrong" transition. Symbolic bisimulation on the other hand allows us to ignore to some extent the way data is distributed across transitions. As long as the same set of data is used in total, then we can find a matching transition in the other process. In the logic FULL, this translates to ensuring that the data quantifier *precedes* the associated transition quantifier to maintain the link with symbolic bisimulation. This is done by tightly associating quantifiers over data and those over transitions in the semantics.

FULL* allows the data quantifier to follow the transition quantifier (in fact, quantifiers may occur in any order). The final Buffer example above illustrates

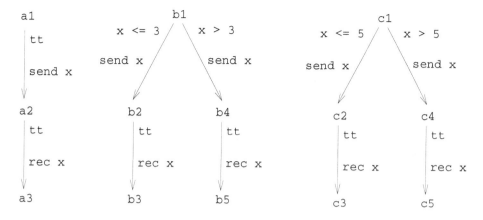

Fig. 3. Bisimilar processes A, B and C respectively

why this is necessary. If repetition patterns are used with data, and the data is required to change each time through the pattern, then it has to be possible to have the data quantifiers inside the modal operators. Therefore we have traded expressivity of the logic against adequacy with symbolic bisimulation.

6   Conclusion

We have presented an extension to the FULL logic with iteration operators. This allows properties over paths whose length is unknown to be expressed. The form of these operators was derived from XTL [12], extended with explicit quantifiers over data and interpreted over symbolic transition systems. In FULL* fundamental verification properties such as liveness, safety and response can be defined, as illustrated in Sect. 4. This is important since the uptake of formal methods for description of systems relies on the expressivity and ease of use of the language for realistic problems. Another key factor is automated analysis, and one of our goals in developing the symbolic framework is to make analysis more tractable by reducing the state space of the system.

Further work will validate this logic using "real-life" examples of LOTOS specification using data, aided by a prototype model checking implementation in CADP. The symbolic nature of the transitions here is not reflected since CADP is built on Binary Coded Graph representation, but the prototype is useful for gaining confidence in the logic. A larger example has been completed using the benchmark RPC case study [16]. The properties expressed concern the ability of the memory to respond appropriately to a call, and are rather similar to those given for the communications protocol above. We will carry out further case studies to demonstrate the expressivity and useful features of FULL*.

References

1. Holzmann, G.J.: The model checker SPIN. IEEE Transactions on Software Engineering **23** (1997) 279–295
2. International Organisation for Standardisation: Information Processing Systems – Open Systems Interconnection – LOTOS – A Formal Description Technique Based on the Temporal Ordering of Observational Behaviour. (1988)
3. Calder, M., Shankland, C.: A Symbolic Semantics and Bisimulation for Full LOTOS. In Kim, M., Chin, B., Kang, S., Lee, D., eds.: Proceedings of FORTE 2001, 21st International Conference on Formal Techniques for Networked and Distributed Systems, Kluwer Academic Publishers (2001) 184–200
4. Calder, M., Maharaj, S., Shankland, C.: A Modal Logic for Full LOTOS based on Symbolic Transition Systems. The Computer Journal **45** (2002) 55–61
5. Calder, M., Maharaj, S., Shankland, C.: An Adequate Logic for Full LOTOS. In Oliveira, J., Zave, P., eds.: Formal Methods Europe'01. LNCS 2021, Springer-Verlag (2001) 384–395
6. Fernandez, J.C., Garavel, H., Kerbrat, A., Mateescu, R., Mounier, L., Sighireanu, M.: CADP (CAESAR/ALDEBARAN Development Package): A Protocol Validation and Verification Toolbox. In Alur, R., Henzinger, T., eds.: Proceedings of CAV'96. Number 1102 in Lecture Notes in Computer Science, Springer-Verlag (1996) 437–440
7. Mateescu, R., Garavel, H.: XTL: A Meta-Language and Tool for Temporal Logic Model-Checking. In: Proceedings of the International Workshop on Software Tools for Technology Transfer STTT'98 (Aalborg, Denmark). (1998)
8. Mateescu, R., Sighireanu, M.: Efficient On-the-Fly Model-Checking for Regular Alternation-Free Mu-Calculus. In: Proceedings of the 5th International Workshop on Formal Methods for Industrial Critical Systems FMICS'2000 (Berlin, Germany). (2000) Full version available as INRIA Research Report RR-3899.
9. Cornejo, M.A., Garavel, H., Mateescu, R., de Palma, N.: Specification and Verification of a Dynamic Reconfiguration Protocol for Agent-Based Applications. Technical Report 4222, INRIA (2001)
10. Sighireanu, M., Mateescu, R.: Verification of the Link Layer Protocol of the IEEE-1394 Serial Bus (FireWire): an Experiment with E-LOTOS. Springer International Journal on Software Tools for Technology Transfer (STTT) **2** (1998) 68–88
11. Hennessy, M., Milner, R.: Algebraic Laws for Nondeterminism and Concurrency. Journal of the Association for Computing Machinery **32** (1985) 137–161
12. Mateescu, R.: Vérification des propriétés temporelles des programmes parallèles. PhD thesis, Institut National Polytechnique de Grenoble (1998)
13. Hennessy, M., Lin, H.: Symbolic Bisimulations. Theoretical Computer Science **138** (1995) 353–389
14. Eertink, H.: Simulation Techniques for the Validation of LOTOS Specifications. PhD thesis, University of Twente (1994)
15. Garavel, H., Lang, F.: NTIF: A General Symbolic Model for Communicating Sequential Processes with Data. In Peled, D., Vardi, M., eds.: Formal Techniques for Networked and Distributed Systems - FORTE 2002. LNCS 2529, Springer-Verlag (2002) 276–291
16. Broy, M., Merz, S., Spies, K., eds.: Formal Systems Specification: The RPC-Memory Specification Case Study. Number 1169 in Lecture Notes in Computer Science. Springer-Verlag (1996)

Extending Separation Logic with Fixpoints and Postponed Substitution

Élodie-Jane Sims[1,2]

[1] CNRS & École Polytechnique, STIX, 91128 Palaiseau, France
[2] Kansas State University, CIS, Manhattan KS 66506, USA
Elodie-Jane.Sims@polytechnique.fr

Abstract. We are interested in static analysis of programs which use shared mutable data structures. We introduce a backward and a forward analyses with a separation logic called $BI^{\mu\nu}$. This logic is an extension of BI logic [7], to which we add fixpoint connectives and a postponed substitution. This allows us to express recursive definitions within the logic as well as the axiomatic semantics of `while` statements. Unlike the existing rule-based approach to program proof using separation logic, our approach does not have syntactical restrictions on the use of rules.

1 Introduction

In this paper we address the problem of doing static analysis of programs [2] which use shared mutable data structures. The final goal of our work is to detect errors in a program (problems of dereferencing, aliasing, etc.) or to prove that a program is correct (with respect to these problems) in an automatic way. John Reynolds, Peter O'Hearn and others have developed [7,10] an extension of Hoare logic called separation logic (also known as *BI* logic) that permits reasoning about such programs. The classical definition of predicates on abstract data structures is extended by introducing a "separating conjunction", denoted $*$, which asserts that its sub-formulae hold for disjoint parts of the heap, and a closely related "separating implication", denoted $-\!\!*$. This extension permits the concise and flexible description of structures with controlled sharing.

We extend this logic with fixpoint connectives to define recursive properties and to express the axiomatic semantics of a `while` statement. We present forward and backward analyses *(sp (strongest postcondition), wlp (weakest liberal precondition)* expressed for all statements and all formulae).

Organization of the paper In Sect. 2 we describe the commands language we analyze and in Sect. 3 we present our logic $BI^{\mu\nu}$. In Sect. 4, we provide a backward analysis with $BI^{\mu\nu}$ in terms of "weakest liberal preconditions". We express the *wlp* for the composition, if $-$ then $-$ else and `while` commands. In Sect. 5, we provide a forward analysis with $BI^{\mu\nu}$ in terms of "strongest postconditions".

C. Rattray et al. (Eds.): AMAST 2004, LNCS 3116, pp. 475–490, 2004.
© Springer-Verlag Berlin Heidelberg 2004

Background Hoare logic [6] and Dijkstra-style weakest-precondition logics [4] are well known. It is also well known that these logics disallow *aliasing*, that is, the logics require that each program variable names a distinct storage location. Therefore, it is difficult to reason about programs that manipulate pointers or heap storage.

Through a series of papers [9,7,10], Reynolds and O'Hearn have addressed this foundationally difficult issue. Their key insight is that a command executes within a *region* of heap storage: they write

$$s, h \models \phi$$

to denote that property ϕ holds true within heap subregion h and local-variable stack s. One could also say that a formula describes some properties of the memories it represents. For example, ϕ might be:

emp   means that the heap is empty

$E \mapsto a, b$ means that there is exactly one cell in the heap, the one
          containing the values of a and b and that E points to it.

$E \hookrightarrow a, b$ is the same except that the heap can contain additional
          cells

With the assistance of a new connective, the "separating conjunction", denoted $*$, Reynolds and O'Hearn write

$$s, h_1 \cdot h_2 \models \phi_1 * \phi_2$$

to assert that both ϕ_1 and ϕ_2 hold but use *disjoint* heap subregions to justify their truth — there is no aliasing between the variables mentioned in ϕ_1 and ϕ_2. For example, consider the two cases below.

If $s = \begin{bmatrix} x \to l_1 \\ y \to l_2 \end{bmatrix}$, $h = \begin{bmatrix} l_1 \to \langle 3, 4 \rangle \\ l_2 \to \langle 1, 2 \rangle \end{bmatrix}$ then $s, h \models (x \mapsto 3, 4) * (y \mapsto 1, 2)$ or also $s, h \models (x \hookrightarrow 3, 4)$ but $s, h \not\models (x \mapsto 3, 4)$.

If $s = \begin{bmatrix} x \to l_1 \\ y \to l_1 \end{bmatrix}$, $h = [l_1 \to \langle 3, 4 \rangle]$ then $s, h \models (x \mapsto 3, 4) \wedge (y \mapsto 3, 4)$ but $s, h \not\models (x \mapsto 3, 4) * (y \mapsto 3, 4)$.

Adjoint to the separating conjunction is a "separating implication":

$$s, h \models \phi_1 \twoheadrightarrow \phi_2$$

asserts, "if heap region h is augmented by h' such that $s, h' \models \phi_1$, then $s, h \cdot h' \models \phi_2$". For example: if $s = \begin{bmatrix} x \to l_1 \\ y \to l_2 \end{bmatrix}$, $h = [l_1 \to \langle 3, 4 \rangle]$ then $s, h \models (y \mapsto 1, 2) \twoheadrightarrow *((x \mapsto 3, 4) * (y \mapsto 1, 2))$.

Istiaq and O'Hearn [7] showed how to add the separating connectives to a classical logic, producing a *bunched implication* logic (*BI* or *separation logic*) in which Hoare-logic-style reasoning can be conducted on while-programs that manipulate temporary-variable stacks and heaps.

A Hoare triple, $\{\phi_1\} C \{\phi_2\}$, uses assertions ϕ_i, written in separation logic; the semantics of the triple is stated with respect to a stack-heap storage model.

Finally, there is an additional inference rule, the *frame rule*, which formalizes compositional reasoning based on disjoint heap regions:

$$\frac{\{\phi_1\}C\{\phi_2\}}{\{\phi_1 * \phi'\}C\{\phi_2 * \phi'\}}$$

where ϕ''s variables are not modified by C.

The reader interested in the *set-of-inference-rules* approach for separation logic is invited to read [7], and [11] for details on the frame rule. The rules could also be found in a survey on separation logics [10]. We do not present the set of rules in this paper.

Our contribution Istiaq and O'Hearn's efforts were impressive but incomplete: the weakest-precondition and strongest postcondition semantics for their while-language were absent, because these require *recursively defined* assertions, which were not in their reach.

The primary accomplishment of this paper is to add least- and greatest-fixed-point operators to separation logic, so that pre- and post-condition semantics for the while-language can be wholly expressed within the logic. As a pleasant consequence, it becomes possible to formalize recursively defined properties on inductively (and co-inductively) defined data structures, e.g.,

$$\text{nonCircularList}(x) =$$
$$\mu X_v.\,(x = \text{nil}) \vee \exists x_1, x_2.(\text{isval}(x_1) \wedge (x \mapsto x_1, x_2 \; * \; X_v[x_2/x]))$$

asserts that x is a linear, non-circular list ($\text{isval}(x_1)$ insures that x_1 is a value, this predicate is defined later).

The addition of the recursion operators comes with a price: the usual definition of syntactic substitution and the classic substitution laws become more complex; the reasons are related both to the semantics of stack- and heap-storage as well as the inclusion of the recursion operators; details are given later in the paper.

2   Commands and Basic Domains

We consider a simple "while"-language with Lisp-like expressions for accessing and creating cons cells.

2.1   Command Syntax

The commands we consider are as follows.

$$C ::= x := E \mid x := E.i \mid E.i := E' \mid x := \text{cons}(E_1, E_2) \mid \text{dispose}(E)$$
$$\mid C_1; C_2 \mid \text{if } E \text{ then } C_1 \text{ else } C_2 \mid \text{skip} \mid \text{while } E \text{ do } C_1$$
$$i ::= 1 \mid 2$$
$$E ::= x \mid 42 \mid \text{nil} \mid True \mid False \mid E_1 \; op \; E_2$$

An expression can denote an integer, an atom, or a cons cell. Here x is a variable in Var and op is an operator in $(Val \times Val) \to Val$ like $+ : (Int \times Int) \to Int$, $Or : (Bool \times Bool) \to Bool$ (about Var and Val see Sect. 2.2).

The second and third assignment statements read and update the heap, respectively. The fourth creates a new cons cell in the heap, and places a pointer to it in x.

Notice that in our language we do not handle two dereferencings in a simple statement (no $x.i.j$, no $x.i := y.j$), this restriction is just for simplicity and does not limit the expressivity of the language (it will just require the addition of intermediate variables).

2.2   Semantic Domains

$$Val = Int \cup Bool \cup Atoms \cup Loc$$
$$S = Var \to_{fin} Val$$
$$H = Loc \to_{fin} Val \times Val$$

Here, $Loc = \{l_1, l_2, ...\}$ is an infinite set of locations, $Var = \{x, y, ...\}$ is an infinite set of variables, $Atoms = \{\texttt{nil}, a, ...\}$ is a set of atoms, and $\to_{fin}$ is for finite partial functions. We call an element $s \in S$ a stack, and $h \in H$ a heap. We also sometimes call an element $(s, h) \in S \times H$ a state or a memory.

We use $dom(h)$ to denote the domain of definition of a heap $h \in H$, and $dom(s)$ to denote the domain of a stack $s \in S$.

An expression is interpreted as a heap-independent value: $[\![E]\!]s \in Val$. For example, $[\![x]\!]s = s(x)$, $[\![42]\!]s = 42$, $[\![\mathbf{true}]\!]s = true$, $[\![E_1 + E_2]\!]s = [\![E_1]\!]s + [\![E_2]\!]s$

2.3   Small-Step Semantics

The semantics of the statements are given in small-step semantics defined by the relation $\leadsto$ on configurations. The configurations include triples C, s, h and terminal configurations s, h for $s \in S$ and $h \in H$.

In the following rules, we use r for elements of $Val \times Val$, $\pi_i r$ with $i \in \{1, 2\}$ for the first or second projection, and $(r|i \to v)$ for the pair like r except that the i'th component is replaced with v, $[s \mid x \to v]$ for the stack like s except that it maps x to v, $(h - l)$ for $h_{\restriction dom(h) \setminus \{l\}}$.

$$\frac{[\![E]\!]s = v}{x := E, s, h \leadsto [s|x \to v], h} \qquad \frac{[\![E]\!]s = l \quad h(l) = r}{x := E.i, s, h \leadsto [s|x \to \pi_i r], h}$$

$$\frac{[\![E]\!]s = l \quad h(l) = r \quad [\![E']\!]s = v'}{E.i = E', s, h \leadsto s, [h|l \to (r|i \to v')]} \qquad \frac{l \in dom(h) \quad [\![E]\!]s = l}{\texttt{dispose}(E), s, h \leadsto s, (h - l)}$$

$$\frac{l \in Loc \quad l \notin dom(h) \quad [\![E_1]\!]s = v_1, [\![E_2]\!]s = v_2}{x := \texttt{cons}(E_1, E_2), s, h \leadsto [s|x \to l], [h|l \to \langle v_1, v_2 \rangle]}$$

$$\frac{C_1, s, h \leadsto C', s', h'}{C_1; C_2, s, h \leadsto C'; C_2, s', h'} \qquad \frac{C_1, s, h \leadsto s', h'}{C_1; C_2, s, h \leadsto C_2, s', h'} \qquad \frac{}{\texttt{skip}, s, h \leadsto s, h}$$

$$\frac{\llbracket E \rrbracket s = True}{\text{if } E \text{ then } C_1 \text{ else } C_2, s, h \rightsquigarrow C_1, s, h} \qquad \frac{\llbracket E \rrbracket s = False}{\text{if } E \text{ then } C_1 \text{ else } C_2, s, h \rightsquigarrow C_2, s, h}$$

$$\frac{\llbracket E \rrbracket s = False}{\text{while } E \text{ do } C, s, h \rightsquigarrow s, h} \qquad \frac{\llbracket E \rrbracket s = True}{\text{while } E \text{ do } C, s, h \rightsquigarrow C; \text{while } E \text{ do } C, s, h}$$

The location l in the $\mathtt{cons}$ case is not specified uniquely, so a new location is chosen non-deterministically.

Let the set of error configurations be: $\Omega = \{C, s, h \mid \nexists K.\ C, s, h \rightsquigarrow K\}$. We say that:

"C, s, h is safe" if and only if $\forall K.\ (C, s, h \rightsquigarrow^* K \Rightarrow K \notin \Omega)$
"C, s, h is stuck" if and only if $C, s, h \in \Omega$

For instance, an error state can be reached by an attempt to dereference $\mathtt{nil}$ or an integer. Note also that the semantics allows dangling references, as in stack $[x \to l]$ with empty heap $[]$.

The definition of safety is formulated with partial correctness in mind: with loops, C, s, h could fail to converge to a terminal configuration but not get stuck.

3 $BI^{\mu\nu}$

In this section, we present the logic $BI^{\mu\nu}$. It is designed to describe properties of the memory. Typically, for analysis it will be used in Hoare triples of the form $\{P\}C\{Q\}$ with P and Q formulae of the logic and C a command.

We present in Sect. 3.1 the syntax of the logic and in Sect. 3.2 its formal semantics. In Sect. 3.3, we give the definition of a *true triple* $\{P\}C\{Q\}$. At the end, in Sect. 3.4, we discuss the additions to separation logic (fixpoints and postponed substitution).

3.1   Syntax of Formulae

$P, Q, R ::=$	$E = E'$	Equality		$E \mapsto E_1, E_2$	Points to
	$\mathtt{false}$	Falsity	$\mid$	$P \Rightarrow Q$	Classical Imp.
$\mid$	$\exists x.P$	Existential Quant.	$\mid$	$\mathtt{emp}$	Empty Heap
$\mid$	$P * Q$	Spatial Conj.	$\mid$	$P \mathrel{-\!\!*} Q$	Spatial Imp.
$\mid$	X_v	Formula Variable	$\mid$	$P[E/x]$	Postponed Substitution
$\mid$	$\nu X_v.P$	Greatest Fixpoint	$\mid$	$\mu X_v.P$	Least Fixpoint

We have an infinite set of variables, $Formula\_Variables$, used for the variables bounded by μ and ν and disjoint from the set Var. They range over sets of states, the others ($x,y,...$) are variables which range over values. For the sake of clarity, uppercase variables with the subscript $_v$ are used to define recursive formulae. We use the term "closed" for the usual notion of closure of variables in Var (closed by $\exists$ or $\forall$) and the term "v-closed" for closure of variables in $Formula\_Variables$ (v-closed by μ or ν).

Our additions to Reynolds and O'Hearn's separation logic are the fixed-point operators $\mu X_v.\ P$ and $\nu X_v.\ P$ and the substitution construction $P[E/x]$.

We can define various other connectives as usual, rather than taking them as primitives:

$$\neg P \triangleq P \Rightarrow \mathtt{false} \qquad \mathtt{true} \triangleq \neg(\mathtt{false})$$
$$P \vee Q \triangleq (\neg P) \Rightarrow Q \qquad P \wedge Q \triangleq \neg(\neg P \vee \neg Q)$$
$$\forall x.P \triangleq \neg(\exists x.\neg P) \qquad E \hookrightarrow a,b \triangleq \mathtt{true} * (E \mapsto a,b)$$
$$x = E.i \triangleq \exists x_1, x_2. (E \hookrightarrow x_1, x_2) \wedge (x = x_i)$$

The set $FV(P)$ of free variables of a formula is defined as usual. The set $Var(P)$ of variables of a formula is defined as usual with $Var(P[E/x]) = Var(P) \cup Var(E) \cup \{x\}$.

3.2   Semantics of Formulae

We use the following notation in formulating the semantics:

- $h \sharp h'$ indicates that the domains of heaps h and h' are disjoint;
- $h \cdot h'$ denotes the union of disjoint heaps (i.e., the union of functions with disjoint domains).

We express the semantics of the formulae in an environment ρ mapping formula variables to set of states: $\rho : Formula\_Variables \rightharpoonup \mathcal{P}(S \times H)$. The semantics of a formula in an environment ρ is the set of states which satisfy it, and is expressed by: $\Gamma_\rho : BI^{\mu\nu} \rightharpoonup \mathcal{P}(S \times H)$

We call $\gamma(P)$ the semantics of a formula P in an empty environment $\gamma(P) = \Gamma_\emptyset(P)$. We also define a forcing relation of the form:

$$s, h \models P \text{ if and only if } s, h \in \gamma(P)$$

and an equivalence: $P \equiv Q$ if and only if $\forall \rho.\Gamma_\rho(P) = \Gamma_\rho(Q)$. We write $\top$ for $S \times H$.

$$
\begin{aligned}
\Gamma_\rho(E = E') &= \{s, h \mid \llbracket E \rrbracket s = \llbracket E' \rrbracket s\} \\
\Gamma_\rho(E \mapsto E_1, E_2) &= \{s, h \mid dom(h) = \{\llbracket E \rrbracket s\} \\
&\quad \text{and } h(\llbracket E \rrbracket s) = \langle \llbracket E_1 \rrbracket s, \llbracket E_2 \rrbracket s \rangle\} \\
\Gamma_\rho(\mathtt{false}) &= \emptyset \\
\Gamma_\rho(P \Rightarrow Q) &= (\top \setminus \Gamma_\rho(P)) \cup \Gamma_\rho(Q) \\
\Gamma_\rho(\exists x.P) &= \{s, h \mid \exists v \in Val.[s|x \to v], h \in \Gamma_\rho(P)\} \\
\Gamma_\rho(\mathtt{emp}) &= \{s, h \mid h = []\} \\
\Gamma_\rho(P * Q) &= \{s, h \mid \exists h_0, h_1 \; h_0 \sharp h_1, \; h = h_0 \cdot h_1 \\
&\quad s, h_0 \in \Gamma_\rho(P) \text{ and } s, h_1 \in \Gamma_\rho(Q)\} \\
\Gamma_\rho(P -\!\!* Q) &= \{s, h \mid \forall h', \text{ if } h \sharp h' \text{ and } s, h' \in \Gamma_\rho(P) \text{ then} \\
&\quad s, h \cdot h' \in \Gamma_\rho(Q)\} \\
\Gamma_\rho(X_v) &= \rho(X_v), \text{ if } X_v \in dom(\rho) \\
\Gamma_\rho(\mu X_v . P) &= \mathrm{lfp}_{\overline{\emptyset}}^{\subseteq} \lambda X. \, \Gamma_{[\rho|X_v \to X]}(P) \\
\Gamma_\rho(\nu X_v . P) &= \mathrm{gfp}_{\overline{\emptyset}}^{\subseteq} \lambda X. \, \Gamma_{[\rho|X_v \to X]}(P) \\
\Gamma_\rho(P[E/x]) &= \{s, h \mid s[x \to \llbracket E \rrbracket s], h \in \Gamma_\rho(P)\}
\end{aligned}
$$

In both cases μ and ν, the X in λX is a fresh variable over sets of elements in $S \times H$ which does not already occur in ρ.

Notice that Γ_ρ is only a partial function. In definitions above, $\mathrm{lfp}_{\bar\emptyset}^{\subseteq}\phi$ $(\mathrm{gfp}_{\bar\emptyset}^{\subseteq}\phi)$ is the least fixpoint *(greatest fixpoint)* of ϕ on the poset $\langle \mathcal{P}(S \times H), \subseteq \rangle$, if it exists. Otherwise, $\Gamma_\rho(\mu X_v.P)$ $(\Gamma_\rho(\nu X_v.P))$ is not defined. For example, this is the case for $\mu X_v . (X_v \Rightarrow \mathtt{false})$.

We do not give a syntactical criterion for formulae with defined semantics (like parity of negation under a fixpoint, etc.), they are the usual ones knowing that in terms of monotonicity, $\rightarrow\!*$ acts like $\Rightarrow$, $*$ acts like $\wedge$, and $[\, / \,]$ do not interfere.

3.3   Interpretation of Triples

Hoare triples are of the form $\{P\}C\{Q\}$, where P and Q are assertions in $BI^{\mu\nu}$ and C is a command. The interpretation adopted ensures that well-specified commands do not get stuck (by this, it differs from the usual interpretation of Hoare triples).

$\{P\}C\{Q\}$ is a true triple iff $\forall s, h$ if $s, h \models P$ and $FV(Q) \subseteq dom(s)$ then
- C, s, h is safe
- if $C, s, h \rightsquigarrow^* s', h'$ then $s', h' \models Q$

This is a partial correctness interpretation; with looping, it does not guarantee termination. This is the reason of expressing "weakest liberal preconditions" for our backward analysis and not "weakest preconditions". However, the safety requirement rules out certain runtime errors and, as a result, we do not have that $\{\mathtt{true}\}C\{\mathtt{true}\}$ holds for all commands. For example, $\{\mathtt{true}\}x := \mathtt{nil}; x.1 := 3\{\mathtt{true}\}$ fails.

3.4   Adding of Fixpoints and Postponed Substitution

In this section, we discuss our motivations for those additions. We show that the postponed substitution connective $[\, / \,]$ is not the classical substitution $\{ \, / \, \}$ and that the usual variable renaming theorem does not hold for $\{ \, / \, \}$.

First motivation Our first motivation for adding fixed-point operators to separation logic came from the habit of the separation logic community to define informally recursive formulae and to use them in proofs of correctness.

Since we have added fixed-point operators to the logic, we can formally and correctly express, for example, that x is a non-cyclic finite list as

$$\mathtt{nclist}(x) = \mu X_v. \, (x = \mathtt{nil}) \vee \exists x_1, x_2.(\mathtt{isval}(x_1) \wedge (x \mapsto x_1, x_2 \ * \ X_v[x_2/x]))$$

and that x is non-cyclic finite or infinite list

$$\mathtt{nclist}(x) = \nu X_v. \, (x = \mathtt{nil}) \vee \exists x_1, x_2.(\mathtt{isval}(x_1) \wedge (x \mapsto x_1, x_2 \ * \ X_v[x_2/x]))$$

where $\mathtt{isval}(x) = (x = \mathtt{true}) \vee (x = \mathtt{false}) \vee (\exists n.n = x + 1)$

In earlier papers [8], Reynolds and O'Hearn use the definition,

$$\mathtt{nclist}(x) = (x = \mathtt{nil}) \vee \exists x_1, x_2.(\mathtt{isval}(x_1) \wedge (x \mapsto x_1, x_2 \ * \ \mathtt{nclist}(x_2)))$$

which is not within the syntax of separation logic.

Second motivation The second motivation was the formulations of the wlp ($\{\ ?\ \}C\{P\}$) and sp ($\{P\}C\{\ ?\ \}$) in the case of while commands, which was not possible earlier. This problem is nontrivial: For separation logic without fixed-points, we might express sp as

$$sp(P, \text{while } E \text{ do } C) = (\text{lfp} \overset{\models}{\text{false}} \lambda X.sp(X \wedge E = \text{true}, C) \vee P) \wedge (E = \text{false})$$

with $\text{lfp} \overset{\models}{\text{false}} \lambda X.F(X)$ defined, if it exists, as a formula P which satisfies:

- $P \equiv F(P)$
- for any formula Q, $(Q \equiv F(Q)$ implies $P \models Q$)

with $P \equiv Q$ if and only if $P \models Q$ and $Q \models P$.

But this implies that during the computation of the sp, each time a while loop would occur, we must find a formula in existing separation logic that was provably the fixpoint, so that we could continue the computation of the sp. In an other sense, this "work" could be seen as the "work" of finding the strongest loop invariant in the application of the usual rule for while loop.

Our addition of fixpoints (and the related postponed substitution) allows us to express the sp directly within the logic: $sp(P, \text{while } E \text{ do } C) = (\mu X_v.sp(X_v \wedge E = \text{true}, C_1) \vee P) \wedge (E = \text{false})$.

Although the definitions of the wlp and sp for the while loop are simple and elegant, the "work" of finding loop invariants is not skipped, however it is now postponed for when we have a specific proof to undertake. For example, we are working on translations of formulae into some other domains, and we have to find an approximation of the translation of fixpoints which is precise and not too expensive to compute. The advantage here is that this work of building the translation is done once and for all, then the analysis can be fully automated while the methodology of a proof system and finding loop invariant implies hand work.

$[\ /\ ]$ is not $\{\ /\ \}$ In this paper, we use the notation $P\{E/x\}$ for capture-avoiding syntactical substitution (that is, the usual substitution of variables). Recall that $[\ /\ ]$ is a connective of the logic (called *postponed substitution*) and is not equivalent to $\{\ /\ \}$.

The distinction between $[\ /\ ]$ and $\{\ /\ \}$ can be viewed in this example:

$$\{\text{true}\}x := y\{\text{true}\} \text{ is false}$$
(the command will be stuck from a state that has no value in its stack for y)

which implies that the axiom for assignment, $\{P\{y/x\}\}x := y\{P\}$, is unsound.

In other versions of separation logic [10], $\{P\{y/x\}\}x := y\{P\}$ was correct, since the definition of a true triple requires $FV(C, Q) \subseteq dom(s)$ and not merely $FV(Q) \subseteq dom(s)$ like here and also because there was no recursion. We believe that our definition is better since it does not require variables of the program to have a default value in the stack and it checks whether a variable has been assigned before we try to access its value.

Unfolding Notice that as usual, we have:
$$\mu X_v.P \equiv P\{\mu X_v.P/X_v\}$$
$$\nu X_v.P \equiv P\{\nu X_v.P/X_v\}$$

$\{/\}$: no variable renaming theorem $\exists y.P \not\equiv \exists z.P\{z/y\}$ with $z \notin Var(P)$
(when $y \neq z$)
Here are two counterexamples to the equivalence:

Example 1: $\gamma(\nu X_v.y = 3 \wedge \exists y.(X_v \wedge y = 5)) \not\equiv \gamma(\nu X_v.y = 3 \wedge \exists z.(X_v \wedge z = 5))$
because (*) let $A \triangleq \gamma(a) \triangleq \gamma(\nu X_v.y = 3 \wedge \exists y.(X_v \wedge y = 5)) = \emptyset$
and (**) let $B \triangleq \gamma(\nu X_v.y = 3 \wedge \exists z.(X_v \wedge z = 5)) = \gamma(y = 3)$
Explanation:
(*) the "$y = 3 \wedge$" part says that all the states in A should satisfy $y = 3$,
the "$\exists y.X_v \wedge y = 5$" part says that for all states in A, we can put a value to y
such that the state satisfies $y = 5$ and is still in A, which brings a contradiction
with the previous condition, so we have $A = \emptyset$.
(**) the "$y = 3 \wedge$" part says that all the states in B should satisfy $y = 3$,
the "$\exists z.X_v \wedge z = 5$" part says that for all states in B, we can put a value to
z such that the states satisfies $z = 5$ and is still in B which is possible for any
state of B, so we have $B = \gamma(y = 3)$.

Example 1 shows that variable renaming has a special behavior when applied
to a formula which is not v-closed.

Example 2: $\gamma(\exists y.\nu X_v.y = 3 \wedge \exists y.(X_v \wedge y = 5)) \not\equiv \gamma(\exists z.\nu X_v.z = 3 \wedge \exists y.(X_v \wedge y = 5))$
because (***) let $C \triangleq \gamma(c) \triangleq \gamma(\exists y.\nu X_v.y = 3 \wedge \exists y.(X_v \wedge y = 5)) = \emptyset$
and (****) let $D \triangleq \gamma(d) \triangleq \gamma(\exists z.\nu X_v.z = 3 \wedge \exists y.(X_v \wedge y = 5)) = S \times H$
Explanation:
(***) $c = \exists y.a$,
the result comes directly from the explanation of the case of A
(****) if we take the explanation of B, we get that $E \triangleq \gamma(e) \triangleq \gamma(\nu X_v.z = 3 \wedge \exists y.(X_v \wedge y = 5)) = \gamma(z = 3)$, then since $d = \exists z.e$ we have that the states in
D are such that we can find a value to give to z such that $z = 5$, this is trivially
true and then any state is in D, so we have $D = S \times H$.

Example 2 shows that variable renaming has a special behavior in case of μ
if the variable renamed is bounded inside the μ .

Full substitution The previous Example 2 leads to the definition of a new
substitution.
Let $\{[\ / \ ]\}$ be a full syntactical variable substitution: $P\{[z/y]\}$ is P in which all
y are replaced by z wherever they occur, for example:
$$(\exists y.P)\{[z/y]\} \triangleq \exists z.(P\{[z/y]\}), \quad (P[E/x])\{[z/y]\} \triangleq (P\{[z/y]\})[E\{z/y\}/x\{z/y\}]$$

The variable renaming theorem for $BI^{\mu\nu}$ If P is v-closed and $z \notin Var(P)$, $y \notin FV(P)$ then $P \equiv P\{[z/y]\}$ (in particular $\exists y.P \equiv \exists z.(P\{[z/y]\})$).

Equivalences on [/]

- If P is v-closed and if $x_1 \notin Var(E)$ and $x_1 \neq x_2$, then:
  $(\exists x_1.P)[E/x_2] \equiv \exists x_1.(P[E/x_2])$.

- $(\exists x.P)[E/x] \equiv (\exists x.P) \wedge is(E)$.

- $(A \vee C)[E/x] \equiv (A[E/x]) \vee (C[E/x])$

- If $y \notin Var(P)$ then
  $(\mu X_v.P)[y/x] \equiv (\mu X_v.P\{[y/x]\}) \wedge is(y)$
  $(\nu X_v.P)[y/x] \equiv (\nu X_v.P\{[y/x]\}) \wedge is(y)$
  One would want this for E instead of y but this is not possible in this way since $(P[E'/x])\{[E/x]\}$ is not defined because $P[E'/E]$ is not defined (what has to be substituted must be a variable). This is the point of the existence of [/] as a connective.
  To understand the last rule, we can come back to the program point of view seeing fixpoints as while loops and [/] as assignments, the precondition for $x := w$; while $x = y$ do $x := x + 1$ is the same as the one for while $w = y$ do $w := w+1$ (if you have seen Sect. 4 this would be $(\nu X_v.(x \neq y) \vee ((x = y) \wedge X_v[x+1/x]))[w/x] \equiv is(w) \wedge (\nu X_v.(w \neq y) \vee ((w = y) \wedge X_v[w+1/w])))$.

Example of unfolding `nclist42`

Let $\texttt{nclist42}(x) \triangleq \mu X_v.(x = \texttt{nil}) \vee \exists x_2.((x \mapsto 42, x_2) * X_v[x_2/x])$ with $x_2 \neq x$.

One would expect there $X_v[x_2/x]$ to be equivalent to $\texttt{nclist42}(x_2)$.

Remark that $\texttt{nclist42}(x_2) \triangleq \mu X_v.(x_2 = \texttt{nil}) \vee \exists x_3.((x_2 \mapsto 42, x_3) * X_v[x_3/x_2])$ with $x_3 \neq x_2$.

Let's prove that $X_v[x_2/x]$ is equivalent to $\texttt{nclist42}(x_2)$.

$$
\begin{aligned}
\texttt{nclist42}(x) \quad &\triangleq \quad \mu X_v.(x = \texttt{nil}) \vee \exists x_2.((x \mapsto 42, x_2) * X_v[x_2/x]) \\
(unfolding) \quad &= \quad (x = \texttt{nil}) \vee \exists x_2.((x \mapsto 42, x_2)* \\
&\quad ((\mu X_v.(x = \texttt{nil}) \vee \exists x_2.((x \mapsto 42, x_2) * X_v[x_2/x]))[x_2/x])) \\
(\text{variable renaming for } BI^{\mu\nu}) = \quad & (x = \texttt{nil}) \vee \exists x_2.((x \mapsto 42, x_2)* \\
&\quad ((\mu X_v.(x = \texttt{nil}) \vee \exists x_3.((x \mapsto 42, x_3) * X_v[x_3/x]))[x_2/x])) \\
(\text{simplification } [/] \text{ case } \mu) = \quad & (x = \texttt{nil}) \vee \exists x_2.((x \mapsto 42, x_2)* \\
&\quad (\mu X_v.(x_2 = \texttt{nil}) \vee \exists x_3.((x_2 \mapsto 42, x_3) * X_v[x_3/x_2]))) \\
&\triangleq \quad (x = \texttt{nil}) \vee \exists x_2.((x \mapsto 42, x_2) * \texttt{nclist42}(x_2))
\end{aligned}
$$

So we have $\texttt{nclist42}(x_2) \equiv (\texttt{nclist42}(x))[x_2/x]$ as expected.

Why $BI + \mu + \nu \neq BI^{\mu\nu}$? Or, why do we need to add $[\,/\,]$ to the syntax ?

Informally said, one can see the fixpoints as a while loop and $[\,/\,]$ as an assignment, then if we have a while loop followed by an assignment, we cannot include the assignment within the loop. So, if an analysis postponed the computation of while loop/fixpoints, then it also has to postpone the computation of assignment/$[\,/\,]$.

This need of $[\,/\,]$ is not surprising. In [3], de Bakker proved that for his simple logic with only fixpoints, there was no sp for the while loop statements.

We define $is(E)$ as a shortcut for $E = E$, which is just a formula insuring that E has a value in the current memory.

For P without any μ, ν, X_v in it, we have $P[E'/x] \equiv P\{E'/x\} \wedge is(E')$. But for $BI^{\mu\nu}$ without the connective $[\,/\,]$, there is no formula in the logic equivalent to $P[E'/x]$, which means that $[\,/\,]$ has to be in the logic syntax.

For example, $(\exists y.P)[E/x] \not\equiv \exists y.(P[E/x])$ when $y \neq x$ but $y \in Var(E)$ but the renaming theorem: $\exists y.P \equiv \exists z.P\{z/y\}$ with $z \notin Var(P)$ does not hold, so the attempt to find an equivalent formula for $(\exists y.P)[E/x]$ will fail.

4   Backward Analysis

We now define the *weakest liberal precondition* (wlp) semantics of the while-loop language with pointers.

If we can establish $\{P\}C\{\texttt{true}\}$ then we will know that execution of C is safe in any state satisfying P. So for our backward analysis, we express wlp such that

$$\{wlp(P,C)\}C\{P\} \text{ is true}$$

Most of the clauses are from Istiaq and O'Hearn [7], but our definition for while E do C is new and crucial.

$$
\begin{aligned}
&wlp(P, && x := E) && = P[E/x] \\
&wlp(P, && x := E.i) && = \exists x_1 \exists x_2.(P[x_i/x] \wedge (E \hookrightarrow x_1, x_2)) \\
& && && \quad\text{with } x_i \notin FV(E, P) \\
&wlp(P, && E.1 := E') && = \exists x_1 \exists x_2.(E \mapsto x_1, x_2) * ((E \mapsto E', x_2) \mathbin{-\!\!*} P) \\
& && && \quad\text{with } x_i \notin FV(E, E', P) \\
&wlp(P, && E.2 := E') && = \exists x_1 \exists x_2.(E \mapsto x_1, x_2) * ((E \mapsto x_1, E') \mathbin{-\!\!*} P) \\
& && && \quad\text{with } x_i \notin FV(E, E', P) \\
&wlp(P, && x := \mathsf{cons}(E_1, E_2)) && = \forall x'.(x' \mapsto E_1, E_2) \mathbin{-\!\!*} P[x'/x] \\
& && && \quad\text{with } x' \notin FV(E_1, E_2, P) \\
&wlp(P, && \mathsf{dispose}(E)) && = P * (\exists a \exists b.(E \mapsto a, b)) \\
& && && \quad\text{with } a, b \notin FV(E) \\
&wlp(P, && C_1; C_2) && = wlp(wlp(P, C_2), C_1)
\end{aligned}
$$

$$wlp(P, \text{if } E \text{ then } C_1 \text{ else } C_2) =$$
$$(E = \texttt{true} \wedge wlp(P, C_1)) \vee (E = \texttt{false} \wedge wlp(P, C_2))$$

$$
\begin{aligned}
&wlp(P, && \mathsf{skip}) && = P \\
&wlp(P, && \text{while } E \text{ do } C_1) && = \nu X_v.((E = \texttt{true} \wedge wlp(X_v, C_1)) \\
& && && \qquad\qquad \vee (E = \texttt{false} \wedge P)) \\
& && && \quad\text{with } X_v \text{ not in } P
\end{aligned}
$$

4.1 Proof of the Backward Analysis

Definition of wlp_o To prove that our definition indeed defines the *wlp*, we will formally relate it to the inverse state-transition function defined by the operational semantics in the domain $\mathcal{P}(S \times H)$:

$wlp_o(\Delta, C) = \{s, h \mid C, s, h \text{ is safe} \wedge (\text{if } C, s, h \leadsto^* s', h' \text{ then } s', h' \in \Delta\}$

We can instantiate the definition of a true triple as:

$$\{wlp(P, C)\}C\{P\} \text{ true}$$
$$\text{if and only if}$$
$$\left(\begin{array}{c} \gamma(wlp(P, C)) \\ \cap\{s, h \mid FV(P) \subseteq dom(s)\} \end{array} \right) \subseteq wlp_o(\gamma(P), C)$$

$$
\begin{array}{ccc}
BI & \xleftarrow{\;wlp\;} & \boxed{BI} \\
\downarrow{\scriptstyle\gamma} & & \downarrow{\scriptstyle\gamma} \\
op & \xleftarrow{\;wlp_o\;} & op \\
\subseteq & &
\end{array}
$$

To prove that our analysis is correct, we express wlp_o for each command, and prove by induction on the syntax of C that for each C and P, we have $\gamma(wlp(P, C)) \subseteq wlp_o(\gamma(P), C)$.

To prove that those preconditions are the weakest we have established that

$$\gamma(wlp(P, C)) = wlp_o(\gamma(P), C)$$

5 Forward Analysis

In the previous section, we have defined *wlp* for each C and P such that $\{wlp(P, C)\}C\{P\}$ is true.

The strongest postcondition semantics $sp(P, C)$ is not always defined, i.e., we can find C and P such that there exists no Q that makes $\{P\}C\{Q\}$ true. This is due to the fact that a true triple requires C to be executable from all states satisfying P *(and also such that $FV(Q) \subseteq dom(s))$* which is obviously not the case for some C and P. (For example, $\{\text{true}\}x = nil; y = x.1; \{?\}$ has no solution, since all states satisfy P but the command can never be executed - nil.1 is not defined).

We therefore have to split the analysis into two steps. The first step checks whether C is executable from all states satisfying P or not. The second step gives $sp(P, C)$ that makes the triple $\{P\}C\{sp(P, C)\}$ true if C is executable from all states satisfying P.

5.1 Step 1: $wlp(\text{true}, C)$

C is executable from any state satisfying P if and only if $P \models wlp(\text{true}, C)$. So for the first step we express the $wlp(\text{true}, C)$ formulae. These are just the formulae given in Sect. 4, instantiated for $P = \text{true}$.

5.2   Step 2: $sp(P, C)$ in Case $P \models wlp(\mathtt{true}, C)$

The postcondition formulae are:

$$
\begin{aligned}
sp(P, \quad x := E) \quad &= \exists x'.\, P[x'/x] \wedge x = E\{x'/x\} \\
&\quad \text{with } x' \notin FV(E, P) \\
sp(P, \quad x := E.i) \quad &= \exists x'.\, P[x'/x] \wedge x = (E\{x'/x\}).i \\
&\quad \text{with } x' \notin FV(E, P) \\
sp(P, \quad E.1 := E') \quad &= \exists x_1 \exists x_2.(E \mapsto E', x_2) * ((E \mapsto x_1, x_2) \twoheadrightarrow P) \\
&\quad \text{with } x_i \notin FV(E, E', P) \\
sp(P, \quad E.2 := E') \quad &= \exists x_1 \exists x_2.(E \mapsto x_1, E') * ((E \mapsto x_1, x_2) \twoheadrightarrow P) \\
&\quad \text{with } x_i \notin FV(E, E', P) \\
sp(P, \quad x := \mathtt{cons}(E_1, E_2)) \quad &= \exists x'.(P[x'/x] * (x \mapsto E_1\{x'/x\}, E_2\{x'/x\})) \\
&\quad \text{with } x' \notin FV(E_1, E_2, P) \\
sp(P, \quad \mathtt{dispose}(E)) \quad &= \exists x_1, x_2.\,((E \mapsto x_1, x_2) \twoheadrightarrow P) \\
&\quad \text{with } x_1, x_2 \notin FV(E, P) \\
sp(P, \quad C_1; C_2) \quad &= sp(sp(P, C_1), C_2) \\
sp(P, \mathtt{if}\ E\ \mathtt{then}\ C_1\ \mathtt{else}\ C_2) &= sp(P \wedge E = \mathtt{true}, C_1) \\
&\quad \vee sp(P \wedge E = \mathtt{false}, C_2) \\
sp(P, \quad \mathtt{skip}) \quad &= P \\
sp(P, \quad \mathtt{while}\ E\ \mathtt{do}\ C_1) \quad &= (\mu X_v.sp(X_v \wedge E = \mathtt{true}, C_1) \vee P) \\
&\quad \wedge (E = \mathtt{false}) \\
&\quad \text{with } X_v \text{not in } P
\end{aligned}
$$

5.3   If $P \not\models wlp(\mathtt{true}, C)$

If $P \not\models wlp(\mathtt{true}, C)$, we conclude that C cannot be executable from all states satisfying P, but for those states from which C is executable, the final states satisfy $sp(P \wedge wlp(\mathtt{true}, C), C)$.

5.4   Proof of the Forward Analysis

We want to prove that: If $P \models wlp(\mathtt{true}, C)$ then $\{P\}C\{sp(P, C)\}$ true.

Definition of sp_o We define the strongest postcondition in the operational domain:
$$sp_o(\Delta, C) = \{s', h' \mid \exists s, h \in \Delta.\ C, s, h \leadsto^* s', h'\}$$

Now, we can instantiate the definition of a true triple as:

$$\{P\}C\{Q\} \text{ true if and only if } P \models wlp(\mathtt{true}, C) \wedge sp_o(\gamma(P), C) \subseteq \gamma(Q)$$

To prove that our analysis is correct, we express sp_o for each command, and prove by induction on the syntax of C that for each C, and P, we have

$$BI \xrightarrow{\quad sp \quad} BI$$

If $P \models wlp(\texttt{true}, C)$
then $sp_o(\gamma(P), C) \subseteq \gamma(sp(P, C))$

$$op \xrightarrow{\quad sp_o \quad} \subseteq op$$

But since sp_o is defined such that it only collects the final states of successful computations, we must only prove that for each command C:

$sp_o(\gamma(P), C) \subseteq \gamma(sp(P, C))$

We have proved that our postconditions are the strongest by proving that:

$$sp_o(\gamma(P), C) = \gamma(sp(P, C))$$

5.5   Remark on Using the Result of Our Analyses

Our analyses always give the most precise and correct result. However, most of the time the result is not suited for humans. Hence, a theorem prover may be necessary. For example, the user will express some initial condition before running a program, the analysis will give the safe initial condition ($wlp(\texttt{true}, C)$) and the user will have to use a theorem prover to check whether the initial condition implies the safety initial condition (or do it by hand). In the other way, the user can express some postcondition and check if the postcondition provided by the analysis implies the postcondition that was expected.

Of course no theorem prover can always answer those questions since the validity of a formula is undecidable [1]. The satisfiability of the μ and ν may also be a problem in practice.

Even if the analyses here are not of direct use to automatically check programs, they can still be useful. We plan to use the logic as an interface language for other point-to analyses . For this we would translate the domain of those analyses into the logic (and conversely). Furthemore, we would also like to prove the correctness of some of those analyses, and then the expressions of our wlp and sp could be helpful. Those wlp and sp could also be useful as an intermediate step for an analysis which would consist in first going from a program to a formula of $BI^{\mu\nu}$, then to another more abstract domain. Then the usual challenge of finding a loop invariant would result in finding an approximation for the translation of μ and ν in this last domain. This last approach is our current and future work.

6   Conclusion

Checking programs dealing with pointers is an old challenge but still alive. There are many pointer analyses but none of them would be elected as the one which is precise enough, efficient and entirely formally proved. We do not get into the challenge in this paper but we think separation logics could become a good help to work towards this aim.

We have proposed an extension of separation logic, with fixpoint connectives and postponed substitution. First, this allows us to express formulae of recursive definitions within the logic. Secondly, it allows us to express the sp and wlp in the case of `while` statements (in our knowledge, there is no forward analysis using separation logic in the literature). We expressed the sp and wlp operators for any statements without any syntactical restrictions on the formulae provided as pre- or post-conditions. This leads to automatic analyses which is a quite different approach from the set of rules analyses which usually include a big part of human work for each program.

Towards real program analyses, the remaining difficulty is to interpret the results of our analyses.

This could be solved by finding some human understandable formulae that are equivalent to the one provided by our analyses, or proving that the result implies (or is implied by) another formula. This would need use of theorem provers, or approximation technics (our analyses are the most precise). (D. Galmiche and D. Méry are doing some work on building a theorem prover [5] but not for our syntax of separation logic. Indeed Calcagno and others proved in [1] that deciding the validity of an assertion in separation logic is not decidable.).

Another approach, which is an ongoing work, is to translate the formulae into some other (more specific or approximating) domains and reversely. This could lead to a two step analysis: first step translate the program into formulae with the wlp or sp, second step translate the formulae into the chosen domain. This approach is parallel to the use of this logic as an interface language between other analyses (this is motivated by the simplicity for human to write useful primitives with this logic, for example to point to a non-cyclic list). As an example, we have illustrated our approach for a partition analysis.

Acknowledgments I am grateful to Patrick Cousot for giving me guidance and advice in this research and to Radhia Cousot and David Schmidt for their helpful discussions. I thank Charles Hymans, Francesco Logozzo, Damien Massé, David Monniaux, Alexander Serebrenik and the anonymous referees for comments and suggestions.

References

1. C. Calcagno, H. Yang, and P. W. O'Hearn. Computability and complexity results for a spatial assertion language for data structures. In *LNCS 2245*, page 108, 2001.
2. P. Cousot and R. Cousot. Systematic design of program analysis frameworks. In *POPL'79*, pages 269–282, San Antonio, Texas, 1979. ACM Press, New York, NY.
3. J. W. de Bakker. *Mathematical Theory of Program Correctness*. Prentice Hall, Englewood Cliffs, NJ, 1980.
4. E. W. Dijkstra. *A Discipline of Programming*. Prentice Halll, Englewood Cliffs, NJ, 1976.
5. D. Galmiche and D. Méry. Connection-based proof search in propositional BI logic. In *CADE'02*, pages 111–128, Denmark, 2002.

6. C. A. R. Hoare. An axiomatic basis for computer programming. *Comm. ACM*, 12:576–580, 1969.
7. S. Ishtiaq and P. O'Hearn. BI as an assertion language for mutable data structures. In *POPL'01*, pages 14–26, 2001.
8. H. Yang P. O'Hearn and J. Reynolds. Syntactic control of interference. In *POPL'04*, Italy, 2004. ACM Press, New York, NY.
9. J. C. Reynolds. Syntactic control of interference. In *POPL'78*, pages 39–46. ACM Press, New York, NY, 1978.
10. J. C. Reynolds. Separation logic : A logic for shared mutable data structures. In *LICS'02*, pages 55–74, Denmark, 2002. IEEE Computer Society.
11. H. Yang and P. O'Hearn. A semantic basis for local reasoning. In *FoSSaCS'02*, Lecture Notes in Computer Science, pages 402–416. Springer, 2002.

A Formally Verified Calculus for Full Java Card

Kurt Stenzel

Lehrstuhl für Softwaretechnik und Programmiersprachen
Institut für Informatik, Universität Augsburg
86135 Augsburg, Germany
stenzel@informatik.uni-augsburg.de

Abstract. We present a calculus for the verification of sequential Java programs. It supports all Java language constructs and has additional support for Java Card. The calculus is formally proved correct with respect to a natural semantics. It is implemented in the KIV system and used for smart card applications.

1 Introduction

The Java language has received a formal treatment for a couple of years now, see e.g. [1][2][9][11][15][24]. While issues like type safety have been solved [22], most people perceive that the problem of proving Java programs correct is not yet solved satisfactorily. Java combines a number of features that pose difficulties for a verification calculus: objects and object updates, static initialization, a class hierarchy and inheritance, exceptions, and threads. Additionally, Java has a large number of language constructs. It is not clear how to integrate these features into a proof system that is fast and (relatively) simple to use.

However, that may depend on the problem domain, i.e. what kind of Java programs are verified. Consider smart cards: they are distributed in large numbers (in credit and health cards or mobile phones), and they are usually security critical (money or privacy is lost if the security is broken). And, they can be programmed in Java Card, but this is difficult and error prone. So smart cards are a very useful area for Java program verification, and, as it turns out, a tractable one.

In this paper we present a calculus for sequential Java that is formally proved correct and implemented in the KIV system [12][3][23]. Section 2 describes the intended problem domain, smart card programming. Section 3 introduces the semantics that is the basis for the calculus and the correctness proofs in section 4. All of them are only presented in examples due to the complexity of Java. Section 5 shows an example proof, section 6 compares with other Java proof systems, and section 7 summarizes.

2 Java Card

Java Card [18] is a variation of Java that is tailored for smart cards. A smart card is a plastic card containing a small processor. Smart cards are used in

C. Rattray et al. (Eds.): AMAST 2004, LNCS 3116, pp. 491–505, 2004.
© Springer-Verlag Berlin Heidelberg 2004

mobile phones, as credit cards, for authentication etc. These processors have very limited speed and memory. Typically they are 8-bit processors running at 5 Mhz. The most advanced cards have 128 KBytes ROM, 64 KBytes EEPROM, and 5 KBytes RAM. Often a cryptographic co-processor is included (e.g. with 1024 Bit RSA and 3-DES). This means that the complete JVM and all applications must fit into some dozen kilobytes of memory.

Java Card has the same language constructs (i.e. expressions and statements) as Java, but omits all features that make the JVM big and slow. It does not support threads, garbage collection, streams, floating point arithmetic, characters and strings, and integers. Essentially, Java Card is sequential Java with fewer primitive data types and a much smaller API. However, Java Card programs often use expressions like postfix increment or compound assignments, and rely heavily on byte and short arithmetic. This means that these constructs must be supported and modeled precisely.

An interesting smart card application is the storage of electronic tickets (e.g. railway tickets). If the smart card contains only genuine tickets they can be inspected and invalidated offline, i.e. without access to a central data base. This goal (only genuine tickets) can be achieved by access control and cryptographic protocols. In this scenario the smart card owner must be considered hostile because he has an interest to load forged or copied tickets onto the card. So the owner may not modify or read the code or data (the code itself is not secret, but cryptographic keys are). This is achieved by loading the program in a secure environment and by setting suitable access rights. The Java Card program on the smart card manages the tickets, and performs the steps of the cryptographic protocols. A typical program (including loading of tickets via internet and offline transfer of tickets) has about 1200 lines of code and about 40 methods. This does not include cryptographic operations like RSA encryption, digital signatures etc. that come from a predefined library (and are defined in the Java Card API).

3   A Natural Semantics for Sequential Java

Several different Java semantics exist, e.g. [5][8][26]. They differ in the used formalism, the number of supported Java constructs, in the assumptions made, and in the level of detail for auxiliary operations[1]. However, the basis for every formal semantics is the Java Language Specification [19]. Different formalisms can be used for the semantics: denotational semantics, abstract state machines (ASMs), Structural Operational Semantics (SOS), natural semantics, and others. (There is also the further notion of small-step and big-step semantics.) Maybe the "best" semantics is one that is closest to the informal descriptions in the Java Language Specification, so that people who are not experts in formal methods have a possibility to understand it. For this reason we choose a natural (big-step) semantics.

---

[1] Perhaps one should distinguish between a *formal* semantics – one written in a logic with a precise semantics and syntax, parsed by a computer, and a *mathematical* semantics that uses mathematical notation, but does not adhere to a precise syntax.

The semantics of a Java statement is a ternary relation: a statement α transforms an initial state $st = v \times h$ consisting of a variable mapping v (as in classical predicate logic) and a heap h into a final state $st' = v' \times h'$ consisting of a new variable mapping v' and a new heap h', $st[\![\alpha]\!]_{tds} st'$. In SOS this is often written as $\langle \alpha, st \rangle \longrightarrow st'$. If α does not terminate there is no final state st'. If α terminates there is exactly one new state st', because Java Card is deterministic. (The semantics can also be viewed as a transition system that (stepwise) transforms states.) The relation $.[\![.]\!]_{tds}.$ is annotated with the list of class and interface (type) declarations tds that comprise the Java program. The semantics of a Java expression e additionally includes a result value, $st[\![e]\!]_{tds} st' \times val$ (or written as $\langle e, st \rangle \longrightarrow \langle val, st' \rangle$). The variable mapping v contains values for the local variables of the program; they are identified with logical variables. The heap h contains the object fields (indexed by a reference and a field name) and a 'reserved' reference with 'reserved' fields for static fields, the initialization state of classes, and whether currently a jump (*abrupt transfer of control*, e.g. when an exception is thrown) happens or whether evaluation proceeds in the normal manner. No other ingredients are needed to describe the state. The heap is specified algebraically, so that the semantics of Java is defined relative to all models of this specification. Evaluation of statements and expressions is described inductively by reduction rules, the semantics is then the smallest set of triples $.[\![.]\!]_{tds}.$ closed under the rules for the relations. The construction guarantees that the smallest set is well defined (no negative occurrences of the relations).

$$\frac{h[mode] \neq normal}{v \times h[\![e]\!] v \times h \times \bot} \ (1) \qquad\qquad \frac{h[mode] = normal}{v \times h[\![l]\!] v \times h \times l_v} \ (2)$$

$$\frac{v \times h[\![e]\!] v_0 \times h_0 \times r_0 \quad r_0 = null \lor h_0[mode] \neq normal}{v_0 \times h_0[\![\textbf{throw new } NullPointerException();]\!] v_1 \times h_1} \ (3)$$
$$\frac{}{v \times h[\![e.f]\!] v_1 \times h_1 \times \bot}$$

$$\frac{v \times h[\![e]\!] v_0 \times h_0 \times r_0 \quad r_0 \neq null \land h_0[mode] = normal}{v \times h[\![e.f]\!] v_0 \times h_0 \times h_0[r_0 \times f]} \ (4)$$

Fig. 1. Semantics rules for jump (1), literal (2), and instance field access (3, 4).

Figure 1 shows four simple rules. The jump rule (1) states that in case of an abrupt transfer of control (when an exception is thrown, or a `return`, `break`, or `continue` statement is encountered) any expression is skipped. The heap h stores Java values that can be accessed with keys (normally a pair of an object reference and a field name, but there are some special keys and special values: one special key *mode* is used to record the evaluation mode, and $h[mode]$ looks up the value in the heap. So $h[mode] \neq normal$ means that no normal evaluation takes place. The expression is not evaluated, the variable mapping and heap remain unchanged, and a dummy value $\perp$ is returned. A literal (2) is evaluated if the mode is normal. The value is the literal evaluated under the current variable mapping l_v (in our case literals may contain algebraic expressions with variables), everything else remains unchanged. In case of an instance field access (3) a NullPointerException is thrown if the invoking expression is null (if initially the mode was not normal the evaluation of e is skipped due to the jump rule; if during evaluation of e an exception occurs the **throw** statement will be skipped due to the throw rule). Otherwise the value is looked up in the heap using the computed reference r_0 together with the field name ($h_0[r_0 \times f]$) (4).

We consider another example: the instance method invocation (see Fig. 2). It has three rules, the first in case the invoking expression is null, the second for exceptions during evaluation of the arguments or the body, or in case the body completes without a **return** (this is possible for **void** methods). Only the third rule is shown. First, the invoker and the arguments are evaluated, then the arguments are bound and the method body α is evaluated. If the body completes with a **return** ($is\_return\_mode(h_{n+1}[mode])$) the mode is set back to normal ($h_{n+1}[mode, normal]$) and the value ($h_{n+1}[mode].val$) returned. The main point is that the real problem of the rule, the dynamic method lookup, is completely hidden in the definition of *getMethod* (an algebraically specified function). Every formal Java semantics must specify precisely how this method lookup works (e.g. what happens in case of a cyclical class hierarchy). The same holds for all other definitions. So the semantics is much more (complicated, longer) than just the rules. This makes it very difficult to compare two semantics that are formally defined in different proof systems.

$$\frac{v \times h[\![e]\!]v_0 \times h_0 \times r_0 \quad v_0 \times h_0[\![e_1]\!]v_1 \times h_1 \times r_1 \ldots v_{n-1} \times h_{n-1}[\![e_n]\!]v_n \times h_n \times r_n}{v \times h[\![e.m(e_1,\ldots,e_n)]\!]v_n \times h_{n+1}[mode, normal] \times h_{n+1}[mode].val}$$
$$r_0 \neq null \wedge h_n[mode] = normal$$
$$(v_n)^{r_1,\ldots,r_n,r_0}_{x_1,\ldots,x_n,this} \times h_n[\![\alpha]\!]v_{n+1} \times h_{n+1} \quad is\_return\_mode(h_{n+1}[mode])$$

$$m(x_1,\ldots,x_n) \{\alpha\} = \text{getMethod}(\text{classOf}(r_0), m, tds),$$

Fig. 2. The third rule for instance method invocation.

All in all 24 expressions and 23 statements are specified. In other semantics the count may differ due to decisions what are considered different constructs. Seven of the 23 statements are not part of Java. They are introduced because they are used in the calculus. **static** and **endstatic** are used for the initialization of super classes and to capture exceptions during initialization; **target** and **targetexpr** define a target for **return** statements, **finally** and **endfinally** are used to describe the beginning and end of a finally block, and **catches** holds a list of catch clauses. The full semantics of the language constructs is described in 123 rules. This is a large number, but every rule describes exactly one case that may occur during evaluation. About 50 nontrivial definitions with several hundred axioms are specified algebraically, plus the heap, and the primitive types for bytes, shorts, and the bitwise integer operations. A pretty printing has more than 50 pages. Obviously it is a nontrivial task to make sure that the specification captures the intended Java semantics in every little detail.

Comparison to other formal semantics. David von Oheimb [25] defines a natural semantics for a subset of Java (9 statements and 12 expressions) in Isabelle with a deep embedding using higher order logic. It is interesting to note that the notion of a 'state' is more complicated than our state, even though a reduced language is used. The state contains a store for local variables, and an explicit exception status; the environment contains not only the Java classes, but also local variables. Börger and Schulte [5] (later expanded in [24]) give an ASM semantics that also includes threads, but is not formalized in a proof system. Usage of ASMs requires more notational overhead, because stacks for local variables and a pointer into the program code is needed to model program evaluation. Both use a very short notation that is sometimes difficult to read, and both aim more at meta properties of the Java language than at source code verification of concrete programs. Therefore they omit e.g. most operations on primitive data types. The LOOP project (Jacobs et. al. [17][13]) uses a coalgebraic approach (and a formalization by a shallow embedding in PVS) where the state includes besides the heap also a stack for method calls and a memory for static data. So compared to others, this semantics seems to be more simple (concerning the state and the used formalism, algebraic specifications) and may be easier to understand for non-experts.

4   A Calculus for Java Card

4.1   Sequents and Dynamic Logic

The calculus is a sequent calculus for dynamic logic [10]. Dynamic logic extends predicate logic with two modal operators, box $[\,.\,]$. and diamond $\langle\,.\,\rangle$. The intuitive meaning of $\langle \text{H}; \alpha \rangle \; \varphi$ is: with initial heap H (a variable) the Java statement α terminates, and afterwards φ holds. φ is again a formula of dynamic logic, i.e. it may contain boxes or diamonds. The meaning of $[\text{H}; \alpha] \; \varphi$ is: if α terminates then afterwards φ holds. The formal semantics is

$$\mathcal{A}, tds, v \models \langle \text{H}; \alpha \rangle \; \varphi :\Leftrightarrow \exists \, v', h'.v \times v(\text{H}) [\![\alpha]\!]_{tds} v' \times h' \text{ and } \mathcal{A}, (v')_{\text{H}}^{h'} \models \varphi$$

$\langle \text{H}; \alpha \rangle \; \varphi$ holds in an model $\mathcal{A}$ with Java type declarations tds and variable mapping v iff there exists a new variable mapping v' and a heap h' such that α with initial mapping v and initial heap $v(\text{H})$ computes v' and h' (this means that α terminates) and afterwards φ holds under the new variable mapping where the new heap h' is bound to the variable H. A sequent $\varphi_1, \ldots, \varphi_m \vdash \psi_1, \ldots, \psi_n$ consists of two lists of formulas (often abbreviated by Γ and Δ) divided by $\vdash$ and is equivalent to the formula $\varphi_1 \wedge \ldots \wedge \varphi_m \rightarrow \psi_1 \vee \ldots \vee \psi_n$. $\varphi_1, \ldots \varphi_m$ can be thought of as preconditions, while one of $\psi_1, \ldots, \psi_n$ must be proved. A Hoare triple $\{\varphi\}\alpha\{\psi\}$ can be expressed as $\varphi \vdash [\text{H}; \alpha]\psi$ or $\varphi \vdash \langle \text{H}; \alpha \rangle \; \psi$ if termination is included. However, more things can be expressed (e.g. program equivalence).

4.2   Some Example Rules

The calculus essentially has one rule for every Java expression and statement, plus some generic rules. It works by symbolic execution of the Java program from its beginning to its end (i.e. computation of strongest postcondition). This means it follows the natural execution of the program, and is much more intuitive than inventing intermediate values (as in a Hoare calculus) or computing weakest preconditions (by evaluating a program from the end to the start). Nested expressions and blocks are flattened to a sequence of simple expressions and statements that can be executed directly. Obviously, this flattening must obey the evaluation order of Java. The additional Java statements mentioned in the last section are used to mark the end of a block. A box or diamond never contains isolated expressions but only statements (e.g. an expression statement). The result of an expression is 'stored' with an assignment to a local variable. Most of the rules are applicable if the program occurs on the left or on the right hand side of the turnstile $\vdash$, and they are applicable for boxes and diamonds.

An instance field assignment has three premises:

$$
\begin{array}{ll}
1. & \Gamma, H[mode] \neq normal \vdash \varphi, \Delta \\
2. & \Gamma, H[mode] = normal, e = \texttt{null} \\
& \vdash \langle H; \texttt{throw new NullPointerException();} \rangle\, \varphi, \Delta \\
3. & \Gamma, H[mode] = normal, e \neq \texttt{null}, H_0 = H[e \times f, e_0] \\
& \vdash \langle H_0; x = e_0; \rangle\, \varphi[H/H_0], \Delta \\
\hline
& \Gamma \vdash \langle H; x = (e.f = e_0); \rangle\, \varphi, \Delta
\end{array}
$$

f is the field name, e the invoking expression. The rule is only applicable if e and e_0 are *basic* expressions, either local variables or literals. If the mode is not normal the expression is skipped (first premise). If the invoking expression is `null` a `NullPointerException` is raised (second premise). Otherwise the heap H is modified by setting the field $e \times f$ to the value of e_0 ($H[e \times f, e_0]$). Then the computation continues with the new heap (bound to H_0, a new variable). Since e and e_0 are local variables or literals they require no further evaluation, but can be taken directly as values. If they are other Java expressions they have to be flattened first.

The flattening rule works as follows:

1. For an expression $x = e$ select the immediate subexpressions $e_1, \ldots, e_n$ of e.
2. Find the first e_i that is not a local variable or a literal, and that does not cause a variable conflict (see case 4).
3. Replace e_i in e by a new variable y yielding e' and add the assignment $y = e_i$ before $x = e'$.
4. A variable conflict occurs if e_i contains an assignment to a variable that occurs in $e_1, \ldots, e_{i-1}$, e.g. in x * (x = 3). In this case a renaming is necessary.

After a finite number of applications the algorithm will return a list of assignments where every subexpression is either a local variable or a literal. Then a rule for this expression is applicable. The test of an `if`, the expression of a `switch`, `return`, or `throw` can also be flattened in this manner (but not the test of a `while`, `do` or `for`). A block is also flattened:

$$\frac{1.\ \Gamma, H[mode] \neq normal \vdash \varphi, \Delta \qquad 2.\ \Gamma, H[mode] = normal \vdash \langle H; \alpha'_1[\underline{x}/\underline{y}]\rangle \ ...\langle H; \alpha'_n[\underline{x}/\underline{y}]\rangle\ \varphi, \Delta}{\Gamma \vdash \langle H; \{\alpha_1...\alpha_n\}\rangle\ \varphi, \Delta}$$

α_i are the top level statements of the block, $\underline{x}$ the local variables declared in the block, and $\underline{y}$ new variables. α'_i is α_i except for a local variable declaration. $ty\ x = e$ becomes an assignment $x = e$. Note that this is not legal Java if e is an array initializer, but we treat it as a normal expression. The replacement with new variables ensures that the variables really behave as local variables. A similar flattening happens for the `try` statement:

$$\frac{1.\ \Gamma, H[mode] \neq normal \vdash \varphi, \Delta \qquad 2.\ \Gamma, H[mode] = normal \vdash \langle H; \alpha\ \texttt{catches}(catches)\ \texttt{finally}(\beta)\rangle\ \varphi, \Delta}{\Gamma \vdash \langle H; \texttt{try}\ \alpha\ catches\ \texttt{finally}\ \beta\rangle\ \varphi, \Delta}$$

The list of catch clauses *catches* is transformed into an additional Java statement `catches(`*catches*`)`, and the finally clause is transformed into an additional Java statement `finally(`β`)`. In this manner expressions and statements can be flattened. Loops can be unwound and treated by induction.

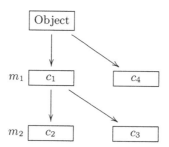

1. $\Gamma, mode(st) \neq normal \vdash \varphi, \Delta$
2. $\Gamma, e = null, mode(st) = normal \vdash \langle H; \textbf{throw } new\ NPE()\rangle\ \varphi, \Delta$
3. $\Gamma, e \neq null, mode(st) = normal \vdash classOf(e, st) \in \{c_1, c_2, c_3\}, \Delta$
4. $\Gamma, e \neq null, mode(st) = normal, classOf(e, st) \in \{c_1, c_3\},$
   $this' = e, \underline{z} = es \vdash \langle H; \alpha'_1\rangle\ \langle H; \textbf{targetexpr}(x)\rangle\ \varphi, \Delta$
5. $\Gamma, e \neq null, mode(st) = normal, classOf(e, st) \in \{c_2\},$
   $this' = e, \underline{z} = es \vdash \langle H; \alpha'_2\rangle\ \langle H; \textbf{targetexpr}(x)\rangle\ \varphi, \Delta$

$$\overline{\Gamma \vdash \langle H; x = e.m(es)\rangle\ \varphi, \Delta}$$

Fig. 3. Example class hierarchy and rule for instance method invocation.

As a last example we show the rule for an instance method invocation. This rule has a variable number of premises depending on the number of possible methods that can be invoked. Figure 3 shows a class hierarchy where class c_1 contains a method declaration m with a body m_1 that is overwritten in class c_2 with another body m_2. If $e.m(e_1, \ldots, e_n)$ is the method invocation the correct method body to invoke depends on the runtime type of e. e must be a local variable or literal and should be a reference $\neq$ null. If e is a reference to an object with type c_1 or c_3 then method m_1 is invoked; if the type of e is c_2 then method m_2 is invoked; the third premise ensures that the type of e is either c_1, c_2, or c_3. Note that there is no possibility to omit this premise. A type correct Java program guarantees this if the heap conforms to the program, but the calculus does not enforce the latter. The heap can contain arbitrary entries that have nothing to do with the classes and declarations of the program. Therefore it is necessary to *prove* that the runtime type of e is 'legal'. Together with a 'jump' case and a null invoker the rule has five premises in this instance (Fig. 3). In premises 4. and 5. the parameters $e_1, \ldots, e_n$ (that must be local variables or literals) are bound to new variables $\underline{z}$, a new variable $this'$ is introduced for this and bound to e, in the method body the formal parameters are replaced with the new variables yielding α_i', and a new statement **targetexpr**(x) is added that will catch a return statement and bind x to the returned value.

4.3   Interactive Proofs

One of the most important features of the calculus is that it is well suited for interactive proofs, much better – we feel – than a Hoare or wp calculus. Because a formula containing a Java statement is only one formula among others, the user can mix predicate logic rules (case distinctions, quantifier instantiations, cut, etc., or advanced rules like rewriting or simplification) freely with Java rules. This helps to reduce the size of non-Java formulas during the proof. Furthermore, a method call (or any other statement) can be either evaluated, or one or several lemmas for the method call can be applied at different points in the proof. The reason is that programs can appear on both sides of the turnstyle $\vdash$ so that it is possible to argue that if program α computes x then program β computes y. The following proof rule is valid:

$$\frac{\langle H; \alpha \rangle \ \underline{x} = \underline{y}, \varphi[\underline{x}/\underline{y}], \Gamma \vdash \psi[\underline{x}/\underline{y}], \Delta}{\langle H; \alpha \rangle \ \varphi, \Gamma \vdash \langle H; \alpha \rangle \ \psi, \Delta}$$

We know $\langle H; \alpha \rangle \ \varphi$, i.e. α terminates and afterwards φ holds (this can be a lemma about α). α can only modify a number of variables $\underline{x}$ (the assigned variables of α and the heap H) and these variables describe the state exactly. Therefore we can introduce new variables $\underline{y}$ that hold the result and then we know that $\varphi[\underline{x}, \underline{y}]$ holds (φ with $\underline{x}$ replaced with $\underline{y}$). Now we want to prove $\langle H; \alpha \rangle \ \psi$ and we know that α computes $\underline{y}$. Therefore we can discard α on the right hand side of the turnstyle $\vdash$ and it remains to prove that $\psi[\underline{x}, \underline{y}]$ holds. This is the typical situation if a lemma is used. But the program α does not disappear from the sequent. It remains in the antecedent and later the user can apply another lemma that introduces another property. Nothing similar is possible in a Hoare calculus. Another advantage is that induction can be used for loops or recursive methods. This means the user has more freedom to structure proofs.

4.4   Correctness of the Calculus

The proof rules are specified in KIV and their correctness with respect to the semantics has been proved. Since the calculus introduces several new operations (the notion of *free* and *assigned* variables, replacement of variables, flattening of expressions, formulas, sequents etc.) the complete specification is considerably larger than just the Java semantics. Especially operations that are defined for all expressions or statements have lots of axioms, and proofs for these operations require the most time. Furthermore, the semantics' relations for expressions, expression lists, statements, and statement lists are mutually recursive and must be formulated and proved for all four relations at once. The most time consuming proofs are concerned with the correct definition of free and assigned variables (the semantics of a Java statement depends only on the values of the free variables, and only assigned variables can be changed) because the definitions should not include superfluous variables. Other important proofs concern the correctness of variable replacement and flattening. In comparison the rule for instance method invocation is quite simple to prove because in the semantics and in the proof rule the method lookup just travels up in the class hierarchy. All 57 rules have been proved correct. The specification and verification effort required several months of work. Currently the calculus is not (relatively) complete for two reasons: First, there is no possibility to argue about the number of recursive method calls. Experience shows that usually induction on the used data structure is sufficient (a counter that becomes smaller, or the length of a list that decreases). However, it is not possible to prove that a method that just calls itself (void m() {m(); }) does not terminate. Second, the semantics does not depend on type correct programs (or programs that pass a Java compiler – the compiler checks more than just type correctness). But for verification we are interested only in type correct programs, and for type correct programs more efficient rules can be designed. (They are correct but incomplete for type incorrect programs.)

   As can be imagined several errors were found during verification. Most of them are errors only for type incorrect programs. For example, the definition of free variables must handle the case that a local variable declaration occurs outside a block or that a local variable occurs outside the scope of a local variable declaration. Other errors were more serious because they concerned type correct programs:

1. The third premise in the instance method invocation was missing (that checks that the runtime type of the invoker is correct, see Fig. 3). As mentioned above this is important even for type correct programs because the heap may not conform to the program.
2. The flattening rule contained no check for variable conflicts. y = x * (x = 3); was flattened to z = (x = 3); y = x * z; which is wrong. (A renaming is needed: x0 = x; z = (x = 3); y = x0 * z;).

3. A similar problem occurred for the postfix increment: x = y++ was transformed into x = y; y = y + 1; which is wrong for x = x++; (the right hand side is fully evaluated before the assignment occurs).

These errors were found because the verification of the proof rules failed. However, some errors were found in the semantics as well.

1. *first active use* was not handled correctly in the semantics rules, i.e. whether static initialization occurs before or after the arguments are evaluated. (compound assignment to static field and new class: before evaluation of arguments, simple assignment to static field and static method invocation: after evaluation of arguments.)
2. The semantics rules for compound assignment failed to cast the result back to byte or short if the right hand side was byte or short. (In byte b = 127; b += 1; the result is (byte)−128.)

These errors were found because the verification of the proof rules also failed, and analysis revealed the errors to be in the semantics. However, both semantics and calculus could be wrong. It is possible to validate the semantics by 'running' test programs in KIV (automatically applying the proof rules) and comparing the output with a run of a Java compiler and JVM (currently 150 examples), and this certainly increases confidence in the semantics, but who would think about writing programs like x = x++;?

5   An Example Program

The aim of the example is to show the 'look and feel' of the calculus for a small example that involves a for loop, exceptions and abrupt termination of the loop. It is typical for Java Card programs to use this programming style. We consider the Java Card application for storing tickets mentioned in section 2. Since the available memory for a Java Card program (an *applet*) is severely limited the maximal number of tickets that can be stored must be fixed in advance. The missing garbage collector means that storage cells cannot be reclaimed. Therefore it is usual programming practice to allocate all objects when the applet is loaded onto the smart card, and to reuse these objects. Our applet has a capacity of 20 tickets that are stored in an array. A field free indicates if the entry is free or not. If a ticket is loaded the value is set to false, if it is deleted after usage the entry is set back to true.

```
class Ticket {boolean free = true; ⟨rest of class⟩}

public class Cardlet extends Applet {
    final static byte MAX = 20;
    Ticket[] tickets = new Ticket[MAX];
    Cardlet() {
        for(byte c=0;c<MAX; c++) tickets[c] = new Ticket();}
    ⟨rest of class⟩}
```

A Java Card applet works as follows: An `install` method is called once when the applet is loaded onto the card. This method typically calls the constructor. All data (fields, objects, and arrays) is stored permanently in the EEPROM of the smart card. When the smart card is inserted in a reader and the applet is selected, a method `select` is called once. Afterwards all communication takes place through a `process` method. The JVM receives the input, a sequence of bytes, stores them in a byte array and calls the `process` method with this array. `process` normally computes an answer that it stores in the byte array, and that will be sent back to the terminal by the JVM. Another possibility is to throw an exception that is converted by the JVM into an error message.

The following method can be used somewhere in the applet to find a free position:

```
byte findFree () {
    for(byte c=0; c < MAX; c++)
        if (tickets[c].free) return c;
    ISOException.throwIt(SW_FILE_FULL);
}
```

If no free position is available an exception is thrown (without creating a new exception object!) from the predefined method `ISOException.throwIt`. This exception ends the `process` method and results in an error message to be sent to the terminal (SW_FILE_FULL). If the `findFree` method is used several times in the code it is good proving practice to formulate some lemmas about its behaviour and re-use them wherever possible. For example,

findFree-install : install(H) $\vdash$ $\langle$H; by = cardlet.findFree();$\rangle$ install(H)

findFree-throw :
$\#$tickets(H) = b2i(MAX), install(H), H[mode] = normal
$\vdash$ $\langle$H; by = cardlet.findFree();$\rangle$ ISOException(SW_FILE_FULL, H)

findFree-ok :
$\#$tickets(H) < b2i(MAX), H = H$_0$, install(H), H[mode] = normal
$\vdash$ $\langle$H; by = cardlet.findFree();$\rangle$ (H = H$_0$ $\wedge$ free_ticket(b2i(by), H))

The method assumes that the array entries are not `null`. This is the case after installation. Hence an invariant *install*(H) is needed for the applet. This invariant will contain other properties of the applet that should hold before and after every communication step (i.e. before and after every call of `process`), including logical properties related to the cryptographic protocols. Finding this invariant is not trivial, but essential for the correctness of the applet. *install*(H) is a user defined predicate for the heap. The second property, *findFree-throw*, states that the method will throw an exception SW_FILE_FULL if the number of tickets stored in the heap is already MAX ($\#$tickets(H) = b2i(MAX), b2i converts a byte into an integer). Here we can assume that the invariant holds, and we must assume that no abrupt transfer of control happens initially (H[mode] = normal) because then the method call will be skipped. Finally, *findFree-ok* states that the method will return a free position if there is one.

We show the proof for *findFree-ok*. Method call and initialization of the `for` loop results in

$H = H_0$, this = cardlet, c = 0, H[mode] = normal,
#tickets(H) < b2i(MAX), install(H)
$\vdash \langle H;$`for`$(c < MAX; c ++)$ `if` $(this.tickets[c].free)$ `return` $c;$ `else` $\{ \})$
$\langle H;$ISOException.throwIt(SW_FILE_FULL);$\rangle$
$\langle H;$`targetexpr`$(by)\rangle$ (free_ticket(b2i(by), H) $\wedge$ H = H_0)

The `for` loop contains no initialization so it can be unwound; *cardlet* is a reference to an object of type `Cardlet` that becomes the value of `this` inside the method. Now we use induction on $|MAX - c|$ and generalize the goal by replacing $c = 0$ with $0 \leq c \wedge c \leq MAX$ (this is done automatically), and add the formula #tickets(H[cardlet – tickets].refval, b2i(c)−1, H) = b2i(c) stating that the number of tickets from 0 to $c - 1$ in the array is c (this means that all tickets below c are not free). H[cardlet – tickets].refval returns the reference that is stored in the `tickets` field of the cardlet object. This property is needed to prove that the loop counter c can never reach MAX. Then we unwind the `for` loop once and obtain

Ind-Hyp, ... $\langle other\ preconditions \rangle$ ...
$\vdash \langle H;$`if` $(b2i(c) < b2i(MAX))$
           `if` $(this.tickets[c].free)$ `return` $c;$ `else` $\{ \}$ $c ++;\rangle$
$\langle H;$`for`$(c < MAX; c ++)$ `if` $(this.tickets[c].free)$ `return` $c;$ `else` $\{ \})$
$\langle H;$ISOException.throwIt(SW_FILE_FULL);$\rangle$
$\langle H;$`targetexpr`$(by)\rangle$ (free_ticket(b2i(by), H) $\wedge$ H = H_0)

If the first `if` test is true and the second one is false we obtain after the postfix increment `c++`

Ind-Hyp, $c_0 = i2b(b2i(c) + 1)$, $\neg$ H[r_0 – free].boolval
... $\langle other\ preconditions \rangle$ ...
$\vdash \langle H;$`for`$(c_0 < MAX; c_0++)$ `if` $(this.tickets[c_0].free)$ `return` $c_0;$ `else`$\{\})$
$\langle H;$ISOException.throwIt(SW_FILE_FULL);$\rangle$
$\langle H;$`targetexpr`$(by)\rangle$ (free_ticket(b2i(by), H) $\wedge$ H = H_0)

The formula on the right hand side of $\vdash$ (with the `for` loop) is identical to the formula where the induction started, except that c is replaced by c_0. This means we can apply the induction hypothesis (This requires proving that no overflow occurs for the **byte** value c.), and obtain a program formula one the left hand side of $\vdash$. The result is an axiom:

$\langle H;$`for`$(c_0 < MAX; c_0 ++) ... \vdash \langle H;$`for`$(c_0 < MAX; c_0 ++) ...$

Aside from the rather longish sequents the proof proceeds as a proof done on paper. The same principle (induction and unwinding of the loop) can be used for `while` or `do` loops. In other proofs only the lemmas for the `findFree` method are used. This allows a nice structuring of the proofs.

6   Comparison to Other Proof Systems for Java

Java Card verification in KIV seems to be unique in that an existing prover was extended to incorporate a Java Card calculus. In the KeY project [16][20] a new prover is developed from scratch; Oheimb [27][25] models a formal (operational) semantics and a Hoare calculus (for a subset of Java) in Isabelle; in the LOOP project [16][17][14] Java together with JML [6] annotations are translated into a formal (denotational) semantics enriched by proof rules for a Hoare- and weakest precondition calculus in PVS; the Krakatoa tool [21] translates into Coq; Jack [7] into a prover for the B method. The KIV approach has two advantages compared to the others: First, an already good prover can be used for the non-Java parts; second, the prover can be tailored to the goals that arise in Java verification. This includes simple things like pretty printing to make the goals more readable, but also special heuristics and simplification strategies (e.g. a special treatment of the heap). The drawback is, of course, that access to the internals of the prover is necessary.

KeY also uses a dynamic logic, but the calculus is not proved correct w.r.t. a formal semantics. The three main differences are: exceptions are modeled as non-termination, blocks are not flattened (though expressions are), and there is no explicit heap. The last two lead to more complex formulas: programs contain a 'prefix' representing nested blocks (including try catch blocks) and 'updates' to objects to cope with aliasing. On the other hand, omitting the heap may help to prove that a method does *not* modify some objects. Oheimb uses a pure Hoare calculus that is tailored for backward reasoning (i.e. computes weakest preconditions), and has proved its correctness and completeness in Isabelle. One specialty of the calculus is that the state must always conform to the (type correct) program. This may require unnecessary proof work when the consequence rule is used (the Java rules preserve conformity), and it is not clear how theorems that are proved for a library or API class can be reused (because the new state contains new objects that do not conform to the original classes). We do not require this conformity. The LOOP project also uses a Hoare calculus and weakest preconditions (formally proved correct). As mentioned earlier, we feel that a dynamic logic allows more flexibility. It is interesting to note that Oheimb uses quadruples as pre- and postconditions, while LOOP uses 8-tuples. So it seems that the calculus presented here has the most simple structure (a heap and two modal operators for the programs).

Another difference to most other approaches is that in KIV the user plays an important role and is expected (and encouraged) to interact with the system to keep the proofs manageable. In KeY, LOOP, Krakatoa, and Jack the prover is viewed as a back end system that is best used fully automatically. Ideally, the prover is fed with some Java goals, generates proof obligations that contain no longer Java statements, and proves these goals automatically. However, this works only up to a given size of the formulas. Better support for interactive proofs can help to reduce the size while proving, but this requires (currently) a rather experienced user.

7 Conclusion

We presented (in excerpts) a formal semantics and calculus for Java Card (essentially sequential Java) that supports all language constructs. The semantics is a natural (big-step) semantics, which is adequate for deterministic sequential programs. The semantics is defined relatively to an algebraic specification of a heap and Java's primitive types. The complete specification is big, but Java is a complex language. The calculus is a sequent calculus for dynamic logic that is more expressive than a Hoare calculus and – we feel – better suited for interactive proofs. The calculus has been proved correct formally with KIV, and is also implemented in KIV. Currently, KIV seems to be the only existing prover that was extended for a Java Card calculus.

The main application area is in the context of smart cards where interesting and security critical (because money and privacy is involved) e-commerce applications like electronic ticketing make a formal verification very desirable. It must be proved that the program correctly implements a cryptographic protocol that guarantees the security of the application. Without formal methods it is very difficult to assess the correctness and security of an interesting smart card application[2]. The programming style in Java Card requires that certain features (byte and short arithmetic, arrays, `for` loops, exceptions etc.) are modeled precisely and handled efficiently in the prover. Experience shows that the main difficulties are reasoning about bytes, shorts, byte arrays, cryptographic methods, and invariants that hold between communications.

One can ask why a prover should support all the complex features of sequential Java. There are three reasons. First, Java Card programs typically use these features; second, to show people who are critical towards formal methods that the field has reached at least that maturity; third, because formal methods do not scale up by themselves. It is not clear if a double sized program requires double effort or more. And even if the increase is linear every prover will eventually fail (see e.g. [17]: "PVS can run for hours without completing the proof, or it can crash because the proof state becomes too big."). So work has still to be done to reduce the complexity of formal proofs.

References

1. J. Alves-Foss, editor. *Formal Syntax and Semantics of Java.* Springer LNCS 1523, 1999.
2. I. Attali and T. Jensen, editors. *Java on Smart Cards: Programming and Security.* Springer LNCS 2041, 2001.
3. M. Balser, W. Reif, G. Schellhorn, K. Stenzel, and A. Thums. Formal system development with KIV. In T. Maibaum, editor, *Fundamental Approaches to Software Engineering*, Springer LNCS 1783, 2000.

---

[2] The author had to inspect (without formal methods) 35.000 lines of Java Card code written by students in a practical course, so this statement comes from own experience.

4. B. Beckert. A dynamic logic for the formal verification of java card programs. In I. Attali and T. Jensen, editors, *Java on Smart Cards*. Springer LNCS 2041, 2000.
5. E. Börger and W. Schulte. A Programmer Friendly Modular Definition of the Semantics of Java. In [1].
6. L. Burdy, Y. Cheon, D. Cok, M. Ernst, J. Kiniry, G. Leavens, K. Leino, and E. Poll. An overview of JML tools and applications. In T. Arts and W. Fokkink, editors, *Proceedings FMICS '03*. Volume 80 of Electronic Notes in Theoretical Computer Science, Elsevier, 2003.
7. N. Burdy, A. Requet, and J.-L. Lanet. Java applet correctness: A developer-oriented approach. In *Formal Methods Europe (FME)*, Springer LNCS, 2003.
8. S. Drossopoulou and S. Eisenbach. Describing the Semantics of Java and Proving Type Soundness. In [1].
9. S. Drossopoulou, S. Eisenbach, B. Jacobs, G. T. Leavens, P. Müller, and A. Poetzsch-Heffter, editors. *Formal Techniques for Java Programs, Proceedings ECOOP2000 Workshop*. Technical Report 269, 5/2000, Fernuniversität Hagen.
10. D. Harel. *First Order Dynamic Logic*. LNCS 68. Springer, Berlin, 1979.
11. P. Hartel and L. Moreau. Formalizing the safety of Java, the Java virtual machine, and Java card. *ACM Computing Surveys (CSUR)*, 33(4), December 2001.
12. KIV home page. http://www.informatik.uni-augsburg.de/swt/fmg/.
13. M. Huisman. *Reasoning about JAVA programs in higher order logic with PVS and Isabelle*. PhD thesis, University of Nijmegen, IPA dissertation series, 2001-03, 2001.
14. M. Huisman and B. Jacobs. Java Program Verification via a Hoare Logic with Abrupt Termination. In T. Maibaum, editor, *Fundamental Approaches to Software Engineering (FASE'00)*. Springer LNCS 1783, 2000.
15. B. Jacobs, G. T. Leavens, P. Müller, and A. Poetzsch-Heffter, editors. *Formal Techniques for Java Programs*. Technical Report 251, Fernuniversität Hagen, 1999.
16. B. Jacobs and E. Poll. A logic for the java modeling language JML. In *Proceedings FASE 2001*, Genova, Italy, 2001. Springer LNCS 2029.
17. B. Jacobs and E. Poll. Java program verification at nijmegen: Developments and perspective. Technical Report NIII-R0318, University of Nijmegen, 2003.
18. *Java Card 2.2 Specification*, 2002. http://java.sun.com/products/javacard/.
19. Bill Joy, Guy Steele, James Gosling, and Gilad Bracha. *The Java (tm) Language Specification, Second Edition*. Addison-Wesley, 2000.
20. KeY project homepage. http://i12www.ira.uka.de/~key.
21. Krakatoa home page. http://krakatoa.lri.fr/.
22. T. Nipkow and D. von Oheimb. Java light is Type-Safe – Definitely. In *25th ACM Symposium on Principles of Programming Languages*. ACM, 1998.
23. W. Reif, G. Schellhorn, K. Stenzel, and M. Balser. Structured specifications and interactive proofs with KIV. In W. Bibel and P. Schmitt, editors, *Automated Deduction—A Basis for Applications*. Kluwer Academic Publishers, 1998.
24. R.F. Stärk, J. Schmid, and E. Börger. *Java and the Java Virtual Machine: Definition, Verification, Validation*. Springer, 2001.
25. D. von Oheimb. *Analyzing Java in Isabelle/HOL: Formalization, Type Safety and Hoare Logic*. PhD thesis, Technische Universität München, 2001.
26. D. von Oheimb and T. Nipkow. Machine-checking the Java Specification: Proving Type-Safety. In [1].
27. D. von Oheimb. Axiomatic semantics for Java^{light} in Isabelle/HOL. In [9].

On Refinement of Generic State-Based Software Components

Sun Meng[1] and Luís S. Barbosa[2]

[1] LMAM, School of Mathematical Science
Peking University, China
sunmeng@water.pku.edu.cn
[2] Department of Informatics
Minho University, Portugal
lsb@di.uminho.pt

Abstract. This paper characterizes refinement of state-based software components modelled as pointed coalgebras for some Set endofunctors. The proposed characterization is parametric on a specification of the underlying behaviour model introduced as a strong monad. This provides a basis to reason about (and transform) state-based software designs.

Keywords: Components, refinement, coalgebraic models.

1 Introduction

Component-based software development [15, 16] emerged as a promising paradigm to deal with the ever increasing need for mastering complexity, software evolution and reuse. From object-orientation it retains the basic principle of encapsulation of data and code. The emphasis, however, is shifted from (class) inheritance to (object) composition to avoid interference between the former and encapsulation and, thus, paving the way to a development methodology based on *third-party assembly* of components. In [3, 2], the authors proposed a coalgebraic characterization of software components as specifications of *state-based* modules, encapsulating a number of services through a public interface and providing limited access to an internal state space. Component persist and evolve in time, being able to interact with the environment during their overall computation. This piece of research has been driven by two key ideas: first, the 'black-box' characterization of software components favors an *observational* semantics; secondly, the proposed constructions should be *generic* in the sense that they should not depend on a particular notion of component behaviour. This led both to the adoption of coalgebra theory [14] to capture observational semantics and to the abstract characterization of possible behaviour models (*e.g.*, partiality or (different degrees of) non-determinism) by strong monads acting as parameters in the resulting calculus.

Within this approach, briefly reviewed in section 2, a set of component connectors have been identified and their properties established as *bisimilarity* equations with respect to a generic behaviour model. Actually, the corner stone of

C. Rattray et al. (Eds.): AMAST 2004, LNCS 3116, pp. 506–520, 2004.
© Springer-Verlag Berlin Heidelberg 2004

our 'components as coalgebras' approach is the use of coinduction to prove $\sim$-results, where $\sim$ is the appropriate bisimilarity relation, as a basis for reasoning and transforming component-based designs. This paper provides a basis to extend the approach toward the *inequational* side through the discussion of suitable notions of *refinement*.

In broad terms *refinement* can be defined as a *transformation* of an 'abstract' into a more 'concrete' design, entailing a notion of *substitution*, but not necessarily *equivalence*. There is, however, a diversity of ways of understanding both what *substitution* means, and what such a *transformation* should seek for. In *data refinement*, for example, after Hoare's landmark paper [8], the 'concrete' model is required to have *enough redundancy* to represent all the elements of the 'abstract' one. This is captured by the definition of a surjection from the former into the latter (the *retrieve map*). Also *substitution* is regarded as 'complete' in the sense that the (concrete) operations accept all the input values accepted by the corresponding abstract ones, and, for the same inputs, the results produced are also the same, up to the retrieve map. This means that, if models are specified, as they usually are in *model-oriented* design methods like VDM[10], in terms of pre and post-conditions, the former are weakened and the latter strengthened, under refinement. In *object-orientation*, on the other hand, *substitution* is expressed in terms of *behaviour subtyping* [11] capturing the idea that 'concrete' objects behave similarly to objects in the 'abstract' class. Finally, refinement in *process algebras* is usually discussed in terms of several 'observation' preorders (see, for example, [7]), most of them justifying transformations entailing *reduction of nondeterminism*.

In general, refinement correctness means that the usage of a system according to its 'abstract' description is still valid if it is actually built according to the 'concrete' one. What is commonly understood by being a *valid usage* is that the corresponding observable consequences are still the same, or, in a less strict sense, a subset thereof. The exact definition, however, depends on the underlying *behaviour model*, which, in our approach to component modelling, is taken as a specification parameter. Therefore, the main contribution of this paper is a semantic characterization of refinement for state-based components, parametric on a strong monad intended to capture components' behavioural models.

After a brief review of the component calculus, in section 2, the paper discusses two levels of component refinement: the *interface* level, concerned with what one may call *plugging compatibility*, in section 3, and the *behavioural* one in section 4, which introduces *forward* and *backward* morphisms as refinement 'witnesses', and section 5 which builds on them to propose a family of refinement preorders. Section 6 proves soundness of *simulations* to establish behavioural refinement. A few examples, along with some prospects for future work, are presented in section 7.

2   Components as Coalgebras

In [3, 2] software components and connectors have been characterised as dynamic systems with a public interface and a private, encapsulated state. A typical example is LBuff: a connector modelling a buffered channel which occasionally looses messages, as represented below:

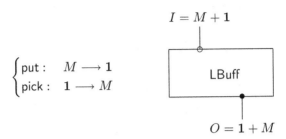

$$I = M + 1$$

$$\begin{cases} \text{put}: & M \longrightarrow 1 \\ \text{pick}: & 1 \longrightarrow M \end{cases}$$

LBuff

$$O = 1 + M$$

The put and pick operations are regarded as 'buttons' or 'ports', whose signatures are grouped together in the diagram (M stands for a message parameter type, 1 for the nullary datatype and $+$ for 'datatype sum'). One might capture LBuff dynamics by a function $a_{\mathsf{LBuff}} : U \times I \longrightarrow \mathcal{P}(U \times O)$ where U denotes the space state. This describes how LBuff reacts to input stimuli, produces output data (if any) and changes state. It can also be written in a curried form as $\overline{a}_{\mathsf{LBuff}} : U \longrightarrow \mathcal{P}(U \times O)^I$ that is, as a *coalgebra* [14] of signature $U \longrightarrow \mathsf{T}\, U$ where functor T captures the transition 'shape':

$$\mathsf{T} = \mathcal{P}(\mathsf{Id} \times O)^I \tag{1}$$

Built in this 'shape' is the possibility of non deterministic evolution captured by the use of $\mathcal{P}$, the finite powerset monad. Concretely, LBuff is defined over $U = M^*$, with nil as the initial state, and dynamics given by

$$a_{\mathsf{LBuff}}\langle u, \mathsf{put}\ m \rangle = \{\langle u, \iota_1 * \rangle, \langle m : u, \iota_1 * \rangle\}$$
$$a_{\mathsf{LBuff}}\langle u, \mathsf{pick} \rangle = \{\langle \mathsf{tail}\ u, \iota_2\ (\mathsf{head}\ u) \rangle\}$$

where put m and pick abbreviates $\iota_1\, m$ and $\iota_2\, *$, respectively.

Non determinism, capturing the occasional loss of messages, is a possible behavioural pattern for this buffer, but, by no means, the only one. Other components will exhibit different *behaviour models*: actually *genericity* is achieved by replacing the *powerset* monad above, by an arbitrary *strong monad*[1] B. In the general case, a component $p : I \longrightarrow O$ is specified as a (pointed) coalgebra in Set

$$\langle u_p \in U_p, \overline{a}_p : U_p \longrightarrow \mathsf{B}(U_p \times O)^I \rangle \tag{2}$$

where point u_p is taken as the 'initial' or 'seed' state. Therefore, the computation of an action will not simply produce an output and a continuation state,

---

[1] A *strong monad* is a monad $\langle \mathsf{B}, \eta, \mu \rangle$ where B is a strong functor and both η and μ strong natural transformations. B being strong means there exist natural transformations $\tau_r^\mathsf{T} : \mathsf{T} \times - \Longrightarrow \mathsf{T}(\mathsf{Id} \times -)$ and $\tau_l^\mathsf{T} : - \times \mathsf{T} \Longrightarrow \mathsf{T}(- \times \mathsf{Id})$ called the right and left strength, respectively, subject to certain conditions. Their effect is to *distribute* the free variable values in the context "$-$" along functor B.

but a B-structure of such pairs. The monadic structure provides tools to handle such computations. Unit (η) and multiplication (μ), provide, respectively, a value embedding and a 'flatten' operation to reduce nested behavioural annotations. Strength, either in its right (τ_r) or left (τ_l) version, will cater for context information.

In such a framework, components become *arrows* in a (bicategorical) universe Cp whose objects are sets, which provide types to input/output parameters (the components' *interfaces*), and component morphisms $h : p \longrightarrow q$ are functions relating the state spaces of p and q and satisfying the following *seed preservation* and *coalgebra* conditions:

$$h\,u_p = u_q \qquad \text{and} \qquad \overline{a}_q \cdot h = \mathsf{B}\,(h \times O)^I \cdot \overline{a}_p \qquad (3)$$

For each triple of objects $\langle I, K, O \rangle$, a composition law is given by functor $;_{I,K,O}$: $\mathsf{Cp}(I,K) \times \mathsf{Cp}(K,O) \longrightarrow \mathsf{Cp}(I,O)$ whose action on objects p and q is

$$p\,;q = \langle \langle u_p, u_q \rangle \in U_p \times U_q, \overline{a}_{p;q} \rangle \qquad \text{with}$$

$$
\begin{aligned}
a_{p;q} &= U_p \times U_q \times I \xrightarrow{\;\cong\;} U_p \times I \times U_q \xrightarrow{\;a_p \times \mathsf{id}\;} \mathsf{B}(U_p \times K) \times U_q \\
&\xrightarrow{\;\tau_r\;} \mathsf{B}(U_p \times K \times U_q) \xrightarrow{\;\cong\;} \mathsf{B}(U_p \times (U_q \times K)) \\
&\xrightarrow{\;\mathsf{B}(\mathsf{id} \times a_q)\;} \mathsf{B}(U_p \times \mathsf{B}(U_q \times O)) \xrightarrow{\;\mathsf{B}\tau_l\;} \mathsf{BB}(U_p \times (U_q \times O)) \\
&\xrightarrow{\;\cong\;} \mathsf{BB}(U_p \times U_q \times O) \xrightarrow{\;\mu\;} \mathsf{B}(U_p \times U_q \times O)
\end{aligned}
$$

Similarly, for each object K, an identity law is given by a functor $\mathsf{copy}_K : 1 \longrightarrow \mathsf{Cp}(K,K)$ whose action is the constant component $\langle * \in 1, \eta_{1 \times K} \rangle$. Note that the definitions above rely solely on the monadic structure of B.

In [3,2] a set of component *combinators* have been defined upon Cp in a similar parametric way and their properties studied. In particular it was shown that any function $f : A \longrightarrow B$ can be lifted to Cp as $\ulcorner f \urcorner = \langle * \in 1, \eta_{(1 \times B)} \cdot (\mathsf{id} \times f) \rangle$. Also defined were both a *wrapping* mechanism $p[f, g]$ which encodes the pre- and post-composition of a component with Cp-lifted functions, and three tensors, capturing, respectively, *external choice* ($\boxplus : I + J \longrightarrow O + R$), *parallel* ($\boxtimes$: $I \times J \longrightarrow O \times R$) and *concurrent* ($\boxast : I+J+I \times J \longrightarrow O+R+O \times R$) composition, given components $p : I \longrightarrow O$ and $q : J \longrightarrow R$. When interacting with $p \boxplus q$: $I + J \longrightarrow O + R$, the environment chooses either to input a value of type I or one of type J, which triggers the corresponding component (p or q, respectively), producing the relevant output. In its turn, *parallel* composition corresponds to a synchronous product: both components are executed simultaneously when triggered by a pair of legal input values. Note, however, that the behaviour effect, captured by monad B, propagates. For example, if B can express component failure and one of the arguments fails, the product will fail as well. Finally, *concurrent* composition combines choice and parallel, in the sense that p and q can be executed independently or jointly, depending on input. Generalized interaction is catered through a 'feedback' mechanism on a subset of the inputs.

3   Interface Refinement

Component *interface refinement* is concerned with type compatibility. The question is whether a component can be transformed, by suitable wiring, to replace another component with a different interface. As the structure of components interface types encodes the available operations, this may capture situations of *extension of functionality*, in the sense that the 'concrete' component may introduce new operations. In the context of object-orientation, this is often called *design sophistication* (rather than *refinement*) and it is known not to be a congruence with respect to typical process combinators (see *e.g.*, [17]). If we structure component input and output parameters as an operations' signature, interface refinement can also be seen as induced by a signature morphism, as in *e.g.*, [13].

To motivate our own approach, consider, from [3], the following law expressing commutativity of *choice*:

$$p \boxplus q \ \sim \ (q \boxplus p)[\mathsf{s}_+, \mathsf{s}_+] \tag{4}$$

where $\mathsf{s}_+ : I + J \longrightarrow J + I$ is a natural isomorphism capturing $+$ commutativity. The law states that $p \boxplus q$ and $q \boxplus p$ are bisimilar *up to isomorphic wiring*. This means that the observational effect of component $p \boxplus q$ can be achieved by $q \boxplus p$, providing the interface of the latter is converted to the interface of the former. Such a conversion is achieved by composition with the appropriate *wires*, leading to a notion of *replaceability*.

Definition 1 *Let p and q be components. We say that $p : I \longrightarrow O$ is* replaceable by $q : I' \longrightarrow O'$, *or q is a* replacement *of p, and write $p \lessdot q$ if there exist functions $w_1 : I \longrightarrow I'$ and $w_2 : O' \longrightarrow O$, to be referred to as the* replacement *witnesses, such that*

$$p \ \sim \ q[w_1, w_2] \tag{5}$$

Furthermore, components p and q are interchangeable *if each of them is a replacement of the other. Formally,*

$$p \doteq q \quad iff \quad p \lessdot q \ \wedge \ q \lessdot p \tag{6}$$

Clearly, $p \boxplus q \doteq q \boxplus p$, using isomorphism s_+ as a wire in both cases. In general, $p \doteq q$ whenever w_1 and w_2 in (5) are isomorphisms.

Lemma 1. *Replaceability $(\lessdot)$ is a preorder on components*

Proof. Clearly, $\lessdot$ is reflexive because $p \lessdot p$ is witnessed by $p \sim p[\mathsf{id}, \mathsf{id}]$. On the other hand, if $p \lessdot q$ and $q \lessdot r$ hold, there exist w_1, w_2, w_3 and w_4 such that $p \sim q[w_1, w_2]$ and $q \sim r[w_3, w_4]$. Thus, a composition result on wrapping [2] and transitivity of $\sim$, entails $p \sim r[w_1 \cdot w_3, w_4 \cdot w_2]$, *i.e.*, $p \lessdot r$. □

Using $\lessdot$ and $\doteq$, some component laws in [2], as (4) above, can be presented in a 'wiring free' form. As another example consider the law relating *concurrent* composition with *choice*,

$$\ulcorner \iota_1 \urcorner \,;\, (p \boxtimes q) \ \sim \ (p \boxplus q) \,;\, \ulcorner \iota_1 \urcorner$$

which gives rise to two replacement inequations:

$$\ulcorner \iota_1 \urcorner ; (p \boxtimes q) \; \prec \; p \boxplus q \quad \text{and} \quad (p \boxplus q) ; \ulcorner \iota_1 \urcorner \; \prec \; p \boxtimes q$$

Finally, the statement that every component p can replace an *inert* component can be expressed as an interface refinement situation: inert $\prec p$.

Relation $\prec$, however, fails to be a pre-congruence with respect to the component operators introduced in [3]. It is easy to check that $\boxplus$, $\boxtimes$ and $\boxasterisk$, as well as *wrapping* are preserved, *i.e.*, if $p \prec p'$ then, for any q, f and g, $p[f, g] \prec p'[f, g]$, $p \boxplus q \prec p' \boxplus q$ and, similarly, for the other two tensors. But things are different with respect to sequential composition and feedback. In these cases, the replaced expression may even become wrongly typed.

What $p \prec p'$ means is that component p can be replaced in *any* context by $p'[w_1, w_2]$, for any functions w_1, w_2 witnessing the fact. The explicit reference to them is actually required, something which is not completely satisfactory in a refinement situation, although common in similar settings (*cf.* [17]).

4 Forward and Backward Morphisms

Interface refinement is essentially concerned with plugging adjustment. Behaviour refinement, on the other hand, affects the internal dynamics of a component while leaving unchanged its external interface: it takes place inside each hom-category of Cp. Intuitively component p is a *behavioural refinement* of q if the behaviour patterns observed from p are a structural restriction, with respect to the *behavioural model* captured by monad B, of those of q. To make precise such a 'definition' we shall first describe behaviour patterns concretely as *generalized transitions*.

Actually, just as transition systems can be coded back as coalgebras, any coalgebra $\langle U, p : U \longrightarrow \mathsf{T}U \rangle$ specifies a (T-shaped) transition structure over its carrier U. For extended polynomial Set endofunctors[2] such a structure may be expressed as a binary relation $\longrightarrow_p \subseteq U \times U$, defined in terms of the *structural membership* relation $\in_\mathsf{T} \subseteq U \times \mathsf{T}\, U$, *i.e.*,

$$u \longrightarrow_p u' \quad \text{iff} \quad u' \in_\mathsf{T} p\, u$$

where $\in_\mathsf{T}$ is defined by induction of the structure of T:

$$
\begin{aligned}
x \in_{\mathsf{Id}} y \quad &\text{iff} \quad x = y \\
x \in_{\underline{K}} y \quad &\text{iff} \quad \text{false} \\
x \in_{\mathsf{T}_1 \times \mathsf{T}_2} y \quad &\text{iff} \quad x \in_{\mathsf{T}_1} \pi_1\, y \; \vee \; x \in_{\mathsf{T}_2} \pi_2\, y \\
x \in_{\mathsf{T}_1 + \mathsf{T}_2} y \quad &\text{iff} \quad \begin{cases} y = \iota_1\, y' & \Rightarrow x \in_{\mathsf{T}_1} y' \\ y = \iota_2\, y' & \Rightarrow x \in_{\mathsf{T}_2} y' \end{cases} \\
x \in_{\mathsf{T}\underline{K}} y \quad &\text{iff} \quad \exists_{k \in K}.\ x \in_\mathsf{T} y\, k \\
x \in_{\mathsf{P}\mathsf{T}} y \quad &\text{iff} \quad \exists_{y' \in y}.\ x \in_\mathsf{T} y'
\end{aligned}
$$

---

[2] The class inductively defined as the least collection of functors containing the identity Id and constant functors $\underline{K}$ for all objects K in the category, closed by functor composition and finite application of product, coproduct, covariant exponential and finite powerset functors.

Notice that, given $x \in U$, $X \in \mathsf{T}U$ and a function $h : U \longrightarrow V$, if $x \in_\mathsf{T} X$ then $h\, x \in_\mathsf{T} \mathsf{T}h\, X$, as it may be shown by induction on the polynomial structure, resorting to the definition of $\in_\mathsf{T}$ and functoriality. Similarly, the dynamics of $p : I \longrightarrow O$, based on functor $\mathsf{B}(\mathsf{Id} \times O)^I$, can be expressed in terms of the following transition relation:

$$u \xrightarrow{\langle i,o \rangle}_p u' \quad \text{iff} \quad \langle u',o \rangle \in_\mathsf{B} (pu)\, i$$

In this setting, a possible (and intuitive) way of regarding component p as a behavioural refinement of q is to consider that p transitions are simply *preserved* in q. For non deterministic components this is understood simply as set inclusion. But one may also want to consider additional restrictions. For example, to stipulate that if p has no transitions from a given state, q should also have no transitions from the corresponding state(s). Or one may adopt the dual point of view requiring transition *reflection* instead of preservation. In any case the same basic question arises: how can such a refinement situation be identified?

In data refinement, as mentioned above, there is a 'recipe' to identify a refinement situation: look for a *retrieve function* to witness it. I.e., a morphism in the relevant category, from the 'concrete' to the 'abstract' model such that the latter can be *recovered* from the former up to a suitable notion of equivalence, though, typically, not in a unique way.

In our components' framework, however, things do not work this way. The reason is obvious: component morphisms are (seed preserving) coalgebra morphisms which are known (*e.g.*, [14]) to entail bisimilarity. Therefore we have to look for a somewhat *weaker* notion of a morphism between coalgebras.

First of all recall that a component morphism from p to q is a seed preserving function $h : U_p \longrightarrow U_q$ such that

$$\mathsf{B}(h \times \mathsf{id}) \cdot a_p = a_q \cdot (h \times \mathsf{id}) \tag{7}$$

In terms of transitions, equation (7) is translated into the following two requirements (by a straightforward generalization of an argument in [14]):

$$u \xrightarrow{\langle i,o \rangle}_p u' \;\Rightarrow\; h\, u \xrightarrow{\langle i,o \rangle}_q h\, u' \tag{8}$$

$$h\, u \xrightarrow{\langle i,o \rangle}_q v' \;\Rightarrow\; \exists_{u' \in U}.\; u \xrightarrow{\langle i,o \rangle}_p u' \wedge v' = h\, u' \tag{9}$$

which jointly state that, not only p dynamics, as represented by the induced transition relation, is *preserved* by h (8), but also q dynamics is *reflected* back over the same h (9). Is it possible to weaken the morphism definition to capture only one of these aspects? The answer is given as follows:

An order $\leq$ on a Set endofunctor T is defined in [9] as a functor $\leq$ which makes the following diagram to commute:

This means that for any function $h : X \longrightarrow Y$, $\mathsf{T}h$ preserves the order, *i.e.*

$$x_1 \leq_{TX} x_2 \quad \Rightarrow \quad (\mathsf{T}h)\, x_1 \leq_{TY} (\mathsf{T}h)\, x_2 \tag{10}$$

In the sequel $\leq$ will be referred to as a *refinement preorder*. Then,

Definition 2 *Let* T *be an extended polynomial functor on* Set *and consider two* T*-coalgebras* $\alpha : U \longrightarrow \mathsf{T}U$ *and* $\beta : V \longrightarrow \mathsf{T}V$. *A forward morphism* $h : \alpha \longrightarrow \beta$ *with respect to a refinement preorder* $\leq$, *is a function from* U *to* V *such that*

$$\mathsf{T}\, h \cdot \alpha \;\leq\; \beta \cdot h$$

Dually, h is called a backwards *morphism if*

$$\beta \cdot h \;\leq\; \mathsf{T}\, h \cdot \alpha$$

The following lemma shows that such morphisms compose and can be taken as witnesses of refinement situations:

Lemma 2. *For* T *an endofunctor in* Set, T*-coalgebras and forward (respectively, backward) morphisms define a category.*

Proof. In both cases, identities are the identities on the carrier and composition is inherited from Set. What remains to be shown is that the composition of forward (respectively, backward) morphisms yields again a forward (respectively, backward) morphism. So, let $h : \alpha \longrightarrow \beta$ and $k : \beta \longrightarrow \gamma$ be two forward (respectively, backward) morphisms. Then

(*forward* case)	(*backward* case)

$$
\begin{array}{ll}
& \mathsf{T}(k \cdot h) \cdot \alpha \\
= & \quad \{\ \mathsf{T} \text{ functor } \} \\
& \mathsf{T}k \cdot (\mathsf{T}h \cdot \alpha) \\
\leq & \quad \{\ h \text{ forward and (10) } \} \\
& \mathsf{T}k \cdot (\beta \cdot h) \\
= & \quad \{\ \cdot \text{ associative } \} \\
& (\mathsf{T}k \cdot \beta) \cdot h \\
\leq & \quad \{\ k \text{ forward } \} \\
& (\gamma \cdot k) \cdot h \\
= & \quad \{\ \cdot \text{ associative } \} \\
& \gamma \cdot (k \cdot h)
\end{array}
\qquad
\begin{array}{ll}
& \gamma \cdot (k \cdot h) \\
= & \quad \{\ \cdot \text{ associative } \} \\
& (\gamma \cdot k) \cdot h \\
\leq & \quad \{\ k \text{ backward } \} \\
& (\mathsf{T}k \cdot \beta) \cdot h \\
= & \quad \{\ \cdot \text{ associative } \} \\
& \mathsf{T}k \cdot (\beta \cdot h) \\
\leq & \quad \{\ h \text{ backward and (10) } \} \\
& \mathsf{T}k \cdot \mathsf{T}h \cdot \alpha \\
= & \quad \{\ \mathsf{T} \text{ functor } \} \\
& \mathsf{T}(k \cdot h) \cdot \alpha
\end{array}
$$

$\square$

Such a split of a coalgebra morphism, witnessing a bisimulation equation, into two conditions, makes it possible to capture separately transition *preservation* and *reflection*. To prove the next result, however, it is necessary to impose an extra condition on the refinement preorder $\leq$ expressing its compatibility with $\in_T$: for all $x \in X$ and $x_1, x_2 \in TX$,

$$x \in_T x_1 \ \wedge\ x_1 \leq x_2 \ \Rightarrow\ x \in_T x_2 \tag{11}$$

Lemma 3. *Let* T *be an extended polynomial functor in* Set, *and* α *and* β *two* T-*coalgebras as above. Let* $\longrightarrow_\alpha$ *and* $\longrightarrow_\beta$ *denote the corresponding transition relations. A forward (respectively, backward) morphism* $h : \alpha \longrightarrow \beta$ *preserves (respectively, reflects) such transition relations.*

Proof. Preservation follows from

$$u \longrightarrow_\alpha u'$$

$\equiv \qquad \{\ \longrightarrow\ \text{definition}\ \}$

$$u' \in_T \alpha\, u$$

$\Rightarrow \qquad \{\ \in_T\ \text{definition}\ \}$

$$h\, u' \in_T (T h \cdot \alpha)\, u$$

$\equiv \qquad \{\ h\ \text{forward and (11)}\ \}$

$$h\, u' \in_T (\beta \cdot h)\, u$$

$\equiv \qquad \{\ \cdot\ \text{associative and}\ \longrightarrow\ \text{definition}\ \}$

$$h\, u \longrightarrow_\beta h\, u'$$

To establish reflection suppose that $h\, u \longrightarrow_\beta v'$, *i.e.*, $v' \in_T (\beta \cdot h)\, u$. As h is a backward morphism we have $\beta \cdot h \leq T\, h \cdot \alpha$, which, together with requirement (11), entails $v' \in_T (T\, h \cdot \alpha)\, u$. This implies the existence of a state $u' \in U$ such that $v' = h\, u'$ and $u' \in_T \alpha\, u$, *i.e.*, $u \longrightarrow_\alpha u'$. $\qquad\square$

5   Behaviour Refinement

The existence of a *forward* (*backward*) morphism connecting two components p and q witnesses a refinement situation whose symmetric closure coincides, as expected, with bisimulation. In the sequel we will restrict ourselves to *forward* refinement[3] and define *behaviour refinement* as the existence of a forward morphism *up to bisimulation*. Formally,

Definition 3 *Component* p *is a behaviour refinement of* q, *written* $q \trianglelefteq p$, *if there exist components* r *and* s *such that* $p \sim r$, $q \sim s$ *and a (seed preserving) forward morphism from* r *to* s.

The exact meaning of a refinement assertion $q \trianglelefteq p$ depends, of course, on the concrete refinement preorder $\leq$ adopted. Let us consider a few possibilities.

---

[3] A similar study can be made about *backward* refinement, although the underlying intuition seems less familiar.

- T-*structural* inclusion, defined by $x \leq y$ iff $\forall_{e \in T_x}$. $e \in_T y$, seems inadequate because the transition relation preserved by a forward morphism is not $\xrightarrow{\langle i, o \rangle}_p$, but simply $\longrightarrow_p$, and, therefore, blind to the outputs produced. This suggests an additional requirement on refinement preorders for Cp components: their definition on a constant functor $\underline{K}$ is equality on set K, i.e., $x \leq_K y$ iff $x =_K y$ so that transitions with different O-labels could not be related.
- Building on this idea, we arrive to a first (good) example:

$$x \subseteq_{\mathsf{Id}} y \quad \text{iff} \quad x = y$$

$$x \subseteq_{\underline{K}} y \quad \text{iff} \quad x =_K y$$

$$x \subseteq_{\mathsf{T}_1 \times \mathsf{T}_2} y \quad \text{iff} \quad \pi_1\, x \subseteq_{\mathsf{T}_1} \pi_1\, y \;\wedge\; \pi_2\, x \subseteq_{\mathsf{T}_2} \pi_2\, y$$

$$x \subseteq_{\mathsf{T}_1 + \mathsf{T}_2} y \quad \text{iff} \quad \begin{cases} x = \iota_1\, x' \wedge y = \iota_1\, y' \;\Rightarrow\; x' \subseteq_{\mathsf{T}_1} y' \\ x = \iota_2\, x' \wedge y = \iota_2\, y' \;\Rightarrow\; x' \subseteq_{\mathsf{T}_2} y' \end{cases}$$

$$x \subseteq_{\mathsf{T}\underline{K}} y \quad \text{iff} \quad \forall_{k \in K}.\; x\, k \subseteq_{\mathsf{T}} y\, k$$

$$x \subseteq_{\mathcal{P}\mathsf{T}} y \quad \text{iff} \quad \forall_{e \in x} \exists_{e' \in y}.\; e \subseteq_{\mathsf{T}} e'$$

A *forward* refinement of non deterministic components based on $\subseteq_{\mathsf{T}}$ captures the classical notion of *nondeterminism reduction*.

- However, this preorder can be tuned to more specific cases. For example, the following 'failure forcing' variant – $\subseteq_{\mathsf{T}}^{E}$, where E stands for 'emptyset' – guarantees that the concrete component fails no more than the abstract one. It is defined as $\subseteq_{\mathsf{T}}$ by replacing the clause for the powerset functor by

$$x \subseteq_{\mathcal{P}\mathsf{T}}^{E} y \quad \text{iff} \quad (x = \emptyset \Rightarrow y = \emptyset) \;\wedge\; \forall_{e \in x} \exists_{e' \in y}.\; e \subseteq_{\mathsf{T}} e'$$

- Relation $\subseteq_{\mathsf{T}}$ is inadequate for partial components: refinement would collapse into bisimilarity, instead of entailing increasing definition in the implementation. An alternative is relation $\subseteq_{\mathsf{T}}^{F}$ (F standing for 'failure') which replaces the sum clause in $\subseteq_{\mathsf{T}}$ by

$$x \subseteq_{\mathsf{T}_1 + \mathsf{T}_2}^{F} y \quad \text{iff} \quad \begin{cases} x = \iota_1\, x' \wedge y = \iota_1\, y' \;\Rightarrow\; x' \subseteq_{\mathsf{T}} y' \\ x = \iota_2\, * & \Rightarrow y = \iota_2\, * \\ y = \iota_2\, * & \Rightarrow \text{true} \end{cases}$$

To illustrate behaviour refinement, consider the lossy buffer LBuff introduced in section 2, and a deterministic buffered channel Buff specified as a coalgebra $M^* \longrightarrow (M^* \times (1 + M))^{M+1}$ with nil as the initial state, and dynamics given by

$$a_{\mathsf{Buff}} \langle u, \mathsf{put}\ m \rangle \;=\; \langle m : u, \iota_1\, * \rangle$$

$$a_{\mathsf{Buff}} \langle u, \mathsf{pick} \rangle \;=\; \langle \mathsf{tail}\ u, \iota_2\ (\mathsf{head}\ u) \rangle$$

To establish LBuff $\trianglelefteq$ Buff it is required first to embed the latter into the space of non-deterministic systems. This is achieved by a (natural) transformation from

$(\mathsf{Id} \times O)^I$ to $\mathcal{P}(\mathsf{Id} \times O)^I$ canonically extending function $\mathsf{sing}\, x = \{x\}$ which is a monad morphism from the *identity* to the *powerset* monads – the behaviour models underlying Buff and LBuff, respectively. Then, it is immediate to verify that the identity function on state space M^* is a *forward* morphism, with respect to the first preorder given above, *i.e.*,

$$(\mathsf{id} \times O) \cdot a_{\mathsf{Buff}} \subseteq a_{\mathsf{LBuff}}$$

Another behaviour refinements of LBuff would arise by choosing different strategies for delivering elements from the buffer. Here are some possibilities, each of them is witnessed by a forward morphism:

– the queuing strategy, leading to the specification Buff as above;
– the stack strategy (LIFO deliver);
– the priority strategy (in which elements carry some probability information);
– the lift strategy (a linear order on the elements is served in alternating increasing and decreasing order).

In the priority strategy, for example, elements are labelled with a 'show-up' probability, introducing an elementary form of probabilistic nondeterminism. As detailed in [3], the corresponding behaviour monad is generated by a monoid $M = \langle [0,1], \min, \times \rangle$ with the additional requirement that for each $m \in M$, $\sum(\mathcal{P}\pi_2)m = 1$. 'Probabilistic' components can be embedded into the space of 'plain nondeterministic' ones where behaviour refinement, wrt $\subseteq_\mathsf{T}$, is discussed.

6   Simulations

In this section we prove that behaviour refinement, as characterized above, can be established by a *simulation* relation $R \subseteq U_p \times U_q$ on the state spaces of the 'concrete' (p) and the 'abstract' (q) components. Again, the notion of a simulation depends on the adopted refinement preorder $\leq$. To proceed in a *generic* way, we adopt an equally generic definition of simulation due to Jacobs and Hughes in [9]:

Definition 4 *Given a* Set *endofunctor* T *and a refinement preorder* $\leq$, *a lax relation lifting is an operation* $Rel_\leq(\mathsf{T})$ *mapping relation* R *to* $\leq \circ Rel(\mathsf{T})(R) \circ \leq$, *where* $Rel(\mathsf{T})(R)$ *is the lifting of* R *to* T *(defined, as usual, as the* T*-image of inclusion* $\langle r_1, r_2 \rangle : R \longrightarrow U \times V$, *i.e.,* $\langle \mathsf{T}r_1, \mathsf{T}r_2 \rangle : \mathsf{T}R \longrightarrow \mathsf{T}U \times \mathsf{T}V$*).*

Given T*-coalgebras* α *and* β, *a simulation is a* $Rel_\leq(\mathsf{T})$*-coalgebra over* α *and* β, *i.e., a relation* R *such that, for all* $u \in U, v \in V$, $\langle u, v \rangle \in R \Rightarrow \langle \alpha\, u, \beta\, v \rangle \in Rel_\leq(\mathsf{T})(R)$.

In order to prove that simulations are a sound proof technique to establish behaviour refinement we consider separately functional and non functional simulations. In any case, however, simulations are assumed to be left total relations[4] as we do not consider *partial* refinements.

---

[4] A relation $R \subseteq U \times V$ is *functional* if every $u \in U$ is related to at most one $v \in V$ and *left total* if for all $u \in U$, there exists some $v \in V$ such that $\langle u, v \rangle \in R$.

Lemma 4. *Let p and q be T-components over state spaces U and V, respectively. For a given refinement preorder $\leq$, if there exists a simulation $R \subseteq U \times V$ which is both functional and left total, then p is a (forward) refinement of q.*

Proof. By assumption, simulation R is the graph of a function. Now, define a forward morphism $h : U \to V$ as $h\,u = v$ iff $\langle u, v \rangle \in R$. Because R is a simulation, for every pair $\langle u, v \rangle \in R$, there should exist $x \in TU$, $y \in TV$, such that $\alpha\,u \leq_{TU} x$, $y \leq_{TV} \beta\,v$, and $\langle x, y \rangle \in Rel(T)(R)$, i.e., $y = Th(x)$. By (10) and $\alpha\,u \leq_{TU} x$ we get $Th(\alpha\,u) \leq_{TV} Th(x)$, and thus $Th(\alpha\,u) \leq_{TV} \beta\,v$. Since R is left total, h is defined for all $u \in U$, making the following diagram to commute:

$$
\begin{array}{ccc}
u & \xrightarrow{\ \ h\ \ } & h\,u = v \\[2pt]
{\scriptstyle \alpha}\downarrow & & \downarrow{\scriptstyle \beta} \\[2pt]
\alpha\,u\,(\leq_{TU} \alpha\,u) & \xrightarrow{\ Th\ } Th(\alpha\,u) \leq_{TV} & \beta\,v
\end{array}
$$

$\square$

Consider, now, the non-functional case (*e.g.* whenever two bisimilar but not equal abstract states are represented by a single concrete state). To prove soundness in this case, the state space of the 'concrete' component p is artificially inflated with an auxiliary state space such that a forward morphism can be found.

Definition 5 *Given a coalgebra $(U, \alpha : U \to TU)$ and a set W, define the extension of α to W as any coalgebra $\widehat{\alpha}$ over $\widehat{U} = U \times W$ such that $T\pi_1 \circ \widehat{\alpha} = \alpha \circ \pi_1$.*

Clearly this auxiliary state space does not interfere with the behaviour of α: π_1 being a coalgebra morphism, the two coalgebras are bisimilar.

Given components p and q and a non-functional simulation R an auxiliary coalgebra $\widehat{p}$ can be defined taking R as the state space (which, because R is left total, is just an extension of p in the sense of the definition above) and the rule $(u', v') \in_T \widehat{\alpha}(u, v)$ iff $u' \in_T a_p u \wedge v' \in_T a_q v$ as its dynamics. With this construction we prove that

Lemma 5. *(Soundness) To prove $q \trianglelefteq p$ it is sufficient to exhibit a left total simulation R relating components p and q.*

Proof. If R is functional the result follows from lemma 4. Otherwise construct $\widehat{p}$ as above: clearly p is bisimilar to $\widehat{p}$ and the graph of projection π_2 from its state space to V defines a simulation between $\widehat{p}$ and q. By definition, $p \sim \widehat{p}$ and the existence of a (seed-preserving) forward morphism from $\widehat{p}$ to q entails $q \trianglelefteq p$. $\square$

Finally notice that, although $\trianglelefteq$ is transitive, it is not always the case that simulations are closed under (relational) composition. This would be a consequence of $Rel_\leq(T)$ preserving composition, but, in general, only the following weaker result holds:

Lemma 6. *Any refinement preorder $\leq$ verifies*

$$Rel_\leq(T)(R \circ S) \subseteq Rel_\leq(T)(R) \circ Rel_\leq(T)(S) \quad and \quad =_{TU} \subseteq Rel_\leq(T)(=_U)$$

Proof. For the first statement note that $\langle u, w \rangle \in Rel_\leq(\mathsf{T})(R \circ S)$ equivales

$$\exists u', w'.(u \leq u' \wedge \langle u', w' \rangle \in Rel(\mathsf{T})(R \circ S) \wedge w' \leq w)$$
$$\{\text{because } Rel(\mathsf{T})(R \circ S) = Rel(\mathsf{T})(R) \circ Rel(\mathsf{T})(S)\}$$
$$\Leftrightarrow \exists u', w'.(u \leq u' \wedge (\exists v'.(\langle u', v' \rangle \in Rel(\mathsf{T})(R) \wedge \langle v', w' \rangle \in Rel(\mathsf{T})(S))) \wedge w' \leq w)$$
$$\Leftrightarrow \exists u', w', v'.(u \leq u' \wedge \langle u', v' \rangle \in Rel(\mathsf{T})(R) \wedge \langle v', w' \rangle \in Rel(\mathsf{T})(S) \wedge w' \leq w)$$
$$\{\text{introducing } v = v'\}$$
$$\Rightarrow \exists u', w', v, v'.(u \leq u' \wedge \langle u', v' \rangle \in Rel(\mathsf{T})(R) \wedge v' \leq v) \wedge$$
$$(v \leq v' \wedge \langle v', w' \rangle \in Rel(\mathsf{T})(S) \wedge w' \leq w)$$
$$\Rightarrow \exists v.\langle u, v \rangle \in Rel_\leq(\mathsf{T})(R) \wedge \langle v, w \rangle \in Rel_\leq(\mathsf{T})(S)$$
$$\Leftrightarrow \langle u, w \rangle \in Rel_\leq(\mathsf{T})(R) \circ Rel_\leq(\mathsf{T})(S)$$

Then consider

$$=_{\mathsf{T}U} \subseteq \leq_{\mathsf{T}U}$$
$$= \leq_{\mathsf{T}U} \circ =_{\mathsf{T}U} \circ \leq_{\mathsf{T}U} = \leq_{\mathsf{T}U} \circ Rel(\mathsf{T})(=_U) \circ \leq_{\mathsf{T}U} = Rel_\leq(\mathsf{T})(=_U)$$

$\square$

7   Discussion and Future Work

In this paper, two levels of refinement for (state-based) components have been introduced. In particular, the notion of *behavioural* refinement parametric on a model of behaviour captured by a strong monad B is, to the best of our knowledge, new. It is *generic* enough to capture a number of situations, depending on both B and the refinement preorder adopted. Nondeterminism reduction is just one possibility among many others. Also note that Poll's notion of *behavioural subtyping* in [13], at the model level, emerges as a particular instantiation.

As mentioned in the introduction, the main motivation underlying this work is the development of *inequational* laws in the context of the component calculus proposed in [3]. Even though there is not enough space in this paper to introduce the derived laws, let us take a brief glimpse. As a first example consider equation

$$\ulcorner !_I \urcorner \sim p ; \ulcorner !_O \urcorner \tag{12}$$

which does not hold for non trivial behaviour models. In fact the Cp lifting of the final arrow (as the lifting of any other function) cannot fail, whereas the the right hand side may fail (whenever p does). Function $! : U_p \times 1 \longrightarrow 1$ is, however, a forward morphism, with respect to $\subseteq_\mathsf{T}^F$ for partial components, or to both $\subseteq_\mathsf{T}$ and $\subseteq_\mathsf{T}^E$ for non deterministic ones. For this last case, note that $\overline{a}_{\ulcorner !_O \urcorner} \cdot ! = \lambda\, i \in I.\, \{*\}$, whereas $\mathsf{B}(! \times \mathrm{id})^I \cdot \overline{a}_{p;\ulcorner !_O \urcorner} \langle u, * \rangle$ equals

$$\lambda\, i \in I.\ \begin{cases} \{*\} & \text{iff } (\overline{a}_p\, u)\, (i) \neq \emptyset \\ \emptyset & \text{iff } (\overline{a}_p\, u)\, (i) = \emptyset \end{cases}$$

Therefore, $\ulcorner !_I \urcorner \trianglelefteq p ; \ulcorner !_O \urcorner$. Similarly, the *cancellation* law for parallel composition $\boxtimes$, which involves a *split*-like construction for components (which, differently from the split of functions [4], is not an universal arrow), is, in general, a refinement result:

$$p \trianglelefteq \langle p, q \rangle ; \ulcorner \pi_1 \urcorner \tag{13}$$

witnessed by projection $\pi_1 : U_p \times U_q \times 1 \to U_p$ as a forward morphism. Yet another example is given by the (pseudo) naturality of $\ulcorner \triangle \urcorner$, where $\triangle$ is the diagonal function, which could be written as

$$\ulcorner \triangle \urcorner ; (p \boxtimes p) \trianglelefteq p ; \ulcorner \triangle \urcorner \tag{14}$$

Finally, *monotonicity* of $\trianglelefteq$ with respect to both *pipeline* composition and the tensor products can be proved by defining a simulation in terms of the argument simulations: if $q \trianglelefteq p$ and $t \trianglelefteq r$ are witnessed by R_1 and R_2, respectively, refinement $q \Box t \trianglelefteq p \Box r$, with $\Box$ standing for $;, \boxtimes, \boxplus$ or $\boxast$ is witnessed by simulation $R = \{((u_p, u_r), (u_q, u_t)) \mid (u_p, u_q) \in R_1 \wedge (u_r, u_t) \in R_2\}$.

Currently we are working on the full development of the refinement calculus and, in particular, in its application to the proof of consistency between static and dynamic diagrams in UML in the context of [12]. Whether this approach scales up to be useful in the classification and transformation of software *architectures* [1] remains a research question. Further comparison with refinement theories in both process algebra (as in, *e.g.*, [5]) and state-based systems (for example in [6]) is also in order.

Acknowledgements

This piece of research was carried on in the context of the PURE Project (*Program Understanding and Re-engineering*) supported by FCT (the Portuguese Foundation for Science and Technology) under contract POSI/ICHS/44304/2002. The work of Sun Meng was further supported by the National Natural Science Foundation of China under grant no. 60273001.

References

1. R. Allen and D. Garlan. A formal basis for architectural connection. *ACM TOSEM*, 6(3):213–249, 1997.
2. L. S. Barbosa. Towards a Calculus of State-based Software Components. *Journal of Universal Computer Science*, 9(8):891–909, August 2003.
3. L. S. Barbosa and J. N. Oliveira. State-based components made generic. In H. P. Gumm, editor, *CMCS'03, Elect. Notes in Theor. Comp. Sci.*, volume 82.1, 2003.
4. R. Bird and O. Moor. *The Algebra of Programming*. Series in Computer Science. Prentice-Hall International, 1997.
5. P. Degano, R. Gorrieri, and G. Rosolini. A categorical view of process refinement. In J. de Bakker, G. Rozenberg, and J. Rutten, editors, *Proc. REX Workshop on Semantics*, pages 138–154. Springer Lect. Notes Comp. Sci. (666), 1992.

6. J. Derrick and E. Boiten. Calculating upward and downward simulations of state-based specifications. *Information and Software Technology*, 41:917–923, July 1999.
7. M. Fokkinga and R. Eshuis. Comparing refinements for failure and bisimulation semantics. Technical report, Faculty of Computing Science, Enschede, 2000.
8. C. A. R. Hoare. Proof of correctness of data representations. *Acta Informatica*, 1:271–281, 1972.
9. B. Jacobs and J. Hughes. Simulations in coalgebra. In H. P. Gumm, editor, *CMCS'03, Elect. Notes in Theor. Comp. Sci.*, volume 82.1, Warsaw, April 2003.
10. C. B. Jones. *Systematic Software Development Using VDM*. Series in Computer Science. Prentice-Hall International, 1986.
11. B. Liskov. Data abstraction and hierarchy. *SIGPLAN Notices*, 23(3), 1988.
12. S. Meng and B. Aichernig. Towards a Coalgebraic Semantics of UML: Class Diagrams and Use Cases. Technical Report 272, UNU/IIST, January 2003.
13. E. Poll. A coalgebraic semantics of subtyping. *Theorectical Informatica and Aplli-cations*, 35(1):61–82, 2001.
14. J. Rutten. Universal coalgebra: A theory of systems. *Theor. Comp. Sci.*, 249(1):3–80, 2000. (Revised version of CWI Techn. Rep. CS-R9652, 1996).
15. C. Szyperski. *Component Software, Beyond Object-Oriented Programming*. Addison-Wesley, 1998.
16. P. Wadler and K. Weihe. Component-based programming under different paradigms. Technical report, Dagstuhl Seminar 99081, February 1999.
17. J. Woodcock and J. Davies. *Using Z: Specification, Refinement and Proof*. Prentice-Hall International, 1996.

Techniques for Executing and Reasoning about Specification Diagrams

Prasanna Thati[1], Carolyn Talcott[2], and Gul Agha[1]

[1] University of Illinois at Urbana Champaign
{thati,agha}@cs.uiuc.edu
[2] SRI International
clt@csl.sri.com

Abstract. Specification Diagrams (SD) [19] are a graphical notation for specifying the message passing behavior of open distributed object systems. SDs facilitate specification of system behaviors at various levels of abstraction, ranging from high-level specifications to concrete diagrams with low-level implementation details. We investigate the theory of *may testing equivalence* [15] on SDs, which is a notion of process equivalence that is useful for relating diagrams at different levels of abstraction. We present a semantic characterization of the may equivalence on SDs which provides a powerful technique to relate abstract specifications and refined implementations. We also describe our prototypical implementation of SDs and of a procedure that exploits the characterization of may testing to establish equivalences between finitary diagrams (without recursion).

Keywords: Graphical specification languages, π-calculus, may testing, trace equivalence, rewriting logic.

1 Introduction

Smith and Talcott introduced Specification Diagrams (SD) [19] as a graphical notation for specifying message passing behaviors of open distributed object systems. SDs not only have an intuitive appeal as other graphical specification languages such as UML [18] and MSC [17], but also have a formal underpinning which makes them amenable to rigorous analysis. SDs draw upon concepts from various formalisms for concurrency; they allow dynamic name generation and name passing as in the π-calculus [14], they have asynchronous communication and enforce the locality discipline on use of names as in concurrent object-based models such as the Actor model [1], they are equipped with imperative notions such as variables, environments, and assignments, and they also incorporate logical features such as assertions and constrains which are appropriate for specification languages.

The language of SDs is designed to be useful at various stages of the software cycle. In the initial stages, one can abstractly express the desired system behavior and its properties without having to switch to another logic, and then progressively refine the abstract specifications into concrete diagrams with implementation details. An important task to be accomplished in this process is to

C. Rattray et al. (Eds.): AMAST 2004, LNCS 3116, pp. 521–536, 2004.

be able to formally prove that a refined diagram is indeed a correct implementation of an abstract specification. The framework of *may testing* [15] is useful for formalizing such a semantic correspondence between diagrams. It is known that may testing is useful for reasoning about safety properties of implementations; specifically, it formalizes the criteria for a refined diagram to be a safe implementation of an abstract specification.

Relating diagrams according to may testing is in general a difficult task. In this paper, we present a characterization of may testing on SDs that provides a powerful technique for relating diagrams. We also present an executable implementation of SDs by modeling the language as a theory in rewriting logic [12]. The Maude tool [5] which supports specifications in rewriting logic can then be used to execute diagrams. Finally, we describe the implementation in Maude of a procedure that exploits the characterization of may testing to relate finitary diagrams that do not involve recursion.

SDs are more of a specification language rather than a programming language in that not every SD is executable. For instance SDs are equipped with the **constraint** construct that is analogous to Dijkstra's **assume** predicate [7]. A constrain specifies a predicate that should hold during a computation; failure of the predicate indicates that such a computation never happens, i.e the entire computation "is cancelled in between the computation" as though it never happened. It is clear that the constrain construct is not implementable in general. SDs are also equipped with certain fairness notions that are not implementable [20] (see the end of Section 3). In this paper, we will consider only the executable fragment of SDs; in particular we discard the constrain construct and the fairness conditions. Although the language we consider is only a fragment of Smith and Talcott's language, from now on we will refer to it as the language of SDs.

A central theme of our work is that we present SDs as an extension of asynchronous π-calculus [3, 9] with certain imperative and logical constructs. We will exploit this connection both to obtain a characterization of may testing and an executable implementation of SDs. Specifically, we will adapt our characterization of may testing for asynchronous π-calculus with locality [24] to obtain a characterization of may testing on SDs. Similarly, we will extend our implementation of asynchronous π-calculus described in [22] to obtain an implementation of SDs. In summary, this paper has three main contributions. It presents SDs as an extension of asynchronous π-calculus and exploits this connection to obtain both a characterization of may testing and an implementation of SDs.

Following is the layout of the rest of this paper. In Sections 2 and 3, we present SDs as an extension of asynchronous π-calculus. In Section 4, we instantiate the framework of may testing on SDs and present an alternate characterization of it. In Section 5, we describe our implementation of SDs in the Maude tool. We conclude the paper in Section 6 with comments on possible directions of future work.

2   Specification Diagram Syntax

We assume a set of values *Val*, which is not specified completely, but is assumed to include booleans and an infinite set of names $\mathcal{N}$. We assume an infinite set of variables *Var*, which can take on values from *Val*. The sets *Var* and *Val* are assumed to be disjoint. We let u, v, w range over *Val*, a, b, c over $\mathcal{N}$, ρ, ξ over sets of names, and x, y, z over *Var*. An environment is a partial function from *Var* to *Val* that is defined for only finitely many variables. We let γ range over environments. We denote an environment as a subset of $Var \times Val$ in the usual way. We assume a set of operations on *Val* that is not specified completely, but is assumed to contain the equality operator $=$, and the boolean operators $\neg, \vee$ and $\wedge$. We also assume a function $n(\cdot)$ on values such that $n(v)$ is the set of all names that are used in constructing the (possibly composite) value v; we assume that this set is always finite. We lift the function $n(\cdot)$ from *Val* to environments in the expected way. We let e, f, g range over expressions and ϕ over boolean expressions. Expressions can contain free variables, and are evaluated in an environment that assigns values to these variables. We assume an evaluation function $\text{eval}(e, \gamma)$ that evaluates an expression e in an environment γ that assigns values to all free variables in e. From now on, we use the words diagrams and processes interchangeably.

SDs are defined by the following context-free grammar. We assume a set of process variables *PrVar* that is disjoint from *Var* and *Val*, and let $X, Y, Z, \ldots$ range over it.

$$D := 0 \mid \bar{a}e \mid a(x).D \mid D_1 \mid D_2 \mid (\nu a)D \mid recX.D \mid X \qquad \textbf{(asynch } \pi\textbf{)}$$
$$\mid \ D_1; D_2 \mid \ D_1 \oplus D_2 \mid \text{fork}(D) \qquad \textbf{(control)}$$
$$\mid \ \{|\gamma : D|\} \mid x := e \qquad \textbf{(imperative)}$$
$$\mid \ \text{pick}(x).D \mid \text{wait}(\phi) \qquad \textbf{(logical)}$$

Following is an informal description of each of these constructs - (i) 0 (*nil*): Trivial behavior that does nothing. (ii) $\bar{a}e$ (*output*): Send an asynchronous message to a with the result of evaluating e as the content. The name a is said to be the *subject* of the output. (iii) $a(x).D$ (*input*): Receive an input u at a and continue as $D\{u/x\}$ (substitution). All occurrences of x in D are bound by the input argument. The name a is said to be the subject of the input. (iv) $D_1 \mid D_2$ (*parallel composition*): Execute D_1 and D_2 parallely (possibly involving interactions between the two). (v) $(\nu a)D$ (*restriction*): Privatize the name a to D. All occurrences of a in D are bound by the restriction. (vi) $recX.D$ (*recursion*): Behave as $D\{recX.D/X\}$. All occurrences of X in D are bound by the recursion operator. (vii) $D_1; D_2$ (*sequential composition*): Execute D_1, and then execute D_2. (viii) $D_1 \oplus D_2$ (*choice*): Execute exactly one of D_1 and D_2. (ix) $\text{fork}(D)$ (*fork*): Make a copy of the current environment and execute D with this copy as its environment. D is to be executed concurrently with the (parent) diagram performing this fork. Specifically, note that the forked diagram and the parent diagram do not share their environments. (x) $\{|\gamma : D|\}$ (*scope*): Execute D in the environment γ. (xi) $x := e$ (*assignment*): Assign to x the result of evaluating e. (xii) $\text{pick}(x).D$ (*pick*): Pick any value v such that $n(v)$ contains only the names that are currently in use, and execute $D\{v/x\}$. In particular, this construct does not generate fresh names. All occurrences of x in D are bound by the pick. (xiii) $wait(\phi)$ (*wait*): Wait until the environment is such that ϕ evaluates to *true*.

The reader is referred to [20] for a graphical representation of these constructs and a description of how many other constructs such as conditionals and loops can be encoded using the these constructs.

SDs impose the discipline of *locality* in the use of names, where the recipient of a name communicated in a message is only allowed to use the name for sending messages; in particular, the recipient does not have the capability to listen to messages targeted to the received name. Locality is a common feature of concurrent object-based systems and it was first formally investigated in the setting of π-calculus by Merro and Sangiorgi [11]. The SD syntax automatically enforces the locality discipline since an input subject is always a name (constant) and names can only be bound by a restriction. In particular, an input subject cannot be bound by the argument of another (enclosing) input, and hence the recepient cannot listen to messages targeted to the names it receives.

In addition to locality, SDs also enforce *uniqueness* of names. Specifically, let $rcp(D)$ be the set of all free names in D that occur as an input subject. This set contains all the free names at which D can currently receive a message. For a *top-level* diagram D, the uniqueness property states that no other process besides D can receive messages at a name in $rcp(D)$. Therefore, in particular, if D_1, D_2 are two top-level diagrams, then $rcp(D_1) \cap rcp(D_2) = \emptyset$. Note that locality guarantees that the uniqueness property is an invariant during execution; the set $rcp(D)$ may expand during the computation as private names of D are exported in outputs, and locality ensures the uniqueness property for these exported names.

We end this section with a few definitions and notational conventions. As usual, we do not distinguish between α-equivalent processes, i.e. processes that differ only in the use of bound names, bound variables, or bound process variables. The functions $fn(\cdot)$ and $bn(\cdot)$ which return the set of all free names and bound names that occur in a process (respectively), are defined as expected. Further, we define $n(D) = fn(D) \cup bn(D)$. A *value substitution* is a partial function from Var to Val that is defined only for finitely many variables. We write $\{\tilde{v}/\tilde{x}\}$ to denote the (value) substitution that maps x_i to v_i and is undefined for all other variables, where x_i and v_i are the ith components of the tuples $\tilde{x}$ and $\tilde{v}$. We write $D\{\tilde{v}/\tilde{x}\}$ to denote the result of simultaneously substituting all occurrences of x_i in D with v_i. As usual, substitution is defined only modulo α-equivalence with the usual renaming of bound names to avoid captures. Similarly, a *process substitution* is a partial function from $PrVar$ to processes, that is defined only for finitely many process variables. The notations $\{\tilde{D}/\tilde{X}\}$ and $D'\{\tilde{D}/\tilde{X}\}$ have the expected meaning.

3   Operational Semantics

We define the SD semantics using a labeled transition system in the SOS style introduced by Plotkin [16]. The transition labels are of five kinds.

i. τ: An internal action.
ii. $(\xi)av$: An input of value v at name a. ξ is the set of names in $n(v)$ that are fresh with respect to the diagram D performing the input.
iii. $(\xi)\bar{a}v$: An output of value v to name a. ξ is the set of names in $n(v)$ that are private to the diagram performing the output. These names will no longer be private to the diagram after the output.
iv. $(\xi)pick(v)$: Execution of a *pick* construct that picks a value v. ξ is the set of all names in $n(v)$ that are private to the diagram performing this action.
v. $(\xi)fork(D, \gamma)$: Execution of a *fork* construct that forks a diagram D with environment γ. ξ is the set of all names in $n(D) \cup n(\gamma)$ that are private to the diagram performing this action.

The functions $fn(\cdot)$ and $bn(\cdot)$ over actions are defined as expected; in particular ξ is the set of bound names for the actions above . We let α range ove the set of all actions, and define $n(\alpha) = fn(\alpha) \cup bn(\alpha)$. For environments γ_1, γ_2, we define $\gamma = \gamma_1; \gamma_2$ as $\gamma(x) = \gamma_1(x)$ if $\gamma_1(x)$ is defined, and $\gamma_2(x)$ otherwise. We write $\gamma[\tilde{x} \to \tilde{u}]$, where $\tilde{x} = x_1.....x_n$ and $\tilde{u} = u_1.....u_n$, as a shorthand for $\{(x_n, u_n)\}; \ldots ; \{(x_1, u_1)\}; \gamma$. We say that D is *trivial* if its syntax does not contain the *input, output, fork, assign, pick,* or *wait* constructs; such a process has the same behavior as 0. For $\xi = \{a_1, \ldots, a_n\}$, we write $(\nu\xi)D$ as an abbreviation for $(\nu a_1) \ldots (\nu a_n)D$; note that this notational convention is defined only modulo the ordering of the names $a_1, \ldots, a_n$ which in any case is irrelevant.

The transition system is defined at two levels - an *inner* level, and an *outer* level. The *inner* level transitions $\overset{\alpha}{\rightarrowtail}$ are between pairs consisting of a diagram and an environment in which the diagram is executed (see Table 1). The *outer* level transitions $\overset{\alpha}{\longrightarrow}$ are defined between *closed* diagrams (see Table 2), i.e. diagrams in which every variable occurrence is bound by an *input, scope* or *pick* construct. The main reason for defining transitions at two levels, besides accounting for environments, is that the execution of *pick* and *fork* constructs is context sensitive. For instance, executing a *pick* can only return a value v such that every name in $n(v)$ is currently in use, and the set of names in use is determined by the entire top-level diagram that contains the *pick* construct. Similarly, in case of the *fork* construct, the forked diagram is to be instantiated in parallel with entire top-level diagram. Using two types of transitions facilitates the definition of a transition system in the SOS style despite the context sensitive nature of *pick* and *fork* constructs.

The transitions in Tables 1 and 2 are all defined modulo α-equivalence on diagrams, i.e. if D_1 and D_2 are α-equivalent then D_1 and D_2 have the same transitions, and so do the pairs $\langle D_1, \gamma \rangle$ and $\langle D_2, \gamma \rangle$. The rules *INP, OUT, REC, BINP, PAR, RES, OPEN* and *COM* are all analogous to the corresponding transition rules for asynchronous π-calculus [2]. The rules *PAR, COM* and *SUM* have symmetric versions that are not shown. We now elaborate on the rules concerned with *pick* and *fork* constructs; the others are self-explanatory.

The transition label of the *PICK* rule includes the value v that is being picked. All the names in $n(v)$ are progressively accounted for by the *RES-PICK* and *TOP-PICK* rules; these names should either be private or already occur

Table 1. Rules for inner level transitions.

$INP \quad \langle a(x).D, \gamma \rangle \xrightarrow{av} \langle D\{v/x\}, \gamma \rangle$

$OUT \quad \langle \overline{a}e, \gamma \rangle \xrightarrow{\overline{a}v} \langle 0, \gamma \rangle \quad eval(e,\gamma) = v$

$$REC \quad \frac{\langle D\{recX.D/X\}, \gamma \rangle \xrightarrow{\alpha} \langle D', \gamma' \rangle}{\langle recX.D, \gamma \rangle \xrightarrow{\alpha} \langle D', \gamma' \rangle}$$

$$BINP \quad \frac{\langle D, \gamma \rangle \xrightarrow{(\xi)av} \langle D', \gamma' \rangle}{\langle D, \gamma \rangle \xrightarrow{(\xi\cup\{b\})av} \langle D', \gamma' \rangle} \quad b \in n(v) \setminus (fn(D) \cup n(\gamma))$$

$$PAR \quad \frac{\langle D_1, \gamma \rangle \xrightarrow{\alpha} \langle D_1', \gamma' \rangle}{\langle D_1|D_2, \gamma \rangle \xrightarrow{\alpha} \langle D_1'|D_2, \gamma' \rangle} \quad bn(\alpha) \cap fn(D_2) = \emptyset$$

$$RES \quad \frac{\langle D, \gamma \rangle \xrightarrow{\alpha} \langle D', \gamma' \rangle}{\langle (\nu b)D, \gamma \rangle \xrightarrow{\alpha} \langle (\nu b)D', \gamma' \rangle} \quad b \notin n(\gamma) \cup n(\alpha)$$

$$COM \quad \frac{\langle D_1, \gamma \rangle \xrightarrow{(\xi)\overline{a}v} \langle D_1', \gamma' \rangle \quad \langle D_2, \gamma \rangle \xrightarrow{(\xi)av} \langle D_2', \gamma' \rangle}{\langle D_1|D_2, \gamma \rangle \xrightarrow{\tau} \langle (\nu\xi)(D_1'|D_2'), \gamma' \rangle}$$

$$OPEN \quad \frac{\langle D, \gamma \rangle \xrightarrow{(\xi)\overline{a}v} \langle D', \gamma' \rangle}{\langle (\nu b)D, \gamma \rangle \xrightarrow{(\xi\cup\{b\})\overline{a}v} \langle D', \gamma' \rangle} \quad \begin{array}{l} b \neq a, b \in n(v), \\ b \notin \xi, b \notin n(\gamma) \end{array}$$

$$OPEN\text{-}FORK \quad \frac{\langle D, \gamma \rangle \xrightarrow{(\xi)fork(D_1,\gamma_1)} \langle D', \gamma' \rangle}{\langle (\nu b)D, \gamma \rangle \xrightarrow{(\xi\cup\{b\})fork(D_1,\gamma_1)} \langle D', \gamma' \rangle} \quad \begin{array}{l} b \in n(D_1) \cup n(\gamma_1) \\ b \notin \xi, b \notin n(\gamma) \end{array}$$

$$SEQ1 \quad \frac{\langle D_1, \gamma \rangle \xrightarrow{\alpha} \langle D_1', \gamma' \rangle}{\langle D_1; D_2, \gamma \rangle \xrightarrow{\alpha} \langle D_1'; D_2, \gamma' \rangle} \qquad SEQ2 \quad \frac{\langle D_2, \gamma \rangle \xrightarrow{\alpha} \langle D_2', \gamma' \rangle}{\langle D_1; D_2, \gamma \rangle \xrightarrow{\alpha} \langle D_2', \gamma' \rangle} \quad D_1 \text{ is trivial}$$

$ASSGN \quad \langle x := e, \gamma \rangle \xrightarrow{\tau} \langle 0, \gamma[x \to v] \rangle \quad eval(e,\gamma) = v$

$SUM \quad \langle D_1 \oplus D_2, \gamma \rangle \xrightarrow{\tau} \langle D_1, \gamma \rangle$

$PICK \quad \langle pick(x).D, \gamma \rangle \xrightarrow{pick(v)} \langle D\{v/x\}, \gamma \rangle$

$FORK \quad \langle fork(D), \gamma \rangle \xrightarrow{fork(D,\gamma)} \langle 0, \gamma \rangle$

$WAIT \quad \langle wait(\phi), \gamma \rangle \xrightarrow{\tau} \langle 0, \gamma \rangle \quad eval(\phi,\gamma) = true$

$$SCOPE \quad \frac{\langle D, \gamma_1;\gamma_2 \rangle \xrightarrow{\alpha} \langle D', \gamma_1[\tilde{x} \to \tilde{v}];\gamma_2' \rangle}{\langle \{|\gamma_1 : D|\}, \gamma_2 \rangle \xrightarrow{\alpha} \langle \{|\gamma_1[\tilde{x} \to \tilde{v}] : D'|\}, \gamma_2' \rangle}$$

$$RES\text{-}PICK \quad \frac{\langle D, \gamma \rangle \xrightarrow{(\xi)pick(v)} \langle D', \gamma' \rangle}{\langle (\nu b)D, \gamma \rangle \xrightarrow{(\xi\cup\{b\})pick(v)} \langle (\nu b)D', \gamma' \rangle} \quad b \in n(v)$$

free in the top-level diagram. This ensures that every name in $n(v)$ is currently in use. The transition label of the *FORK* rule contains both the diagram that is being forked, and the environment in which the forked diagram will be executed. The *TOP-FORK* rule instantiates this diagram along with the environment, in parallel with the top-level diagram. This ensures that the newly forked diagram executes concurrently with the current diagram, and that the two diagrams do not share their environments. Finally, a note on the *TOP-OUT* rule. This rule accounts for asynchrony in message exchanges. A message emitted by the *OUT* rule can either be immediately exported by the *TOP* rule, or it can be buffered by the *TOP-OUT* rule. Note that the arguments of a buffered message have already been evaluated by the *OUT* rule that created the message.

Table 2. Rules for outer level transitions.

$$TOP \quad \frac{\langle D, \emptyset \rangle \xrightarrow{\alpha} \langle D', \emptyset \rangle}{D \xrightarrow{\alpha} D'} \quad \alpha \neq \begin{array}{l} pick, \\ fork \end{array} \qquad TOP\text{-}FORK \quad \frac{\langle D_1, \emptyset \rangle \xrightarrow{(\xi)fork(D_2, \gamma)} \langle D'_1, \emptyset \rangle}{D_1 \xrightarrow{\tau} (\nu\xi)(D'_1|\{|\gamma : D_2|\})}$$

$$TOP\text{-}OUT \quad \frac{\langle D_1, \emptyset \rangle \xrightarrow{(\xi)\bar{a}v} \langle D'_1, \emptyset \rangle}{D_1 \xrightarrow{\tau} (\nu\xi)(D'_1|\bar{a}v)} \qquad TOP\text{-}PICK \quad \frac{\langle D, \emptyset \rangle \xrightarrow{(\xi)pick(v)} \langle D', \emptyset \rangle}{D \xrightarrow{\tau} D'} \quad n(v) \subseteq \xi \cup fn(D)$$

Let $\mathcal{L}$ denote the set of all input and output actions; these are the visible actions. Note that every top-level transition is labeled with a τ or an action in $\mathcal{L}$. We let s, r, t range over the set of *traces* $\mathcal{L}^*$. The functions $fn(.), bn(.)$ and $n(.)$ are extended to $\mathcal{L}^*$ the obvious way. We define a complementation function on $\mathcal{L}$ as $\overline{(\xi)xy} = (\xi)\bar{x}y$, $\overline{(\xi)\bar{x}y} = (\xi)xy$, and extend this to $\mathcal{L}^*$ the obvious way. The α-equivalence over traces is defined as expected, and α-equivalent traces are not distinguished. For example, the traces $(b)ab.\bar{b}a$ and $(c)ac.\bar{c}a$ are α-equivalent; we do not distinguish between the bound names b and c. Since we work modulo α-equivalence on traces, for convenience we assume the following *normality* condition on any trace s we consider – if $s = s_1.\alpha.s_2$ then $(n(s_1) \cup fn(\alpha)) \cap bn(\alpha.s_2) = \emptyset$.

We define the relation $\Longrightarrow$ as the reflexive transitive closure of $\xrightarrow{\tau}$, and $\overset{\beta}{\Longrightarrow}$ as $\Longrightarrow \overset{\beta}{\longrightarrow} \Longrightarrow$. For $s = l.s'$, we write $D \overset{l}{\Longrightarrow} \overset{s'}{\Longrightarrow} Q$ compactly as $D \overset{s}{\Longrightarrow} Q$. We write the assertion $D \overset{s}{\Longrightarrow} D'$ for some D', as $D \overset{s}{\Longrightarrow}$. We define $[D] = \{s \mid D \overset{s}{\Longrightarrow}\}$. Not every trace produced by the transition system corresponds to a valid computation. For example, we have $(\nu a)(a(x).D|\bar{a}v|\bar{b}a) \overset{(a)\bar{b}a}{\longrightarrow} a(x).D|\bar{a}v \overset{\bar{a}v}{\longrightarrow}$. But the message $\bar{a}v$ is not observable due to the *locality* property of SDs (see Section 2); the locality property prevents the recipient of the private name a from listening to messages targeted to a. Further, due to the uniqueness property the message $\bar{a}v$ in the top-level diagram $a(x).D|\bar{a}v$ is not observable, although we have the transition $a(x).D|\bar{a}v \overset{\bar{a}v}{\longrightarrow}$. To account for this, we define for a set of names ρ, the notion of a ρ-well-formed trace such that only ρ-well-formed traces can be exhibited by a diagram D with $rcp(D) = \rho$.

Definition 1. *We define $rcp(\rho, s)$ inductively as $rcp(\rho, \epsilon) = \rho$, $rcp(\rho, s.(\xi)av) = rcp(\rho, s)$, and $rcp(\rho, s.(\xi)\bar{a}v) = rcp(\rho, s) \cup \xi$. We say s is ρ-well-formed if $s = s_1.(\xi)\bar{a}v.s_2$ implies $a \notin rcp(\rho, s_1)$. We say s is well-formed if it is $\emptyset$-well-formed.*

For convenience we adopt the following hygiene condition on traces (in addition to the normality condition). Whenever we consider a ρ-well-formed trace s, we have $\rho \cap bn(s) = \emptyset$. The following lemma captures the intuition behind Definition 1.

Lemma 1. *Let $rcp(D_2) \cap \rho = \emptyset$. Then the computation $D_1|D_2 \overset{s}{\Longrightarrow}$ can be unzipped into $D_1 \overset{s}{\Longrightarrow}$ and $D_2 \overset{\bar{s}}{\Longrightarrow}$ such that s is ρ-well-formed.* □

The original definition of SDs by Smith and Talcott [19] also accounts for certain fairness conditions. For instance, it is required that every message that is sent during the course of a computation is eventually received. Such fairness conditions are in general not implementable, making SDs more of a specification language rather than a programming language. For instance, it is in general impossible to decide if a diagram can eventually evolve to a state where it can receive a certain message. Since our intention is to focus on an executable fragment (or variant) of SDs, we drop these fairness conditions.

4   May Testing on Specification Diagrams

The may testing equivalence [15] is a notion of process equivalence which is useful to relate specifications at different levels of abstraction. It is an instance of the general notion of behavioral equivalence where, roughly, two processes are said to be equivalent if they are indistinguishable in all contexts of use. Specifically, the context consists of an observing process that runs in parallel and interacts with the process being tested. The observer can in addition signal a success by emitting a special event. The process being tested is said to pass the test proposed by the observer if there *exists* a run in which the observer signals a success; note that due to possible non-determinism the observer and the process can take one of many possible computation paths. Two process are said to be may equivalent if they pass exactly the same set of tests.

We consider a generalized version of the usual may equivalence, where the equivalence is parameterized with a set of names that determines the set of observers that can be used to decide the equivalence. We originally introduced this generalized notion in the context of asynchronous π-calculus with locality [24].

Definition 2 (may testing). *Observers are diagrams that can emit a special message $\overline{\mu}\mu$. We let O range over the set of observers. We say D may O if $D|O \overset{\overline{\mu}\mu}{\Longrightarrow}$. We say $D_1 \sqsubseteq_\rho D_2$ if for every O with $rcp(O) \cap \rho = \emptyset$, we have D_1 may O implies D_2 may O. We say $D_1 \simeq_\rho D_2$ if $D_1 \sqsubseteq_\rho D_2$ and $D_2 \sqsubseteq_\rho D_1$.*

Thus, only observers that do not listen at names in ρ are used to decide the preorder $\sqsubseteq_\rho$; the larger the parameter ρ the smaller the observer set that is used to decide $\sqsubseteq_\rho$.

May testing is known to be useful for reasoning about safety properties of concurrent systems. Specifically, by viewing the observer's success as something bad happening, $D_1 \sqsubseteq_\rho D_2$ can be interpreted as D_1 is a safe implementation of the specification D_2, because if the specification D_2 is guaranteed to not cause anything bad to happen in a given context (that does not listen to names in ρ), then the implementation D_1 would also not cause anything bad to happen in the same context.

The universal quantification over contexts in the definition of may testing makes it very hard to prove equalities. Specifically, to prove an equivalence, one has to consider all possible interactions between the given processes and all

Table 3. A preorder relation on traces.

(drop)	$s_1.(\xi)s_2 \preceq s_1.(\xi)av.s_2$	if $(\xi)s_2 \neq \perp$
(delay)	$s_1.(\xi)(\alpha.av.s_2) \preceq s_1.(\xi)av.\alpha.s_2$	if $(\xi)(\alpha.av.s_2) \neq \perp$
(annihilate)	$s_1.(\xi)s_2 \preceq s_1.(\xi)av.\bar{a}v.s_2$	if $(\xi)s_2 \neq \perp$

possible observers. The typical approach to circumvent this problem is to find an alternate characterization of the equivalence that involves only the process being tested [2, 4, 8]. For SDs, a variant of the trace semantics characterizes the parameterized may preorder. In fact, it turns out that the characterization is similar to the one for asynchronous π-calculus with locality that we presented in [24]; the only difference arising due to the fact that unlike SDs the calculus in [24] is not equipped with the mismatch operator on names. The characterization in [24] is in turn an adaptation of the characterization for asynchronous π-calculus [2]. We skip the proofs of all the propositions in this section as they are relatively simple extensions of the proofs in [2] and [24]. The main difference arises due to the mismatch capability on names in SDs (this capability is absent in the formalisms in [2] and [24]), which can be handled using the techniques we presented in [23].

The trace based characterization of $\sqsubseteq_\rho$ over SDs follows. We define a preorder $\preceq$ on traces as the reflexive transitive closure of the laws shown in Table 3, where $(\xi)\cdot$ is defined as

$$(\xi)s = \begin{cases} s & \text{if } \xi = \emptyset \text{ or } \xi \cap fn(s) = \emptyset \\ (\xi \setminus \{b\})s_1.(\xi' \cup \{b\})av.s_2 & \text{if } b \in \xi, b \in n(v) \text{ and there are } s_1, s_2, a, \xi' \\ & \text{s.t. } s = s_1.(\xi')av.s_2 \text{ and } b \notin fn(s_1) \cup \{a\} \\ \perp & \text{otherwise} \end{cases}$$

The expression $(\xi)s$ returns $\perp$, if there is $b \in \xi$ such that b is used in s before it is received for the first time, i.e. the first free occurrence of b in s is *not* in the argument of an input. Otherwise, the expression binds the first such occurrence (in an input in s) of every $b \in \xi$, and returns the resulting trace. The intuition behind the preorder $\preceq$ is that if a process leads an observer to a success by exhibiting a trace s, then it can also lead the observer to a success by exhibiting any trace $r \preceq s$. Specifically, inputs are not observable since they are asynchronous, and hence they can be dropped, delayed, or annihilated with complementary output actions.

Lemma 2. *If* $O \overset{s.\bar{\mu}\mu}{\Longrightarrow}$, *then* $r \preceq s$ *implies* $O \overset{r.\bar{\mu}\mu}{\Longrightarrow}$. $\qquad\square$

Definition 3. *We say* $[D_2] \preceq_\rho [D_1]$ *if for every ρ-well-formed trace s, $D_1 \overset{s}{\Longrightarrow}$ implies there is $r \preceq s$ such that $D_2 \overset{r}{\Longrightarrow}$.*

Theorem 1. $D_1 \sqsubseteq_\rho D_2$ *if and only if* $[D_2] \preceq_\rho [D_1]$. $\qquad\square$

So far, we have allowed a given pair of diagrams D_1 and D_2 to be compared with $\sqsubseteq_\rho$ for arbitrary ρ. But if D_1 and D_2 are top-level diagrams, due to the uniqueness property of names (see Section 2) it makes sense to compare D_1 and D_2 with $\sqsubseteq_\rho$ only if $fn(D_1), fn(D_2) \subseteq \rho$. In this case, we can in fact strengthen Theorem 1 by dropping the *annihilation* law. Specifically, for ρ such that $fn(D_1), fn(D_2) \subseteq \rho$, we have $D_1 \sqsubseteq_\rho D_2$ if and only if for every ρ-well-formed trace s, $D_1 \overset{s}{\Longrightarrow}$ implies $D_2 \overset{r}{\Longrightarrow}$ for some $r \preceq s$ using only the laws *delay* and *drop*. The reason behind this is that since s is ρ-well-formed and $rcp(D) \subseteq \rho$, s cannot contain complimentary input and output actions.

5 Executable Specification in Rewriting Logic

We now specify the language of SDs as a theory in rewriting logic [12]. The Maude tool [5] which supports specifications in rewriting logic can then be used to execute SDs. We also present a procedure implemented in Maude, that exploits Theorem 1 to decide the may preorder relation between finitary diagrams (without recursion). To simplify matters we assume that the set of values *Val* only contains names. Extending this to arbitrary value sets will need sophisticated symbolic techniques [10, 25], which is out of the scope of this paper.

Since we have represented specification diagrams as an extension of asynchronous π-calculus we can smoothly extend the specification of asynchronous π-calculus in rewriting logic that we introduced in [22] to obtain an executable specification of SDs. The main idea behind specifying SDs in rewriting logic is to represent an (inner or outer) transition rule of form

$$\frac{P_1 \to Q_1 \quad \dots \quad P_n \to Q_n}{P_0 \to Q_0}$$

as a *conditional* rewrite rule of the form $P_0 \longrightarrow Q_0 \quad \text{if} \quad P_1 \longrightarrow Q_1 \wedge \dots \wedge P_n \longrightarrow Q_n$, where the condition includes rewrites. This was first introduced by Verdejo et al. [26] for implementing CCS [13]. Such conditional rules with rewrite conditions are executable in version 2.0 of the Maude language and system [5]; the rewrite conditions are solved by means of an implicit search process. The reader is referred to [5, 26] for further details.

In the Maude specification of the SD syntax that follows, the sorts Chan, Var, and PrVar are used to represent names, variables and process variables, and the sort Env is used to represent environments. Following are operator declarations for a few of the SD constructs.

```
sorts Chan Var PrVar Env Diag .
op  _(_)._  : Chan Qid Diag -> Diag .    op {|_:_|} : Env Diag -> Diag .
op _|_ : Diag Diag -> Diag .  ops new[_]_  rec[_]_  : Qid Diag -> Diag .
```

The sort Qid represents quoted identifiers. To manage name and variable bindings in specification diagrams, we use CINNI as a calculus for explicit substitutions [21] which has been implemented in Maude. CINNI gives a first-order

Table 4. The CINNI operations.

[a := x]	[shiftup a]	[shiftdown a]	[lift a S]
a{0} ↦ x	a{0} ↦ a{1}	a{0} ↦ a{0}	a{0} ↦ [shiftup a] (S a{0})
a{1} ↦ a{0}	a{1} ↦ a{2}	a{1} ↦ a{0}	a{1} ↦ [shiftup a] (S a{1})
...	...	...	...
a{n+1} ↦ a{n}	a{n} ↦ a{n+1}	a{n+1} ↦ a{n}	a{n} ↦ [shiftup a] (S a{n})
b{m} ↦ b{m}	b{m} ↦ b{m}	b{m} ↦ b{m}	b{m} ↦ [shiftup a] (S b{m})

representation of terms with bindings and capture-free substitutions, instead of going to the metalevel to handle identifiers and bindings. The main idea in such a representation is to keep the bound identifier inside the binders as it is, but to replace its use by the identifier followed by an index which is a count of the number of binders with the same identifier it jumps before it reaches the place of use. This combines the best of the approaches based on standard variables and de Bruijn indices [6]. Following this idea, we define terms of sorts Chan, Var and PrVar as indexed identifiers as follows.

```
op  _{_} : Qid Nat -> Chan .          op  _{_} : Qid Nat -> Var .
op  _{_} : Qid Nat -> PrVar .
```

Note that the operator _{_}_ is (adhoc) overloaded. Following are the constructors for environments.

```
op  emptyEnv : -> Env .          op  (__) : Var Chan -> Env .
op  _;_ : Env Env -> Env .       eq  {| emptyEnv : D |} = D .
eq  {|(X CX);E : D|} = {|E : {|(X CX) : D |} |} .
```

As a result of the equations above, from now on we can assume that the environment γ in $\{|\gamma : D|\}$ is of form $(x\ a)$. For name substitutions we introduce the sort ChanSubst along with the following operations. The intuitive meaning of these operations is described in Table 4 (see [21] for more details).

```
op [_:=_] : Qid Chan -> ChanSubst .  op [shiftdown_] : Qid -> ChanSubst .
op [shiftup_] : Qid -> ChanSubst .
op [lift__] : Qid ChanSubst -> ChanSubst .
```

We introduce the sort PrSubst for process substitutions, along with similar operations as above. Using these, explicit substitutions for SDs can be defined equationally. Following are some interesting equations. Note how the substitution is lifted as it moves across a binder.

```
Var NS : ChanSubst .  Var PS : PrSubst .
eq  NS (CX(Y) . D ) = (NS CX)(Y) . ([lift Y NS] D) .
eq  NS (D1 | D2) = (NS D1) | (NS D2) .
eq  NS ({|(X CX) : D |} = {|(X (NS CX)) : [lift X NS] D |} .
```

We now describe the specification of the transition system. As mentioned earlier, the transition rules are represented as conditional rewrite rules with the premises as conditions of the rule. Since rewrites do not have labels unlike the labeled transitions, we make the label a part of the resulting term; thus these rewrites are of the form $P \Rightarrow \{\alpha\}Q$.

A problem to overcome in giving an executable specification of the transition system is that the transitions of a term can be *infinitely branching* because of the *INP* and *OPEN* rules. In case of the *INP* rule, there is one branch for every possible name that can be received in the input (recall that we have assumed names to be the only values). In case of the *OPEN* rule, there is one branch for every name that is chosen to denote the private channel that is being emitted (recall that the transition rules are defined only modulo α-equivalence).

To overcome this problem, we define transitions relative to an execution environment[1]. The environment is represented abstractly as a set of free names CS that it may use while interacting with the process, and both the inner and outer level transitions are modeled as rewrite rules over terms of the form [CS] P. The set CS expands during bound input and output interactions when private names are exchanged between the process and its environment. The infinite branching due to the *INP* rule is avoided by allowing only the names in CS to be received in free inputs. Since CS is assumed to contain all the free names in the environment, an input argument that is not in CS would be a private name of the environment. Now, since the identifier chosen to denote the fresh name is irrelevant, all bound input transitions can be identified to a single input. With these simplifications, the number of input transitions of a term becomes finite. Similarly, in the *OPEN* rule, since the identifier chosen to denote the private name emitted is irrelevant, instances of the rule that differ only in the chosen name are not distinguished.

Following is the specification of a few of the inner level transitions (see Table 1).

```
sorts Action VisAction VisActionType EnvProc TraceProc .
subsort VisAction < Action .       subsort EnvProc < TraceProc .
ops i o : -> VisActionType .
op  f : VisActionType Chan Chan -> VisAction .
op  b : VisActionType Chan Qid -> VisAction .
op  [_]<_,_> : ChanSet Diag Env -> EnvProc .
op  {_}_ : Action TraceProc -> TraceProc [frozen] .

rl [INP] : [CY CS] < CX(X) . D , E > =>
           {f(i,CX,CY)} [CY CS] < [X := CY] D , E > .
crl [BINP] : [CS] < D, E > =>
           {[shiftdown 'U] b(i,CX,'U)} ['U{0} [shiftup 'U] CS] < D1 , E1 >
              if (not flag in CS) /\ CS1 := flag 'U{0} [shiftup 'U] CS /\
                 [CS1] < [shiftup 'U] D , [shiftup 'U] E > =>
                                     {f(i,CX,'U{0})} [CS1] < D1 , E1 > .
```

---

[1] We have overloaded the word environment. So far we have used it to denote variable bindings. We now also use it to refer to the external process with which the process under consideration is interacting. It should be clear from the context as to which of these we mean.

```
crl [OPEN] : [CS] <new [X] D , E> =>
        {[shiftdown X] b(o,CY,X)} [X{0} CS1] <D1 , E1>
        if CS1 := [shiftup X] CS /\ E2 := [shiftup X] E /\
        [CS1] <D,E2> => {f(o,CY,X{0})} [CS1] <D1,E1> /\ X{0} =/= CY .
```

We have shown the constructors for only the sort VisAction that represents visible actions, i.e input and output actions. Since names are assumed to be the only values, these actions are of form $(\hat{b})ab$ or $(\hat{b})\bar{a}b$, where the metavariable $\hat{b}$ ranges over $\{\emptyset, \{b\}\}$. The operators f and b are used to construct free and bound actions respectively. Name substitutions on actions are defined equationally as expected. The implementation of the INP, BINP and OPEN rules is similar to that of the corresponding rules for asynchronous π-calculus [22]. We explain only the BINP rule in detail, and refer the reader to [22] for further details.

In the BINP rule, since the identifier chosen to denote the bound argument is irrelevant, we use the constant 'U for all bound inputs, and thus 'U{0} denotes the fresh name received. Note that in contrast to the BINP rule of Table 1, we do not check if 'U{0} is in the free names of the process performing the input or its variable bindings, and instead we shift up the channel indices appropriately in CS, D, and E in the righthand side and condition of the rule. This is justified because the transition target is within the scope of the bound name in the input action. Note also that the channel CX in the action is shifted down because it is now out of the scope of the bound argument. The set CS is expanded by adding the received channel 'U{0} to it. Finally, we use a special constant flag of sort Chan, to ensure termination. The constant flag is used to prevent the BINP rule from being fired again while evaluating the condition. Without this check, we will have a non-terminating execution in which the BINP rule is repeatedly fired.

Following is the implementation of one of the outer level transitions (see Table 2).

```
sorts  EnvDiag TraceDiag .    subsort EnvDiag < TraceDiag .
ops tau bpick : -> Action .   op fpick : Chan -> Action .
op  [_]_ : ChanSet Diag -> EnvDiag .
op  {_}_ : Action TraceDiag -> TraceDiag [frozen] .
crl [TOP-PICK] : [CS] D => {tau} [CS1] D1
                 if [CS] < D , emptyEnv > => {A} [CS1] < D1 , emptyEnv > /\
                 (A == bpick \/ (A == pick(CX) /\ CX in freenames(D))) .
```

The constant bpick is used to represent a pick action that picks a bound name, while fpick is used to denote an action that picks a free name. Note that the operator {_}_ above is declared frozen. This forbids rewriting of its arguments; otherwise rewrites can be applied to any subterm. We use the frozen attribute because otherwise for a recursive process D the term [CS] D may have a non-terminating rewrite sequence [CS]D => {A1} [CS]D1 => {A1}{A2} [CS]D2 => .... But since {_}_ is declared frozen a term [CS] D can be rewritten only once. To compute all possible successors of a term, we explicitly generate the transitive closure of one step transitions as follows (the dummy operator [_] declared below is used to prevent infinite loops).

```
sort TEnvDiag .
op  [_] : EnvDiag -> TEnvDiag [frozen] .
crl [reflx] : [ P ] => {A} Q if P => {A} Q .
crl [trans] : [ P ] => {A} R  if P => {A} Q /\ [ Q ] => R .
```

Now, the set of all traces exhibited by [CS] D can be computed by finding all the one step successors of [[CS] D]. The traces appear (with tau actions) as the prefix of these one step successors. Of course, there can be infinitely many one step successors if D is recursive, but using the meta-level facilities of Maude we can compute only as many of them as needed. To represent traces, we introduce the sort Trace as follows.

```
subsort  VisAction < Trace .
op  epsilon : -> Trace .     op  [_] : Trace -> TTrace .
op  _._ : Trace Trace -> Trace [assoc id: epsilon] .
ceq [TR1 . b(IO,CX,Y) . TR2] =
                [TR1 . b(IO,CX,'U) . [Y := 'U{0}] [shiftup 'U] TR2]
                                    if Y =/= 'U .
```

The equation above defines α-equivalence on traces the expected way. The function rwf which checks if a trace is ρ-well-formed can be defined along the lines of Definition 1.

We encode the relation $\preceq$ of Table 3 as rewrite rules on terms of sort TTrace. Specifically $r \preceq s$ *if cond* is encoded as s => r if cond.

```
rl  [Delay] : [ ( TR1 . f(i,CX,CY) . b(IO,CU,V) . TR2 ) ] =>
               [ ( TR1 . b(IO,CU,V) . ([shiftup V] f(i, CX , CY)) . TR2 ) ] .
crl [Delay] : [ ( TR1 . b(i,CX,Y) . f(IO,CU,CV) . TR2 ) ] =>
                 [ ( TR1 . bind(Y , f(IO,CU,CV) . f(i,CX,Y{0}) . TR2) ) ]
                 if bind(Y , f(IO,CU,CV) . f(i,CX,Y{0}) . TR2) =/= bot .
```

The operator bind implements the function $(\hat{y})\cdot$ on traces. Note that in the first Delay rule, the channel indices of the free input action are shifted up when it is delayed across a bound action, since it gets into the scope of the bound argument. Similarly, in the second Delay rule, when the bound input action is delayed across a free input/output action, the channel indices of the free action will be shifted down by the bind operation. The other two subcases of the Delay rule, namely, where a free input is to be delayed across a free input or output, and where a bound input is to be delayed across a bound input or output, are not shown as they are similar.

To decide $D_1 \sqsubseteq_\rho D_2$ for finitary diagrams D_1, D_2 without recursion, we exploit the alternate characterization of $\sqsubseteq_\rho$ given by Theorem 1. But a problem with this approach is that finiteness of D only implies that the length of traces in $[D]$ is bounded, but the number of traces in $[D]$ can be infinite (even modulo α-equivalence) because the INP rule is infinitely branching. To avoid the problem of having to compare infinite sets, we observe that

$$[D_2] \precsim_\rho [D_1] \quad \text{if and only if} \quad [D_2]_{fn(D_1,D_2)} \precsim_\rho [D_1]_{fn(D_1,D_2)},$$

where for a set of traces S we define $S_\xi = \{s \in S \mid fn(s) \subseteq \xi\}$. Now, since the traces in $[D_1]$ and $[D_2]$ are finite in length, it follows that the sets of traces

$[D_1]_{fn(D_1,D_2)}$ and $[D_2]_{fn(D_1,D_2)}$ are finite modulo α-equivalence. In fact, the set of traces generated for `[[fn(D1,D2)] D1]` by our implementation, contains exactly one representative from each α-equivalence class of $[D_1]_{fn(D_1,D_2)}$.

Given processes D_1 and D_2, we generate the set of all traces (modulo α-equivalence) of `[[fn(D1,D2)] D1]` and `[[fn(D1,D2)] D2]` using the metalevel facilities of Maude. Then for each ρ-well-formed trace T in $[D_1]_{fn(D_1,D_2)}$, we compute the reflexive transitive closure of T with respect to the rewrite rules for the laws in Table 3. We then use the fact that $[D_2]_{fn(D_1,D_2)} \precsim_\rho [D_1]_{fn(D_1,D_2)}$ if and only if for every ρ-well-formed trace T in $[D_1]_{fn(D_1,D_2)}$ the closure of T and $[D_2]_{fn(D_1,D_2)}$ have a common element. We skip the details of the implementation using metalevel facilities of Maude, as they are the same as that for asynchronous π-calculus [22].

6   Conclusion

We have presented the executable fragment of SDs as an extension of asynchronous π-calculus. We have exploited this relation to both obtain an implementation of SDs, and to develop a theory of may testing for SDs. An interesting direction for future work is to investigate the theory of may testing for the entire SD language. Features such as fairness conditions and the **constraint** predicate, which we have not considered here, change the characterization of may testing in a non-trivial way. Another problem of interest is to extend the implementation of may testing for the case where there are other infinite value domains besides names, such as integers and lists. This would involve the use of sophisticated symbolic techniques to handle these infinite value domains [10, 25]. The resulting implementation of may testing over the full fledged SD language with a rich value set can be used for reasoning about practical examples; such case studies are also a topic of interest.

References

1. G. Agha. *Actors: A Model of Concurrent Computation in Distributed Systems.* MIT Press, 1986.
2. M. Boreale, R. De Nicola, and R. Pugliese. Trace and testing equivalence on asynchronous processes. *Information and Computation*, 172(2):139–164, 2002.
3. G. Boudol. Asynchrony and the π-Calculus. Technical Report 1702, INRIA Technical Report, May 1992.
4. I. Castellani and M. Hennessy. Testing theories for asynchronous languages. In *FSTTCS '98*, volume 1530 of *Lecture Notes in Computer Science*, pages 90–101, 1998.
5. M. Clavel, F. Durán, S. Eker, P. Lincoln, N. Martí-Oliet, J. Meseguer, and J. F. Quesada. Towards Maude 2.0. In *International Workshop on Rewriting Logic and its Applications*, volume 36 of *Electronic Notes in Theoretical Computer Science*, pages 297–318, 2000.
6. N. G. de Bruijn. Lambda calculus with nameless dummies, a tool for automatic formula manipulation, with application to the Church-Rosser theorem. *Proc. Kninkl. Nederl. Akademie van Wetenschappen*, 75:381–392, 1972.

7. E. W. Dijkstra and C. S. Scholten. *Predicate Calculus and Program Semantics.* Springer Verlag, 1990.
8. M. Hennessy. *Algebraic Theory of Processes.* MIT Press, 1988.
9. K. Honda and M. Tokoro. An Object Calculus for Asynchronous Communication. In *Fifth European Conference on Object-Oriented Programming*, July 1991. LNCS 512, 1991.
10. A. Ingolfsdottir and H. Lin. A symbolic approach to value passing processes. In *Handbook of Process Algebra*, pages 427–478. Elsevier Publishing, 2001.
11. M. Merro and D. Sangiorgi. On Asynchrony in Name-Passing Calculi. In *International Colloquium on Automata Languages and Programming.* Springer-Verlag, 1998. LNCS 1443.
12. José Meseguer. Rewriting as a unified model of concurrency. In J. C. M. Baeten and J. W. Klop, editors, *CONCUR'90*, volume 458 of *Lecture Notes in Computer Science*, pages 384–400, 1990.
13. R. Milner. *Communication and Concurrency.* Prentice Hall, 1989.
14. R. Milner, J. Parrow, and D. Walker. A calculus of mobile processes (Parts I and II). *Information and Computation*, 100:1–77, 1992.
15. R. De Nicola and M. Hennessy. Testing equivalence for processes. *Theoretical Computer Science*, 34:83–133, 1984.
16. G. D. Plotkin. A structural approach to operational semantics. Technical Report DAIMI FN-19, Computer Science Dept., Aarhus University, September 1981.
17. ITU-T Recommendation Z.120. Message sequence charts, 1996.
18. J. Rumbaugh, I. Jacobson, and G. Booch. *Unified Modeling Language Reference Manual.* Addison-Wesely, 1998.
19. S. Smith and C. Talcott. Modular reasoning for actor specification diagrams. In *Formal Methods in Object-Oriented Distributed Systems.* Kluwer Academic Publishers, 1999.
20. S. Smith and C. Talcott. Specification diagrams for actor systems. *Higher-Order and Symbolic Computation*, 2002. To appear.
21. M. O. Stehr. CINNI — A generic calculus of explicit substitutions and its application to λ-, ς- and π-calculi. In *International Workshop on Rewriting Logic and its Applications*, volume 36 of *Electronic Notes in Theoretical Computer Science*, pages 71–92, 2000.
22. P. Thati, K. Sen, and N. Martí-Oliet. An executable specification of asynchronous π-calculus and may testing in Maude 2.0. In *International Workshop on Rewriting Logic and its Applications*, 2002. Electronic Notes in Theoretical Computer Science, vol. 71.
23. P. Thati, R. Ziaei, and G. Agha. A theory of may testing for actors. In *Formal Methods for Open Object-based Distributed Systems*, March 2002.
24. P. Thati, R. Ziaei, and G. Agha. A theory of may testing for asynchronous calculi with locality and no name matching. In *AMAST '02*, volume 2422 of *Lecture Notes in Computer Science*, pages 222–238. Springer Verlag, 2002.
25. A. Verdejo. Building tools for lotos symbolic semantics in maude. In *International Conference on Formal Techniques for Networked and Distributed Systems*, volume 2529 of *Lecture Notes in Computer Science*, pages 292–307. Springer Verlag, 2002.
26. A. Verdejo and N. Martí-Oliet. Implementing CCS in Maude 2. In *International Conference on Rewriting Logic and its Applications*, 2002. Electronic Notes in Theoretical Computer Science, vol. 71.

Formalising Graphical Behaviour Descriptions

Kenneth J. Turner

Computing Science and Mathematics, University of Stirling, Stirling FK9 4LA, UK

Abstract. CRESS (Chisel Representation Employing Systematic Specification) is used for graphical behaviour description, underpinned by formal and implementation languages. Plug-in frameworks adapt it for particular application domains such as Intelligent Networks, Internet Telephony and Interactive Voice Response. The CRESS notation and its syntax are explained. The semantics of CRESS is discussed with reference to its interpretation in LOTOS.

Keywords: Graphical Specification, LOTOS (Language Of Temporal Ordering Specification), SDL (Specification and Description Language), Voice Service

1 Introduction

1.1 Background

Diagrammatic representations abound in science and engineering. For example, software engineering uses flowcharts, entity-relationship diagrams, data-flow diagrams, state diagrams, and various UML diagrams. In general, diagrams are valued because they give a clear overview of a system. In industry, graphical representations are regarded as more accessible than textual ones (especially for more formal specifications).

This paper describes the basis of CRESS (Chisel Representation Employing Systematic Specification). Although CRESS was inspired by the need to represent voice services, it is a general-purpose way to represent behaviour graphically. CRESS can therefore be used for a variety of other applications. Unlike many diagrammatic forms, CRESS is precise in that diagrams are interpreted by an underlying formal model. CRESS diagrams can also be translated to implementation languages. CRESS supports:

- graphical behaviour description, open to non-specialists and industrial engineers
- a precise interpretation that allows rigorous analysis and development
- a portable toolset that facilitates specification, implementation, analysis and testing.

The same service diagrams can be used for multiple purposes. CRESS is neutral with respect to the target language. For formal analysis, CRESS diagrams are automatically translated to LOTOS (Language Of Temporal Ordering Specification [4]) and to SDL (Specification and Description Language [6]). For implementation, CRESS diagrams are automatically translated to Perl (for Internet Telephony services) or to VoiceXML (for Interactive Voice Response services).

CRESS was initially based on the industrial notation Chisel developed by BellCore [1]. However, CRESS has been considerably extended since its beginnings. For example, it now supports the notion of plug-in domains: the vocabulary and concepts required

C. Rattray et al. (Eds.): AMAST 2004, LNCS 3116, pp. 537–552, 2004.

for each application area are defined separately. CRESS has been used in the domains of Intelligent Networks, Internet Telephony and Interactive Voice Response.

Although other papers by the author have described the *applications* of CRESS, the present paper considers the *foundations* of CRESS, namely the composition, syntax and semantics of CRESS diagrams. The CRESS philosophy focuses on two aspects:

- The graphical notation is of most interest and value to domain experts such as communications engineers. Such users require a convenient and pragmatic way of describing services.
- The rigorous analysis of formal specifications derived from diagrams is of most interest and value to formalists. Such users require precise expression with the ability to reason formally about the specifications.

The translation from diagrams to specifications is of limited interest to both categories of user. Although the paper gives some indication of how translation is achieved, it is a secondary issue. In particular, the translation procedure is not formalised though the results of the translation are formal. The important point is that specifications generated from CRESS are precise and can be reasoned about.

1.2   Relationship to Other Work

Diagrammatic notations, e.g. visual programming languages, are common in software engineering. However few graphical approaches have a formal basis. Statecharts [3], LSCs (Live Sequence Charts [7]), and UML all have graphical representations with a formal basis. The following techniques are perhaps closest to CRESS:

- SDL has a graphical syntax and a formal interpretation. However SDL lacks composition mechanisms that are appropriate for building services. It is a specialised notation that needs expert knowledge. SDL also does not have support for specific application domains.
- MSCs (Message Sequence Charts [5]) are higher-level and more straightforward. Several authors have given formal meaning to MSCs. Like SDL, MSCs are also rather generic and lack composition mechanisms suitable for services.
- UCMs (Use Case Maps [2]) are useful for high-level descriptions of requirements. UCMs have been represented using LOTOS and MSCs. However the formalisation of UCMs is not complete, and they lack support for specific application domains.

CRESS aims to circumvent these problems:

- CRESS accepts plug-in domains that allow it to be readily adapted for use in a variety of applications.
- CRESS supports the needs of voice services, though it is more widely applicable. Specific mechanisms are provided in CRESS for service composition.
- CRESS is intended as a front-end for formal representations. That is, CRESS is translated into a formal language into order to give it precise meaning. The implied semantic model of CRESS is reasonably straightforward, so there can be confidence in the equivalence of different formal models.

Formally-based language translation has been studied for many decades. [9] is an example of the state-of-the-art. Although such approaches might in principle be used for CRESS, they would be a side interest. In addition, the scale and challenges of translating CRESS to very different target languages cast doubt on the practicability of formally-based translation. CRESS includes concurrent, nondeterministic and event-driven behaviour. CRESS descriptions may be cyclic and may contain context-sensitive expressions. All these aspects would be challenging for a formal approach to translator design. Even if the problems could be overcome, proving the translation correct would be very difficult and time-consuming. The outcome would also be of little benefit for the intended purposes of CRESS (graphical description and formal analysis).

1.3   Overview of the Paper

As motivation for CRESS, section 2 briefly overviews the role of CRESS and how it has been applied. Section 3 gives a condensed summary of the CRESS notation. An overview is given in section 4 of the syntax and static semantics of diagrams. Section 5 deals with the semantics of diagrams by considering their interpretation in LOTOS.

2   CRESS Application

2.1   The Role of CRESS

Why not define system behaviour using some implementation technique used for program development? Such a definition would naturally be rather low-level, language-dependent, and possibly platform-dependent. It would be hard to analyse the definition rigorously. Historically, this is how communications services were defined. The approach suffers from major problems such as incompatibility among vendors, inconsistency among features, lack of portability, and cost of testing.

A formally-based approach is therefore an obvious choice. Why not then just specify behaviour using a selected formal method? There are a number of problems that CRESS aims to circumvent:

Acceptability: Formal methods have achieved only limited penetration into industry. In general, engineers are not trained in formal methods and are hesitant to use them. CRESS exploits the benefits of formal methods 'behind the scenes' without forcing them on the user. As a graphical notation, CRESS is more accessible to the non-specialist. Indeed, CRESS benefits from its origins in Chisel as a notation that can be used by all stakeholders in system development.

Architecture: CRESS provides a framework and vocabulary for specifying applications in various domains. CRESS is therefore close to the architectural level at which a domain specialist would normally think. If a plain formal language is used, it is necessary to describe behaviour in terms of *language* concepts rather than in terms of *architectural* concepts. This leads to specifications that are low-level (in architectural terms) and verbose.

Neutrality: CRESS is not oriented towards any particular domain, target language or platform. It can therefore act as a front-end for creating formal specifications. CRESS diagrams are currently translated into LOTOS and SDL, and could be translated into many constructive formal languages.

Implementation: Formal specifications are usually rather distant from implementation. As a result, the effort put into specification is often not exploited when implementation takes place – there is a discontinuity in development that risks introducing inconsistencies. CRESS is unusual in that the *same* diagrams can be translated into both formal specifications and into implementations (in selected languages).

So, when is CRESS applicable? In general it can be used to give a constructive description of a system that has inputs, outputs, actions and events. That is, CRESS is useful for reactive systems. Although CRESS can be used to describe the behaviour of an entire system, its specialised capabilities come into their own when the system can be considered as a base behaviour (service) modified by optional additional capabilities (features). Such a situation is common in voice applications such as telephony, but is also common in many other cases. For example, software applications frequently have plug-in modules that are used to extend their abilities. CRESS is therefore most appropriate for modular systems.

CRESS cannot be used for non-constructive (e.g. declarative, logical) descriptions. Although CRESS descriptions are hierarchical in the sense of diagrams invoking other diagrams, CRESS is not able to describe a system at multiple levels of abstraction.

CRESS scales reasonably well. Since systems are usually described by multiple diagrams, a complex system can be handled as long as its behaviour can be broken down into manageable diagrams. When used for verification (proof), CRESS has the same characteristics as the formal language in which CRESS is interpreted. In practice, this usually places severe limits on what can be verified. When used for validation (testing), CRESS is much more practical. As long as tests can be defined fairly independently, a complex system can be validated without much regard to scale.

2.2 Domain Frameworks

In themselves, CRESS diagrams have no meaning. In fact, CRESS is deliberately open-ended as far as applications are concerned. Although CRESS defines a structure for diagrams, diagram content is governed by plug-in domains that define the syntax and semantics of diagram contents. As will be seen, CRESS has been used in three domains with four target languages.

Some aspects are domain-specific but language-independent. For example, a domain defines the names and parameter types of input signals, output signals, actions and event conditions. The sources and destinations are given for signals. To resolve certain kinds of composition problems, diagrams may be given priorities that control the order in which they are applied.

Some aspects are both domain-specific and language-specific. A specification architecture gives an outline specification in the chosen target language. This identifies the communicating subsystems and the domain-independent data types. When CRESS diagrams in this domain are translated to this language, the generated code is embedded in the specification architecture. Most of the generated code deals with behaviour, but some of it defines domain-specific data types.

2.3   Tool Support

The CRESS toolset is written largely in Perl for portability. About 14,000 lines of code in numerous modules are used to support seven main tools. The major CRESS tool resembles a conventional compiler. However CRESS is unusual in interpreting a graphical description. In addition, the possibly cyclic nature of graphs requires special treatment.

The CRESS lexer reads each diagram and turns it into a directed graph. The CRESS parser combines all the graphs into one, and checks the syntax and static semantics of the resulting graph. A CRESS code generator for each target language translates the graph into this language. These tools are invoked by a preprocessor that is used with a specification architecture. This determines the configuration and deployment of diagrams, and embeds their translation into the specification framework.

When CRESS is interpreted in a formal language, verification and validation are obvious choices. [15] describes how these can be performed. In general, CRESS is open to any verification technique that would normally be used with the formal language. However, CRESS is accompanied by its culinary counterpart MUSTARD (Multiple-Use Scenario Test And Refusal Description) as a practical means of validating behaviour. The MUSTARD tool translates validation scenarios into the target language and automatically checks them against the system specification. This is used to establish confidence in the CRESS description, and also to check for incompatibilities among features.

As far as possible, the toolset is automated so that the user is unaware of it. For example, a LOTOS user issues a TOPO toolset command to process a given specification. This invokes the CRESS toolset, creates a new specification from the CRESS diagrams, and uses MUSTARD to validate it. Similarly, an SDL user clicks on a button in the Telelogic TAU toolset to perform comparable actions.

2.4   Applications of CRESS

The following is a brief overview of CRESS in selected domains. Refer to the cited papers for more details.

IN: The Intelligent Network is an architecture for telephony networks that separates call routing (service switching) from call processing (service control). In particular, dedicated network elements handle complex services. The IN supports a flexible range of services such as Call Forwarding, Call Screening, Credit Card Calling, and FreePhone. IN services are complemented by ones that reside in switches (exchanges), e.g. Call Waiting or Conference Calling. The major issues in defining IN services are vendor-independence and mutual compatibility.
CRESS has been used to model a range of typical services from the IN [10, 12]. Such descriptions are independent of vendor, platform and language. The CRESS descriptions are interpreted in LOTOS or SDL, allowing rigorous analysis in isolation (checking correctness) and in combination (checking mutual compatibility).

Internet Telephony: Voice calls can be supported over an Internet-like network. This may be the Internet proper, though voice traffic is increasingly being carried over private networks that follow Internet standards. H.323 is a well-established set of standards for digital telephony. However SIP (Session Initiation Protocol [8]) is a simpler and more flexible alternative that is rapidly gaining in popularity. For example, SIP is being widely used to replace conventional telephony, and has been adopted for the new generation (3G) of mobile telephones.

CRESS has been used to model SIP and its associated services. In fact, SIP is sufficiently new that there is not yet a consensus on what a SIP service is. An important aspect of the CRESS work has therefore been to define a SIP service architecture [11–13]. For formal analysis, a range of SIP services has been described in CRESS and interpreted in LOTOS and SDL. This part of the work has had similar goals to the IN study. To gain practical benefits, the same service diagrams are also translated into Perl for use with SIP CGI (Common Gateway Interface).

IVR: Interactive Voice Response systems are used to provide callers with a voice interface. Speech recognition allows natural language enquiries, and speech synthesis gives fixed or generated voice responses. IVR systems are a major growth area, largely because of customer dissatisfaction with touch-tone enquiry systems. Among competing standards, VoiceXML [16] has gained a dominant position.

CRESS has been used to model VoiceXML and its associated services [12–15]. In fact, VoiceXML does not recognise the concepts of service or feature. Part of the CRESS work has therefore been to investigate the value of these ideas in an IVR setting. For formal analysis, a range of IVR services has been described in CRESS and interpreted in LOTOS and SDL. This part of the work has had similar goals to the IN study. For practical deployment, the same service diagrams are also translated into VoiceXML for deployment in an IVR server.

3   CRESS Notation

CRESS may appear to define state diagrams. However, state is intentionally implicit in CRESS because this allows more abstract descriptions to be given. CRESS has explicit support for defining and composing features. Plug-in domains adapt the notation for selected application areas. Sample diagrams from the domains mentioned earlier are used to illustrate the notation.

3.1   Diagram Expressions

CRESS expressions offer the usual logical, arithmetic and comparison operators. String and set operators are also supported. Assignment is indicated by '<−'. Since CRESS acts as a front-end for a variety of target languages, it does not itself define operator precedences. Complex expressions are therefore parenthesised as necessary.

3.2   Diagram Types

A CRESS diagram is a directed, possibly cyclic, graph. Ultimately, CRESS deals with a single diagram. However it is convenient to construct diagrams from smaller pieces. A multi-page diagram, for example, is linked through connectors. More usefully, CRESS supports the notion of feature diagrams that extend or modify other diagrams. This is especially useful for systems that have a base functionality plus additional capabilities. CRESS supports three similar kinds of diagrams:

Root Diagrams describe the fundamental behaviour of a system. An extract from a sample root diagram appears in figure 1. This describes POTS (Plain Old Telephone Service, i.e. an ordinary telephone call). For example the subscriber may go off-hook and dial a number, resulting in both telephones ringing.

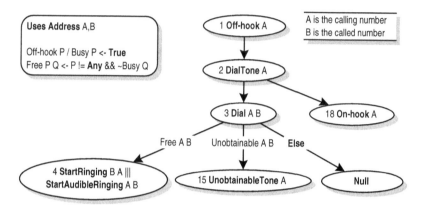

Fig. 1. CRESS Root Diagram (*Incomplete*) for the Plain Old Telephone Service

Spliced (Cut-and-Paste) Feature Diagrams are applied to matching behaviour in another diagram. The feature is copied and used to replace the matching behaviour. Note that this is a static, syntactic operation. A modified diagram is created by the process of composition. A spliced feature has a unique entry node. It has one or more exit nodes that indicate how it flows back to the diagram being modified. A spliced feature can be used to replace, extend or modify behaviour in the original diagram. A sample spliced feature appears in figure 2 that describes the Calling Number Delivery feature in Intelligent Networks. This matches node POTS 4 that follows node POTS 3 via the *Free A B* arc. If the called number *B* has subscribed to the *CallingNumber* feature, the caller's number *A* is displayed. The feature continues from nodes POTS 5 or POTS 13.

Template (Macro) Feature Diagrams are similar but are parameterised. The actual parameters are determined by pattern matching the triggering node to the template. A template feature may have several exit nodes, but only one of these (**Finish**) continues with the original behaviour. A template is instantiated and applied statically when diagrams are composed. Sample template features appear in figures 3, 4 and 5. Figure 3 describes an Intelligent Networks feature to check if the called number *Q* is busy; if so and the callee has subscribed to call forwarding, the call is diverted to the selected *ForwardBusy* number. Figure 4 describes an Internet Telephony feature that blocks a caller *P* who appears in the screening list *ScreenIn* for callee *Q*.

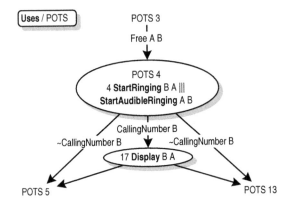

Fig. 2. CRESS Spliced Feature Diagram for Calling Number Delivery

Figure 5 describes an Interactive Voice Response feature for charity donations. It asks the user whether to start again after a request to clear all inputs. Unless the user re-confirms clearing, the donation details are submitted to a server.

It is normally preferable to use template features rather than spliced features. This is because the latter often have to repeat large pieces of the matching diagram. If a feature loops back to its initial node, this is interpreted as meaning a loop back to the triggering node. This allows a chain of features to be triggered by the same node; the instantiated features are combined in sequence.

3.3   Diagram Nodes

Diagrams contain several kinds of nodes, distinguished by shapes that are arbitrary but known: comments, behaviour nodes, labels and rule boxes.

Comments (parallel lines, figure 1) contain explanatory text. Some diagram editors used with CRESS allow hyperlinks to be added as comments, e.g. links to an audio commentary, document or figure.

Behaviour Nodes (ovals) contain behaviours and their parameters. A node is identified by a number, optionally followed by one or more symbols to indicate its kind:

- '<', '>' (figure 4 nodes 1 and 2) means the node contains input, output signals (used if a signal can be sent in both directions)
- '+', '−', '=' (figure 3 node 1, figure 5 node 1) means a template is appended, prefixed, substituted (relative to the triggering node)
- '!' (figure 5 node 3) means a node will not be matched by a template (used to prevent unintended or recursive substitution).

A behaviour node contains signals (inputs or outputs, figure 1 nodes 1 and 2) or actions (like programming language statements, figure 5 node 3). Behaviours may carry variables or expressions as parameters (figure 1 node 1, figure 5 node 2) Several behaviours may occur in parallel ('|||', figure 1 node 4). Each behaviour may be followed by variable assignments separated by '/'.

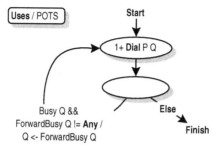

Fig. 3. CRESS Template Feature Diagram for Call Forward on Busy Line

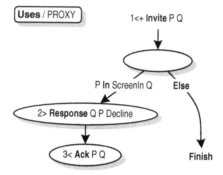

Fig. 4. CRESS Template Feature Diagram for Call Screening

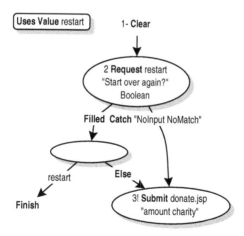

Fig. 5. CRESS Template Feature Diagram for Restarting A Charity Donation

Labels (plain text or ovals) are used as connectors to join parts of a diagram. A source label cites the diagram name (optional, meaning the same diagram) and node number (figure 2 node POTS 3). A destination label (figure 2 node POTS 5) may define variable assignments much as a behaviour node does.

Special labels are also used. A **Start** node (figure 3) is the initial one of a graph. It is required when a graph is cyclic so the initial node is ambiguous; it is implicit when the initial node is well-defined (e.g. figure 1). A **Null** node (figure 1) does nothing. It may be used as a shorthand to join a number of nodes to a number of other nodes. A **Finish** node (figure 3) is used to indicate the exit of a template. In fact, labels can be left empty for these special nodes (e.g. a **Null** label is normally omitted, figures 3 and 4).

Rule Boxes (rounded rectangles) have multiple purposes. They give a variety of rules to define variables, diagram interdependencies, assignments, macros, signal transformations, and configurations. A **Uses** clause may begin by optionally declaring the variables local to a diagram (figure 1). This is optionally followed by '/' and a list of other diagrams that the diagram depends on (figure 3). For example, a feature diagram lists the root (or other) diagrams that it may modify. The explicit dependencies among diagrams are used to determine the full set of diagrams needed.

A variable initialisation (not illustrated) assigns a value at the start of behaviour. Variables may also be assigned when a certain behaviour occurs. For example in the first rule of figure 1, the occurrence of *Off-hook* for any telephone P causes that telephone to be marked as busy. Parameterised macros and their expansion may be defined. In the second rule of figure 1, *Free P Q* expands to a check that P is a defined telephone number and Q is not busy. Signal transformations (not illustrated) are macros that allow one signal and its parameters to be replaced by another.

Rule boxes may also be used to define a system configuration (not illustrated). A **Deploys** clause lists the behaviour diagrams that apply. In addition, feature parameters may be defined (e.g. the forwarding number for a telephone).

3.4   Diagram Arcs

The arcs between nodes indicate the flow of control. Arcs may be unlabelled (figure 1 node 1 to 2) or may be labelled with guards. These may define value conditions (imposing a restriction on progress, figure 1 node 3 to 4). **Else** means the complement of other value conditions (figure 1 node 3 to **Null**). The use of **Else** is not obligatory, but if it is omitted then a dynamic check can be performed for all guards being unfulfilled.

Guards may also define event conditions that are activated by dynamic occurrence of an event (figure 5 node 2 to 3). Event conditions are distinguished by their names (e.g. **Filled**, **Catch**). Since events may be intercepted at several levels in a CRESS description, there is no equivalent of **Else** for event conditions.

A guard may be followed by assignments separated by '/' (figure 3 empty node to 1). This can be necessary to change system state without executing other behaviour. Sometimes it is convenient to have an empty guard condition but associated assignments, i.e. the progression to a new behaviour node changes the system state.

4   CRESS Syntax and Static Semantics

The syntax and static semantics of CRESS are handled by a preprocessor, a lexical analyser and a parser.

4.1 Specification Architecture Preprocessor

CRESS is driven by a specification architecture for a given domain and target language. This may contain preprocesssor calls that invoke the CRESS tools:

Cress(*Options,Diagrams*): generates code for the diagrams to be included. Translator options may optionally be given. The diagrams may be listed explicitly, but normally the keyword **Features** is used to mean those diagrams defined by the system configuration. To deal with the situation that a domain may have multiple root diagrams (e.g. Internet Telephony), all diagrams with a certain prefix may be included. For example, **Cress**(PROXY) will include all SIP Proxy Server diagrams.

Cress(Profiles): generates code for domain-specific user profiles, e.g. the features and their parameters that subscribers have selected.

Cress(Types): generates code for domain-specific data types.

The preprocessor calls the CRESS lexical analyser and subsequent tools.

4.2 Diagram Lexical Analysis

A graph editor is the most obvious tool to create CRESS diagrams. In fact, the requirements of CRESS are modest. The graph editor must be able to create a small number of node shapes, and must be able to associate multi-line labels with nodes and arcs. For practical reasons, the graph editor must also be multi-platform. A surprising number of graph editors fail to meet these criteria (e.g. Graphlet, GraphMaker, JGraphPad, Open-JGraph, W3Pal). One of the few appropriate graph editors is yEd from yWorks GmbH; this is free and, being Java-based, is multi-platform. It is hoped to use the DiaGen tool in future to create a CRESS-specific editor. Graph editors usually provide analysis features that are irrelevant to CRESS (e.g. finding the minimum spanning tree of a graph). They also tend to draw rather visually plain graphs.

An obvious alternative is a drawing package, but again many of these do not meet the CRESS criteria (e.g. Corel Draw, Dia, jfig, xfig). One of the few suitable drawing packages is Diagram! from Lighthouse Design; the figures in this paper were drawn using this tool. Although Diagram! runs on four different platforms and is free, it requires NEXTSTEP/OPENSTEP which limits its portability.

Another issue is the file representation of graphs. There are many competing formats, a number of which fail to represent the basic graph topology required by CRESS. GML (Graph Modeling Language) and XGMML (eXtensible Graph Markup and Modeling Language) are both suitable. In principle, GraphML (Graph Markup Language) would also be suitable but in practice it is used with proprietary extensions. Support for standard formats among graph editors is rather patchy. CRESS therefore accepts graphs in GML or XGMML format as well as the format created by the Diagram! tool.

The CRESS lexical analyser parses the file representation of a diagram and reduces it to a common internal format. In fact this is tricky, partly because nodes and arcs may appear in any order in the file, and partly because the graph may be cyclic. The generated graph is handed off to the CRESS parser for syntax and static semantic checking.

4.3   Diagram Syntactic Analysis

A list of CRESS diagrams is considered at the same time. There must be at most one root diagram. The other diagrams constitute a hierarchy of features that modify each other or the root diagram. Feature interaction is a well-known problem whereby independently defined features may interfere with each other. A common solution is to prioritise features. In CRESS, feature diagrams have priorities that ensure a well-defined order of application. Each feature in turn is combined with another or with the root diagram. If it is a spliced feature, the feature nodes are cut and pasted into the root diagram. If it is a template feature, the current root diagram is scanned for matching nodes. Each of these is then modified by the instantiated template. The final result is a single diagram. Various manipulations are carried out as this diagram is created:

- Source and destination labels are paired up, joining subgraphs.
- Event names are normalised. CRESS allows some latitude in naming for readability (e.g. *Start Audible Ringing* vs. *StartAudibleRinging*) and for British vs. American English (e.g. *Dialling* vs. *Dialing*).
- '<' and '>' labels on nodes are used to disambiguate signal directions and are then removed. A '!' label preventing template matching is removed after all templates substitutions have been handled.
- The successor nodes of each node are ordered by signal name. This simplifies interpretation in some languages (e.g. SDL).
- **Null** nodes are removed where possible to reduce the size of the graph. This cannot always be done (e.g. in a recursive loop, figure 3).
- Nodes that are reached via an **Else** or an event condition are moved to the end of the node successor list. This simplifies error-checking.
- Since a graph may be cyclic, nodes that appear earlier or later in the graph are specially marked.

The consistency of the final diagram is then checked for syntactic and static semantic correctness. More than 50 checks are applied, including the following as examples:

- A graph must have at most one **Start** node and one **Finish** node.
- Behaviour node labels must be unique.
- A behaviour node must contain events of the same type (input, output, action).
- A branch cannot lead to multiple output behaviour nodes. Non-deterministic inputs are allowed, but not non-deterministic outputs.
- A **Null** node cannot cycle back to itself.
- Expressions must be well-formed and have the correct parameter types for events and operators.
- At most one **Else** may appear in a list of value conditions, and **Else** cannot appear as the only such condition.

The CRESS parser performs the diagram composition, manipulation and checking described above. The result is a graph stored in an internal format. This is handed off to a CRESS code generator for translation to some specification or implementation language.

5   CRESS Dynamic Semantics

For space reasons, the formal interpretation of CRESS is explained here with reference to just one language – LOTOS. The interpretation using SDL is broadly similar, though restrictions on SDL inputs and outputs considerably complicate translation. Code for SDL is generated in SDL/PR (program-like) format.

5.1   Specification Architecture Using LOTOS

As an example of what a specification architecture looks like, the following is an outline for Intelligent Network services and LOTOS. There are comparable LOTOS specifications for Internet Telephony and Interactive Voice Response. Domain-defined specification architectures are also defined for SDL, Perl and VoiceXML as appropriate.

```
Specification INSystem [User] : NoExit                          (* Intelligent Network *)
  Library                                                          (* library types *)
  Type Address                                                 (* address operations *)
  Type Addresses                                                      (* addresses *)
  Type BooleanOperations                                       (* boolean operations *)
  Type Digit                                                            (* digits *)
  Type Number                                                          (* numbers *)
  Type Statuses                                                   (* call statuses *)
  Type StatusResult                                          (* call status results *)
  Cress(Types)                                                    (* generate types *)
  Behaviour INStructure [User]                                 (* overall behaviour *)
Where                                                            (* local definition *)
  Process INStructure [User]                                 (* IN network structure *)
    Hide Bill,Stat,Scp In                                      (* hide internal signals *)
      (
        (
          (
            CallInstances [Bill,Scp,Stat,User]                     (* call instances *)
          |[User,Stat]|                                (* synchronise user/status messages *)
            CallCoordinator [User,Stat] ({})                      (* call coordinator *)
          )
        |[Scp]|                                     (* synchronise service control messages *)
          ServiceControl [Scp,Stat]                           (* service control point *)
        )
      |[Stat]|                                         (* synchronise status messages *)
        StatusManager [Bill,Stat] (0, Cress(Profiles))   (* generate subscriber profiles *)
      )
    |[Bill]|                                         (* synchronise on billing messages *)
      BillingSystem [Bill]                                        (* billing system *)
  Process CallInstances [Bill,Scp,Stat,User]                       (* call instances *)
    Cress(Features)                                      (* generate feature behaviour *)
  Process CallCoordinator [User,Stat] (Addresses)                (* call coordinator *)
  Process ServiceControl [Scp,Stat]                        (* service control point *)
  Process StatusManager [Bill,Stat] (Time, Statuses)            (* status manager *)
  Process BillingSystem [Bill]                                    (* billing system *)
```

Node Type	Node Visited Once	Node Visited Earlier	Node Visited Later
Graph Start	instantiate top-level process, whose definition is then begun		
Action	translation is domain-specific		
Input/Output	event, interleaved if events in parallel	call already defined process	start new process definition, then translate input/output as normal
Null	start new process	call already defined process	start new process definition
Value Guard	guard	guard before call of already defined process	guard before call of new process, whose definition is then begun
Event Guard	start new event handler process	not allowed	not allowed
Graph End	close off all process definitions		

Fig. 6. Outline CRESS Denotation in LOTOS

5.2 Interpretation Using LOTOS

The dynamic semantics of CRESS is handled by code generators for each target language. Since graphs may be cyclic, a distinction is made between a behaviour node that is visited once, one that is visited earlier in the graph, and one that is visited later. When a code generator walks the graph, it recognises when it has already visited a node. Different code is usually generated for the first and subsequent visits to a node.

A potential difficulty with any automatic code generation is relating the generated code to the original source. In the case of CRESS, the code generators go to a lot of trouble to create human-readable, well laid out code. In addition, virtually every line of the generated code has automatically produced comments. These relate the code directly to the diagram that created it. It is possible to use the generated code without having to be aware of this. However for some purposes (e.g. simulation or verification), the comments are useful for the expert to relate the code and the CRESS diagrams.

Fixed data types are defined by the specification architecture. Type definitions are also automatically generated for signals and events defined by the plug-in domain. Expressions are translated to their LOTOS equivalents. If the domain requires user profiles, these are translated into LOTOS from the CRESS system configuration diagram.

The specification architecture includes fixed process definitions. Diagram-defined processes are generated from nodes according to the outline strategy in figure 6. This gives an idea of the denotations (code skeletons) for various CRESS constructs; it is not possible to define the full mapping here. In the main the translation to LOTOS is straightforward except for the following points:

- Although LOTOS does not really distinguish inputs and outputs, they are translated slightly differently since CRESS outputs must use only constants and variables with defined values. The CRESS toolset performs a data-flow analysis to determine this. A behaviour parameter with an undefined value becomes a '?' event parameter in LOTOS (like input), while a behaviour parameter with a defined value is prefixed with '!' (like output).
- Normally a behaviour node is translated as the corresponding LOTOS event. If the node contains parallel behaviours, these are translated as sequential or concurrent events as defined by a code generator option.

- Behaviour nodes along a path usually become a sequence of LOTOS events. However if several paths lead to a node, a new LOTOS process is defined for the behaviour from that node. A branch to the node then becomes a call of this process.
- Value guards are translated into their direct LOTOS equivalents. An **Else** becomes the logical complement of all the accumulated guards. If an **Else** is omitted, a code generator option produces a dynamic check for the guards being unfulfilled.
- Event guards are very complex to translate into LOTOS. The problem is that the name of a CRESS event may be constructed only dynamically. Where the event is handled is also determined dynamically, as events may be caught at several levels of a CRESS description. Event handling in a LOTOS specification must, however, be defined statically. Fortunately it is possible to statically determine the scope of all event handlers. This allows the translator to define a static process that dispatches events according to their context. A node that follows an event guard will start a new LOTOS process definition. When a CRESS event occurs dynamically, the event dispatcher calls the appropriate process according to its context. More about the event model can be found in [15].

6   Conclusion

The role of CRESS has been seen as a graphical notation for describing system behaviour, particularly for voice services but also for reactive systems generally. The notation, syntax, static semantics and dynamic semantics of CRESS have been discussed. The notion of plug-in application domains makes CRESS very adaptable. CRESS is also powerful in that the same diagrams can be given a formal interpretation or be used for implementation. The use of CRESS has been briefly reviewed for Intelligent Networks, Internet Telephony and Interactive Voice Response.

References

1. A. V. Aho, S. Gallagher, N. D. Griffeth, C. R. Schell, and D. F. Swayne. SCF3/Sculptor with Chisel: Requirements engineering for communications services. In K. Kimbler and W. Bouma, editors, *Proc. 5th. Feature Interactions in Telecommunications and Software Systems*, pages 45–63. IOS Press, Amsterdam, Netherlands, Sept. 1998.
2. D. Amyot, L. Charfi, N. Gorse, T. Gray, L. M. S. Logrippo, J. Sincennes, B. Stepien, and T. Ware. Feature description and feature interaction analysis with use case maps and LOTOS. In M. H. Calder and E. H. Magill, editors, *Proc. 6th. Feature Interactions in Telecommunications and Software Systems*, pages 274–289. IOS Press, Amsterdam, Netherlands, May 2000.
3. D. Harel and E. Gery. Executable object modeling with Statecharts. In *Proc. 18th International Conference on Software Engineering*, pages 246–257. Institution of Electrical and Electronic Engineers Press, New York, USA, 1996.
4. ISO/IEC. *Information Processing Systems – Open Systems Interconnection – LOTOS – A Formal Description Technique based on the Temporal Ordering of Observational Behaviour.* ISO/IEC 8807. International Organization for Standardization, Geneva, Switzerland, 1989.
5. ITU. *Message Sequence Chart (MSC)*. ITU-T Z.120. International Telecommunications Union, Geneva, Switzerland, 2000.

6. ITU. *Specification and Description Language*. ITU-T Z.100. International Telecommunications Union, Geneva, Switzerland, 2000.

7. R. Marelly, D. Harel, and H. Kugler. Multiple instances and symbolic variables in executable sequence charts. In *Proc. 17th ACM Object-Oriented Programming, Systems, Languages and Applications*, pages 83–100, New York, USA, 2002. ACM Press.

8. J. Rosenberg, H. Schulzrinne, G. Camarillo, A. Johnson, J. Peterson, R. Sparks, M. Handley, and E. Schooler, editors. *SIP: Session Initiation Protocol*. RFC 3261. The Internet Society, New York, USA, June 2002.

9. T. Rus. A unified language processing methodology. *Theoretical Computer Science*, 281(1–2):499–536, June 2002.

10. K. J. Turner. Formalising the CHISEL feature notation. In M. H. Calder and E. H. Magill, editors, *Proc. 6th. Feature Interactions in Telecommunications and Software Systems*, pages 241–256, Amsterdam, Netherlands, May 2000. IOS Press.

11. K. J. Turner. Modelling SIP services using CRESS. In D. A. Peled and M. Y. Vardi, editors, *Proc. Formal Techniques for Networked and Distributed Systems (FORTE XV)*, number 2529 in Lecture Notes in Computer Science, pages 162–177. Springer-Verlag, Berlin, Germany, Nov. 2002.

12. K. J. Turner. Formalising graphical service descriptions using SDL. In R. Reed and J. Reed, editors, *SDL 2003*, number 2708 in Lecture Notes in Computer Science, pages 183–202. Springer-Verlag, Berlin, Germany, July 2003.

13. K. J. Turner. Representing new voice services and their features. In D. Amyot and L. Logrippo, editors, *Proc. 7th. Feature Interactions in Telecommunications and Software Systems*, pages 123–140. IOS Press, Amsterdam, Netherlands, June 2003.

14. K. J. Turner. Specifying and realising interactive voice services. In H. König, M. Heiner, and A. Wolisz, editors, *Proc. Formal Techniques for Networked and Distributed Systems (FORTE XVI)*, number 2767 in Lecture Notes in Computer Science, pages 15–30. Springer-Verlag, Berlin, Germany, Sept. 2003.

15. K. J. Turner. Analysing interactive voice services. *Computer Networks*, Jan. 2004. In press.

16. VoiceXML Forum. *Voice eXtensible Markup Language*. VoiceXML Version 2.0. VoiceXML Forum, Jan. 2003.

Model-Checking Distributed Real-Time Systems with States, Events, and Multiple Fairness Assumptions*

Farn Wang

Dept. of Electrical Engineering, National Taiwan University
1, Sec. 4, Roosevelt Rd., Taipei, Taiwan 106, ROC
+886-2-23635251 ext. 435; FAX:+886-2-23671909
farn@cc.ee.ntu.edu.tw, http://cc.ee.ntu.edu.tw/~farn

Model-checker/simulator **Red** 5.1 will be available at http://cc.ee.ntu.edu.tw/~val

Abstract. At this moment, there lacks a specification language for distributed real-time system properties involving states and events. There also lacks a language for fairness assumptions in dense-time systems. We have defined a new temporal logic, $TECTL^f$, for the flexible specification of distributed real-time systems with constraints involving events, states, and fairness assumptions. Then we have also designed algorithms for model-checking $TECTL^f$ formulas. Finally, we have endeavored to implement and experiment the ideas in our tool, **Red** 5.1, and shown that the ideas could be used in practice.

Keywords: Distributed, real-time, model-checking, verification, events, fairness.

1   Introduction

It has long been argued that neither pure state-based nor pure event-based languages quite support the natural expressiveness desirable for the specification of real-world systems [11,17,18]. The inadequacy of such pure languages is even more salient in the setting of distributed real-time systems, where multiple clocks do not tick at the same times. For example, to specify that an event must eventually happen using dense-time state-based language $TCTL$ (a branching-time temporal logic) [1], one common style is to use an artificial state (or urgent state) that immediately follows this event. But for distributed real-time systems, this style looks cumbersome and unnatural since there is no perfect way to say how long this urgent state should last or whether we allow other transitions to happen in this state. Although there were many works in combining state-based and event-based languages [11, 16–18, 20], they were all developed for untimed

---

* The work is partially supported by NSC, Taiwan, ROC under grants NSC 92-2213-E-002-103, NSC 92-2213-E-002-104, and by the System Verification Technology Project of Industrial Technology Research Institute, Taiwan, ROC (2004).

C. Rattray et al. (Eds.): AMAST 2004, LNCS 3116, pp. 553–567, 2004.

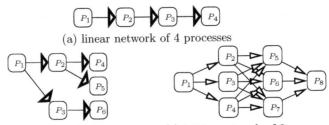

(a) linear network of 4 processes

(b) binary tree network of 6 processes   (c) lattice network of 8 processes

Fig. 1. Network configurations for diffusive computations

systems. Facing the undecidability of verification problems of discrete-time extensions of linear temporal logics [3, 4], we have chosen to extend TCTL[1] [1] to a specification language for dense-time properties involving both states and events.

Another challenge arises in specifying that "something good will eventually happen" in distributed real-time systems. Such properties are called "liveness" in linear-time logics and "inevitability" [27] in branching-time logics. For example, after turning on their computers, people *"naturally"* expect that operating systems will start correctly. But such a property can be impossible to verify with the models of nondeterministic automata unless we assume that each component of the system has a *fair* share of execution. The concept means that we do not have to worry about those "unnatural" behaviors as long as the systems work fine with those "natural" behaviors. For example, we may have the linear, binary tree, and lattice networks in figures 1(a), (b), and (c) for diffusive computing [10]. An arrow from a process P_i to another P_j means that the finishing of P_i may lead to the finishing of P_j. For convenience, we let $\mathtt{src}(i)$ be the index set of processes with an outgoing arrow to P_i. P_1 will nondeterministically make transition, with event σ_1, to its *finished* state. For each $i > 1$, process P_i will nondeterministically check if any process with index in $\mathtt{src}(i)$ has *finished* and make transition, with event σ_i, to its *finished* state. Note here, we add a dummy cycle transition with events $\sigma_1, \sigma_2, \sigma_3, \sigma_4$ to their respective *finished* states. This is the traditional way to model fairness in finite-execution systems. The automata do not guarantee that all processes will enter their *finished* states. But it will be odd that if some process does not execute infinitely often. Thus with the fairness assumption that "every process executes infinitely often," we can deduce that all processes will enter their *finished* states.

Two commonly accepted concepts of fairness assumptions [13] are:

- **Correctness with strong fairness assumption:** The concept of *strong fairness* means that "something will happen infinitely often." For the networks in figure 1 with four processes, in order to verify the eventual finishing of all processes, we may want to assume that "for each i, transitions with event σ_i execute infinitely often."

---

[1] TCTL model checking probelm against timed automatas is PSPACE-complete.

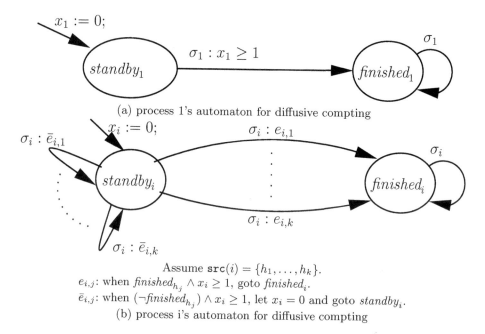

$x_1 := 0;$

$\sigma_1 : x_1 \geq 1$

$standby_1$ $finished_1$ σ_1

(a) process 1's automaton for diffusive compting

$\sigma_i : \bar{e}_{i,1}$

$x_i := 0;$ $\sigma_i : e_{i,1}$

$standby_i$ $finished_i$ σ_i

$\sigma_i : e_{i,k}$

$\sigma_i : \bar{e}_{i,k}$

Assume $\mathtt{src}(i) = \{h_1, \ldots, h_k\}$.

$e_{i,j}$: when $finished_{h_j} \wedge x_i \geq 1$, goto $finished_i$.

$\bar{e}_{i,j}$: when $(\neg finished_{h_j}) \wedge x_i \geq 1$, let $x_i = 0$ and goto $standby_i$.

(b) process i's automaton for diffusive compting

Fig. 2. Diffusive propagations with fairness assumptions

- **Correctness with weak fairness assumption:** The concept of *weak fairness* means that "something will eventually stabilize to a state." In figure 2, one example is that "eventually all processes finish their tasks."

  Motivated by the discussion in previous paragraphs, we have extended TCTL [1] with event constraints and fairness assumptions and defined a new specification language *TECTL^f* (acronym for *Timed Event CTL with Fairness assumptions*) in section 3. For example, for the systems in figure 2, the goal of diffusive computing is that all processes will eventually enter their *"finished"* states. This specification can be expressed with strong fairness assumptions as

$$\forall^{[\sigma_1 \geq 1, \sigma_2 \geq 1, \sigma_3 \geq 1, \sigma_4 \geq 1]} \Diamond \bigwedge_{0 \leq i \leq m} finished_i \qquad (A)$$

Here the strong fairness assumptions are written in square brackets as event predicates to the superscript of the universal path quantifier $\forall$. Strong fairness assumption $\sigma_i \geq 1$ means that transitions with one or more events σ_i will happen infinitely often. Such a style of fairness assumptions as event predicates allows for flexible characterizations of sets of transitions.

We shall present algorithms for model-checking *TECTL^f* formulas in section 4. We have used our ideas to extend the functionality of our TCTL model-checker **Red** [21–25]. Our implementation, **Red** 5.1, uses CRD (Clock-Restriction Diagram), a BDD-like structure [5, 8], with symbolic manipulation algorithms. We shall report our experiments to check the performance of our techniques in section 5. The tool and benchmarks can be downloaded for free at

http://cc.ee.ntu.edu.tw/~val

2   Timed Automata

We use the widely accepted model of *timed automata* [2] to describe the transitions in dense-time state-spaces. A *timed automaton* is a finite-state automaton equipped with a finite set of clocks which can hold nonnegative real-values. At any moment, the timed automaton can stay in only one *mode* (or *control location*). Each mode is labeled with an invariance condition on clocks. In its operation, one of the transitions can be triggered when the corresponding triggering condition is satisfied. Upon being triggered, the automaton instantaneously transits from one mode to another and resets some clocks to zero. In between transitions, all clocks increase their readings at a uniform rate.

For convenience, given a set P of atomic propositions and a set X of clocks, we use $B(P, X)$ as the set of all Boolean combinations of atoms of the forms p and $x \sim c$, where $p \in P$, $x \in X \cup \{0\}$, "$\sim$" is one of $\leq, <, =, >, \geq$, and c is an integer constant.

Definition 1. (timed automata) A timed automaton A is given as a tuple $\langle \Sigma, X, Q, I, \mu, E, \gamma, \lambda, \tau, \pi \rangle$ with the following restrictions. Σ is a finite set of event names. X is a finite set of clocks. Q is a finite set of modes. $I \in B(Q, X)$ is the initial condition. $\mu : Q \mapsto B(\emptyset, X)$ defines the conjunctive invariance condition of each mode. E is the finite set of transitions. $\gamma : E \mapsto (Q \times Q)$ defines the source and destination modes of each transition. $\lambda : (E \times \Sigma) \mapsto \mathcal{N}$ defines the number of instances of an event type that happen on each transition. Such a general scheme allows for the modeling of broadcasting and multicasting of many generic transmission events. $\tau : E \mapsto B(\emptyset, X)$ and $\pi : E \mapsto 2^X$ respectively defines the triggering condition and the clock set to reset of each transition. ∎

Definition 2. (states) A state of $A = \langle \Sigma, X, Q, I, \mu, E, \gamma, \lambda, \tau, \pi \rangle$ is a pair like (q, ν) with $q \in Q$ and $\nu : X \mapsto \mathcal{R}^+$, the set of nonnegative real numbers.   ∎

We say a state (q, ν) satisfies a state predicate $\eta \in B(P, X)$, where P is either $\emptyset$ or Q, iff the following inductive conditions are satisfied.

- $(q, \nu) \models q'$ iff $q' = q$;
- $(q, \nu) \models x \sim c$ iff $\nu(x) \sim c$;
- $(q, \nu) \models \eta_1 \vee \eta_2$ iff $(q, \nu) \models \eta_1$ or $(q, \nu) \models \eta_2$;
- $(q, \nu) \models \neg \eta_1$ iff it is not the case that $(q, \nu) \models \eta_1$.   ∎

For any $t \in \mathcal{R}^+$, $\nu + t$ is a valuation identical to ν except that for every $x \in X$, $\nu(x) + t = (\nu + t)(x)$. Given $\bar{X} \subseteq X$, $\nu \bar{X}$ is a new valuation identical to ν except that for every $x \in \bar{X}$, $\nu \bar{X}(x) = 0$.

Definition 3. (runs) Given a timed automaton $A = \langle \Sigma, X, Q, I, \mu, E, \gamma, \lambda, \tau, \pi \rangle$, a *run* is an infinite computation of A along which time diverges. Formally speaking, a run is an infinite sequence of state-time pairs $((q_0, \nu_0), t_0)((q_1, \nu_1), t_1)$ $\ldots ((q_k, \nu_k), t_k) \ldots \ldots$ such that

- $t_0 t_1 \ldots t_k \ldots \ldots$ is a monotonically increasing divergent real-number sequence, i.e., $\forall c \in \mathcal{N}, \exists h > 1, t_h > c$, and
- **Invariance condition:** for all $k \geq 0$, for all $t \in [0, t_{k+1} - t_k]$, $(q_k, \nu_k + t) \models \mu(q_k)$; and

- **Transitions:** for all $k \geq 0$, either
  - **a null transition:** $q_k = q_{k+1}$ and $\nu_k + (t_{k+1} - t_k) = \nu_{k+1}$; or
  - **a discrete transition e:** denoted $q_k \xrightarrow{e} q_{k+1}$. The constraints is that there is an $e \in E$ such that $\gamma(e) = (q_k, q_{k+1})$, $(q_k, \nu_k + t_{k+1} - t_k) \models \tau(e)$, and $(\nu_k + t_{k+1} - t_k)\pi(e) = \nu_{k+1}$. ∎

3 $TECTL^f$ (Timed Event CTL with Fairness Assumptions)

3.1 Syntax

In our model of timed automata, a transition can be labeled with many instances of an event type. Thus, we devise *event predicates*, to characterize complex event constraints on transitions, with the following syntax.

$$\epsilon ::= \sum_i a_i \sigma_i \sim c \mid \neg\epsilon_1 \mid \epsilon_1 \vee \epsilon_2$$

where σ_i is an event name and a_i's and c are integer constants. For example, for figure 1, $\sigma_i \geq 1$ represents the set of transitions with at least one σ_i event.

$TECTL^f$ is an extension to TCTL [1] with the following syntax rules.

$$\phi ::= q \mid x \sim c \mid \phi_1 \vee \phi_2 \mid \neg\phi_1 \mid x.\phi_1 \mid \exists^{[\alpha_1,\dots,\alpha_m]}_{\langle\beta_1,\dots,\beta_n\rangle} \phi_1 \mathcal{U}^\epsilon \phi_2 \mid \exists^{[\alpha_1,\dots,\alpha_m]}_{\langle\beta_1,\dots,\beta_n\rangle} \Box^\epsilon \phi_1$$

Here $q \in Q$, $x \in X$, $c \in \mathcal{N}$. $\alpha_1, \dots, \alpha_m, \beta_1, \dots, \beta_n$ are either event predicates or $TECTL^f$ formulas. ϵ is either null (not specified) or an event predicate. ϕ_1 and ϕ_2 are $TECTL^f$ formulas. When ϵ is not specified in a path formula, then we may simply write $\phi_1\mathcal{U}\phi_2$ and $\Box\phi_1$. The modal operators are intuitively explained in the following.

- $x.\phi$ means that "if there is a clock x with reading zero now, then ϕ is satisfied."
- $\exists^{[\alpha_1,\dots,\alpha_m]}_{\langle\beta_1,\dots,\beta_n\rangle}$ means "there exists a run satisfying *strong fairness assumptions* $\alpha_1, \dots, \alpha_m$ and *weak fairness assumptions* $\beta_1, \dots, \beta_n$."
- $\phi_1\mathcal{U}^\epsilon\phi_2$:
  - When ϵ is null, it works as a classical until-formula and means that along a computation, ϕ_1 is true until ϕ_2 happens.
  - When ϵ is specified, it means that along a computation, ϕ_1 is true until a transition, satisfying ϵ, happens and immediately makes ϕ_2 true.
- $\Box^\epsilon\phi_1$:
  - When ϵ is null, it works as a classical always-formula and means that along a computation, ϕ_1 is true in every state.
  - When ϵ is specified, it means that along a computation, whenever a transition satisfying ϵ happens, immediately after the transition, ϕ_1 is true.

Also we adopt the following standard shorthands :
- *true* for $0 = 0$ and *false* for $\neg true$;
- $\phi_1 \wedge \phi_2$ for $\neg((\neg\phi_1) \vee (\neg\phi_2))$

- $\phi_1 \to \phi_2$ for $(\neg\phi_1) \vee \phi_2$
- $\exists^{[\,]}$ and $\exists_{\langle\,\rangle}$ can both be abbreviated as $\exists$.
- $\exists^{[\alpha_1,...,\alpha_m]}_{\langle\beta_1,...,\beta_n\rangle}\Diamond^\epsilon\phi_1$ for $\exists^{[\alpha_1,...,\alpha_m]}_{\langle\beta_1,...,\beta_n\rangle}true\,\mathcal{U}^\epsilon\phi_1$,
- $\forall^{[\alpha_1,...,\alpha_m]}_{\langle\beta_1,...,\beta_n\rangle}\Box^\epsilon\phi_1$ for $\neg\exists^{[\alpha_1,...,\alpha_m]}_{\langle\beta_1,...,\beta_n\rangle}\Diamond^\epsilon\neg\phi_1$
- $\forall^{[\alpha_1,...,\alpha_m]}_{\langle\beta_1,...,\beta_n\rangle}\phi_1\mathcal{U}^\epsilon\phi_2$ for $\neg((\exists^{[\alpha_1,...,\alpha_m]}_{\langle\beta_1,...,\beta_n\rangle}(\neg\phi_2)\mathcal{U}^\epsilon\neg(\phi_1\vee\phi_2))\vee(\exists^{[\alpha_1,...,\alpha_m]}_{\langle\beta_1,...,\beta_n\rangle}\Box^\epsilon\neg\phi_2))$
- $\forall^{[\alpha_1,...,\alpha_m]}_{\langle\beta_1,...,\beta_n\rangle}\Diamond^\epsilon\phi_1$ for $\forall^{[\alpha_1,...,\alpha_m]}_{\langle\beta_1,...,\beta_n\rangle}true\,\mathcal{U}^\epsilon\phi_1$

Example 1. For the system in figure 2, we have specification (A) already with strong fairness assumptions in page 555. Another specification is *"If process 1 eventually stabilizes to the finished state and process 2 executes infinitely many times, then process 2 will eventually finish."* The specification with strong and weak fairness assumptions is

$$\forall^{[\sigma_2\geq 1]}_{\langle finished_1\rangle}\Diamond finished_2 \tag{B}$$

One last specification is *"Every time after process 2 executes a transition and ends in the* standby *mode, it will stay there for at least one more time units."* In $TECTL^f$, this specification is

$$\forall\Box^{\sigma_2\geq 1}(standby_2 \to x.\Box(x\leq 1 \to standby_2)) \tag{C}$$

$\blacksquare$

3.2   Semantics

Given an event predicate ϵ and a transition $e\in E$, we say e satisfies ϵ, in symbols $e\models\epsilon$, according to the following inductive rules.

- $e\models\sum_i a_i\sigma_i \sim c$ iff $\sum_i a_i\lambda(e,\sigma_i)\sim c$;
- $e\models\neg\epsilon$ iff it is not the case that $e\models\epsilon$; and
- $e\models\epsilon_1\vee\epsilon_2$ iff either $e\models\epsilon_1$ or $e\models\epsilon_2$.

For convenience of presentation, we have the following definition.

Definition 4. (runs with timed fairness assumption) Given event predicates or $TECTL^f$ formulas $\alpha_1,...,\alpha_m,\beta_1,...,\beta_n$, a run

$$\rho = ((q_0,\nu_0),t_0)((q_1,\nu_1),t_1)...((q_k,\nu_k),t_k)......$$

satisfies *strong fairness assumptions* $\alpha_1,...,\alpha_m$ and *weak fairness assumptions* $\beta_1,...,\beta_n$, denoted $\rho\models\overset{\infty}{\Diamond}\{\alpha_1,...,\alpha_m\}\wedge\overset{\infty}{\Box}\{\beta_1,...,\beta_n\}$, iff along the run,

- *states or transitions satisfying* $\alpha_1,...,\alpha_m$ *respectively happen infinitely often; and*
- *the system eventually stabilizes to states or transitions satisfying* $\beta_1,...,\beta_n$.

Formally speaking, $\rho\models\overset{\infty}{\Diamond}\{\alpha_1,...,\alpha_m\}\wedge\overset{\infty}{\Box}\{\beta_1,...,\beta_n\}$, iff

- for each $1\leq i\leq m$ and $c\in\mathcal{N}$,
  - if α_i is a $TECTL^f$ formula, there are a $k>1$ and a $t\in[0,t_{k+1}-t_k]$ such that $t_k+t>c\wedge(q_k,\nu_k+t)\models\alpha_i$;
  - if α_i is an event predicate, there are a $k>1$ and an $e\in E$ such that $t_k>c\wedge q_k\overset{e}{\to}q_{k+1}\wedge e\models\alpha_i$;

- for each $1 \leq j \leq n$, there is a $c \in \mathcal{N}$,
  - if β_j is a $TECTL^f$ formula, then for every $k > 1$ and $t \in [0, t_{k+1} - t_k]$ such that $t_k + t > c$, $(q_k, \nu_k + t) \models \beta_j$.
  - if β_j is an event predicate, then for every $k > 1$ and $e \in E$ such that $t_k > c$ and $q_k \xrightarrow{e} q_{k+1}$, $e \models \beta_j$. ∎

Note that we bind the concept of fairness to the divergence of time. For example in the condition of strong fairness, we require that α_i happens at infinitely and divergently many clock readings. This is quite different from the traditional fairness concepts in untimed systems [13].

Definition 5. (Satisfaction of $TECTL^f$ formulas): We write in notations $A, (q_1, \nu_1) \models \phi$ to mean that ϕ is satisfied at state (q_1, ν_1) in timed automaton A. The satisfaction relation is defined inductively as follows.

- When $\epsilon \in B(Q, X)$, $A, (q_1, \nu_1) \models \epsilon$ iff $(q_1, \nu_1) \models \epsilon$, which was previously defined in the beginning of subsection 3.2;
- $A, (q_1, \nu_1) \models \phi_1 \vee \phi_2$ iff either $A, (q_1, \nu_1) \models \phi_1$ or $A, (q_1, \nu_1) \models \phi_2$
- $A, (q_1, \nu_1) \models \neg \phi_1$ iff $A, (q_1, \nu_1) \not\models \phi_1$
- $A, (q_1, \nu_1) \models x.\phi$ iff $A, (q_1, \nu_1\{x\}) \models \phi$, where $\nu_1\{x\}$ is a valuation identical to ν_1 except that $x = 0$.
- $A, (q_1, \nu_1) \models \exists_{\langle \beta_1, \ldots, \beta_n \rangle}^{[\alpha_1, \ldots, \alpha_m]} \phi_1 \mathcal{U}^\epsilon \phi_2$ iff there exists a run $\rho = ((q_1, \nu_1), t_1)((q_2, \nu_2), t_2) \ldots$, an $i \geq 1$, and a $\delta \in [0, t_{i+1} - t_i]$ such that
  - $\rho \models \overset{\infty}{\Diamond} \{\alpha_1, \ldots, \alpha_m\} \wedge \overset{\infty}{\Box} \{\beta_1, \ldots, \beta_n\}$;
  - if ϵ is null, then $A, (q_i, \nu_i + \delta) \models \phi_2$ and for all j, δ' such that either $(0 \leq j < i) \wedge (\delta' \in [0, t_{j+1} - t_j])$ or $(j = i) \wedge (\delta' \in [0, \delta))$, $A, (q_j, \nu_j + \delta') \models \phi_1$;
  - if ϵ is an event predicate, then there is an $e \in E$ with $q_i \xrightarrow{e} q_{i+1} \wedge e \models \epsilon$ such that $A, (q_{i+1}, \nu_{i+1}) \models \phi_2$ and for all j, δ' with $(0 \leq j \leq i) \wedge (\delta' \in [0, t_{j+1} - t_j])$, $A, (q_j, \nu_j + \delta') \models \phi_1$.
- $A, (q_1, \nu_1) \models \exists_{\langle \beta_1, \ldots, \beta_n \rangle}^{[\alpha_1, \ldots, \alpha_m]} \Box^\epsilon \phi_1$ iff there exists a run $\rho = ((q_1, \nu_1), t_1)(q_2, \nu_2), t_2) \ldots$ such that
  - $\rho \models \overset{\infty}{\Diamond} \{\alpha_1, \ldots, \alpha_m\} \wedge \overset{\infty}{\Box} \{\beta_1, \ldots, \beta_n\}$;
  - if ϵ is null, then for every $i \geq 1$ and $\delta \in [0, t_{i+1} - t_i]$, $A, (q_i, \nu_i + \delta) \models \phi_2$.
  - if ϵ is an event predicate, then for every $i \geq 1$ and $e \in E$ such that $q_i \xrightarrow{e} q_{i+1} \wedge e \models \epsilon$ implies $A, (q_{i+1}, \nu_{i+1}) \models \phi_2$.

A timed automaton $A = \langle \Sigma, X, Q, I, \mu, E, \gamma, \lambda, \tau, \pi \rangle$ satisfies a $TECTL^f$ formula ϕ, in symbols $A \models \phi$, iff for every state $(q_0, \nu_0) \models I$, $A, (q_0, \nu_0) \models \phi$. ∎

4   Model-Checking Algorithm

The key component in the algorithm is for the construction of state-space representations for the following formula:

$$\exists_{\langle \beta_1, \ldots, \beta_n \rangle}^{[\alpha_1, \ldots, \alpha_m]} \Box^\epsilon \phi_1 \qquad \text{(NZF: non-Zeno Fairness)}$$

for any given $\alpha_1, \ldots, \alpha_m, \beta_1, \ldots, \beta_n, \epsilon$ (specified or not), and ϕ_1. Then in subsection 4.2, we present a symbolic algorithm for the construction of state-space representations characterized by NZF formulas. In subsection 4.3, we present our symbolic evaluation algorithm for $TECTL^f$ formulas. Finally in subsection 4.4, we discuss an alternative NZF evaluation algorithm to the one in subsection 4.2. We shall compare the performance of these two algorithms in section 5.

4.1 Basic Reachability Procedures

We need two basic procedures, one for the computation of weakest preconditions of discrete transitions and the other for those of backward time-progressions. Details about the two procedures can be found in [15,21–24,26]. Given a state-space representation η and a discrete transition e, the first procedure, xtion_bck(η, e) with $\gamma(e) = (q, q')$, computes the weakest precondition
  • in which every state satisfies the invariance condition $\mu(q)$; and
  • from which we can transit to states in η through e.
η can be represented as a DBM set [12] or as a BDD-like data-structures [21, 23, 25]. Our algorithms are independent of the representation scheme of η. The second procedure, time_bck(η), computes the space representation of states
  • from which we can go to states in η simply by time-passage; and
  • every state in the time-passage also satisfies the invariance condition imposed by $\mu()$ for whatever modes the states are in.
With the two basic procedures, we can construct the event version of the symbolic backward reachability procedure, denoted reachable$^\epsilon$_bck$_\beta(\eta_0, \eta_1, \eta_2)$ for convenience, as in [15, 21–24, 26]. Intuitively, reachable$^\epsilon$_bck$_\beta(\eta_0, \eta_1, \eta_2)$ characterizes the state-space for $\exists \eta_1 \mathcal{U} \eta_2$ except for the following three differences.
  • Only transitions satisfying β, a state predicate, are permitted in the fixpoint calculation; and
  • When ϵ is null, all states constructed in the least fixpoint iterations must satisfy η_0; and
  • When ϵ is an event predicate, the postcondition immediately after a transition satisfying ϵ must satisfy η_0.
The design of this procedure is for the evaluation of formulas like $\exists \Box^\epsilon \phi_1$. Note that the semantics of $\exists \Box^\epsilon \phi_1$ says that immediately after a discrete transition satisfying ϵ, the destination state must satisfy ϕ_1; in all other cases, the states are not required to satisfy ϕ_1.

Computationally, when ϵ is null, reachable$^\epsilon$_bck$_\beta(\eta_0, \eta_1, \eta_2)$ can be defined as the least fixpoint of equation:

$$Y = \eta_2 \vee \left(\eta_0 \wedge \eta_1 \wedge \text{time\_bck}(\eta_0 \wedge \eta_1 \wedge \bigvee_{e \in E; e \models \beta} \text{xtion\_bck}(Y, e)) \right).$$

i.e., reachable$^\epsilon$_bck$_\beta(\eta_0, \eta_1, \eta_2) \equiv$

$$\text{lfp} Y. \left(\eta_2 \vee \left(\eta_0 \wedge \eta_1 \wedge \text{time\_bck}(\eta_0 \wedge \eta_1 \wedge \bigvee_{e \in E; e \models \beta} \text{xtion\_bck}(Y, e)) \right) \right).$$

The monotonicity of F in fixpoint equation $Y = F(Y)$ ensures the computability of the least fixpoint. Given an event predicate ϵ, we let E^ϵ be the set of discrete transitions satisfying ϵ while letting $E^{\neg \epsilon}$ be the set of those discrete transitions not satisfying ϵ. When ϵ is an event predicate, reachable$^\epsilon$_bck$_\beta(\eta_0, \eta_1, \eta_2) \equiv$

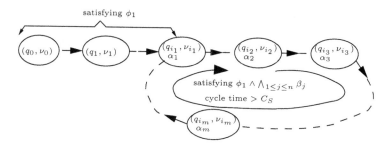

Fig. 3. The run segment for NZF $\exists^{[\alpha_1,\ldots,\alpha_m]}_{\langle\beta_1,\ldots,\beta_n\rangle}\Box^\epsilon\phi_1$ and $q_{i_m} = q_k$ and $\nu_{i_m} = \nu_k$

$$\mathtt{lfp}Y.\eta_2 \vee \eta_1 \wedge \mathtt{time\_bck}\left(\eta_1 \wedge \left(\begin{array}{c}\bigvee_{e\in E^\epsilon;e\models\beta}\mathtt{xtion\_bck}(Y \wedge \eta_0, e) \\ \vee \bigvee_{e\in E^{\neg\epsilon};e\models\beta}\mathtt{xtion\_bck}(Y, e)\end{array}\right)\right)$$

Note that while calculating the fixpoint, we use
- formula $\bigvee_{e\in E^\epsilon;e\models\beta}\mathtt{xtion\_bck}(Y \wedge \eta_0, e)$ to enforce that states immediately after transitions satisfying ϵ must satisfy η_0; and
- formula $\bigvee_{e\in E^{\neg\epsilon};e\models\beta}\mathtt{xtion\_bck}(Y, e)$ to enforce that states immediately after transitions not satisfying ϵ have no obligation.

4.2   Evaluation of NZF

To maintain the non-Zeno run quantification of TCTL, it means that when we say something like "a run along which all fairness assumptions are honored," we really mean *"a run along which all fairness assumptions are honored AND TIME DIVERGES."* Considering that time-divergence means *"time increases by at least $c \geq 1$ infinitely often,"* non-Zenoness can actually be considered as a strong fairness condition.

An NZF formula $\exists^{[\alpha_1,\ldots,\alpha_m]}_{\langle\beta_1,\ldots,\beta_n\rangle}\Box^\epsilon\phi_1$ is satisfied at a state (q_0, ν_0) with a cycle of states and a path leading from (q_0, ν_0) to the cycle such that
- if ϵ is null, all states along the path and cycle satisfy ϕ_1;
- if ϵ is an event predicate, all states immediately following a transition in E^ϵ along the path and cycle must satisfy ϕ_1;
- all states along the cycle satisfy the $TECTL^f$ formulas in $\beta_1,\ldots,\beta_n$;
- all transitions along the cycle satisfy the event predicates in $\beta_1,\ldots,\beta_n$;
- for each of $1 \leq i \leq m$, α_i is satisfied at least once along the cycle; and
- the cycle time is greater than $c \geq 1$.

In our experience [27], we found that the biggest constant, denoted C_S, used in the timed automaton and the $TECTL^f$ formula is usually a proper choice for c for efficient evaluation. A picture for such a run segment from (q_0, ν_0) is in figure 3. For convenience, we need the following notations. Let $L[\phi_1]$ be the state-predicate characterizing the set of states satisfying ϕ_1 and $\mathtt{gfp}X.F(X)$ denote the greatest solution to the the equation of $X = F(X)$. Similarly, the monotonicity of F in fixpoint equation $Y = F(Y)$ ensures the computability of the greatest fixpoint.

Also, let $B_0^\epsilon(X) \equiv x > C_S$ for all X. Finally for conciseness, let β and $\check{\beta}$ be the shorthands for $\bigwedge_{1 \le j \le n; \beta_j \text{ is an event predicate}} \beta_j$ and $\bigwedge_{1 \le j \le n; \beta_j \text{ is an event predicate}} \beta_j$ respectively. The state space satisfying $\exists_{\langle \beta_1, \dots, \beta_n \rangle}^{[\alpha_1, \dots, \alpha_m]} \Box^\epsilon \phi_1$ can then be characterized as follows.

$$\texttt{reachable}^\epsilon\texttt{\_bck}_{true}(L[\phi_1], true, \texttt{gfp}X.(x.B_m^\epsilon(X))) \qquad \text{(FX\_NZF)}$$

where for all $1 \le i \le m$,

- if α_i is a $TECTL^f$ formula, $B_i^\epsilon(X)$ is defined as
  $$\texttt{reachable}^\epsilon\texttt{\_bck}_\beta \left(L[\phi_1], L[\check{\beta}], L[\alpha_i] \wedge B_{i-1}^\epsilon(X) \right)$$
- if ϵ is null and α_i is an event predicate, $B_i^\epsilon(X)$ is defined as
  $$\texttt{reachable}^\epsilon\texttt{\_bck}_\beta \left(L[\phi_1], L[\check{\beta}], \bigvee_{e \in E; e \models \alpha_i \wedge \beta} \texttt{xtion\_bck}(B_{i-1}^\epsilon(X), e) \right)$$
- if ϵ and α_i are both event predicates, $B_i^\epsilon(X)$ is defined as

$$\texttt{reachable}^\epsilon\texttt{\_bck}_\beta \left(L[\phi_1], L[\check{\beta}], \left(\begin{array}{c} \bigvee_{e \in E^\epsilon; e \models \alpha_i \wedge \beta} \texttt{xtion\_bck}(B_{i-1}^\epsilon(X) \wedge L[\phi_1], e) \\ \vee \bigvee_{e \in E^{\neg\epsilon}; e \models \alpha_i \wedge \beta} \texttt{xtion\_bck}(B_{i-1}^\epsilon(X), e) \end{array} \right) \right)$$

The cycle is embodied through the greatest fixpoint $\texttt{gfp}X.(x.B_m^\epsilon(X))$. Note that the weak fairness assumptions as event predicates are enforced through β as the restriction to transitions. The other weak ones are enforced through $L[\check{\beta}]$ as the second argument to $\texttt{reachable}^\epsilon\texttt{\_bck}_\beta()$. The cycle is composed of m segments, each of which fulfills a strong fairness assumption and is constructed differently according to whether the corresponding strong assumption is for a $TECTL^f$ formula or an event predicate. In the 2nd and 3rd items, for an event predicate strong assumption α_i, we start constructing the corresponding segment from the precondition of the event, i.e., $\bigvee_{e \in \dots; e \models \alpha_i \wedge \beta} \texttt{xtion\_bck}(\dots, e)$. The fulfillment of the assumptions are checked with the existence of such cycles. The following lemma is for the correctness of this formulation.

Lemma 1. : *formula (FX_NZF) characterizes the set of states which satisfy* $\exists_{\langle \beta_1, \dots, \beta_n \rangle}^{[\alpha_1, \dots, \alpha_m]} \Box^\epsilon \phi_1$. ∎

Suppose that we have already calculated the state-space representations $L[\phi']$ for all the proper subformulas ϕ' of $\exists_{\langle \beta_1, \dots, \beta_n \rangle}^{[\alpha_1, \dots, \alpha_m]} \Box^\epsilon \phi_1$. We can use the following algorithm to compute state-predicates for NZFs.

$NZF([\alpha_1, \dots, \alpha_m], \langle \beta_1, \dots, \beta_n \rangle, \epsilon, W)$
/* where W is $L[\phi_1]$, i.e., the state-predicate for states satisfying ϕ_1. */ {
    let $\check{\beta}$ be $\bigwedge_{1 \le j \le n; \beta_j \text{ is not an event predicate}} \beta_j$;
    let β be $\bigwedge_{1 \le j \le n; \beta_j \text{ is an event predicate}} \beta_j$;
    $R := W \wedge L[\check{\beta}]$; $R' = false$;
    Repeat until $R = R'$, {
        $R' := R$; $R := R \wedge x > C_S$, where x is never used anywhere else.   (1)
        for $i := m$ to 1, if α_i is not an event predicate,
            $R := \texttt{reachable}^\epsilon\texttt{\_bck}_\beta(W, L[\check{\beta}], R \wedge L[\alpha_i])$;   (2)
        else if ϵ is null,

$$R := \mathtt{reachable}^\epsilon\_\mathtt{bck}_\beta(W, L[\breve{\beta}], \bigvee\nolimits_{e \in E; e \models \beta \wedge \alpha_i} \mathtt{xtion\_bck}(R, e)); \tag{3}$$

else

$$R := \mathtt{reachable}^\epsilon\_\mathtt{bck}_\beta \left(W, L[\breve{\beta}], \left(\begin{array}{c} \bigvee_{e \in E^\epsilon; e \models \beta \wedge \alpha_i} \mathtt{xtion\_bck}(R \wedge L[\phi_1], e) \\ \vee \bigvee_{e \in E^{\neg \epsilon}; e \models \beta \wedge \alpha_i} \mathtt{xtion\_bck}(R, e) \end{array} \right) \right); \tag{4}$$

$$R := R' \wedge \mathtt{var\_eliminate}(\mathtt{bypass}(x = 0 \wedge R, x), \{x\}); \tag{5}$$

}

return $\mathtt{reachable}^\epsilon\_\mathtt{bck}_{true}(W, true, R)$;

}

The repeat-loop in NZF() calculates state-space representation of $\mathtt{gfp}X.(x. B_m(X))$. The inner for-loop iterates to check the fulfillment of $\alpha_m, \alpha_{m-1}, \ldots, \alpha_1$ in sequence. Through the backward reachability analysis steps from statements (1) to (4), we expect the cycle to go backward from states with $x > C_S$ (when $C_S \geq 1$) to states with $x = 0$. The cyclic execution time $> C_S$ is enough for non-Zenoness. The return statement uses one final least fixpoint procedure on the result of the repeat-loop to calculate the state-space representatoin for formulation (FX_NZF).

In procedure NZF(), we use procedure $\mathtt{var\_eliminate}(\eta, \{x\})$ which partially implements the Fourier-Motzkin elmination [14] and will eliminate all information in state-predicate η related to x. But before the application of this procedure, we first apply procedure $\mathtt{bypass}(\eta, x)$ which will add to η any transitivity information deducible from x. For example,

$$\mathtt{var\_eliminate}(\mathtt{bypass}(x < 3 \wedge y - x \leq -2, x), \{x\})$$
$$= \mathtt{var\_eliminate}(x < 3 \wedge y - x \leq -2 \wedge y < 1, \{x\})$$
$$= y < 1$$

Note that in the first step, new constraint $y < 1$ is deduced. Details of procedures $\mathtt{var\_eliminate}()$ and $\mathtt{bypass}()$ can be found in [25].

4.3   Evaluation of $TECTL^f$ Formulas

The following evaluation algorithm uses the procedures presented in the last two subsections as basic blocks to evaluate $TECTL^f$ formulas.

Eval($A, \bar{\phi}$) {

   switch ($\bar{\phi}$) {

case (*false*):	return *false*;
case (q):	return q;
case ($x \sim c$):	return $x \sim c$; ;
case ($\phi_1 \vee \phi_2$):	return Eval(A, ϕ_1) $\vee$ Eval(A, ϕ_2);
case ($\neg\phi_1$):	return $\neg$Eval(A, ϕ_1);
case ($x.\phi_1$):	return $\mathtt{var\_eliminate}(\mathtt{bypass}(x = 0 \wedge \mathrm{Eval}(A, \phi_1), x), \{x\})$;
case ($\exists_{\langle \beta_1, \ldots, \beta_n \rangle}^{[\alpha_1, \ldots, \alpha_m]} \phi_1 \mathcal{U}^\epsilon \phi_2$):	$Y_1 := \mathrm{Eval}(A, \phi_1)$;
	$Y_2 := \mathrm{Eval}(A, \phi_2) \wedge \mathrm{NZF}([\alpha_1, \ldots, \alpha_m], \langle \beta_1, \ldots, \beta_n \rangle, \mathtt{NULL}, true)$;
	if ϵ is an event predicate, $Y_2 := \bigvee_{e \in E^\epsilon} \mathtt{xtion\_bck}(Y_2, e)$;
	return $\mathtt{reachable\_bck}_{true}(true, Y_1, Y_2)$;
case ($\exists_{\langle \beta_1, \ldots, \beta_n \rangle}^{[\alpha_1, \ldots, \alpha_m]} \Box^\epsilon \phi_1$):	$W := \mathrm{Eval}(A, \phi_1)$;
	return $\mathrm{NZF}([\alpha_1, \ldots, \alpha_m], \langle \beta_1, \ldots, \beta_n \rangle, \epsilon, W)$;

   }

}

Theorem 1. *Given a timed automaton A, a $TECTL^f$ formula ϕ, and a state (q, ν), $(q, \nu) \models \phi$ iff $(q, \nu) \models Eval(A, \phi)$.*

Proof: The framework of the proof is a standard induction on the structure of ϕ. Other than that, we also need to check that instead of evaluating $\exists^{[\alpha_1,...,\alpha_m]}_{\langle\beta_1,...,\beta_n\rangle}\phi_1\mathcal{U}^\epsilon$ ϕ_2), we evaluate $\exists\phi_1\mathcal{U}^\epsilon\left(\phi_2 \wedge \exists^{[\alpha_1,...,\alpha_m]}_{\langle\beta_1,...,\beta_n\rangle}\square true\right)$. The NZF is asserted additionally when the liveness property ϕ_2 is fulfilled. Finally, when ϵ is not null, we start the least fixpoint evaluations for $\exists\mathcal{U}^\epsilon$-formulas from the preconditions of those transitions that satisfying ϵ. This is compatible with the semantics of $\mathcal{U}^\epsilon$.   ∎

Given a timed automaton A and a $TECTL^f$ formula ϕ, A satisfies ϕ iff $Eval(A, \neg\phi)$ is *false*.

4.4   Another Algorithm for NZF Evaluation

There can be other algorithms for the valuation of NZF. We have experimented with an algorithm that uses m auxiliary variables $a_1, \ldots, a_m$ for the bookkeeping of strong fairness assumption fulfillment in NZF. The auxiliary variables serve the purpose to flag whether along the computation, corresponding strong fairness assumptions have been fulfilled. The advantage of this algorithm is that inside the greatest fixpoint evaluation, we only need one iteration of least fixpoint evaluation with the new procedure `reachable_aux_vars`$^\epsilon$`_bck`$_\beta$`()` instead of m iterations of least fixpoint evaluation with `reachable`$^\epsilon$`_bck`$_\beta$`()`. The idea is that initially we start `reachable_aux_vars`$^\epsilon$`_bck`$_\beta$`()` from states where $a_1, \ldots, a_m$ are all false. Then in the execution of `reachable_aux_vars`$^\epsilon$`_bck`$_\beta$`()`, each time we compute a weakest precondition from a basic time-progression (`time_bck()`) or from a discrete transition (`xtion_bck()`), we check whether a strong assumption α_i has just been fulfilled and set the corresponding a_i to true if it has. In the end of this single iteration of least fixpoint evaluation, the states in the cycles that have fulfilled all strong fairness assumptions are those with $a_1 \wedge \ldots \wedge a_m$ true. Then the state space satisfying $\exists^{[\alpha_1,...,\alpha_m]}_{\langle\beta_1,...,\beta_n\rangle}\square^\epsilon\phi_1$ can also be characterized with the following formulation.

```
reachableᵉ_bck_true(
    L[φ₁], true,
    gfpX.var_eliminate(
        (x.reachable_aux_varsᵉ_bck_β(L[φ₁], L[β̃], X ∧ x > C_S ∧ ⋀_{1≤i≤m} ¬a_i)) ∧ ⋀_{1≤i≤m} a_i,
        {a₁,...,a_m}
    )
)
```

We shall report the performance comparison in the next section.

5   Implementation and Experiment

We have implemented the ideas in our model-checker/simulator, **Red** version 5.1, for communicating timed automata [19]. The events are interpreted as input-output event pairs through communication channels. **Red** uses the new BDD-like data-structure, CRD (Clock-Restriction Diagram) [23–25], and supports both

Table 1. Performance data of scalability w.r.t. number of processes

spec.	# proc's	Linear networks		Tree networks		Lattice networks	
		sequential	aux. var.s	sequential	aux. var.s	sequential	aux. var.s
(A)	2	0.03s/5k	0.01s/5k	0.02s/5k	0.01s/5k	0.02s/5k	0.00s/5k
	4	0.10s/14k	0.09s/14k	0.08s/14k	0.13s/14k	0.14s/16k	0.17s/18k
	6	0.53s/28k	0.47s/28k	0.47s/27k	1.00s/35k	0.81s/34k	1.93s/58k
	8	1.86s/46k	1.71s/46k	1.99s/46k	4.82s/70k	5.42s/64k	15.0s/192k
	10	5.16s/70k	4.74s/70k	5.45s/70k	19.1s/145k	17.0s/93k	99.8s/697k
	12	11.7s/100k	11.0s/100k	15.0s/99k	68.7s/301k	64.3s/141k	781s/2452k
	14	23.7s/136k	23.4s/136k	34.4s/135k	239s/635k	219s/198k	6289s/9203k
	16	44.4s/179k	45.4s/179k	82.1s/179k	837s/1315k	1046s/377k	52383s/35356k
	18	79.1s/229k	82.5s/229k	163s/229k	3098s/2801k	2074s/601k	Not
	20	127s/288k	143s/288k	341s/287k	12532s/5961k	5381s/1011k	Available
(B)	2	0.00s/5k	0.00s/5k	0.01s/5k	0.02s/5k	0.00s/5k	0.00s/5k
	4	0.01s/14k	0.02s/14k	0.04s/14k	0.03s/14k	0.01s/16k	0.02s/16k
	6	0.05s/27k	0.08s/27k	0.10s/27k	0.12s/27k	0.18s/33k	0.21s/33k
	8	0.15s/46k	0.17s/46k	0.39s/46k	0.41s/46k	0.85s/64k	0.85s/64k
	10	0.29s/70k	0.36s/70k	1.12s/70k	1.22s/70k	3.11s/93k	3.29s/93k
	12	0.59s/99k	0.59s/99k	2.83s/99k	3.14s/99k	10.7s/141k	10.9s/141k
	14	0.99s/136k	1.05s/136k	6.95s/135k	7.43s/145k	34.1s/198k	34.8s/198k
	16	1.56s/179k	1.69s/179k	15.5s/179k	16.4s/179k	107s/363k	108s/363k
	18	2.38s/230k	2.48s/230k	34.0s/229k	35.3s/229k	280s/570k	287s/570k
	20	3.42s/288k	3.57s/288k	70.5s/288k	73.2s/288k	789s/1010k	792s/1010k
(C)	2	0.02s/5k	Not	0.02s/5k	Not	0.01s/5k	Not
	4	0.19s/14k	Available	0.16s/14k	Available	0.17s/16k	Available
	6	1.14s/28k		0.92s/28k		0.89s/33k	
	8	4.85s/46k		3.70s/46k		4.76s/64k	
	10	15.6s/70k		12.5s/70k		14.2s/93k	
	12	42.1s/99k		37.7s/103k		50.0s/141k	
	14	100s/136k		103s/174k		174s/259k	
	16	212s/178k		282s/306k		604s/629k	
	18	433s/229k		689s/474k		1358s/906k	
	20	791s/288k		1878s/877k		4096s/1630k	

data collected on a Pentium 4 Mobile 1.6GHz with 256MB memory running LINUX;
s: seconds; k: kilobytes of memory in data-structure; N/A: not available;

forward and backward analyses, full $TECTL^f$ model-checking with non-Zeno computations, deadlock detection, and counter-example generation. Users can also declare global and local (to each process) variables of type clock, integer, and pointer (to identifier of processes). Boolean conditions on variables can be tested and variable values can be assigned. **Red** 5.1 also accepts $TECTL^f$ formulas with quantifications on process identifiers for succinct specification.

We use the diffusive computing algorithm [10] in figure 2, with the three network configurations in figures 1, to demonstrate the performance of our techniques. The $TECTL^f$ specifications in our experiment include properties (A), (B) and (C) in example 1 in page 558. In table 1, please find the performance data.

Data in the "sequential" columns were collected with the NZF evaluation algorithm in subsection 4.2 while those in the "aux. var.s" columns were with the algorithm with auxiliary variables. Specification (C) does not have fairness assumptions and does not use options for choosing NZF algorithms.

Our techniques seem quite efficient for the benchmarks. When we carefully check the data, we find that **Red** 5.1 can actually handle much higher concurrencies since the memories used at 20 clocks are still not very much in the "sequential" columns. It is also interesting to see that the NZF evaluation algorithm in subsection 4.2 performs better than the alternative in subsection 4.4. Though the latter needs less iterations of least fixpoint evaluations inside the greates fixpoint evaluation, the single iteration of $\mathtt{reachable\_aux\_vars}^\epsilon\_\mathtt{bck}_\beta()$ incurs huge memory complexities and usually performs worse than the algorithm in subsection 4.2. This is understandable since the complexity of BDD-based algorithms is usually exponential to the number of variables.

6 Conclusion

We investigate how to add the concepts of events and fairness assumptions to TCTL model-checking and define the new language of $TECTL^f$. Our framework allows for the specification and verification of punctual event properties and multiple strong and weak fairness assumptions. Our implementation and experiments have shown that the ideas could be of practical use. More research are expected to develop useful specification languages for distributed real-time systems.

References

1. R. Alur, C. Courcoubetis, D.L. Dill. Model Checking for Real-Time Systems, IEEE LICS, 1990.
2. R. Alur, D.L. Dill. Automata for modelling real-time systems. ICALP' 1990, LNCS 443, Springer-Verlag, pp.322-335.
3. R. Alur, T.A. Henzinger. A really temporal logic, 30th IEEE FOCS, pp.164-169, 1989.
4. R. Alur, T.A. Henzinger. Real-Time Logics: Complexity and Expressiveness. Information and Computation **104**, 35-77, 1993.
5. J.R. Burch, E.M. Clarke, K.L. McMillan, D.L.Dill, L.J. Hwang. Symbolic Model Checking: 10^{20} States and Beyond, IEEE LICS, 1990.
6. M. Bozga, C. Daws. O. Maler. Kronos: A model-checking tool for real-time systems. 10th CAV, June/July 1998, LNCS 1427, Springer-Verlag.
7. J. Bengtsson, K. Larsen, F. Larsson, P. Pettersson, Wang Yi. UPPAAL - a Tool Suite for Automatic Verification of Real-Time Systems. Hybrid Control System Symposium, 1996, LNCS, Springer-Verlag.
8. R.E. Bryant. Graph-based Algorithms for Boolean Function Manipulation, IEEE Trans. Comput., C-35(8), 1986.
9. E. Clarke and E.A. Emerson. Design and Synthesis of Synchronization Skeletons using Branching-Time Temporal Logic, Proceedings of Workshop on Logic of Programs, Lecture Notes in Computer Science 131, Springer-Verlag, 1981.
10. Chandy, Misra. Parallel Program Design - A Foundation, Addison-Wesley, 1988.
11. S. Chaki, E.M. Clarke, J. Ouaknine, N. Sharygina, N. Sinha. State/Event-based Software Model Checking. IFM 2004, LNCS 2999, Springer-Verlag.

12. D.L. Dill. Timing Assumptions and Verification of Finite-state Concurrent Systems. CAV'89, LNCS 407, Springer-Verlag.
13. E.A. Emerson, C.-L. Lei. Modalities for Model Checking: Branching Time Logic Strikes Back, Science of Computer Programming 8 (1987), pp.275-306, Elsevier Science Publishers B.V. (North-Holland).
14. J.B. Fourier. (reported in:) Analyse des travaux de l'Académie Royale des Sciences pendant l'année 1824, Partie Mathématique, 1827.
15. T.A. Henzinger, X. Nicollin, J. Sifakis, S. Yovine. Symbolic Model Checking for Real-Time Systems, IEEE LICS 1992.
16. M. Huth, R. Jagadeesan, D. Schmidt. Modal transition systems: A foundation for three-valued program analysis. ESOP 2001, LNCS 2028, Springer Verlag.
17. E. Kindler, T. Vesper. ESTL: A Temporal Logic for Events and States. ATPN 1998, LNCS 1420, Springer-Verlag.
18. D. Kozen. Results on the propositional mu-calculus. Theoretical Computer Science, 27:333-354, 1983.
19. A. Shaw. Communicating Real-Time State Machines. IEEE Transactions on Software Engineering 18(9), September, 1992.
20. R. De Nicola, F. Vaandrager. Three Logics for Branching Bisimulation. Journal of the ACM (JACM), 42(2):458-487, 1995.
21. F. Wang. Efficient Data-Structure for Fully Symbolic Verification of Real-Time Software Systems. TACAS'2000, LNCS 1785, Springer-Verlag.
22. F. Wang. Region Encoding Diagram for Fully Symbolic Verification of Real-Time Systems. the 24th COMPSAC, Oct. 2000, Taipei, Taiwan, ROC, IEEE press.
23. F. Wang. RED: Model-checker for Timed Automata with Clock-Restriction Diagram. Workshop on Real-Time Tools, Aug. 2001, Technical Report 2001-014, ISSN 1404-3203, Dept. of Information Technology, Uppsala University.
24. F. Wang. Symbolic Verification of Complex Real-Time Systems with Clock-Restriction Diagram, Proceedings of FORTE, August 2001, Cheju Island, Korea.
25. F. Wang. Efficient Verification of Timed Automata with BDD-like Data-Structures, to appear in special issue of STTT (Software Tools for Technology Transfer, Springer-Verlag) for VMCAI'2003, LNCS 2575, Springer-Verlag.
26. F. Wang, P.-A. Hsiung. Efficient and User-Friendly Verification. IEEE Transactions on Computers, Jan. 2002.
27. F. Wang, G.-D. Huang, F. Yu. TCTL Inevitability Analysis of Dense-Time Systems. 8th CIAA (Conference on Implementation and Application of Automata), July 2003, Santa Barbara, CA, USA; LNCS 2759, Springer-Verlag.

Author Index